Table of Contents

• • • • • • • • • • • • • • • • • • • •

Student Solutions Manual, Study Guide, and Problems Book

Biochemistry

FIFTH EDITION

Reginald H. Garrett
University of Virginia

Charles M. Grisham
University of Virginia

Prepared by

David K. Jemiolo
Vassar College

Steven M. Theg
University of California, Davis

BROOKS/COLE
CENGAGE Learning·

Australia · Brazil · Japan · Korea · Mexico · Singapore · Spain · United Kingdom · United States

For product information and technology assistance, contact us at **Cengage Learning Customer & Sales Support, 1-800-354-9706**

For permission to use material from this text or product, submit all requests online at **www.cengage.com/permissions** Further permissions questions can be emailed to **permissionrequest@cengage.com**

ISBN-13: 978-1-133-10851-1
ISBN-10: 1-133-10851-2

Brooks/Cole
20 Davis Drive
Belmont, CA 94002-3098
USA

Cengage Learning is a leading provider of customized learning solutions with office locations around the globe, including Singapore, the United Kingdom, Australia, Mexico, Brazil, and Japan. Locate your local office at: **www.cengage.com/global**

Cengage Learning products are represented in Canada by Nelson Education, Ltd.

To learn more about Brooks/Cole, visit **www.cengage.com/brookscole**

Purchase any of our products at your local college store or at our preferred online store **www.cengagebrain.com**

Printed in the United States of America
1 2 3 4 5 6 7 16 15 14 13 12

Preface

• •

In one scene in the movie Stripes (Columbia Picture Corporation 1981), privates John Winger and Russell Zissky (played by Bill Murray and Harold Ramis) attempt to persuade their platoon to an all night training session to prepare for the next day's final parade. The troops are skeptical of the plan; however, Zissky wins them over by his testimony of the importance of cramming. He proudly reports that he had, in fact, once learned two semesters of geology in a single three-hour all nighter.

It would seem unlikely that this approach would work well with biochemistry (or even geology). Rather a steady diet of reading, problem solving, and reviewing might be a better plan of attack. This study guide was written to accompany *"Biochemistry"* by Garrett and Grisham. It includes chapter outlines, guides to key points covered in the chapters, in-depth solutions to the problems presented in the textbook, additional problems, and detailed summaries of each chapter. In addition, there is a glossary of biochemical terms and key text figures.

Several years ago I spent part of a sabbatical in Italy and in preparation took a year- long course in elementary Italian. I had not been on the student-end of an academic interaction for several years and taking a language course was an excellent opportunity to be reminded of the difficulties of learning something for the first time. Memorization is part and parcel to the study of any language and so I found myself committing to memory nouns, verbs, adverbs, adjectives, and complex, irregular verb conjugations. The study of biochemistry has parallels to language studies in that memorization is necessary. What makes the study of biochemistry somewhat easier, however, are the common themes, the interconnections between various facets of biochemistry, and the biological and chemical principles at work. The authors have done a marvelous job in presenting these aspects of biochemistry and I have attempted to highlight them here. Biochemistry is a demanding discipline but one well worth the effort for any student of the sciences. *Buona fortuna.*

Acknowledgments

It is often stated that teaching a subject is the best way to learn it. In teaching my one-semester biochemistry course at Vassar College, because there is never enough time to cover all the topics, I used to worry about forgetting certain aspects of biochemistry. Thanks to Charles Grisham and Reginald Garrett, this fear is no longer with me. I thank both authors for the marvelous text and the opportunity to relearn all of biochemistry. I also thank my co-author Steven Theg. To my wife Kristen I give special thanks for putting up with me during this project.

David K. Jemiolo
Poughkeepsie, NY August
2011

Every time I work on this project I am grateful for the chance to learn and relearn aspects of biochemistry from Reginald Garrett and Charles Grisham through their scholarly and readable text. My co-author Dave Jemiolo displays the same vast knowledge of biochemistry, and I am grateful for the opportunity to work with him on this book. I am especially thankful for Jill, Chris, Alex and Sam for providing the context in which all this makes sense.

Steven M. Theg Davis,
CA August 2011

Why study biochemistry?

This excerpt from *Poetry and Science* by the Scottish poet Hugh MacDiarmid (1892-1978), which first appeared in *Lucky Poet* (1943), might help with an answer.

Poetry and Science

Wherefore I seek a poetry of facts. Even as The
profound kinship of all living substance Is made clear
by the chemical route.
Without some chemistry one is bound to remain
Forever a dumbfounded savage
In the face of vital reactions. The
beautiful relations Shown only by
biochemistry
Replace a stupefied sense of wonder With
something more wonderful Because natural
and understandable. Nature is more
wonderful
When it is at least partly understood. Such an
understanding dawns
On the lay reader when he becomes
Acquainted with the biochemistry of the glands
In their relation to diseases such as goitre
And their effects on growth, sex, and reproduction. He will
begin to comprehend a little
The subtlety and beauty of the action
Of enzymes, viruses, and bacteriophages, These
substances which are on the borderland Between the
living and the non-living.
He will understand why the biochemist
Can speculate on the possibility
Of the synthesis of life without feeling
That thereby he is shallow or blasphemous. He will
understand that, on the contrary, He finds all the
more
Because he seeks for the endless
---'Even our deepest emotions
May be conditioned by traces
Of a derivative of phenanthrene!'

*Science is the Differential
Calculus of the mind,
Art is the Integral Calculus;
they may be Beautiful apart, but are great
only when combined.*

Sir Ronald Ross

In this poem, MacDiarmid argues strongly for the importance of studying biochemistry to understand and appreciate Nature itself. The poem was published in 1943, well before the molecular revolution in biochemistry, well before the first protein structure was solved or the first gene cloned yet MacDiarmid seems to have appreciated the importance of enzyme kinetics and enzyme catalysis and to anticipate the value of recombinant DNA technology: "The subtlety and beauty of the action of enzymes, viruses, and bacteriophages...." He even suggests that a fundamental understanding of life itself might be possible through biochemistry.

It is interesting to see how biochemists are portrayed in movies and films in this electronic age. In the 1996 film *The Rock* staring Sean Connery and Nicholas Cage, Cage plays a biochemist enlisted by the FBI to deal with a threat involving VX gas warheads. (VX is a potent acetylcholinesterase inhibitor.) Cage's character, Stanley Goodspeed, delivers this memorable line, which informs the audience of his expertise: " Look, I'm just a biochemist. Most of the time, I work in a little glass jar and lead a very uneventful life. I drive a Volvo, a beige one. But what I'm dealing with here is one of the most deadly substances the earth has ever known, so what say you cut me some friggin' slack!" Perhaps Stanley is overstating the danger inherent in his work but he is surely understating the importance of his occupation.

Chapter 1

The Facts of Life: Chemistry Is the Logic of Biological Phenomena

• •

Chapter Outline

❖ Properties of living systems
  ⋏ Highly organized - Cells > organelles > macromolecular complexes > macromolecules (proteins, nucleic acids, polysaccharides)
  ⋏ Structure/function correlation: Biological structures serve functional purposes
  ⋏ Energy transduction: ATP and NADPH –energized molecules
  ⋏ Steady state maintained by energy flow: Steady state not equilibrium
  ⋏ Self-replication with high, yet not perfect, fidelity
❖ Biomolecules
  ⋏ Elements: Hydrogen, oxygen, carbon, nitrogen (lightest elements of the periodic table capable of forming a variety of strong covalent bonds)
    • Carbon -4 bonds, nitrogen -3 bonds, oxygen –2 bonds, hydrogen -1 bond
  ⋏ Compounds: Carbon-based compounds –versatile
  ⋏ Phosphorus- and sulfur-containing compounds play important roles
❖ Biomolecular hierarchy
  ⋏ Simple compounds: H_2O, CO_2, NH_4^+, NO_3^-, N_2
  ⋏ Metabolites: Used to synthesize building block molecules
  ⋏ Building blocks: Amino acids, nucleotides, monosaccharides, fatty acids, glycerol
  ⋏ Macromolecules: Proteins, nucleic acids, polysaccharides, lipids
  ⋏ Supramolecular complexes: Ribosomes, chromosomes, cytoskeleton
❖ Membranes: Lipid bilayers with membrane proteins
  ⋏ Define boundaries of cells and organelles
  ⋏ Hydrophobic interactions maintain structures
❖ Organelles: Mitochondria, chloroplasts, nuclei, endoplasmic reticulum Golgi, etc.
❖ Cells: Fundamental units of life
  ⋏ Living state: Growth, metabolism, stimulus response and replication
❖ Properties of biomolecules
  ⋏ Directionality or structural polarity
    • Proteins: N-terminus and C-terminus
    • Nucleic acids: 5'- and 3'- ends
    • Polysaccharides: Reducing and nonreducing ends
  ⋏ Information content: Sequence of monomer building blocks and 3-dimensional architecture
❖ 3-Dimensional architecture and intermolecular interactions (via complementary surfaces) of macromolecules are based on weak forces
  ⋏ Van der Waals interactions (London dispersion forces)
    • Induced electric interactions that occur when atoms are close together
    • Significant when many contacts form complementary surfaces
  ⋏ Hydrogen bonding
    • Donor and acceptor pair: Direction dependence

1

- Donor is hydrogen covalently bonded to electronegative O or N
- Acceptor is lone pair on O or N
  - Ionic interactions
    - Stronger than H bonds
    - Not directional
    - Strength influenced by solvent properties
  - Hydrophobic interactions: Occur when nonpolar groups added to water
    - Water molecules hydrogen bond
    - Nonpolar groups interfere with water H-bonding and to minimize this nonpolar groups aggregate
- ❖ Structural complementarity
  - Biomolecular recognition depends on structural complementarity
  - Weak chemical forces responsible for biomolecular recognition
  - Life restricted to narrow range of conditions (temperature, pH, salt concentration, etc.) because of dependence on weak forces. Denaturation: Loss of structural order in a macromolecule
- ❖ Enzymes: Biological catalysts capable of being regulated
- ❖ Cell types
  - Prokaryotes: Bacteria and archaea: Plasma membrane but no internal membrane-defined compartments
    - Archaea include thermoacidophiles, halophiles and methanogens
  - Eukaryotes: Internal membrane-defined compartments: Nuclei, endoplasmic recticulum, Golgi, mitochondria, chloroplasts, vacuoles, peroxisomes
  - Viruses and bacteriophages: Incomplete genetic systems

Chapter Objectives

Understand the basic chemistry of H, O, N and C.
   H forms a single covalent bond. When bound to an electronegative element, like O or N, the electron pair forming the covalent bond is not equally shared, giving rise to a partial positive charge on the hydrogen (this is the basis of H bonds which will be covered in the next chapter). In extreme cases the H can be lost as a free proton.
   O forms two covalent bonds and has two lone pairs of electrons. It is an electronegative element and when bound to hydrogen it will cause H to be partially positively charged. O is highly reactive due to its high electronegativity.
   N forms up to three covalent bonds and has a single lone pair of electrons. It is an electronegative element and will create a partial positive charge on a hydrogen bonded to it.
   C forms four covalent bonds. With four single bonds, tetrahedral geometry is predominant. With one double bond, carbon shows trigonal planar geometry, with an additional pair of electrons participating in a pi bond.

Macromolecules and subunits
   Proteins are formed from amino acids composed of C, H, O, N, and in some instances S.
Nucleic acids are formed from nucleotides that are composed of phosphate, sugar and nitrogenous base components. (Nucleosides lack phosphate).
Polysaccharides are made of carbohydrates or sugar molecules.
Lipids are a class of mostly nonpolar, mostly hydrocarbon molecules.

Macromolecular structures
   Macromolecular structures are composed of complexes of macromolecules (i.e., proteins, nucleic acids, polysaccharides and lipids). The ribosome, made up of protein and ribonucleic acid, is a prime example.

Organelles
   Organelles are subcellular compartments defined by lipid bilayer membranes.

Cell types
   There are two fundamental cell types: eukaryotic, having organelles and a defined nuclear region, and prokaryotic, lacking organelles and a membrane-enclosed region of genetic material. The archaea and bacteria comprise the prokaryotes.

Problems and Solutions

1. The Biosynthetic Capacity of Cells
The nutritional requirements of Escherichia coli *cells are far simpler than those of humans, yet the macromolecules found in bacteria are about as complex as those of animals. Because bacteria can make all their essential biomolecules while subsisting on a simpler diet, do you think bacteria may have more biosynthetic capacity and hence more metabolic complexity than animals? Organize your thoughts on this question, pro and con, into a rational argument.*

Answer: Although it is true that *Escherichia coli* are capable of producing all of their essential biomolecules (e.g. there is no minimum daily requirement for vitamins in the world of wild-type *E. coli*), they are rather simple, single-cell organisms capable of a limited set of responses. They are self sufficient, yet they are incapable of interactions leading to levels of organization such as multicellular tissues. Multicellular organisms have the metabolic complexity to produce a number of specialized cell types and to coordinate interactions among them.

2. Cell Structure
Without consulting figures in this chapter, sketch the characteristic prokaryotic and eukaryotic cell types and label their pertinent organelle and membrane systems.

Answer: Prokaryotic cells lack the compartmentation characteristic of eukaryotic cells and are devoid of membrane bound organelles such as mitochondria, chloroplasts, endoplasmic reticulum, Golgi apparatus, nuclei, peroxisomes and vacuoles. Both cell types are delimited by membranes and contain ribosomes.

3. The Dimensions of Prokaryotic Cells and Their Constituents
Escherichia coli *cells are about 2 μm (microns) long and 0.8 μm in diameter.*
a. How many E. coli cells laid end to end would fit across the diameter of a pinhead? (Assume a pinhead diameter of 0.5 mm.)

Answer:

$$E.\ coli \text{ per pinhead} = \frac{0.5\ mm\ \text{dia.}}{\dfrac{2\ \mu m}{E.\ coli}}$$

$$= \frac{0.5\times10^{-3}\text{m dia.}}{\dfrac{2\times10^{-6}\text{m}}{E.\ coli}}$$

$$= 250\ E.\ coli \text{ per pinhead}$$

b. What is the volume of an E. coli cell? (Assume it is a cylinder, with the volume of a cylinder given by $V=\pi r^2 h$, where $\pi = 3.14$.)

Answer:

$$V = \pi \times r^2 \times h$$

$$= 3.14 \times \left(\frac{0.8\ um}{2}\right)^2 \times 2\ um$$

$$= 3.14 \times \left(0.4\times10^{-6}\text{m}\right)^2 \times 2\times10^{-6}\text{m}$$

$$= 1\times10^{-18}\text{m}^3$$

But, $1\ m^3 = \left(100\ cm\right)^3 = 10^6 cm^3 = 10^6 ml = 10^3 L$

$$V = 1\times10^{-18}\text{m}^3 = 1\times10^{-15}\text{L} = 1\ fL \text{ (femtoliter)}$$

c. What is the surface area of an E. coli cell? What is the surface-to-volume ratio of an E. coli cell?

Answer:

$$\text{Surface Area} = 2 \times \pi \times r^2 + \pi \times d \times h$$

$$\text{Surface Area} = 2 \times 3.14 \times \left(0.4 \times 10^{-6}\,\text{m}\right)^2 + 3.14 \times \left(0.8 \times 10^{-6}\,\text{m}\right) \times \left(2 \times 10^{-6}\,\text{m}\right)$$

$$\text{Surface Area} = 6.03 \times 10^{-12}\,\text{m}^2$$

$$\frac{\text{Surface Area}}{\text{Volume}} = \frac{6 \times 10^{-12}\,\text{m}^2}{1 \times 10^{-18}\,\text{m}^3 \text{ (from b)}}$$

$$\text{Surface Area per volume} = 6 \times 10^{6}\,\text{m}^{-1}$$

d. Glucose, a major energy-yielding nutrient, is present in bacterial cells at a concentration of about 1 mM. What is the concentration of glucose, expressed as mg/ml? How many glucose molecules are contained in a typical E. coli cell? (Recall that Avogadro's number = 6.023 x 10²³.)

Answer:

$$\left[\text{Glucose}\right] = 1\,\text{mM} = 1 \times 10^{-3}\,\frac{\text{mol}}{\text{L}}$$

$$\text{Glucose} = C_6H_{12}O_6$$

$$M_r = 6 \times 12 + 12 \times 1.0 + 6 \times 16$$

$$M_r = 180$$

$$\left[\text{Glucose}\right] = 1 \times 10^{-3}\,\frac{\text{mol}}{\text{L}} \times 180\,\frac{\text{g}}{\text{mol}}$$

$$\left[\text{Glucose}\right] = 0.18\,\frac{\text{g}}{\text{L}} = 0.18\,\frac{\text{mg}}{\text{ml}}$$

$$\text{moles of glucose} = \text{concentration} \times \text{volume}$$

$$\text{moles of glucose} = 1 \times 10^{-3}\,\frac{\text{mol}}{\text{L}} \times 1 \times 10^{-15}\,\text{L (from b)}$$

$$\text{moles of glucose} = 1 \times 10^{-18}$$

$$\#\text{ molecules} = 1 \times 10^{-18}\,\text{mol} \times 6.023 \times 10^{23}\,\frac{\text{molecules}}{\text{mol}}$$

$$\#\text{ molecules} = 6 \times 10^{5}\,\text{molecules}$$

e. A number of regulatory proteins are present in E. coli at only one or two molecules per cell. If we assume that an E. coli contains just one molecule of a particular protein, what is the molar concentration of this protein in the cell? If the molecular weight of this protein is 40 kD, what is its concentration, expressed as mg/ml?

Answer:

$$\frac{1\,\text{molecule}}{6.023 \times 10^{23}\,\frac{\text{molecules}}{\text{mol}}} = 1.66 \times 10^{-24}\,\text{mol}$$

$$\text{Molar Concentration} = \frac{\text{moles}}{\text{volume (in liters)}} = \frac{1.66 \times 10^{-24}\,\text{mol}}{1 \times 10^{-15}\,\text{L (from b)}}$$

$$\text{Molar Concentration} = 1.66 \times 10^{-9}\,\text{M} = 1.7\,\text{nM}$$

$$\left[\text{Protein}\right] = 1.66 \times 10^{-9}\,\frac{\text{mol}}{\text{L}} \times 40{,}000\,\frac{\text{g}}{\text{mol}}$$

$$\left[\text{Protein}\right] = 6.6 \times 10^{-5}\,\frac{\text{g}}{\text{L}} = 6.6 \times 10^{-5}\,\frac{\text{mg}}{\text{ml}} \text{ or } 66\,\frac{\text{ug}}{\text{L}} \text{ or } 66\,\frac{\text{ng}}{\text{ml}}$$

f. An E. coli cell contains about 15,000 ribosomes, which carry out protein synthesis. Assuming ribosomes are spherical and have a diameter of 20 nm (nanometers), what fraction of the E. coli cell volume is occupied by ribosomes?

Answer:

$$\text{Volume of 1 ribosome} = \frac{4}{3} \times \pi \times r^3$$

$$\text{Volume of 1 ribosome} = \frac{4}{3} \times 3.14 \times \left(\frac{20 \times 10^{-9}\,\text{m}}{2}\right)^3$$

$$\text{Volume of 1 ribosome} = 4.2 \times 10^{-24}\,\text{m}^3$$

$$\text{Volume of 15,000 ribosomes} = 15,000 \times 4.2 \times 10^{-24}\,m^3 = 6.3 \times 10^{-20}\,m^3$$

$$\text{Fractional volume} = \frac{\text{Volume ribosomes}}{\text{Volume cell}}$$

$$\text{Fractional volume} = \frac{6.3 \times 10^{-20}\,m^3}{1 \times 10^{-18}\,m^3\ \text{(from b)}}$$

$$\text{Fractional volume} = 0.063 \text{ or } 6.3\%$$

g. The E. coli *chromosome is a single DNA molecule whose mass is about 3.0 x 10⁹ daltons. This macromolecule is actually a circular array of nucleotide pairs. The average molecular weight of a nucleotide pair is 660 and each pair imparts 0.34 nm to the length of the DNA molecule. What is the total length of the E. coli *chromosome? How does this length compare with the overall dimensions of an E. coli *cell? How many nucleotide pairs does this DNA contain? The average E. coli *protein is a linear chain of 360 amino acids. If three nucleotide pairs in a gene encode one amino acid in a protein, how many different proteins can the E. coli *chromosome encode? (The answer to this question is a reasonable approximation of the maximum number of different kinds of proteins that can be expected in bacteria.)*

Answer: The number of moles of base pairs in 3.0 x 10⁹ Da dsDNA is given by

$$= \frac{3.0 \times 10^9\ \dfrac{g}{\text{mol dsDNA}}}{660\ \dfrac{g}{\text{mol bp}}}$$

$$= 4.55 \times 10^6\ \frac{\text{mol bp}}{\text{mol dsDNA}}$$

$$\text{Length} = 4.55 \times 10^6\ \frac{\text{mol bp}}{\text{mol dsDNA}} \times 0.34\ \frac{\text{nm}}{\text{bp}}$$

$$\text{Length} = 4.55 \times 10^6 \times 0.34 \times 10^{-9}\,\text{m}$$

$$\text{Length} = 1.55 \times 10^{-3}\,\text{m} = 1.55\ \text{mm} = 1,550\ \mu\text{m}$$

$$\text{Length } E.\,coli\ =\ 2\ \mu\text{m}$$

$$\frac{\text{Length DNA}}{\text{Length } E.\,coli} = \frac{1,550\ \mu\text{m}}{2\ \mu\text{m}} = 775$$

To calculate the number of different proteins that would be encoded by the *E. coli* chromosome:

$$360\ \frac{\text{aa}}{\text{protein}} \times 3\ \frac{\text{bp}}{\text{aa}} = 1080\ \frac{\text{bp}}{\text{protein}}$$

$$\text{\# different proteins} = \frac{4.55 \times 10^6\,\text{bp}}{1080\ \dfrac{\text{bp}}{\text{protein}}} = 4,213\ \text{proteins}$$

The exact number can be found at NCBI (http://www.ncbi.nlm.nih.gov/). The genomes of a number of strains of *E. coli* have been sequenced but the first one was K-12 strain MG1655. At NCBI, search for MG1655 and view hits in the genome database. There should be 16 of them and NC_00913 should be one of them. Activate this link (or search for NC_00913 directly and then activate it). The returned page should indicate that this strain of *E. coli* has 4,145 protein coding genes.

4. The Dimensions of Mitochondria and Their Constituents
Assume that mitochondria are cylinders 1.5 μm in length and 0.6 μm in diameter.
a. What is the volume of a single mitochondrion?

Answer :

$$V = \pi \times r^2 \times h$$

$$V = 3.14 \times \left(3 \times 10^{-7}\,m\right)^2 \times \left(1.5 \times 10^{-6}\,m\right)$$

$$V = 4.24 \times 10^{-19}\,m^3$$

But $1\,m^3 = 10^3\,L$

$$V = 4.24 \times 10^{-19}\,m^3 \times \frac{10^3\,L}{m^3} = 4.24 \times 10^{-16}\,L = 0.424\,fL$$

b.   Oxaloacetate is an intermediate in the citric acid cycle, an important metabolic pathway localized in the mitochondria of eukaryotic cells. The concentration of oxaloacetate in mitochondria is about 0.03 μM. How many molecules of oxaloacetate are in a single mitochondrion?

Answer:

$$\# \text{ molecules} = \text{Molar concentration} \times \text{volume} \times 6.023 \times 10^{23}\,\frac{\text{molecules}}{\text{mol}}$$

$$\# \text{ molecules} = 0.03 \times 10^{-6}\,\frac{\text{mol}}{L} \times 4.24 \times 10^{-16}\,L\ \text{(from a)} \times 6.023 \times 10^{23}\,\frac{\text{molecules}}{\text{mol}}$$

$$\# \text{ molecules} = 7.66 \text{ molecules (less than 8 molecules)}$$

5. The Dimensions of Eukaryotic Cells and Their Constituents
Assume that liver cells are cuboidal in shape, 20 μm on a side.
a. How many liver cells laid end to end would fit across the diameter of a pinhead? (Assume a pinhead diameter of 0.5 mm.)

Answer:

$$\# \text{ liver cells} = \frac{0.5\,\dfrac{\text{mm}}{\text{pinhead}}}{20\,\dfrac{\mu m}{\text{cell}}}$$

$$\# \text{ liver cells} = \frac{0.5 \times 10^{-3}\,\dfrac{m}{\text{pinhead}}}{20 \times 10^{-6}\,\dfrac{m}{\text{cell}}}$$

$$\# \text{ liver cells} = 25\,\frac{\text{cells}}{\text{pinhead}}$$

b. What is the volume of a liver cell? (Assume it is a cube.)

Answer:

$$\text{Volume of cubic liver cell} = \text{length}^3 = \left(20 \times 10^{-6}\,m\right)^3$$

$$\text{Volume of cubic liver cell} = 8 \times 10^{-15}\,m^3 \times \left(\frac{100\,cm}{m}\right)^3 \times \left(\frac{1\,L}{1000\,cm^3}\right)$$

$$\text{Volume of cubic liver cell} = 8 \times 10^{-12}\,L = 8\,pL$$

c. What is the surface area of a liver cell? What is the surface-to-volume ratio of a liver cell? How does this compare to the surface-to-volume ratio of an E. coli cell? (Compare this answer to that of problem 3c.) What problems must cells with low surface-to-volume ratios confront that do not occur in cells with high surface-to-volume ratios?

Answer:

$$\text{Surface Area} = 6 \times \left(20 \times 10^{-6} \text{m}\right) \times \left(20 \times 10^{-6} \text{m}\right) = 2.4 \times 10^{-9} \text{m}^2$$

$$\frac{\text{Surface Area}}{\text{Volume}} = \frac{2.4 \times 10^{-9} \text{m}^2}{8 \times 10^{-15} \text{m}^3 \text{(from b)}}$$

$$\frac{\text{Surface Area}}{\text{Volume}} = 3.0 \times 10^5 \text{m}^{-1}$$

The surface-to-volume ratio of liver to that of *E. coli* is given by:

$$\frac{3.0 \times 10^5 \text{m}^{-1}}{6 \times 10^6 \text{m}^{-1} \text{ (from 3c)}} = 0.05 \ (1/20^{\text{th}})$$

The volume of a cell sets or determines the cell's maximum metabolic activity while the surface area defines the surface across which nutrients and metabolic waste products must pass to meet the metabolic needs of the cell. Cells with a low surface-to-volume ratio have a high metabolic capacity relative to the surface area for exchange.

d. A human liver cell contains two sets of 23 chromosomes, each set being roughly equivalent in information content. The total mass of DNA contained in these 46 enormous DNA molecules is 4 x 10^{12} daltons. Because each nucleotide pair contributes 660 daltons to the mass of DNA and 0.34 nm to the length of DNA, what is the total number of nucleotide pairs and the complete length of the DNA in a liver cell? How does this length compare with the overall dimensions of a liver cell?

Answer:

$$\text{\# base pairs} = \frac{4.0 \times 10^{12} \text{Da}}{660 \dfrac{\text{Da}}{\text{base pair}}}$$

$$\text{\# base pairs} = 6.1 \times 10^9 \text{bp}$$

$$\text{length} = 0.34 \frac{\text{nm}}{\text{bp}} \times 6.1 \times 10^9 \text{bp}$$

$$\text{length} = 2.06 \text{ m}$$

$$\text{length relative to liver cell} = \frac{2.06 \text{ m}}{20 \ \mu\text{m}} = \frac{2.06 \text{ m}}{20 \times 10^{-6} \text{ m}}$$

$$\text{length relative to liver cell} = 1.03 \times 10^5 \text{ or about 100,000 times greater!}$$

The maximal information in each set of liver cell chromosomes should be related to the number of nucleotide pairs in the chromosome set's DNA. This number can be obtained by dividing the total number of nucleotide pairs calculated above by 2. What is this value?

Answer: The maximal information is 3.0 x 10^9 bp.

If this information is expressed in proteins that average 400 amino acids in length and three nucleotide pairs encode one amino acid in a protein, how many different kinds of proteins might a liver cell be able to produce? (In reality livers cells express at most about 30,000 different proteins. Thus, a large discrepancy exists between the theoretical information content of DNA in liver cells and the amount of information actually expressed.)

Answer:

$$\# \text{ proteins} = 400\,\frac{\text{aa}}{\text{protein}} \times 3\,\frac{\text{bp}}{\text{aa}} = 1{,}200\,\frac{\text{bp}}{\text{protein}}$$

$$\# \text{ proteins} = \frac{3.0 \times 10^9\,\text{bp}}{1{,}200\,\dfrac{\text{bp}}{\text{protein}}} = 2.5 \times 10^6\,\text{proteins}$$

6. The Principle of Molecular Recognition Through Structural Complementarity
Biomolecules interact with one another through molecular surfaces that are structurally complementary. How can various proteins interact with molecules as different as simple ions, hydrophobic lipids, polar but uncharged carbohydrates, and even nucleic acids?

Answer: The amino acid side chains of proteins can participate in a number of interactions through hydrogen bonding, ionic bonding, hydrophobic interactions, and van der Waals interactions. For example, the polar amino acids, acidic amino acids and their amides, and the basic amino acids all have groups that can participate in hydrogen bonding. Those amino acid side chains that have net charge can form ionic bonds. The hydrophobic amino acids can interact with nonpolar, hydrophobic surfaces of molecules. Thus, amino acids are capable of participating in a variety of interactions. A protein can be folded in three dimensions to organize amino acids into surfaces with a range of properties.

7. The Properties of Informational Macromolecules
What structural features allow biological polymers to be informational macromolecules? Is it possible for polysaccharides to be informational macromolecules?

Answer: Biopolymers, like proteins and nucleic acids, are informational molecules because they are vectorial molecules, composed of a variety of building blocks. For example, proteins are linear chains of some 20 amino acids joined head-to-tail to produce a polymer with distinct ends. The information content is the sequence of amino acids along the polymer. Nucleic acids (DNA and RNA) are also informational molecules for the same reason. Here, the biopolymer is made up of 4 kinds of nucleotides. Monosaccharides can be linked to form polymers. When a polymer is formed from only one kind of monosaccharide, as for example in glycogen, starch, and cellulose, even though the molecule is vectorial (i.e., it has distinct ends) there is little information content. There are, however, a variety of monosaccharides and monosaccharide derivatives that are used to form polysaccharides. Furthermore, monosaccharides can be joined in a variety of ways to form branch structures. Branched polysaccharides composed of a number of different monosaccharides are rich in information.

8. The Importance of Weak Forces in Biomolecular Recognition
Why is it important that weak forces, not strong forces, mediate biomolecular recognition?

Answer: Life is a dynamic process characterized by continually changing interactions. Complementary interactions based on covalent bonding would of necessity produce static structures that would be difficult to change and slow to respond to outside stimuli.

9. Interatomic Distances in Weak Forces versus Chemical Bonds
What is the distance between the centers of two carbon atoms (their limit of approach) that are interacting through van der Waals forces? What is the distance between the centers of two carbon atoms joined in a covalent bond? (See Table 1.4)

The limit of approach of two atoms is determined by the sum of their van der Waals radii, which are given in Table 1.4. For two carbon atoms the limit of approach is (0.17 nm + 0.17 nm) 0.34 nm. The distance between the centers of two carbon atoms joined in a covalent bond is the sum of the covalent radii of the two carbons or (0.077 nm + 0.077 nm) 0.154 nm. Clearly, two carbons sharing electrons in a covalent bond are closer together than are two carbons interacting through van der Waals forces.

10. The Strength of Weak Forces Determines the Environmental Sensitivity of Living Cells
Why does the central role of weak forces in biomolecular interactions restrict living systems to a narrow range of environmental conditions?

Answer: The weak forces such as hydrogen bonds, ionic bonds, hydrophobic interactions, and van der Waals interactions can be easily overcome by low amounts of energy. Slightly elevated temperatures are sufficient to break hydrogen bonds. Changes in ionic strength, pH, concentration of particular ions, etc., all potentially have profound effects on macromolecular structures dependent on the weak forces.

11. Cells As Steady-State Systems
Describe what is meant by the phrase "cells are steady-state systems".

Answer: Life is characterized as a system through which both energy and matter flow. The consequence of energy flow in this case is order, the order of monomeric units in biopolymers, which in turn produce macromolecular structures that function together as a living cell.

12. A Simple Genome and Its Protein-Encoding Capacity
The genome of the Mycoplasma genitalium consists of 523 genes, encoding 484 proteins, in just 580,074 base pairs (Table 1.6). What fraction of the M. genitalium genes encodes proteins? What do you think the other genes encode? If the fraction of base pairs devoted to protein-coding genes is the same as the fraction of the total genes that they represent, what is the average number of base pairs per protein-coding gene? If it takes 3 base pairs to specify an amino acid in a protein, how many amino acids are found in the average M. genitalium protein? If each amino acid contributes on average 120 daltons to the mass of a protein, what is the mass of an average M. genitalium protein?

What fraction of the M. genitalium genes encodes proteins?

Answer:

$$f_{protein} = \frac{484}{523} = 0.925 \text{ or } (92.5\%)$$

What do you think the other genes encode?

Answer: The other genes likely code for ribosomal RNAs and transfer RNAs. To make a functional ribosome it takes at least three ribosomal RNAs, a small subunit rRNA, a large subunit rRNA and a 5S rRNA. To decode 61 triplet codons requires a minimum of 32* tRNAs. So, a minimum set of tRNAs and rRNAs is 35 (32 + 3). Of the 523 genes, 484 are proteins leaving 39 genes to code for RNAs.

*Essentially 2 tRNA's for each XXN triplet set except for TAN, which only requires one. This is because TAA and TAG are stop codons that require proteins for recognition. This would give 31 tRNAs but an extra one should be included for initiation of protein synthesis. In bacteria a methionine codon starts a protein-coding region and it is decoded by a special initiator tRNA, which is different from the one used at internal methionine codons.

Of the few RNAs that we are missing by this accounting one is the RNA portion of RNase P a ribonuclease involved in tRNA processing. Another is the so-called 10Sa RNA, a tRNA like RNA that is involved in decoding faulty mRNAs. The 4.5S RNA of the signal recognition particle, a complex involved in synthesis of membrane and secreted proteins is also coded in the genome. This leaves perhaps one or two RNAs unaccounted for whose functions are still unknown.

A complete listing of genes for *M. genitalium* may be found by doing a search at the NCBI web site (http://www.ncbi.nlm.nih.gov/) for this organism. You can either restrict your search to "Genome" using the pull down search menu or do a search on all databases and then inspect hits for the genome database. Information for *M. genitalium* G37 is in NC_000908. In August of 2011 the number of genes listed in this organism was 524, encoding 475 proteins. That these numbers are slightly different than those listed above emphasizes the dynamic nature of the interpretation of the genomic information.

If the fraction of base pairs devoted to protein-coding genes is the same as the fraction of the total genes that they represent, what is the average number of base pairs per protein-coding gene?

Answer: Assuming no overlap of genes, and using the numbers in the original problem:

$$\text{Amount of genome devoted to proteins} = \frac{484}{523} \times 580,074 = 536,818 \text{ bp}$$

The average number of base pairs per protein-coding gene is found by dividing this number by the number of protein genes. Thus,

$$\text{Average size of gene coding for protein} = \frac{536{,}818}{484} = 1{,}109 \ \frac{\text{bp}}{\text{protein}}$$

Note: This number is simply the genome size divided by the total number of genes.

If it takes 3 base pairs to specify an amino acid in a protein, how many amino acids are found in the average M. genitalium protein?

Answer:

$$\text{Average number of amino acids} = \frac{1{,}109}{3} = 370 \ \frac{\text{amino acids}}{\text{protein}}$$

To calculate the actual average number of amino acids in *M. genitalium* proteins, visit NC_000908 at NCBI. You will find a table summarizing this organism's genome. Activating the "Protein coding: 475" link will direct you to a table of all the proteins for *M. genitalium*. At the bottom of the page use the "Send to" pull down menu to select "Text". This will return a tab delimited text file of the information in the table. Simply copy all of it except the very first line and paste this information into an Excel spread sheet. The information presented in the "length" column is the length in codons or amino acids for all the proteins. The average of this column is 369, which is in very good agreement with the average calculated above.

If each amino acid contributes on average 120 daltons to the mass of a protein, what is the mass of an average M. genitalium protein?

Answer:

$$\text{Average protein size} = 370 \times 120 = 44{,}400 \ \text{daltons}$$

13. An Estimation of Minimal Genome Size for a Living Cell
Studies of existing cells to determine the minimum number of genes for a living cell have suggested that 206 genes are sufficient. If the ratio of protein-coding genes to non–protein-coding genes is the same in this minimal organism as the genes of Mycoplasma genitalium, how many proteins are represented in these 206 genes?

Answer: For *M. genitalium* we determined in question 12 that 92.5% of the genes of this organism are protein-coding genes. Assuming the same percentage applies to a minimum set of genes then 191 of the 206 genes are protein-coding genes.

$$\text{Protein-coding genes} = 0.925 \times 206 = 190.6 = 191$$

How many base pairs would be required to form the genome of this minimal organism if the genes are the same size as M. genitalium genes?

Answer: In question 12 we were told that 580,074 base pairs code for 523 genes. The genome size required to code for 206 genes is calculated as follows:

$$\frac{580{,}074}{523} = \frac{x}{206}$$

$$x = 206 \times \frac{580{,}074}{523} = 228{,}480$$

Note: This calculation assumes that genes essentially do not overlap. A smaller genome size could be possible by allowing overlapping, but this would constrain the protein sequences.

14. An Estimation of the Number of Genes in a Virus
Virus genomes range in size from approximately 3500 nucleotides to approximately 280,000 base pairs. If viral genes are about the same size as M. genitalium genes, what is the minimum and maximum number of genes in viruses?

Answer: In question 12 we determined that the average gene size in *M. genitalium* is 1109 (the genome size - 580,074- divided by the number of genes -523). Applying this average gene size to viral genomes we find:

$$\text{Minimum number of viral genes} = \frac{3,500}{1,109} = 3.15 \approx 3 \text{ genes}$$

$$\text{Maximum number of viral genes} = \frac{280,000}{1,109} = 252 \text{ genes}$$

15. Intracellular Transport of Proteins

The endoplasmic reticulum (ER) is a site of protein synthesis. Proteins made by ribosomes associated with the ER may pass into the ER membrane or enter the lumen of the ER. Devise a pathway by which:
a. a plasma membrane protein may reach the plasma membrane.
b. a secreted protein may be deposited outside the cell.

Protein synthesis starts out on ribosomes located in the cytoplasm of cells. Proteins destined to be excreted or to become membrane proteins are synthesized with a signal sequence located near the N-terminus of the protein. (Protein synthesis begins at the N-terminus.) This signal sequence directs the ribosome to the endoplasmic reticulum where the ribosome docks with the reticular membrane. Endoplasmic reticulum studded with ribosomes is called rough endoplasmic reticulum. The signal-sequence-containing protein is synthesized by rough endoplasmic reticulum-bound ribosomes that synthesize the protein and simultaneously export it into the lumen of the endoplasmic reticulum. For a protein to be transported to the plasma membrane it must be packaged into membrane vesicles in the endoplasmic reticulum since the reticular membrane is separate from the plasma membrane. Vesicles from the endoplasmic reticulum containing membrane proteins do not, however, move directly to the plasma membrane. Rather they are routed to the Golgi apparatus where a variety of post-translational modifications occur. Once proteins move through the Golgi they are repackaged into vesicles that are directed to the plasma membrane. Secreted proteins follow the same pathway. Both membrane proteins and excreted proteins contain signal sequences that get them into the endoplasmic reticulum. Membrane proteins contain an additional domain or domains that are hydrophobic in nature and anchor the proteins into the reticular membrane.

Preparing for the MCAT® Exam

16. Biological molecules often interact via weak forces (H bonds, van der Waals interactions, etc.). What would be the effect of an increase in kinetic energy on such interactions?

Answer: Weak forces are easily disrupted by increases in the kinetic energies of the interacting components. Thus, slight increases in temperature can disrupt weak forces. Biological molecules, like proteins whose three-dimensional structures are often determined by weak force interactions, may undergo conformational changes even with modest changes in temperature leading to inactivation or loss of function.

17. Proteins and nucleic acids are informational macromolecules. What are the two minimal criteria for a linear informational polymer?

Answer: Informational macromolecules must be directional (vectorial) and they must be composed of unique building blocks. Both nucleic acids and proteins are directional polymers. The directionality of a single nucleic acid is 5' to 3' whereas that of a protein is N-terminus to C-terminus. The repeat units in nucleic acid polymers are four different nucleoside monophosphates. The repeat units in proteins are 20 amino acids. The information content of a nucleic acid, especially dsDNA, is its linear sequence. The same is true for proteins; however, proteins typically fold into unique three-dimensional structures, which show biological activity.

Additional Problems

1. Silicon is located below carbon in the periodic chart. It is capable of forming a wide range of bonds similar to carbon yet life is based on carbon chemistry. Why are biomolecules made of silicon unlikely?

2. Identify the following characters of the Greek alphabet: $\alpha, \beta, \gamma, \delta, \Delta, \varepsilon, \zeta, \theta, \kappa, \lambda, \mu, \nu, \pi, \rho, \sigma, \Sigma, \tau, \chi, \phi, \psi$ and ω.

3. Give a common example of each of the weak forces at work.

4. On a hot dry day, leafy plants may begin to wilt. Why?

Abbreviated Answers

1. Covalent silicon bonds are not quite as strong as carbon covalent bonds because the bonding electrons of silicon are shielded from the nucleus by an additional layer of electrons. In addition, silicon is over twice the weight of carbon. Also, silicon oxides (rocks, glass) are extremely stable and not as reactive as carbon.

2. These Greek letters are commonly used in biochemistry but this set is not the complete Greek alphabet. alpha (α), beta (β), gamma (γ), delta (δ), capital delta (Δ), epsilon (ε), zeta (ζ), theta (θ), kappa (κ), lambda (λ), mu (μ), nu (ν), pi (π), rho (ρ), sigma (σ), capital sigma (Σ), tau (τ), chi (χ), phi (ϕ), psi (ψ), and omega (ω), the last letter of the Greek alphabet.

3. Ice is an example of a structure held together by hydrogen bonds. Sodium and chloride ions are joined by ionic bonds in table salt crystals. A stick of butter is a solid at room temperature because of van der Waals forces. The energetically unfavorable interactions between water and oil molecules cause the oil to coalesce.

4. The tonoplast loses water and begins to shrink causing the plant cell membrane to exert less pressure on the cell wall.

Summary

The chapter begins with an outline of the fundamental properties of living systems: complexity and organization, biological structure and function, energy transduction, and self-replication. What are the underlying chemical principles responsible for these properties? The elemental composition of biomolecules is dominated by hydrogen, carbon, nitrogen and oxygen. These are the lightest elements capable of forming strong covalent bonds. In particular, carbon plays a key role serving as the backbone element of all biomolecules. It can participate in as many as four covalent bonds arranged in tetrahedral geometry and can produce a variety of structures including linear, branched, and cyclic compounds.

The four elements are incorporated into biomolecules from precursor compounds: CO_2, NH_4^+, NO_3^- and N_2. These precursors are used to construct more complex compounds such as amino acids, sugars, and nucleotides, which serve as building blocks for the biopolymers; proteins, polysaccharides, and nucleic acids, as well as fatty acids and glycerol which are the building blocks of lipids. These complex macromolecules are organized into supramolecular complexes such as membranes and ribosomes that are components of cells, the fundamental units of life.

Proteins, nucleic acids and polysaccharides are biopolymers with structural polarity due to head-to-tail arrangements of asymmetric building block molecules. In these biopolymers, the building blocks are held together by covalent bonds, but they assume an elaborate architecture due to weak, noncovalent forces such as van der Waals interactions, hydrogen bonds, ionic bonds and hydrophobic interactions. The three-dimensional shape is important for biological function, especially for proteins. At extreme conditions such as high temperature, high pressure, high salt concentrations, extremes of pH, and so on, the weak forces may be disrupted, resulting in loss of both shape and function in a process known as denaturation. Thus, life is confined to a narrow range of conditions.

Life demands a flow of energy during which energy transductions occur in the organized, orderly, small, manageable steps of metabolism, each step catalyzed by enzymes.

The fundamental unit of life is the cell. There are two types: eukaryotic cells with a nucleus and prokaryotic cells without a nucleus. Prokaryotes are divided into two groups, eubacteria and archaea. All cells contain ribosomes, which are responsible for protein synthesis; however, prokaryotic cells contain little else in the way of subcellular structures. Eukaryotic cells, found in plants, animals, and fungi, contain an array of membrane-bound compartments or organelles, including a nucleus, mitochondria, chloroplasts, endoplasmic reticulum, Golgi apparatus, vacuoles, lysosomes, and perixosomes. Organelles are internal compartments in which particular metabolic processes are carried out.

Chapter 2

Water: The Medium of Life

• •

Chapter Outline

❖ Properties of water: High boiling point, high melting point, high heat of vaporization, high surface tension, high dielectric constant, maximum density as a liquid: All due to ability of water to hydrogen bond
❖ Water structure
  ⌁ Electronegative oxygen, two hydrogens: Nonlinear arrangement: Dipole
  ⌁ Two lone pairs on oxygen: H-bond acceptors
  ⌁ Partially positively charged hydrogens: H-bond donors
❖ Ice
  ⌁ Lattice with each water interacting with 4 neighboring waters
  ⌁ H-bonds: Directional, straight and stable
❖ Liquid: H-bonds present but less than 4 and transient
❖ Solvent properties of water
  ⌁ High dielectric constant decreases strength of ionic interactions between other molecules
    · Force of ionic interaction, $F = e_1e_2/Dr^2$, inversely dependent on D
    · Salts dissolve in water
  ⌁ Interaction with polar solutes through H-bonds
  ⌁ Hydrophobic interactions: Entropy-driven process minimizes solvation cage
  ⌁ Amphiphilic molecules: Polar and nonpolar groups
❖ Colligative properties: Freezing point depression, boiling point elevation, lowering of vapor pressure, osmotic pressure effects: Depend on solute particles per volume
❖ Ionization of water
  ⌁ Ions: hydrogen ion H^+ (protons), hydroxyl ion OH^-, hydronium ion H_3O^+ (protonated water)
  ⌁ Ion product: $K_w = [H_2O] \times K_{eq} = 55.5 \times K_{eq} = 10^{-14} = [H^+][OH^-]$
  ⌁ $pH = -\log_{10}[H^+]$, $pOH = -\log_{10}[OH^-]$, $pH + pOH = 14$
❖ Strong electrolytes: Completely dissociate: Salts, strong acids, strong bases
❖ Weak electrolytes: Do not fully dissociate: Hydrogen ion buffers
❖ Buffers
  ⌁ Henderson-Hasselbalch equation: $pH = pK_a + \log_{10}([A^-]/[HA])$
  ⌁ Biological buffers: Phosphoric acid ($pK_1 = 2.15$, $pK_2 = 7.2$, $pK_3 = 12.4$); histidine ($pK_a = 6.04$); bicarbonate ($pK_{overall} = 6.1$)
  ⌁ "Good" buffers: pK_a's in physiological pH range and not influenced by divalent cations

Chapter Objectives

Water

Its properties arise because of the ability of water molecules to form H bonds and to dissociate to H^+ and OH^-. Thus, water is a good solvent, has a high heat capacity and a high dielectric constant.

Acid-Base Problems

For acid-base problems the key points to remember are:

Henderson-Hasselbalch: $pH = pK_a + \log([A^-]/[HA])$

13

Conservation of acid and conjugate base:

$$[A^-] + [HA] = \text{Total concentration of weak electrolyte added.}$$

Conservation of charge:

$$\sum [\text{cations}] = \sum [\text{anions}] \text{ i.e., the sum of the cations must equal the sum of the anions.}$$

In many cases, simplifications can be made to this equation. For example, $[OH^-]$ or $[H^+]$ may be small relative to other terms and ignored in the equation. For strong acids, it can be assumed that the concentration of the conjugate base is equal to the total concentration of the acid. For example, an x M solution of HCl is x M in Cl^-. Likewise for x M NaOH, the $[Na^+]$ is x M. In polyprotic buffers (e.g., phosphate, citrate, etc.), the group with the pK_a closest to the pH under study will have to be analyzed using the Henderson-Hasselbalch equation. For groups with pK_as 2 or more pH units away from the pH, they are either completely protonated or unprotonated.

The solution to a quadratic equation of the form, $y = ax^2 + bx + c$ is:

$$x = \frac{-b \pm \sqrt{b^2 - 4ac}}{2a}$$

Problems and Solutions

1. Calculating pH from [H⁺]
Calculate the pH of the following.
a. 5 x 10⁻⁴ M HCl

Answer: HCl is a strong acid and fully dissociates into $[H^+]$ and $[Cl^-]$.
Thus, $[H^+] = [Cl^-] = [HCl]_{total\ added}$

$$pH = -\log_{10}[H^+] = -\log_{10}[HCl_{total}]$$
$$= -\log_{10}(5 \times 10^{-4}) = 3.3$$

b. 7 x 10⁻⁵ M NaOH

Answer: For strong bases like NaOH and KOH,

$$pH = 14 + \log_{10}[\text{Base}]$$
$$= 14 + \log_{10}(7 \times 10^{-5}) = 9.85$$

c. 2 μM HCl

Answer:

$$pH = -\log_{10}[H^+] = -\log_{10}[HCl_{total}]$$
$$= -\log_{10}(2 \times 10^{-6}) = 5.70$$

d. 3 x 10⁻² M KOH

Answer:

$$pH = 14 + \log_{10}(3 \times 10^{-2}) = 12.5$$

e. 0.04 mM HCl

Answer:

$$pH = -\log_{10}[H^+] = -\log_{10}[HCl_{total}]$$
$$= -\log_{10}(0.04 \times 10^{-3}) = 4.4$$

f. 6 x 10⁻⁹ M HCl= 0.06 x 10⁻⁷ M HCl

Answer: Beware! Naively one might fall into the trap of simply treating this like another strong acid problem and solving it like so:

$$pH = -\log_{10}[H^+] = -\log_{10}[HCl_{total}]$$
$$= -\log_{10}(6 \times 10^{-9}) = 8.22$$

However, something is odd. This answer suggests that addition of a small amount of a strong acid to water will give rise to a basic pH! What we have ignored is the fact that water itself will contribute H^+ into solution so we must consider the ionization of water as well. There are two approaches we can take in solving this problem. As a close approximation we can assume that:

$$[H^+] = 10^{-7} + [HCl] \text{ or,}$$
$$[H^+] = 10^{-7} + 6 \times 10^{-9} = 10^{-7} + 0.06 \times 10^{-7} = 1.06 \times 10^{-7}$$
$$pH = -\log_{10}(1.06 \times 10^{-7}) = 6.97$$

The exact solution uses the ion product of water.

$$[H^+][OH^-] = K_W = 10^{-14} \quad \text{(the ion product of water)} \quad (1)$$

In solution HC1 fully dissociates into $H^+ + C1^-$. For any solution the sum of negative and positive charges must be equal. Since we are dealing with monovalent ion we can write:

$$[H^+] = [Cl^-] + [OH^-] \quad (2)$$

Now because HC1 is fully dissociated:

$$[Cl^-] = [HCl_{total}] = 6 \times 10^{-9} \quad (3)$$

Substituting this into equation (2), solving equation (2) for $[OH^-]$ and substituting in equation (1) we have a quadratic equation in H^+:

$$[H^+]^2 - 6 \times 10^{-9}[H^+] - 10^{-14} = 0$$

whose general solution is given by

$$[H^+] = \frac{-b \pm \sqrt{b^2 - 4ac}}{2a} = \frac{6 \times 10^{-9} \pm \sqrt{(6 \times 10^{-9})^2 + 4 \times 10^{-14}}}{2}$$

Before firing up the calculator a little reflection suggests that the argument under the square root is dominated by 4×10^{-14} whose root is 2×10^{-7}. Furthermore, of the two solutions (i.e., ±) it must be the + solution (- given rises to a negative $[H^+]$!)
Therefore,

$$[H^+] = \frac{6 \times 10^{-9} + 2 \times 10^{-7} + [OH^-]}{2}$$

and , $pH = -\log_{10}[H^+] = 6.99$

2. Calculating [H⁺] from pH
Calculate the pH or pOH of the following.
a. [H⁺] in vinegar

Answer:

From Table 2.3 we find that pH=2.9
since pH=$-\log_{10}[H^+]$
$[H^+] = 10^{-pH} = 10^{-2.9} = 1.26 \times 10^{-3}M = 1.26$ mM

b. [H⁺] in saliva

Answer:

From Table 2.3 we find that in saliva pH=6.6
since pH=$-\log_{10}[H^+]$
$[H^+] = 10^{-pH} = 10^{-6.6} = 2.5 \times 10^{-7}M = 0.25$ μM

c. [H⁺] in household ammonia

Answer:

The pH of ammonia is 11.4 thus
$$[H^+]=10^{-pH}=10^{-11.4}=4 \times 10^{-12}M = 4 \text{ pM}$$

d. [OH⁻] in milk of magnesia

Answer:

The pH of milk of magnesia is 10.3.
From pH + pOH = 14,
pOH=14 - 10.3 = 3.7
$$[OH^-]=10^{-pOH}=10^{-3.7}=2\times10^{-4}M = 0.2 \text{ mM}$$

e. [OH⁻] in beer

Answer:

The pH of beer is 4.5.
From pH + pOH=14,
pOH=14 - 4.5=9.5
$$[OH^-]=10^{-pOH}=10^{-9.5}=3.16\times10^{-10}M = 0.316 \text{ nM}$$

f. [H⁺] inside a liver cell

Answer:

The pH of a liver cell is 6.9 thus
$$[H^+]=10^{-pH}=10^{-6.9}=1.26\times10^{-7}M=0.126 \text{ μM}$$

3. Calculating [H⁺] and pKₐ from the pH of a solution of weak acid
The pH of a 0.02 M solution of an acid was measured at 4.6.
a. What is the [H⁺] in this solution?

Answer:

$$[H^+]=10^{-pH}=10^{-4.6}=2.5\times10^{-5}M = 25 \text{ μM}$$

b. Calculate the acid dissociation constant Kₐ and pKₐ for this acid.

Answer:

$$HA \rightleftharpoons H^+ + A^-$$
$$K_a = \frac{[H^+][A^-]}{[HA]}$$

Assume that a small amount, *x*, of HA dissociates into equal molar amounts of H⁺ and A⁻. We then have:

$$HA \rightarrow H^+ + A^- \text{ or,}$$
$$0.02 - x \leftrightarrow x + x$$

From (a) we know that $[H^+] = 25 \text{ μM} = x = A^-$

And, $[HA] = 0.02\text{-}25\times10^{-6} \approx 0.02$

Thus, $K_a = \dfrac{(25\times10^{-6})^2}{0.02} = \dfrac{625\times10^{-12}}{0.02} = 3.13\times10^{-8} \text{ M}$

$$pK_a = -\log_{10}(3.13\times10^{-8}) = 7.5$$

4. Calculating the pH of a solution of a weak acid; calculating the pH of the solution after the addition of strong base
The Kₐ for formic acid is 1.78 × 10⁻⁴ M.
a. What is the pH of a 0.1 M solution of formic acid?

Answer:

$$pH = pK_a + \log_{10} \frac{[A^-]}{[HA]} \text{ or } [H^+] = K_a \frac{[HA]}{[A^-]} \text{ (1)}$$

For formic acid, $[H^+] \approx [A^-]$ (2)

and $[HA] + [A^-] = 0.1$ M or

$[HA] = 0.1 - [A^-]$ (3)

and, using equation (2) we can write

$[HA] = 0.1 - [H^+]$

Substituting this equation and (2) into (1) we find:

$$[H^+] = K_a \frac{0.1 - [H^+]}{[H^+]} \text{ or}$$

$[H^+]^2 + K_a[H^+] - 0.1K_a = 0$, a quadratic whose solutions are

$$[H^+] = \frac{-K_a \pm \sqrt{K_a^2 + 0.4K_a}}{2}$$

The argument under the square root sign is greater than K_a. Therefore, the correct solution is the positive root. Further K_a^2 is small relative to $0.4K_a$ and can be ignored.

$$[H^+] = \frac{-K_a + \sqrt{0.4K_a}}{2} = \frac{1.78 \times 10^{-4} + \sqrt{0.4 \times 1.78 \times 10^{-4}K_a}}{2}$$

$[H^+] = 0.00413$ M

$pH = -\log_{10}[H^+] = 2.38$

b. 150 ml of 0.1 M NaOH is added to 200 ml of 0.1 M formic acid, and water is added to give a final volume of 1 L. What is the pH of the final solution?

Answer: The total concentration of formic acid is:

$$[HA_{total}] = \frac{0.2 \text{ L} \times 0.1 \text{ M}}{1 \text{ L}} = 0.02 \text{ M}$$

When 150 ml of 0.1 M NaOH is added its total concentration is:

$$[NaOH] = \frac{0.15 \text{ L} \times 0.1 \text{ M}}{1 \text{ L}} = 0.015 \text{ M}$$

Since NaOH is a strong base it will fully dissociate into equal amounts of Na^+ and OH^-. The OH^- will react with an equivalent number of free protons and to compensate for loss of protons the protonated form of formic acid, i.e., HA, will dissociate. The final concentration of HA is found as follows:

$[HA] = [HA_{total}] - [OH^-] = 0.02 \text{ M} - 0.015 \text{ M} = 0.0005 \text{ M}$ and,

$[A^-] = 0.015$ M

From the Henderson-Hasselbalch equation we have:

$$pH = pK_a + \log_{10} \frac{[A^-]}{[HA]}$$

$$pH = -\log_{10}(1.78 \times 10^{-4}) + \log_{10} \frac{0.015}{0.005}$$

$$pH = 3.75 + \log_{10} \frac{0.015}{0.005}$$

$$pH = 4.23$$

5. Prepare a buffer by combining a solution of weak acid with a solution of the salt of the weak acid Given 0.1 M solutions of acetic acid and sodium acetate, describe the preparation of 1 L of 0.1 M acetate buffer at a pH of 5.4.

Answer: From the Henderson-Hasselbalch equation, i.e.,

$$pH = pK_a + \log_{10} \frac{[A^-]}{[HA]}$$

$$\frac{[A^-]}{[HA]} = 10^{(pH-pK)} = 10^{(5.4-4.76)} = 10^{0.64} = 4.37 \ (1)$$

Further, we want

$$[HA] + [A^-] = 0.1 \ M \ (2)$$

Solving equation (1) for $[A^-]$ and substituting into (2) we find:

$$[HA] + 4.37 \times [HA] = 5.37 \times [HA] = 0.1 \ M$$

$$[HA] = 0.0186 \ M$$

Substituting this value of $[HA]$ into (2), we find that $[A^-] = 0.0814$

Therefore, combine 186 ml 0.1 M acetic acid with 814 ml 0.1 M sodium acetate.

6. Calculate the $HPO_4^{2-}/H_2PO_4^-$ in a muscle cell from the pH
If the internal pH of a muscle cell is 6.8, what is the $[HPO_4^{2-}]/[H_2PO_4^-]$ ratio in this cell?

Answer: The dissociation of phosphoric acid proceeds as follows:

$$H_3PO_4 \ \square \ H^+ + H_2PO_4^- \ \square \ H^+ + HPO_4^{2-} \ \square \ H^+ + PO_4^{3-}$$

Each dissociation has the following pK_a values: 2.15, 7.20 and 12.40.

Now, at pH = 6.8, we expect the first equilibrium to be completely to the right and the last equilibrium to be to the left (i.e., we expect phosphoric acid to be in the doubly or singly protonated forms).

From the Henderson-Hasselbalch equations i.e.,

$$pH = pK_a + \log_{10} \frac{[A^-]}{[HA]} \text{ where}$$

$$[A^-] = [HPO_4^{2-}]; [HA] = [H_2PO_4^-]$$

$$\log_{10} \frac{[HPO_4^{2-}]}{[H_2PO_4^-]} = pH - pK_a$$

$$\frac{[HPO_4^{2-}]}{[H_2PO_4^-]} = 10^{(pH-pK_a)} = 10^{(6.8-7.2)} = 0.398$$

7. Given 0.1 M solutions of Na_3PO_4 and H_3PO_4, describe the preparation of 1 L of a phosphate buffer at a pH of 7.5. What are the molar concentrations of the ions in the final buffer solution, including Na^+ and H^+?

Answer: (See preceding two problems)

$$\frac{[HPO_4^{2-}]}{[H_2PO_4^-]} = 10^{(pH-pK_a)} = 10^{(7.5-7.2)} = 2.0 \text{ and,}$$

$$[HPO_4^{2-}] + [H_2PO_4^-] = 0.1 \ M \text{ so,}$$

$$[H_2PO_4^-] = 0.0333 \ M \text{ and } [HPO_4^{2-}] = 0.0667 \ M$$

For charge neutrality

$$[H^+] + [Na^+] = 2 \times [HPO_4^{2-}] + [H_2PO_4^-] \text{ or}$$

$$[Na^+] \approx 2 \times [HPO_4^{2-}] + [H_2PO_4^-]$$

$$[Na^+] = 2 \times 0.0667 \ M + 0.0333 \ M = 0.1667 \ M$$

Na^+ comes from the Na_3PO_4 solution, where it is 0.3 M. (Note: The solution is 0.1 M in Na_3PO_4, which dissociates into 3 Na^+ and 1 PO_4^{3-}.) To make the solution we need to add enough Na_3PO_4 to result in 0.1667 M Na^+. This is calculated as follows:

$$0.3 \ M \times x = 0.1667 \ M \times 1000 \ ml$$

$$x = 555.7 \ ml$$

Where x is the volume in ml of tribasic sodium phosphate. This will contribute 0.05557 M of phosphate to the solution. The final phosphate concentration must be 0.1 M and so the remainder must come from phosphoric acid. Therefore, 444.3 mL of phosphoric acid must be added.

The final solution will be:

$$0.0667 \text{ M in } [HPO_4^{2-}]$$

$$0.0333 \text{ M in } [H_2PO_4^-]$$

$$0.1667 \text{ M in } [Na^+]$$

$$3.16 \times 10^{-8} \text{M in } [H^+]$$

8. Polyprotic acids: Phosphate species abundance at different pHs
What are the approximate fractional concentrations of the following phosphate species at pH values of 0, 2, 4, 6, 8, 10, and 12?
a. H_3PO_4 b. $H_2PO_4^-$ c. HPO_4^{2-} d. PO_4^{3-}

Answer: For phosphoric acid the following equilibria apply:

$$H_3PO_4 \; \rightleftarrows \; H^+ + H_2PO_4^- \; \rightleftarrows \; H^+ + HPO_4^{2-} \; \rightleftarrows \; H^+ + PO_4^{3-}$$

$$2.15 \qquad\qquad 7.20 \qquad\qquad 12.40$$

The Henderson-Hasselbalch equation may be used to calculate the ratio of any two species that differ by a proton. Thus,

$$pH = pK_a + \log_{10} \frac{[A^-]}{[HA]} \quad \text{or}$$

$$\frac{[A^-]}{[HA]} = 10^{(pH - pK_a)}$$

By applying this equation at a pH value and at the 3 pKa's we can calculate the following ratios:

$$\frac{[H_2PO_4^-]}{[H_3PO_4]} = 10^{(pH-2.15)}, \frac{[HPO_4^{2-}]}{[H_2PO_4^-]} = 10^{(pH-7.20)}, \frac{[PO_4^{3-}]}{[HPO_4^{2-}]} = 10^{(pH-12.40)}$$

or

$$\frac{[H_2PO_4^-]}{[H_3PO_4]} = x, \frac{[HPO_4^{2-}]}{[H_2PO_4^-]} = y, \frac{[PO_4^{3-}]}{[HPO_4^{2-}]} = z$$

The fraction of any one species, at a particular pH, is its concentration divided by the sum of the concentrations of all of the species. For example,

$$f_{H_3PO_4} = \frac{[H_3PO_4]}{[H_3PO_4] + [H_2PO_4^-] + [HPO_4^{2-}] + [PO_4^{3-}]}$$

This fraction can be written as a function of x, y and z as follows:

$$f_{H_3PO_4} = \frac{[H_3PO_4]}{[H_3PO_4] + x[H_3PO_4] + x \times y[H_3PO_4] + x \times y \times z[H_3PO_4]}$$

or

$$f_{H_3PO_4} = \frac{1}{1 + x + x \times y + x \times y \times z}$$

Using this equation we can evaluate the fraction of the fully protonated species at each of the pH values given. For the other species, we can take a similar approach to find expressions for their fractional value as a function of x, y and z.

$$f_{H_3PO_4} = \frac{1}{1 + x + (x \times y) + (x \times y \times z)}$$

$$f_{H_2PO_4^-} = \frac{1}{(1/x) + 1 + y + (y \times z)}$$

$$f_{HPO_4^{2-}} = \frac{1}{1/(x \times y) + 1/y + 1 + z}$$

$$f_{PO_4^{3-}} = \frac{1}{1/(x \times y \times z) + 1/(x \times y) + 1/z + 1}$$

The values of x, y and z may be calculated by hand. A more efficient approach would be to use a spreadsheet to evaluate x, y and z at the pH values given and then to calculate the corresponding fractions at each of the pH values. The following two tables give this information (to three decimal places).

pH	x	y	z
0	0.007	0.000	0.000
2	0.708	0.000	0.000
4	70.795	0.001	0.000
6	7079.458	0.063	0.000
8	707945.784	6.310	0.000
10	70794578.438	630.957	0.004
12	7079457843.841	63095.734	0.398

pH	f_{H3P}	f_{H2P}	f_{HP}	f_P
0	0.993	0.007	0.000	0.000
2	0.585	0.415	0.000	0.000
4	0.014	0.985	0.001	0.000
6	0.000	0.941	0.059	0.000
8	0.000	0.137	0.863	0.000
10	0.000	0.002	0.994	0.004
12	0.000	0.000	0.715	0.285

Note: This is the exact solution. A good approximation is to evaluate the Henderson-Hasselbalch for species whose pK_a are nearest the pH under consideration. For example, for pH 0, 2 and 4, evaluation of the Henderson-Hasselbalch for $pK_a = 2.15$ would give answers close to those presented above.

9. Polyprotic acids: Citric acid species at various pHs
Citric acid, a tricarboxylic acid important in intermediary metabolism, can be symbolized as H_3A. Its dissociation reactions are

$$H_3A \rightleftharpoons H^+ + H_2A^- \quad pK_{a1} = 3.13$$
$$H_2A^- \rightleftharpoons H^+ + HA^{2-} \quad pK_{a2} = 4.76$$
$$HA^{2-} \rightleftharpoons H^+ + A^{3-} \quad pK_{a3} = 6.40$$

If the total concentration of the acid and its anion forms is 0.02 M, what are the individual concentrations of H_3A, H_2A^-, HA^{2-}, and A^{3-} at pH 5.2?

Answer: For citric acid

$$H_3A \rightleftharpoons H^+ + H_2A^- \rightleftharpoons H^+ + HA^{2-} \rightleftharpoons H^+ + A^{3-}$$
$$3.13 \qquad 4.76 \qquad 6.40$$

At pH = 5.2 the predominant equilibrium will involve H_2A^- and HA^{2-}.

$$pH = pK_a + \log_{10}\frac{[HA^{2-}]}{[H_2A^-]}$$

$$\frac{[HA^{2-}]}{[H_2A^-]} = 10^{(pH-pK_a)} = 10^{(5.2-4.76)} = 2.754 \ (1)$$

And, considering the other two equilibria we have

$$\frac{[A^{3-}]}{[HA^{2-}]} = 10^{(pH-pK_a)} = 10^{(5.2-6.4)} = 0.063 \ (2) \text{ and}$$

$$\frac{[H_2A^-]}{[H_3A]} = 10^{(pH-pK_a)} = 10^{(5.2-3.13)} = 117.5 \ (3)$$

These terms are related as follows:

$$[H_3A] + [H_2A^-] + [HA^{2-}] + [A^{3-}] = 0.02 \ (4)$$

Using equations (1), (2) and (3) we can relate the concentration of any one species to any other.

$$[HA^{2-}] = 2.754 \times [H_2A^-] \text{ from (1), and}$$

$$[A^{3-}] = 0.063 \times [HA^{2-}] \text{ from (2), or substituting from above}$$

$$[A^{3-}] = 0.063 \times 2.754 \times [H_2A^-] = 0.174 \times [H_2A^-]$$

And,

$$[H_3A] = \frac{[H_2A^-]}{117.5} = 0.00851 \times [H_2A^-] \text{ from (3)}$$

Substituting these expressions into (4), we find:

$$0.00851 \times [H_2A^-] + [H_2A^-] + 2.754 \times [H_2A^-] + 0.174 \times [H_2A^-] = 0.02 \text{ M or}$$

$$3.937 \times [H_2A^-] = 0.02 \text{ M or}$$

$$[H_2A^-] = \frac{0.02}{3.937} = 0.00508 \text{ M} = 5.08 \text{ mM}$$

And, substituting this value back into equations (1), (2), and (3) we find:

From (1) $[HA^{2-}] = 2.754 \times [H_2A^-] = 2.754 \times 5.08 \text{ mM} = 14.00 \text{ mM}$

From (2) $[A^{3-}] = 0.174 \times [H_2A^-] = 0.174 \times 0.0051 \text{ M} = 0.88 \text{ mM}$

From (3) $[H_3A] = 0.009 \times [H_2A^-] = 0.00851 \times 0.0051 \text{ M} = 4.34 \times 10^{-5} \text{ M} = 0.04 \text{ mM}$

We could have anticipated these results because the pH is far from two of the pK$_a$s. Only the equilibrium between H$_2$A$^-$ and HA^{2-} with a pK$_a$ = 4.76 will be significant at pH = 5.2.

10. Calculate the pH change in a phosphate buffer when acid or base is added
a. If 50 ml of 0.01 M HCl is added to 100 ml of 0.05 M phosphate buffer at pH 7.2, what is the resultant pH? What are the concentrations of H₂PO₄⁻ and HPO₄²⁻ in the final solution?

Answer: The relevant pK$_a$ for phosphoric acid is 7.2 governing the following equilibrium,

$$H_2PO_4^- \rightleftarrows H^+ + HPO_4^{2-}$$

From the Henderson-Hasselbalch equation we find that

$$pH = pK_a + \log_{10}\frac{[HPO_4^{2-}]}{[H_2PO_4^-]} \text{ or}$$

$$[HPO_4^{2-}] = [H_2PO_4^-] \times 10^{(pH-pK_a)} = [H_2PO_4^-] \times 10^{(7.20-7.20)} = [H_2PO_4^-]$$

And, since $[HPO_4^{2-}] + [H_2PO_4^-] = 0.05 \text{ M}$

$$[HPO_4^{2-}] = [H_2PO_4^-] = 0.025 \text{ M}$$

In 100 ml we have $100 \text{ ml} \times \frac{1 \text{ L}}{1000 \text{ ml}} \times 0.025 \text{ M} = 0.0025 \text{ mol of each.}$

Now, addition of 50 ml of 0.1 M HCl accomplishes two things:
(1) It dilutes the solution; and,
(2) It introduces protons that will convert HPO₄²⁻ to H₂PO₄⁻.
The moles of protons is given by:

$$50 \text{ ml} \times \frac{1 \text{ L}}{1000 \text{ ml}} \times 0.01 \text{ M} = 0.0005 \text{ mole}$$

Thus, 0.0005 mol of HPO_4^{2-} will be converted to $H_2PO_4^-$ or

There will be 0.0025-0.0005 mole HPO_4^{2-} and 0.0025 + 0.0005 mole $H_2PO_4^-$

$$pH = pK_a + \log_{10}\frac{[HPO_4^{2-}]}{[H_2PO_4^-]} = 7.2 + \log_{10}\frac{0.0025 - 0.0005}{0.0025 + 0.0005} = 7.02$$

And

$$[HPO_4^{2-}] = \frac{(0.0025 - 0.0005) \text{ mol}}{100 \text{ ml} + 50 \text{ ml}} \times \frac{1000 \text{ ml}}{1 \text{ L}} = 0.0133 \text{ M}$$

$$[H_2PO_4^-] = \frac{(0.0025 + 0.0005) \text{ mol}}{100 \text{ml} + 50 \text{ ml}} \times \frac{1000 \text{ ml}}{1 \text{ L}} = 0.0200 \text{ M}$$

Note: This amount of acid added to 100 ml of water gives pH = 2.5.

b. If 50 ml of 0.01 M NaOH is added to 100 ml of 0.05 M phosphate buffer at pH 7.2, what is the resultant pH? What are the concentrations of H₂PO₄⁻ and HPO₄²⁻ in this final solution?

Answer: For NaOH, the same equations apply with one important difference: HPO_4^{2-} is increased and $H_2PO_4^-$ is decreased. Thus,

$$pH = pK_a + \log_{10} \frac{[HPO_4^{2-}]}{[H_2PO_4^-]} = 7.2 + \log_{10} \frac{0.0025 + 0.0005}{0.0025 - 0.0005} = 7.38$$

And

$$[H_2PO_4^-] = \frac{(0.0025 - 0.0005) \text{ mol}}{100 \text{ ml} + 50 \text{ ml}} \times \frac{1000 \text{ ml}}{1 \text{ L}} = 0.0133 \text{ M}$$

$$[HPO_4^{2-}] = \frac{(0.0025 + 0.0005) \text{ mol}}{100 \text{ ml} + 50 \text{ ml}} \times \frac{1000 \text{ ml}}{1 \text{ L}} = 0.0200 \text{ M}$$

Note: If added to water instead, this amount of NaOH would result in a solution with pH = 11.5.

11. Explore the bicarbonate/carbonic acid buffering system of blood plasma
At 37°C, if the plasma pH is 7.4 and the plasma concentration of HCO₃⁻ is 15 mM, what is the plasma concentration of H₂CO₃? What is the plasma concentration of CO₂(dissolved)? If metabolic activity changes the concentration of CO₂ (dissolved) to 3 mM, and [HCO₃⁻] remains at 15 mM, what is the pH of the plasma?

Answer: Given the pH and the concentration of bicarbonate, we can use the following equation to calculate the amount of H₂CO₃:

$$pH = 3.57 + \log_{10} \frac{[HCO_3^-]}{[H_2CO_3]} \quad \text{(From page 43 of the textbook.)}$$

$$\frac{[HCO_3^-]}{[H_2CO_3]} = 10^{(pH-3.57)}$$

$$[H_2CO_3] = [HCO_3^-] \times 10^{(3.57-pH)}$$

$$= 15 \text{ mM} \times 10^{(3.57-7.4)}$$

$$= 2.22 \times 10^{-6}$$

$$= 2.22 \text{ } \mu M$$

For the bicarbonate buffer system:
The concentration of CO₂(dissolved) is calculated by first solving for this term:

$$pH = pK_{overall} + \log_{10} \frac{[HCO_3^-]}{[CO_{2(d)}]}$$

where $K_{overall} = K_a K_h$

and K_a = acid dissociation constant for H_2CO_3,

K_h = equilibrium constant for hydration of CO_2

$pK_{overall} = 6.1$ (See page 43.)

$$\log_{10} \frac{[HCO_3^-]}{[CO_{(d)}]} = pH - pK_{overall} = 7.4 - 6.1 = 1.3$$

$$\frac{[HCO_3^-]}{[CO_{(d)}]} = 10^{1.3} = 19.95$$

$$[CO_{(d)}] = \frac{[HCO_3^-]}{19.95} = \frac{15 \text{ mM}}{19.95} = 0.75 \text{ mM}$$

If $[CO_{(d)}] = 3$ mM and $[HCO_3^-] = 15$ mM

$$pH = 6.1 + \log_{10} \frac{15 \text{ mM}}{3 \text{ mM}} = 6.1 + 0.7 = 6.8$$

12. How to prepare a buffer solution: an anserine buffer

Draw the titration curve for anserine (Figure 2.16). The isoelectric point of anserine is the pH where the net charge on the molecule is zero; what is the isoelectric point for anserine? Given a 0.1 M solution of anserine at its isoelectric point and ready access to 0.1 M HCl, 0.1 M NaOH and distilled water, describe the preparation of 1 L of 0.04 M anserine buffered solution, pH 7.8?

Answer:

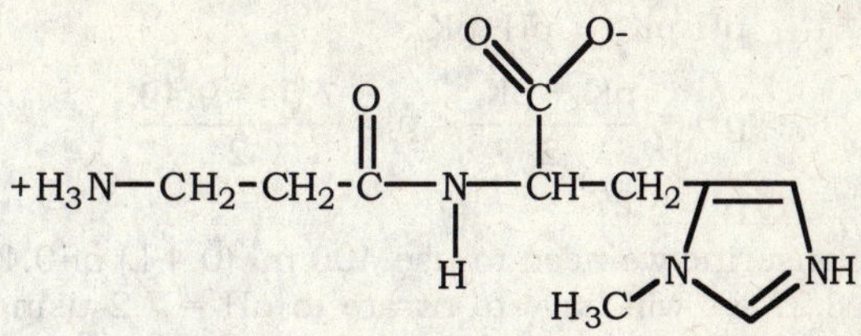

The structure of anserine is shown above. It has three ionizable groups: a carboxyl group pK_{a1} = 2.64, imidazole nitrogen pK_{a2} = 7.04 and amino group pK_{a3} = 9.49. Starting at acidic pH, all three groups will be protonated and thus the molecule will have a +2 charge. As base is added, the carboxyl group will be the first to deprotonate with a midpoint at 2.64. When fully deprotonated at about 4.64, anserine will have a +1 charge. As the pH approaches 7.0, the imidazole group will deprotonate leaving anserine uncharged. Finally, as the pH passes 9.49, the amino group will deprotonate and by about pH 11.5 anserine will have a –1 charge.

The titration curve is shown below with the pK_a's labeled. The isoelectric point, pI, is the pH at which the molecule is uncharged. Clearly, this will happen at a pH at which the carboxyl group's –1 charge is balanced by positive charges from both the imidazole group and the amino group. The isoelectric point must be between the pK_a's of the imidazole and amino groups. Thus, the sum of the protonated imidazole group and the protonated amino group must equal to one equivalent of charge.

Let I and IH^+ be the unprotonated and protonated imidazole groups respectively.
Let A and AH^+ be the unprotonated and protonated amino groups.
$$[IH^+] + [AH^+] = \text{one equivalent}$$
That is, the sum of the positively charged species for the imidazole and the amino groups must sum to the one equivalent of negative charge from the carboxylate. (Remember, there is one of each group.) The concentrations of the imidazole species, protonated and unprotonated, must sum to one equivalent. The same is true for the amino species. Thus,
$$[I] + [IH^+] = [A] + [AH^+] \text{ and so}$$
$$[I] + \cancel{[IH^+]} = \cancel{[IH^+]} + [AH^+] \quad (1)$$
$$[A] + \cancel{[AH^+]} = [IH^+] + \cancel{[AH^+]} \quad (2)$$
From equations (1) and (2) we can see that:
$$[AH^+] = [I] \text{ or } [IH^+] = [A] \quad (3)$$

The Henderson-Hasselbalch equations for each are:
$$pH = pK_2 + \log\frac{[I]}{[IH^+]} = pK_3 + \log\frac{[A]}{[AH^+]}$$

Substituting (3) into these we have:
$$pH = pK_2 + \log\frac{[AH^+]}{[IH^+]} \text{ and}$$
$$pH = pK_3 + \log\frac{[IH^+]}{[AH^+]}$$

Solving these two equations for the log terms, which are inversely related, and setting them equal we have:

$$pH - pK_2 = \log \frac{[AH^+]}{[IH^+]}$$

$$-pH + pK_3 = -\log \frac{[IH^+]}{[AH^+]} = \log \frac{[AH^+]}{[IH^+]}$$

Thus,

$$pH - pK_2 = -pH + pK_3$$

$$pH = \frac{pK_2 + pK_3}{2} = pI = \frac{7.04 + 9.49}{2}$$

$$pI = 8.27$$

In order to prepare 1 L of 0.04 M anserine we need to use 400 ml (0.4 L) of 0.1 M anserine stock. (1 L×0.04 M = 0.1 M×0.4 L). Since the pH = 8.27 we will have to titrate to pH = 7.2 using 0.1 M HCl. At pH = 8.27 the amino group is nearly fully protonated whereas the imidazole group is nearly unprotonated. Using the Henderson-Hasselbalch equation for the imidazole group, we can determine what the ratio of unprotonated to protonated form at pH 8.27 and 7.20.

$$pH = 7.04 + \log \frac{[I]}{[IH^+]}$$

At 7.2, $\dfrac{[I]}{[IH^+]} = 10^{(7.2-7.04)} = 1.445$

And since $[I] + [IH^+] = 0.04$

$$1.445 \times [IH^+] + [IH^+] = 0.04$$

$$[IH^+]_{7.2} = \frac{0.04}{2.445} = 0.0164$$

At 8.27, $\dfrac{[I]}{[IH^+]} = 10^{(8.27-7.04)} = 16.98$

And since $[I] + [IH^+] = 0.04$

$$16.98 \times [IH^+] + [IH^+] = 0.04$$

$$[IH^+]_{8.27} = \frac{0.04}{17.98} = 0.0022$$

$$[IH^+]_{7.2} - [IH^+]_{8.27} = 0.0164 - 0.0022 = 0.0142$$

Thus, we will need to add 142 mL of 0.1 M HCl to adjust the imidazole group. We will need additional HCl to titrate the amino group from pH 8.27 to 7.20.

$$pH = 9.49 + \log \frac{[N]}{[NH^+]}$$

At 7.2, $\dfrac{[N]}{[NH^+]} = 10^{(7.2-9.49)} = 0.00513$

And since $[N] + [NH^+] = 0.04$

$$0.00513 \times [NH^+] + [NH^+] = 0.04$$

$$[NH^+]_{7.2} = \frac{0.04}{1.00513} = 0.0398$$

At 8.27, $\dfrac{[N]}{[NH^+]} = 10^{(8.27-9.49)} = 0.06026$

And since $[N] + [NH^+] = 0.04$

$$0.06026 \times [NH^+] + [NH^+] = 0.04$$

$$[IH^+]_{8.27} = \frac{0.04}{1.06026} = 0.03773$$

$$[IH^+]_{7.2} - [IH^+]_{8.27} = 0.0398 - 0.03773 = 0.00207$$

Thus, we will need to add 21 mL of 0.1 M HCl to titrate the amino group. The total HCl need will be 163 ml.

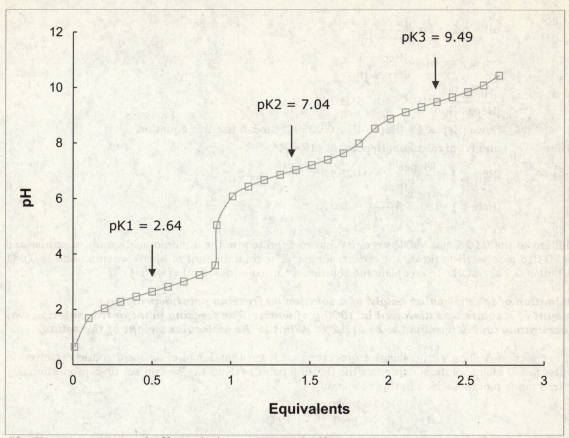

13. How to prepare a buffer solution: a HEPES buffer
Given a solution of 0.1 M HEPES in its fully protonated form, and ready access to 0.1 M HCl, 0.1 M NaOH and distilled water, describe the preparation of 1 L of 0.025 M HEPES buffer solution at pH 7.8?

Answer: The structure of HEPES is shown below.

The sulfonic acid group has a pK_a of around 3 and the ring nitrogen has a pK_a of 7.55. HEPES in its fully protonated form would have its tertiary nitrogen protonated and hence positively charged and a protonated, uncharged sulfonic acid group. To achieve this state a strong acid, like HCl must be added making the chloride salt of HEPES. To bring the pH to 7.8 would require a strong base like NaOH. One equivalent of NaOH would have to be added to deprotonate the sulfonic acid group. Additional NaOH would be required to deprotonate the tertiary nitrogen. To make 1 L of 0.025 M HEPES using 0.1 M stock we need:

$$0.025\frac{mole}{L} \times 1\ L = 0.1\frac{mole}{L} \times x$$

Solving for x we find that x = 0.25 L or 250 mL

The amount of base needed to adjust the pH to 7.8 can be calculated using the Henderson-Hasselbalch equation. Let [Hepes^{1-}] = the concentration of the unprotonated tertiary amine form (whose charge would be 1- due to the sulfonate group) and let [Hepes-H] be the protonated tertiary amine form (which is uncharged because the protonated tertiary nitrogen's positive charge balances the sulfonate's negative charge).

$$pH = pK_a + \log \frac{[Hepes^{1-}]}{[Hepes\text{-}H]}$$

$$7.8 = 7.55 + \log \frac{[Hepes^{1-}]}{[Hepes\text{-}H]}$$

$$\frac{[Hepes^{1-}]}{[Hepes\text{-}H]} = 10^{0.25} = 1.78 \ \ (1)$$

Since $[Hepes^{1-}] + [Hepes\text{-}H] = 0.025$ M, we can use this equation
and (1) to calculate $[Hepes^{1-}]$ at pH = 7.8.

$$[Hepes^{1-}] + \frac{[Hepes^{1-}]}{1.78} = 0.025 \text{ M}$$

$$[Hepes^{1-}] = 0.025 \times \frac{1.78}{2.78} = 0.016 \text{ M}$$

Thus, in addition to the 0.025 mol NaOH we must add to deprotonate the sulfonic acid group, we must add an additional 0.016 mol NaOH to titrate the tertiary amine. The total amount of NaOH we must add is 0.041 moles or 410 ml of 0.1 M NaOH. To complete the solution we must add 340 ml of water.

14. Determination of the molecular weight of a solution by freezing point depression
A 100-g amount of a solute was dissolved in 1000 g of water. The freezing point of this solution was measured accurately and determined to be –1.12°C. What is the molecular weight of the solute?

Answer: In the section dealing with colligative properties we learn that 1 mol of an ideal solute dissolved in 1000 g of water (a 1.0 molal solution) depresses the freezing point by 1.86°C. We can set up a proportionality to calculate how many mol was added in this problem.

$$\frac{1.0 \text{ molal}}{-1.86°C} = \frac{x}{-1.12°C}$$

$$x = \frac{1.12}{1.86} \times 1.0 \text{ molal}$$

$$x = 0.60 \text{ molal}$$

The 0.6 molal solution was made by adding 100 g solute, which represents 0.6 mol. Therefore, the solute's molecular weight is:

$$\frac{100 \text{ g}}{0.6 \text{ mol}} = 167 \text{ Da}$$

15. How to prepare a buffer solution: a triethanolamine buffer
Shown here is the structure of triethanolamine in its fully protonated form:

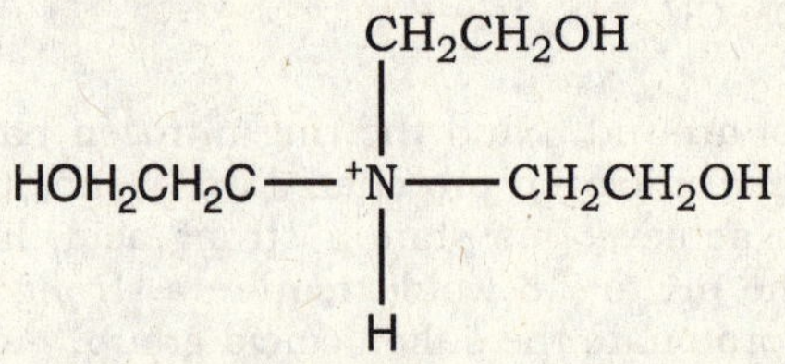

$$HOH_2CH_2C - \overset{\overset{\displaystyle CH_2CH_2OH}{|}}{\underset{\underset{\displaystyle H}{|}}{{}^+N}} - CH_2CH_2OH$$

Its pKₐ is 7.8. You have available at your lab bench 0.1 M solutions of HCl, NaOH, and the uncharged (free base) form of triethanolamine, as well as ample distilled water. Describe the preparation of a 1 L solution of 0.05 M triethanolamine buffer, pH 7.6.

Answer: The free base form of triethanolamine is the molecule shown above but with its tertiary nitrogen unprotonated. The solution we are asked to make must be 0.05 M triethanolamine. So, the first step is to calculate the amount of triethanolamine that is needed to make 1 L of 0.05 M solution using a 0.1 M solution.

$$0.05 \text{ M} \times 1 \text{ L} = 0.1 \text{ M} \times x$$

$$x = 0.5 \text{ L} = 500 \text{ ml}$$

The pH of the free base solution is basic because when added to water triethanolamine protonates and depletes the solution of free protons. To adjust the pH to 7.6 we will have to add HCl. The amount of HCl needed is determined by application of the Henderson-Hasselbalch equation using 7.8 for pK_a and 7.6 for pH.

$$pH = pK_a + \log\frac{[\text{Triethanolamine}]}{[\text{Triethanolamine} \cdot H^+]}$$

$$\frac{[\text{Triethanolamine}]}{[\text{Triethanolamine} \cdot H^+]} = 10^{(pH-pK_a)}$$

$$\frac{[\text{Triethanolamine}]}{[\text{Triethanolamine} \cdot H^+]} = 10^{(7.6-7.8)} = 10^{(-0.2)} = 0.6310$$

At pH 7.6 the ratio of free base to protonated triethanolamine is 0.631 and we know that the sum of the concentrations of these species is 0.05. Or,

$$\frac{[\text{Triethanolamine}]}{[\text{Triethanolamine} \cdot H^+]} = 0.6310$$

And,

$$[\text{Triethanolamine}] + [\text{Triethanolamine} \cdot H^+] = 0.05 \text{ M}$$

There are two equations with two unknowns. Solving for one unknown and substituting gives:

$$[\text{Triethanolamine}] = 0.6310 \times [\text{Triethanolamine} \cdot H^+]$$

And,

$$[\text{Triethanolamine}] + [\text{Triethanolamine} \cdot H^+] = 0.05 \text{ M}$$

Or,

$$0.6310 \times [\text{Triethanolamine} \cdot H^+] + [\text{Triethanolamine} \cdot H^+] = 0.05 \text{ M}$$

And,

$$1.6310 \times [\text{Triethanolamine} \cdot H^+] = 0.05 \text{ M}$$

$$[\text{Triethanolamine} \cdot H^+] = \frac{0.05 \text{ M}}{1.6310} = 0.0307 \text{ M}$$

To adjust the pH to 7.6, which we should recognize is below the pK_a, we will have to add 0.0307 moles of HCl. (We are making up 1L.) Using 0.1 M HCl, we will need

$$\frac{0.0307 \text{ mole}}{0.1 \text{ M}} = 0.307 \text{ L} = 307 \text{ ml}$$

So, the solution is made by mixing 500 ml of 0.1 M triethanolamine (free base) with 0.1 M HCl using 307 ml to drop the pH to 7.6. Then adjust the final volume to 1 L.

16. How to prepare a buffer solution: a Tris buffer solution
Tris-hydroxymethyl aminomethane (TRIS) is widely used for the preparation of buffers in biochemical research. Shown here is the structure TRIS in its protonated form:

$$\begin{array}{c} +NH_3 \\ | \\ HOH_2C - C - CH_2OH \\ | \\ CH_2OH \end{array}$$

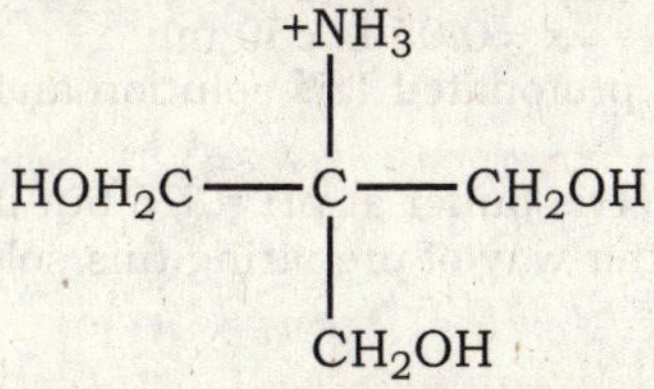

Its acid dissociation constant, K_a, is 8.32×10^{-9}. You have available at your lab bench a 0.1 M solution of TRIS in its protonated form, 0.1 M solutions of HCl and NaOH, and ample distilled water. Describe the preparation of a 1 L solution of 0.02 M TRIS buffer, pH 7.8.

Answer: Using the 0.1 M TRIS solution the amount needed to make 1 L of 0.02 M is determined as follows:

$$\frac{x \times 0.1 \text{ M}}{1 \text{ L}} = 0.02 \text{ M}$$

Where x is the volume of 0.1 M to be used.

Solving for x we find:

$$x = \frac{0.02 \text{ M} \times 1 \text{ L}}{0.1 \text{ M}} = 0.2 \text{ L} = 0.2 \text{ L} \times \frac{1000 \text{ ml}}{\text{L}} = 200 \text{ ml}$$

Next let's use the Henderson-Hasselbalch equation to calculate the ratio of TRIS base to protonated TRIS at pH = 7.6. Note: We are given the value for K_a, 8.32 x 10^{-9}. The pK_a is calculated as follows:

$$pK_a = -\log\left(8.32 \times 10^{-9}\right) = 8.0799$$

Use the Henderson-Hasselbalch equation as follows:

$$pH = pK_a + \log\frac{[\text{Tris base}]}{[\text{Tris} \cdot \text{H}^+]}$$

$$7.6 = 8.0799 + \log\frac{[\text{Tris base}]}{[\text{Tris} \cdot \text{H}^+]}$$

$$\frac{[\text{Tris base}]}{[\text{Tris} \cdot \text{H}^+]} = 10^{(7.6-8.0799)} = 10^{-0.4799} = 0.331$$

But,

$$[\text{Tris base}] + [\text{Tris} \cdot \text{H}^+] = 0.02 \text{ M}$$

Use the last two equations to solve for the concentration of each species.

$$[\text{Tris base}] = 0.331 \times [\text{Tris} \cdot \text{H}^+]$$

But,

$$0.331 \times [\text{Tris} \cdot \text{H}^+] + [\text{Tris} \cdot \text{H}^+] = 0.02 \text{ M}$$

$$1.331 \times [\text{Tris} \cdot \text{H}^+] = 0.02 \text{ M}$$

$$[\text{Tris} \cdot \text{H}^+] = \frac{0.02 \text{ M}}{1.331} = 0.0150$$

And,

$$[\text{Tris base}] = 0.02 \text{ M} - [\text{Tris} \cdot \text{H}^+] = 0.02 \text{ M} - 0.0150 = 0.005 \text{ M}$$

The protonated Tris solution was likely made using Tris·HCl, which is the chloride salt of Tris base. Thus, it contains equal amounts of chloride and Tris distributed between its protonated and unprotonated forms but mainly as the protonated form. The 200 ml represents 0.02 mole of protonated Tris. To adjust the pH to 7.6 we will need to lower the protonated Tris from 0.02 moles to 0.015 mole and so we will need to add NaOH in the amount of 0.005 mole. This corresponds to the following volume of 0.1 M NaOH:

$$x \times 0.1 \text{ M} = 0.005 \text{ mol}$$

$$x = 0.05 \text{ L} = 50 \text{ ml}$$

The final recipe is to use 200 ml of 0.1 M protonated Tris solution and add 50 ml of 0.1 M NaOH.

The final solution will actually be 0.02 M Tris buffer at pH = 7.6 but it will contain 5 mM NaCl produced when Tris HCl was adjusted with NaOH. A better way of preparing this solution is to use Tris base and then titrate with HCl to pH = 7.6.

17. Plot the titration curve for Bicine and calculate how to prepare a pH 7.5 Bicine buffer solution
Bicine (N, N–bis (2-hydroxyethyl) glycine) is another commonly used buffer in biochemistry labs. The structure of bicine in its fully protonated form is shown below:

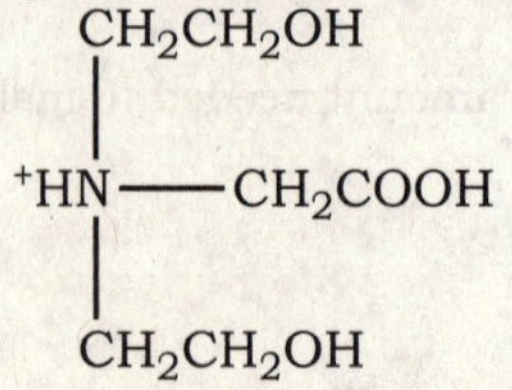

$$\text{CH}_2\text{CH}_2\text{OH}$$
$$|$$
$$^+\text{HN} \longrightarrow \text{CH}_2\text{COOH}$$
$$|$$
$$\text{CH}_2\text{CH}_2\text{OH}$$

a. Draw the titration curve for Bicine, assuming the pKₐ for its free COOH group is 2.3 and the pKₐ for its tertiary amino group is 8.3.

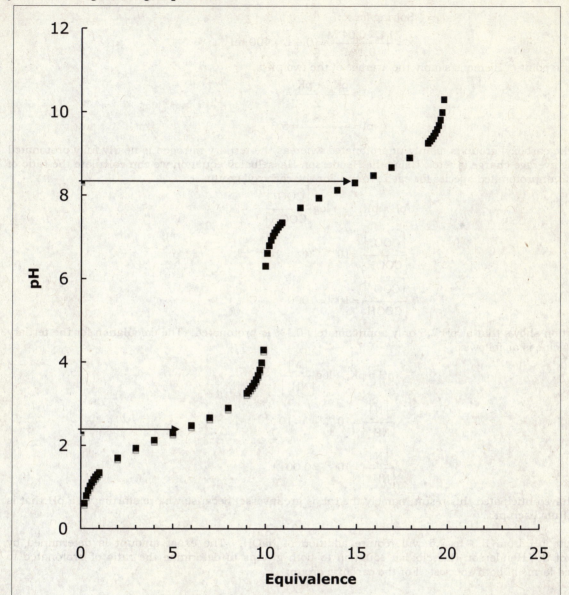

b. Draw the structure of the fully deprotonated form (completely dissociated form) of bicine.

$$CH_2CH_2OH$$

$$:N\!\!-\!\!CH_2COO^-$$

$$CH_2CH_2OH$$

c. You have available a 0.1 M solution of Bicine at its isoelectric point (pH$_I$), 0.1 M solutions of HCl and NaOH, and ample distilled H₂O. Describe the preparation of 1 L of 0.04 M Bicine buffer, pH 7.5.

Answer: The volume of 0.1 M bicine needed is:

$$\frac{x \times 0.1 \text{ M}}{1 \text{ L}} = 0.04 \text{ M}$$

Solving for x :

$$x = \frac{1 \text{ L} \times 0.04 \text{ M}}{0.1 \text{ M}} = 0.4 \text{ L} = 400 \text{ ml}$$

The isoelectric point of Bicine is simply the average of the two pKₐs.

$$pI = \frac{pK_{a1} + pK_{a2}}{2}$$

$$pI = \frac{2.3 + 8.3}{2} = 5.3$$

At this pH the carboxyl group is mainly unprotonated whereas the tertiary nitrogen is nearly fully protonated and thus the average charge is zero. Using the Henderson-Hasselbalch equation we can calculate the ratio of protonated to unprotonated species for each group. For the carboxyl group:

$$pH = pK_{COOH} + \log\frac{\left[COO^-\right]}{\left[COOH\right]}$$

$$\frac{\left[COO^-\right]}{\left[COOH\right]} = 10^{(pH - pK_{COOH})} = 10^{(5.3 - 2.3)}$$

$$\frac{\left[COO^-\right]}{\left[COOH\right]} = 10^{(3)} = 1000$$

This calculation shows that at pH 7.5 only approximately 0.1% is protonated. The calculation for the tertiary nitrogen of bicine is as follows:

$$pH = pK_N + \log\frac{\left[N\right]}{\left[NH^+\right]}$$

$$\frac{\left[N\right]}{\left[NH^+\right]} = 10^{(pH - pK_N)} = 10^{(5.3 - 8.3)}$$

$$\frac{\left[N\right]}{\left[NH^+\right]} = 10^{(-3)} = 0.001$$

We should have anticipated this result, namely the ratios are inverse, because we are starting at a pH that is equidistant from each pKₐ.

To adjust the pH from 5.3 to 7.5 will require addition of NaOH. The exact amount is determined by application of the Henderson-Hasselbalch equation to both groups to determine the ratio of protonated to unprotonated forms of both species. For the carboxyl group:

$$pH = pK_{COOH} + \log\frac{\left[COO^-\right]}{\left[COOH\right]}$$

$$\frac{\left[COO^-\right]}{\left[COOH\right]} = 10^{(pH - pK_{COOH})} = 10^{(7.5 - 2.3)}$$

$$\frac{\left[COO^-\right]}{\left[COOH\right]} = 10^{(5.2)} = 1.58 \times 10^5$$

For the amino group:

$$pH = pK_N + \log\left[\frac{N}{NH^+}\right]$$

$$\left[\frac{N}{NH^+}\right] = 10^{(pH-pK_N)} = 10^{(7.5-8.3)}$$

$$\left[\frac{N}{NH^+}\right] = 10^{(-0.8)} = 0.158$$

Using these equations and remembering that the total sum of COO- and COOH is equal to 0.04 mol (0.04 M times 1 L) and that the same is true for the protonated and unprotonated nitrogen we can calculate the moles of each species at the starting pH (i.e., the pI) and at pH = 7.5. The results are shown to five places. The column labeled "delta" is the change in the species from pI to pH = 7.5.

	pI = 5.3	pH = 7.5	delta
COO-	0.03996	0.04000	-0.00004
COOH	0.00004	0.00000	0.00004
N	0.00004	0.00546	-0.00542
NH+	0.03996	0.03454	0.00542
		Sum =	0.00546

Using 0.1 M NaOH we need 54.6 ml. The final solution is then adjusted to a final volume of 1L with 545.4 ml of water.

d. What is the concentration of fully protonated form of Bicine in your final buffer solution?

Answer: At any pH there will be four possible forms of bicine shown in the chart below. The fraction of each species is simply the fraction of the carboxyl species times the fraction of the nitrogen species.

For example, COO-/NH refers to Bicine with unprotonated carboxyl group and protonated nitrogen. The value 0.863 under the fraction column was calculated using data in the chart in part c. The fraction of carboxyl group that is unprotonated is calculated by dividing the value for COO- in the above chart by the sum of COOH and COO-. The same is done for the nitrogen. The molar amount is the value under fraction times 0.04. The sum shows that all species are accounted for. There are only two species at significant levels, both have the carboxyl group unprotonated. Fully protonated Bicine is only at 0.2 µM.

Species	Fraction	Molar Amount
COO-/N	0.1364413	0.0054577
COO-/NH	0.8635524	0.0345421
COOH/N	0.0000009	0.0000000
COOH/NH	0.0000055	0.0000002
		Sum = 0.04

18. Calculate the concentration of Cl⁻ in gastric juice
Hydrochloric acid is a significant component of gastric juice. If chloride is the only anion in gastric juice, what is its concentration if pH =1.2?

Answer: Strong acids like HCl fully dissociate in solution and so the pH, which is $-\log[H^+]$, is closely approximated by $-\log[HCl]$

$$HCl \rightarrow H^+ + Cl^-$$

Thus, pH = –log[HCl] = 1.2 giving

[HCl] = $10^{-1.2}$ = 0.0631 or 63.1 mM
So the chloride concentration is 63.1 mM.

19. Calculate the concentration of lactate in blood plasma at pH 7.4 if [lactic acid] = 15 μM. From the pK$_a$ for lactic acid given in Table 2.4, calculate the concentration of lactate in blood plasma (pH = 7.4) if the concentration of lactic acid is 1.5 μM.

Answer: From Table 2.4 the pK$_a$ of lactic acid is 3.86. To determine the concentration of lactate at pH = 7.4 we need to use the Henderson-Hasselbalch equation.

$$pH = pK_a + \log \frac{\left[\text{lactate}\right]}{\left[\text{lactic acid}\right]}$$

$$pH = 7.4,$$
$$pK_a = 3.86,$$
$$\left[\text{lactic acid}\right] = 1.5 \ \mu M = 1.5 \times 10^{-6} M$$
$$\left[\text{lactate}\right] = \left[\text{lactic acid}\right] \times 10^{(7.4-3.86)} = 1.5 \times 10^{-6} M \times 10^{3.54}$$
$$\left[\text{lactate}\right] = 0.0052 \ M = 5.2 \ mM$$

Note: The statement that the concentration of lactic acid is 1.5 μM was taken literally. It is possible that the concentration refers to the sum of lactate and its protonated, lactic acid form. In this case the solution is as follows:

$$pH = pK_a + \log \frac{\left[\text{lactate}\right]}{\left[\text{lactic acid}\right]}$$

$$pH = 7.4,$$
$$pK_a = 3.86,$$
$$\left[\text{lactic acid}\right] = 1.5 \ \mu M = 1.5 \times 10^{-6} M = \left[\text{lactate}\right] + \left[\text{lactic acid}\right]$$
$$\left[\text{lactate}\right] = \left[\text{lactic acid}\right] \times 10^{(7.4-3.86)} = \left[\text{lactic acid}\right] \times 10^{3.54} = 3,467 \times \left[\text{lactic acid}\right]$$
$$3,467 \times \left[\text{lactic acid}\right] + \left[\text{lactic acid}\right] = 1.5 \times 10^{-6} M$$
$$\left[\text{lactic acid}\right] = \frac{1.5 \times 10^{-6} M}{3,468} = 4.32 \times 10^{-10} M, \text{ and}$$
$$\left[\text{lactate}\right] = 3,467 \times \left[\text{lactic acid}\right] \approx 1.5 \times 10^{-6} M$$

That is, the pH is so far from the pKa that essentially all the lactic acid is in the unprotonated lactate form.

20. Draw the titration curve for a weak acid and determine its pK$_a$ from the titration curve
When a 0.1 M solution of a weak acid was titrated with base, the following results were obtained:

Equivalence of base added	pH Observed
0.05	3.4
0.15	3.9
0.25	4.2
0.40	4.5
0.60	4.9
0.75	5.2
0.85	5.4
0.95	6.0

Plot the results of this titration and determine the pK$_a$ of the weak acid from your graph.

Answer: In the chart shown below the values for pH$_{calculated}$ were determined using the Henderson-Hasselbalch equation with a guess for pK$_a$ shown. Values of [A⁻] were assumed to be equal to the amount of base added while the values of [HA] were 1- [A-].

equivalence	pH	pH$_{calculated}$	pK$_a$	A-	HA
0.05	3.4	3.42	4.7	0.05	0.95
0.15	3.9	3.95		0.15	0.85
0.25	4.2	4.22		0.25	0.75
0.4	4.5	4.52		0.4	0.6
0.6	4.9	4.88		0.6	0.4
0.75	5.2	5.18		0.75	0.25
0.85	5.4	5.45		0.85	0.15
0.95	6	5.98		0.95	0.05

In the graph shown below pH is plotted against equivalence for the observed data (closed circles) and for the pH$_{calculated}$ (×). There is very close agreement between the two data sets indicating that the guess for 4.7 is close to the true value. We know we are dealing with an acid (something with pK$_a$ lower that 7.0) because the initial pH is 3.42.

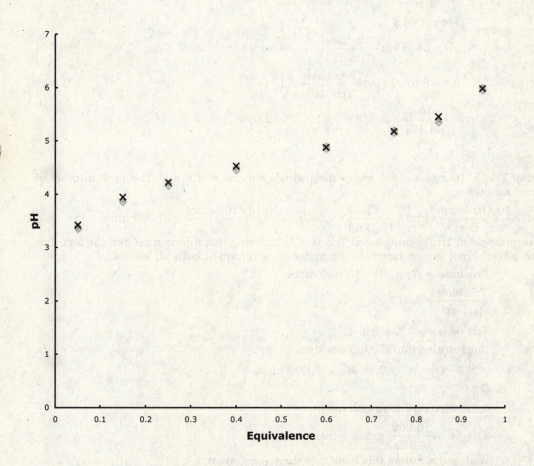

Preparing for the MCAT® Exam

21. The enzyme alcohol dehydrogenase catalyzes the oxidation of ethyl alcohol by NAD+ to give acetaldehyde plus NADH and a proton:

$$CH_3CH_2OH + NAD^+ \rightarrow CH_3CHO + NADH + H^+$$

The rate of this reaction can be measured by following the change in pH. The reaction is run in 1 ml 10 mM TRIS buffer at pH 8.6. If the pH of the reaction solution falls to 8.4 after ten minutes, what is the rate of alcohol oxidation, expressed as nanomoles of ethanol oxidized per sec per ml of reaction mixture?

Answer: The number of protons produced after 10 minutes can be determined using the Henderson-Hasselbalch equation. The reaction is being conducted using Tris buffer at an initial pH of 8.6. In problem 16 we were given the following information about Tris:

$$pK_a = -\log\left(8.32 \times 10^{-9}\right) = 8.0799$$

We will use the Henderson-Hasselbalch equation to determine the ratio of protonated and unprotonated Tris at pH 8.6 and at pH 8.4.

For pH = 8.6

$$pH = pK_a + \log\frac{[\text{Tris base}]}{[\text{Tris} \cdot \text{H}^+]}$$

$$8.6 = 8.0799 + \log\frac{[\text{Tris base}]}{[\text{Tris} \cdot \text{H}^+]}$$

$$\frac{[\text{Tris base}]}{[\text{Tris} \cdot \text{H}^+]} = 10^{(8.6-8.0799)} = 10^{0.5201} = 3.312$$

For pH = 8.4

$$pH = pK_a + \log\frac{[\text{Tris base}]}{[\text{Tris} \cdot \text{H}^+]}$$

$$8.4 = 8.0799 + \log\frac{[\text{Tris base}]}{[\text{Tris} \cdot \text{H}^+]}$$

$$\frac{[\text{Tris base}]}{[\text{Tris} \cdot \text{H}^+]} = 10^{(8.4-8.0799)} = 10^{0.3201} = 2.090$$

The total concentration of Tris is 10 mM and we are dealing with a volume of 1.0 ml. The total number of moles of Tris is calculated as follows:

$$10 \text{ mM} \times 1 \text{ ml} = \frac{10 \times 10^{-3} \text{ mol}}{\text{L}} \times 1 \text{ ml} \times \frac{1 \text{ L}}{1000 \text{ ml}} = 1.0 \times 10^{-5} \text{ mol} \times \frac{1 \times 10^9 \text{ nmol}}{\text{mol}} = 10,000 \text{ nmol}$$

At pH 8.6, the ratio of unprotonated Tris to protonated Tris is 3.312. Using this information and the fact that both forms must sum to 10,000 nmol we can calculate the moles of each form at both pH values.

$$\text{Tris base} + \text{Tris} \cdot \text{H}^+ = 10,000 \text{ nmol}$$

$$\frac{\text{Tris base}}{\text{Tris} \cdot \text{H}^+} = 3.312$$

$$\text{Tris base} = 3.312 \times \text{Tris} \cdot \text{H}^+$$

Substituting into the top equation

$$3.312 \times \text{Tris} \cdot \text{H}^+ + \text{Tris} \cdot \text{H}^+ = 10,000 \text{ nmol}$$

Or,

$$4.312 \times \text{Tris} \cdot \text{H}^+ = 10,000 \text{ nmol}$$

$$\text{Tris} \cdot \text{H}^+ = \frac{10,000}{4.312} \text{ nmole} = 2319 \text{ nmol}$$

And, substituting this back into the top equation

$$\text{Tris base} + 2319 \text{ nmol} = 10,000 \text{ nmol}$$

$$\text{Tris base} = 7681 \text{ nmol}$$

The moles of each species at the two pH values are shown below

	pH		
	8.6	8.4	Delta
Tris base	7681	6764	917
TrisH+	2319	3236	917
Total	10000	10000	

The column headed "Delta" is the amount of acid in nmoles that must have been produced to change the ratio of Tris species. To be correct we should calculate the amount of acid needed to lower the pH from 8.6 to 8.4 without regard to buffering. This value, however, is very small. It is calculated as follows:

$$pH = -\log[H^+]$$

$$[H^+] = 10^{-pH}$$

At pH 8.6

$$[H^+] = 10^{-8.6} = 2.51 \times 10^{-9} M$$

At pH 8.4

$$[H^+] = 10^{-8.4} = 3.98 \times 10^{-9} M$$

The change in protons is the difference inconcentrations times the volume

$$\text{moles produced} = \left(3.98 \times 10^{-9} M - 2.51 \times 10^{-9} M\right) \times 1 \times 10^{-3} L \times \frac{1 \times 10^9 \, \text{nmol}}{\text{mol}} = 0.0015 \, \text{nmol}$$

So, we have determined that 917 nmol of proton were produced in 10 minutes. The rate is calculated as follows:

$$\text{rate} = \frac{917 \, \text{nmole}}{10 \, \text{min}} \times \frac{1 \, \text{min}}{60 \, \text{sec}} = 1.53 \, \frac{\text{nmole}}{\text{sec}}$$

22. In light of the Human Biochemistry box, what would be the effect on blood pH if cellular metabolism produced a sudden burst of carbon dioxide?

Answer: The enzyme carbonic anhydrase rapidly hydrates carbon dioxide to form carbonic acid, which dissociates into a proton and bicarbonate as shown below.

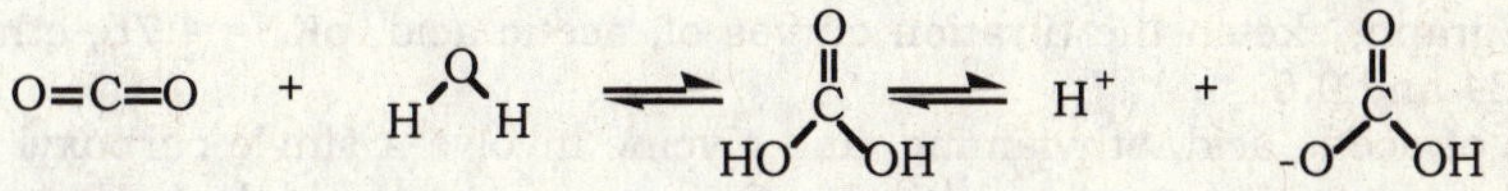

Thus, production of carbon dioxide should lead to additional protons and hence a decrease in pH (acidosis). The body can compensate by increasing exchange of carbon dioxide in the lungs. In the lungs, since carbon dioxide is being eliminated the above equilibrium shifts to the left causing a decrease in proton concentration and an increase in pH.

23. On the basis of Figure 2.12, what will the pH of acetate-acetic acid solution when the ratio of [acetate]/[acetic acid] is 10?
   a. 3.76
   b. 4.76
   c. 5.76
   d. 11.24

Answer: The pK_a of acetic acid is 4.76. To realize a ratio of acetate to acetic acid of 10 nearly an equivalent of base (NaOH) must be added and from Figure 2.12 we see that this corresponds to a pH of around 6. So, the correct answer must be "c", 5.76. One can easily calculate this value using the Henderson-Hasselbalch equation, $pH = pK_a + \log[A^-]/[HA]$. The term $[A^-]/[HA]$ is the ratio of acetate to acetic acid, which is given as 10. The log 10 = 1.0, thus the pH = 4.76 +1 = 5.76.

Additional Problems

1. Tris (Tris[hydroxymethyl]aminomethane) is a commonly used buffer with a pK_a = 8.06. In making up a Tris solution, an appropriate amount of the free base is dissolved in water and the pH of the solution is adjusted, often with HCl. The M_r of Tris is 121.1.
(a). Describe the preparation of 1 L of 25 mM solution, pH = 8.4, using solid Tris base and 1 M HCl.
(b). Draw the structure of Tris in its protonated and unprotonated forms.

2. Many proteins are sensitive to divalent cations and often chelating agents are included in protein isolation solutions to bind divalent cations, lowering their concentration, and removing them from proteins. Two common agents used for this purpose are EDTA (Ethylenediaminetetraacetic acid) and EGTA (Ethyleneglycol-bis-(β-aminoethyl ether) N,N,N',N'-tetraacetic acid).
(a). Draw the structures of EDTA and EGTA
(b). These agents function by binding divalent cations. EDTA binds both Mg^{2+} and Ca^{2+} tightly while EGTA has a greater specificity for Ca^{2+} (Both will bind a number of other divalent cations as well.) However, cation binding is pH-dependent. In making a solution of EDTA, it is often suggested that the pH of the solution be adjusted to around 8. The pK_as of the four acetic acid groups on EDTA are 2.0, 2.67, 6.16 and 10.26. At pH = 8.0, what is the predominant ionic species of EDTA?
(c). Divalent cations bind avidly to the fully ionized forms of both EDTA and EGTA. What fault do you find with using a solution of 10 mM Tris (pH = 7.0) and 25 mM EDTA to both buffer a protein solution and chelate divalent cations?

3. To make up 1 L of a 0.5 M solution of EDTA starting with the free acid, approximately how much 10 M NaOH will have to be added to adjust the pH to 7.0? Do you expect this solution to have a pH-buffering capacity? Explain.

4. The "Good" buffers were developed to provide buffers that would not interact strongly with divalent cations such as Mg^{2+} and Ca^{2+}. One such buffer is PIPES, (Piperazine-N,N'-bis[2-ethanesulfonic acid]), pK_a = 6.8. You are asked to make 1L of 25 mM solution of PIPES, pH = 7.0. In preparing the solution starting from the free acid, you notice that PIPES is not very soluble in water. You find that the solution is acidic (around 2), and so you begin to add NaOH. After addition of approximately 3.75 mL of 10 M NaOH, the solution clears. The pH is around 7. Explain.

5. (a). On the same graph, sketch the titration curves of; acetic acid, pK_a = 4.76; ethylamine, pK_a = 10.63; and, glycine, pK_a = 2.34 and 9.6.
    (b). The titrations of acetic acid, ethylamine, and glycine involve a single carboxyl group, a single amino group, and a carboxyl and amino group respectively. Can you suggest why the carboxyl group and the amino groups on glycine have lower pK_as than the same groups on acetic acid and ethylamine?

Abbreviated Answers

1. (a) Add 3.028 g Tris to approximately 900 mL of water. Carefully titrate the pH to 8.4 using approximately 7.8 ml of 1 M HCl. Finally, adjust the volume of the solution to 1 L.

(b)   protonated   unprotonated

$$CH_2OH$$
$$HOH_2C-C-CH_2OH$$
$$NH_3^+$$

$$CH_2OH$$
$$HOH_2C-C-CH_2OH$$
$$NH_2$$

2. (a)
EDTA

$$HOOC-CH_2 \quad\quad\quad\quad CH_2-COOH$$
$$N-CH_2-CH_2-N$$
$$HOOC-CH_2 \quad\quad\quad\quad CH_2-COOH$$

EGTA

$$HOOC-CH_2$$
$$N-CH_2-CH_2-O-CH_2-CH_2-O-CH_2-CH_2-N$$
$$HOOC-CH_2 \quad\quad\quad\quad\quad\quad\quad\quad\quad\quad\quad CH_2-COOH$$
$$CH_2-COOH$$

(b) $EDTA^{3-}$ (i.e., the singly-protonated species).

(c) Divalent cation binding to EDTA will result in release of a mole of proton per mole of cation bound to EDTA. The Tris functions as a proton buffer but it is one pH unit below its pK_a. Thus, it will not buffer the solution very effectively.

3. To adjust the solution to pH = 7.0, sufficient NaOH must be added to fully titrate the two carboxyl groups with pK_as = 2.0 and 2.67. This will require 100 mL of 10 M NaOH. Further, an additional 43.7 mL will be required to titrate the carboxyl group whose pK_a = 6.16. This is calculated using the Henderson-Hasselbalch equation. The total amount of NaOH is 143.7 mL.
   The solution will not be a good buffer at pH = 7.0 because the pH is 0.8 of a pH unit away from the nearest pK_a.

4. To bring the pH of the solution to its pK_a normally requires approximately one-half of an equivalent of (in this case) base. The 3.75 mL of NaOH represents (3.75 mL x 10 M =) 0.0375 equivalents of OH⁻. One liter of 25 mM Pipes contains 0.025 moles of Pipes and 0.0375 equivalents of OH⁻ corresponds to 1.5 equivalents of Pipes. Therefore, there must be two titratable groups on PIPES.

5.. The pK_as of the carboxyl group and the amino group of glycine are shifted to lower pH values relative to the same groups on acetic acid and ethylamine. Clearly, the groups are influencing each other's pK_as. The pK_a of the carboxyl group is lowered because the positively charged amino group influences it. The unprotonated carboxyl group is negatively charged. This state is favored by having a positively charged amino group nearby. Thus, the carboxyl group becomes a slightly stronger acid and will give up its proton at lower pH values. Similarly, the amino group becomes a slightly stronger acid because in its protonated state it is positively charged.

Summary

The most abundant molecule in living systems is H_2O. What physical properties make water such an important component of life? The two hydrogens in a water molecule make covalent bonds with a single O. However, there is something special about these covalent bonds. Oxygen is an electronegative element with a high affinity for electrons, surpassed only by F. (This high attraction for electrons is a consequence of a positively-charged nucleus rather unshielded by electron clouds.) In covalent bonds with H, the electrons will be attracted to O, giving rise to charge separation or a dipole moment. The hydrogen becomes partially positively-charged whereas the oxygen becomes partially negatively-charged. The four electron pairs in H_2O form a distorted tetrahedron, with two lone pairs separated by an angle greater than 109° and the two shared electron pairs (shared between O and H) at 104.5°. The partially positively-charged hydrogen can interact with lone-pair electrons to form hydrogen bonds. We will see that hydrogen bonds, although weak compared to covalent bonds, are important interactions in life.
   Water molecules can interact with each other to form hydrogen-bonded structures. Ice is a hydrogen-bonded structure, with each water molecule bonded to four other water molecules. In two of these bonds, the hydrogens interact with lone-pair electrons on two other water molecules. In these cases, water is serving as a hydrogen-bond donor. The two electron pairs make hydrogen bonds, participating as hydrogen-bond acceptors, with positively-charged hydrogens on two other water molecules. The thermal properties of water are a consequence of hydrogen bonds. For example, the density of water decreases with decreasing temperature down to 4°C; as water molecules lose kinetic energy, they can approach each other more closely. As the temperature moves to the freezing point of water, more and more hydrogen bonds form, causing water

molecules to move apart and density to decrease. Thus, ice floats. The environmental consequence is that ponds and lakes freeze from the surface, allowing life to continue in the liquid interior. The transition from solid to liquid to gas is accompanied by disruption of hydrogen bonds. A great many hydrogen bonds remain in liquid water, as evidenced by the difference in the heat of melting versus the heat of sublimation. Water has a high heat capacity (the ability to absorb heat without a large increase in temperature), because energy in the form of heat is used to break hydrogen bonds rather than increase temperature (increase kinetic energy). Finally, the rather large amount of energy required to vaporize water means that water can absorb energy with little increase in temperature.

Water is a good solvent for ionic and polar substances, because it forms hydrogen bonds with these substances. Salts dissolve and dissociate because the ionic components become hydrated with water molecules. The high dielectric constant of water is responsible for decreasing the attractive force between two ionic components of a salt. (The force between two charges, e_1 and e_2, is given by $F = e_1e_2/Dr^2$; it is inversely dependent on the dielectric constant.)

The attraction between water molecules is strong and the tendency of water molecules to make these interactions is great. For example, at an air/water interface, interaction of water molecules on the surface is responsible for surface tension. But surface tension comes at a price. Water molecules at an interface are forced to make irregular hydrogen bonds, often with less than perfect geometry. In fact, this occurs at any interface, and when nonpolar molecules are put into water, each molecule in effect represents an interface. To minimize the interface area, there is a tendency, driven by hydrogen bonding of H_2O, for nonpolar substances to coalesce, to minimize their contact with water molecules. This tendency is termed hydrophobic interactions. We will see that biological membranes and proteins rely on hydrophobic interactions to maintain an ordered structure.

There are two other properties of water that are of extreme importance to biological systems: osmotic pressure and ionization of water. Osmotic pressure arises when a semipermeable membrane through which water can pass separates two aqueous compartments. If a solute is dissolved in one of the aqueous compartments (and the solute is not freely permeable to the membrane), water will move from the solute-free compartment into the compartment containing solute. The amount of pressure necessary to prevent this movement of water is the osmotic pressure. Osmotic pressure is a colligative property and as such its magnitude is directly proportional to concentration.

Water is capable of ionizing in solution to form hydrogen ions and hydroxyl ions. The pH is a measure of the hydrogen ion concentration: $pH = -\log[H^+]$. The pH scale in aqueous solution is set by the magnitude of the ion product. In aqueous solutions, $pH + pOH = 14$. Using this formula, the pH of solutions of strong acids or bases can be calculated. For weak acids and bases, the Henderson-Hasselbalch equation must be used: $pH = pK_a + \log([A^-]/[HA])$. Several of the problems illustrate the use of this equation.

Chapter 3

Thermodynamics of Biological Systems

• •

Chapter Outline

❖ Thermodynamic concepts
  ⋏ Systems
    ✦ Isolated systems cannot exchange matter or energy with surroundings
    ✦ Closed systems exchange energy but not matter
    ✦ Open systems exchange both energy and matter
  ⋏ First Law: $\Delta E = q + w$
    ✦ ΔE = change in internal energy (state function), q = heat absorbed, and w = work done on the system
      • Work = generalized force × generalized distance
    ✦ H = E + PV: H = enthalpy (energy transferred at P = constant): $\Delta H = q$ when work limited to $P\Delta V$
    ✦ $\Delta H° = -Rd(\ln K_{eq})/d(1/T)$: van't Hoff plot $R\ln K_{eq}$ vs $1/T$ (R = gas constant = 8.314J/mol K)
  ⋏ Second Law: Disorder or randomness
    ✦ $S = k\ln W$ where k = Botzmann's constant = 1.38×10^{-23} J/K
    ✦ W = number of ways of arranging the components of a system at fixed internal energy
    ✦ $dS = dq/T$ for reversible process
    ✦ $\Delta S > 0$ for irreversible process, $\Delta S = 0$ for reversible process
  ⋏ Third law: Entropy of a perfectly ordered, crystalline array at 0 K is exactly zero
    ✦ $S = \int_0^T C_p\, d\ln T$, C_p = heat capacity = dH/dT for constant P process
    ✦ Third law allows for calculation of S at any temperature provided C_p is known
  ⋏ Gibbs free energy: G = H -TS
    ✦ $\Delta G = \Delta H - T\Delta S$ for constant T processes
    ✦ $\Delta G = \Delta G° + RT\ln([P]/[R])$ and $\Delta G° = -RT\ln([P]_{eq}/[R]_{eq})$
    ✦ When protons involved in process: $\Delta G°' = \Delta G° \pm RT\ln[H^+]$
      • ("+" if reaction produces protons; "-" if protons consumed)
❖ Coupled processes: Enzymatic coupling of a thermodynamically unfavorable reaction with a thermodynamically favorable reaction to drive the unfavorable reaction. Thermodynamically favorable reaction is often hydrolysis of high-energy molecule
❖ Energy transduction: high-energy phosphate in ATP and reduced cofactor NADPH
  ⋏ Phototrophs use light energy to produce ATP and NADPH
  ⋏ Chemotrophs use chemical energy to produce ATP and NADPH
❖ High-energy molecules
  ⋏ Phosphoric anhydrides (ATP, ADP, GTP, UTP, etc.)
  ⋏ Enol phosphates (phosphoenolpyruvate a.k.a. PEP)
  ⋏ Phosphoric-carboxylic anhydrides (1,3-bisphosphoglycerate)
  ⋏ Guanidino phosphates (creatine phosphate)
❖ Group transfer reactions
  ⋏ Transfer of proton: $pK_a = \Delta G/2.303RT$

    ⅄   Transfer of electron: $\Delta\mathscr{E}_0 = -\Delta G/n\mathscr{F}$
- ❖ Why is hydrolysis of high-energy bonds favorable
  - ⅄ Destabilization of reactant due to electrostatic repulsion
  - ⅄ Product isomerization and resonance stabilization
  - ⅄ Entropy factors
- ❖ Thermodynamics of ATP hydrolysis influenced by: pH, cation concentration, reactant and product concentrations

Chapter Objectives

Laws of Thermodynamics

    The first law of thermodynamics is simply a conservation of energy statement. The internal energy changes if work and/or heat are exchanged. In biological systems, we are usually dealing with constant-pressure processes and in this case the term enthalpy, **H**, is used. Enthalpy is the heat exchanged at constant pressure. Since enthalpy is heat, it is readily measured using a calorimeter or from a plot of $R(\ln K_{eq})$ versus $1/T$, a van't Hoff plot. (To get ahead of the story, the point is that if ΔG and ΔH are known, ΔS can be calculated.)

    The second law of thermodynamics introduces the term entropy, **S**, which is a measure of disorder or randomness in a system. A spontaneous reaction is accompanied by an increase in disorder. For a reversible reaction, $dS_{reversible} = dq/T$. Also, $S = k \ln W$ where k = Boltzmann's constant, and W = the number of ways to arrange the components of a system without changing the internal energy.

Gibbs Free Energy

    The change is Gibbs free energy for a reaction is the amount of energy available to do work at constant pressure and constant volume. This is an important concept and should be understood. For a general reaction of the type

$$A + B \rightleftharpoons C + D$$

$$\Delta G = \Delta G° + \ln\frac{[C][D]}{[A][B]}$$

When a reaction is at equilibrium, clearly it can do no work (nor have work done on it) without change so it must be true that $\Delta G = O$ at equilibrium. Thus, $\Delta G° = -RT \ln K_{eq}$, and by measuring the equilibrium concentrations of reactants and products, $\Delta G°$ can be evaluated. Knowing the initial concentrations of reactants and products and $\Delta G°$ allows a calculation of ΔG. Why is this important? The sign on ΔG tell us in which direction the reaction will proceed. A negative ΔG indicates that the reaction has energy to do work and will be spontaneous in the direction written. A positive ΔG indicates work must be done on the reaction for it to proceed as written, otherwise it will run in reverse. The magnitude of ΔG is the amount of energy available to do work when the reaction goes to equilibrium. Finally, the relationship $\Delta G = \Delta H - T\Delta S$ can be used to evaluate ΔS, the change in disorder, if ΔG and ΔH are known.

High Energy Compounds

    ATP is the energy currency of cells and its hydrolysis is used to drive a large number of reactions. For ATP and other high-energy biomolecules, you should understand the properties that make them energy-rich compounds. These include: destabilization due to electrostatic repulsion, stabilization of hydrolysis products by ionization and resonance, and entropy factors. Examples of high-energy compounds (from Table 3.3 in Garrett and Grisham) include: phosphoric acid anhydrides (e.g., ATP, ADP, GTP, UTP, CTP, and PP_i), phosphoric-carboxylic anhydrides (acetyl phosphate and 1,3-bisphosphoglycerate), enol phosphates (PEP), and guanidinium phosphates (creatine and arginine phosphate). ATP is the cardinal high-energy compound. Hydrolysis of GTP is important in signal transduction and protein synthesis. UTP and CTP are used in polysaccharide and phospholipid synthesis, respectively. We will encounter 1,3-bisphosphoglycerate and PEP in glycolysis. These high-energy compounds, along with creatine and arginine phosphates, are used to replenish ATP from ADP. Other important high-energy compounds include coenzyme A derivatives such as acetyl-CoA, and succinyl-CoA important in the citric acid cycle. Aminoacylated-tRNAs, the substrates used by the ribosome for protein synthesis, are high-energy compounds.

Chapter 3 · Thermodynamics of Biological Systems

Problems and Solutions

1. Calculating K_{eq} and ΔG from concentrations
An enzymatic hydrolysis of fructose-1-P

$$\text{Fructose-1-P} + H_2O \rightleftharpoons \text{fructose} + P_i$$

was allowed to proceed to equilibrium at 25°C. The original concentration of fructose-1-P was 0.2 M, but when the system had reached equilibrium the concentration of fructose 1-P was only 6.52 x 10⁻⁵ M. Calculate the equilibrium constant for this reaction and the standard free energy of hydrolysis of fructose 1-P.

Answer: For Fructose-1-P + H₂O $\rightleftharpoons$ fructose + P_i the equilibrium constant, K_{eq}, is given by

$$K_{eq} = \frac{[\text{fructose}]_{eq}[P]_{eq}}{[\text{fructose}-1-P]_{eq}}$$

At 25°C or 298 K (273 + 25), [fructose-1-P]$_{eq}$ = 6.52 x 10⁻⁵ M. Initially [fructose-1-P] = 0.2 M. The amount of the fructose produced is given by: 0.2 M - 6.52 x 10⁻⁵ M.
And, since an equal amount of [P_i] is produced, K_{eq} may be written as follows:

$$K_{eq} = \frac{(0.2 \text{ M} - 6.52 \times 10^{-5})(0.2 \text{ M} - 6.52 \times 10^{-5})}{6.52 \times 10^{-5}}$$

$$K_{eq} = 613 \text{ M}$$

$$\Delta G° = -RT\ln K_{eq}$$

$$\Delta G° = -(8.314 \text{ J/mol K}) \times 298 \text{ K} \times \ln 613$$

$$\Delta G° = -15.9 \text{ kJ/mol}$$

2. Calculating $\Delta G°$ and $\Delta S°$ from $\Delta H°$
The equilibrium constant for some process A $\rightleftharpoons$ B is 0.5 at 20°C and 10 at 30°C. Assuming that $\Delta H°$ is independent of temperature, calculate $\Delta H°$ for this reaction. Determine $\Delta G°$ and $\Delta S°$ at 20° and at 30°C. Why is it important in this problem to assume that $\Delta H°$ is independent of temperature?

Answer:

At 20°C

$$\Delta G° = -RT\ln K_{eq}$$

$$\Delta G° = -(8.314 \text{ J/mol K}) \times (273 + 20) \text{ K} \times \ln 0.5$$

$$= 1.69 \text{ kJ/mol}$$

At 30°C

$$\Delta G° = -RT\ln K_{eq}$$

$$\Delta G° = -(8.314 \text{ J/mol K}) \times (273 + 30) \text{K} \times \ln 10$$

$$= -5.80 \text{ kJ/mol}$$

From the equation $\Delta G° = \Delta H° - T\Delta S°$, we see that $\Delta G°$ is linearly related to T when $\Delta H°$ is independent of temperature. If this is the case, then $d\Delta H°/dT = 0$ (i.e., the heat capacity is zero). A plot of $\Delta G°$ versus T will be linear with a slope $= -\Delta S°$ and a y intercept $= \Delta H°$.

$$-\Delta S° = \text{slope} = \frac{\Delta G°_{30°C} - \Delta G°_{20°C}}{T_{30°C} - T_{20°C}} = \frac{-5.8 - 1.69}{303 - 293}$$

$$\Delta S° = 0.75 \text{ kJ/mol K}$$

$\Delta H°$ can be calculated using $\Delta H° = \Delta G° + T\Delta S°$.

For 20°C, $\Delta H° = 1.69 \text{ kJ/mol} + 293 \text{ K} \times 0.75 \text{ kJ/mol K} = 221.5 \text{ kJ/mol}$

For 30°C, $\Delta H° = -5.80 \text{ kJ/mol} + 303 \text{ K} \times 0.75 \text{ kJ/mol K} = 221.5 \text{ kJ/mol}$

Therefore, $\Delta H° = 221.5$ kJ/mol and $\Delta S° = 0.75$ kJ/mol K at both temperatures, and, $\Delta G° = 1.69$ kJ/mol at 20° C and $\Delta G° = -5.80$ kJ/mol at 30° C.

3. Calculating ΔG from ΔG°′
The standard state free energy of hydrolysis of acetyl phosphate is ΔG° = -42.3 kJ/mol.

$$Acetyl\text{-}P + H_2O \rightarrow acetate + P_i$$

Calculate the free energy change for the acetyl phosphate hydrolysis in a solution of 2 mM acetate, 2 mM phosphate and 3 nM acetyl phosphate.

Answer: Since temperature is not specified in this problem, we are free to choose one. Both 20°C and 25°C are common choices, and 37°C makes sense if we are talking about reactions in warm-blooded mammals. The answer given in the back of the text for this answered was calculated using 20°C.

$$\Delta G = \Delta G° + RT \ln \frac{[Product]_{initial}}{[Reactant]_{initial}}$$

$$\Delta G = \Delta G° + 8.314 \text{ J/mol K} \times (273+20) \text{ K} \times \ln \frac{[acetate][P_i]}{[acetyl\text{-}P]}$$

$$\Delta G = -42,300 \text{ J/mol} + 8.314 \text{ J/mol K} \times (273+20) \text{ K} \times \ln \frac{(2 \times 10^{-3})(2 \times 10^{-3})}{(3 \times 10^{-9})}$$

$$\Delta G = -24.8 \text{ kJ/mol}$$

4. Understanding state functions
Define a state function. Name three thermodynamic quantities that are state functions and three that are not.

Answer: A state quantity is a variable that describes the condition of a system but is independent of the past history of the system. Variables like temperature, pressure and volume are state quantities. When a state quantity, a variable, can be mathematically related to another variable, the relationship is a state function. Functions are mathematical expressions that relate a variable to another variable such that each value of one variable is determined by or related to a single value of the other variable. A state function then is a variable quantity of a system that is related mathematically to another variable but is independent of the history of the system. It is a characteristic of the condition of a system. Equations that relate state quantities are known as equations of state. We know, for example, that the ideal gas law relates temperature, pressure, volume and number of moles of a gas. Each of these is a state quantity and the ideal gas law is a state function and an equation of state.

Temperature is a state function that depends on the energy of molecules of an object. An object's temperature is independent of how energy was transferred to the object. Heating the object or cooling it or doing work on it could have all resulted in energy transfer to the object. The internal energy of an object is a state function. From the first law of thermodynamics we know that a change in internal energy occurs by heat or by work or by a combination of both. But, the internal energy does not depend on the events that brought it to its present value. The object could have been cooled, heated, worked on, allowed to do work or any combination to bring it to some internal energy value.

Another example of a state function is gravitational potential energy of a closed system -an object that cannot exchange matter. The gravitational potential energy depends on the location an object is in a gravitational field and the object's mass. For the case of an object on earth its potential energy is a function of the relative height of the object above some reference height (i.e., E = mgh). Gravitational potential energy is thus a state function. Changing the object's height or altitude can change its gravitational potential energy. To accomplish this energy must be expended either by the object or on the object and this energy is not a state function. Suppose we decide to increase the gravitational potential energy of the object by raising its height. The energy expended in doing this will depend on path. If, for example, we drive it to the top of a mountain, energy will be expended as heat (from the car's engine) and friction and only a portion of the energy released by combustion of say gasoline will be conserved as gravitational potential energy. The amount of energy expended will clearly depend on the route we take but the gravitational potential energy of the object will depend only on the final height of the object. The change in altitude experienced by the object is also a state function whereas the distance traveled by the object to reach its new altitude is not.

In the example above if we change an object's gravitational potential energy by an adiabatic process then the object's internal energy is increased by an amount equal to its gravitational potential energy increase. The first law of thermodynamics is a state function but heat and work are not.

To summarize, a list of state functions includes internal energy, temperature, volume, pressure, entropy and Gibbs free energy. Variables that are not state functions include work and heat, distance and time.

5. Calculating the effect of pH on $\Delta G°$

ATP hydrolysis at pH 7.0 is accompanied by release of a hydrogen ion to the medium

$$ATP^{4-} + H_2O \rightleftharpoons ADP^{3-} + HPO_4^{2-} + H^+$$

If the $\Delta G°'$ for this reaction is -30.5 kJ/mol, what is $\Delta G°$ (that is, the free energy change for the same reaction with all components, including H^+, at a standard state of 1 M)?

Answer: The reaction produces H^+ and we can use the following equation to calculate $\Delta G°$:

$\Delta G° = \Delta G°' - RT \ln [H^+]$ where $\Delta G°' = -30.5$ kJ/mol, T = 298 K, R = 8.314 J/mol K and $[H^+] = 10^{-7}$. Thus,

$$\Delta G° = -30.5 \text{ kJ/mol} - 8.314 \times 10^{-3} \text{kJ/mol K} \times (273+25) \text{ K} \times \ln 10^{-7.0}$$

$$\Delta G° = -30.5 + 39.9 = 9.4 \text{ kJ/mol}$$

6. Calculating K_{eq} and $\Delta G°$ for coupled reactions

For the process $A \rightleftharpoons B$, $K_{eq}(AB)$ is 0.02 at 37°C. For the process $B \rightleftharpoons C$, $K_{eq}(BC)$ is 1000 at 37°C.
a. Determine $K_{eq}(AC)$, the equilibrium constant for the overall process $A \rightleftharpoons C$ from $K_{eq}(AB)$ and $K_{eq}(BC)$.
b. Determine standard state free energy changes for all three processes, and use $\Delta G°(AC)$ to determine $K_{eq}(AC)$. Make sure that this value agrees with that determined in part a, of this problem.

Answer: For $A \rightleftharpoons B$ and $B \rightleftharpoons C$,

$$K_{eq}(AB) = \frac{[B]_{eq}}{[A]_{eq}}, \text{ and } K_{eq}(BC) = \frac{[C]_{eq}}{[B]_{eq}}$$

By solving for $[B]_{eq}$ in the above two equations we find:

$$[B]_{eq} = K_{eq}(AB) \times [A]_{eq} = \frac{[C]_{eq}}{K_{eq}(BC)}$$

This equation can be rearranged to give:

$$\frac{[C]_{eq}}{[A]_{eq}} = K_{eq}(AC) = K_{eq}(AB) \times K_{eq}(BC) = 0.02 \times 1000 \text{ or,}$$

$$K_{eq}(AC) = 20$$

The standard free energy change is calculated as follows:

$$\Delta G°(AB) = -RT \ln K_{eq}(AB) = -8.314 \text{ J/mol K} \times 310 \text{ K} \times \ln 0.02$$

$$\Delta G°(AB) = 10.1 \text{ kJ/mol}$$

$$\Delta G°(BC) = -RT \ln K_{eq}(BC) = -8.314 \text{ J/molK} \times 310 \text{ K} \times \ln 1000$$

$$\Delta G°(BC) = -17.8 \text{ kJ/mol}$$

$$\Delta G°(AC) = -RT \ln K_{eq}(AC) = -8.314 \text{ J/mol K} \times 310 \text{ K} \times \ln 20$$

$$\Delta G°(AC) = -7.72 \text{ kJ/mol}$$

or

$$\Delta G°(AC) = \Delta G°(AB) + \Delta G°(BC)$$

$$\Delta G°(AC) = 10.1 \text{ kJ/mol} - 17.8 \text{ kJ/mol} = 7 -.70 \text{ kJ/mol}$$

7. Understanding resonance structures

Draw all possible resonance structures for creatine phosphate and discuss their possible effects on resonance stabilization of the molecule.

8. Calculating K_{eq} from $\Delta G°'$

Write the equilibrium constant, K_{eq}, for the hydrolysis of creatine phosphate and calculate a value for K_{eq} at 25°C from the value of $\Delta G°'$ in Table 3.3.

Answer: For the reaction:

$$\text{creatine phosphate} + H_2O \rightarrow \text{creatine} + P_i, \Delta G°'=-43.3 \text{ kJ/mol}$$

$$K_{eq} = \frac{[\text{creatine}]_{eq}[P_i]_{eq}}{[\text{creatine phosphate}]_{eq}}$$

From $\Delta G°'=-RT\ln K_{eq}$, we can write:

$$K_{eq} = e^{\frac{-\Delta G°'}{RT}} = e^{\frac{-(-43.3\times10^3)}{8.314\times298}}$$

$$K_{eq} = 3.89\times10^7$$

9. Imagining creatine phosphate and glycerol-3-phosphate as energy carriers

Imagine that creatine phosphate, rather than ATP, is the universal energy carrier molecule in the human body. Repeat the calculation presented in section 3.8, calculating the weight of creatine phosphate that would need to be consumed each day by a typical adult human if creatine phosphate could not be recycled. If recycling of creatine phosphate were possible, and if the typical adult human body contained 20 grams of creatine phosphate, how many times would each creatine phosphate molecule need to be turned over or recycled each day? Repeat the calculation assuming that glycerol-3-phosphate is the universal energy carrier, and that the body contains 20 grams of glycerol-3-phosphate.

Answer: The calculation presented in section 3.8 determined the number of moles of ATP that must be hydrolyzed under cellular conditions to provide 5,860 kJ of energy. Under standard conditions, ATP hydrolysis yields 35.7 kJ/mol (See Table 3.3) whereas under cellular conditions the value is approximately 50 kJ/mol. In order to repeat the calculation using creatine phosphate, we must estimate the free energy of hydrolysis under cellular conditions. Alternatively, we can assume that the same number of moles of creatine phosphate must be hydrolyzed as ATP. The energy of hydrolysis of creatine phosphate is larger (i.e., more negative) than that of ATP. So, hydrolysis of an equivalent molar amount of creatine phosphate will release considerably more energy. Let us try both solutions.

First, let us assume that an equivalent number of moles of creatine phosphate is hydrolyzed. (The reason for considering this is that metabolic energy is in effect quantized as high-energy bonds.) From section 3.8 we see that 117 moles of ATP are required. An equal number of moles of creatine phosphate weigh:

$$117 \text{ mol} \times 180 \text{ g/mol} = 21,060 \text{ g}$$

And, the turnover of creatine phosphate is

$$\frac{21,060 \text{ g}}{20 \text{ g}} = 1,052 \text{ times}$$

To calculate the free energy of hydrolysis of creatine phosphate under cellular conditions, let us assume that in resting muscle, creatine phosphate is approximately 20 mM, the concentration of P_i is approximately 5 mM (See Problem 10), and approximately 10% of creatine phosphate or 2 mM is as creatine. Using these values and the standard free energy of hydrolysis of creatine phosphate (from Table 3.3) we calculate the energy of hydrolysis under cellular conditions as follows:

$$\Delta G = \Delta G° + RT\ln \frac{[\text{creatine}][P_i]}{[\text{creatine phosphate}]}$$

$$\Delta G = -43.3 \text{ kJ/mol} + 8.314\times10^{-3} \text{ kJ/mol K} \times (273+37) \text{ K} \times \ln \frac{(2\times10^{-3})(5\times10^{-3})}{20\times10^{-3}}$$

$$\Delta G = -62.9 \text{ kJ/mol}$$

The number of moles of creatine phosphate is given by

$$\frac{5860 \text{ kJ}}{62.9 \text{ kJ/mol}} = 93.2 \text{ mol}$$

The molecular weight of creatine phosphate is 180. Therefore

$$93.2 \text{ mol} \times 180 \text{ g/mol} = 16,780 \text{ g is required.}$$

The turnover of creating phosphate is $\frac{16,780 \text{ g}}{20 \text{ g}} = 839$ times.

If one uses -43.3 kJ/mol for the ΔG of hydrolysis of creatine phosphate, the number of moles required is 5,860 kJ ÷ 43.3 kJ/mol = 135.3 mol. The M_r of creatine is 131.1 and so it would require 131.1 g/mol × 135.3 mol = 17,738 g.

The solution to the problem using glycerol-3-phosphate is slightly more complicated. From Table 3.3 we see that the standard free energy of hydrolysis of the compound is only -9.2 kJ/mol, considerably lower than the -35.7 kJ/mol listed for ATP hydrolysis. So, we cannot simply assume that an equivalent number of moles of glycerol-3-phosphate will substitute for ATP hydrolysis as we did for creatine phosphate above because hydrolysis of an equivalent number of moles of glycerol-3-phosphate will not supply sufficient energy. To solve the problem, we must estimate the energy of hydrolysis of glycerol-3-phosphate under cellular conditions as we did for creatine phosphate hydrolysis. Let us assume that $[P_i]$ = 5 mM, and that the ratio of [glycerol]:[glycerol-3-phosphate] is 1:10. Under these conditions,

$$\frac{5860 \text{ kJ}}{9.2 \text{ kJ/mol}} = 637 \text{ mol or } 637 \text{ mol} \times 172.08 \text{ g/mol} = 109,615 \text{ g}$$

The turnover of creating phosphate is $\frac{109,615 \text{ g}}{20 \text{ g}}$ = 5,480 times.

$$\Delta G = -9.2 \text{ kJ/mol} + 8.314 \times 10^{-3} \text{ kJ/mol K} \times (273+37) \text{ K} \times \ln\frac{5 \times 10^{-3}}{10}$$

$\Delta G = -28.8$ kJ/mol

The number of moles of glycerol-3-phosphate is given by

$$\frac{5860 \text{ kJ}}{28.8 \text{ kJ/mol}} = 203 \text{ mol}$$

The molecular weight of glycerol-3-phosphate is 172.08. Therefore

203 mol × 172.08 g/mol = 35,013 g is required.

The turnover of glycerol-3-phosphate is $\frac{35,013 \text{ g}}{20 \text{ g}}$ = 1,751 times.

10. Calculating ΔG in a rat liver cell
Calculate the free energy of hydrolysis of ATP in a rat liver cell in which the ATP, ADP, and P_i concentrations are 3.4, 1.3, and 4.8 mM, respectively.

Answer: For [ATP] = 3.4 mM, [ADP] = 1.3 mM, and $[P_i]$ = 4.8 mM, calculate the ΔG of hydrolysis of ATP.

$$\Delta G = \Delta G^{\circ\prime} + RT\ln\frac{[ADP][P_i]}{[ATP]}$$

$$\Delta G = -30.5 \text{ kJ/mol (From Table 3.3)} + 8.314 \times 10^{-3} \times (293+25) \text{ K} \times \ln\frac{(1.3 \times 10^{-3})(4.8 \times 10^{-3})}{3.4 \times 10^{-3}}$$

$\Delta G = -46.1$ kJ/mol

At 37°C, $\Delta G = -46.7$ kJ/mol

11. Calculating ΔG°′ and K_{eq} from cellular concentrations
Hexokinase catalyzes the phosphorylation of glucose from ATP, yielding glucose-6-P and ADP. Using the values of Table 3.3, calculate the standard-state free energy change and equilibrium constant for the hexokinase reaction.

Answer: Hexokinase catalyzes the following reaction:
$$\text{glucose + ATP} \rightleftharpoons \text{glucose-6-P + ADP}$$
This reaction may be broken down into the following two reactions:
$$\text{glucose} + P_i \rightleftharpoons \text{glucose-6-P} + H_2O$$
$$\text{ATP} + H_2O \rightleftharpoons \text{ADP} + P_i$$
From Table 3.3, we find that $\Delta G^{\circ\prime}$ = -13.9 kJ/mol for glucose-6-P hydrolysis.
Thus, the reverse reaction, namely reaction (1), must have $\Delta G^{\circ\prime}$ = +13.9 kJ/mol.

From Table 3.3, we also find that ATP hydrolysis has $\Delta G^{\circ\prime}$ = - 30.5 kJ/mol.
The overall $\Delta G^{\circ\prime}$ for phosphoryl transfer from ATP to glucose is:

$$\Delta G^{\circ\prime} = +13.9 + (-30.5) = -16.6 \text{ kJ/mol and,}$$

$$K_{eq} = e^{\frac{\Delta G^{\circ\prime}}{RT}} = e^{\frac{(16.6 \times 10^3)}{8.314 \times 310}} = 626.9$$

12. Evaluating the reactivity of high-energy molecules
Would you expect the free energy of hydrolysis of acetoacetyl-coenzyme A (see diagram) to be greater than, equal to, or less than that of acetyl-coenzyme A? Provide a chemical rationale for your answer.

$$H_3C-\overset{\overset{O}{\|}}{C}-CH_2-\overset{\overset{O}{\|}}{C}-S-CoA$$

Answer: Hydrolysis of acetyl-coenzyme A produces free coenzyme A and acetate whereas hydrolysis of acetoacetyl-coenzyme A releases acetoacetate and coenzyme A. Acetate is relatively stable; however, acetoacetate is unstable and will break down to acetone and CO_2. Thus, the instability of one of the products of hydrolysis of acetoacetyl-coenzyme A, namely acetoacetate, will make the reverse reaction (i.e., production of acetoacetyl-coenzyme A) unlikely. Furthermore, the terminal acetyl group of acetoacetyl-coenzyme A is electron-withdrawing in nature and will destabilize the thiol ester bond of acetoacetyl-CoA. Thus, the free energy of hydrolysis of acetoacetyl-coenzyme A is expected to be greater than that of acetyl-coenzyme A and in fact, the free energy of hydrolysis is -43.9 kJ/mol for acetoacetyl-CoA and -31.5 kJ/mol for acetyl-CoA.

13. Assessing the reactivity of carbamoyl phosphate
Consider carbamoyl phosphate, a precursor in the biosynthesis of pyrimidines:

$$^{+}H_3N \underset{}{\overset{\overset{O}{\overset{\|}{C}}}{}} O-PO_3^{-}$$

Based on the discussion of high-energy phosphates in this chapter, would you expect carbamoyl phosphate to possess a high free energy of hydrolysis? Provide a chemical rationale for your answer.

Answer: Is carbamoyl phosphate destabilized due to electrostatic repulsion? The carbonyl oxygen will develop a partial-negative charge causing charge-repulsion with the negatively charged phosphate. So, charge destabilization exists. Are the products stabilized by resonance or by ionization? Without regard to resonance states of the phosphate group, there are two possible resonance structures for carbamoyl phosphate:

$$H_2N-\overset{\overset{O}{\|}}{C}-O-PO_3^{2-} \longleftrightarrow {}^{+}H_2N=\overset{\overset{O^{-}}{\|}}{C}-O-PO_3^{2-}$$

However, we expect that the carbonyl-carbon must pass through a positively charged intermediate and in doing so, affect phosphate resonance. Thus, the products are resonance stabilized. Finally, are there entropy factors? The products of hydrolysis are phosphate and carbamic acid. Carbamic acid is unstable and decomposes to CO_2 and NH_3 unless stabilized as a salt by interacting with a cation. With these considerations in mind, we expect carbamoyl phosphate to be unstable and therefore a high energy compound. The free energy of hydrolysis of carbamoyl phosphate is -51.5 kJ/mol, whereas for acetyl phosphate it is only -43.3 kJ/mol.

14. Assessing the effect of temperature on equilibrium
You are studying the various components of the venom of a poisonous lizard. One of the venom components is a protein that appears to be temperature sensitive. When heated, it denatures and is no longer toxic. The process can be described by the following simple equation:

$$T \text{ (toxic)} \ \square \ N \text{ (nontoxic)}$$

There is only enough protein from this venom to carry out two equilibrium measurements. At 298 K, you find that 98% of the protein is in its toxic form. However, when you raise the temperature to 320 K, you find that only 10% of the protein is in its toxic form.
a. Calculate the equilibrium constants for the T to N conversion at these two temperatures.

Answer: For the reaction $T \rightarrow N$

$$K_{eq} = \frac{[N]_{eq}}{[T]_{eq}}$$

At 298 K, [T] = 98% of the total protein concentration and [N] = 2%. Thus,

$$K_{eq} = \frac{[N]_{eq}}{[T]_{eq}} = \frac{2}{98} = 0.0204$$

At 320 K, [T] = 10% and [N] = 90%. Thus,

$$K_{eq} = \frac{[N]_{eq}}{[T]_{eq}} = \frac{90}{10} = 9$$

b. Use the data to determine the $\Delta H°$, $\Delta S°$, and $\Delta G°$ for this process.

Answer: Using Figure 3.2 as a guide, the following chart calculates $1/T$ and $R\ln K_{eq}$ for the data. (R = 8.314 J/mol·K)

T (°K)	K_{eq}	1/T	$R\ln K_{eq}$
298	0.0204	0.0034	-32.3599
320	9	0.0031	18.2677

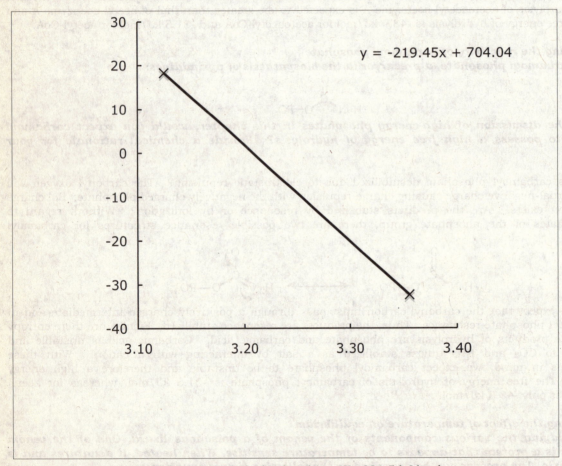

$$y = -219.45x + 704.04$$

The slope of this van't Hoff plot is $-\Delta H°$ in kJ/mol and so $\Delta H° = 291.5$ kJ/mol.

In general,

$$\Delta G^\circ = -RT \ln K_{eq}$$

And,

$$\Delta G^\circ = \Delta H^\circ - T\Delta S^\circ$$

Or,

$$\Delta S^\circ = \frac{\Delta H^\circ - \Delta G^\circ}{T}$$

Values of ΔG° at both temperatures were calculated above.

T (°K)	K_{eq}	$\Delta G^\circ =$ -$RT\ln K_{eq}$ (kJ/mol)	ΔH° (kJ/mol)	ΔS° (kJ/mol·K)
298	0.0204	9.64	291.5	0.95
320	9	-5.85	291.5	0.93

15. Analyzing the thermodynamics of protein denaturation
Consider the data in Figures 3.3 and 3.4. Is the denaturation of chymotrypsinogen spontaneous at 58°C? And what is the temperature at which the native and denatured forms of chymotrypsinogen are in equilibrium?

Answer: Figure 3.3 is a plot of ΔG° versus temperature and from it we see that at 58°C, ΔG° is approximately –3 kJ/mol. Thus, we expect denaturation to be spontaneous at this temperature if we start with a solution of chymotrypsinogen in its native conformation. The data presented in Figure 3.5 do not address the question of spontaneity of the reaction at 58°C. The ΔS° is positive at this temperature, suggesting that disorder in the denatured state is greater than that in the native state. This is consistent with an unfolding process but not predictive of the process occurring.

To calculate the temperature at which the process is at equilibrium we must find the temperature at which $K_{eq} = 1$ and $\Delta G^\circ = 0$. We can use the data used to construct Figure 3.3 to make this calculation. From Figure 3.3 we see that a plot of $R\ln K_{eq}$ versus $1000/T$ is linear and with a slope of magnitude 533 kJ/mol. A linear plot implies the following:

$$R \ln K_{eq} = -533 \times \frac{1000}{T} + b$$

We can solve the equation for b, the x-intercept, and can evaluate b by using any one data point in Figure 3.3 that is in the linear portion of the plot. Once b is evaluated, we can substitute it into the above equation and find T at which $K_{eq} = 1.0$. Remembering that T is in degrees Kelvin, we find that the protein is in equilibrium at T = 56.6°C.

16. Assessing the meaning of heat capacity changes
Consider Tables 3.1 and 3.2, as well as the discussion of Table 3.2 in the text, and discuss the meaning of a positive ΔC_P in Table 3.1.

Answer: In each case of protein denaturation presented in Table 3.1, we see that ΔC_p is positive. The corresponding ΔG° values in each case are also positive, implying that the native state is favored over the denatured state i.e., denaturation is unfavorable for these proteins in the conditions specified. Table 3.2 presents us with a reaction also accompanied by a positive ΔC_p, namely, transfer of toluene into water. This process is unfavorable because hydrophobic toluene will order water molecules due to its inability to hydrogen bond with water. A similar process occurs when proteins denature: their hydrophobic cores become exposed to water.

17. Assessing the effect of pH on metabolic reactions
The difference between ΔG and $\Delta G^{\circ\prime}$ was discussed in Section 3.3. Consider the hydrolysis of acetyl phosphate (Figure 3.10) and determine the value of ΔG° for this reaction at pH 2, 7, and 12. The value for $\Delta G^{\circ\prime}$ for the enolase reaction (Figure 3.11) is 1.8 kJ/mol. What is the value of ΔG° for enolase at pH 2, 7, and 12? Why is this case different from that of acetyl phosphate?

Answer: When protons are involved in process: $\Delta G^{\circ\prime} = \Delta G^\circ \pm RT\ln[H+]$ where the plus sign is used when protons are produced and the minus sign when protons are consumed. Hydrolysis of acetyl phosphate produces acetate, phosphate, and a proton and hence the plus sign is used. The $\Delta G^{\circ'}$ for acetyl phosphate is given as -43.3 kJ/mol.

$$\Delta G^{\circ'} = \Delta G^{\circ} + RT\ln[H^+] \quad (1)$$

Solving for ΔG°

$$\Delta G^{\circ} = \Delta G^{\circ'} - RT\ln[H^+]$$

$$\Delta G^{\circ} = -43{,}300 \text{ J/mol} - (8.314 \text{ J/mol} \cdot \text{K}) \times (273+25)\text{K} \times \ln[10^{-7}]$$

$$\Delta G^{\circ} = -33{,}700 \text{ J/mol} = -33.7 \text{ kJ/mol}$$

We were given the $\Delta G^{\circ'}$ value, which is for a reaction at pH 7. The value of ΔG° is the same for pH 2, 7, and 12 because it is the value when $[H^+] = 1$ M. We could, however, use equation (1) to calculate $\Delta G^{\circ'}$ at pH 2.0 and pH 12. Remembering that $[H^+] = 10^{-pH}$ we find $\Delta G^{\circ'}$ values of -14.78 kJ/mol and -71.82 kJ/mol for pH 2.0 and 12.0 respectively (and of course -43.3 kJ/mol for pH 7.0).

The enolase reaction does not involve a proton and so $\Delta G^{\circ} = \Delta G^{\circ'}$.

18. Analyzing the energetics of coupled reactions
What is the significance of the magnitude of $\Delta G^{\circ'}$ for ATP in the calculations in the box on page 67? Repeat these calculations for the case of coupling of a reaction to 1,3-bisphosphoglycerate hydrolysis to see what effect this reaction would have on the equilibrium ratio for components A and B under the conditions stated on this page.

Answer: The value of $\Delta G^{\circ'}$ for hydrolysis of ATP, a highly favorable reaction, is -30.5 kJ/mol. Conversion of A to B under standard conditions has $\Delta G^{\circ'}$ of +13.8 kJ/mol, which is unfavorable. By coupling the two reactions the overall $\Delta G^{\circ'}$ is -16.7 kJ/mol, which indicates that under standard conditions the coupled reaction will proceed toward hydrolysis of ATP and conversion of A to B. But, net conversion of A to B is exaggerated in the example given in the textbook because the concentrations of ATP, ADP and P_i are likely not equilibrium values. If they were and ATP hydrolysis was not coupled to anything the values would predict a $\Delta G^{\circ'}$ of approximately +17 kJ/mol. It is likely the case that the concentrations are being maintained by coupling of favorable catabolic processes to maintain ATP concentrations high. Given this, the ratio of B to A would be greatly skewed in favor of production of B.

To make the same calculation but using 1,3-bisphophoglycerate hydrolysis one would have to know the concentrations of 1,3-BPG and 3-phosphoglycerate for *E. coli*. Table 18.2 gives values of 0.001 mM and 0.12 mM in erythrocytes for 1,3-BPG and 3-PG and 1 mM for P_i. The overall $\Delta G^{\circ'}$ for coupling of A to B with 1,3-BPG to 3-PG would be +13.8 kJ/mol -49.6 kJ/mol =-35.8 kJ/mol. The corresponding K_{eq} is:

$$K_{eq} = e^{\frac{35{,}800}{8.314 \times 298}} = 1.89 \times 10^6$$

The reaction is $A + 1{,}3 \text{ BPG} + H_2O \rightarrow B + 3\text{-PG} + P_i$

$$K_{eq} = \frac{[B_{eq}][3-PG][P_i]}{[A_{eq}][1{,}3-BPG]}$$

$$\frac{[B_{eq}]}{[A_{eq}]} = \frac{K_{eq}[1{,}3-BPG]}{[3-PG][P_i]} = \frac{1.89 \times 10^6 \times 0.00001}{0.00012 \times .001}$$

$$\frac{[B_{eq}]}{[A_{eq}]} = 1.57 \times 10^9$$

Preparing for the MCAT® Exam

19. Analyzing resonance effects in hydrolysis of high-energy phosphates
The hydrolysis of 1,3-bisphosphoglycerate is favorable, due in part to the increase resonance stabilization of the products of the reaction. Draw resonance structures for the reactant and the products of this reaction to establish that this statement is true.

Answer: Hydrolysis of 1,3-bisphosphate produces 3-phosphoglycerate and inorganic phosphate. The structures are shown below.

Additional resonance of electrons is available in the products as indicated by the curved arrows.

20. Analyzing the energetics of coupled reactions
The acyl-CoA synthetase reaction activates fatty acids for oxidation in cells:
$$R\text{-}COO^- + CoASH + ATP \rightarrow R\text{-}COSCoA + AMP + pyrophosphate$$
The reaction is driven forward in part by hydrolysis of ATP to AMP and pyrophosphate. However, pyrophosphate undergoes further cleavage to yield two phosphate anions. Discuss the energetics of this reaction both in the presence and absence of pyrophosphate cleavage.

Answer: On the reactant side ATP is the least stable molecule due to charge repulsion of the negatively-charged triphosphate group. This is relieved, somewhat, by hydrolysis to form AMP and pyrophosphate, but pyrophosphate is still unstable due to charge repulsion. The carboxylate of the fatty acid (R-COO⁻) loses resonance possibilities when it forms the thioester bond in fatty acyl Coenzyme A (the COS part of R-COSCoA). It is clear that further hydrolysis of pyrophosphate to two phosphates would relieve charge repulsion in pyrophosphate and increase resonance possibilities in the two phosphates. Furthermore, additional numbers of products would contribute to entropy, which would help drive the reaction.

Questions for Self Study

1. True of False.
   a. The internal energy of an isolated system is conserved. _____
   b. A closed system can exchange matter but not energy with the surroundings. _____
   c. An open system includes a system and its surroundings. _____
   d. An open system can exchange matter with another open system. _____
   e. The internal energy of an open system is always constant. _____

2. The first law of thermodynamics states that there are only two ways to change the internal energy of any system. What are they?

3. Enthalpy, H, is defined as $H = E + PV$ and $\Delta E = q + w$. Under what conditions is $\Delta H = q$?

4. Define the terms in the following expression: $S = k \ln W$.

5. Match the items in the two columns.
   a. $\Delta G = \Delta H - T\Delta S$ 1. Reaction spontaneous as written.
   b. $\Delta G = \Delta G° + RT \ln ([P]/[R])$ 2. Used to determine standard Gibbs free energy change.
   c. $\Delta G° = 0$ 3. Reaction unfavorable.
   d. $\Delta G = 0$ 4. Used to calculate amount of free energy released when
   reaction proceeds to equilibrium.
   e. $\Delta G > 0$ 5. Definition of change in Gibbs free energy.
   f. $\Delta G < 0$ 6. System at equilibrium.
   g. $\Delta G° = - RT \ln K_{eq}$ 7. $K_{eq} = 1$.

6. The compounds shown below include ATP, pyrophosphate, phosphoenolpyruvate, creatine phosphate, and 1,3-bisphosphoglycerate. They are all examples of high-energy compounds. Identify each, locate the high-energy bond and list the products of hydrolysis of this bond.

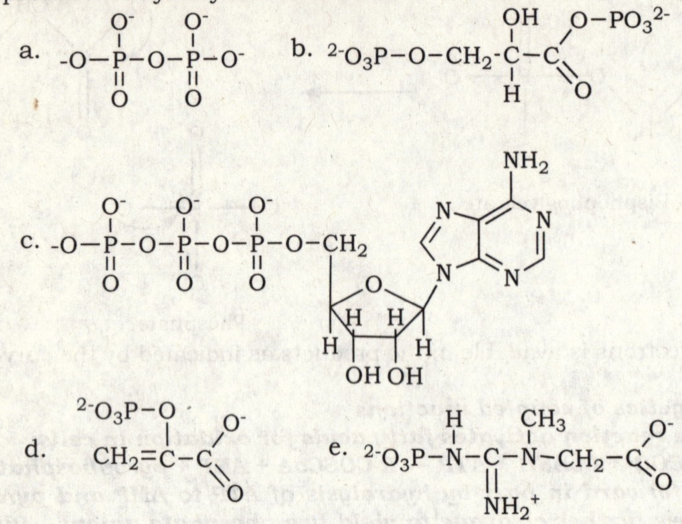

a.
$$\text{-O-P-O-P-O}^-$$

b.
$$^{2-}O_3P\text{-O-CH}_2\text{-C-C}$$
with OH, $O\text{-PO}_3^{2-}$

c.
$$\text{-O-P-O-P-O-P-O-CH}_2$$
(attached to adenine/ribose structure with NH_2, $OH\ OH$)

d.
$$^{2-}O_3P\text{-O}$$
$$CH_2=C\text{-C}$$

e.
$$^{2-}O_3P\text{-N-C-N-CH}_2\text{-C}$$
with H, CH_3, NH_2^+, O^-

7. What are the three chemical reasons for the large negative energy of hydrolysis of phosphoric acid anhydride linkage for compounds such at ATP, ADP, and pyrophosphate?

8. Creatine phosphate is an example of a guanidinium phosphate, a high-energy phosphate compound. This compound is abundant in muscle tissue. What is its function?

9. During protein synthesis amino acids are joined in amide linkage on the ribosome. The substrates for this reaction are not free amino acids but rather amino acids attached via their carboxyl groups to the 3' hydroxyl group of tRNAs. What kind of bond is formed between the amino acid and the tRNA? Is this a high-energy bond? Given the fact that aminoacyl-tRNA formation is accompanied by hydrolysis of ATP to AMP and PP_i and that PP_i is subsequently hydrolyzed to $2\ P_i$, how many high-energy phosphates are consumed to produce an aminoacyl-tRNA?

10. Although the standard free energy of hydrolysis of ATP is around -30 kJ/mol the cellular free energy change is even more negative. What factors contribute to this?

Answers

1. a.T; b.F; c.F; d.T; e.F.

2. Energy flow in the form of heat or work.

3. In general, $\Delta H = \Delta E + P\Delta V + V\Delta P = q + w + P\Delta V + V\Delta P$. When the pressure of the system remains constant (i.e., $\Delta P = 0$) and work is limited to only mechanical work (i.e., $w = -P\Delta V$) then $\Delta H = q$.

4. S is the entropy, k is Boltzmann's constant, ln W is the natural logarithm of the number of ways, W, of arranging the components of a system.

5. a.5; b.4; c.7; d.6; e.3; f.1; g.2.

6. a./pyrophosphate, hydrolysis products $2\ P_i$. b./1,3-bisphosphoglycerate, hydrolysis products 3-phosphoglycerate and P_i. c./ATP, hydrolysis products either ADP and P_i or AMP and pyrophosphate. d./phosphoenolpyruvate, hydrolysis products pyruvate and P_i. e./creatine phosphate, hydrolysis products creatine and P_i. The location of high-energy bonds is shown below.

a. $-O-\overset{O^-}{\underset{O}{P}}-\overset{\downarrow}{O}-\overset{O^-}{\underset{O}{P}}-O^-$ b. $^{2-}O_3P-O-CH_2-\overset{OH}{\underset{H}{C}}-\overset{O-PO_3^{2-}}{\underset{O}{C}}\nwarrow$

c. $-O-\overset{O^-}{\underset{O}{P}}-\overset{\downarrow}{O}-\overset{O^-}{\underset{O}{P}}-\overset{\downarrow}{O}-\overset{O^-}{\underset{O}{P}}-O-CH_2$ [adenosine structure with NH2 adenine base and ribose OH OH]

d. $\overset{^{2-}O_3P-O}{\underset{CH_2=C-C}{\underset{O}{|}}}\overset{O^-}{}$ e. $^{2-}O_3P-\overset{H}{N}-\overset{CH_3}{\underset{\underset{NH_2^+}{||}}{C}}-\overset{}{N}-CH_2-\overset{O^-}{\underset{O}{C}}$

7. Bond strain due to electrostatic repulsion, stabilization of products by ionization and resonance, and entropy factors.

8. Creatine phosphate is used to replenish supplies of ATP by transferring phosphate to ADP in a reaction catalyzed by creatine kinase.

9. Amino acid ester bonds in aminoacyl-tRNAs are high-energy bonds produced at the expense of two high-energy phosphate bonds.

10. The presence of divalent and monovalent cations and the maintenance of low levels of ADP and P_i and high levels of ATP are responsible for the large negative free energy change of ATP under cellular conditions.

Additional Problems

1. Show that $\Delta G° = -RT \ln K_{eq}$.

2. The term $\Delta G°$ may be evaluated by measuring ΔG for a reaction in which reactants and products start out at 1 M concentration. The ΔG measured in this case is the standard-state free energy and is equal to $\Delta G°$. Prove this.

3. If a particular reaction is allowed to reach equilibrium, is the equilibrium concentration of product ever dependent on the initial concentrations of reactant and product?

4. Confusion reigns supreme when work and energy expenditure are discussed. Define the term work. Are work and energy expenditure synonymous?

5. There are two statements about spontaneity of reactions: $\Delta S > 0$, and $\Delta G < 0$. Justify that these statements are in fact true descriptions of spontaneity.

6. DNA ligase catalyzes formation of a phosphodiester bond between a 5'-phosphate and a 3'-hydroxyl group on the ends of two DNAs to be joined. Many ligases use hydrolysis of ATP to drive phosphodiester bond synthesis; however, the *E. coli* ligase uses NAD^+ as a high-energy compound. Explain why NAD^+ is considered a high-energy compound. What type of reaction is required to release energy from NAD^+ and what are the products?

7. You find yourself in the laboratory, late at night, working on an important assay and you discover that the last of the ATP stock is used up. After frantically searching everywhere, you find a 10 mg bottle of 2'-deoxyadenosine 5'-triphosphate. Is this a high-energy compound? Will it substitute for ATP in your assay?

8. Stock solutions of ATP are usually adjusted to around neutral pH to stabilize them. Why?

9. Intense muscle activity depletes ATP and creatine phosphate stores and produces ADP, creatine, AMP, and P_i. ADP is produced by ATP hydrolysis catalyzed by myosin, a component of the contractile apparatus of muscle. Creatine is a product of creatine kinase activity, and AMP is produced by adenylate kinase from two ADPs. Given this information, explain how high-energy phosphate compounds in muscle are interconnected.

10. The standard free energy of hydrolysis of glucose-1-phosphate is -21 kJ/mol whereas it is only -13.9 kJ/mol for glucose-6-phosphate. Provide an explanation for this difference.

Abbreviated Answers

1. At equilibrium $\Delta G = 0$ and the concentration of reactants and products are at their equilibrium values. Using

$$\Delta G = \Delta G^{\circ\prime} + RT \ln \frac{[C][D]}{[A][B]} \text{ we see that}$$

$$\Delta G = 0 = \Delta G^{\circ} + RT \ln \frac{[C_{eq}][D_{eq}]}{[A_{eq}][B_{eq}]}$$

. or, solving for ΔG° we find that:

$$\Delta G^{\circ} = -RT \ln \frac{[C_{eq}][D_{eq}]}{[A_{eq}][B_{eq}]} = -RT \ln K_{eq}$$

2. Using

$$\Delta G = \Delta G^{\circ\prime} + RT \ln \frac{[C][D]}{[A][B]} \text{ we see that}$$

$$\Delta G = \Delta G^{\circ} + RT \ln \frac{1M \times 1M}{1M \times 1M} = \Delta G^{\circ} + RT \ln 1$$

But, $\ln 1 = 0$, and

$$\Delta G = \Delta G^{\circ}$$

3. The equilibrium constant is independent of the initial concentrations of reactants and products but the equilibrium constant is the ratio of the product of the concentration of products to the product of the concentration of reactants. This ratio is independent of initial concentrations. Clearly, the absolute amount of product depends on the initial concentration of reactant.

4. Work is application of a force through some distance the consequence of which is a change in the energy of the object upon which the force is acting. Application of a force on an object at rest will cause the object to accelerate and in the absence of friction this will impart kinetic energy to the object. In the presence of friction, energy will be dissipated as heat raising the temperature of the object. However, one can have energy expenditure without doing work. As an example, holding an object at some position in a gravitational field requires a force equal and opposite to the force of gravity and creation of this force requires expenditure of energy yet no work is done if the object does not move.

5. Entropy is a measure of disorder and for a reaction to be spontaneous $\Delta S > 0$ or since $S_f - S_i > 0$, $S_f > S_i$. A spontaneous reaction results in an increase in disorder. Gibbs free energy is the amount of energy available to do work at constant pressure and temperature. Using G as a criterion for spontaneity requires that $\Delta G < 0$ or since $G_f - G_i < 0$, $G_f < G_i$. The amount of energy available to do work decreases for a spontaneous reaction.

6. NAD^+ contains a single high-energy phosphoric anhydride linkage between AMP and nicotinamide monophosphate nucleotide. Hydrolysis will release AMP and nicotinamide nucleoside monophosphate.

7. 2'-Deoxyadenosine 5'-triphosphate or dATP contains two high-energy phosphoric anhydride bonds. However, if the assays being performed are enzymatic assays (as opposed to some chemical assay), then it is out-of-the-question to even think about substituting dATP for ATP. ATP is a ribonucleotide; dATP is a deoxyribonucleotide. dATP is used for synthesis of DNA and little else.

8.   At first thought it might might seem reasonable to make the ATP solution slightly acidic. This will neutralize the negatively-charged phosphates and should lead to stabilization of the phosphoric anhydride linkages. The problem with this idea is that the N-glycosidic linkage is acid labile. This bond is more stable in alkaline solution but the anhydride bonds are readily cleaved by hydroxide attack.

9.   The key here is to recognize that kinases are enzymes that transfer the γ-phosphate of ATP to a target molecule. Thus, creatine kinase phosphorylates creatine and adenylate kinase phosphorylates AMP. The following scheme outlines how ATP, ADP, AMP, creatine phosphate, creatine, and P_i are related.

$$\text{ATP + AMP} \quad \leftarrow \quad \text{ADP + ADP (adenylate kinase)}$$
$$\downarrow \qquad\qquad\qquad \uparrow$$
$$\text{ATP + H}_2\text{O} \quad \rightarrow \quad \text{ADP + P}_i \quad \text{(myosin)}$$
$$\nearrow \qquad\qquad\qquad \downarrow$$
$$\text{ATP + creatine} \quad \leftarrow \quad \text{ADP + creatine phosphate (creatine kinase)}$$

10. Phosphate is an electron-withdrawing group that is attached to quite different carbons in glucose-6-phosphate versus glucose-1-phosphate. In the latter, carbon-1 is already bonded to an electronegative oxygen in the pyranose form.

Summary

Thermodynamics - a collection of laws and principles describing the flows and interchanges of heat, energy and matter - can provide important insights into metabolism and bioenergetics. Thermodynamics distinguishes between closed systems, which cannot exchange matter or energy with the surroundings; isolated systems, which may exchange heat, but not matter, with the surroundings; and open systems, which may exchange matter, energy or both with the surroundings.

The first law of thermodynamics states that the total energy of an isolated system is conserved. E, the internal energy function, which is equal to the sum of heat absorbed and work done on the system, is a useful state function, which keeps track of energy transfers in such systems. In constant pressure processes, it is often more convenient to use the enthalpy function (H = E + PV) for analysis of heat and energy exchange. For biochemical processes, in which pressure is usually constant and volume changes are small, enthalpy (H) and internal energy (E) are often essentially equal. Enthalpy changes for biochemical processes can often be determined from a plot of R(ln K_{eq}) versus 1/T (a van't Hoff plot).

Entropy is a measure of disorder or randomness in the system. The second law of thermodynamics states that systems tend to proceed from ordered (low entropy) states to disordered (high entropy) states. The entropy change for any reversible process is simply the heat transferred divided by the temperature at which the transfer occurs: $dS_{rev} = dq/T$.

The third law of thermodynamics states that the entropy of any crystalline, perfectly ordered substance must approach zero as the temperature approaches 0 K, and at 0 K, entropy is exactly zero. The concept of entropy thus invokes the notion of heat capacity - the amount of heat any substance can store as the temperature of that substance is raised by one degree K. Absolute entropies may be calculated for any substance if the heat capacity can be evaluated at temperatures from 0 K to the temperature of interest, but entropy changes are usually much more important in biochemical systems.

The free energy function is defined as G = H - TS and, for any process at constant pressure and temperature (i.e., most biochemical processes), $\Delta G = \Delta H - T\Delta S$. If ΔG is negative, the process is exergonic and will proceed spontaneously in the forward direction. If ΔG is positive, the reaction or process is endergonic and will proceed spontaneously in the reverse direction. The sign and value of ΔG, however, do not allow one to determine how fast a process will proceed. It is convenient to define a standard state for processes of interest, so that the thermodynamic parameters of different processes may be compared. The standard state for reactions in solution is 1 M concentration for all reactants and products. The free energy change for a process A + B ⇌ C + D at concentrations other than standard state is given by:

$$\Delta G = \Delta G° + RT \ln([C][D]/[A][B])$$

and the standard state free energy change is related to the equilibrium constant for the reaction by $\Delta G° = -RT$ ln K_{eq}. This states that the equilibrium established for a reaction in solution is a function of the standard state free energy change for the process. In essence, $\Delta G°$ is another way of writing an equilibrium constant. Moreover, the entropy change for a process may be determined if the enthalpy change and free energy change calculated from knowledge of the equilibrium constant and its dependence on temperature are known.

The standard state convention of 1 M concentrations becomes awkward for reactions in which hydrogen ions are produced or consumed. Biochemists circumvent this problem by defining a modified standard state of 1 M concentration for all species except H^+, for which the standard state is 1×10^{-7} M or pH 7.

Thermodynamic parameters may provide insights about a process. As an example, consider heat capacity changes: Positive values indicate increase freedom of movement whereas negative values indicate less freedom of motion.

Many processes in living things must run against their thermodynamic potential, i.e., in the direction of positive ΔG. These processes are driven in the thermodynamically unfavorable direction via coupling with highly favorable processes. Coupled processes are vitally important in intermediary metabolism, oxidative phosphorylation, membrane transport and many other processes essential to life.

There is a hierarchy of energetics among organisms: certain organisms capture solar energy directly, whereas others derive their energy from this group in subsequent chemical processes. Once captured in chemical form, energy can be released in controlled exergonic reactions to drive life processes. High energy phosphate anhydrides and reduced coenzymes mediate the flow of energy from exergonic reactions to the energy-requiring processes of life. These molecules are transient forms of stored energy, rapidly carrying energy from point to point in the organism. One of the most important high energy phosphates is ATP, which acts as an intermediate energy shuttle molecule. The free energy of hydrolysis of ATP (for hydrolysis to ADP and phosphate) is less than that of PEP, cyclic-AMP, 1,3-BPG, creatine phosphate, acetyl phosphate and pyrophosphate, but is greater than that of the lower energy phosphate esters, such as glycerol-3-phosphate and the sugar phosphates. Group transfer potential - the free energy change that occurs upon hydrolysis (i.e., transfer of a chemical group to water) - is a convenient parameter for quantitating the energetics of such processes. The phosphoryl group transfer potentials for high energy phosphates range from -31.9 kJ/mole for UDP-glucose to -35.7 kJ/mole for ATP to -62.2 kJ/mole for PEP.

The large negative ΔG°' values for the hydrolysis of phosphoric acid anhydrides (such as ATP, GTP, ADP, GDP, sugar nucleotides and pyrophosphate) may be ascribed to 1) destabilization of the reactant due to bond strain caused by electrostatic repulsion, 2) stabilization of the products by ionization and resonance, and 3) increases in entropy upon product formation. Mixed anhydrides of phosphoric and carboxylic acids - known as acyl phosphates - are also energy-rich. Other classes of high energy species include enol phosphates, such as PEP, guanidinium phosphates, such as creatine phosphate, cyclic nucleotides, such as 3',5'-cyclic AMP, amino acid esters, such as aminoacyl-tRNA and thiol esters, such as coenzyme A. Other biochemically important high energy molecules include the pyridine nucleotides (NADH, NADPH), sugar nucleotides (UDP-glucose) and S-adenosyl methionine, which is involved in the transfer of methyl groups in many metabolic processes.

Though the hydrolysis of ATP and other high energy phosphates are often portrayed as simple processes, these reactions are far more complex in real biological systems. ATP, ADP and similar molecules can exist in several different ionization states, and each of these individual species can bind divalent and monovalent metal ions, so that metal ion complexes must also be considered. For example, ATP has five dissociable protons. The adenine ring amino group exhibits a pK_a of 4.06, whereas the last proton to dissociate from the triphosphate chain possesses a pK_a of 6.95. Thus, at pH 7, ATP is a mixture of ATP^{4-} and $HATP^{3-}$. If equilibrium constants for the hydrolysis reactions are re-defined in terms of total concentrations of the ionized species, then such equilibria can be considered quantitatively, and fractions of each ionic species in solution can be determined. The effects of metal ion binding equilibria on the hydrolysis equilibria can be quantitated in a similar fashion.

The free energy changes for hydrolysis of high energy phosphates such as ATP are also functions of concentration. In the environment of the typical cell, where ATP, ADP and phosphate are usually 5 mM or less, the free energy change for ATP hydrolysis is substantially larger in magnitude than the standard state value of -30.5 kJ/mole.

High energy molecules such as ATP are rapidly recycled in most biological environments. The typical 70 kg human body contains only about 50 grams of ATP/ADP. Each of these ATP molecules must be recycled nearly 2,000 times each day to meet the energy needs of the body.

Chapter 4

Amino Acids and the Peptide Bond

• •

Chapter Outline

❖ Amino acids: Tetrahedral alpha (α) carbon with H, amino group, carboxyl group, and side chain
❖ Amino acids in proteins joined by peptide bonds
  ⅄ Amino group reacted with carboxyl group and removal of water
  ⅄ Peptide bonds join amino acid residues in polypeptides
❖ Classification
  ⅄ Nonpolar: Ala, val, leu, ile, pro, met, phe, trp
  ⅄ Polar uncharged: Gly, ser, thr, tyr, asn, gln, cys
  ⅄ Acidic: Asp, glu
  ⅄ Basic: His, lys, arg
❖ Amino acids come in a variety of forms: There are far more than 20 amino acids
  ⅄ The 20 amino acids presented above are coded for in the genetic code
  ⅄ Selenocysteine and pyrrolysine are the 21st and 22nd amino acids to be coded for
  ⅄ Modified amino acids: Hydroxylysine, hydroxproline, thyroxin, methyl histidine, methyl arginine, methyl lysine, γ–carboxyglutamic acid, pyroglutamic acid, aminoadipic acid
❖ Small molecules derived from amino acids: GABA, histamine, serotonin, β–alanine, epinephrine, penicillamine, ornithine, citrulline, betaine, homocysteine, homoserine
❖ Ionic properties of amino acids
  ⅄ α–Carboxyl group: pK_a's range from 2.0 to 2.4
  ⅄ α–Amino group: pK_a's range from 9.0 to 9.8
  ⅄ Side chains
    • Acidic pK_a's for asp, glu
    • Basic pK_a's for his, arg, lys
❖ Amino acid chemistry: Edman reagent
  ⅄ Used to modify N-terminus and to sequence peptides
❖ Chirality: Asymmetric carbon has four different groups attached
  ⅄ Optically active compounds
    • Dextrorotatory (+), clockwise rotation
    • Levorotatory (-), counterclockwise rotation
  ⅄ Asymmetric carbon compounds
    • One asymmetric carbon: Mirror image isomers or enantiomers
    • More than one asymmetric carbon: enantiomers and diastereomers
  ⅄ D,L System: Typically naturally occurring amino acids: L-amino acids; related to L-glyceraldehyde
  ⅄ (R,S) System: Assign priorities to groups about the chiral carbon
    • $SH > OH > NH_2 > COOH > CHO > CHOH > CH_3$
      • First four above: Priority order is M_r of heaviest element
      • Last four: Priority is oxidation state of carbon
❖ Ultraviolet spectral properties of amino acids: Only phe, try, trp absorb UV light above 250 nm
❖ NMR signal from amino acids sensitive to magnetic environment
❖ Separation of amino acids

57

      ⅄   Ion exchange
- Anion exchangers: Matrix is positively charged so anions bind
- Cation exchangers: Matrix is negatively charged so cations bind

      ⅄   Gas chromatography and high-performance liquid chromatography (HPLC)

❖ Peptide bonds: Carboxyl group + amino group – water = peptide (amide) bond
      ⅄   Trans arrangement (usually) of carbonyl oxygen and amide hydrogen

❖ Double-bond character
      ⅄   Pi electron delocalization prevents free rotation about peptide bond
      ⅄   Amide plane includes carbonyl, amide and alpha carbons

❖ Peptide classification
      ⅄   Up to 12 amino acid residues, dipeptide, tripeptide, tetrapeptide, etc.
- Numerical prefix refers to number of amino acid residues not peptide bonds

      ⅄   Oligopeptides: 12 to 20 residues
      ⅄   Polypeptides: Several dozen residues
      ⅄   Proteins: Polypeptides
- Monomeric proteins have one chain
- Multimeric proteins have more than one chain
  - Homomultimeric: Identical chains
  - Hetromultimeric: Different chains

Chapter Objectives

It is imperative to learn the structures and one-letter codes for the amino acids. Amino acids are key molecules, both as components of proteins and as intermediates to important biomolecules. The structures are shown in Figure 4.3 of *Biochemistry*. Figure 4.3 is presented on the following two pages.

Classification of Amino Acids

Here is an alternative classification scheme.

Acidic amino acids and their amides:
> **aspartic acid, asparagine, glutamic acid, glutamine**.

Basic amino acids:
> **histidine, lysine, arginine**.

Aromatic amino acids:
> **phenylalanine, tyrosine, tryptophan**.

Sulfur containing amino acids:
> **cysteine, methionine**.

Imino acid:
> **proline**.

Hydrophobic side chains:
> **glycine, alanine, valine, leucine, isoleucine**.

Hydroxylic amino acids:
> **serine, threonine, (tyrosine)**.

To memorize the structures of the 20 amino acids, here are some aids:

The side chain of **glycine** is **H** and that of **alanine** is a methyl group, CH_3.

The side chain of **valine** is just the methyl side chain of alanine with two methyl groups substituting for two Hs. (In effect it is just dimethylalanine.)

The side chain of **leucine** is just valine with a $-CH_2-$ group between the valine side chain and the α-carbon (or, leucine is alanine with the valine side chain attached).

Isoleucine is an isomer of leucine with one of the methyl groups exchanged with a H on the carbon attached to the α-carbon.

Several of the amino acids have alanine as a base.

Phenylalanine has a phenyl group (i.e., benzene ring) attached to alanine.

Tyrosine is hydroxylated phenylalanine.

Serine has a hydroxyl group substituting for a H on the methyl group of alanine.

Cysteine can be thought of as the sulfur variety of serine i.e., with a sulfhydryl group in place of the hydroxyl group of serine.

Threonine, a hydroxylic amino acid, can be thought of as methylated serine.

The acidic amino acids and their amides are quite easy to remember. **Aspartic acid** and **glutamic acid** have carboxylic acid groups attached to $-CH_2-$ and $(-CH_2-)_2$, respectively. The amides of these amino acids, namely **asparagine** and **glutamine**, have nitrogen attached to the side chain carboxyl groups in amide linkage.

The basic amino acids all have six carbons, counting the carboxyl carbon and the α–carbon. **Lysine's** side chain is a linear chain of four $-CH_2-$ groups with a terminal amino group. **Arginine** has a linear side chain of three $-CH_2-$ groups to which a guanidinium group is attached. (The sixth carbon is the central C of the guanidinium group.) Guanidinium is like urea but with a nitrogen group replacing the carbonyl oxygen. **Histidine** has an imidazole ring attached to alanine. The ring contains two nitrogens and three carbons, a five-membered ring.

An easy way to remember the structure of **proline** is to know that it derives from glutamic acid. The carboxyl side chain forms a C-N bond to the amino group to form this imino acid. (In the process, the carbon is reduced so the side chain carboxyl oxygens do not show up in proline.)

Methionine is a sulfur-containing amino acid. Biologically, it is also used as a methyl group donor in the molecule S-adenosylmethionine (SAM). These two facts indicate that the side chain ends in a $S-CH_3$, which is attached to $(-CH_2-)_2$. The remaining amino acid is **tryptophan**. Its side chain is an indole ring attached to a $-CH_2-$ group. In effect it can be thought of as alanine with a indole ring. Substituted indole rings are widely used in biochemistry. The basic structure is a six-membered benzene ring fused to a five-membered ring containing a single N.

Nomenclature

The one-letter code for the amino acids must be memorized. In journals, the one letter codes are used to show amino acid sequences. See the answer to Problem 2 for a discussion of how the one letter codes are organized.

Chirality

The α–carbons of amino acids, excepting Gly, have four different groups attached to them, and hence they are chiral carbons. Two different arrangements of the four groups about the chiral carbon can be made, giving rise to two structural isomers. Structural isomers that are mirror images of each other are called enantiomers. When two or more chiral centers exist on a single molecule, 2^n structural isomers are possible, where n = the number of chiral centers. The various structural isomers include enantiomers (mirror-image-related isomers) and non mirror-image-related isomers known as diastereomers.

The structural isomers can be distinguished from each other using the R and S prefixes assigned to each chiral center of a molecule. The R and S prefixes are assigned according to a set of sequence rules. The four groups about a chiral center are assigned priority numbers with the highest priority given to the atom with the highest atomic number, thus S > O > N > C > H (for elements found in amino acids). When two or more groups have the same primary element, then elements attached to the primary element are considered. So for example, SH > OR > OH > NH_2 > COOH > CHO > CH_2OH > CH_3. Once priority assignments are made, the molecule is viewed with the group having the lowest priority away from the reader. Moving from highest to lowest priority, for the three remaining groups, clockwise movement indicates R and counterclockwise indicates S configuration.

UV-absorbing Properties

Only three amino acids, phenylalanine, tyrosine, and tryptophan, absorb light in the near UV range (i.e., 230 nm to 310 nm). These amino acids will dominate the UV absorption spectra of proteins. The wavelength maxima for tyrosine and tryptophan are around 280 nm. In addition, these two amino acids have extinction coefficients quite a bit larger than the extinction coefficient of phenylalanine, whose λ max = 260 nm. Thus very many proteins have maximum absorbance around 280 nm. In contrast, nucleic acids absorb light with a maximum around 260 nm due to the nucleotides. The extinction coefficients of the nucleotides are larger than those for the aromatic amino acids. In general, nucleic acid contamination of a protein sample will contribute to UV absorption at 260 nm. Often the 260/280 ratio is used as a criterion for purity.

Figure 4.3 The amino acids that are the building blocks of most proteins can be classified as (a) nonpolar (hydrophobic), (b) polar, neutral, (c) acidic, or (d) basic. Also shown are the one-letter and three-letter codes used to denote amino acids. For each amino acid, the ball-and-stick (left) and space-filling (right) models show only the side chain.

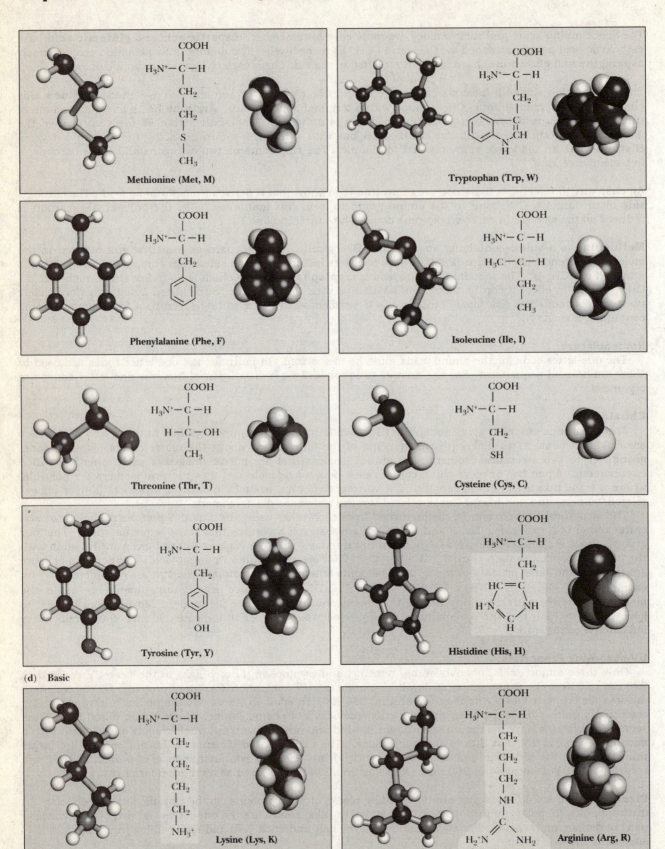

Methionine (Met, M)

Tryptophan (Trp, W)

Phenylalanine (Phe, F)

Isoleucine (Ile, I)

Threonine (Thr, T)

Cysteine (Cys, C)

Tyrosine (Tyr, Y)

Histidine (His, H)

(d) Basic

Lysine (Lys, K)

Arginine (Arg, R)

(a) Nonpolar (hydrophobic)

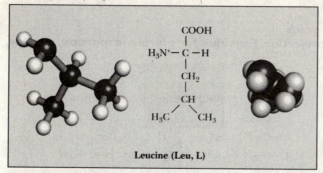

Leucine (Leu, L)

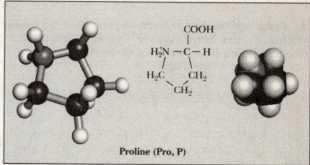

Proline (Pro, P)

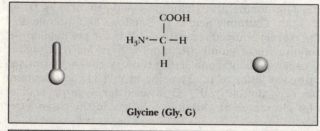

Alanine (Ala, A)

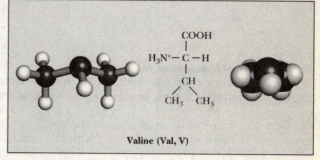

Valine (Val, V)

(b) Polar, uncharged

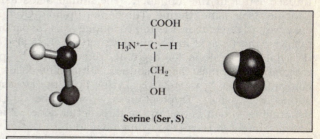

Glycine (Gly, G)

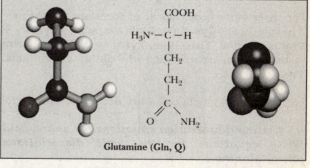

Serine (Ser, S)

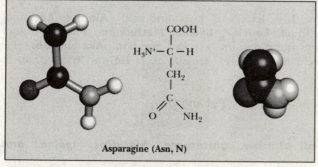

Asparagine (Asn, N)

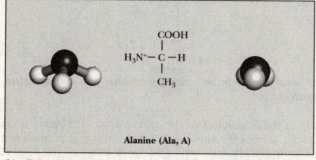

Glutamine (Gln, Q)

(c) Acidic

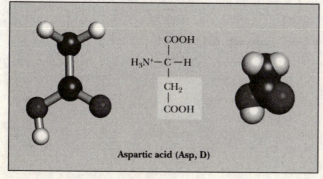

Aspartic acid (Asp, D)

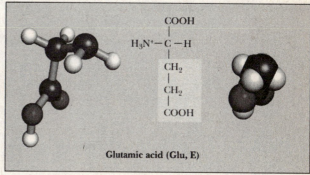

Glutamic acid (Glu, E)

Problems and Solutions

1. Drawing Fischer projection formulas for amino acids
Without consulting chapter figures, draw Fisher projection formulas for glycine, aspartate, leucine, isoleucine, methionine, and threonine.

Answer:

| Glycine | Aspartate | Leucine | Isoleucine | Methionine | Threonine |

2. Knowing abbreviation for amino acids
Without reference to the text, give the one-letter and three-letter abbreviations for asparagine, arginine, cysteine, lysine, proline, tyrosine, and tryptophan.

Answer: For many of the amino acids the one letter code is the first letter of the amino acid. Problems arise when two or more amino acids start with the same letter. Whenever this occurs, the amino acid with the lowest molecular weight is assigned the letter. The other amino acids are given a phonetic letter or the closest unused letter in the alphabet. The amino acids assigned their first letters are: Glycine, Alanine, Valine, Leucine, Isoleucine, Serine, Threonine, Proline, Cysteine, Methionine and Histidine. Aspartic acid is D, an easy way to remember it is to pronounce it as "asparDic" acid. Glutamic acid is E, E follows D, glutamic acid follows aspartic acid in chemical structure. E is also the closest unused letter to G. F is for phenylalanine, (Fenylalanine). R is for arginine, remembered by pronouncing arginine (Rginine). N is for asparagine; remember to stress the N sound. Glutamine follows asparagine, its one letter code is the next unused consonant after N, namely Q. Lysine is K, the closest unused letter to L. Tyrosine is Y; Y is the dominant sound in tyrosine. Tryptophan is W; remember "double ring, double you". B is used for aspartic acid or asparagine, which is Asx in the three-letter code. Z is for glutamine or glutamic acid (glx in the three letter code). Finally, X refers to any amino acid.

For three-letter codes, most of them are simply the first three letters of the amino acid. **Ala**nine, **Arg**inine, **Asp**artic acid, **Cys**teine, **Glu**tamic acid, **Gly**cine, **His**tidine, **Leu**cine, **Lys**ine, **Met**hionine, **Phe**nylalanine, **Pro**line, **Ser**ine, **Thr**eonine, **Tyr**osine, **Val**ine. Asparagine and glutamine are **Asn** and **Gln**. **Asx** and **Glx** refer to asparagine/aspartic acid and glutamine/glutamic acid respectively. Isoleucine is **Ile** and tryptophan is **Trp**.

For selenocysteine its codes are **Sec** and U and pyrrolysine's codes are **Pyl** and O.

3. Writing dissociation equations for amino acids
Write equations for the ionic dissociations of alanine, glutamate, histidine, lysine, and phenylalanine.

Answer: Alanine dissociation:

Glutamate Dissociation:

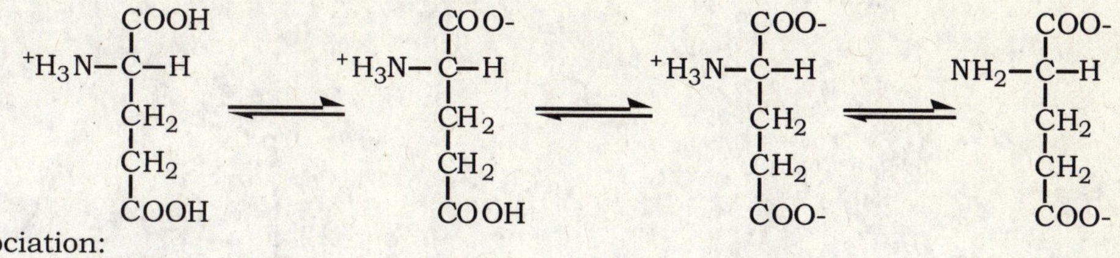

Histidine dissociation:

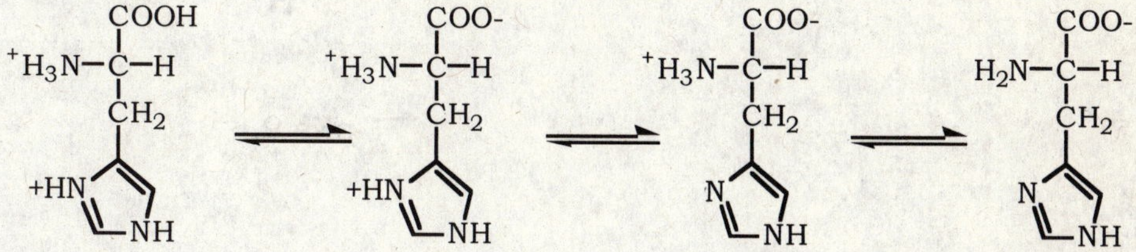

Lysine dissociation:

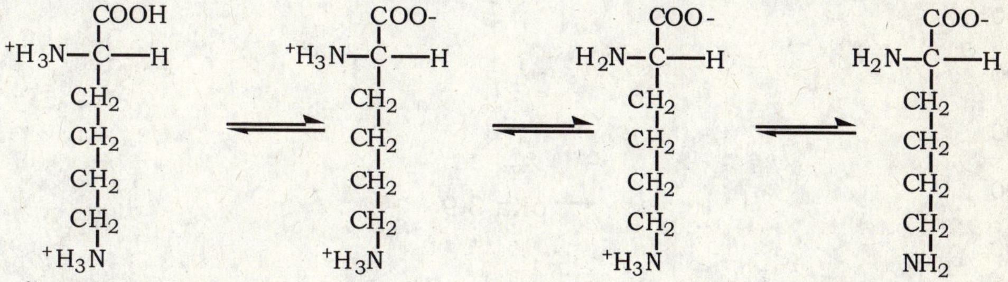

Phenylalanine dissociation:

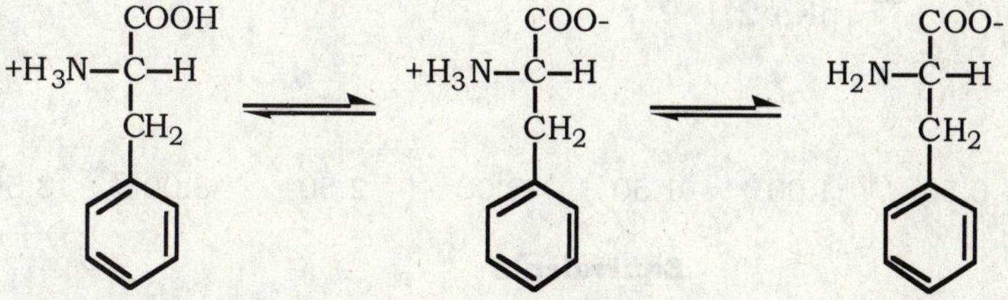

4. Understanding chemical effects on amino acid pK_a values
How is the pK_a of the α-NH_3^+ group affected by the presence on an amino acid of the α-COOH?

Answer: The pK_a on an isolated amino group is around 10 (as for example, the side chain amino group of lysine whose pK_a = 10.5 (see Table 4.1). In general, the pK_as of α-amino groups in the amino acids are around 9.5. Thus, the proximity of the α-carboxyl group lowers the pK_a of the α-amino group.

5. Drawing titration curves for the amino acids
(Integrates with Chapter 2) Draw an appropriate titration curve for aspartic acid, labeling the axis and indicating the equivalence points and the pK_a values.

Answer: For a titration curve the independent variable is the amount of acid or base used to titrate the amino acid. The dependent variable is pH.

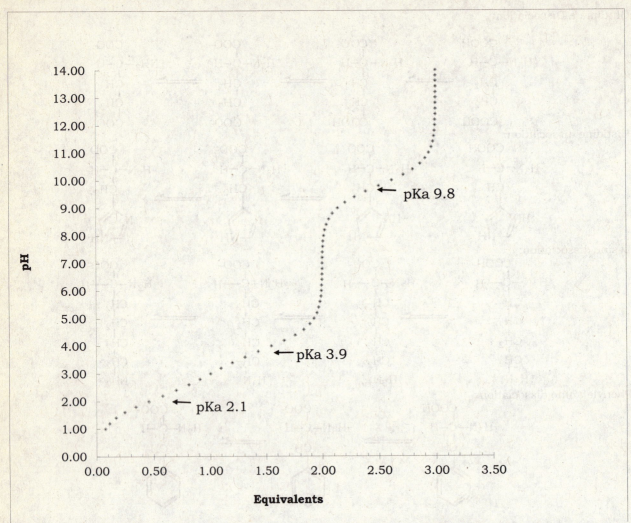

The equivalence point is a point on the titration curve reached after a molar equivalent of acid or base is added to the amino acid. For aspartic acid there are actually three equivalence points at around 3.0, 6.2 and 12.2.

6. Calculating concentrations of species in amino acid solutions
(Integrates with Chapter 2) Calculate the concentrations of all ionic species in a 0.25 M solution of histidine at pH 2, pH 6.4, and pH 9.3.

Answer: The pK$_a$s for histidine are 1.8, 6.0, and 9.2. At pH 2.0, we can ignore the two higher pK$_a$s. At this pH, histidine's side chain and the α-amino group are protonated and thus both are positively charged. So, the ionic species of histidine at pH = 2.0 depend critically on the ionization state of the carboxyl group. When the carboxyl group is in the carboxylate form, histidine will be in the 1+ ionic form. When the carboxyl group is protonated, and hence uncharged, histidine will carry a 2+ charge. For the α-carboxyl group:

$$pH = pK_a + \log \frac{[COO^-]}{[COOH]} \text{ or,}$$

$$\frac{[COO^-]}{[COOH]} = 10^{pH-pK_a} = 10^{2.0-1.8} = 1.58 \text{ or } [COO^-] = 1.58 \times [COOH] \text{ and,}$$

$$[COO^-] + [COOH] = 0.25 \text{ M}$$

Thus, $1.58 \times [COOH] + [COOH] = 2.58 \times [COOH] = 0.25$ M and,

$$[COOH] = \frac{.25 \text{ M}}{2.58} = 0.097 \text{ M and,}$$

$$[COO^-] = 1.58 \times [COOH] = 0.153 \text{ M}$$

So, at pH=2.0, $[His^{2+}]=0.097$ M and $[His^{1+}]=0.153$ M

The concentration of uncharged histidine and negatively charged histidine are both small but can be estimated as follows. The concentration of uncharged histidine will be approximately equal to the concentration of the unprotonated side-chain species. Using the Henderson-Hasselbalch equation with $pK_a = 6.0$, pH = 2.0, and $[His^{1+}] = 0.153$ M, we find that

$$pH = 6.0 + \log \frac{[His^0]}{[His^{1+}]} \text{ or,}$$

$$\frac{[His^0]}{[His^{1+}]} = 10^{pH-6.0} = 10^{2.0-6.0} \text{ and,}$$

$$[His^0] = [His^{1+}] \times 10^{-4} = 1.53 \times 10^{-5} \text{M}$$

The concentration of negatively charged histidine is calculated using the Henderson-Hasselbalch equation with pKa = 9.2, pH = 2.0, and $[His^0] = 1.53 \times 10^{-5}$ M. In this case

$$pH = 9.2 + \log \frac{[His^{1-}]}{[His^0]}, \text{ or}$$

$$\frac{[His^{1-}]}{[His^0]} = 10^{pH-9.2} = 10^{2.0-9.2} \text{ and,}$$

$$[His^{1-}] = [His^0] \times 10^{-7.2} = 1.53 \times 10^{-5} \times 10^{-7.2} = 9.6 \times 10^{-13} \text{M}$$

At pH = 2.0 we have: $\quad [His^{2+}] = 0.097$ M , $\qquad [His^0] = 1.53 \times 10^{-5}$ M,

$\qquad\qquad\qquad\qquad\quad [His^{1+}] = 0.153$ M, $\qquad\quad [His^{1-}] = 9.6 \times 10^{-13}$ M.

At pH = 6.4, we can assume that the carboxyl group is fully unprotonated and carries a -1 charge while the α-amino group is fully protonated and carries a +1 charge. Thus, the ionic species of histidine depends critically on the ionic state of the imidazole side chain. When the imidazole group is protonated, histidine carries a net +1 charge. When it is unprotonated, histidine is uncharged.
So, using

$$pH = 6.0 + \log \frac{[His^0]}{[His^{1+}]} \text{ and } [His^0]+[His^{1+}]=0.25 \text{ M },$$

we find that the ionic species of histidine at pH=6.4 are:

$$[His^0]=0.179 \text{ M and } [His^{1+}] = 0.071 \text{ M}$$

$[His^{2+}]$ and $[His^{1-}]$ are calculated using the Henderson-Hasselbalch equation with pH = 6.4, pK_a = 1.8 and 9.2 and $[His^{1+}] = 0.071$ M and $[His^0] = 0.179$ M. We find $[His^{2+}] = 2.8 \times 10^{-6}$ M and $[His^{1-}] = 2.25 \times 10^{-4}$ M.

At pH = 6.4 we have: $\quad [His^{2+}] = 1.78 \times 10^{-6}$ M , $\qquad [His^0] = 0.179$ M,

$\qquad\qquad\qquad\qquad\quad [His^{1+}] = 0.071$ M, $\qquad\qquad\quad [His^{1-}] = 2.84 \times 10^{-4}$ M.

At pH = 9.3, the carboxyl group is unprotonated and negatively charged, the imidazole group is unprotonated and uncharged, and only the protonation state of the α-amino group needs to be determined. Histidine, at pH = 9.3 will be portioned between the His^0 and His^{1-} states. From

$$pH = 9.2 + \log \frac{[His^{1-}]}{[His^0]} \text{ and } [His^0] + [His^{1-}] = 0.25 \text{ M} ,$$

we find that the ionic species of histidine at pH=9.3 are:

$[His^0] = 0.111$ M, $\qquad [His^{1-}] = 0.139$ M and

$[His^{2+}] = 1.75 \times 10^{-12}$M, $[His^+] = 5.5 \times 10^{-5}$M

There is an alternate solution. Consider any one single ionizable group on histidine. For example, consider the carboxyl group. One can use the Henderson-Hasselbalch equation to evaluate the fraction of carboxyl groups that are protonated and the fraction that are unprotonated. This is done as follows:

$$pH = pK_{COOH} + \log \frac{[A^-]}{[HA]}$$

$$\frac{[A^-]}{[HA]} = 10^{(pH-pK_{COOH})}$$

$$[A^-] = [HA] \times 10^{(pH-pK_{COOH})}$$

$$f_{COO^-} = \frac{[A^-]}{[HA]+[A^-]} = \frac{[HA] \times 10^{(pH-pK_{COOH})}}{[HA]+[HA] \times 10^{(pH-pK_{COOH})}} = \frac{10^{(pH-pK_{COOH})}}{1+10^{(pH-pK_{COOH})}}$$

And since $f_{COO^-} + f_{COOH} = 1$,

$$f_{COOH} = 1 - f_{COO^-}$$

These fractional terms can be easily evaluated for the amino group and the imidazole side chain by substituting the appropriate pK$_a$ value in the above equations. Doing this we would be left with six values, two each for each of the ionizable groups on histidine. The fraction of histidines with any particular arrangement of protonated or unprotonated groups is simply the product of the appropriate fractions. For example, the fraction of molecules with the carboxyl group unprotonated, side chain and amino group both protonated is:

$$f_{COO^-} \times f_{IH+} \times f_{NH_3^+}$$

Practically speaking our first approach to solving this problem was a practical one in that it evaluated only the more likely species to be encountered.

7. Calculating pH in amino acid solutions
(Integrates with Chapter 2) Calculate the pH at which the γ-carboxyl group of glutamic acid is two-thirds dissociated.

Answer: The pK$_a$ of γ-carboxyl group of glutamic acid is 4.3.

Using the Henderson-Hasselbalch equation

$$pH = pK_{COOH} + \log \frac{[C_\gamma OO^-]}{[C_\gamma OOH]}$$

When the γ-carboxyl group is two-thirds dissociated:

$$\frac{[C_\gamma OO^-]}{[C_\gamma OO^-]+[C_\gamma OOH]} = \frac{2}{3} \text{ or, } [C_\gamma OO^-] = \frac{2}{3}\left([C_\gamma OO^-]+[C_\gamma OOH]\right)$$

Thus, $[C_\gamma OOH] = \frac{1}{2}[C_\gamma OO^-]$

Upon substituting into the Henderson-Hasselbalch equation we find:

$$pH = 4.3 + \log \frac{[C_\gamma OO^-]}{[C_\gamma OOH]} = 4.3 + \log \frac{[C_\gamma OO^-]}{\frac{1}{2}[C_\gamma OO^-]} = 4.3 + \log 2$$

$$pH = 4.6$$

8. Calculating pH in amino acid solutions
(Integrates with Chapter 2) Calculate the pH at which the ε-amino group of lysine is 20% dissociated.

Answer: For the lysine side chain, pK$_a$ = 10.5.

66

Using the Henderson-Hasselbalch equation

$$pH = pK_{N_\varepsilon H_3^+} + \log\frac{[N_\varepsilon H_2]}{[N_\varepsilon H_3^+]}$$

When the ε-amino group is 20% dissociated:

$$\frac{[N_\varepsilon H_2]}{[N_\varepsilon H_2]+[N_\varepsilon H_3^+]} = 0.2 \ (20\%) \text{ or, } [N_\varepsilon H_2] = 0.2\big([N_\varepsilon H_2]+[N_\varepsilon H_3^+]\big)$$

Thus, $[N_\varepsilon H_2] = \frac{1}{4}[N_\varepsilon H_3^+]$

Upon substituting into the Henderson-Hasselbalch equation we find:

$$pH = 10.5 + \log\frac{[N_\varepsilon H_2]}{[N_\varepsilon H_3^+]} = 10.5 + \log\frac{\frac{1}{4}[N_\varepsilon H_3^+]}{[N_\varepsilon H_3^+]} = 10.5 - \log 4$$

$$pH = 9.9$$

9. Calculating the pH of amino acid solutions
(Integrates with Chapter 2) Calculate the pH of a 0.3 M solution of (a) leucine hydrochloride, (b) sodium leucinate, and (c) isoelectric leucine.

Answer: The solution 0.3 M leucine HCl is composed of 0.3 M leucine and 0.3 M HCl. Thus, we expect it to have an acidic pH and only the α-carboxyl group of leucine is involved in proton equilibrium.

The pKa of the α-carboxyl of leucine is 2.4. Thus,

$$pH = pK_a + \log\frac{[A^-]}{[HA]} = 2.4 + \log\frac{[COO^-]}{[COOH]} \ (1) \text{ and,}$$

$$[COO^-]+[COOH] = 0.3 \text{ M } (2)$$

For charge neutrality:

$$[Cl^-]+[COO^-] = [H^+]+[NH_3^+]$$

If we recognize that $[Cl^-]=[NH_3^+]$ (because Cl⁻ and the amino acid are in equimolar amounts)

The charge neutrality equation may be simplified to:

$[COO^-]=[H^+]$ and combined with (1) and (2) we have:

$$pH = 2.4 + \log\frac{[H^+]}{0.3-[H^+]}$$

If we make the assumption that $[H^+]\ll 0.3$

$$pH = 2.4 + \log\frac{[H^+]}{0.3} = 2.4 + \log[H^+] - \log 0.3 = 2.4 - \log 0.3 + \log[H^+]$$

Recognizing that $\log[H^+]=-pH$ we see that

$pH = 2.4 - \log 0.3 - pH$ or

$$pH = \frac{2.4 - \log 0.3}{2} = 1.46$$

Note: If we did not make the simplification that $[H^+]<0.3$ we would have had to solve a quadratic and would have found that pH = 1.49.

(b) sodium leucinate.
Here we assume the solution is basic; therefore, only the α-amino group is involved. The mathematical solution is similar to that given for (a) with pKa = 9.6.

$$pH = pK_a + \log\frac{[NH_2]}{[NH_3^+]} = 9.6 + \log\frac{[NH_2]}{[NH_3^+]} \ (1) \ \text{and,}$$

$[NH_2] + [NH_3^+] = 0.3M \ (2)$

For charge neutrality:

$[OH^-] + [COO^-] = [NH_3^+] + [Na^+]$

If we recognize that $[COO^-] = [Na^+]$ (because Na^+ and the amino acid are in equimolar amounts)

The charge neutrality equation may be simplified to:

$[OH^-] = [NH_3^+]$ and combined with (1) and (2) we have:

$$pH = 9.6 + \log\frac{0.3 - [OH^-]}{[OH^-]}$$

If we make the assumption that $[OH^-] << 0.3$

$$pH = 9.6 + \log\frac{0.3}{[OH^-]} = 9.6 + \log 0.3 - \log[OH^-]$$

Recognizing that $pOH \equiv -\log[OH^-]$ and that $pH + pOH = 14$ we have

$pH = 9.6 + \log 0.3 + pOH = 9.3 + \log 0.3 + 14 - pH$ or

$$pH = \frac{9.6 + 14 + \log 0.3}{2} = 11.5$$

(c) isoelectric leucine

Isoelectric leucine is a leucine solution at a pH at which there is no net charge on the molecule. For this to occur the carboxyl group must be unprotonated to the same extent as the amino group is protonated. This must occur at a pH half way between the pK_as of the two groups

At the pI, $[COO^-] = [NH_3^+]$

Using the Henderson-Hasselbalch equations

$$pH = pK_{COOH} + \log\frac{[COO^-]}{[COOH]} \ \text{and} \ pH = pK_{NH_3^+} + \log\frac{[NH_2]}{[NH_3^+]}$$

and recognizing that $[COOH] = [\text{Amino acid}]_{total} - [COO^-]$

and that $[NH_2] = [\text{Amino acid}]_{total} - [NH_3^+]$

upon substituting, solving for $[COO^-]$ and $[NH_3^+]$, and setting these terms

equal at pH=pI we find that

$$pI = \frac{pK_{COOH} + pK_{NH_3^+}}{2} \ \text{or}$$

$$pI = \frac{2.4 + 9.6}{2} = 6.0$$

10. Understanding stereochemical transformations of amino acids
Absolute configurations of the amino acids are referenced to D- and L-glyceraldehyde on the basis of chemical transformations that can convert the molecule of interest to either of these reference isomeric structures. In such reactions, the stereochemical consequences for the asymmetric centers must be understood for each reaction step. Propose a sequence of reactions that would demonstrate that L(–)-serine is stereochemically related to L(–)-glyceraldehyde.

Answer: The aldehyde group of L(-)-glyceraldehyde is converted to L-glyceric acid by oxidation and L-glyceric acid is converted to 2-hydroxypropanoic acid. 2-Hydroxypropanoic acid is converted to 2-bromopropanoic acid by nucleophilic substitution using concentrated HBr. The reaction results in alteration in stereochemistry. 2-Bromopropanoic acid is subsequently converted to L-(-)-alanine by ammonolysis and alteration in stereochemistry. Finally, L-(-)-alanine is convert serine. The reaction sequence is shown below.

L-(-)-Glyceraldehyde   L-Glyceric acid   2-Hydroxypropanoic acid

$$\begin{array}{ccccc}
\text{CHO} & & \text{COOH} & & \text{COOH} \\
| & & | & & | \\
\text{HO-C-H} & \rightarrow & \text{HO-C-H} & \rightarrow & \text{HO-C-H} \\
| & & | & & | \\
\text{CH}_2\text{OH} & & \text{CH}_2\text{OH} & & \text{CH}_3
\end{array}$$

$$\begin{array}{ccccc}
\text{COOH} & & \text{COOH} & & \text{COOH} \\
| & & | & & | \\
\text{H-C-Br} & \rightarrow & \text{H}_2\text{N-C-H} & \rightarrow & \text{H}_2\text{N-C-H} \\
| & & | & & | \\
\text{CH}_3 & & \text{CH}_3 & & \text{CH}_2\text{OH}
\end{array}$$

2-Bromopropanoic acid   L-(-)-Alanine   L-(-)-Serine

11. Understanding amino acid stereochemistry
Describe the stereochemical aspects of the structure of cystine, the structure that is a disulfide-linked pair of cysteines.

Answer: Cystine (dicysteine) has two chiral carbons, the two α-carbons of the cysteine moieties. Since each chiral center can exist in two forms there are 4 stereoisomers of cystine. However, it is not possible to distinguish the difference between L-cysteine/D-cysteine and D-cysteine/L-cysteine dimers. So three distinct isomers are formed.

$$\begin{array}{cccc}
\text{H} & \text{H} & \text{H} & \text{H} \\
| & | & | & | \\
{}^-\text{OOC-C-NH}_3{}^+ & {}^+\text{H}_3\text{N-C-COO}^- & {}^-\text{OOC-C-NH}_3{}^+ & {}^+\text{H}_3\text{N-C-COO}^- \\
| & | & | & | \\
\text{CH}_2 & \text{CH}_2 & \text{CH}_2 & \text{CH}_2 \\
| & | & | & | \\
\text{S} & \text{S} & \text{S} & \text{S} \\
| & | & | & | \\
\text{S} & \text{S} & \text{S} & \text{S} \\
| & | & | & | \\
\text{CH}_2 & \text{CH}_2 & \text{CH}_2 & \text{CH}_2 \\
| & | & | & | \\
{}^+\text{H}_3\text{N-C-COO}^- & {}^+\text{H}_3\text{N-C-COO}^- & {}^-\text{OOC-C-NH}_3{}^+ & {}^-\text{OOC-C-NH}_3{}^+ \\
| & | & | & | \\
\text{H} & \text{H} & \text{H} & \text{H} \\
\text{L-cysteine-} & \text{L-cysteine-} & \text{D-cysteine-} & \text{D-cysteine-} \\
\text{L-cysteine} & \text{D-cysteine} & \text{L-cysteine} & \text{D-cysteine}
\end{array}$$

The two middle structures are meso-cystine and they can be identified as such because the top half and bottom half are mirror images. The two halves of L-cystine are composed of identical molecules. Remember in Fischer projections bonds drawn horizontally are coming out of the plane of the paper whereas vertical bonds are into the plane. Fischer projections can only be rotated by 180° to retain their stereochemistry. Thus, the top half of L-cystine is L-cysteine. Taking L-cysteine and rotating it by 180° will give the bottom half of the structure and joining these gives L-cystine.

12. Understanding amino acid reaction mechanisms
Draw a simple mechanism for the reaction of a cysteine sulfhydryl group with iodoacetamide.

Answer:

At slightly basic pH the cysteine sulfhydryl will be unprotonated and thus negatively charged. Nucleophilic attack by the sulfur with iodine functioning as a good leaving group will complete the reaction.

13. Determining tyrosine content of an unknown protein

A previously unknown protein has been isolated in your laboratory. Others in your lab have determined that the protein sequence contains 172 amino acids. They have also determined that this protein has no tryptophan and no phenylalanine. You have been asked to determine the possible tyrosine content of this protein. You know from your study of this chapter that there is a relatively easy way to do this. You prepare a pure 50 μM solution of the protein, and you place it in a sample cell with a 1 cm path length, and you measure the absorbance of this sample at 280 nm in a UV-visible spectrophotometer. The absorbance of the solution is 0.372. Are there tyrosines in this protein? How many? (Hint: You will need to use Beer's Law, which is described in any good general chemistry or physical chemistry textbook. You will also find it useful to know that the units of molar absorptivity are M^{-1}cm^{-1}.)

Answer: Beer's law relates the absorbance of a solution to the path length (l) solute concentration [C] and light-absorbing properties of the solute.

For a substance, the absorbance is defined as follows:

$$A_\lambda = -\log \frac{I}{I_0}$$

Where I_0 is the intensity of incident light and I is the intensity of light after it passes through the substance. The ratio of I to I_0 is the transmission, T. The subscript, λ, specifies the wavelength of light and this is often the wavelength at which absorbance is maximum for the substance. Beer's law is:

$$A_\lambda = -\log \frac{I}{I_0} = \varepsilon \times [C] \times \ell$$

In our case Beer's law is being applied to a solution of protein. The term [C] refers to the concentration of light-absorbing compound in solution. We are told that the protein lacks phenyalanine and tryptophan and so the only other amino acid that absorbs light at 280 nm is tyrosine*.

We are only interested in measuring the absorption of light due to the protein and not, for example to the cuvette or solvent used to dissolve the protein. To accomplish this the cuvette is first filled with solvent only and the instrument electronically zeroed. This, in effect, sets the transmitted light intensity for a cuvette filled with solvent-only equal to the incident light intensity. (The absorbance is zero when the transmission is one and this occurs when $I = I_0$.) The cuvette is then filled with the solution under consideration and the absorbance recorded. It is standard practice to use cuvettes with 1 cm path length. Thus, the magnitude of the absorbance measured is linearly dependent on the concentration, [C], and on the molar absorptivity, ε.

The absorbance is given as 0.371. Using Beer's Law we can determine the concentration provided we know the molar absorptivity. The molar absorptivity of tyrosine is 1,490 M^{-1}cm^{-1}.

$$A_{280nm} = 0.371 = \varepsilon \times [C] \times \ell$$

$$[C] = \frac{0.371}{1,490 \text{ M}^{-1}\text{cm}^{-1} \times 1 \text{ cm}} = 2.49 \times 10^{-4} \text{M}$$

The molar concentration calculated above is the concentration of the chromophore, namely tyrosine, in the protein solution, which is at 50 μM. If we assume that there are n moles of tyrosine per mole of protein then the molarity of tyrosine is simply n × 50 μM. Equating these two and solving for n we find the number of tyrosines in the protein.

$$2.49 \times 10^{-4} M = n \times 50 \times 10^{-6} M$$

$$n = \frac{2.49 \times 10^{-4} M}{50 \times 10^{-6} M} = 4.98 \approx 5$$

*The UV absorption spectra of proteins are determined by the aromatic amino acids and to a very small extent by cystine (essentially di-cysteine, two cysteines linked by a disulfide bond). Phenylalanine does not absorb at 280 nm but tryptophan, tyrosine and cystine do. (See "How to measure and predict the molar absorption coefficient of a protein" by Pace, CN, Vajdos F, Fee L, Grimsley G and Gray T in Proteins Science (1995) 4:2411-23.) The protein being studied was likely treated with a reducing agent, like dithiothreitol or β-mercaptoethanol to break disulfide bonds and so there should be no contribution to UV absorbance by cystine at 280 nm. Another problem that can arise when measuring light absorbing properties of proteins is that the absorptivity of the amino acids is influenced by the local environment in which the amino acid exists in the protein. Denaturing agents like 8 M urea or 6 M guanidinium HCl are often used in conjunction with a reducing agent to convert the protein into a so-called random coil whose UV-light absorbing properties more closely mimic those of its amino acid composition.

14. A rule of thumb for amino acid content of proteins
The simple average molecular weight of the 20 common amino acids is 138, but most biochemists use 110 when estimating the number of amino acids in a protein of known molecular weight. Why do you suppose this is? (Hint: there are two contributing factors to the answer. One of them will be apparent from a brief consideration of amino acid compositions of common proteins. See for example Figure 5.16 of this text.)

Answer: The first thing to realize is that a protein is composed of amino acid residues, not complete amino acids. When a peptide bond is formed between a carboxyl group and an amino group through a condensation reaction a water molecule is lost. Thus, the average molecular weight must be corrected for water loss (138 - 18 =) to 120. Figure 5.16 shows the frequency of amino acids in the SWISS-PROT protein knowledgebase. This information can be found at http://ca.expasy.org/sprot/relnotes/relstat.html for the 2011 release. It is clear from the data that the frequencies of finding amino acids in proteins are not all equal. Leucine, alanine, and serine are the most frequently-found amino acids whereas histidine, cysteine and tryptophan are found least frequently. Leucine is approximately 9 times more abundant in the data base than is tryptophan.

The data to construct Figure 5.16 can be found at the SWISS-PROT URL. The data were transferred to a spread (see below) and sorted. The molecular weights were then entered and used along with the percentage (converted to frequency) to calculate a relative molecular weight. The column headed Percent should sum to 100 but it does not likely due to rounding. When percent was converted to frequency the percent value for each amino acid was divided by the sum (99.88). The sum of the numbers in the right-most column is the average molecular weight of an amino acid found in the data base. This number, 129.13, is then corrected for loss of water (129.13 -18 =) to 111.13 indicating that the rule of thumb is pretty good.

Amino acid	Percent	Mr	Mr x frequency
L	9.66	131.17	12.69
A	8.26	89.09	7.37
G	7.08	75.07	5.32
V	6.87	117.15	8.06
E	6.75	147.13	9.94
S	6.54	105.09	6.88
I	5.97	131.17	7.84
K	5.85	146.19	8.56
R	5.53	174.20	9.64
D	5.45	133.10	7.26
T	5.33	119.12	6.36
P	4.69	115.13	5.41
N	4.06	132.12	5.37
Q	3.93	146.15	5.75
F	3.86	165.19	6.38
Y	2.92	181.19	5.30
M	2.42	149.21	3.62

H	2.27	155.16	3.53
C	1.36	121.16	1.65
W	1.08	204.23	2.21
SUM	99.88		129.13

15. Understanding the chemistry of the dipeptide sweetener aspartame
The artificial sweeteners Equal and NutraSweet contain aspartame, which has the structure:

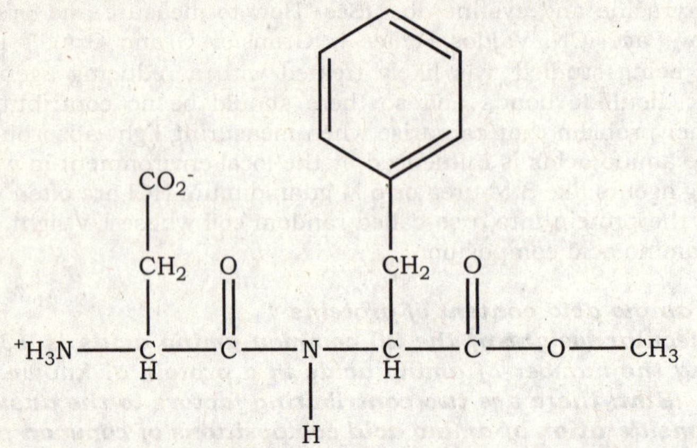

What are the two amino acids that are components of aspartame? What kind of bond links these amino acids? What do you suppose might happen if a solution of aspartame was heated for several hours at a pH near neutrality? Suppose you wanted to make hot chocolate sweetened only with aspartame, and you stored it in a thermos for several hours before drinking it. What might it taste like?

Answer: The drawing of aspartame is slightly modified from the one shown in the text. The two amino acids are aspartic acid and phenylalanine. Phenylalanine is actually a methyl ester (with a methyl group attached to its carboxyl group in ester linkage). The two amino acids are joined by a peptide bond and this bond, while quite stable, will slowly hydrolyze upon heating. Hydrolysis will release aspartic acid and phenylalanine. What will the consequences be to the taste of the hot chocolate? Amino acids have a flavor and perhaps the best example is the savory flavor of the monosodium salt of glutamic acid or MSG. The taste is called umami and it is set off by a membrane-bound taste receptor. Aspartic acid is also detected by umami receptors but not as strongly as glutamic acid. Several amino acids, like glycine, alanine, and others are sweet and phenylalanine, tyrosine, the branched chain amino acids and a few others are bitter tasting. Depending on the sensitivity of your taste buds to aspartic acid and phenylalanine, hot chocolate with hydrolyzed aspartame could range from bitter to bitter/savory.

16. Understanding a defect of amino acid metabolism
Individuals with phenylketonuria must avoid dietary phenylalanine because they are unable to convert phenylalanine to tyrosine. Look up this condition and find out what happens if phenylalanine accumulates in the body. Would you advise a person with phenylketonuria to consume foods sweetened with aspartame? Why or why not?

Answer: Phenylketonuria is a metabolic disorder caused by a deficiency of the enzyme phenylalanine hydroxylase. The deficiency arises when inactivating mutations occur in the gene for phenylalanine hydroxylase. The gene is located on human chromosome 12 and the disease is an autosomal recessive genetic disease. The enzyme converts phenylalanine into tyrosine using molecular oxygen and tetrahydrobiopterin. Tyrosine is 4-hydroxyphenylalanine and thus the reaction is a hydroxylation. It is interesting that phenylalanine is listed as an essential amino acid in humans but tyrosine is not. This is because tyrosine can be made from phenylalanine but humans lack the ability to synthesize the phenyl ring in both amino acids. The hydroxylation of phenylalanine serves two purposes, it produces tyrosine and it is in the pathway for degradation of excess phenylalanine. In general an amino acid is metabolized by removing the amino group and then catabolizing the carbon skeleton. A group of enzymes called amino transferases transfer the amino group from an amino acid to an alpha keto acid acceptor. This converts the amino acid into an alpha-keto acid and the alpha keto acceptor into an amino acid. The alpha keto acid derived from phenylalanine is phenylpyruvate. When phenylalanine hydroxylase is missing, phenylalanine cannot be converted into tyrosine but phenylalanine is converted into phenylpyruvate, which builds up in the blood and is passed into the urine hence the name of the disease is phenylketonuria. The consequence of elevated phenylpyruvate is mental retardation but this can be avoided by early detection of the disease and a diet low in phenylalanine. Neonatal screening checks for PKU among a long list of other metabolic diseases. Clearly, foods with aspartame are a very rich source of phenylalanine and should be avoided.

17. Distinguishing prochiral isomers
In this chapter, the concept of prochirality was discussed. Citrate (see Figure 19.2) is a prochiral molecule. Describe the process by which you would distinguish between (R-) and (S-) portions of this molecule and how an enzyme would discriminate between similar but distinct moieties.

Answer: For an asymmetric carbon R and S are determined by first assigning priority scores to each of the groups bound to the carbon. The priority scores are based on molecular weight of the element forming the bond but if two are identical then by their respective oxidation states. Once priorities for the four groups are assigned the molecule is rotated such that the group with the lowest priority is facing into the page (i.e., pointing away from the viewer). This will place the other three groups in a plane. If movement from highest to lowest priority among these groups is clockwise the configuration is R and if counterclockwise it is S. A prochiral carbon has 4 groups bounded to it but two of the groups are identical. The central carbon in citric acid is prochiral. Groups bounded to it include an hydroxyl group, a carboxyl group, and two carboxymethyl groups. If the priority of one of these carboxymethyl groups were increased (say by substituting a deuterium for one hydrogen) then the central carbon becomes chiral and the chirality rules for determining R and S apply. For an enzyme to produce a chiral compound from citrate the binding site for citrate would simply have to recognize the hydroxyl group and the carboxyl group attached to the central carbon of citrate and have an active site to one side or the other of these. In effect, the three sites on the enzyme would be in an R or S arrangement.

18. Understanding buffering capacity of amino acids
Amino acids are frequently used as buffers. Describe the pH range of acceptable buffering behavior for the amino acids alanine, histidine, aspartic acid, and lysine.

Answer: Buffers work best roughly within ±1 pH unit of a pK_a. The amino acids listed all have alpha amino groups with pK_a's around 9.0 and alpha carboxyl groups with pK_a's around 2.0. So, all of the amino acids would be expected to be reasonably good buffers in the ranges of 1.0 to 3.0 and 8.0 to 10.0. This ignores completely the side chains. Alanine's side chain does not have a dissociable hydrogen so no further buffering is expected but the other three amino acids do. Histidine's side chain has a pK_a around 6.0 and so it will serve as an effective buffer in the range 5.0 to 7.0. Aspartic acid's side chain pK_a at around 4.0 will buffer from 3.0 to 5.0. Finally, lysine will buffer through its side chain amino group from about 9.0 to 10.0.

Preparing for the MCAT® Exam

19. Understanding the chemistry of cysteine
Although the other common amino acids are used as buffers, cysteine is rarely used for this purpose. Why?

Answer: Cysteine has the nasty habit of undergoing oxidation/reduction reactions in which two cysteines will react to form cystine (effectively dicysteine). As this reaction proceeds cysteines are consumed. Now this is not expected to change the buffering capacity around pH = 2.0 and 10.8 (pK_a of the alpha amino group) but it will lower the buffering capacity of the side chain whose pK_a is 8.3.

20. Understanding the stereochemistry of amino acids
Draw all the possible isomers of threonine and assign (R,S) nomenclature to each.

Answer: The first thing to recognize is that threonine has two chiral carbons, the alpha carbon and the beta carbon. Thus, there are ($2^n = 2^2 =$) 4 different stereoisomers with two pairs of enantiomers (mirror images) and two pairs of diastereomers. Starting with L-threonine, which is shown below one can make D-threonine by changing the configuration about both the alpha and beta carbons (e.g., switch the amino and hydrogen on the alpha carbon and then switch the hydrogen and hydroxyl on the beta carbon. This action produces D-threonine, the mirror image of L-threonine (enantiomers). L-allothreonine, a diastereomer of L-threonine, is made by switching the configuration about only the beta carbon and D-allothreonine is L-allothreonine's mirror image. These are all shown below.

To assign (R,S) nomenclature we have to assign priorities to each of the groups about the chiral carbons. Let's start with L-threonine.

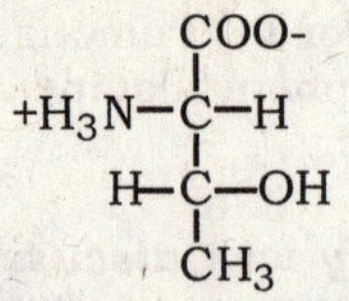

$$
\begin{array}{c}
\text{COO-} \\
| \\
{}^+\text{H}_3\text{N}-\text{C}-\text{H} \\
| \\
\text{H}-\text{C}-\text{OH} \\
| \\
\text{CH}_3
\end{array}
$$

The groups about the α-carbon of L-threonine are assigned the following priority: $NH_3^+ > COO^- > C_\beta > H$. In Fisher projections, two groups joined to the α-carbon (i.e., NH_3^+ and H) are coming out of the plane while the other two groups are into the plane. By rotating the molecule such that the group with the lowest priority, namely H, is away from the reader, the NH_3^+ group will be located on the right side of the α-carbon, the carboxyl group above and to the left, and the remaining β-carbon to the left and below the α-carbon:

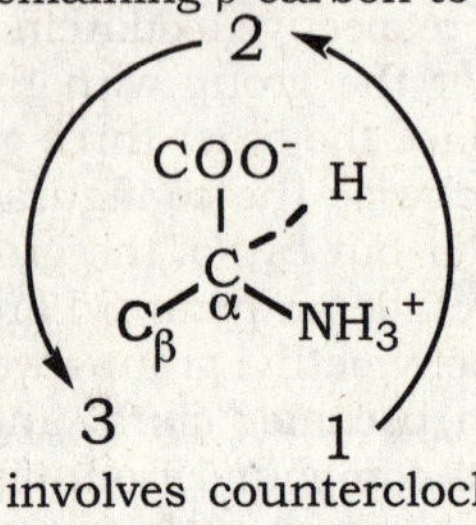

Since moving from highest to lowest priority involves counterclockwise movement the α-carbon is in the *S* configuration.

For the β-carbon, the priority is $OH > C\alpha > CH_3 > H$. (The orientation of the $C\alpha$-$C\beta$ bond was already defined for the α-carbon as being into the plane therefore it must be out of the plane for the β-carbon, as must be the bond to -CH$_3$. Thus, bonds to OH and H are into the plane. By rotating the molecule such that H is behind the β-carbon, we find the OH to the left, CH$_3$ below and C$_\alpha$ above:

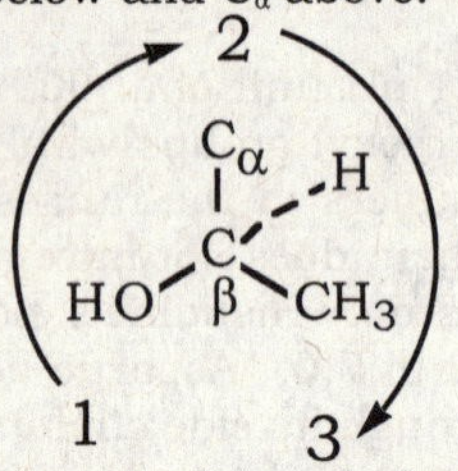

Since moving from highest to lowest priority involves clockwise movement the α-carbon is in the *R* configuration. The β-carbon is in the *R* configuration.

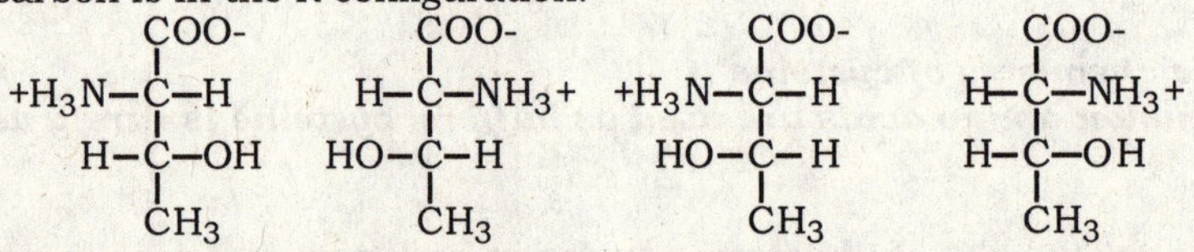

$$
\begin{array}{cccc}
\text{COO-} & \text{COO-} & \text{COO-} & \text{COO-} \\
| & | & | & | \\
{}^+\text{H}_3\text{N}-\text{C}-\text{H} & \text{H}-\text{C}-\text{NH}_3^+ & {}^+\text{H}_3\text{N}-\text{C}-\text{H} & \text{H}-\text{C}-\text{NH}_3^+ \\
| & | & | & | \\
\text{H}-\text{C}-\text{OH} & \text{HO}-\text{C}-\text{H} & \text{HO}-\text{C}-\text{H} & \text{H}-\text{C}-\text{OH} \\
| & | & | & | \\
\text{CH}_3 & \text{CH}_3 & \text{CH}_3 & \text{CH}_3 \\
\text{L-Threonine} & \text{D-Threonine} & \text{L-Allothreonine} & \text{D-Allothreonine}
\end{array}
$$

By inspection we see that

L-threonine is (2S, 3R) threonine
D-threonine is (2*R*,3*S*) threonine
L-allothreonine is (2*S*,3*S*) threonine
D-allothreonine is (2*R*,3*R*) threonine

Questions for Self Study

1. Fill in the blanks: A ____ bond is an amide bond between two ____. The bond is formed by a reaction between the ___ of one amino acid and the ___ of a second amino acid with the elimination of the elements of ___ . The result is a linear chain with an ___ end and a ___ end. The chain can be extended into a polymer with additional amino acids. The polymer formed is called either a ___ or a ___ depending on the number of amino acids joined. The amide bond can be broken by the addition of the elements of water across it in a ___ reaction.

2. List the four classes of amino acids.

3. Indicate which class of amino acids best represents the statements presented below:
   a. Their side chain may contain a hydroxyl group. ___
   b. Negatively charged at basic pH.___
   c. Relatively poorly soluble in aqueous solution.___
   d. Relatively soluble in aqueous solution.___
   e. Positively charged at acidic pH.___
   f. Side chains are capable of hydrogen bonding.___
   g. Includes the amino acids arginine, histidine and lysine.___
   h. Side chains contain predominantly hydrocarbons.____
   i. Side chains are responsible for the hydrophobic cores of globular proteins.___
   j. The amides of glutamic and aspartic acid belong to this group.___

4. From the list of amino acids presented indicate which one best fits the description:
   a. The smallest amino acid (It also lacks a stereoisomer).
       alanine, glycine, phenylalanine, glutamine
   b. An aromatic amino acid.
       glutamic acid, tyrosine, isoleucine, proline
   c. A sulfur-containing amino acid.
       methionine, aspartic acid, arginine, leucine
   d. An amino acid capable of forming sulfur-sulfur bonds.
       methionine, proline, cysteine, tryptophan
   e. A branched chain amino acid.
       methionine, leucine, aspartic acid, asparagine

5. Fill in the blanks. Most of the amino acids are chiral compounds because they have at least one carbon atom with ___ different groups attached. Thus, there are ___ ways of arranging the four groups about the carbon atom. For the case of an amino acid with a single asymmetric carbon, the isomers pairs that can be formed are called ___; they are nonsuperimposable, ___ images of each other. Using the D,L system of nomenclature, the commonly occurring amino acids are all ___ amino acids.

6. Which three amino acids absorb light in the ultraviolet region of the spectrum above 250 nm?

7. Of the following techniques which are commonly used to separate amino acids: ion exchange, NMR, electroporation, HPLC, UV spectroscopy, Edman reaction?

8. DTNB (Ellman's reagent), iodoacetate, and N-ethylmaleimide are three agents that are used to modify which amino acid? What is the reactive group on the amino acid with which all three agents react?

9. Although there are twenty common amino acids, there are hundreds of different amino acids found in nature, some of which derive from the common amino acids by simple chemical (biochemical) modification. Name one common modification of amino acids found in nature.

10a. Of the twenty common amino acids which amino acid has a pK_a around neutrality.
b. What amino acid lacks an amino group?

Answers

1. peptide; amino acids; amino group; carboxyl group; water; N-terminal (or amino); C-terminal (or carboxyl); polypeptide; protein; hydrolysis.

2. Nonpolar or hydrophobic; neutral polar; acidic; basic.

3. a. neutral polar; b. acidic; c. nonpolar; d. neutral polar, acidic, and basic; e. basic;
f. neutral polar, acidic, and basic; g. basic; h. nonpolar; i. nonpolar; j. neutral polar.

4. a. glycine; b. tyrosine; c. methionine; d. cysteine; e. leucine.

5. Four; two; enantiomers (or mirror image isomers); mirror; L.

6. Phenylalanine, tyrosine, and tryptophan.

7. ion exchange, HPLC

8. cysteine; sulfhydryl group.

9. Hydroxylation, methylation, carboxylation, phosphorylation.

10. a. Histidine; b. proline.

Additional Problems

1. Sketch the titration curve of histidine.

2. In the protein hemoglobin, a number of salt bridges (ionic bonds) are broken in going from oxygenated to deoxygenated hemoglobin. One in particular involves an aspartic acid/histidine ionic interaction. (a). At pH = 7.0, would you expect this salt bridge to form? Explain. (b). Under certain conditions (low oxygen tension) the salt bridge does form. Explain how the influence of aspartic acid on histidine is similar to the influence of the α-carboxyl group on the pKa of the α-amino group (and vice versa).

3. How might the presence of calcium ions effect the apparent pK$_a$s of the side chains of aspartic acid and glutamic acid?

4a. The ultraviolet spectra of proteins are largely determined by what three amino acids?
b. In working with nucleic acids it is often useful to measure the optical densities at 260 nm and 280 nm. Nucleic acids (DNA and RNA) absorb more strongly at 260 nm than 280 nm such that the ratio A$_{260}$: A$_{280}$ ranges from 1.6 to 2.0. What might happen to the A$_{260}$: A$_{280}$ ratio of a solution of nucleic acid contaminated with protein?
c. Proteins typically have a A$_{260}$: A$_{280}$ ratio less than one. Estimate the A$_{260}$: A$_{280}$ ratio of a protein whose amino acid composition includes two moles of Trp and one mole of Tyr per mole of protein. For appropriate extinction coefficients (molar absorptivity) use information given at http://omlc.ogi.edu/spectra/PhotochemCAD/html/index.html. Activate the link to the appropriate amino acids and then under the absorption spectrum activate the link to "Extinction Data".
d. Certain proteins lack both Trp and Tyr yet still absorb ultraviolet light in the range of 240 nm to 270 nm. What amino acid is responsible for this? How would the A$_{260}$: A$_{280}$ ratio of a protein rich in this amino acid differ from a typical protein containing Trp and Tyr? Use the information at the URL given above to answer this.

5. Poetry comes in many forms. Can you write a short, two line poem, using the single-letter abbreviations for the amino acids?

Abbreviated Answers

1. Histidine is the only amino acid with a pK$_a$ around neutrality. The pK$_a$s of histidine are 1.8, 6.0, and 9.2. The titration curve will be similar to the one shown in the answer to problem 5 but with an additional plateau around pH = 6.0.

2. (a) The pK$_a$s of the side chains of aspartic acid and histidine are 3.9 and 6.0 respectively. At pH = 7.0, the side chain of aspartic acid is unprotonated and therefore negatively charged, whereas the side chain of histidine is also unprotonated and therefore uncharged. We might not expect this interaction to occur.

   (b) The fact that the salt bridge forms indicates that the two amino acid side chains are close together in the three dimensional structure of the protein. The proximity of the negatively charged aspartic acid must therefore raise the pK$_a$ of the histidine side group.

3. Calcium (Ca^{2+}) may bind to carboxylate ions. Therefore, the presence of calcium may lower the apparent pK$_a$s of aspartic acid and glutamic acid.

4. a./Phe, Tyr, and Trp.; b./ A$_{260}$: A$_{280}$ ratio may be lower.
c./The molar absorptivity for Trp at 280 nm and 260 nm are 5,502 and 3,765, respectively. The molar absorptivity for Tyr at 280 nm and 260 nm are approximately 1,209 and 612, respectively. The A$_{260}$: A$_{280}$ ratio is given by:

$$\text{Ratio } 260{:}280 = \frac{(2 \times 3{,}756) + 612}{(2 \times 5{,}502) + 1206} = 0.67$$

d./Phe.   The $A_{260} : A_{280}$ ratio would be much greater than 1.

$$\text{Ratio } 260{:}280 = \frac{144}{2} = 72$$

5.   Here is a start that is by no means original. "I think that I shall never see a ...."* Try composing one with the nucleoside one-letter code (A,T,G and C) if you really want a challenge.

* The very next word of this poem by Joyce Kilmer is "poem". Before the discovery of pyrrolysine we would have had to stop here because "o" was not used for an amino acid. With pyrrolysine and selenocysteine, the amino acid alphabet has been extended to 25 (20 amino acids, pyrrolysine, selenocysteine, b and z for asx and glx, respectively, and x for any amino acid) of the 26 letters in the English alphabet. The only letter to still unused is j. Here is the complete poem, Trees by Joyce Kilmer.

> I think that I shall never see
> A poem lovely as a tree.
> A tree whose hungry mouth is prest
> Against the sweet earth's flowing breast;
>
> A tree that looks at God all day,
> And lifts her leafy arms to pray;
> A tree that may in summer wear
> A nest of robins in her hair;
>
> Upon whose bosom snow has lain;
> Who intimately lives with rain.
> Poems are made by fools like me,
> But only God can make a tree.

Summary

Amino acids are the building blocks of proteins. There are 20 amino acids commonly found in proteins. The general structure of an amino acid is based on a central α-carbon, to which is attached a carboxyl group, a hydrogen, an amino group, and a variable side chain or R group. Amino acids can be classified according to the nature of their R groups into four classes. The nonpolar, hydrophobic amino acids include **alanine**, **valine**, **leucine** and **isoleucine** in addition to **proline**, **methionine**, **phenylalanine** and **tryptophan**. Polar, uncharged amino acids include **glycine**, **serine**, **threonine**, **tyrosine**, **asparagine**, **glutamine,** and **cysteine**. The acidic amino acids are **aspartic acid** and **glutamic acid** and the basic amino acids include **histidine**, **lysine**, and **arginine**. The amino acids are weak polyprotic acids. The carboxyl group is a rather strong carboxylic acid with pK_a around 2.0, whereas the amino group is a base with pK_a around 9.5. At neutral pH, the carboxyl group is unprotonated and negatively charged whereas the amino group is protonated and positively charged. Considering only these two groups at pH = 7.0, amino acids are zwitterionic. Several of the side chains have titratable groups including aspartic acid, glutamic acid, cysteine, tyrosine, arginine, histidine, and lysine.

In proteins, peptide (or amide) bonds join amino acids. The carboxyl group of one amino acid is joined to the amino group of another amino acid, with elimination of the elements of water.

In addition to the twenty common amino acids, there are a large number of modified amino acids found in nature. Examples include hydroxylation of lysine and proline, methylation of lysine, carboxylation of glutamic acid, phosphorylation of serine, threonine and tyrosine. Metabolites derived from amino acids play key roles in cells. Knowing the structures of ornithine, citrulline, aspartic acid and arginine will serve as a strong basis for understanding the urea cycle. Transamination of glutamic acid and aspartic acid produce α-ketoglutarate and oxaloacetate, respectively, two citric acid cycle intermediates. The same reaction with alanine produces pyruvate. A great many small molecules acting as neurotransmitters, hormones, antibiotics, and others are directly derived from amino acids.

The twenty common amino acids are coded for genetically as are selenocysteine and pyrrolysine. The latter two amino acids are coded for by stop codons.

The α-carbons of all of the amino acids except glycine are chiral. Thus amino acids can exist as stereoisomers. In nature, the L amino acids predominate. We will consider stereochemistry again when sugar structures are discussed.

The UV light absorbing properties of unconjugated proteins are dominated by the aromatic amino acids, phenylalanine, tyrosine and tryptophan. NMR can characterize amino acids and this technique is used to determine the three-dimensional structure of peptides and small proteins.

There are several methods for separating and analyzing amino acid mixtures. These included ion exchange chromatography, gas chromatography and HPLC. It is often the case that the amino acids are derivatized on their amino end with a hydrophobic group and the modified amino acids separated by reverse phase chromatography.

The fundamental structure of a protein is a series of amide planes joining alpha carbons. The amide plane contains the peptide bond, a polar bond with partial double-bond character. Amino acid chains are composed of amino acid residues and are classified as peptides (12 residues), oligopeptides (up to a few dozen peptides) and polypeptides. Proteins are polypeptides with one or more peptide chains.

Chapter 5

Proteins: Their Primary Structure and Biological Functions

• •

Chapter Outline

❖ Protein architecture
  ⊼ Fibrous proteins
  ⊼ Globular proteins
  ⊼ Membrane proteins
❖ Protein structure
  ⊼ Primary structure: Sequence of amino acids
  ⊼ Secondary structure: Helices and sheets formed by interactions between peptide planes
  ⊼ Tertiary structure: 3-D shape
  ⊼ Quaternary structure: How chains interact in multimeric protein
  ⊼ Protein conformation determined by primary structure and weak force interactions
  ⊼ Configuration: Actual arrangement of atoms held together by covalent bonds
❖ Protein purification
  ⊼ Size distinction: Size exclusion chromatography, ultracentrifugation, ultrafiltration
  ⊼ Solubility distinction: Ammonium sulfate (salts), organic solvent, pH (solubility minimum at isoelectric point)
  ⊼ Differential interactions: Ion exchange chromatography, hydrophobic interaction chromatography
❖ Amino acid composition
  ⊼ 6N HCl, 110°C, 24, 48, 72 hr
    • Ser/Thr extrapolate to zero time, Val/Ile extrapolate to infinity
    • Trp destroyed, Asn/Gln converted to Asp/Glu
❖ Protein Sequencing
  ⊼ Separation of polypeptide chains: Reduction of disulfide bonds and denaturation
  ⊼ Inactivation of sulfhydryl groups: Performic acid oxidation or reduction with β–mercaptoethanol followed alkylation
  ⊼ Amino acid composition
  ⊼ N- and C-terminus identification
    • N-terminus chemistry: Edman degradation -phenylisothiocyanate to product PTH derivative of N-terminal amino acid
    • C-terminus: Carboxypeptidase C or Y cleaves any C-terminal amino acid, carboxypeptidase B cleaves only Arg or Lys, carboxypeptidase A does not cleave Arg, Lys, or Pro
  ⊼ Polypeptide chain fragmentation: Sets of short, overlapping peptides
    • Trypsin: Carbonyl side of Arg, Lys
    • Chymotrypsin: Carbonyl side of Phe, Tyr, Trp
    • Clostripain: Carbonyl side of Arg
    • Endopeptidase Lys-C: Carbonyl side of Lys
    • Staphyloccal protease: Carbonyl side of Asp, Glu
    • Cyanogen Bromide: Carbonyl side of Met

- Hydroxylamine: Asn/Gly bond
- Acid hydrolysis: Asp/Pro
  - ⅄ Sequence determination
    - Edman degradation
    - Mass Spectrometry
  - ⅄ Sequence reconstruction: Match sequence of members of set of overlapping peptides
  - ⅄ Location of disulfide bonds: Analysis of fragments from nonreduced protein by diagonal electrophoresis (first dimension nonreducing -second dimension reducing)
- ❖ Mass spectrometry
  - ⅄ Electrospray Ionization
    - Often used in tandem to sequence peptides
      - Filter peptides with first mass spectrometer
      - Selected peptide fragmented and fragments analyzed by second mass spectrometer
  - ⅄ Matrix-Assisted Laser Desorption Ionization-Time of Flight: MALDI-TOF
    - High sensitivity and accuracy used to determine mass of peptides
    - Peptide Mass Fingerprinting measures mass of proteolytic fragments of protein
      - Comparison to database allows for protein identification
- ❖ Amino acid sequence analysis
  - ⅄ Homologous proteins
    - Proteins that perform the same function in different organisms
    - Proteins related evolutionarily and detected by sequence similarities
  - ⅄ Homologous proteins can be orthologous or paralogous
    - Orthologous: Proteins in different species but with similar function and sequence: Arose by gene duplication before speciation
    - Paralogous: Proteins is single species but with similar function and related function: Arose by gene duplication after speciation
  - ⅄ Computer programs to discover sequence similarities
    - BLAST (Basic Local Alignment Search Tool)
      - Align sequences, compare amino acids using scoring matrix, insert gaps to account for insertions/deletions (indels)
      - BLOSUM (Blocks Substitution Matrix) used to score observed substitutions
  - ⅄ Phylogenetic studies: Comparison of identical protein from many species
    - Identifies invariant residues that are crucial to function
    - Used to construct phylogenetic trees to establish evolutionary relationships among proteins
- ❖ Tertiary structure analysis identifies proteins with similar structure yet different function
  - ⅄ Amino acid analysis reveals many differences indicating distant relationship
- ❖ Merrifield solid phase synthesis: Chemical synthesis of polypeptides
- ❖ Protein complexity
  - ⅄ Simple proteins: Composed of amino acids but no other chemical groups
    - Amino acid residues may be modified by post-translational modifications
    - RESID: Database of modifications (http://www.ncifcrf.gov/RESID)
  - ⅄ Conjugated proteins: Protein with prosthetic group: glyco- sugar, lipo- lipids, nucleo- nucleic acid, phospho- phosphorylated amino acid, metallo- metal atom, hemo- protoporphyrin, flavo- FMN or FAD
- ❖ Biological function of proteins
  - ⅄ Enzymes: Biological catalysts
  - ⅄ Regulatory proteins: Hormones, DNA binding proteins
  - ⅄ Transport proteins: Hemoglobin, serum albumin
  - ⅄ Storage proteins: Zein, ovalbumin, phaseolin, ferritin
  - ⅄ Contractile and Motile proteins: Actin, myosin, dynein, kinesin, tubulin
  - ⅄ Structural proteins: α-Keratins, collagen, elastin, B-keratin, proteoglycans
  - ⅄ Scaffold or Adapter proteins: SH2, SH3, PH modules, PDZ-containing proteins
  - ⅄ Protective proteins: Antibodies, blood clotting factors, antifreeze proteins, toxins
  - ⅄ Exotic proteins: Monellin, glue protein

Chapter Objectives

Protein Structure

The four levels of protein structure are important to understand. The primary structure refers to the sequence of amino acids. The secondary structure refers to regular structures formed by hydrogen bond interactions between peptide bonds within a polypeptide. These include helices, sheet structures, and turns. The tertiary structure is the three-dimensional shape of the protein. Finally, the quaternary structure, reserved for multimeric proteins, describes how subunits are arranged.

Protein Sequencing

The basic strategy of protein sequencing is to first purify the protein and separate it into its component polypeptide chains. End group analysis identifies the N- and C-terminal residues. The chains are cleaved to produce smaller more manageable pieces whose amino acid sequence is determined using Edman degradation or mass spectrometry. Understand the basic idea behind Edman chemistry as applied to amino acid sequencing. It is often easier to sequence a protein by cloning its gene. However, knowing the sequence of even short oligopeptides from the protein gives information to search for the protein's gene.

In fragmenting a protein, remember: trypsin cuts after lysine and arginine; clostripain cuts after arginine; chymotrypsin cleaves after the aromatic amino acids, phenylalanine, tyrosine and tryptophan; and, cyanogen bromide cleaves after methionine.

Amino Acid Sequence Analysis

The primary structure of proteins may be analyzed to identify sequence similarities. These similarities reveal evolutionary relatedness among proteins and functional relatedness. Scoring matrices are used to identify related proteins. Invariant positions indicate functional importance of the particular position (and amino acid residue).

Biological Functions

Proteins are the most diverse category of biomolecules responsible for an array of biological functions. There are three classes of proteins: fibrous proteins, globular proteins, and membrane proteins. Protein functions include: enzyme catalysis, regulation, transportation, storage, contractility and motility, structure, connectivity (scaffold) and as natural defense mechanisms.

Problems and Solutions

1. Calculating the molecular weight and subunit organization of a protein from its metal content
The element molybdenum (atomic weight 95.95) constitutes 0.08% of the weight of nitrate reductase. If the molecular weight of nitrate reductase is 240,000, what is its likely quaternary structure?

Answer: If molybdenum (95.95 g/mole) is 0.08% of the weight of nitrate reductase, and nitrate reductase contains at a minimum 1 atom of Mo per molecule, then

$$95.95 = 0.08\% \times M_r$$

$$M_r = \frac{95.95}{0.0008} = 119,937 \approx 120,000$$

Since the actual molecular weight is 240,000, the quaternary structure must be a dimer.

2. Solving the sequence of an oligopeptide from sequence analysis data
Amino acid analysis of an oligopeptide 7 residues long gave
> Asp Leu Lys Met Phe Tyr

The following facts were observed:
> a. *Trypsin treatment had no apparent effect.*
> b. *The phenylthiohydantoin released by Edman degradation was*

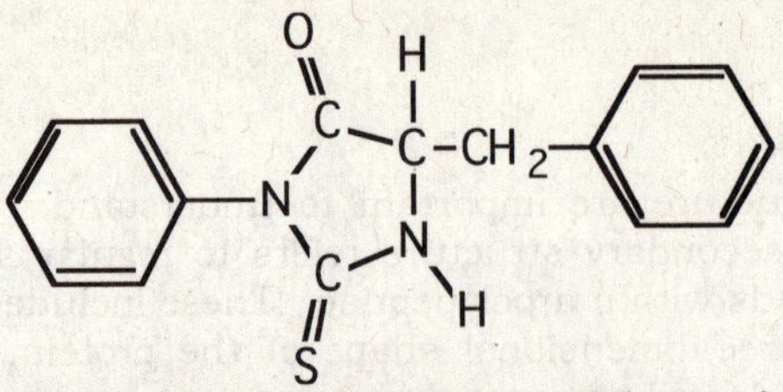

c. **Brief chymotrypsin treatment yielded several products including a dipeptide and a tetrapeptide. The amino acid composition of the tetrapeptide was Leu, Lys, and Met.**
d. Cyanogen bromide treatment yielded a dipeptide, a tetrapeptide, and free Lys.
What is the amino acid sequence of this heptapeptide?

Answer: From (a) we suspect that the C terminal amino acid is lysine. Trypsin cleaves on the carboxyl side of lysine or arginine. There is no arginine in this protein and trypsin digestion leaves the protein undigested. A lysine on the C-terminus would be consistent with this result.
From (b) we can conclude that the heptapeptide has phenylalanine as the N terminal amino acid because the PTH derivative shown in (b) that of phenylalanine.

                    Phe _ _ _ _ _ Lys

Chymotrypsin cleaves strongly at phenylalanine, tyrosine (and tryptophan). It must have released the N-terminal Phe in addition to a dipeptide and tetrapeptide. Since the tetrapeptide does not contain Phe or Tyr it must have derived from the C terminus by cleavage at position 3. An aromatic amino acid cannot be located at position 2 so we are left to conclude that Tyr is at position 3.

                    Phe _ Tyr _ _ _ Lys

From (d) we realize that there must be two methionines, which occupy two of the four unassigned positions. One Met must be located at position 6 to account for Lys production upon cyanogen bromide cleavage.

                    Phe _ Tyr _ _ Met Lys

We still have to account for one Met, one Leu, and one Asp. For cyanogen bromide to produce a tetrapeptide, M must be located at position 4.

                    Phe _ Tyr Met _ Met Lys

Leu must occupy position 5 because it is found in the tetrapeptide produced by chymotrypsin. This leaves Asp in position 2.

                    Phe Asp Tyr Met Leu Met Lys

3. Solving the sequence of an oligopeptide from sequence analysis data
Amino acid analysis of a decapeptide revealed the presence of the following products:
    NH₄⁺   Asp   Glu   Tyr   Arg   Met   Pro   Lys   Ser   Phe
The following facts were observed:
    a. Neither carboxypeptidase A or B treatment of the decapeptide had any effect.
    b. Trypsin treatment yielded two tetrapeptides and free Lys.
    c. Clostripain treatment yielded a tetrapeptide and a hexapeptide.
    d. Cyanogen bromide treatment yielded an octapeptide and a dipeptide of sequence NP (using the one-letter codes).
    e. **Chymotrypsin treatment yielded two tripeptides and a tetrapeptide. The N-terminal chymotryptic peptide had a net charge of -1 at neutral pH and a net charge of -3 at pH 12.**
    f. One cycle of Edman degradation gave the PTH derivative:

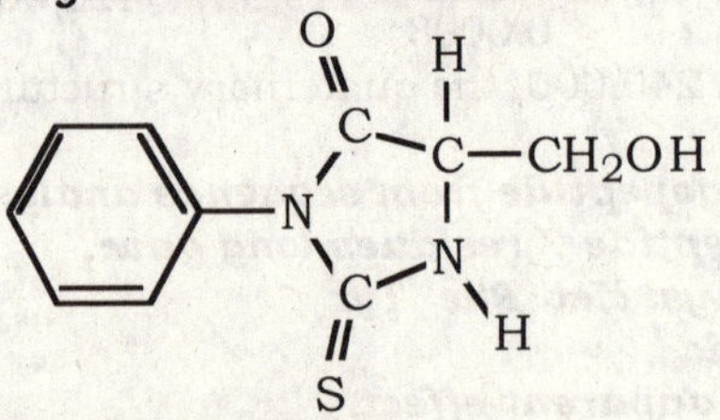

What is the amino acid sequence of this decapeptide?

Answer: Carboxypeptidase A will cleave at the C-terminus, except for Pro, Lys, and Arg whereas carboxypeptidase B cleavage occurs at C-terminal Arg and Lys. No action by either indicates either a circular peptide or proline as C-terminus. One cycle of Edman degradation identifies serine as the N-terminal amino acid. Thus,

                    S _ _ _ _ _ _ _ _ P

Cyanogen bromide treatment cleaves after M and releases two products that account for 10 amino acid residues. Thus, there must be only one M and it must be located near the C-terminus to release the dipeptide NP.

$$S _____ M N P$$

Trypsin cleavage yielded two tetrapeptides and free lysine. Looking at the sequence we constructed so far the C-terminus must be in one of the tetrapeptides and so either an R or K must be at position 6. Positions 2 and 3 cannot be either R or K because the trypsin products do not include a di- or tripeptide. So, there must be an R or K at position 4. Since 4 and 6 both contain R or K, the free lysine must come from position 5. Thus,

$$S \_ \_ (R/K) K (R/K) \_ M N P$$

Since the peptide contains both R and K the possibilities for the sequence of positions 4, 5 and 6 are RKR, RKK and KKR and only RKK would generate free lysine and two moles of it. This would put R at position 4, which is consistent with clostripain treatment that yielded a tetrapeptide and a hexapeptide. So,

$$S \_ \_ R K K \_ M N P$$

F or Y must be located at position 7 to insure that chymotrypsin cleavage produces a tripeptide from the C-terminus and nothing longer. In addition an aromatic must also be located at position 3. The only amino acid unaccounted for is either Glu (E) or Gln (Q), which is located at position 2. This gives:

$$(1) \ S \ (E/Q) \ (F/Y) \ R \ K \ K \ (F/Y) \ M \ N \ P$$

For the N-terminal chymotryptic peptide, S (E/Q) (F/Y), to have a net charge of -1 at neutral pH and -3 at pH = 12 it must contain glutamic acid (E) and tyrosine (Y) to account for the charge.

$$S \ E \ Y \ R \ K \ K \ F \ M \ N \ P$$

4. Solving the sequence of an oligopeptide from sequence analysis data
Analysis of the blood of a catatonic football fan revealed large concentrations of a psychotoxic octapeptide. Amino acid analysis of this octapeptide gave the following results:

2 Ala   1 Arg   1 Asp   1 Met   2 Tyr   1 Val   1 NH_4^+

The following facts were observed:
a. Partial acid hydrolysis of the octapeptide yielded a dipeptide of the structure

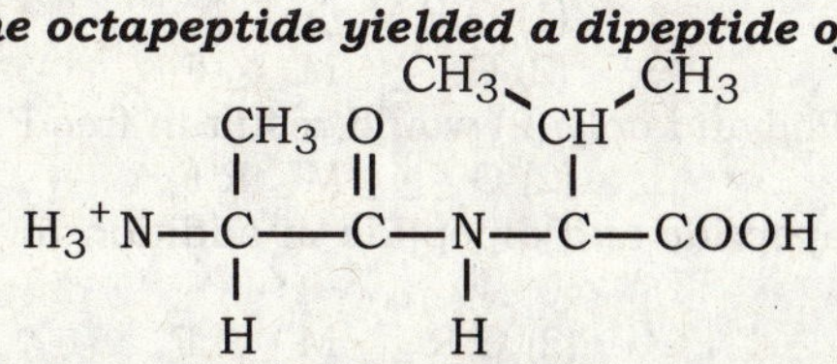

b. Chymotrypsin treatment of the octapeptide yielded two tetrapeptides, each containing an alanine residue.
c. Trypsin treatment of one of the tetrapeptides yielded two dipeptides.
d. Cyanogen bromide treatment of another sample of the same tetrapeptide yielded a tripeptide and free Tyr.
e. End group analysis of the other tetrapeptide gave Asp.
What is the amino acid sequence of this octapeptide?

Answer: The dipeptide produced by partial acid hydrolysis is Ala Val, AV.
Chymotrypsin cleaves at aromatics like Y. To produce two tetrapeptides from an octapeptide with two tyrosines, one must be the C-terminus and the other at position 4. Thus:

$$\_ \_ \_ Y \_ \_ \_ Y$$

Since trypsin cleaves, in this case, after R from (c) we infer that R is at position 2 or 6.

$$\_ R \_ Y \_ \_ \_ Y$$
$$\_ \_ \_ Y \_ R \_ Y$$

That CNBr cleavage of the tetrapeptides produces free Y indicates the sequence MY adjacent to R. Thus,

$$\_ R M Y \_ \_ \_ Y$$
$$\_ \_ \_ Y \_ R M Y$$

End group analysis of the second chymotrypsin tetrapeptide indicates Asp = D, but the presence of NH_4^+ suggests asparagine, N. (The end in question must be the N-terminus.) Thus:

$$\_ R M Y N \_ \_ Y$$
$$N \_ \_ Y \_ R M Y$$

We can now place AV (from (a)) into these sequences:

$$\_ R M Y N A V Y$$
$$N A V Y \_ R M Y$$

The only amino acid unaccounted for is an additional A giving us:

$$A R M Y N A V Y$$
$$N A V Y A R M Y$$

5. Solving the sequence of an oligopeptide from sequence analysis data
Amino acid analysis of an octapeptide revealed the following composition:

        2 Arg   1 Gly   1 Met   1 Trp   1 Tyr   1 Phe   1 Lys

The following facts were observed:
   a. *Edman degradation gave*

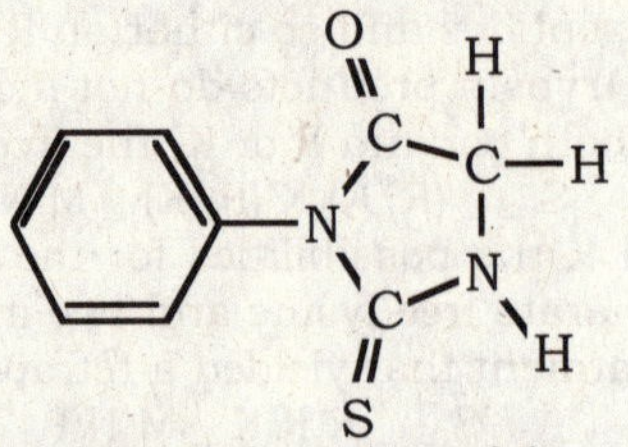

   b. *CNBr treatment yielded a pentapeptide and a tripeptide containing phenylalanine.*
   c. *Chymotrypsin treatment yielded a tetrapeptide containing a C-terminal indole amino acid and two dipeptides.*
   d. *Trypsin treatment yielded a tetrapeptide, a dipeptide, and free Lys and Phe.*
   e. *Clostripain yielded a pentapeptide, a dipeptide, and free Phe.*
What is the amino acid sequence of this octapeptide?

Answer: Edman degradation reveals that the N-terminal amino acid is G.
CNBr tells us that methionine is located at either position 3 or 5. Thus:

        (1) G _ M _ _ _ _ _
        (2) G _ _ _ M _ _ _

For CNBr cleavage to give F in the tripeptide, F must be located either at position 2 in (1) or on the C terminus of (2).

        (1) G F M _ _ _ _ _
        (2) G _ _ _ M _ _ F

Clostripain cleavage releases free Phe and only (2) would result in free F, if R precedes F. Thus,

        (2) G _ _ _ M _ R F

For clostripain to produce a pentapeptide and dipeptide in addition to free F, a second R must be located at position 2.

        (2) G R _ _ M _ R F

Next let's turn to the trypsin results. For trypsin digestion to release free K, K must follow R in the sequence. Thus,

        (2) G R K _ M _ R F

We are only left with W and Y. Tryptophan (W) is an indole amino acid and for it to be on the C terminus of a tetrapeptide produced by chymotrypsin, it must be located at position 4 leaving Y at 6. The peptide must be

        G R K W M Y R F

6. Solving the sequence of an oligopeptide from sequence analysis data
Amino acid analysis of an octapeptide gave the following results:

        1 Ala   1 Arg   1 Asp   1 Gly   3 Ile   1 Val   1 NH_4^+

The following facts were observed:
   a. *Trypsin treatment yielded a pentapeptide and a tripeptide.*
   b. *Chemical reduction of the free α-COOH and subsequent acid hydrolysis yielded 2-aminopropanol.*
   c. *Partial acid hydrolysis of the tryptic pentapeptide yielded, among other products, two dipeptides, each of which contained C-terminal isoleucine. One of these dipeptides migrated as an anionic species upon electrophoresis at neutral pH.*
   d. *The tryptic tripeptide was degraded in an Edman sequenator, yielding first A, then B:*

A.

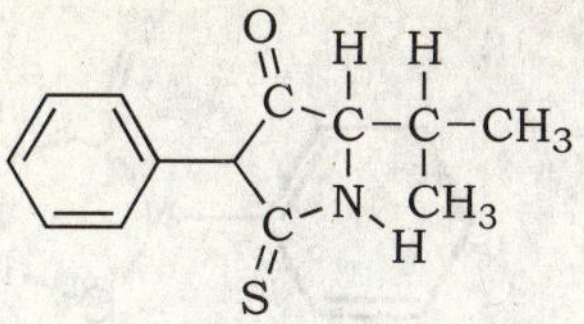

B.

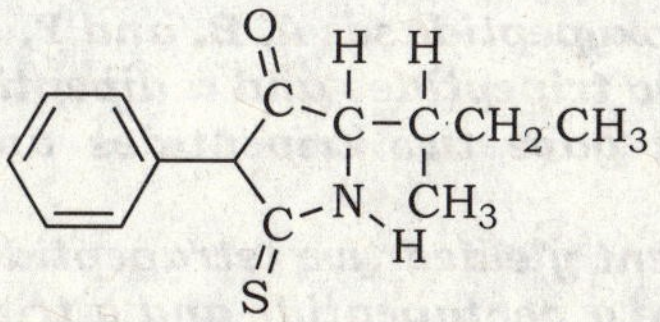

What is an amino acid sequence of the octapeptide? Four sequences are possible, but only one suits the authors. Why?

Answer: Trypsin digestion indicates R at position 3 or 5. Thus,

(1) _ _ R _ _ _ _ _

(2) _ _ _ _ R _ _ _

The product, 2-aminopropanol, must derive from reduction of a carboxyl terminal alanine. The structures of both are shown below.

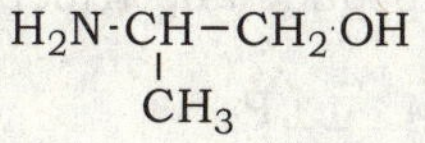

$H_2N \cdot CH - CH_2 OH$
|
CH_3 2-aminopropanol

$H_2N - CH - C - OH$
|
CH_3 alanine

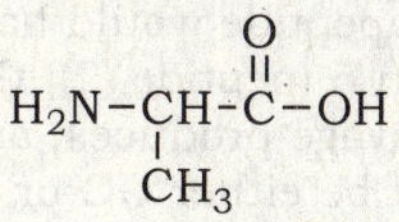

Our possibilities so far are:

(1) _ _ R _ _ _ _ A

(2) _ _ _ _ R _ _ A

From (d) we identify the PTH-derivatized amino acids as valine and isoleucine. So, the tryptic tripeptides must contain the sequence VI. For VI to be released by trypsin digestion it must precede R in (1) or it must follow R in (2).

(1) V I R _ _ _ _ A

(2) _ _ _ _ R V I A

Ammonium production and the finding that 1 Asp is present indicate that Asn (N) was in the original octapeptide. So, in the above two peptides we still need to account for G, N and two I's.

From (c) we know that _ I and _ I are located in the tryptic pentapeptide and these must be GI and NI. So, we know that GI and NI are in the remaining places in (1) and (2) but we do not know their order. Placing them into (1) and (2) produces four candidates:

(1a) V I R G I N I A

(1b) V I R N I G I A

(2a) G I N I R V I A

(2b) N I G I R V I A

GI is not charged at neutral pH but NI would be converted to DI by acid hydrolysis and DI is anionic at neutral pH.

Given the fact that both authors are at the University of Virginia we could guess which they prefer!

7. *Solving the sequence of an oligopeptide from sequence analysis data*
An octapeptide consisting of 2 Gly, 1 Lys, 1 Met, 1 Pro, 1 Arg, 1 Trp, and 1 Tyr was subjected to sequence studies. The following was found:
   a. Edman degradation yielded

b. *Upon treatment with carboxypeptidases A, B, and Y, only carboxypeptidase Y had any effect.*
c. *Trypsin treatment gave two tripeptides and a dipeptide.*
d. *Chymotrypsin treatment gave two tripeptides and a dipeptide. Acid hydrolysis of the dipeptide yielded only Gly.*
e. *Cyanogen bromide treatment yielded two tetrapeptides.*
f. *Clostripain treatment gave a pentapeptide and a tripeptide.*
What is the amino acid sequence of this octapeptide?

Answer: Edman degradation indicates glycine at the N-terminus.
Carboxypeptidase A cuts at all C termini except P, R and K, whereas carboxypeptidase B cuts at R and K. Carboxypeptidase Y cleavage indicates that the peptide is not circular and that the C-terminus is proline (P). Thus we have:

$$G \_ \_ \_ \_ \_ \_ P$$

Cyanogen bromide cleavage reveals M at position 4. Thus,

$$G \_ \_ M \_ \_ \_ P$$

Clostripain cleaves after R and for it to produce a pentapeptide and tripeptide R must be at either position 3 or 5.

$$(1)\ G \_ R\ M \_ \_ \_ P$$
$$(2)\ G \_ \_ M\ R \_ \_ P$$

If R is at position 3 (peptide (1)) then the C terminal pentapeptide would have to have K in it at position 5 or 6 for the trypsin digestion to make sense. If R is at position 5 (peptide (2)) then K must be at position 3 for the trypsin digestion to work as observed. Chymotrypsin cleavage produces, among other things, a dipeptide that yielded only G upon acid hydrolysis. The dipeptide could be either GG or GW because acid hydrolysis of GW would yield free G due to the acid instability of W. GG cannot be correct because it would have to be on the N terminus and in either model peptide GG would not be released as a dipeptide upon chymotrypsin digestion. So, we must have the sequence GW in the peptide. GW at the N terminus of (1) would produce the desired digestion products. And, position 5 in (1) would have to be Y. Thus,

$$(1)\ G\ W\ R\ M\ Y \_ \_ P$$

If this option is correct then K would only fit at position 6 and this leaves a remaining G at 7:

$$(1)\ G\ W\ R\ M\ Y\ K\ G\ P$$

For peptide (2) placement of GW cannot be at positions 2 and 3 because GGW would be released by chymotrypsin digestion and this peptide would yield only G upon acid hydrolysis but it is a tripeptide, not a dipeptide. Placing GW at positions 6 and 7 would give a peptide that, upon digestion with chymotrypsin, would yield free P, which is not observed. So, our first peptide fits the bill.

$$(1)\ G\ W\ R\ M\ Y\ K\ G\ P$$

8. Solving the sequence of an oligopeptide from sequence analysis data
Amino acid analysis of an oligopeptide containing nine residues revealed the presence of the following amino acids:

$$\text{Arg} \quad \text{Cys} \quad \text{Gly} \quad \text{Leu} \quad \text{Met} \quad \text{Pro} \quad \text{Tyr} \quad \text{Val}$$

The following was found:
a. *Carboxypeptidase A treatment yielded no free amino acid.*
b. *Edman analysis of the intact oligopeptide released*

    c. Neither trypsin nor chymotrypsin treatment of the nonapeptide released smaller fragments. However, combined trypsin and chymotrypsin treatment liberated free Arg.
    d. CNBr treatment of the eight-residue fragment left after combined trypsin and chymotrypsin action yielded a six-residue fragment containing Cys, Gly, Pro, Tyr and Val; and a dipeptide.
    e. Treatment of the six-residue fragment with β-mercaptoethanol yielded two tripeptides. Brief Edman analysis of the tripeptide mixture yielded only PTH-Cys. (The sequence of each tripeptide, as read from the N-terminal end, is alphabetical if the one-letter designation for amino acids is used.)
What is the amino acid sequence of this octapeptide?

Answer: The amino acid composition accounts for only eight amino acids and since we are dealing with a nonapeptide there must be two of one of the amino acids. If we focus on the information given in (e) we should conclude that there are two cysteines that are cross-linked as cystine in the original peptide. From (a) we infer that the C-terminus is R or P because carboxypeptidase A cuts at all C termini except P, R and K. And this peptide contains both P and R. From (b) we see that the PHT-derivatized amino acid is leucine thus the N-terminus must be L. We know that trypsin cleaves after R and K and that the peptide contains one R yet digestion with trypsin did not release smaller fragments. It could be that R is on the C terminus, which is consistent with the carboxypeptidase A results. Alternatively, R could be internal such that it is between the two cysteines that are cross-linked. Chymotrypsin digestion also gives no smaller fragments. Since chymotrypsin cleaves after Y (W and F) one conclusion is that Y is on the C terminus. But, this would not be consistent with carboxypeptidase A treatment so it must be internal and it must be between the two cysteines. Combined digestion with trypsin and chymotrypsin yields free R. For this to happen Y must precede R, thus the peptide must contain YR and it cannot be on the C terminus (because it would have released R after chymotrypsin digestion alone). So, YR must be internal and it must be between the two cysteines.

$$L \_ \_ \_ \_ \_ \_ \_ P \text{ with } C - YR - C \text{ somewhere}$$

From (d) we realize that because M is not in the six-residue fragment it must be in the dipeptide. Of the amino acids all but L is accounted for in the six-residue fragment and trypsin/chymotrypsin digestion released R. Thus, the dipeptide must be LM.
This gives us:

$$L M \_ \_ \_ \_ \_ P \text{ with } C - YR - C \text{ somewhere}$$

From (e) we also learn that the two tripeptides have N-terminal C's and that they are joined by a disulfide bond.

$$
\begin{array}{l}
C\_\_ \\
| \\
C\_\_
\end{array}
$$

We reasoned above that YR must have been somewhere between the two C's. But, R is not in the six-residue fragment yet Y is. It must be that Y is on the N terminus of one of the tripeptides. The six-residue fragment also contains P, which we established is on the C terminus so we must have

$$
\begin{array}{l}
C\_YR \\
| \\
C\_P
\end{array}
$$

G and V are still unaccounted for and to fit into the above model in alphabetic order it must be as such

$$
\begin{array}{l}
CVYR \\
| \\
CGP
\end{array}
$$

The final sequence is:

$$
\begin{array}{ccc}
S & \!\!\!\!\!\!\!\!\!\!\text{———} & S \\
| & & | \\
\end{array}
$$
$$L\,M\,C\,V\,Y\,R\,C\,G\,P$$

9. Describe the solid-phase chemical synthesis of a small peptide
Describe the synthesis of the dipeptide Lys-Ala by Merrifield's solid phase chemical method of peptide synthesis. What pitfalls might be encountered if you attempted to add a leucine residue to Lys-Ala to make a tripeptide?

Answer: To synthesize Lys-Ala, a resin with alanine attached via its carboxyl group is used to supply the C-terminal alanine. The N-terminal amino acid, lysine, is introduced with both its α-amino group and its ε-amino group blocked with t-BocCl (tertiary-butyloxycarbonyl chloride). The carboxyl group of the incoming lysine is activated with dicyclohexylcarbodiimide, so that it reacts with amino groups of the resin bound alanine, forming a peptide bond. One concern with lysine is the presence of an amino group in the side chain. This amino group may react with an incoming carboxyl group to produce a branched product.

For peptide synthesis, tBoc was originally used in Merrifield synthesis to block the amino terminus whereas amino-containing side chains were blocked by other groups. The tBoc amino-terminus protecting group is removed using trifluoroacetic acid (TFA). Thus, the peptide is subjected to TFA treatment during each cycle. This treatment also causes some side-chain amino groups to deprotect, which results in production of unwanted products. To avoid this problem Fmoc (9-fluorenyl-methoxycarbonyl) is often used in place of tBoc.

10. Identify proteins using BLAST searches for peptide fragment sequences
Go to the National Center for Biotechnology Information web site at http://www.ncbi.nlm.nih.gov/. From the menu of Popular Resources on the right hand side, click on "BLAST". Under the Basic BLAST heading on the new page that comes up, click on "protein blast". In the Enter Query Sequence box at the top of the page that comes up enter the following sequence: NGMMKSSRNLTKDRCK. Confirm that the database under "Choose Search Set" is set on "nr" (non-redundant protein sequences), then click the BLAST button at the bottom of the page to see the results of your search. Next, enter this sequence from a different protein: SLQTASAPDVYAIGECA. Identify the protein from which this sequence was derived.

Answer: The first sequence returns hits mainly to RNase I (pancreatic RNase) with the best hit (fifth hit) on the list (these change regularly as new information is added to the database) to pancreatic ribonuclease from *Bos taurus*. The alignment is shown below.

```
Query  1    NGMMKSSRNLTKDRCK  16
            N MMKS RNLTKDRCK
Sbjct  53   NQMMKS-RNLTKDRCK  67
```

The top line is the query sequence and the bottom line is the region of pancreatic ribonuclease in which the best alignment is located. The middle line shows the positions that are identical in both sequences. Pancreatic ribonuclease differs from the query sequence in two places, one is a substitution and the other is a deletion. (Note: The alignment is shown in Courier font, which is a non proportional font that gives equal spacing to all characters and allows lines to be in register.) If you change the sequence substituting A for G and deleting an S and repeat the BLAST you will get better matches. Finally, you could repeat but this time looking into the Reference proteins database and get much cleaner hits.

The best hit for the second sequence (at the time of this writing) is nitrite reductase (NADPH) from *Neurospora crassa*. This is an enzyme involved in nitrate assimilation. The alignment is shown below.

```
Query  1    SLQTASAPDVYAIGECA  17
            SLQT SAPDVYAIGECA
Sbjct  317  SLQT-SAPDVYAIGECA  332
```

The two differ by a single deletion of an A. If you repeat the BLAST with the A removed you will get better hits and the BLAST picks out a conserved protein domain.

11. Calculating the mass of a protein from mass spectrometric m/z values
Electrospray ionization mass spectrometry (ESI-MS) of the polypeptide chain of myoglobin yielded a series of m/z peaks (similar to those shown in Figure 5.21 for aerolysin K). Two successive peaks had m/z values of 1304.7 and 1413.2, respectively. Calculate the mass of the myoglobin polypeptide chain from these data.

Answer: In electrospray mass spectrometry, the equation describing the observed m/z peaks is: m/z = [M + n(mass of proton)]/n(charge of proton), where M = the mass of the protein and n = number of positive charges per protein molecule. The two successive peaks at 1304.7 and 1413.2 must differ by a single proton. Let the number of positive charges for the first peak be n. Since the second peak has a higher m/z value it must have one fewer positive charge or n-1 positive charge. The mass, M, could be evaluated at either m/z value if we knew n. To determine n, we will first solve the equation for M as a function of m/z and n.

$$\left(\frac{m}{z}\right) = \frac{[M+n]}{n}$$

$$n\left(\frac{m}{z}\right) = [M+n]$$

$$M = (n)\left(\frac{m}{z} - 1\right)$$

We will now evaluate M at each of the m/z values.

$$M = \left(n_1\right)_1 \left(\left(\frac{m}{z}\right)_1 - 1\right) \text{ and } M = \left(n_2\right)\left(\left(\frac{m}{z}\right)_2 - 1\right)$$

We will then set these equal with n_1 corresponding to m/z = 1304.7 and n_2 corresponding to m/z = 1413.2. The two peaks differ by a single positive charge with n_1 having one more positive charge than n_2. Thus, $n_1 = 1 + n_2$.

$$\left(n_1\right)\left(\left(\frac{m}{z}\right)_1 - 1\right) = \left(n_2\right)\left(\left(\frac{m}{z}\right)_2 - 1\right)$$

$$\left(1 + n_2\right)\left(1304.7 - 1\right) = \left(n_2\right)\left(1413.2 - 1\right)$$

$$n_2 = \frac{\left(1304.7 - 1\right)}{\left(1413.2 - 1\right) - \left(1304.7 - 1\right)}$$

$$n_2 = 12$$

But, $M = \left(n_2\right)\left(\left(\frac{m}{z}\right)_2 - 1\right)$

$$M = 12 \times \left(1413.2 - 1\right)$$

$$M = 16,946$$

Or, $M = \left(n_1\right)\left(\left(\frac{m}{z}\right)_1 - 1\right)$

$$M = 13 \times \left(1304.7 - 1\right)$$

$$M = 16,948$$

Thus, M=16,947±1

12. Phosphorylation of proteins induces new properties
Phosphoproteins are formed when a phosphate group is esterified to an –OH group of a Ser, Thr, or Tyr side chain. A typical cellular pH values, this phosphate group bears two negative charges $-PO_3^{2-}$. Compare this side-chain modification to the 20 side chains of the common amino acids found in proteins and comment on the novel properties that it introduces into side-chain possibilities.

Answer: The only amino acids negatively charged at neutral pH are aspartic acid and glutamic acid, both with minus-one charge distributed over a carboxylate group. Phosphorylated serine, threonine and tyrosine differ from these in that they carry a 2- charge and are considerably bigger. Negative charges have the potential to interact with cations and one might expect quite different interactions of a phosphate group compared to a carboxylate group. The large, bulky phosphate group might be expected to allow for ionic interactions not possible with any of the other amino acids. Modification of serine, threonine and tyrosine side chains by phosphorylation converts polar, uncharged groups into charged groups.

Phosphorylation is often seen as a transient modification that changes the reactivity of proteins. It is often the case that when a particular serine (or threonine) is the target for phosphorylation, the protein is engineered to change the serine to aspartic acid or alanine in an attempt to mimic the phosphorylated and unphosphorylated states of the protein.

13. Using graphical analysis to determine the K_D for the interaction between a protein and its ligand
A quantitative study of the interaction of a protein with its ligand yielded the following results:
Ligand concentration (mM)   1,   2,   3,   4,   5,   6,   9,   12
*v (moles of ligand bound   0.28, 0.45, 0.56, 0.60, 0.71, 0.75, 0.79, 0.83
per mole of protein)*

Plot a graph of [L] versus v. Determine K_D, the dissociation constant for the interaction between the protein and its ligand, from the graph.

Answer: Plot v on the y-axis versus [L] on the x-axis and determine the location at which the $v = 0.5$. The data are plotted below as ×. A horizontal line was drawn at $v = 0.5$ and at the point it appears to intersect the data a perpendicular line was drawn to the [L]-axis. On the plot the open circles are values of v calculated using the equation shown below and 2.4 mM as a guess for K_D.

$$v = \frac{[L]}{K_D + [L]}$$

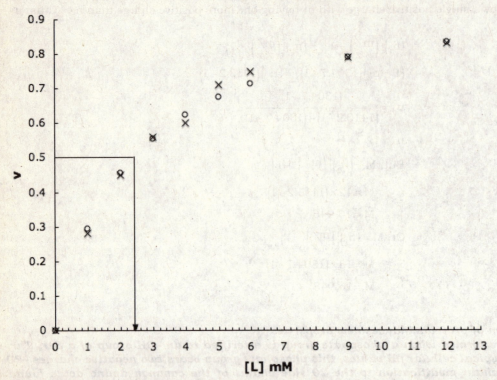

The data (×) and calculated points (∘) are in close agreement suggesting that our estimate of K_D is a good one. But, this was accomplished "by eye" by actually making numerous guesses for K_D and then plotting them to see which value best fit the data. A more sophisticated approach is to transform the equation into a linear form.

$$v = \frac{[L]}{K_D + [L]}$$

$$\left(v\right)^{-1} = \left(\frac{[L]}{K_D + [L]}\right)^{-1}$$

$$\frac{1}{v} = \frac{K_D + [L]}{[L]}$$

$$\frac{1}{v} = K_D \times \frac{1}{[L]} + 1$$

A plot of $1/v$ versus $1/[L]$ should be linear with a slope equal to K_D. By transforming the data (from v to $1/v$ and [L] to $1/[L]$) and then plotting it we can determine the "best fit line" by linear regression. This approach will determine a value of K_D that fits the data by minimizing the difference between data and calculated values.

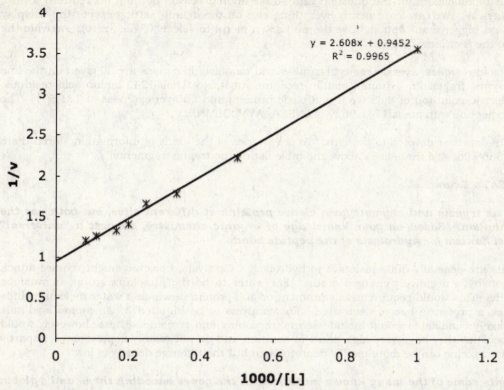

The equation of the "best fit" line is shown above and from it we can see that the value for K_D is 2.61 mM. On the direct plot (*v* versus [L]) if you plot the open circles with data calculated using K_D = 2.61 mM you will not see a dramatic difference from K_D = 2.4 mM. You will see, however, that the calculated values for K_D = 2.61 mM for low values of *v* fit the data better. In the $1/v$ versus $1/[L]$ plot these values are large in magnitude and quite spread out along the graph. Because of this they tend to dominate the best-fit linear regression.

Biochemistry on the Web

14. Exploring the ExPASy proteomics web site
The human insulin receptor substrate-1 (IRS-1) is designated protein P35568 in the protein knowledge base on the ExPASy Web site (http://us.expasy.org/). Go to the PeptideMass tool on this Web site and use it to see the results of trypsin digestion of IRS-1. How many amino acids does IRS-1 have? What is the average molecular mass of IRS-1? What is the amino acid sequence of the tryptic peptide of IRS-1 that has a mass of 1741.9629?

Answer: At the ExPASy Proteomics Server activate the link to "proteomics". On the returned page under "Tools" you should find PeptideMass. (A simple way to locate something on a complicated web page is to search for it using a Find command.) Activating this link will return PeptideMass, a program that performs virtual digestions of proteins. To direct the program to the sequence for the human insulin receptor substrate-1, simply type P35568 in the text box. (P35568 is the protein's accession number is this database.)

The PeptideMass tool is used to predict the size of proteins digested by specific proteases or cleaved after particular residues. It is often the case that this information will be used to compare the results with protein isolated by a technique like 2-dimensional gel electrophoresis and then cleaved with an enzyme like trypsin and analyzed by mass spectrometry.

The program predicts the size of the fragments taking into account the location of cleavage by the protease and other modifications carried out by the investigator. For example, it is often the case that the protein to be digested is treated with a reducing agent like β-mercaptoethanol or dithiothreitol (DTT) to reduce cystine disulfides. Following treatment with reducing agent the protein sulfhydryls (cysteines) may be treated with an agent like iodoacetic acid to carboxymethylate them. This will put an adduct on each cysteine and so this must be taken into account when running PeptideMass to accurately determine the mass of tryptic fragments. Choices are provided in a pull down menu under "The peptide masses are with cysteines treated with:". We

91

are told nothing about modifications in the question and so we should select "nothing (in reduced form)", which is the default anyway. We can leave nearly everything else on the default setting except the "Display the peptides with a mass bigger than" option. Use the pull down menu to select "0" (i.e., zero). Activate the program by clicking on the "Perform" button.

The number of amino acid residues, average molecular mass and monoisotopic mass are all given on the line preceding the list of tryptic fragments. Human insulin receptor substrate-1 has 1242 amino acid residues, has a theoretical isoelectric point (pI) of 8.83 (so it is a basic protein) and an average mass of 131,591. The sequence of the tryptic peptide with mass 1741.9629 is LNSEAAAVVLQLMNIR.

ExPASy is a very useful database devoted to proteins. To get an idea of the kinds of information stored there activate the link to P35568 (located three lines above the table listing the tryptic fragments).

Preparing for the MCAT® Exam

15. Proteases such as trypsin and chymotrypsin cleave proteins at different sites, but both use the same reaction mechanism. Based on your knowledge of organic chemistry, suggest a "universal" protease reaction mechanism for hydrolysis of the peptide bond.

Answer: Peptide bonds are generally quite resistant to hydrolysis. Catalysis, however, might require attack on the carbonyl carbon by a negatively-charged group. For water to be the attacking group, it must be activated to a hydroxide. This would require an environment on the protein in which a water molecule binds and subsequently loses a proton to become activated. For catalysis to be initiated by an amino acid side chain on the protein likely candidates would include serine, threonine and tyrosine. These, however, would have to be activated because they have very high pKa's in solution and do not readily deprotonate. Aspartic acid, glutamic acid and cysteine can be more readily deprotonated but their charge density is low.

16. Table 5.4 presents some of the many known mutations in the genes encoding the α- and β-globin subunits of hemoglobin.
   a. Some of these mutations affect subunit interactions between the subunits. In an examination of the tertiary structure of globin chains, where would you expect to find amino acid changes in mutant globins that affect formation of the hemoglobin α₂β₂ quaternary structure?
   b. Other mutations, such as the S form of the β-globin chain, increase the tendency of hemoglobin tetramers to polymerize into very large structures. Where might you expect the amino acid substitutions to be in these mutants?

Answer: Subunit interactions leading to a stable quaternary structure occur between groups on interacting surfaces of the subunits. Often these groups are hydrophobic. So, amino acid mutations on the surfaces of the subunits leading to removal of hydrophobic amino acids might be good candidates for quaternary structure mutants.

Hemoglobin S is a mutation from glutamic acid to valine, which represents a change from a negatively-charged amino acid to a hydrophobic amino acid. The presence of glutamic acid on the surface of hemoglobin is expected to block polymerization due to charge repulsion. In hemoglobin S, however, the valine allows for hydrophobic inaction leading to polymerization.

Questions for Self Study

1. Fill in the blanks. Proteins are linear chains of ___ ___ held together by covalent ___ ___ bonds. Although these bonds are typically drawn as C-N single bonds, they in fact have partial double-bond characteristics because of delocalization of ___ ___. The result of this delocalization is to constrain four atoms in a single plane termed the ___ plane. The four atoms are ___, ___, ___, and ___. Thus, linear chains of amino acids can be considered as a string of ___ carbons joined together by amide or peptide planes.

2. Match the terms in the first column with terms in the second column
   a. Dipeptide
   b. Protein
   c. Polypeptide
   d. Amino acid residue
   e. Peptide

   1. An amino acid minus the elements of water.
   2. The bond joining amino acids in a protein.
   3. Two amino acids joined in amide linkage.
   4. Subunit of a multimeric protein.
   5. A molecule composed of polypeptide chains.

3. List the four levels of protein structure and give a brief definition of each.

4. Proteins represent an extremely diverse set of biomolecules. Match the functional class of proteins with their appropriate function

Class	Function
a. Enzymes	1. Source of amino acids for developing organisms.
b. Regulatory proteins	2. Move glucose across the cell membrane.
c. Transport proteins	3. Various venoms and toxins.
d. Storage proteins	4. Structural components of motile cellular appendages.
e. Contractile proteins	5. Biological catalysts.
f. Structural proteins	6. Certain extracellular, fibrous elements.
g. Scaffold proteins	7. Certain hormones.
h. Protective proteins	8. Responsible for multiprotein complex formation.

5. Trypsin, chymotrypsin, clostripain, staphylococcal proteins, and cyanogen bromide are agents used in amino acid sequence determination. What is their purpose?

6. Short polypeptides are subjected to Edman degradation for what purpose?

7. What is the difference between amino acid composition and amino acid sequence?

8. Answer True or False
   a. Proteins sharing a significant degree of sequence similarity are homologous. ___
   b. Proteins that perform the same function in different organisms are referred to as homologous. ___
   c. Invariant residues usually are a result of random mutations. ___
   d. Evolutionarily related proteins always have similar biological activities. ___

9. In solid phase synthesis of peptides the amino group of the incoming amino acid is blocked. Why?

10. Give one reason why synthetic peptides may be useful tools.

Answers

1. Amino; acids; amide (or peptide); pi; electrons; amide (or peptide); the carbonyl carbon; the carbonyl oxygen; the amide nitrogen; amide hydrogen; alpha.

2. a. 3; b. 5; c. 4; d. 1; e. 2.

3. The primary structure is the sequence of amino acids of a protein. The secondary structure refers to regular structures such as helices and sheets that form as a result of hydrogen bonding interactions. The tertiary structure of a protein is its three dimensional shape. Finally, the quaternary structure is reserved for proteins composed of multiple subunits and describes how the subunits interact.

4. a. 5; b. 7; c. 2; d. 1; e. 4; f. 6; g. 8; h 3.

5. These agents are used to cleave a polypeptide into short oligopeptides. Each oligopeptide is then sequenced. At least two different agents are used to separately cleave the polypeptide in order to produce overlapping oligopeptides.

6. Edman degradation is used to determine the amino acid sequence of an oligopeptide from the N-terminus to the C-terminus.

7. The amino acid sequence is the primary structure of a polypeptide and refers to the sequence of amino acids, N-terminus to C-terminus. The amino acid composition is the total number of each amino acid of each kind in a polypeptide.

8. a. T; b. T; c. F; d. F.

9. Blocking the amino group of the incoming amino acids prevents them from reacting with themselves either in solution or on the solid phase.

10. Antigenic peptides for antibody production; pharmacologically active peptides; structure-function studies; confirmation of structure of naturally occurring peptides.

Additional Problems

1. Protein purification schemes typically involve several techniques. For the techniques listed below provide a brief explanation for each and indicate at which relative point in a purification procedure the technique might be employed.
   - (a) Molecular Sieve Chromatography
   - (b) Ultracentrifugation
   - (c) Ammonium Sulfate Fractionation
   - (d) Affinity Chromatography
   - (e) Polyacrylamide Gel Electrophoresis
   - (f) Reverse Phase HPLC

2. How are protein purification procedures monitored?

3. What is the role of sulfur in protein structure?

4. The apparent molecular weight of globular proteins is often determined by SDS polyacrylamide gel electrophoresis. Explain.

5. The ultraviolet absorption spectra of proteins rarely looks like the sum of the absorption spectra of the component amino acids but the situation improves when the protein is treated with high concentrations of urea or guanidinium hydrochloride. Explain.

6. Determining the amino acid sequence of even a small portion of a protein is extremely useful. Often this information can be used to identify the protein. Calculate the total number of different amino acid sequences for polypeptide chains of length 5, 10, and 15 amino acids.

Abbreviated Answers

1. (a) Molecular sieve chromatography or molecular exclusion chromatography separates proteins based on size. A porous matrix is employed such as beads of dextrans, agarose, or polyacrylamide. A column of hydrated beads contains two aqueous volumes, the volume within the beads or the internal volume, and the volume outside of the beads, the external volume. The degree to which a protein can enter the internal volume is dependent on the protein's size. Large proteins cannot enter the internal volume at all, whereas smaller proteins may be partially or completely accessible to the internal volume. Separation of proteins of different sizes occurs because for every volume element added to the column, the proteins are displaced that volume element down the column. Large proteins can move only in the external volume whereas smaller proteins move in the external and internal volumes. Molecular sieve chromatography is likely to be used at a later stage in protein purification. It may be performed under conditions favoring the native state of the protein in which case the elution profile is dependent on the native molecular weight of the protein. Alternatively, denaturing conditions may be employed allowing individual polypeptide chains to be separated.

(b). In ultracentrifugation, very high centrifugal forces are employed to separate proteins. Depending on the technique employed, proteins may be separated according to size and shape, or according to density. Very often one of the first steps employed in a purification scheme is to briefly centrifuge a cell lysate to separate cellular debris from soluble proteins. In differential ultracentrifugation, a sample is subjected to ultracentrifugation runs of increasing duration and/or force to separate large proteins from small proteins. Density gradients, either continuous or discontinuous, may be employed to fractionate proteins by size or by density. Centrifugation in general may be used at any stage in a purification scheme.

(c). Ammonium sulfate fractionation operates on the principle that under certain ionic conditions protein charge may be shielded by ionic interactions with solvent components allowing proteins to aggregate and precipitate. Generally, it is used in early stages of protein purification. Variations of this basic technique include the use of other salts as precipitating agents or changing pH. Proteins are least soluble at their isoelectric point, the pH at which a protein's net charge is zero.

(d). Affinity chromatography exploits the fact that proteins have complementary surfaces that exhibit incredible specificity. These surfaces might be a substrate binding site, a cofactor binding site, or a protein-protein interaction surface as examples. The principle is to attach a molecule, that interacts specifically with a particular protein, to an inert, insoluble matrix. A mixture containing the protein of interest is added to the affinity matrix under conditions favoring complementary interactions. The matrix is washed to remove unbound protein and then subjected to conditions that disrupt the complementary interactions. These conditions may be high salt, or high ionic strength, a change in pH, or high concentration of cofactor or substrate. In favorable cases, affinity chromatography may be a one-step purification scheme. Generally, affinity chromatography is employed at a later stage at which the protein of interest has been partially purified and therefore represents a high percentage of the total protein.

(e). Polyacrylamide gel electrophoresis separates proteins according to their charge to mass ratio and their size. The driving force is a voltage gradient. Charged proteins are forced to move through a porous matrix (typically of polyacrylamide). Separation occurs because the degree to which movement of a protein is impeded is proportional to the size of the protein. Under so-called native conditions, protein mixtures may be separated into component proteins retaining biological activity. In the presence of denaturing agents such as urea or SDS (see question 4), individual polypeptide chains may be separated. Gel electrophoresis is employed both preparatively and analytically (see question 2).

(f). HPLC refers to high performance (or high pressure) liquid chromatography. In general, the efficiency of chromatographic procedures depends on the number of interactions or discriminations made per volume or per distance. In high performance chromatography, a large number of interactions are made possible by using chromatography matrix material of very small size (resulting in an increase in the number of theoretical plates). The small size greatly increases resistance to flow and so chromatography must be carried out using high pressure. Reverse phase HPLC is a particular application of HPLC that employs chromatography matrices with hydrophobic surfaces. Initial applications of chromatography employed cellulose, in the form of paper, as a stationary phase. Molecules are moved over this stationary phase in solutions containing organic solvents. Thus, the mobile phase favors hydrophobic compounds whereas the stationary phase favors hydrophilic compounds (because cellulose contains numerous polar groups). In reverse phase, the stationary phase is hydrophobic whereas the mobile phase, at least initially, is an aqueous solution. Proteins that adsorb to the column are removed by addition of increasing amounts of organic solvent (e.g., methanol) or by increasing the ionic strength.

2. After each stage in a protein purification scheme, the total amount of protein is measured and the amount of the protein of interest is quantitated. Quantitation may include determining the biological activity of the protein by performing a specific assay. The specific activity is the activity normalized to total protein concentration. SDS polyacrylamide gel electrophoresis (SDS PAGE) is routinely employed to monitor purification. Using very thin slab gels, large numbers of samples can be examined in only a few hours. SDS PAGE allows a determination of the complexity of samples with respect to the number and kinds of polypeptides present.

3. Of the two sulfur-containing amino acids, cysteine and methionine, the former is capable of an array of interactions that greatly influence protein structure. For example, cysteine (side chain $-CH_2-SH$) by virtue of its polar side chain may influence protein structure by making polar interactions with either the solvent or with other side chains. In addition, sulfur may coordinate iron atoms. Finally, sulfhydryl groups are capable of forming covalent bonds with each other. Two cysteines joined by a disulfide bond is referred to as cystine. Cystine formed by an oxidation reaction between two cysteines represents a covalent bond that is capable of bringing distant portions of a polypeptide chain together or of joining separate polypeptides.

4. Globular proteins contain a core of hydrophobic amino acids that is largely responsible for their globular shape. This core is formed by hydrophobic interactions of amino acids distributed throughout the polypeptide chain. For typical globular proteins, the number of hydrophobic amino acids is a fixed percentage of the total number of amino acids. In solution, SDS forms micelles around each hydrophobic amino acid. This fact is exploited in SDS polyacrylamide gel electrophoresis to produce a protein/SDS complex with a charge dependent on the number of hydrophobic amino acids. The intrinsic charge of a protein contributes

negligibly to the total charge of a protein/SDS complex. Because the total charge is proportional to the number of hydrophobic amino acids, and, the number of hydrophobic amino acids is a fixed percentage of the total number of amino acids, the charge to mass ratio of protein/SDS complexes for globular proteins is constant. This means that protein/SDS complexes will migrate at a constant rate in an electric field, a rate completely independent of protein size. Separation of proteins is accomplished by forcing protein/SDS complexes through a polyacrylamide matrix. Here, migration is inversely dependent on size. By comparing the mobility of a protein of unknown size to the mobility of proteins of known size, an estimate of the molecular weight can be made.

5. The amino acids phenylalanine, tryptophan, and tyrosine absorb ultraviolet light and these amino acids are largely responsible for the UV-light absorption properties of most proteins. However, the physical dimension of a protein may be comparable to the wavelength of ultraviolet light. In this case the protein will exhibit light scattering. Urea and guanidinium HCl are both powerful denaturing agents that will disrupt protein/protein interactions and unfold proteins, thus minimizing light scattering. Also, there are effects due to polarity of the environment in which the aromatic ring is located.

6. The number of arrangements of 20 different amino acids taken n at a time is $(20)^n$.

Polypeptide Length	Number of Polypeptides
5	3.2×10^6
10	1.3×10^{13}
15	3.3×10^{19}

Summary

In describing protein structure, four levels are considered. The primary structure of a protein is the sequence of amino acids in each of the polypeptide chains of a protein. The secondary structure is a collection of regular structures that may be formed by H-bonding interactions with the groups of the peptide bond. The tertiary structure is the actual three-dimensional shape of the protein, which is the result of weak interactions involving interactions of amino acids with each other and with solvent and includes the location and orientation of secondary structure elements, amino acid side chains, and the backbone of the main chain. Finally, the quaternary structure, reserved for proteins composed of more than one polypeptide chain, describes how the polypeptides are arranged in the protein.

Protein characterization begins with purification of the protein. Protein purification is accomplished using a number of techniques including differential solubility, differential centrifugation and gradient centrifugation; ion exchange, molecular sieve, or affinity column chromatography.

Next, the purified protein is separated into its individual polypeptide chains. This is accomplished by treating the protein with a reducing agent to disrupt interchain disulfide bonds and converting the resulting free sulfhydryls into forms that will not readily reform disulfides. This is accomplished with either performic acid or a combination of 2-mercaptoethanol and an alkylating agent (such as iodoacetate). To disrupt weak bonds, denaturing agents such as 8 M urea or 6 M guanidinium HCl are used.

The nature of the N-terminal and C-terminal amino acids of the polypeptide components is determined using techniques specific for each group. For example, phenylisothiocyanate is used during Edman degradation to convert N-terminal amino acids into PTH (phenylthiohydantoin) derivatives. C-terminal residues may be identified after enzymatic cleavage with carboxypeptidase A, B, C or Y (A is blocked by Pro, Arg and Lys; B only cleaves at Arg or Lys; C and Y cleave at any amino acid).

The amino acid composition is determined by first hydrolyzing the polypeptide chains into amino acids using 6 N HCl at 110°C for 24, 48 and 72 hr to insure that total hydrolysis has occurred. A time course is performed because some amino acids (e.g., Ser and Thr) are unstable under these conditions and slowly degrade whereas others (e.g., hydrophobic amino acids) are only slowly released. The time course can be extrapolated to zero time to determine the initial level of unstable amino acids and to infinite time to predict when total hydrolysis will occur. The mixture of amino acids produced by acid hydrolysis is analyzed on an amino acid analyzer to determine the relative amount of each amino acid in the mixture. An amino acid analyzer separates and quantitates amino acids or amino acid derivatives. The separation may be based on the ionic properties of the amino acids (i.e., ion exchange chromatography), or it may be based on the degree of hydrophobicity of derivatized amino acids (i.e., reverse phase HPLC).

The primary structure or amino acid sequence is established by fragmenting polypeptides into short oligopeptides suitable for Edman degradation (see below). Fragmentation is accomplished by either 1) enzymatic cleavage at specific residues in the polypeptide using trypsin (hydrolyzes peptide bonds whose carbonyl group is donated by Arg or Lys), chymotrypsin (Phe, Tyr or Trp), clostripain (Arg), staphylococcal protease (Asp, Glu), or 2) using chemical means such as cyanogen bromide to cleave after Met.

The amino acid sequences of the short oligopeptides are established using Edman degradation. The basic idea is to subject an oligopeptide to multiple rounds of modification and cleavage. In Edman chemistry, the amino-terminal amino acid is modified using phenylisothiocyanate and specifically cleaved to release the derivatized N-terminal residue while producing a new terminus. The amino acid sequences of all of the oligopeptides produced by fragmentation using at least two different and overlapping fragmentation techniques are compared to establish how the individual fragments are represented in the protein. Mass spectrometry is also being used to establish the sequence of short oligonucleotides. In tandem mass spectrometry a protein fragmented by protease digestion is filtered using the first mass spectrometer into its components. Each component can then be directed into a collision chamber and then into a second mass spectrometer. The collision chamber fragments peptides at specific locations and comparison of mass differences between fragments reveals the amino acid residue lost by fragmentation. In peptide mass fingerprinting, proteins isolated, for example, by two-dimensional polyacrylamide gel electrophoresis (isoelectric focusing followed by SDS PAGE) are digested with a proteolytic enzyme like trypsin and the tryptic fragments analyzed by MALDI-TOF. The identity of the protein can be established by comparing the size of tryptic fragments to those predicted by analysis of proteins in a database.

Amino acid sequence information is useful in helping to determine the function of proteins. Proteins that have a high degree of sequence similarity often perform similar functions. Proteins whose genes evolve from a common ancestral gene are said to be homologous and this homology is evidenced by high sequence similarity. Homologous proteins can be orthologous or paralogous. Orthologous proteins are proteins found in different organisms. They arose by gene duplication before the two organisms diverged from each other. Paralogous proteins are homologous proteins found in the same organism and they arose from gene duplication within the species. BLAST programs are used to identify sequence similarities and these programs use scoring matrices to help identify distantly related proteins. BLOSUM substitution matrix derives from sequence information of related proteins and provides a numerical score for the likelihood of observing a particular amino acid substitution.

Chemical synthesis of peptides using solid-phase synthesis allows for production of peptides with a number of uses in the laboratory.

Proteins have a diversity of functions and this diversity is possible because of the diversity in amino acid building blocks, the use of post-translational modifications of amino acids and inclusion of nonprotein constituents. As a crude classification, proteins are cataloged as fibrous, globular, or membrane. Fibrous proteins are elongated in shape; globular proteins are spherical in shape and water soluble; membrane proteins associate with biological membranes. Of the biopolymers, proteins are by far the most diverse group serving a number of biological purposes. The vast majority of biological catalysts are enzymes, proteins capable of increasing the rate of specific reactions. Regulatory proteins modulate physiological responses of cells by binding to specific cell components. Transport proteins facilitate the movement of substances. Storage proteins provide a source of amino acids. Contractile and motile proteins are proteinaceous elements involved in force generation and motility in cells. Structural proteins provide mechanical support. Scaffold proteins recruit different proteins into multiprotein complexes. Protective proteins serve as a defense barrier. Many proteins have additional components including carbohydrates, lipids, nucleic acids and nucleotides, metal ions, and other more complicated organic molecules (coenzymes and prosthetic groups).

Chapter 6

Proteins: Secondary, Tertiary, and Quaternary Structure

• •

Chapter Outline

❖ Protein conformation: Three-dimensional shape
  ↳ Weak forces act on amino acids to stabilize conformation: Hydrogen bonding, hydrophobic interactions, electrostatic interactions, van der Waals forces
    • H bonds: Peptide backbone and side chains H-bond wherever possible
    • Hydrophobic interactions: Nonpolar residues aggregate to form protein core
    • Ionic interactions: Occur between charged amino acids typically on surface
    • Van der Waals: Weak but numerous and involve complementary surfaces
  ↳ Amino acid sequence contains information to fold protein into conformation
❖ Secondary Structure: Structural elements formed by peptide plane interactions
  ↳ Peptide Plane
    • Carbonyl oxygen -H bond acceptor
    • Amide hydrogen -H bond donor
  ↳ ϕ and ψ angles: Describe orientation of peptide planes about alpha carbons
    • ϕ angle: Rotation about C_α-N bond
    • ψ angle: Rotation about C_α-$C_{carbonyl}$ bond
    • Ramachandran plot: Sterically reasonable values of these angles
  ↳ Alpha Helix (Pauling and Corey)
    • Peptide carbonyl of ith residue H bonded to peptide H of i+4th residue
    • 3.6_{13}: 3.6 residues per turn, 13 atoms between carbonyl oxygen and amide hydrogen
    • 0.54 nm per turn (pitch), 0.15 nm per residue along helix (Z) axis
    • H bonds parallel to helix axis
    • Peptide planes point in same direction along helix
      • Helix has dipole moment
    • Proline, glycine helix breakers
  ↳ Beta pleated sheets
    • Polypeptide chain fully extended
    • Chains or strands joined by H bonds
    • Parallel sheets
    • Antiparallel sheets
  ↳ Beta turn
    • Carbonyl of ith residue H bonded to amide nitrogen of i+3
    • Peptide plane in cis
    • Gly or Pro common
❖ Tertiary structure principles: How atoms of polypeptide are arranged in space
  ↳ Four principles
    • Secondary structures formed when possible
    • Secondary structures associate and pack to form layers
    • Elements between secondary structures are short and direct
    • Stability of final fold

- Large number of intramolecular H bonds
- Reduction in surface area
  - Two factors
    - Hydrophobic residues cluster in core
    - Polar backbone N-H and C=O H bond in helices and sheets
- ❖ Fibrous Proteins: Proteins whose polypeptide chains are parallel to a single axis
  - α–Keratin: α helical coiled coils
    - Subunit: 311-314 residue-long α helix-rich region flanked by nonhelical regions
    - Two-stranded coiled coils with helix repeat of 0.51 nm
    - 7-residue quasi repeat with 1st and 4th residues nonpolar
    - Disulfide bonds: Rigid, inextensible, and insoluble structure
  - Fibroin and β-keratin: Stacked β–sheets
    - Stacked antiparallel β–sheets
    - Gly-Ser or Gly-Ala repeats
    - Sheets stack Gly to Gly
  - Collagen: A triple helix: Connective tissue protein in animals
    - Tropocollagen: Basic unit of three intertwined chains
    - Type I $\alpha1(I)_2\alpha2(I)$, Type II $\alpha1(II)_3$, Type III $\alpha1(III)_3$
    - Amino acid composition and amino acid sequence
      - One-third Gly and one-third Pro or Hyp
      - Modified amino acids: Hydroxyproline and hydroxylysine (vitamin C- dependent reaction)
      - Gly-Pro-Hyp repeat with G placed in center of three-stranded coil
    - Collagen fibrils
      - Staggered arrays of 5 tropocollagen (300 nm)
      - 68 nm periodicity with 40 nm hole regions
        - 5×68 nm = 300nm + 40 nm
- ❖ Globular proteins
  - Core composed of hydrophobic amino acids
  - Peptide groups often H bonded as helices or sheets
    - Surface helices amphiphilic
    - Internal helices or sheets are hydrophobic
    - Solvent exposed helices hydrophilic
  - Packing of elements results in formation of small cavities imparting flexibility to structure
  - Ordered regions with well-defined, nonrepetitive structure
  - Disordered, flexible regions
- ❖ Protein domains
  - Proteins of around 250 amino acid residues have simple, compact globular shape
  - Larger proteins made up of domains or modules
    - Domains often impart unique function
    - Domains often coded for on single exons
      - Protein domain: Fundamental unit of evolution
    - Proteins classified on basis of domains
      - SCOP (http://scop.mrc-lmb.cam.ac.uk/scop)
      - CATH (http://www.cathdb.info)
        - Hierarchy based on structural elements, arrangements, function, sequence
- ❖ Loss of 3-D structure by disruption of weak forces: denaturation
  - Sharp temperature dependence indicates two-state transition
  - Temperature, pH, organic solvents, detergents, urea: denaturing agents
  - 3-Dimensional structure determined by amino acid sequence
  - Anfinsen: RNase A denaturation/renaturation with urea and reducing agent
- ❖ Molecular motion: Proteins are dynamic structures
  - Atomic fluctuations: Movement over small distances
  - Collective motions: Movement of groups of atoms
  - Conformational changes: Movement of large domains

- ❖ Chain folding
  - ⋏ Structural stability
  - ⋏ Right-handed structures
    - · Right-handed twists in sheets
    - · Right-handed cross-overs joining secondary structural elements
  - ⋏ β-strand connections
    - · Antiparallel strands connected by hairpins
    - · Parallel strands connected by cross-overs forming βαβ loop
  - ⋏ Hydrophobic amino acids sequestered in interior: Layers
    - · Two layers of backbone form a single hydrophobic core
    - · Three layers of backbone form two hydrophobic cores
    - · Four-layered and five-layered structures
- ❖ Globular protein classification based on secondary structure
  - ⋏ Antiparallel α helix
  - ⋏ Parallel or mixed β–sheet
  - ⋏ Antiparallel β–sheet
  - ⋏ Metal- and Disulfide-rich
- ❖ Molecular chaperones: Protein complexes that catalyze process of protein folding
  - ⋏ Intrinsically unstructured proteins
    - · Do not possess uniform structural properties
    - · Capable of binding multiple ligands
    - · Form large intermolecular interfaces
- ❖ Protein design: Domains: Regions 40-100 amino acids long that form stable tertiary structures
- ❖ Quaternary structure
  - ⋏ Subunits typically fold into independent globular structures
  - ⋏ Subunits interactions
    - · Isologous interactions: Identical faces of identical subunits
    - · Heterologous interactions: Interacting surfaces not identical
  - ⋏ Symmetry
    - · Cyclic symmetry: Single rotational axis
    - · Dihedral symmetry: 2-fold axis perpendicular to n-fold axis
  - ⋏ Subunit association forces
    - · Hydrophobic interactions between faces
    - · Disulfide bond stabilized
  - ⋏ Polymers: Open quaternary structures
- ❖ Structural and functional advantages to quaternary associations
  - ⋏ Protein stabilized by reduction in surface-to-volume ratio
  - ⋏ Genetic economy: Encode self-assembling subunits rather than single complex protein
  - ⋏ Catalytic sites brought together
- ❖ Cooperativity: Influence on catalytic sites by neighboring sites

Chapter Objectives

The key to understanding protein structure is to understand the weak forces responsible for maintaining a protein in the folded state: hydrogen bonds, hydrophobic interactions, electrostatic interactions, and van der Waals forces. It will be helpful to review the structures of the amino acids and to recall the kinds of weak interactions each amino acid side chain is capable of making.

Peptide Bond
Know the characteristics of the peptide bond: the fact that it is planar because of π-bonding and resonance stabilization and the fact that it has a hydrogen bond donor and an acceptor.

Secondary Structure

Protein secondary structures are regular structures formed by hydrogen bonds between amide planes. Understand the structure of the α-helix. The key is to understand how hydrogen bonding occurs in the α-helix. The carbonyl group of the ith residue is H-bonded to the peptide hydrogen of the $1 + 4^{th}$ residue. Other, less common helices are found in nature (2_7 ribbon, 3_{10} helix and π-helix) that form when hydrogen bonds occur between residues closer or further along the polypeptide chain. Sheet structures form between fully extended peptide chains. Beta-turns are tight turns stabilized by a hydrogen bond and requiring *cis* orientation of amide planes. The α-helix and β-sheet are shown in Figures 6.6 (4th edition) and 6.11 (3rd edition).

Tertiary Structure

The tertiary structure of a protein is its three-dimensional structure. Fibrous proteins have secondary structure elements arranged parallel to a single axis. Examples include α-keratin and collagen. Be familiar with the structures of these two proteins. Globular proteins are by far the most abundant class of proteins. As additional protein structures are solved by x-ray crystallography and nuclear magnetic resonance, a greater understanding of the rules governing protein folding and structure stabilization will unfold. Currently a majority of globular proteins are classified into four broad groups: antiparallel α-helix, parallel or mixed β-sheet, antiparallel β-sheet, and the small metal- and disulfide-rich proteins.

Quaternary Structure

Quaternary structure is reserved for proteins composed of multiple subunits. The subunits represent distinct structural domains that interact in specific ways to form the native protein.

Protein Denaturation

The three-dimensional shape of a functional protein is called the native conformation. A conformational change to a state or states lacking activity is known as denaturation. Protein denaturation may occur when the weak forces, responsible for the native conformation, are disrupted. Denaturing agents or conditions include: heat, high concentrations of urea, guanidine HCl, high salt concentrations, low ionic strength solutions, SDS, organic solvents, or extremes in pH.

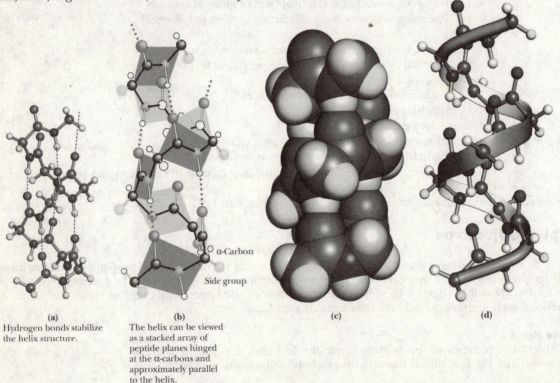

(a)
Hydrogen bonds stabilize the helix structure.

(b)
The helix can be viewed as a stacked array of peptide planes hinged at the α-carbons and approximately parallel to the helix.

α-Carbon

Side group

(c)

(d)

Figure 6.6 Four different graphic representations of the α-helix. (a) A stick representation with H bonds as dotted lines, as originally conceptualized in Pauling's 1960 *The Nature of the Chemical Bond*. (b) Showing the arrangement of peptide planes in the helix. (Illustration: Irving Geis. Rights owned by Howard Hughes Medical Institute. Not to be reproduced without permission.) (c) A space-filling computer graphic presentation. (d) A "ribbon structure" with an inlaid stick figure, showing how the ribbon indicates the path of the polypeptide backbone.

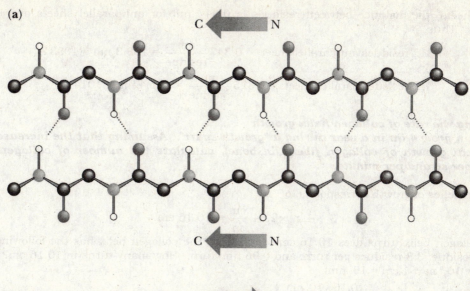

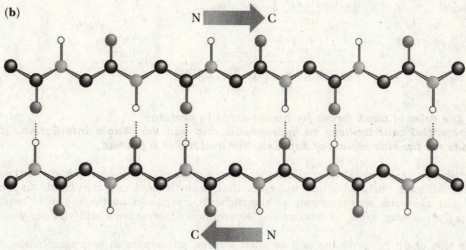

Figure 6.10 The arrangement of hydrogen bonds in (a) parallel and (b) antiparallel β–pleated sheet.

Problems and Solutions

1. Determining the length of a keratin molecule

The central rod domain of a keratin protein is approximately 312 residues in length. What is the length (in Å) of the keratin rod domain? If this same peptide segment were a true α-helix, how long would it be? If the same segment were a β-sheet, what would its length be?

Answer: α-Keratin has extensively distorted α-helical secondary structure. The helix repeats every 3.6 residues but has a pitch of 0.51 nm (compared to 0.54 nm for a true α-helix).

$$\frac{312 \text{ residues}}{3.6 \dfrac{\text{residues}}{\text{turn}}} = 86.7 \text{ turns}$$

$$86.7 \text{ turns} \times 0.51 \frac{\text{nm}}{\text{turn}} = 44.2 \text{ nm} = 442 \text{ Å}$$

For an α-helix of the same number of residues:

$$86.7 \text{ turns} \times 0.54 \frac{\text{nm}}{\text{turn}} = 46.8 \text{ nm} = 468 \text{ Å}$$

For pleated sheets, the distance between residues is 0.347 nm for antiparallel sheets and 0.325 nm for parallel sheets. Thus,

$$312 \text{ residues antiparallel sheet} \times 0.347 \frac{nm}{residue} = 108.3 \text{ nm} = 1083 \text{ Å}$$

$$312 \text{ resisues parallel sheet} \times 0.325 \frac{nm}{residue} = 101.4 \text{ nm} = 1014 \text{ Å}$$

2. Calculating the rate of collagen helix growth

A teenager can grow 4 in. in a year during a "growth spurt". Assuming that the increase in height is due to vertical growth of collagen fibers (in bone), calculate the number of collagen helix turns synthesized per strand per minute.

Answer: Four inches of growth corresponds to

$$4 \text{ in} \times 2.54 \frac{cm}{in} = 10.16 \text{ cm}$$

How many collagen helix turns does 10.16 cm represent? The collagen helix has the following parameters: 0.29 nm per residue; 3.3 residues per turn; and 0.96 nm/turn. How many turns in 10.16 cm? (Note: 1cm = 10^{-2} m, 1 nm = 10^{-9} m ∴ 1 cm = 10^7 nm)

$$\frac{10.16 \times 10^7 \text{nm}}{0.96 \frac{nm}{turn}} = 1.06 \times 10^8 \text{ turns}$$

$$\frac{1.06 \times 10^8 \text{ turns}}{1 \text{ yr} \times 365 \frac{days}{yr} \times 24 \frac{hr}{day} \times 60 \frac{min}{hr}} = 201 \frac{turns}{min}$$

3. Assessing the roles of weak forces for amino acids in proteins

Discuss the potential contributions to hydrophobic and van der Waals interactions and ionic and hydrogen bonds for the side chains of Asp, Leu, Tyr and His in a protein.

Answer: Aspartic acid has a relatively small side chain composed of a -CH_2- group and a carboxyl group. The presence of the ionizable carboxyl group indicates that aspartic acid can participate in ionic bonds. In addition, lone-pair electrons on the oxygen of the carboxyl group can participate in H bonds, as can the hydrogen when the carboxyl group is protonated. Hydrophobic interactions and van der Waals interactions are negligible.

The leucine side chain is an alkane and as such will not participate in hydrogen bonds or ionic bonds. The side chain is hydrophobic and relatively bulky, indicating that it will participate in hydrophobic interactions and is capable of participating in numerous van der Waals interactions.

Tyrosine has a phenolic group attached to the α carbon by a methylene bridge (i.e., -CH_2-). The phenolic group is weakly ionizing with a pK_a of 10. Thus, only under special conditions or environments would it be expected to participate in ionic bonds. The hydroxyl group can both donate and accept H bonds. When protonated, tyrosine is capable of hydrophobic interactions, and, because it is a bulky amino acid, it is expected to participate in numerous van der Waals interactions.

The imidazole side chain of histidine has an ionizable nitrogen and, when protonated, allows histidine to participate in ionic bonds. In addition, hydrogen bond donor and acceptor groups support participation in hydrogen bonds. Thus, the ability to form both hydrogen bonds and ionic bonds precludes hydrophobic bonding. The large side chain is expected to participate in van der Waals interaction.

4. The role of proline residues in β-turns

Pro is the amino acid least commonly found in α-helices but most commonly found in β-turns. Discuss the reasons for this behavior.

Answer: Proline is an imino acid; it has a secondary nitrogen with only one hydrogen. When participating in a peptide bond, the nitrogen no longer has hydrogen bound to it. Thus, peptide bonds in which the nitrogen is supplied by proline are incapable of functioning as hydrogen-bond donors. A prolyl residue will interrupt α-helices when located on the C-terminal end of a helix. If located within 3 residues of the N terminus, proline is capable of hydrogen bonding through its carbonyl oxygen. The β-bend requires a *cis* peptide bond and proline stabilizes the *cis* conformation. Recall a β-bend involves hydrogen bonding of the carbonyl oxygen of a peptide bond with an amide hydrogen three residues away. Usually, adjacent peptide bonds are *trans* but for β-bend formation a *cis* conformation is required.

5. Assessing the crossovers for flavodoxin
For flavodoxin (pdb ID = 5NLL) identify the right-handed cross-overs and the left-handed cross-overs in the parallel β-sheet.

Answer: Visit the protein data bank at http://www.rcsb.org/pdb/home/home.do and download 5NLL. There are a number of options for viewing this protein's structure and these are found under the image. Launch one of the programs or download the pdb file for 5NLL. (MTB Protein Workshop, Rasmol and Swiss-PDB viewer are all reasonable programs to use. The later two are stand alone programs you must download.) View the molecule as a ribbon and locate the N terminus and C terminus.

For a right-handed cross-over, moving in the N-terminal to C-terminal direction, the cross-over moves in a clockwise direction. The reverse is true for a left-handed cross-over, namely, movement from N-terminus to C-terminus is accompanied by counterclockwise rotation.

Various programs have options for coloring secondary structure elements and this is a useful option to identify sheets and helices. Using this option we find that flavodoxin has 5 regions of α-helix and 5 regions involved in sheet formation. The core of the protein is formed by a 5-stranded, curved parallel β-sheet that has two α-helices inside and three α-helices outside (a three layered protein) the curved parallel sheet.

Moving from the N terminus the first secondary structure element we encounter is the second strand of the parallel β-sheet followed by an α-helix, which is inside the curved parallel sheet, and then the outside strand of the parallel β-sheet. A short helix, which runs on the outside curve of the sheet, connects the outside strand of the sheet to its third strand. Another helix again on the outside of the sheet connects to the fourth strand of the sheet. The fifth strand is also connected by a helix running on the outside. Finally, the C-terminus is part of the last helix that is on the inside of the sheet. A diagram of the connectivity of the parallel β-sheet strands is shown below.

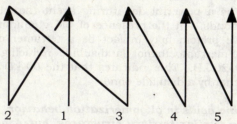

The cross-overs from strand 2 to 3, 3 to 4 and 4 to 5 are easy to identify as right-handed. For example, moving along strand 2 the connection to strand 3 moves in a clockwise direction. The connection between 1 and 2 is also right-handed. To see this, flip the diagram vertically.

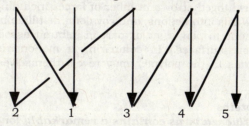

Now, rotate by 180° so that strand 2 is to the right of strand 1.

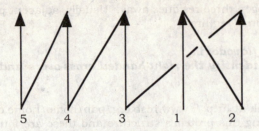

Focus on moving from strand 1 to 2. The connection is clockwise and thus right-handed. All the cross-overs in flavodoxin are right-handed.

6. Assessing the range of φ and ψ angles in proteins
Choose any three regions in the Ramachandran plot and discuss the likelihood of observing that combination of φ and φ in a peptide or protein. Defend your answer using suitable molecular models of a peptide.

Answer: The Ramachandran plot reveals allowable values of φ and φ. The plots consider steric hindrance and will be somewhat specific for individual amino acids. For example, glycine has more allowable φ and φ angles in α-helical conformations than do bulky amino acid like phenylalanine.

7. Protein structure evaluation based on gel filtration data
A new protein of unknown structure has been purified. Gel filtration chromatography reveals that the native protein has a molecular weight of 240,000. Chromatography in the presence of 6 M guanidine hydrochloride yields only a peak for a protein of M_r 60,000. Chromatography in the presence of 6 M guanidine hydrochloride and 10 mM β-mercaptoethanol yields peaks for proteins of M_r 34,000 and 26,000. Explain what can be determined about the structure of this protein from these data.

Answer: Guanidine hydrochloride is a powerful denaturing agent that disrupts tertiary and quaternary structure by disrupting hydrogen bonds. In the presence of 6M guanidine-HCl, only a 60 kD species is observed, indicating that the native protein may in fact be a tetramer of four 60 kD subunits. Upon denaturation in the presence of β-mercaptoethanol (a disulfide reducing agent), two protein peaks are observed, one at 34 kD and one at 26 kD. This indicates that the 60 kD subunit is a heterodimer of two chains, 34 kD and 26 kD, held together by a disulfide bond.

8. Understanding the role of amino acids in oligomerization behavior
Two polypeptides, A and B, have similar tertiary structures, but A normally exists as a monomer, whereas B exists as a tetramer, B₄. What differences might be expected in the amino acid composition of A versus B?

Answer: Oligomeric proteins are held together by a number of forces including hydrogen bonds, ionic bonds, hydrophobic interactions, van der Waals interactions, and covalent, disulfide bonds. Of these interactions, we might expect hydrophobic interactions to play a major part in subunit associations. (Subunit associations involve interactions of two or more surfaces.) In comparing a monomeric protein with a homologous, polymeric protein, interacting regions in the polymer may reveal themselves as sequences of hydrophobic amino acid residues.

9. Evaluation of α-helices in proteins
The hemagglutinin protein in influenza virus contains a remarkably long α-helix with 53 residues.
   a. How long is this α-helix (in nm)?
   b. How many turns does this helix have?
   c. Each residue in an α-helix is involved in two H bonds. How many H bonds are present in this helix?

Answer: The α–helix repeats every 0.54 nm (3.6 amino acid residues) and the distance along the helical axis between residues is 0.15 nm. An α-helix of 53 residues is:

$$\frac{53 \text{ residues}}{3.6 \frac{\text{resisues}}{\text{turn}}} \times 0.54 \frac{\text{nm}}{\text{turn}} = 7.95 \text{ nm, or}$$

$$53 \text{ residues} \times 0.15 \frac{\text{nm}}{\text{residue}} = 7.95 \text{ nm}$$

The number of turns is:

$$\frac{53 \text{ residues}}{3.6 \frac{\text{residues}}{\text{turn}}} = 14.7 \text{ turns}$$

To calculate the number of H bonds, the ith residue is H-bonded to the $(i + 4)^{th}$ residue to form a helix. For a 53-residue helix, the carbonyl groups of the first 49 residues are involved in H bonding ($i + 4 = 53$; $\therefore$, $i = 49$). The amide H for the four residues on the N terminal side are not hydrogen-bonded nor are the carbonyl groups of the last 4 residues, on the C-terminal side. Thus, 49 hydrogen bonds are made.

10. Understanding the role of gly residues in protein secondary and tertiary structure
It is often observed that Gly residues are conserved in proteins to a greater degree than other amino acids. From what you have learned in this chapter, suggest a reason for this observation.

Answer: In considering protein structure we learned that globular proteins are compact structures composed of short helices and sheets. In order to form a compact structure the polypeptide chain must make sharp bends and this is often accomplished by having a glycine residue in the bend. Because glycine has a small side chain, it is easily accommodated in a tight bend.

11. Understanding H-bond formation in proteins
Which amino acids would be capable of forming H bonds with a lysine residue in a protein?

Answer: Lysine has an amino group in its side chain, which at neutral pH is positively charged. Further, the amino hydrogens are capable of functioning as hydrogen-bond donors. So, amino acid residues with hydrogen bond acceptors should be able to form hydrogen bonds with a lysine residue. This would include aspartic and glutamic acid, which could also interact via ionic bonding. Serine and threonine both contain hydroxyl groups, which can function as a H-bond acceptors. Asparagine and glutamine would also be expected to participate in H-bonding as H-bond acceptors through their carbonyl carbon on their side chains. Other possibilities include unprotonated histidine, unprotonated lysine and arginine. The side chains cysteine and tyrosine might also interact as hydrogen bond donors.

12. Assessing the pH dependence of poly-L-glutamate structure
Poly-L-glutamate adopts an α-helical structure at low pH but becomes a random coil above pH 5. Explain this behavior.

Answer: The side chain of glutamic acid contains a carboxylic acid group with a pKa of 4.25. At pH 5.0 glutamic acid's side chain is largely ionized with a 1- charge. Thus, polyglutamic acid at this pH has negative charges, which will repel each other. At pH 2.0, the side chain is protonated and thus uncharged making it possible for the peptide chain to form an α-helix, which brings amino acids closer together than in random coil.

13. Exploring the dimensions of the α-helix and coiled coils
Imagine that the dimensions of the alpha helix were such that there were exactly 3.5 amino acids per turn, instead of 3.6. What would be the consequences for coiled-coil structures?

Answer: In a helix with 3.5 residues per turn, after two turns each additional residue would line up with a residue in the first two turns. Thus, two helices could interact by simply lining up side-by-side, provided the helices are longer than two turns. For the α-helix the helices would have to be longer than 5 turns for this to occur. Thus, significant numbers of interactions between side chains in a coiled-coil made up of α-helices would only occur if the helices wrapped around each other. A helix with a repeat of 3.5 residues per turn could interact by simply lining up to form a structure that may not be as stable as a twisted coil.

14. (Research Problem) The nature and roles of linear motifs in proteins

In addition to domains and modules, there are other significant sequence patterns in proteins – known as linear motifs *– that are associated with a particular function. Consult the biochemical literature to answer the following questions:*
- *What are linear motifs?*
- *How are they different than domains?*
- *What are their functions?*
- *How can they be characterized?*

There are several papers that are good starting points for this problem:
Neduva, V. and Russell, R., 2005. Linear motifs: evolutionary interactions switches. FEBS Letters *579:3342-3345.*
Gibson, T., 2009. Cell regulation: determined to signal discrete cooperation. Trends in Biochemical Sciences *34:471-482.*
Diella, F., Haslam, N., Chica., C. et al., 2009. Understanding eukaryotic linear motifs and their role in cell signaling and regulation. Frontiers of Bioscience *13:6580-6603.*

Answer: Proteins are often composed of domains, which are modular elements that fold into compact structures. Domains are typically responsible for a particular function and to form a stable structure with a particular function at least 30 amino acid residues are required. Analysis of protein sequence reveals that domains are only part of a protein's structure. Often domains are joined by short, disordered sequences that are also often associated with function These short linear sequences (3 to 10 residues) are known as linear motifs. They are called "linear" because long-range folding of the polypeptide chain is not required for their function. In addition, the sequence conservation often involves only a few residues. Linear motifs interact with target proteins with rather low affinity compared to protein-protein interactions that involve large complementary surfaces. Linear motifs can be classified into four types, each with a general function. PTM (post-translational modification) sites are locations that are covalently modified with a variety of groups (glycosylation, phosphorylation, acetylation, etc.). Ligand sites are responsible for binding of ligands or docking of interacting proteins. Processing sites are locations at which cleavage occurs. Finally, subcellular targeting sites direct proteins to a particular cellular location or compartment. Linear motifs are difficult to recognize because of their short sequence in which only a subset of positions is conserved. The Diella paper has a section titled *Methods to predict linear motifs* that explains the variety of techniques for characterizing linear motifs.

15. (Research Problem) The nature of protein-protein interactions
How do proteins interact? When one protein binds to another, one or both changes conformation. Two hypothesis have been proposed to describe such binding: in the induced fit *model, the interaction between a protein and a ligand induces a conformation change (in the protein or ligand) through a stepwise process. In the* conformational selection *model, the unliganded protein (in the absence of the ligand) exists as an ensemble of conformations in a dynamic equilibrium. The binding ligand interacts preferentially with one among many of these conformations and shifts the equilibrium in favor of the selected conformation. Three recent papers shed light on this question:*
Boehr, D. and Wright, R.E., 2008. How do proteins interact? Science *320:1429-1430.*
Gsponer. J. et al. 2008. A coupled equilibrium shift mechanism in calmodulin-mediated signal transduction. Structure *16:736-746.*
Lange, O.F., et al., Recognition dynamics up to microseconds revealed from an RDC-derived ubiquitin ensemble in solution. Science *320:1471-1475.*
- *What proteins were studied in these papers?*
- *What techniques were used, and what time scales of protein motion were studied?*
- *What were the conclusions of these papers, and how do these results illuminate the choice between induced fit and conformational selection in protein-protein interactions.?*

Answer: In the Lange paper the protein under study is ubiquitin, a small (76 amino acid residues), highly conserved protein that is used to modify a number of proteins post-translationally. Modification occurs by amide-bond formation between the C-terminus (a glycine) and a side chain amino group on a lysine on the target protein. The protein was studied because 46 ubiquitin crystal structures had been solved showing structural heterogeneity. Gsponer and coworkers studied the calcium binding protein calmodulin. Calmodulin is about twice as big as ubiquitin. It binds calcium and target proteins to impart calcium regulation to the target proteins and it is involved in a long list of cellular processes.

In both studies nuclear magnetic resonance (NMR) relaxation was used to study conformations of the two proteins with time scales up to microseconds.

The conclusion of both papers is that the two proteins exist in solution in a number of conformations, which supports the conformational selection model. Under specific conditions (temperature, pressure, and solvent conditions) the proteins exist in an ensemble of conformations. For ubiquitin this ensemble includes the 46 ubiquitin structures determined by x-ray crystallography. The ensemble for calmodulin with bound calcium includes a range of structures that resemble the conformation of calmodulin bound to one of its target proteins, myosin light chain kinase.

16. (Research Problem) Conformational transitions in proteins
How do proteins accomplish conformational changes? How is it that proteins convert precisely and efficiently from one conformation to another? Recall from Figure 6.34 that any folding/unfolding transition must involve movement across a free-energy landscape, and try to imagine the nature of a conformational transition. Are bonds formed and broken along the way? What kinds of bonds and interactions might be involved? Suggest how such conformational transitions might occur. One reference that will be useful in this regard is:
Boehr, D., 2009. During transitions proteins make fleeting bonds. Cell 139:1049-1051

Answer: The take home message is that proteins in their native state are dynamic structures. The free-energy landscape shown in Figure 6.34 has a number of energy "valleys" that correspond to different conformations of the native protein. The difference in free energies between valleys determines the relative abundance of conformations whereas the energy barriers (hills) separating valleys determines the kinetics of movement between valleys. Under a specific set of conditions proteins populate a variety of conformational states including active and inactive states. To transition from one state to another the protein must overcome the activation barrier and this seems to be accomplished by transient hydrogen bonds leading to concerted motions. Localized changes in hydrogen bonding as opposed to global changes avoid the protein transitioning through a completely unfolded state to reach another conformation. Proteins have evolved to not only occupy different conformations but to provide efficient mechanisms that allow for transitions between conformations.

17. (Historical Context) The third person on the α-helix publication
Who was Herman Branson? What was his role in the elucidation of the structure of the a-helix? Did he receive sufficient credit and recognition for his contributions? And how did the rest of his career unfold? Do a Google search on Herman Branson to learn about his life, and read the article by David Eisenberg under Further Reading. You may also wish to examine the original paper by Pauling, Corey and Branson, as well as the following web site: http://www.pnas.org/site/misc/classics1.shtml

Pauling, L., Corey, R.B., and Branson, H.R., 1951 The structure of proteins: two hydrogen-bonded helical configurations of the polypeptide chain. Proceedings of the National Academy of Sciences 37:235-240.

Answer: To stimulate your research on Herman Branson you should know that he was discussed in an article titled *Blacks Who Should Have Won Nobel Prizes* published in The Atlanta Post on 28 January 2011. The article discusses six individuals with George Washington Carver included in the list along with Herman Branson. Branson was a physicist who spend some time in Linus Pauling's lab in 1948 using his mathematical skills to explore helical structures in proteins. It was clear that Pauling was on the path to the alpha helix as an important structural component of proteins and Branson played a role in confirming this finding but what role did Branson really play and why was he not included as a Nobel Prize recipient? A Nobel Prize (except for peace) can be awarded to up to three individuals so there was room for Branson.

Preparing for the MCAT® Exam

18. Analyzing peptide sequences and their structural consequences
Consider the following peptide sequences:

>   EANQIDEMLYNVQSLTTLEDTVPW
>   LGVHLDITVPLSWTWTLYVKL
>   QQNWGGLVVILTLVWFLM
>   CNMKHGDSQCDERTYP
>   YTREQSDGHIPKMNCDS
>   AGPFGPDGPTIGPK

Which of the preceding sequences would be likely to be found in each of the following:
a. A parallel β-sheet
b. An antiparallel β-sheet

> **c. A tropocollagen molecule**
> **d. The helical portions of a protein found in your hair?**

Answer: To help identify β-sheets we can look at distributions of hydrophobic amino acids, which are typically distributed on one side of an antiparallel sheet and both sides of a parallel sheet. To help we will display hydrophobic amino acids in bold. (The font is changed to Courier, a non-proportional font, to allow for alignment of the sequence with the residue numbers.)

```
123456789012345678901234
EANQIDEMLYNVQSLTTLEDTVPW
LGVHLDITVPLSWTWTLYVKL
QQNWGGLVVILTLVWFLM
CNMKHGDSQCDERTYP
YTREQSDGHIPKMNCDS
AGPFGPDGPTIGPK
```

The second sequence has hydrophobic amino acids at nearly every other residue. If this sequence were in β-sheet structure it would have a hydrophobic surface and a hydrophilic surface. Thus, sequence two is likely to be an antiparallel sheet.

The first and third sequences also contain significant numbers of hydrophobic amino acids. For the third sequence, if it were in β-sheet structure it would have hydrophobic amino acids on both sides of the sheet consistent with a parallel sheet.

The first sequence also has nearly an equal distribution of hydrophobic amino acids at odd-numbered and even-numbered positions, which would be expected to support parallel sheet formation. However, the first sequence seems to have hydrophobic amino acids distributed evenly throughout its length. If fact, starting at the alanine in position 2, one sees hydrophobic amino acids at positions 5 (an I) and 8 (an M), which is reminiscent of the primary structure of α-keratins with quasi repeating segments, seven-residues long and containing hydrophobic amino acids at positions 1, 4 and 7. If arranged in an α-helix, this would put hydrophobic amino acids on one face of the helix. Thus, the first sequence is a good candidate for the helical portions of a protein found in hair.

Of the remaining sequences, the best fit for a tropocollagen sequence is the sixth sequence. Tropocollagen is a triple helix of a special type formed by intertwining three chains together. The chains have a three-residue repeat, Gly-x-y, with x and y often being proline (and or hydroxyproline).

Programs designed to identify parallel and antiparallel sheets have been developed and several may be found at SwissProt (http://us.expasy.org/sprot/) under Proteomics tools (http://us.expasy.org/tools/).   Under "Primary Structure Analysis" the link to ProtScale gives several useful programs to analyze primary protein structure. The figure below was constructed by analyzing each of the sequences using "beta-sheet /Chou & Fasman". There are also programs listed that will analyze parallel and antiparallel sheets. Sequences are entered into the text box. The programs calculate the probability of a particular structure using a scoring matrix, which gives the likelihood of finding a particular amino acid in a structure. Average scores are calculated over a window of amino acids. Before activating the program, change the window size to the smallest allowable. Because several of the sequences are quite small it may help to add glycines to each end of the sequences.

Data for the figure was generated using ten G's on each end of each of the sequences. After the program was activated the results were reported both graphically and numerically by activating the link at the bottom of the results page. The numerical results were pasted into a Microsoft Word document as unformatted text using "Paste Special". The data were converted from text to table and the values were then transferred to Excel.

Sequences 1, 2 and 3 all score high, relative to 4, 5 and 6 in the Chou and Fasman analysis in addition to antiparallel and parallel sheet programs. Sequence 3 scores highest as a parallel sheet.

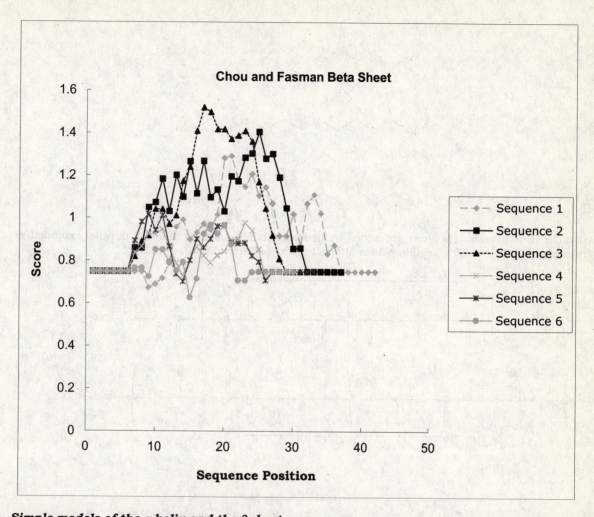

Chou and Fasman Beta Sheet

Legend:
- Sequence 1
- Sequence 2
- Sequence 3
- Sequence 4
- Sequence 5
- Sequence 6

Y-axis: Score
X-axis: Sequence Position

19. Simple models of the α-helix and the β-sheet
To fully appreciate the elements of secondary structure in proteins, it is useful to have a practical sense of their structures. On a piece of paper, draw a simple but large zigzag pattern to represent a β-strand. Then fill in the structure, drawing the locations of the atoms of the chain on this zigzag pattern. Then draw a simple, large coil on a piece of paper to represent an α-helix. Then fill in the structure, drawing the backbone atoms in the correct locations along the coil and indicating the locations of the R groups in your drawing.

Answer: For the beta-pleated sheet, the zigzag below represents alpha carbons of the polypeptide backbone. The lines connecting the alpha carbons contain the elements of the peptide plane with the carbonyl oxygen alternating out of or into the plane of the paper (because the most stable conformation of adjacent peptide planes is trans). The R group and hydrogen are added to the alpha carbons as shown. To make an extended sheet, strands would be aligned side-by-side in parallel or antiparallel fashion.

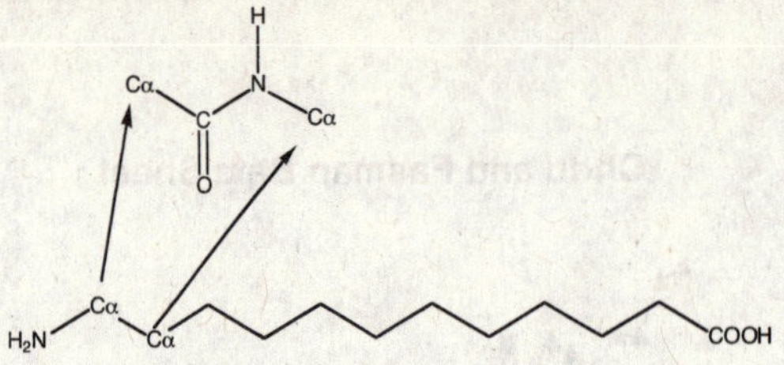

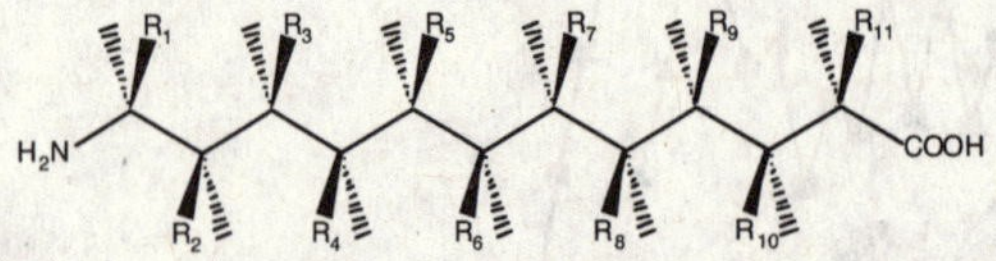

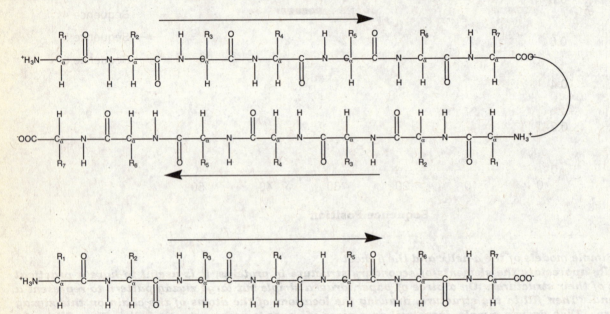

To visualize how strands of a sheet are joined by hydrogen bonds, antiparallel and parallel extended peptide chains are drawn below. The peptide planes alternate (trans conformation).

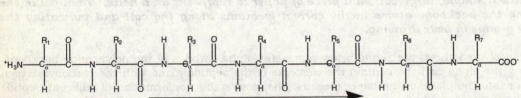

Draw hydrogen bonds between H-bond donors and acceptors to see how the strands of a sheet are held together.

For an alpha helix, the carbonyl oxygen of the ith amino acid residue is hydrogen bonded to the hydrogen on the nitrogen of the $(i + 4)^{th}$ amino residue. Use the template below to show this connection. Then, count atoms moving from the oxygen of the carbonyl group to the hydrogen. You should count 13 atoms. (Note: peptide planes are in *trans* in this "random coil". In an α-helix, peptide plans are in *cis*.)

Do the same on the remaining templates for other helices (i to i + 3, i + 2 and i = 5).

20. Comparing dissociation constants and free energies for protein oligomers
The dissociation constant for a particular protein dimer is 1 micromolar. Calculate the free energy difference for the monomer to dimer transition.

Answer: The situation is described by the following reaction:

$$D \rightleftharpoons M + M$$

The dissociation constant, K_D, is the concentration of monomer, [M], at which the dimer, D, is half dissociated. The standard free energy difference is related to K_D as follows:

$$\Delta G° = -RT \ln K_D$$

Given R = 8.314 J·K^{-1}·mol^{-1}, T = 25 + 273 = 298 K and K_D = 1×10^{-6} M,

$$\Delta G = -8.314 \text{ J K}^{-1} \text{ mol}^{-1} \times 298 \times \ln(1 \times 10^{-6}) = 34.2 \text{ kJ·mol}^{-1}$$

Questions for Self Study

1. List the weak interactions responsible for maintaining protein conformation.

2. For the structures shown below indicate by D, A, D&A, or B if the structure contains only a hydrogen bond donor, only a hydrogen bond acceptor, a donor and separately an acceptor, or a group that functions both as donor or acceptor.

a. |
    C=O
    |

b. —NH$_2$

c. —OH

d.   O
        ||
    —C—N—
        |
        H

e. =N—

3. Fill in the blanks. The α-helix is an example of a ____ structural element. It is a helical structure formed by ____ ____ between groups in peptide bonds. In particular, the ____ ____ of the ith residue of an α–helix is bonded to the ____ ____ of the (i+4)th residue. The result is a helical structure that repeats every ____ residues. The ____ of the helix, or distance along the z-axis per turn, is 0.54 nm. This arrangement orients the amide planes ____ to the helix axis and the side chains ____ to the helix axis.

4. Proteins composed predominantly of β-pleated sheets may be expected to form structures that are flexible yet inextensible. Please explain.

5. What two amino acids are most suited to beta-turns?

6. What posttranslational modification is necessary to produce collagen and what water soluble vitamin is this modification dependent on?

7. Give two examples of fibrous proteins.

8. What are the four globular protein groups found in nature?

9. The agents listed below are all expected to affect protein tertiary structure. For each give a brief explanation of how they affect protein structure.
   a. Urea
   b. SDS
   c. High temperature
   d. β-mercaptoethanol
   e. Distilled water
   f. Organic solvents

10. What are the advantages to quaternary associations?

Answers

1. Hydrogen bonds, hydrophobic interactions, electrostatic interactions, van der Waals interactions.

2. a. A; b. B; c. B; d. D&A; e. A

3. Secondary; hydrogen bonding; carbonyl oxygen; amide nitrogen; 3.6; pitch; parallel; perpendicular.

4. The polypeptide chain in a β-pleated sheet is already fully extended; however, the chains are free to bend.

5. Proline and glycine.

6. Hydroxylation; vitamin C.

7. α-Keratin, silk fibroin proteins, collagen.

8. Antiparallel α-helix; parallel or mixed β-sheet; antiparallel β-sheet; small metal- and disulfide-rich proteins.

9. Urea will disrupt hydrogen bonds. SDS is a detergent that will disrupt the hydrophobic cores of proteins. High temperature will disrupt weak forces. β–Mercaptoethanol will reduce disulfide bonds. Distilled water will perturb ionic interactions. Organic solvents will interact with hydrophobic amino acid side chains.

10. Genetic economy and efficiency; structural stability due to reduction of surface-to-volume ratio; formation of catalytic centers; cooperativity.

Additional Problems

1. For the following conditions or agents, explain how they may denature proteins: heat, urea, guanidine HCl, distilled water, SDS, liquid phenol.

2. An often-cited example of protein denaturation is the change accompanying the heating of egg-whites. Clearly, this process is irreversible yet the classic experiments of Anfinsen on RNase A showed that denaturation is reversible. Rectify these two seemingly conflicting facts.

3. In the movie *Papillon*, a criminal, played by the late Steve MacQueen, is imprisoned on a remote tropical island for many years. Needless to say, conditions in prison are not idyllic; the prisoners are served only enough gruel to keep them alive. In one scene, the main character reaches into his mouth and pulls out a tooth. What disease might the prisoner have suffered from?

4. Gelatin desserts are prepared by adding hot water to a dry powder. This popular dessert is made from collagen from pig skins. Given your understanding of the structure of collagen explain the gel-like consistency of gelatin.

5. Myoglobin was the first globular protein whose structure was solved by x-ray crystallography. The protein contains a hydrophobic core formed by the sides of several α-helices. For human myoglobin, the sequence of

helix E is given below. For this helix, label the hydrophobic residues likely to contribute to the hydrophobic core.

E Helix Sequence: S E D L K K H G A T V L T A L G G I L

Abbreviated Answers

1. Gentle heating will disrupt hydrogen bonds. The structures of urea and guanidine HCl are shown below:

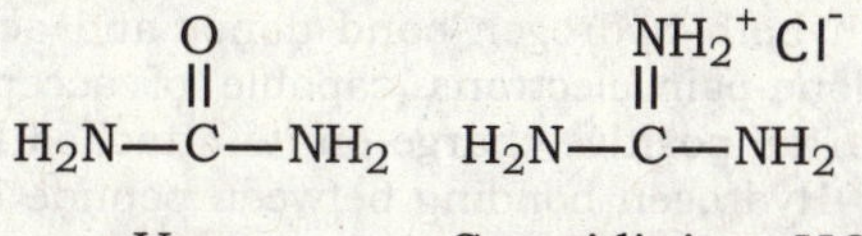

$$\underset{\text{Urea}}{H_2N-\overset{\overset{\displaystyle O}{\|}}{C}-NH_2} \qquad \underset{\text{Guanidinium HCl}}{H_2N-\overset{\overset{\displaystyle NH_2^+\ Cl^-}{\|}}{C}-NH_2}$$

Urea and guanidine HCl are both very soluble in water and may be used to form highly concentrated solutions. These agents denature proteins by disrupting hydrogen bonds. The low ionic strength and high dielectric constant of distilled water cause changes in ionic bonds that may lead to protein denaturation. SDS is an ionic detergent that disrupts hydrophobic interactions. In solution, SDS forms micelle structures by association of their hydrocarbon tails. Micelles will form around hydrophobic amino acid side chains, disrupting the normal hydrophobic interactions these side chains make. Liquid phenol will disrupt hydrophobic interactions.

2. Egg whites contain about 10% by weight of ovalbumin and, while it is true the heat of a hot frying pan is sufficient to denature the protein, the real problem comes subsequently. The denatured protein readily aggregates in this highly concentrated protein solution and the aggregation is for all intents and purposes irreversible. Anfinsen's experiments were done at much lower protein concentrations (and with a considerably smaller protein).

3. The disease was probably scurvy caused by a deficiency of vitamin C. Cross-linking of collagen fibers is catalyzed by prolyl hydroxylase in a vitamin C-dependent reaction. A deficiency of vitamin C results in incomplete cross-linking leading to defective collagen fibers. Collagen fibers are found in connective tissues, which are thus weakened in scurvy.

4. Collagen fibers hydrate and interact to form a tangled mass of crisscrossed fibers.

5. The residues in helix E are plotted in a helical wheel plot to identify the location of amino acids on the helix surface. To construct a helical wheel plot, a projection of the residues in a helix is made along the helical axis (Z-axis) onto the X-Y plane. In an α-helix, the helix repeats every 3.6 residues. Thus, each residue is 100° apart. The helical wheel plot is shown below with hydrophobic amino acids highlighted.

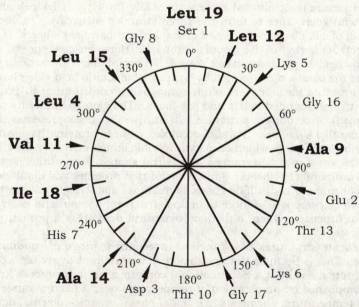

The plot clearly shows that hydrophobic amino acids (in bold face) are concentrated along one side of the helix. This is the side of the helix that faces the hydrophobic core.

115

Summary

Proteins are composed of linear chains of amino acids held together by peptide bonds. The sequence of amino acids is the primary structure of the protein. It contains all the information necessary to fold up the chains into a functional protein. This folding involves weak forces such as hydrogen bonds, electrostatic interactions, hydrophobic interactions and van der Waals interactions. The peptide bond itself is capable of participating in hydrogen bonds with both hydrogen-bond donor and acceptor groups. For example, the carbonyl group has two pairs of lone-pair electrons capable of accepting hydrogen bonds. Also, the electronegative nitrogen induces a partial positive charge on its attached hydrogen, allowing the hydrogen to function as a hydrogen-bond donor. Hydrogen bonding between peptide bond groups is the basis of protein secondary structure elements, namely, helices, pleated sheets and turns.

The side chains of the amino acids are capable of a variety of interactions including, hydrogen bonds, ionic bonds, hydrophobic interactions, and van der Waals interactions. These interactions and secondary structural elements are responsible for the tertiary structure of proteins, their actual three-dimensional shape.

Secondary structure elements include helices, pleated sheets and β-turns. Stable helices are a result of hydrogen bond formation between donor and acceptor groups in peptide bonds. The α-helix is a right-handed helix (in moving from the N terminus to the C terminus, rotation is clockwise) formed by the carbonyl oxygen of the ith residue hydrogen bonded to the amide hydrogen of the $(i + 4)^{th}$ residue. The helix makes one turn every 0.54 nm and has 3.6 amino acid residues per turn. Amino acids are spaced 0.15 nm along the helix axis The hydrogen bonds are parallel to the helical axis, with the peptide-bond planes and the amino acid side chains perpendicular. The α-helix is also referred to as a 3.6_{13} helix; it has 3.6 residues per turn with the hydrogen-bonded groups separated by 13 atoms. Other helical structures are known that involve formation of hydrogen bonds between the carbonyl of the ith amino acid and the amide nitrogen at i + 2, (2_7 ribbon), i + 3, (3_{10}), and i + 5, (4.4_{16} or π helix). When the polypeptide chain is in an extended state, β-pleated sheets may form by hydrogen bonding between parallel chains (parallel β-pleated sheets) or antiparallel chains (β-pleated sheets). The disposition of α-carbons in these structures forms a pleated pattern.

The peptide chain is often required to make sharp turns (as for example when antiparallel β-pleated sheets are formed) and the β-turn is often used to accomplish this. In this secondary structure element, a sharp turn is stabilized by hydrogen bonding between the carbonyl of a peptide bond with the amide hydrogen 2 residues away. In a polypeptide chain, peptide bonds are usually *trans* but in the β-turn a *cis* arrangement is required. Proline and glycine are often found in β-turns because they can assume the *cis* configuration.

Some proteins are dominated by secondary structure. In α-keratin, found in hair, wool, claws, fingernails, and horns of animals, two-stranded coiled coils composed of distorted α-helices predominate. Connective tissue, including tendons, cartilage, bones, teeth, skin, and blood vessels, contains collagen, whose structure is dominated by a triple helix. The collagen helix has 3.3 residues per turn, a fact reflected in its amino acid sequence. The primary structure is dominated by repeats of Gly-Pro-Hyp. Glycines are located on the helix interior, making contact with each other to form the three stranded structure. In contrast to these helical proteins, the fibroin protein of silk fibers is composed of stacked antiparallel β-sheets.

The largest class of proteins is that of the globular proteins. Here, helices, sheets, and turns are used to form a globular structure held together by weak forces. Hydrophobic interactions play a major role in globular protein structure; the core is often formed by hydrophobic amino acid side chains. Structural motifs consisting of secondary structure elements are often components contributing to protein structure, as for example β-sheet arrays with right-handed twists and βαβ-loops. The globular proteins can be classified based on the type and arrangement of secondary structure. In antiparallel helix proteins, the core structure is a bundle composed of antiparallel α–helices. Parallel or mixed β-sheet proteins have an extended sheet of β-structure. The sheet may be internal, in which case hydrophobic amino acids are distributed on both sides of the sheet. Alternatively, the sheet may be arranged in a barrel shape. Antiparallel β-sheet proteins have two sheets, each composed of antiparallel β-sheets. Finally, metal-rich proteins and disulfide-rich proteins rely on either metal binding sites or numerous disulfide bonds to maintain a stable globular structure.

Some proteins appear to be composed of more than one structural domain and each of the domains serves a particular role in protein function. Often, a distinct structural domain is a protein module dedicated to a particular function.

Proteins are dynamic, delicate structures that function in a narrow range of conditions such as pH, ionic strength and temperature. This is because tertiary structure is stabilized by weak forces. When these weak forces are perturbed the protein assumes a nonfunctional conformation in a process known as denaturation. Denaturation can be accomplished by an elevation of temperature over a narrow range giving rise to a sharp transition. In addition to temperature, chemicals like acids, bases, organic solvents, detergents and urea and guanidine hydrochloride are potent denaturing agents.

The experiments of Anfinsen and White showed that the information to fold a peptide chain into a protein is contained in the primary structure. Working with ribonuclease, a small protein capable of hydrolyzing phosphodiester bonds in RNA, they showed that enzymatic activity is dependent on the native conformation of the protein. The protein contains numerous disulfide bridges that stabilize the native state. By reducing these bonds and treating the protein with a denaturing agent, activity is lost. However, the activity is regained if the denaturant is first removed, allowing the chain to fold into its native conformation before disulfide bonds are allowed to reform.

Quaternary structure is reserved for oligomeric proteins, proteins composed of distinct subunits. When the subunits are identical, the protein is a homomultimer; when the subunits are different, the protein is a heteromultimer. In multimeric proteins, the subunits typically fold into independent structures and make isologous (same surface) or heterologous (different surface) interactions with other subunits in the protein. The driving forces for these interactions are hydrophobic in nature and disulfide bonds may stabilize the interactions. Quaternary associations have functional advantages including stability due to a reduction is surface-to-volume ratio, genetic economy, formation of active protomeric enzymes by bringing catalytic sites, located on separate sites on the monomer, together in the polymer and, finally, cooperativity of ligand binding sites and catalytic sites.

Chapter 7

Carbohydrates and Glycoconjugates of the Cell Surfaces

• •

Chapter Outline

- ❖ Carbohydrates $(CH_2O)_n$, $n \geq 3$
- ❖ Nomenclature
    - ⅄ Monosaccharides (simple sugars)
    - ⅄ Olio and polysaccharides: Polymers of simple sugars
- ❖ Classification
    - ⅄ Aldose (aldehyde) and ketose (ketone)
    - ⅄ Triose, tetrose, pentose, hexose, etc.
- ❖ Stereochemistry
    - ⅄ Aldose $n \geq 3$, ketose $n \geq 4$ have asymmetric carbons (chiral centers)
    - ⅄ D- and L- Configuration: Refer to configuration of highest numbered asymmetric carbon
    - ⅄ D- and L- forms: Mirror images: Enantiomers
    - ⅄ With >1 asymmetric carbon: Diastereomers: Configurations that differ at 1 or more chiral carbons but not mirror image molecules
    - ⅄ Epimers: Two molecules that differ in configuration about 1 asymmetric carbon
- ❖ Ring structures
    - ⅄ Pyranoses: Six-membered, oxygen-containing ring
    - ⅄ Furanose: Five-membered, oxygen-containing ring
    - ⅄ Anomeric carbon: Ketone or aldehyde carbon that becomes chiral upon ring formation
    - ⅄ Anomers: α, β differ in configuration about anomeric carbon
        - • α-Configuration: In Fisher projection, OH of anomeric carbon on same side as OH of highest numbered asymmetric carbon
        - • β-Configuration: In Fisher projection, OH of anomeric carbon on opposite side as OH of highest numbered asymmetric carbon
    - ⅄ Haworth projections: Three-dimensional representation: Groups to right in Fisher projection draw down in Haworth projection
    - ⅄ Conformations
        - • Chair and boat conformations due to ring pucker
        - • Axial and equatorial orientation of groups attached to ring
- ❖ Sugar derivatives
    - ⅄ Acids: Oxidation of free anomeric carbon to carboxylate
        - • Aldonic acid (gluconic acid from glucose)
    - ⅄ Acids: At C-6: Uronic acid (glucuronic acid from glucose)
    - ⅄ Alcohols: Alditols: Carbonyl reduced to alcohol
    - ⅄ Deoxysugars: H replaces OH group
    - ⅄ Phosphate esters (ATP and GTP)
    - ⅄ Amino sugars: Amino group in place of OH: Glucosamine
        - • Muramic acid and neuraminic acid: Glucosamine with acids linked to C-1 or to C-3
    - ⅄ Glycosides: Anomeric carbon reacted with alcoholic function.
- ❖ Sugars with free anomeric carbon are reducing sugars

- End of sugar polymer with free anomeric carbon: Reducing end
- Opposite end: Nonreducing end
❖ Oligosaccharides
  - Disaccharides
    - Maltose: Diglucose
    - Lactose: Galactose and glucose
    - Sucrose: Fructose and glucose: Nonreducing sugar
❖ Storage polysaccharides
  - Starch: α–Amylose and amylopectin
    - α–Amylose: Linear chains of α(1→4)D-glucose
    - Amylopectin: Linear chains of α(1→4)D-glucose with α(1→6)D-glucose branches every 12 to 30 residues
    - Plant digestion of starch: Starch phosphorylase: Phosphorolytic cleavage producing glucose-1-phosphate
    - Animal digestion of starch
      - α–Amylase: Animals: Hydrolysis of internal α(1→4) glycosidic linkage
  - Glycogen: α(1→4) D-glucose chains with α(1→6) D-glucose branches every 8 to 12 residues
  - Dextrans: Bacteria: α(1→6) D-glucose polymers
❖ Structural polymers
  - Cellulose: Plant cell wall: Linear polymer of β(1→4) D-glucose
  - Chitin: Exoskeleton and fungi cell wall: β(1→4) N-acetyl-D-glucosamine
  - Agar: Two components: Agarose and agaropectin
    - Agarose: Alternating D-galactose and 3,6-anhydro-L-galactose chains with 6-methyl-D-galactose side chains
    - Agaropectin: Agarose with negatively charged sulfate or carboxylate groups
  - Glycosaminoglycans: Polymers with disaccharide repeat: Negatively charged
    - Heparin: Sulfated glucuronic acid/disulfated glucosamine: Highly charged
    - Hyaluronate, condroitin-4 or -6 sulfate, dermatan sulfate, keratan sulfate
❖ Peptidoglycan of bacterial cell walls
  - Gram-negative bacterial cell wall
    - NAG-NAM repeats crosslinked with tetrapeptide
    - Cell wall sandwiched between inner and outer lipid bilayers
    - Outer lipid bilayer: Lipopolysaccharides: Antigenic
  - Gram-positive bacterial cell walls
    - NAM-NAG repeats crosslinked with tetrapeptide and pentaglycine
    - No outer lipid membrane
    - Teichoic acid: Ribitol phosphate or glycerol phosphate polymers
❖ Glycoproteins
  - Functions: Structural, enzymatic, receptors, transport, immunoglobins
  - Linkage
    - O-linked
      - Amino acids are serine, threonine or hydroxylysine
      - Sugars: N-Acetylgalactosamine
      - Cell surface glycoproteins: Two motifs
        - Glycosylated extracellular domain: Adopts extended conformation
        - Glycosylated extracellular stem: Separates extracellular globular domain from membrane surface
      - Antifreeze glycoproteins: [A-A-T]$_n$-A-A
        - n from 5 to 50
        - T has O-linked disaccharide
    - N-linked
      - Amino acid is asparagine
      - Sugars: Two N-acetylglucosamines linked to branched mannose triad
      - Variety of functions: Protein stabilization, protect against proteolysis, help folding
        - Blood protein age sensed by liver receptor sensitive to charge on protein

oligosaccharides

❖ Proteoglycans: Glycoproteins
  ⋏ Sugar: Glycosaminoglycans: O-linked to Ser of (Ser-Gly) repeats
  ⋏ Function: Interact with a variety of proteins via glycosaminoglycan
    • BBXB and BBBXB (B = basic amino acid) are binding sites on proteins that bind to glycosaminoglycan
  ⋏ Syndecan: Transmembrane proteoglycan
    • Binds to actin cytoskeleton intracellularly
    • Binds to fibronectin extracellularly
  ⋏ Growth-modulating proteoglycans
    • Heparin internalization inhibits cell proliferation
    • Heparin binds fibroblast growth factor and protects it from degradation
    • Proteoglycan synthesis and secretion stimulated by transforming growth factor β
    • Some proteoglycan core proteins have domains that may bind to growth factor receptors
  ⋏ Cartilage flexibility and resilience
    • Matrix: Hyaluronic acid filaments studded with proteoglycan
    • Polyanionic complex coordinates water that is squeezed out and reabsorbed
❖ Carbohydrate code
  ⋏ Glycoconjugates: Information rich protein-sugar interactions
    • Sugar polymers provide information
      • Information organized by enzymes (glycosyltransferases, glycosidases)
    • Sugar-binding proteins like lectins translate code
      • Proteins bind with specificity and high affinity
    • Interactions mediate a host of cell activities: migration, cell interactions, immune response, blood clotting
      • Selectins involved in movement of leukocytes along endothelial cells
      • Galectins regulate cell adhesion, growth, inflammation, immunity and metastasis
      • C-Reactive protein, a pentraxin that limits tissue damage by binding to damaged membranes

Chapter Objectives

It is important to know basic carbohydrate nomenclature, stereochemistry, and chemistry. Monosaccharides, with the general formula, $(CH_2O)_n$, are either aldoses or ketoses. The more important monosaccharides range from n = 3 to n = 7. Because carbohydrates contain a number of chiral centers (carbons with four different groups attached), L- and D- isomers can be formed. Figures 7.2 and 7.3 show D-aldoses and D-ketoses having from three to six carbons.

Cyclic carbohydrates are produced by an intramolecular reaction between an aldehyde or ketone group and a hydroxyl group. In this reaction, the carbonyl carbon of the aldehyde or ketose is converted to a chiral carbon, the anomeric carbon. As a consequence, two structural isomers, α or β anomers, can be formed. Stable five- and six-membered rings, furanose and pyranose, are very common. Sugars with free anomeric carbons are reducing sugars. Oligosaccharides and polysaccharides are composed of monosaccharides held together by glycosidic bonds. If a linear polymer contains a free anomeric carbon at one end, that end is called the reducing end and the other end the nonreducing end.

The following carbohydrates should be known:

trioses:	glyceraldehyde and dihydroxyacetone
pentoses:	ribose, ribulose, and deoxyribose
hexoses:	fructose, galactose, glucose, and mannose
heptoses:	sedulose
disaccharides:	lactose, maltose, and sucrose
polysaccharides:	amylose, amylopectin, glycogen, cellulose, and chitin

Important modified sugars include sugar acids, sugar phosphate esters, deoxy sugars, and amino sugars. Many of these modifications are found in the sugar moieties of important biological compounds. Monosaccharides do not display as wide a range of chemical characteristics as for example the amino acids. Nonetheless, polysaccharides are a diverse group of biomolecules because of the ability to join monosaccharides at a number of different positions to form branch polymers and the availability of numerous modified monosaccharides for polymer construction.

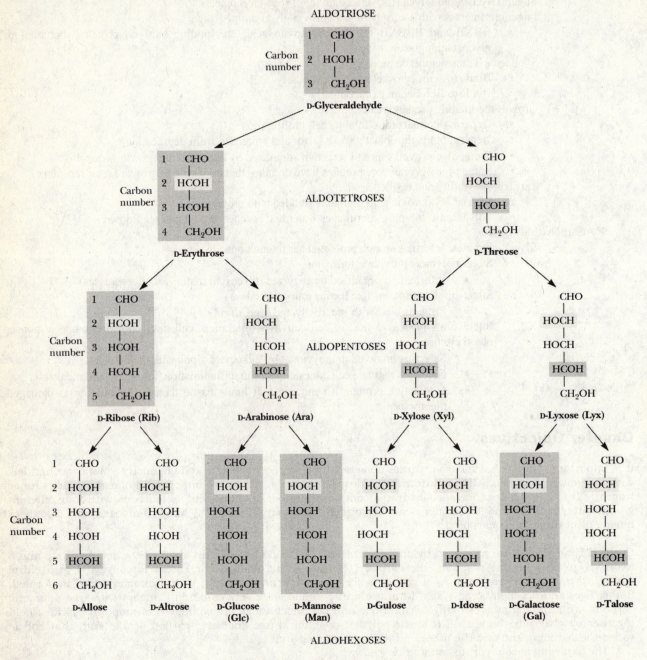

Figure 7.2 The structure and stereochemical relationships of D-aldoses having three to six carbons. The configuration in each case is determined by the highest numbered asymmetric carbon.

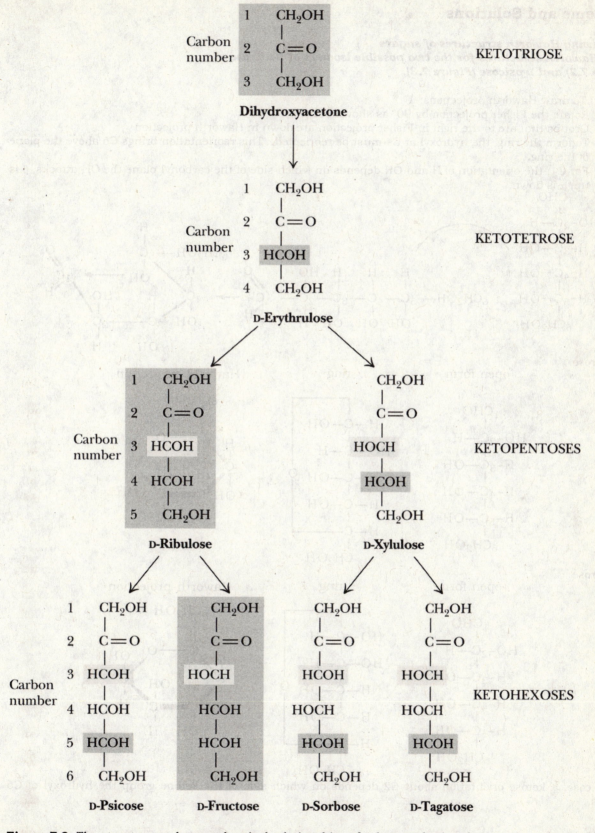

Figure 7.3 The structure and stereochemical relationships of D-ketoses having three to six carbons. The configuration in each case is determined by the highest numbered asymmetric carbon.

Problems and Solutions

1. Drawing Haworth structures of sugars
Draw Haworth structures for the two possible isomers of D-altrose (Figure 7.2) and D-psicose (Figure 7.3).

Answer: To draw Haworth projections:
1. Rotate the Fisher projection by 90° as shown below.
2. Groups that are to the right in Fisher projection are down in Haworth projection.
3. To form the ring, the hydroxyl at C4 must be reoriented. This reorientation brings C6 above the plane of the ring.
4. For C1, the orientation of H and OH depends on which side of the carbonyl plane the OH attacks, β is up, α is down.

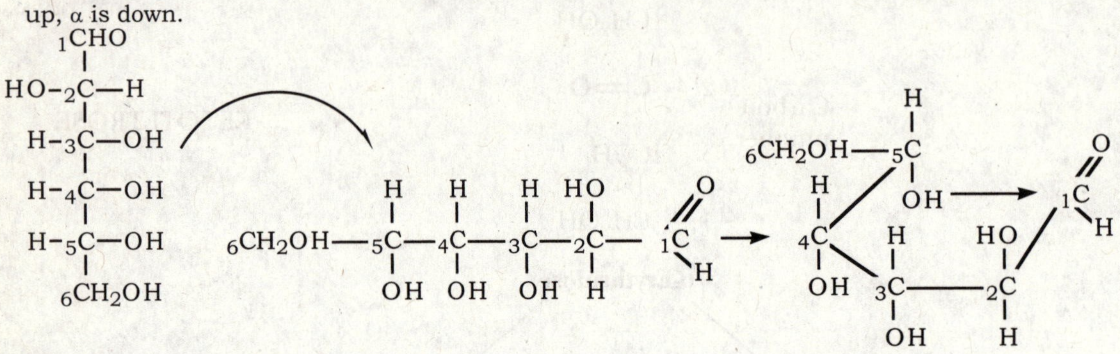

α-D-altrose

open form   ring   Haworth projection

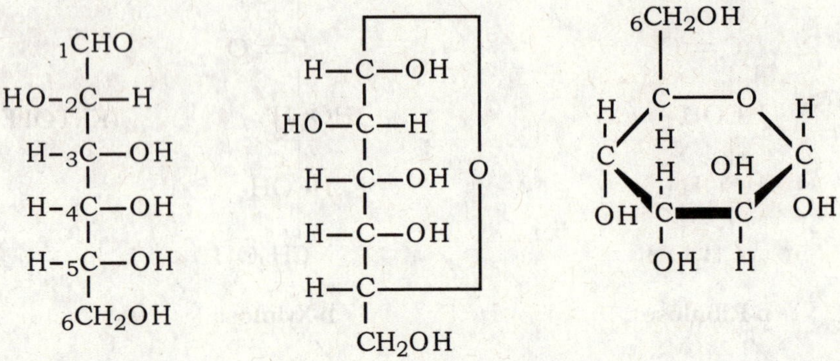

β-D-altrose

open form   ring   Haworth projection

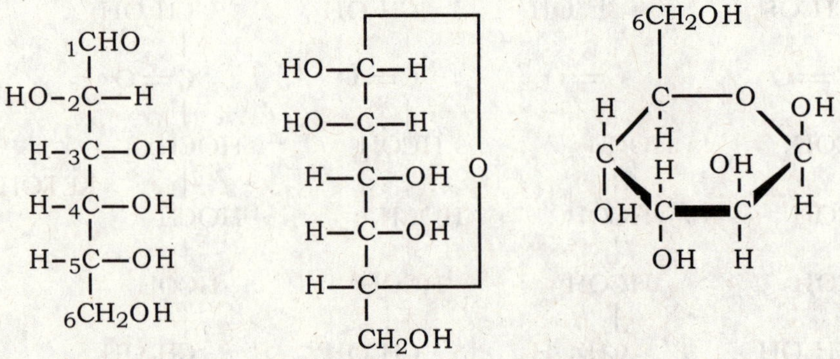

For psicose, a ketose orientation about C2 depends on which side of the ketone group the hydroxyl of C5 attacks.

124

α-D-psicose

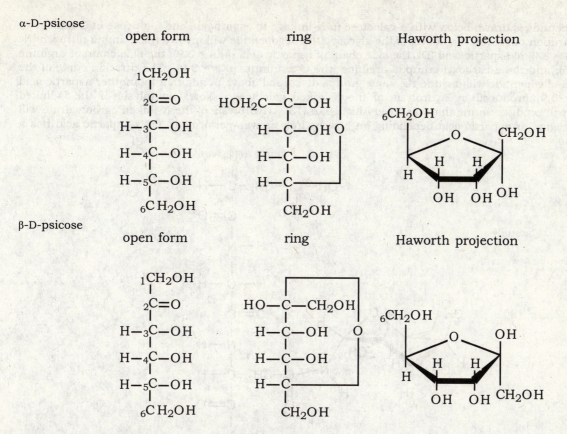

open form   ring   Haworth projection

β-D-psicose

open form   ring   Haworth projection

2. Drawing the structure of a glycoprotein
(Integrates with Chapters 4 and 5.) Consider the peptide DGNILSR, where N has a covalently linked galactose and S has a covalently linked glucose. Draw the structure of this glycopeptide, and also draw titration curves for the glycopeptide and for the free peptide that would result from hydrolysis of the two sugar residues.

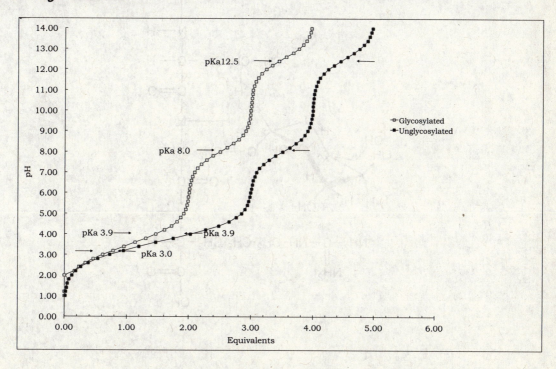

Answer: The peptide is drawn below with α-galactose in N-linkage to asparagine and α-glucose in O-linkage to serine. The titration curve (shown above) of the glycosylated oligopeptide will involve the α-amino nitrogen (N-terminus, pKa = 8.0) of aspartic acid (D), the side chain of aspartic acid (pKa = 3.9), the side chain or arginine (R, pKa = 12.5), and the α-carboxyl group of arginine (the C-terminus, pKa = 3.0). The titration curve of the unglycosylated oligopeptide will include the same groups discussed above in addition to another aspartic acid group (pKa = 3.9) produced by hydrolysis of the N-linked galactose. Note: Hydrolysis of the O-linked carbohydrate will produce serine that has a very high pKa (13). Hydrolysis of the N-linked carbohydrate will produce asparagine or aspartic acid depending on how hydrolysis was performed. Only aspartic acid has a side chain pKa.

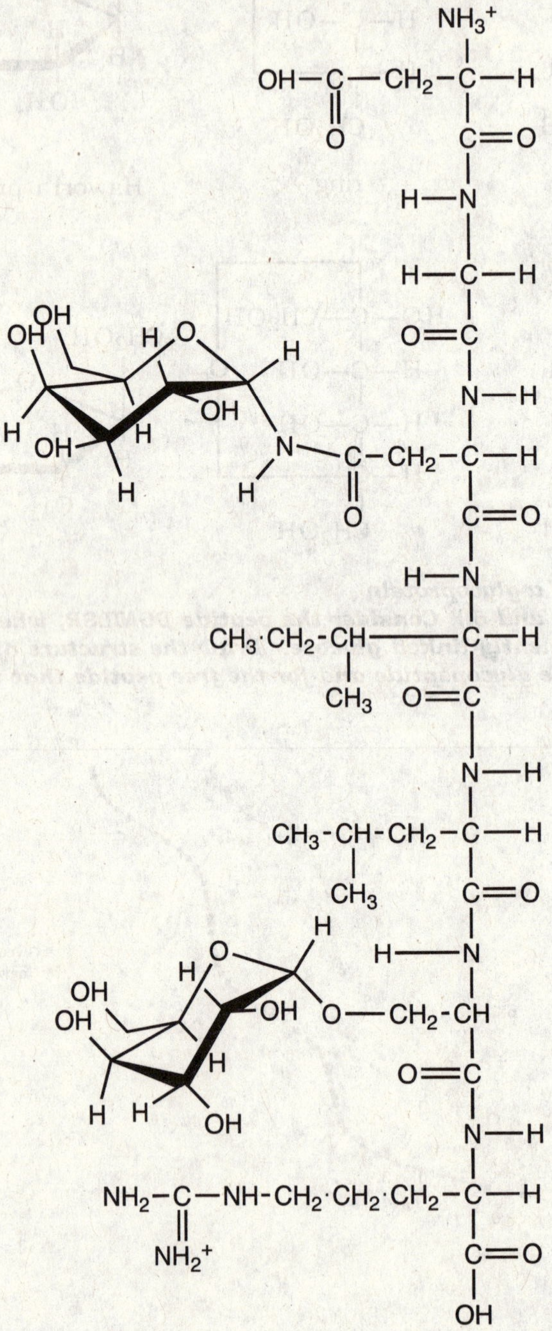

3. Separating glycated Hb from normal Hb
(Integrates with Chapters 5 and 6.) Human hemoglobin can react with sugars in the blood (usually glucose) to form covalent adducts. The α-amino groups of N-terminal valine in the Hb β-subunits react with the C-1 (aldehyde) carbons of monosaccharides to form aldimine adducts, which rearrange to form very stable ketoamine products. Quantitation of this "glycated hemoglobin" is important clinically, especially for diabetic individuals. Suggest at least three methods by which glycated Hb could be separated from normal Hb and quantitated.

Answer: Modification of the N-terminus of a polypeptide will change the ionic properties and separation of modified and unmodified protein may be accomplished by ion exchange. The two forms may also be separated using isoelectric focusing because they have different pI's. Addition of a glucosyl moiety will change the protein's molecular weight; however, this small difference may be difficult to detect unless the protein is cleaved into smaller polypeptides or even individual amino acids. Mass spectrometry may also be used to detect and quantitate this modification.

4. Naming and characterizing a disaccharide
Trehalose, a disaccharide produced in fungi, has the following structure:

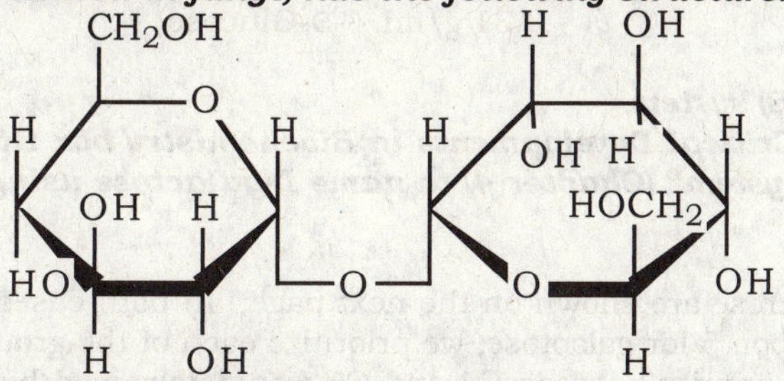

a) What is the systematic name for this disaccharide?
b) Is trehalose a reducing sugar? Explain.

Answer: Trehalose is a disaccharide of α–D-glucose and α–D-glucose, both in the pyranose form. Thus, the systematic name is:

α-D-glucopyranosyl-(1★1)-α-D-glucopyranoside.

A reducing sugar contains a free aldehyde or ketone group. In glucose, carbon 1 is an aldehyde. In trehalose, carbons 1 for both glucosyl moieties are connected by a glycosidic bond. Thus, trehalose is not a reducing sugar.

5. Drawing the Fischer projection of a simple sugar
Draw Fischer projection structures for L-sorbose (D-sorbose is shown in Figure 7.3).

Answer: L-sorbose

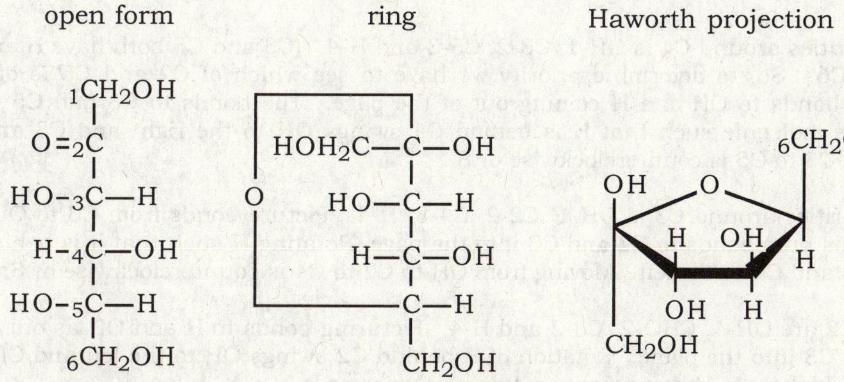

6. Calculating the composition of anomeric sugar mixtures
α-D-glucose has a specific rotation, $[α]_D^{20}$, of +112.2°, whereas β-D-glucose has a specific rotation of +18.7°. What is the composition of a mixture of α-D- and β-D-glucose, which has a specific rotation of 83.0°?

Answer: The specific rotation is the number of degrees through which plane polarized light is rotated in traveling 1 decimeter through a sample at a concentration of 1g/mL, or symbolically:

$$[\alpha]_D^{20} = \frac{\text{rotation (degrees)}}{\text{path length (dm)} \times \text{conc (g/mL)}}$$

A mixture of α-D- and β-D-glucose at 1 g/mL, with a specific rotation of 83°, contains x g/mL of α-D-glucose and y g/mL of β-D-glucose or:

$$(1) \quad x + y = 1 g / mL$$

where x = concentration of α-D-glucose and y = concentration of β-D-glucose. The rotation of the mixture is equal to the sum of the rotations due to the two components. Thus,

rotation (mixture) = rotation (α-D-glucose) + rotation (β-D-glucose)

and,

$$\text{rotation} = [\alpha]_D^{20} \times 1 dm \times \text{conc}$$

$$(2) \quad 83° = 112.2x + 18.7y$$

Solving equations (1) and (2) for x and y, we have:

$$x = 0.69 \text{ g/mL α-D-Glucose}$$
$$y = 0.31 \text{ g/mL β-D-Glucose}$$

7. Naming sugars in the (R,S) system
Use the information in the Critical Developments in Biochemistry box titled "Rules for Description of Chiral Centers in the (R,S) System" (Chapter 4) to name D-galactose using (R,S) nomenclature. Do the same for L-altrose.

Answer: D-galactose and L-altrose are shown on the next page. In both cases, carbons 2, 3, 4, and 5 are all chiral centers. Starting at carbon 5 for galactose, we prioritize each of the groups as follows: OH-1, C4-2, C6-3 and H-4. (The difference in priority between C4 and C6 was determined by recognizing that both carbons have H and OH bonded to them but C6 has an additional H whereas C4 is bonded to C3 making C4 higher priority over C6.) To arrange the molecule such that the group with lowest priority is behind C5, recognize that H and OH are out of the plane and thus C6 and C4 are into the plane. Rotating the molecule such that H is placed behind C5 will flip the OH to the left and C4 and C6 to the right as shown below.

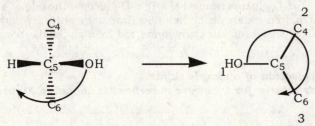

Moving from highest to lowest priority is clockwise and thus R. So, C5 is D-galactose is R.

Assignment of priorities around C4 is OH-1, C3-2, C5-3 and H-4. (C3 and C5 both have H and OH and single bonds to C2 and C6. So, to determine priority we have to see which of C2 and C6 is of higher priority.) Picture C4 having bonds to OH and H coming out of the page. The bonds to C3 and C5 must be into the page. Rotating the molecule such that H is behind C4 swings OH to the right and C3 and C5 to the left. Moving from OH to C3 to C5 is counterclockwise or S.

Assignment of priorities around C3 is OH-1, C2-2, C4-3, H-4. Picture bonds from C3 to OH and H as being out of the page. This puts bonds to C2 and C3 into the page. Rotating H such that it is behind C3 swings OH to the right and C2 and C4 to the left. Moving from OH to C2 to C4 is counterclockwise or S.

Priorities around C2 are OH-1, CHO-2, C3-3 and H-4. Picturing bonds to H and OH as out of the plane puts bonds to CHO and C3 into the plane. Rotation of H behind C2 swings OH to the left and CHO and C3 to the right. Moving from highest to lowest priority moves clockwise or R.

Thus, D-galactose is 2R, 3S, 4S, 5R.

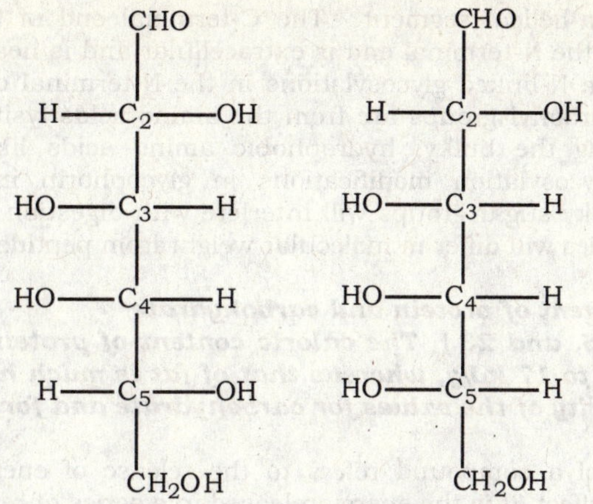

$$\text{CHO}$$

H	C$_2$	OH	
HO	C$_3$	H	
HO	C$_4$	H	
H	C$_5$	OH	
CH$_2$OH			

$$\text{CHO}$$

H	C$_2$	OH	
HO	C$_3$	H	
HO	C$_4$	H	
HO	C$_5$	H	
CH$_2$OH			

D-galactose   L-altrose

An easy way to visualize the molecule is to download its PDB file, which I found at Klotho: Biochemical Compounds Declarative Database at http://www.biocheminfo.org/klotho/. I viewed the PDB file using either RasMol or Swiss PdbViewer.

For L-altrose we would go through a similar exercise. The priorities about each of the carbons would be identical to those described for the carbons of galactose. L-altrose is 2R, 3S, 4S, 5S.

8. Determining the branch points and reducing ends of amylopectin
A 0.2-g sample of amylopectin was analyzed to determine the fraction of the total glucose residues that are branch points in the structure. The sample was exhaustively methylated and then digested, yielding 50 μmol of 2,3-dimethylglucose and 0.4 μmol of 1,2,3,6-tetramethylglucose.
   a. What fraction of the total residues are branch points?
   b. How many reducing ends does this amylopectin have?

Answer: Methylation reactions are expected to modify hydroxyl groups. In amylopectin, 1⋆4 linkage ties up the hydroxyls on carbon 1 and 4. Of the remaining hydroxyls, on carbons 2, 3, 5 and 6, C-5 is involved in pyranose ring formation leaving only carbons 2, 3 and 6 to be modified. The observation that 2,3-dimethylglucose is produced implies that C-6 is unreactive, indicating that these residues are branch points. Thus, the 50 μmol of 2,3-dimethylglucose derives from branch points. The total number of moles of glucose residues in 0.2 g amylopectin is:

$$\frac{0.2\,\text{g}}{162\,\dfrac{\text{g}}{\text{mole}}} = 1.23 \times 10^{-3}\,\text{moles glucose residues}$$

(The molecular weight of a glucose residue is C$_6$H$_{12}$O$_6$ minus H$_2$0 = 162 g/mol.)

$$\text{Fraction of residues at branches} = \frac{50 \times 10^{-6}\,\text{mol}}{1.23 \times 10^{-3}\,\text{mol}}$$
$$= 0.0405 \text{ or } 4\%$$

This analysis is expected to yield predominantly 2,3,6-trimethylglucose, a small amount of 2,3,4,6-tetramethylglucose from non-reducing ends of chains and 1,2,3,6-tetramethylglucose from reducing ends, in addition to 2,3-dimethylglucose from branch points. The number of reducing ends in amylopectin is given by the concentration of 1,2,3,6-tetramethylglucose, which is 0.4 μmol. The number of reducing ends is:

$$0.4 \times 10^{-6}\,\text{mol} \times 6.02 \times 10^{23}\,\frac{\text{molecules}}{\text{mol}} = 2.4 \times 10^{17}\,\text{ends}$$

9. The effect of carbohydrates on proteolysis of glycophorin
(Integrates with Chapters 4, 5, and 9.) Consider the sequence of glycophorin (see Figure 9.10), and imagine subjecting glycophorin and also a sample of glycophorin treated to remove all sugars, to treatment with trypsin and chymotrypsin. Would the presence of sugars in the native glycophorin make any difference in the results?

Answer: Glycophorin A is an intrinsic membrane protein that spans the erythrocyte (red blood cell)

129

membrane using a single alpha-helical segment. The C-terminal end of the protein is intracellular and completely unmodified whereas the N-terminal end is extracellular and is heavily glycosylated. In particular, there are numerous O-linked an N-linked glycosylations in the N-terminal domain of this protein. Trypsin cleaves peptide bonds whose carbonyl groups are from the amino acids lysine and arginine. Chymotrypsin cleaves peptide bonds involving the bulky, hydrophobic amino acids like phenylalanine, tyrosine and tryptophan. Although the glycosylation modifications in glycophorin involve serines, threonines and asparagines the presence of bulky sugar groups will interfere with digestion by these enzymes. In addition, peptides containing sugar moieties will differ in molecular weight from peptides without modifications.

10. Assessing the caloric content of protein and carbohydrate
(Integrates with chapters 4, 5, and 23.) The caloric content of protein and carbohydrate are quite similar, at approximately 16 to 17 kJ/g, whereas that of fat is much higher at 38 kJ/g. Discuss the chemical basis for the similarity of the values for carbohydrate and for protein.

Answer: The caloric content of a compound refers to the release of energy that occurs upon complete oxidation of the compound. In effect, it is the energy released in a series of reactions converting carbon-based compounds to carbon dioxide and water. Thus, the further away a typical carbon of a compound is from carbon dioxide the larger its caloric content. For fats, carbons are highly reduced whereas carbohydrate carbons and protein carbons are more oxidized.

11. Writing a mechanism for starch phosphorylase
Write a reasonable chemical mechanism for the starch phosphorylase reaction (Figure 7.22).

Answer: Starch phosphorylase cleaves glucose units from starch by phosphorlysis using phosphate to release glucose as glucose-1-phosphate. All phosphorylases have pyridoxal phosphate as cofactor and it is thought that the phosphate group of this cofactor plays a role in catalysis. The phosphate group of pyridoxal phosphate may function alternatively as an acid and then a base to promote the cleavage of the glycosidic bond by the substrate phosphate.

12. Assessing the toxicity of laetrile
Laetrile treatment is offered in some countries as a cancer therapy. This procedure is dangerous, and there is no valid clinical evidence of its efficacy. Suggest at least one reason that laetrile treatment could be dangerous for human patients.

Answer: Whenever confronted with an unfamiliar molecule it is extremely useful to look up its structure. Here are two web sites that are extremely useful in this regard: http://chemfinder.cambridgesoft.com/ and http://www.genome.ad.jp/ligand/. The first URL is to ChemFinder, a very useful chemical database. In addition to structural information, results pages from hits in this database provide a number of very informative links to biochemistry, health, physical properties, regulations, online ordering and others in addition to showing structure and listing useful physical properties like molecular weight, chemical formulae etc. However, it requires you to register with the site. The second URL is a ligand database, which is linked to a number of other databases. (For a diagram of these links visit: http://www.genome.ad.jp/dbget/dbget.links.html.) Ligand will conduct searches on several databases including pathways, enzymes and compounds. In 2004 NIH released PubChem, another database of small organic compounds. It is searchable at http://www.ncbi.nlm.nih.gov/.

Laetrile was not found in the Ligand database but was found by ChemFinder and PubChem. Its structure is shown below. It is a substituted glycoside, a diglucose substituted with mandelonitrile, which contains a cyanide group. Metabolism of this compound should produce cyanide, which is highly toxic.

Some plants use so-called cyanogenic glucosides as defense mechanisms against herbivores. Laetrile is a stereoisomeric form of amygdalin (which is found in both Ligand and ChemFinder.). Amygdalin was first isolated from almonds. The famous biochemist Krebs isolated one of the stereoisomeric forms, a levorotary form which was named laetrile.

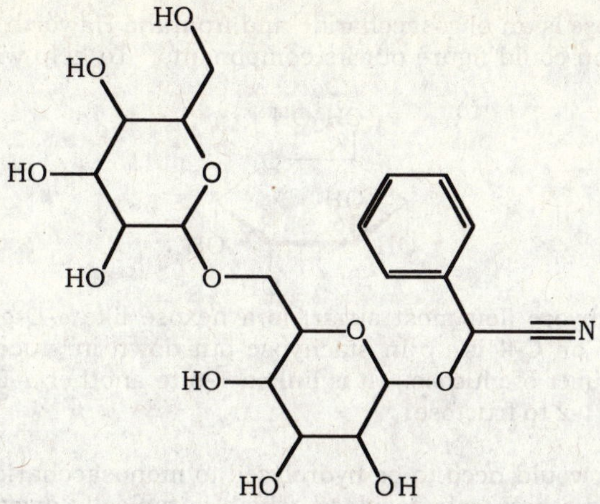

13. Assessing the efficacy of glucosamine and chondroitin for arthritis pain
Treatment with chondroitin and glucosamine is offered as one popular remedy for arthritis pain. Suggest an argument for the efficacy of this treatment, and then comment on its validity , based on what you know of polysaccharide chemistry.

Answer: Chondroitin is a glycosaminoglycan consisting of linear chains of D-glucuronate and N-acetyl-D-galactosamine-4(or 6)-sulfate. Glucuronate is a sugar acid derived from glucose by oxidation of C-6 to a carboxylic acid function. At neutral pH this group has a 1- charge. Sulfated sugars are also negatively charged. Thus, condroitin sulfate is highly negatively-charged. One question to ask is how effectively is condroitin sulfate absorbed by the body and how well is it transported to joints. Glucosamine is used in synthesis of cartilage matrix proteoglycan. In addition to being a component of hyaluronate and keratan sulfate, it is converted into galactosamine, which is found in chondroitin. It is possible that dietary supplements of chondroitin and/or glucosamine help to maintain levels of glycosaminoglycans. Dietary supplements, however, and not subject to regulations covering drugs and pharmaceuticals.

14. Assessing a remedy for flatulence
Certain foods, particularly beans and legumes, contain substances that are indigestible (at least in part) by the human stomach, but which are metabolized readily by intestinal microorganisms, producing flatulence. One of the components of such foods is stachyose, shown below:

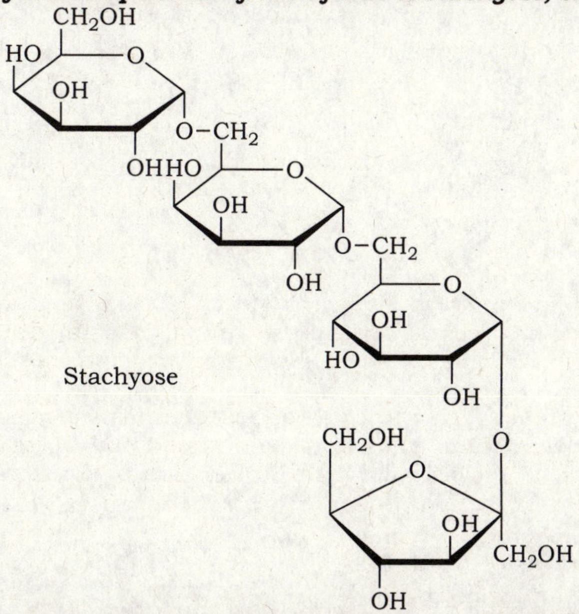

Stachyose

Beano is a commercial product that can prevent flatulence. Describe the likely breakdown of stachyose in the human stomach and intestines and how Beano could contribute to this process. What would be an appropriate name for the active ingredient in stachyose?

Answer: It is clear that stachyose is an oligosaccharide and from the Haworth projection shown and the rules already given in this chapter you could figure out its components. To help with identification α-D-glucose is shown below.

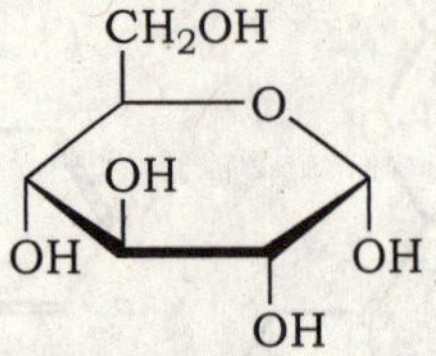

$$CH_2OH$$

The first sugar residue in stachyose (left most sugar) is a hexose like α-D-glucose but with one important differences: the hydroxyl group on C-4 is up in stachyose but down in glucose. This sugar is simply α–D-galactose. (Galactose is a 4' epimer of glucose.) It is linked 1-6 to another α–D-galactose, which is linked 1-6 to α–D-glucose, which is linked 1-2 to fructose.

To metabolize this compound it would need to be hydrolyzed to monosaccharides and would require an alpha galactosidase. (Recall lactose is a beta galactosidase, which is typically digested by beta galactosidases like lactase.) Beano contains an alpha galactosidase, which should cleave the first two glycosidic bonds releasing two moles of galactose and sucrose. The sucrose is readily metabolized by sucrase into glucose and fructose.

Why would failure to metabolize stachyose cause digestive problems? Carbohydrate digestion occurs in the stomach and small intestines where poly- and oligosaccharides are converted to simpler sugars and absorbed by the gut. If this process does not occur, the saccharides move into the large intestine, where resident microorganisms reside. Normally they are maintained on a rather Spartan diet of well digested food stuffs but when given a rich load of carbohydrates they have a burst of metabolic activity, creating among other things a great deal of gas.

(Beans and broccoli are a few of the foods that may cause excessive gas if not well cooked. The reason for refried beans and the practice of cooking Boston baked beans for many hours should be apparent. Long cooking times lead to breakdown of complex carbohydrates making them easier to digest.)

The reaction mechanism of a few alpha-galactosidases has been worked out. For the so-called retaining glycosidases, which conserve stereochemistry about the anomeric carbon, the glycosidic bond is first attacked by a nucleophilic side chain on the enzyme forming an enzyme-modified intermediate. This adduct is then hydrolyzed by water. The following mechanism is taken from an article by Lee and coworkers published in Biochemical Journal (2001) 359, 381-386.

15. Determining the systematic name for a tetrasaccharide
Give the systematic name for stachyose.

Answer: β-D-Fructofuranosyl-O-α-D-galactopyranosyl-(1★6)-O-α-D-galactopyranosyl-(1★6)-O-α-D-glucopyranoside

16. Assessing the formation and composition of limit dextrins
Prolonged exposure of amylopectin to starch phosphorylase yields a substance called a limit dextrin.
Describe the chemical composition of limit dextrins, and draw a mechanism for the enzyme-

133

catalyzed reaction that can begin the breakdown of a limit dextrin.

Answer: Amylopectin is a polymer of α-D-glucose linked (1★4) with (1★6) branches every 12 to 30 residues. Starch phosphorylase (EC 2.4.1.1) is an enzyme that cleaves α(1★4)-linked D-glucose by phosphorylysis starting at the non-reducing end of a glucose polymer and releasing α-D-glucose-1-phosphate. (A search for "starch phosphorylase" at http://ca.expasy.org/, a mirror site in Canada for ExPASy Proteomics Server, returns a number of hits for the enzyme in the Swiss-Prot (UniProtKB) database. Activating one of them (like, for example the one for *Solanum tuberosum*, the enzyme found in potatoes) will return the Swiss-Prot entry for this enzyme. Activating the link to EC 2.4.1.1 returns a "NiceZyme view of the enzyme". On this page you will find a description of the reaction catalyzed by phosphorylases in general.)

Since the enzyme is confronted with a sugar polymer with two types of linkages we can anticipate it having problems with the branches. The enzyme will remove glucose moieties until the end of the polymer approaches an α(1★6) branch to within two to three residues producing a limit dextrin. Thus, a limit dextrin is an amylopectin with short branches.

One way of dealing with the limit dextrin is to hydrolyze the α-(1★6) branches to release either free glucose or maltose using α-1,6 glucosidase. An enzyme mechanism for hydrolysis would involve activation of a water molecule. A number of glucosidases use a mechanism that involves two carboxylic acid amino acid side chains acting as a general acid and a general base. The following mechanism is from "Glycosidase mechanism" by Rye, D.S., and Withers, S.G. in Current Opinion in Chemical Biology 2000, 4:573-580.

17. Assessing the chemistry and enzymology of "Light Beer"
Biochemist Joseph Owades revolutionized the production of beer in the United States by developing a simple treatment with an enzyme that converted regular beer into "light beer," which was marketed aggressively as a beverage that "tastes great," even though it is "less filling." What was the enzyme-catalyzed reaction that Owades used to modify the fermentation process so cleverly, and how is regular beer different from light beer?

Answer: The alcohol in beer is produced by fermentation of carbohydrates by yeast. The carbohydrates derive from starch in barley grain, which includes amylose and amylopectin. Yeast cannot, however, ferment either of these polysaccharides and so to make starch sugars available for yeast fermentation, barley grains are induced to sprout. Soaking barley grains in water induces production of alpha and beta amlyase, two enzymes that digest starch into simpler carbohydrates that would be used by the barley endosperm to begin plant growth. Alpha amylase is an endohydrolase that cleaves α-(1★4) linkage randomly in starch. Beta amylase digests α-(1★4) linkage starting at the nonreducing ends of polysaccharides. Combined action of alpha and beta amylase converts starch into a complex mixture that includes glucose and maltose (di-glucose), both of which are fermentable by yeast, and numerous oligosaccharides. The oligosaccharides are produced in part because digestion of starch by amlyase enzymes is interrupted to prevent utilization of sugar by the endosperm. Additionally, neither enzyme can hydrolyze α-(1★6) linkage and so the oligosaccharides are branched.

A typical "regular" beer contains approximately 100 Calories from alcohol and around 50 Calories from carbohydrate in a 12 ounce serving. Light beers are around 100 Calories per serving. While some "light beers" have slightly lower alcohol content (by around 1%) the big difference between regular beers and "light" beers is the carbohydrate content. Lower carbohydrate content is achieved by treating the ferment with a variety of enzymes including alpha and beta amylase and a debranching enzyme, an α-(1★6) glucosidase that will hydrolyze α-(1★6) branches. The consequence of this is to convert more of the carbohydrates into a form that is fermentable by yeast. Yeast fermentation results in a brew that has a higher alcohol content than normal beer but this is diluted to a value typically below that or regular beer.

The first light beer was produced in 1967 by Rheingold Brewery and marketed as Gablinger's Diet Beer. The ad campaign for Gablinger's beer was a flop apparently because it highlighted the consequence of excessive caloric intake due to beer drinking. Owades gave the light beer formulation to Meister Brau of Chicago. This company sold a "lite" beer for a few years before the company was bought by Miller. Miller brewed their own light beer but sold it as Miller Lite. Their "taste great, less filling" ad campaign was a success that made Miller Lite the first successful low calorie beer.

18. Brewing "Light Beer" on a budget
Amateur brewers of beer, who do not have access to the enzyme described in Problem 17, have nonetheless managed to brew light beers using a readily available commercial product. What is that product, and how does it work?

Answer: There are a few ways of getting low calorie beer. One is to use a sugar source with more glucose and maltose and less complex sugar so the yeast ferment more of the sugar into alcohol. Another is to use diastatic malt, which is malt with active enzymes that would continue to hydrolyze complex carbohydrates during fermentation. Finally, some home beer brewers noticed that using Beano® produces a beer with more alcohol and fewer calories (and one that produces less flatulence). The active ingredient in Beano is alpha galactosidase, an enzyme that hydrolyzes bonds linked to galactose's anomeric carbon in alpha configuration. The enzyme is active on melibiose, which is galactose alpha (1★6) glucose, and thus it will hydrolyze (1★6) linkage. It is used to breakdown complex carbohydrates as described in the answer to problem 13. The problem here, however, is to breakdown alpha (1★6) branches involving glucose. While enzymes are exquisitely tuned to accept their substrates, many can operate on closely related molecules with somewhat lower efficacy. The active ingredient in Beano®, is a case in point, possessing in addition the ability to hydrolyze (1→6)-glucose linkages with reduced efficiency.

19. Assessing the growth potential and enzymology of a prolific plant nuisance
Kudzu is a vine that grows prolifically in the southern and southeastern United States. A native of Japan, China, and India, kudzu was brought to the United States in 1876 at the Centennial Exposition in Philadelphia. During the Great Depression of the 1930s, the Soil Conservation Service promoted kudzu for erosion control, and farmers were paid to plant it. Today, however, kudzu is a universal nuisance, spreading rapidly, and covering and destroying trees in large numbers. Already covering 7 to 10 million acres in the U.S., kudzu grows at the rate of a foot per day. Assume that the kudzu vine consists almost entirely of cellulose fibers, and assume that the fibers lie parallel to the vine axis. Calculate the rate of the cellulose synthase reaction that adds glucose units to the growing cellulose molecules. Use the structures in your text to make a reasonable estimate of the unit length of a cellulose molecule (from one glucose monomer to the next).

Answer: Cellulose is a polymer of β-D-glucose linked (1★4). The glucose chain is extended with alternating glucose residues flipped 180º. Thus, the repeat unit is a di-glucose termed cellobiose. The length from the oxygen on carbon 1 of glucose to the oxygen on carbon 4 of the same glucose is approximately 0.55 nm making the length of cellobiose approximately 1.1 nm. The repeat distance measured by x-ray diffraction is approximately 1.03 nm and we will use this distance in our calculation below.

The velocity of growth is one foot per day. Converting this to ft per min we have:

$$\frac{1 \text{ ft}}{\text{day}} \times \frac{1 \text{ day}}{24 \text{ hr}} \times \frac{1 \text{ hr}}{60 \text{ min}} = 6.94 \times 10^{-4} \frac{\text{ft}}{\text{min}}$$

And, conversion to nm per min we have:

$$6.94 \times 10^{-4} \frac{\text{ft}}{\text{min}} \times \frac{12 \text{ in}}{\text{ft}} \times \frac{2.54 \text{ cm}}{\text{in}} \times \frac{1 \times 10^7 \text{ nm}}{1 \text{ cm}} = 2.12 \times 10^5 \frac{\text{nm}}{\text{min}}$$

For an enzyme, velocity is typically given in units of micromole per min:

$$2.12 \times 10^5 \, \frac{nm}{min} \times \frac{2 \text{ molecules glucose}}{1.03 \, nm} \times \frac{mol}{6.02 \times 10^{23} \text{molecules}} \times \frac{1 \times 10^6 \text{ micromol}}{1 \text{ mol}} = 6.83 \times 10^{-13} \, \frac{\mu mol}{min}$$

This corresponds to about 6,850 glucose molecules per second.

Preparing for the MCAT® Exam

20. Assessing the interaction between antithrombin III and heparin
Heparin has a characteristic pattern of hydroxyl and anionic functions. What amino acid side chains on antithrombin III might be the basis for the strong interactions between this protein and the anticoagulant heparin?

Answer: Heparin is a sulfated sugar polymer containing hydroxyl groups and negatively-charged sulfates. Antithrombin III might interact with the negatively-charged sulfates via ionic interactions using strongly basic amino acids like lysine and arginine and to a lesser extent histidine. Because hydroxyl groups are both hydrogen bond donors and acceptors a number of amino acids may be used to interact with these groups including any amino acid that can function as a H-bond donor or acceptor. Analysis of antithrombin III using tools available at SwissProt reveals that the protein does not have an excess of basic amino acids (pI = 6.32) nor does it have significant charge clusters.

21. Assessing the suitability of cartilage components
What properties of hyaluronate, chondroitin sulfate, and keratan sulfate make them ideal components of cartilage?

Answer: The structures of hyaluronate, chondroitin sulfate and keratan sulfate are shown below. It is clear that these glycosaminoglycans are rich in anionic groups (carboxylate or sulfate) in addition to hydroxyl groups and amide bonds all of which interact strongly with water. These compounds will be heavily hydrated and form a cushion in joints.

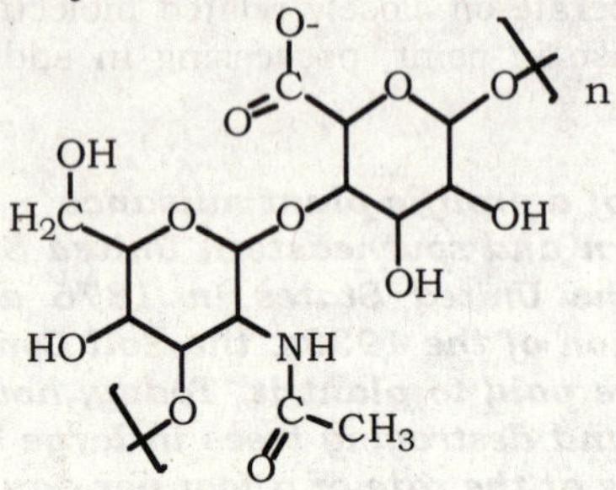

Hyaluronate

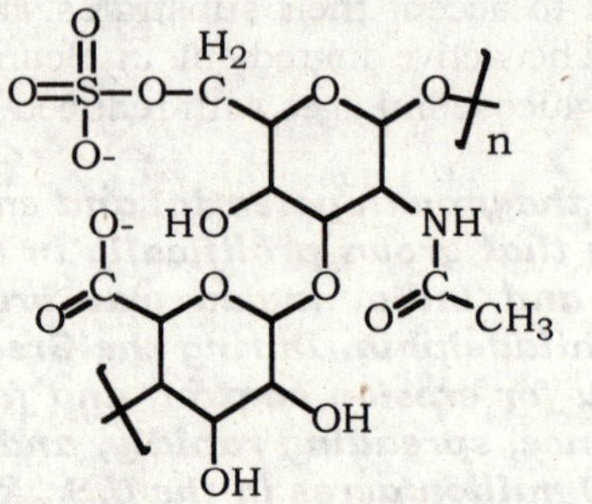

Chondroitin sulfate

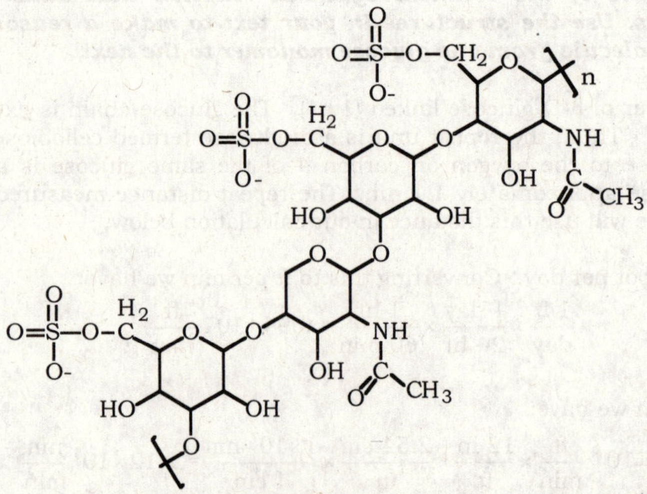

Keratatn sulfate

Questions for Self Study

1. For the compounds shown below identify the following: aldose, ketose, chiral center, potential pyranose, potential furanose, anomeric carbon, enantiomers, epimers, diastereomers.

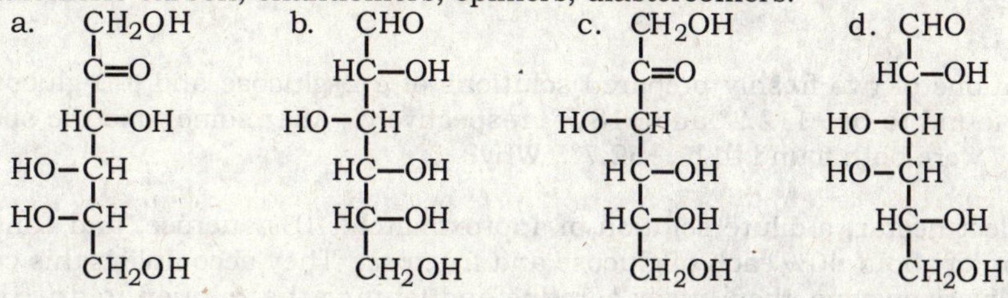

```
   a.  CH₂OH        b.  CHO          c.  CH₂OH        d.  CHO
        |                |                |                |
       C=O             HC—OH            C=O             HC—OH
        |                |                |                |
      HC—OH            HO—CH           HO—CH            HO—CH
        |                |                |                |
      HO—CH            HC—OH            HC—OH            HO—CH
        |                |                |                |
      HO—CH            HC—OH            HC—OH            HC—OH
        |                |                |                |
      CH₂OH            CH₂OH            CH₂OH            CH₂OH
```

2. For each of following disaccharides what are their monosaccharide components? Which are reducing sugars?
   a. lactose, b. sucrose, c. maltose.

3. Starch (S), glycogen (G), and dextrans (D) are storage polysaccharides found in plants, animals, and yeast and bacteria. Which applies the best to the following statements?
   a. Contains α-amylose.
   b. Many α(1★6) branches.
   c. Is used to produce Sephadex, a chromatographic material.
   d. Is cleaved by phosphorylation.
   e. May have 1★2, 1★3, or 1★4 branch points depending on species.

4. Amylose and cellulose are both plant polysaccharides composed of glucose in 1★4 linkage. However, amylose is a storage polysaccharide that is readily digested by animals whereas cellulose is a structural polysaccharide that is not digested by animals. What difference in the two polymers accounts for this?

5. Another structural polysaccharide, found in cell walls of fungi and the exoskeletons of crustaceans, insects, and spiders, is chitin. Chitin is similar to cellulose in that the repeat units are held together by β(1★4) linkage. However, the repeat units are different for the two polymers. What are they?

Answers

1. b. or d.;
   a. or c.;
   carbons 3, 4, 5 of a. and c. and carbons 2, 3, 4, 5 of b and d.;
   b. or d.
   a. or c.
   carbon 2 of a. and c., carbon 1 of b. and d.
   a. and c.
   b. and d.
   b. and d.

2. For lactose: galactose and glucose. For sucrose: glucose and fructose. For maltose: glucose and glucose. Lactose and maltose are reducing sugars.

3. a. S; b. G; c. D; d. S or G; e. D.

4. Amylose is α-(1★4)-linked D-glucose whereas cellulose is β-(1★4)-linked D-glucose. Animals have enzymes that cleave amylose linkage but not cellulose linkage.

5. Cellulose is a polymer of D-glucose; chitin is a polymer of N-acetyl-D-glucosamine.

Additional Problems

1. Protocols for ethanol precipitation of small quantities of DNA often include the addition of glycogen to act as a carrier. Typically, the ethanol precipitation is carried out by adding two volumes of 95% ethanol to a solution of salty DNA at 4°C. Explain why glycogen will precipitate under these conditions. What properties

of the two polymers, DNA and glycogen, make them behave in a similar manner under these conditions?

2. In one orientation, newsprint can be torn to produce a smooth, straight tear; however, in a perpendicular orientation, the tear line is jagged. Given the fact that newsprint is made from cellulose, explain this behavior.

3. The specific rotations of two freshly prepared solutions of α-D-glucose and β-D-glucose, each at 1 g/mL, were measured and found to be +112.2° and +18.7° respectively. After some time the specific rotations were remeasured and they were both found to be +52.7°. Why?

4a. Honey bees collect nectar, a dilute solution of approximately 10% sucrose, and convert it into honey, a concentrated solution of about 40% each of glucose and fructose. They accomplish this conversion by mixing the nectar with a salivary enzyme, then busily aerating and fanning the solution to drive off water. Given the specific rotations, $[\alpha]_D^{20}$, of sucrose (+66.5°), glucose (+52.7°), and fructose (-92°), calculate the rotation that accompanies honey production.
b. The enzyme is known as invertase. Why?

5a. Pectins are highly branched polysaccharides that are an important component of plant cell walls. The predominant fibers in plant cell walls are cellulose microfibrils composed of cellulose molecules in parallel arrays. In the plant cell wall, the cellulose microfibrils are embedded in a network that is composed of, among other things, pectin. Pectin is highly soluble and readily extracted from plant tissue with hot water whereas cellulose is not. Given this information, pectin is responsible for what property of plant cell walls ?
b. Pectins are used to make jams and jellies. Typically, an acidic fruit is boiled with sugar and pectin and the mixture allowed to cool. With luck, a jam or jelly of the proper consistency is produced. The principal monosaccharide in pectin is galacturonic acid. Why is the combination of acid, sucrose, and pectin important for gellation?

Abbreviated Answers

1. Glycogen is a polymer of glucose that is soluble in aqueous solution because of the presence of numerous hydroxyl groups capable of hydrogen bonding with water. It is insoluble in alcohol because this uncharged polymer readily aggregates. DNA is a sugar/phosphate polymer that is also quite soluble in aqueous solution. To precipitate DNA with ethanol, a counter ion must be present in order to form a DNA salt. In the absence of a counter ion, DNA is negatively charged and will not readily precipitate from dilute solutions with ethanol.

2. In newsprint, cellulose fibers are somewhat aligned along one axis of the print. This axis can be determined by ripping the paper in two orientations. A rip along the axis in which the cellulose fibers are aligned will produce a smooth tear whereas a rip perpendicular to this axis will produce a jagged tear.

3. Initially the two solutions start out with only one D-glucose anomer but with time they both equilibrate to a mixture of the two anomeric forms. The addition of alkali will speed up this conversion. Can you calculate the concentration of α-D-glucose and β-D-glucose in the mixture? See the answer to question 5 from Garrett and Grisham for an example of how to solve the problem.

4a. Assuming a 1 dm path length, the rotation of a 10% solution of sucrose (10 g/100 mL) is given by

rotation (degrees) = $[\alpha]_D^{20} \times 1 \text{ dm} \times 0.1 \text{ g/mL}$

$$= +66.5° \times 1 \times 0.1 = +6.65°$$

Similarly, the rotation of a mixture of 40% glucose and 40% fructose is given by

rotation (degrees) = $+52.7° \times 1 \times 0.4 + (-92°) \times 1 \times 0.4 = -15.7°$

b. The conversion of sucrose to a mixture of glucose and fructose is accompanied by inversion of rotation (the sign changes from positive to negative). A mixture of equal parts glucose and fructose is called invert sugar. Invert sugar is made by bees during honey production. Cane sugar is converted to invert sugar by catalysis using invertase in dilute HCl.

138

5a. Anyone who has eaten overcooked green vegetables should know that pectins are responsible for the rigidity of plant cell walls. As vegetables are cooked, pectins are slowly leached out of the plant cell walls and, as a result, the material becomes tender.

b. Because galacturonic acid is a sugar acid, pectin is a charged molecule at neutral pH. Pectin will form a gel-like mass when the acidic groups on galacturonic acid residues are neutralized at slightly acidic pH. Gel formation is also dependent on the presence of sucrose that apparently forms hydrogen bonds with neutralized pectin.

Summary

Carbohydrates are the most abundant class of organic molecules found in nature. As their name implies, they are hydrates of carbon, with the formula $(CH_2O)_n$. The versatile nature of carbohydrates arises from their chemical features, including: 1) the existence of one or more asymmetric centers, 2) the ability to adopt linear or ring structures, 3) the formation of polymer structures through glycosidic bonds, and 4) the ability to form multiple hydrogen bonds.

Carbohydrates include monosaccharides, oligosaccharides and polysaccharides. Monosaccharides include aldoses and ketoses, and may also be named for the number of carbons they contain (i.e., trioses, tetroses, etc.). Aldoses with at least three carbons and ketoses with at least four carbons contain chiral centers - carbon atoms with four different groups. Monosaccharides are usually named using the Fischer nomenclature convention which designates the chiral center farthest from the carbonyl carbon as either D- or L-configuration. The D-forms of monosaccharides predominate in nature, but L-forms are found in specialized roles. D- and L- forms of monosaccharides are mirror images of each other and are designated as enantiomers.

Monosaccharides spontaneously cyclize, forming cyclic hemiacetals with an additional chiral center, and these cyclic furanose and pyranose forms are the preferred structure for monosaccharides in solution. The α- and β-anomers of cyclic monosaccharides may undergo interconversion with intermediate formation of the linear aldehyde or ketone in a process called mutarotation. Pyranose forms are usually favored over furanose forms for aldohexose sugars, and furanose forms are usually more stable for ketohexoses. Furanose and pyranose rings are puckered, not planar, and the most stable conformations place as many bulky groups as possible in equatorial orientations. β-D-glucose, which can adopt a conformation with all its bulky groups in equatorial positions, is the most widely occurring organic group in nature and the central hexose in carbohydrate metabolism. Derivatives of monosaccharides include sugar acids such as gluconic acid, sugar alcohols such as sorbitol (used in sugarless gums), deoxy sugars such as rhamnose (a component of the toxin ouabain), sugar esters such as ATP and GTP, amino sugars, such as muramic and neuraminic acids, and acetals, ketals and glycosides.

Oligosaccharides, the simplest of which are the disaccharides, consist of monosaccharides linked by glycosidic bonds. Oligosaccharides possessing a free, unsubstituted anomeric carbon are referred to as reducing sugars. Sucrose is not a reducing sugar, but maltose and lactose are. Lactose is the principal carbohydrate in milk, but some individuals do not produce the lactase that breaks lactose down to galactose and glucose and cannot tolerate lactose in their diet. Maltose is di-glucose with α(1★4) linkage and cellobiose is di-glucose with β(1★4) linkage.

Polysaccharides function as storage materials, as structural components of organisms, as protective substances, and as mediators of cellular recognition and communication. Starch and glycogen are the principal storage polysaccharides in plants and animals, respectively. Both are α(1★4)-linked chains of glucose units. α-Amylose is a linear starch molecule and amylopectin is a highly branched structure. Glycogen is more highly branched than amylopectin. Dextrans are α(1★6)-linked glucose polymers produced by yeast and bacteria and are components of dental plaques. Structural polysaccharides of note include cellulose, a β(1★4)-linked glucose polymer which is the most abundant carbohydrate polymer in nature. The β-linkage makes cellulose difficult to digest for most animals, but bacteria living symbiotically in the digestive tracts of termites, shipworms and ruminant animals secrete cellulases which effect the breakdown of cellulose. Chitin, a β(1★4)-linked N-acetyl-D-glucosamine polymer, is the principal skeletal material in crustaceans, insects and spiders and is also present in the cell walls of fungi. It can occur in sheets composed of parallel or anti-parallel chains or in structures with mixed parallel and antiparallel sheets. Other structural polysaccharides include the alginates of marine brown algae; agar found in marine red algae; glycosaminglycans, such as heparin, hyaluronates, chondroitins, keratin sulfate and dermatan sulfate; peptidoglycan and lipopolysaccharide in bacterial cell walls; and glycoproteins and proteoglycans.

Bacterial cell walls are made of the peptidoglycan called murin. The backbone of this structure is a polymer of N-acetylglucosamine linked β(1★4) to N-acetylmuramic acid. This polysaccharide is substituted with a tetrapeptide that is attached to the lactic acid moiety of NAM by its N-terminus and crosslinked to the C-terminus of another tetrapeptide via a side chain lysine. Gram-negative bacteria have this structure sandwiched between two plasma membranes, the outer one modified by lipopolysaccharides that are antigenic determinants called O antigens. Gram-positive cell walls have pentaglycine bridges between the side chain lysine and the C-terminus of another peptide. Additionally, polymers of ribitol phosphate or glycerol phosphate are covalently attached to murin. Animal cells have an array of polysaccharides on their cell surface in the form of glycoproteins. Proteoglycans are found in the extracellular matrix. Glycoprotein have either O-linked or N-linked saccharides. O-linkage occurs on hydroxyl groups on serine, threonine or hydroxylysine and the carbohydrate is typically N-acetylgalactosamine. O-Linked membrane glycoproteins have two structural motifs. In one, glycosylation occurs through out the entire extracellular domain of the protein causing it to adopt an extended conformation. In the second, a globular domain is held distal to the membrane surface by a glycosylated domain that adopts an extended conformation. O-linked glycoproteins are used by Antarctic and Artic fish as antifreeze. N-linkage occurs on asparagines with a dimer of N-acetylglucosamine linked to a branched mannose triad, often with additional sugars attached to mannose. N-glycosylation is known to support a number of functions including stabilization of protein conformation, protection from proteolysis and membrane trafficking to intracellular organelles. In blood proteins, degradation of oligosaccharides on N-linked glycoproteins is used as a measure of protein age and functionality. Asialoglycoprotein receptors in the liver bind to proteins not fully glycosylated. The receptor-protein complex is internalized by endocytosis and degraded.

Proteoglycans are glycoproteins with glycosamionglycans O-linked to serine residues in Ser-Gly repeats. Proteoglycans are either soluble components of the extracellular matrix or integral transmembrane proteins. In general proteoglycans, via their glycosaminoglycan groups, bind to specific proteins containing BBXB and BBBXXB (B is basic) sequences. Several proteoglycans are involved in modulation of cell growth.

The complexity of living systems is a consequence in part of an organism's proteome, the proteins expressed by an organism's genome. This complexity also depends on a diverse array of polysaccharides that arise from a number of modified monosaccharides joined in a variety of ways to produced complex linear and branched structures. Cells present themselves to the environment coated with carbohydrate and this coating is used in an array of cell functions including signaling, recognition, growth and regulation of cell differentiation. For example, selectins are adhesion proteins found on the cell surface of leukocytes and on endothelial cells that line the walls of blood vessels. Movement of leukocytes along the vascular wall is achieved by selectin-carbohydrate interactions. Galectins are carbohydrate binding proteins involved in cell adhesion, growth regulation, inflammation, immunity and metastasis.

Chapter 8

Lipids

• •

Chapter Outline

❖ Lipids: Compounds composed largely of reduced carbons and exhibiting low water solubility. Lipids may be completely hydrophobic or, if they contain polar groups, amphipathic
❖ Fatty acids: Carboxyl head group and hydrocarbon tail
  ⅄ Typically even number of carbons (14 to 24)
  ⅄ Saturated fatty acids lack carbon-carbon double bonds
  ⅄ Unsaturated fatty acids contain one or more double bonds in cis configuration
    • Monosaturated: Single carbon-carbon double bond
    • Polyunsaturated: cis Double bonds separated by methylene carbon ($-CH_2-$)
  ⅄ Nomenclature:
    • Systematic: e.g., octadecanoic acid: 18-carbons long
    • Common names: Stearic acid
    • Short hand notation:
      • 18:0 -18 carbons long with no double bonds
      • 18:1(Δ^9) -18 carbons long with one double bond starting at carbon 9
  ⅄ Essential fatty acids: Linoleic and γ-linolenic acid: Plant fatty acids required in diets of animals: used to synthesize arachidonic acid -precursor to eicosanoids (prostaglandins)
❖ Triacylglycerols (triacylglycerides): Fats and oils
  ⅄ Glycerol esterified with three fatty acids
  ⅄ Saponification: Alkali hydrolysis of triacylglycerols: Products: Glycerol and free fatty acids (Salt of free fatty acids -soap)
❖ Glycerophospholipids: A class of phospholipids
  ⅄ 1,2 Diacylglycerol with phosphate ester at carbon 3
  ⅄ Stereospecific numbering system: Number glycerol carbons based on (R,S) system: C1 is carbon that would lead to S-configuration if its priority were increased
  ⅄ Phosphatidic acid: sn-Glycerol-3-phospate with fatty acids esterified to C1 and C2
    • Phosphatidic acid precursor of glycerophospholipids
    • C1 fatty acid typically saturated
    • C2 fatty aid typically unsaturated
  ⅄ Glycerophospholipids: Polar group esterified to phosphate of phosphatidic acid
    • Ethanolamine: Phosphatidylethanolamine
    • Choline: Phosphatidylcholine
    • Glycerol: Phosphatidylglycerol
    • Serine: Phosphatidylserine
    • Inositol: Phosphatidylinositol
    • Diphosphatidyl glycerol (cardiolipin)
  ⅄ Ether glycerophospholipids
    • Ether linkage at carbon 1
    • Alkyl group esterified to carbon 2
  ⅄ Plasmalogens: Ether glycerophospholipids with cis-α,β-unsaturated alkyl group at C-1
❖ Sphingolipids: Backbone of sphingosine: 18-carbon amino alcohol with C-C trans double bond

141

    ⅄   Ceramide: sphingosine with fatty acid in amide linkage

    ⅄   Sphingomyelins: Alcohol esterified to phosphoceramide

        ·   Choline

        ·   Ethanolamine

    ⅄   Glycosphingolipids: Ceramide with sugars in β-glycosidic linkage

        ·   Cerebroside: Sugar is glucose or galactose

        ·   Sulfatide: Sugar is galactose with sulfate esterified at carbon 3 of galactose

        ·   Gangliosides: Ceramide with three or more sugars including sialic acid

❖ Waxes: Esters of long-chain alcohol and long-chain fatty acid

❖ Terpenes: Class of lipids formed from 2-methyl-1,3-butadiene (isoprene)

    ⅄   Monoterpene: Two isoprene units: 10 carbons

    ⅄   Sesquiterpenes: Three isoprene units: 15 carbons

    ⅄   Diterpenes: Four isoprenes: two monoterpenes: 20 carbons

    ⅄   Triterpenes: Six isoprenes: 30 carbons: Cholesterol precursors squalene and lanosterol

    ⅄   Tetraterpenes: Eight isoprenes: 40 carbons: carotenoids

    ⅄   Polyprenols: Long-chain polyisoprenoid alcohols

❖ Steroids

    ⅄   Cholesterol

        ·   Bile salts: Cholic acid and deoxycholic acid

        ·   Steroid hormones

            •   Androgens: Testosterone

            •   Estrogens: Estradiol

            •   Progestins: Progesterone

            •   Glucocorticoids: Cortisol

            •   Mineralocorticoids

❖ Lipid metabolites are biological signal molecules

    ⅄   Phospholipases hydrolyze ester bonds

        ·   Phospholipase A_1 and A_2 release fatty acid from C-1 or C-2

        ·   Phospholipase C and D release phosphorylated head group or unphosphorylated

    ⅄   Inositol phospholipids hydrolyzed by phospholipase C

        ·   Diacylglycerol: Activates protein kinase C

        ·   Inositol trisphosphate: Regulates cellular calcium levels

    ⅄   Acachidonic acid released by phospholipase A_2

        ·   Used to synthesize eicosanoids: Local hormones

    ⅄   Sphingosine-1-phosphate

        ·   Allergic reactions, heart rate, cell movement and migration

Chapter Objectives

Lipids are amphipathic molecules with both polar and nonpolar groups. Understand why this is the case for both simple lipids (like cholesterol) and complex lipids (like phospholipids).

Fatty acids are important components of membrane lipids and triacylglycerols (fats and oils). Fatty acids are most commonly composed of an even number of carbons. Saturated fatty acids have only carbon-carbon single bonds (their carbons are saturated with respect to hydrogens), whereas unsaturated fatty acids have one or more (polyunsaturated) carbon-carbon double bonds in *cis* configuration. When more than one double bond is present, the bonds are not conjugated but rather separated by a -CH2- group.

Complex lipids include triacylglycerols, glycerophospholipids and sphingolipids. They are classified as complex because they contain at least one fatty acid group. For example, triacylglycerols have a glycerol backbone to which three fatty acids are esterified. Glycerophospholipids again have a glycerol backbone but with only two fatty acids (in ester linkage to carbons 1 and 2) and, attached to the third carbon of glycerol, a phosphate group to which a polar alcohol is linked (like ethanolamine, choline, serine or inositol). Sphingolipids are composed of sphingosine, a fatty acid, and a polar head group (like phosphocholine or one or more sugar moieties).

The carbons of simple lipids (terpenes or cholesterol and its derivatives) all derive from isoprene. Know the general structure of isoprene and cholesterol and appreciate the fact that important biomolecules such as steroid hormones and bile salts are derivatives of cholesterol.

Problems and Solutions

1. Drawing structures of triacylglycerols
Draw the structures of (a) all the possible triacylglycerols that can be formed from glycerol from stearic and arachidonic acid and (b) all the phosphatidylserine isomers that can be formed from palmitic and linolenic acids.

Answer: Triacylglycerols have a glycerol backbone to which three fatty acids are esterified. With nonidentical fatty acids at carbons 1 and 3, carbon 2 is chiral. Whereas two stereoisomers are possible, biological triacylglycerols have the L- configuration. Stearic acid is an 18-carbon saturated fatty acid. Arachidonic acid is a 20-carbon fatty acid with four *cis* double bonds at carbons 5, 8, 11 and 14.

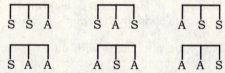

b. Palmitic acid is 16:0; linolenic acid is 18:3($\Delta^{9,12,15}$). The backbone structure of phosphatidylserine is 3-phosphoglycerol with L-serine in phosphate ester linkage. Fatty acids are esterified at carbons 1 and 2. There is a preference for unsaturated fatty acids at carbon 2. The alpha carbon of serine is chiral; L and D serine are possible; however, the L isomer occurs in phosphatidylserine. The central carbon in the glycerol backbone of phosphatidylserine is prochiral and only one isomer is used, the one based on *sn*-glycerol-3-phosphate.

Note: Phosphatidylserine with unsaturated lipids at position 1 are very rare. Unsaturated fatty acids are usually found at position 2.

2. Distinguishing structures of glycerolipids and sphingolipids
Describe in your own words the structural features of
   a. *a ceramide, and how it differs from a cerebroside.*
   b. *a phosphatidylethanolamine, and how it differs from a phosphatidylcholine.*
   c. *an ether glycerophospholipid, and how it differs from a plasmalogen.*
   d. *a ganglioside, and how it differs from a cerebroside.*
   e. *testosterone, and how it differs from estradiol.*

Answer: a.) Ceramide (N-acylsphingosine) is derived from sphingosine, a long-chain amino dialcohol synthesized from palmitic acid (fatty acid 16:0) and serine. (During synthesis of sphingosine, the carboxyl group of serine is lost as carbon dioxide, the carboxyl carbon of palmitic acid is attached to serine's alpha carbon as a ketone and subsequently reduced to an alcohol, and the Cα-Cβ bond of palmitic acid is oxidized to a trans double bond.) Ceramide has a fatty acid attached to sphingosine by an amide bond.
Cerebrosides, 1-β-D-galactoceramide and 1-β-D-glucoceramide, have monosaccharides attached in glycosidic linkage to ceramide at what was serine's side chain.

b.) Phosphatidylethanolamine and phosphatidylcholine are both glycerophospholipids synthesized from phosphatidic acid. Phosphatidic acid is *sn*-glycerol-3-phosphate with fatty acids esterified to carbons 1 and 2. In phosphatidylethanolamine the amino alcohol, ethanolamine, is joined to phosphatidic acid in phosphate ester linkage. The phosphoethanolamine moiety is the head group. In phosphatidylcholine the head group is phosphocholine. Choline is N,N,N-trimethylethanolamine.

c.) As the name implies ether glycerophospholipids are glycerophospholipids with an alkyl chain attached to carbon 1 of glycerol by ether linkage. A fatty acid is esterified to carbon 2. Plasmalogens are ether glycerophospholipids with a cis-α,β-double bond on the ether-linked alkyl chain.

d.) Cerebrosides, as explained in (a.) are glycolipids with either galactose or glucose attached in glycosidic linkage. Gangliosides are synthesized from 1-β-D-glucoceramide and contain additional sugar moieties including galactose and sialic acid (N-acetylneuraminic acid).

e.) Testosterone and estradiol are steroid hormones derived from cholesterol. Testosterone, an androgen, and estradiol, an estrogen, mediate the development of sexual characteristics in animals. Because they are both synthesized from cholesterol they share cholesterol's basic structure (of three fused six-membered rings a one fused five-membered ring) but lack cholesterol's alkyl chain. In its place is a hydroxyl group. The oxygen derived from cholesterol's hydroxyl group is a carbonyl oxygen in testosterone but a hydroxyl group in estradiol (diol implies two hydroxyl groups).

3. Identifying glycerophosphoipids
From your memory of the structures, name
   a. *the glycerophospholipids that carry a net positive charge.*
   b. *the glycerophospholipids that carry a net negative charge.*
   c. *the glycerophospholipids that have zero net charge.*

Answer: a.) Since all the glycerophospholipids derive from phosphatidic acid, in order to form a positively charged lipid, the head group must carry a positive charge. This is true for ethanolamine and choline and for serine at low pH. However, keeping in mind that the phosphodiester bond is negatively charged, phosphatidylethanolamine, phosphatidylcholine, and phosphatidylserine are positively charged only at low pH where the phosphodiester group is protonated and uncharged and the head group carries a single positive charge.

b.)   Negatively charged glycerophospholipids include phosphatidic acid, phosphatidylglycerol, phosphatidylinositol, phosphatidylserine, and diphosphatidylglycerol (cardiolipin). Because glycerol and inositol are uncharged, lipids with these head groups will have -1 charge at all but acidic pH values. The charge on phosphatidic acid will range from 0 to -1 to -2 depending on pH. When phosphotidylserine is negatively charged (at all but acidic values of pH) it ranges from -1 to -2 at basic pH when serine's amino group is not protonated. Cardiolipin typically carries -2 charge.

c.) Phosphatidylethanolamine and phosphatidylcholine are the only uncharged glycerophospholipids at neutral pH.

4. Comparing health effects of cholesterol and plant sterols
Compare and contrast two individuals, one of whose diet consist largely of meats containing high levels of cholesterol, and the other of whose diet is rich in plant sterols. Are their risks of cardiovascular disease likely to be similar or different? Explain your reasoning.

Answer: The American Heart Association identifies high blood cholesterol levels as one of the major risk factors for cardiovascular disease. Fortunately, for many individuals, blood cholesterol levels can be maintained at low levels by avoiding diets high in cholesterol and saturated fatty acids. Foods that contain cholesterol include meat, poultry and seafood, and dairy products. (Lean red meats contain similar amounts of cholesterol as poultry and fish. Plants do not contain cholesterol.)

High blood cholesterol levels lead to atherosclerosis, a thickening and hardening of arteries that is a consequence of plaque formation on arterial walls. Thus, diets high in cholesterol may contribute to plaque formation.

Since plants lack cholesterol, foods derived from plants do not contribute directly to high cholesterol levels. Diets high in triacylglycerols containing saturated fatty acids will raise blood cholesterol. In addition, trans fatty acids may also contribute to increased blood cholesterol. One source of trans fatty acids is from margarine produced by hydrogenation of vegetable oils. In addition, there is evidence that plant sterols (phytosterols) may actually lower blood cholesterol by inhibiting cholesterol absorption.

5. Evaluating organic secretions of trees
James G. Watt, Secretary of the Interior (1981-1983) in Ronald Reagan's first term, provoked substantial controversy by stating publicly that trees cause significant amounts of air pollution. Based on your reading of this chapter, evaluate Watt's remarks.

Answer: During the 1980 presidential campaign Ronald Reagan stated that trees cause more air pollution than do automobiles. Trees in fact do emit large quantities of hydrocarbons principally in the form of isoprenes. These volatile hydrocarbons may react with ozone to form compounds similar to those found in smog. (Trees, however, play a key role in removing pollutants from air.)

(The press had a field day with Reagan's comment, which evoked numerous light-hearted reactions. For example, at Claremont College a tree was draped with a banner that read: "Chop me down before I kill again." (*The Washington Post* Oct. 15, 1980))

6. Evaluating nutritional benefits of foods
In a departure from his usual and highly popular western, author Louis L'Amour wrote a novel in 1987, Last of the Breed (Bantom Press), in which a military pilot of Native American ancestry is shot down over the former Soviet Union and is forced to use the survival skills of his ancestral culture to escape his enemies. On the rare occasions when he is able to trap and kill an animal for food, he selectively eats the fat, not the meat. Based on your reading of this chapter, what was his reasoning for doing so?

Answer: Fats and oils are composed of highly reduced carbons and, therefore, they release large amounts of energy when metabolized aerobically into carbon dioxide and water. L'Amour's hero likely knew of the high caloric content of animal fat. In addition, oxidation of triacylglycerols produces water, which might be of some value in dry climates.

7. Exploring the action of phospholipase A_2
As you read Section 8.8, you might have noticed that phospholipase A_2, the enzyme found in rattlesnake venom, is also the enzyme that produces essential and beneficial lipid signals in most organisms. Explain the differing actions of phospholipase A_2 in these processes.

Answer: Phospholipase A_2 (EC 3.1.1.4) hydrolyzes glycerophospholipids at carbon 2 releasing free fatty acids and monoacylglycerol (lysophosphopholipid). When the fatty acid at position 2 is arachidonic acid this compound can serve as substrate for prostaglandin synthesis. Prostaglandins have a range of physiological activities including response to inflammation and pain. Lysophospholipids are also lipid signaling molecules that act through G-protein-coupled receptors and regulate cellular functions such as migration and chemotaxis, calcium ion homeostasis and cell proliferation and survival.

Rattlesnake venom catalyzes the same reaction producing lysophospholipid and fatty acids. It is likely that these products, acting as detergents, contribute to cell membrane damage initiated by phospholipase activity. Further, there is evidence that venom phospholipase A_2 can exhibit specificity in toxicity by binding to specific membrane target proteins. Target proteins on muscle cells, for example, localize phospholipase activity to muscle.

8. Evaluating the action of warfarin
Visit a grocery store near you, stop by the rodent poison section and examine a container of warfarin or a related product. From what you can glean from the packaging, how much warfarin would a typical dog (40 lbs) have to consume to risk hemorrhages and/or death?

Answer: The MSDS (Material Safety Data Sheet) for a commercially available product containing 98% warfarin lists the LD50 (the lethal dose for 50% of animals tested) as 3 mg/kg. A 40 lb dog (1 pound = 0.454 kilogram) weighs 18.2 kg and needs about 55 mg of warfarin for a dose equivalent to the LD50.

I found the MSDS for a rodenticide in the form of wafarin-coated pellets. The MSDS for rat was listed as 20 g/kg but the product was only 0.025% warfarin by weight. (The LD50 based only on the active ingredient is about 5 mg/kg.) It would take around 360 g (0.8 lb) to be lethal for a 40 lb rat (and presumably for the same sized dog). I found the MSDS for another rodenticide, a "warfarin meal", which listed the LD50 for rats at 200 g/kg but did not state the percentage of the active ingredient. It is likely to be only about 0.0025% by weight. About 8 lbs of this product would be required to do in the dog.

9. Understanding isoprene structures
Refer to Figure 8.13 and draw each of the structures shown and try to identify the isoprene units in each of the molecules. (Note that there may be more than one correct answer for some of these molecules, unless you have the time and facilities to carry out ^{14}C labeling studies with suitable organisms.)

MONOTERPENES

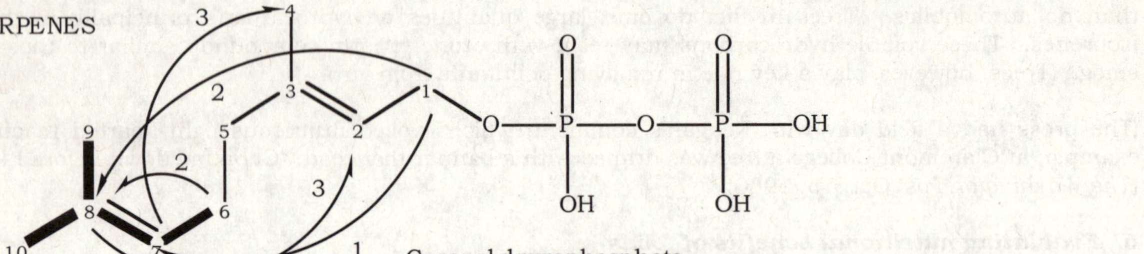

Geranyl dpyrophosphate

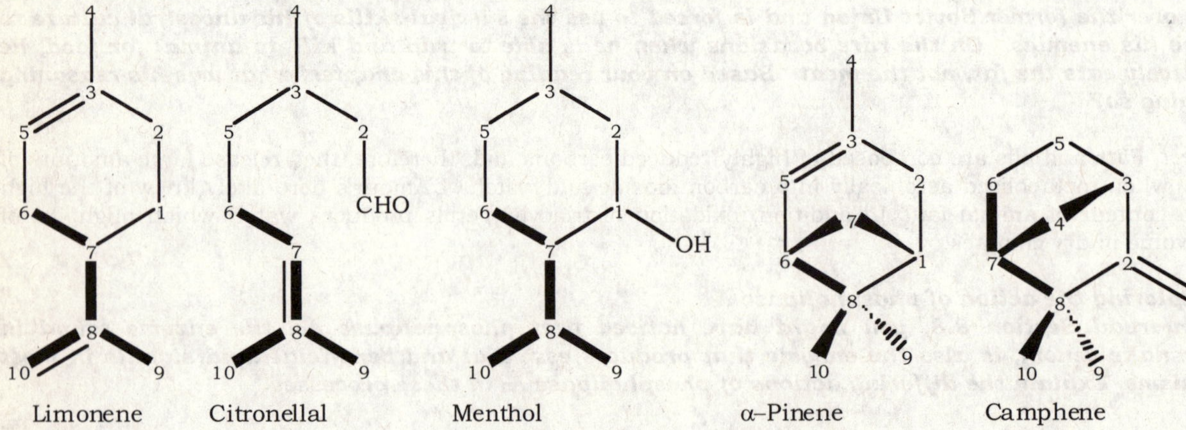

| Limonene | Citronellal | Menthol | α–Pinene | Camphene |
| path 1 | path 1 | path 1 | path 1, 2 | path 3 |

SESQUITERPENES

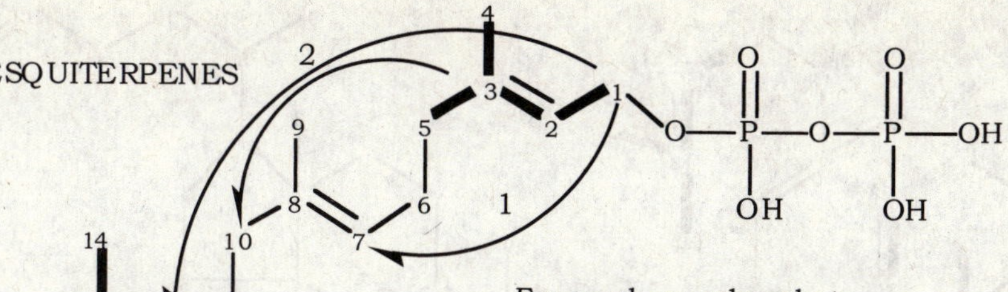

Farnesyl pyrophosphate

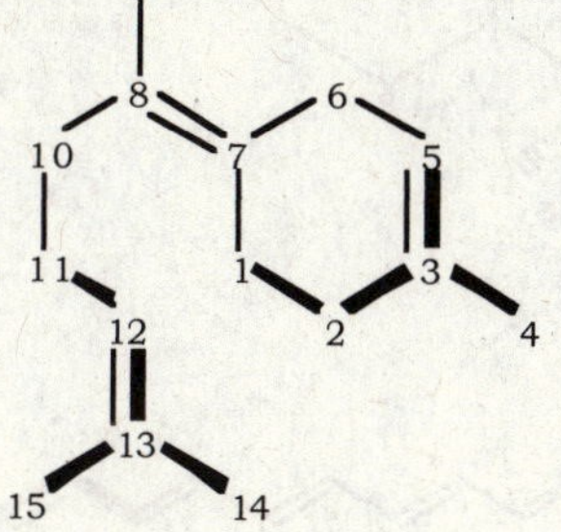

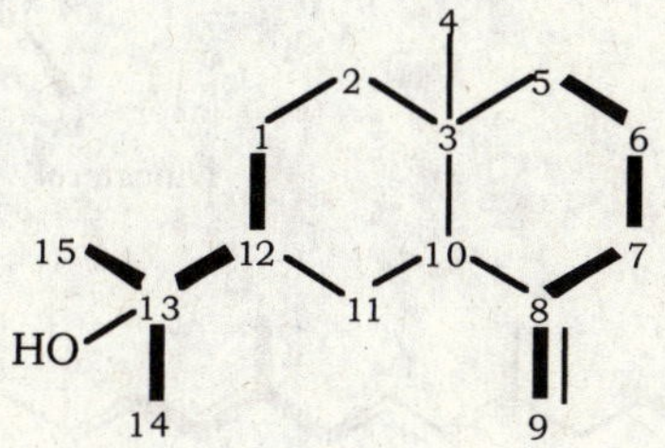

DITERPENES

Bisabolene

Eudesmol

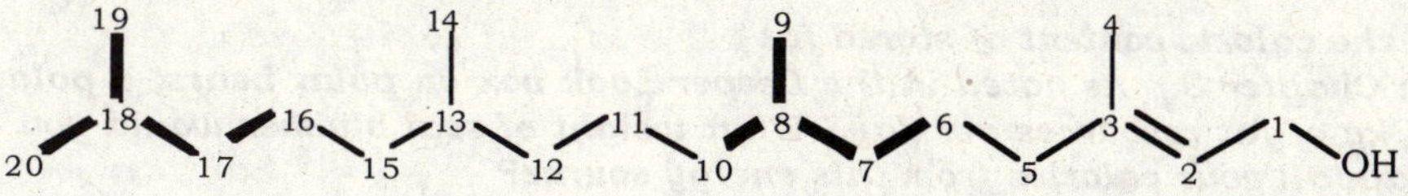

Phytol

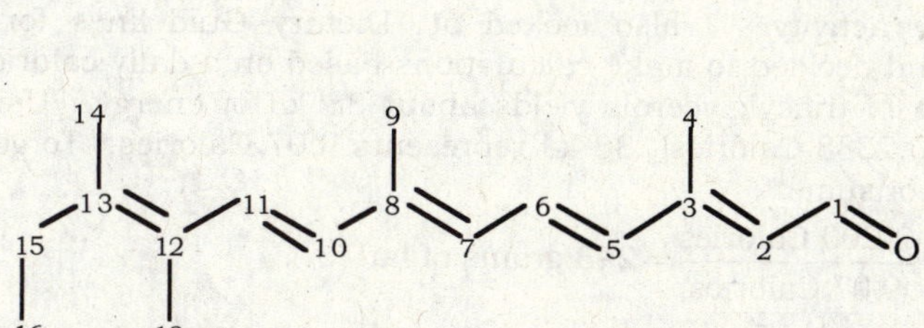

All-trans-retinal

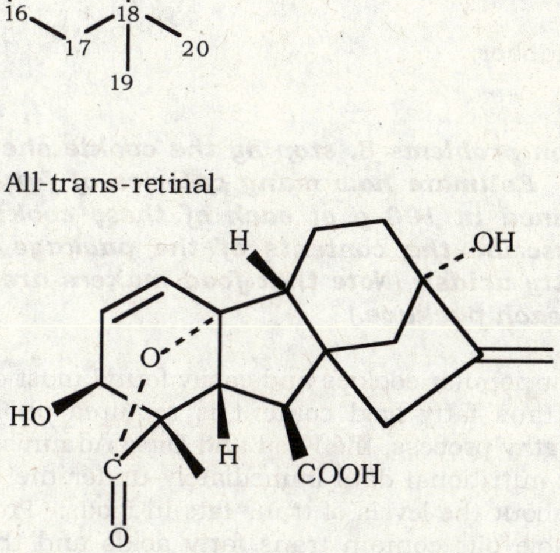

Gibberellic acid

TRITERPENES

Squalene

Lanosterol

TETRAPENES

Lycopene

10. Evaluating the caloric content of stored fat
(Integrates with Chapter 3.) As noted in the Deeper Look box on polar bears, a polar bear may burn as much as 1.5 kg of fat resources per day. What weight of seal blubber would you have to ingest if you were to obtain all your calories from this energy source?

Answer: The amount of seal blubber you would have to ingest really depends on who you are and what you do, in terms of physical activity. I searched for on-line calculators that would help me determine my caloric intake given my age, sex and daily activity. I also looked at "Dietary Guidelines for Americans" (at http://www.health.gov/) and in the end decided to make calculations based on a daily caloric intake of 2,200 Calories (2,200 kcalories). One gram of triacylglycerols yields about 38 kJ of energy. Using the following conversion, 1 kJ = 0.2388 kcalories (0.2388 Calories), 38 kJ represents 9.07 Calories. To get 2,200 Calories from seal blubber you would need to consume

$$\frac{2,200 \text{ Calories}}{\dfrac{9.07 \text{ Calories}}{\text{gram fat}}} = 243 \text{ grams of fat!}$$

This is about a half-a-pound of USDA prime seal blubber.

11. Assessing the fat composition of desserts
If you are still at the grocery store working on problems 8, stop by the cookie shelves and choose your three favorite cookies from the shelves. Estimate how many calories of fat, and how many other calories from other sources, are contained in 100 g of each of these cookies. Survey the ingredients listed on each package, and describe the contents of the package in terms of (a) saturated fat, (b) cholesterol, and (c) trans fatty acids. (Note that food makers are required to list ingredients in order of decreasing amounts in each package.)

Answer: I looked up nutritional information on four popular cookies and easily found most of the information asked for in this question. Information about trans fatty acid content is required on nutritional labels starting 1 January 2006 (21 CFR 101). After a lengthy process, the Food and Drug Administration decided to require this information on a separate line in the nutritional data immediately under the line for saturated fatty acids. Prior to this consumers had no idea about the levels of trans fats in foods. Products containing "hydrogenated" or "partially hydrogenated" vegetable oils contain trans fatty acids and this information is typically found in the ingredients section of nutritional labels. You will see, however, that hydrogenated vegetable oils are often listed as an ingredient despite levels of trans fats listed as 0 (zero). This happens because of the serving size and significant figures. A level of trans fats of 0.49 gram is 0 when recorded to one significant figure. It is interesting to note that there is not a standard serving size. I purposefully avoided listing the brand name of the cookies. The data are presented in two charts. The chart below has information taken directly from the nutritional information label.

Cookie	Serving size (g)	Calories per serving (listed)	Total fat per serving (g)	Total saturated fat (g)	Trans fat (grams)	Cholesterol per serving (mg)	Total carbohydrate (g)	Percent of calories from fat	Percent calories from carbs$	Dietary fiber (g)	Protein (g)	Calculated Total Calories#
Three-layered	34	160	7.0	2.0	0	0	25	37.5	62.5	1.0	2.0	171
Chocolate chip	33	160	8.0	2.5	0*	0	22	43.8	55.0	1.0	2.0	168
Vanilla Wafers	30	140	6.0	2.0	0	0	21	35.7	60.0	1.0	1.0	142
Fruit-based	31	110	2.0	0.0	0*	0	22	18.2	80.0	1.0	1.0	110

$ Total calories from fat are listed on the nutritional facts panel.
# Calculated using 9 Calories per gram for fat, 4 for carbohydrate and 4 for protein.
* The ingredients listed some form of hydrogenated vegetable oil.

The data in the first chart were used to calculate various values presented below.

Cookie	Calories per gram	Calories per 100 grams	Total fat per 100 grams	Total saturated fat per 100 grams	Cholesterol per 100 grams	Total carbohydrates per 100 grams	Percent of calories as fat*	Percent of calories as carbohydrate
Three-layered	4.7	470.6	20.6	5.9	0	73.5	39.4	62.5
Chocolate chip	4.8	484.8	24.2	7.6	0	66.7	45.0	55.0
Vanilla Wafers	4.7	466.7	20.0	6.7	0	70.0	38.6	60.0
Fruit-based	3.5	354.8	6.5	0.0	0	71.0	16.4	80.0

* Calculated using 9 Calories per gram for fat.

Clearly, there are differences. You may notice that the total Calories does not equal to the sum of the Calories from fat and carbohydrate. In some cases, the data presented by the cookie makers are not consistent and they (the cookie makers) attribute this to rounding errors! The zeros under cholesterol should not be taken as Gospel because they were based on a serving size less than 100 grams and likely rounded to 1 significant figure. Finally, as a gentle reminder nutritional Calories (note the capital C) are actually kilocalories.

12. Evaluating the structures of sterols
Describe all of the structural differences between cholesterol and stigmasterol?

Answer: The structures of both compounds are shown below. There are two differences: Stigmasterol has a trans double bond between carbons 22 and 23; and, carbon 24 in stigmasterol is modified with an ethyl group.

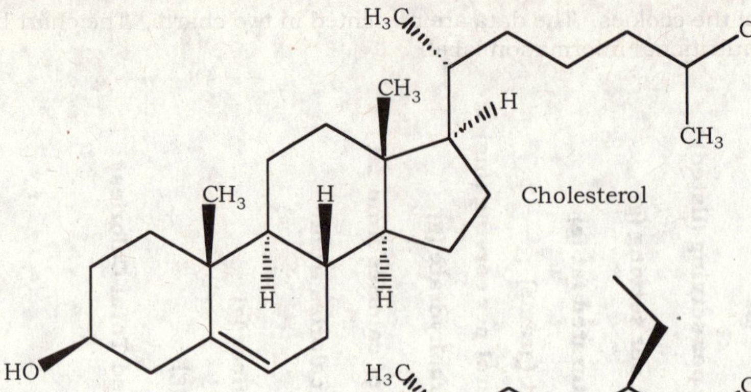

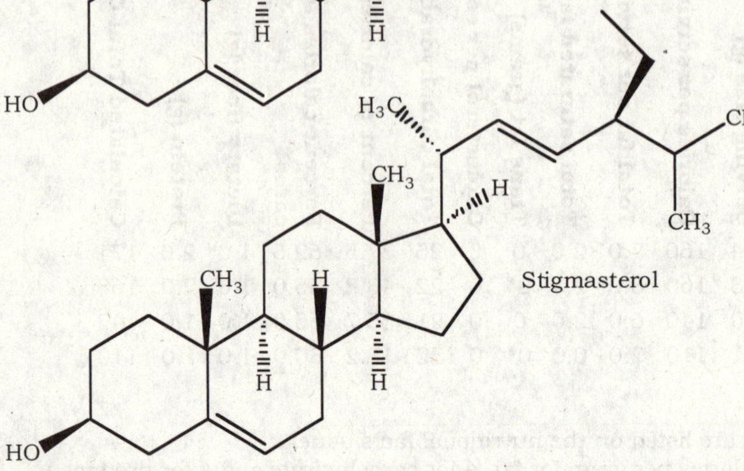

Cholesterol

Stigmasterol

13. Understanding the functions of steroid hormones
Describe in your own words the functions of androgens, glucocorticoids, and mineralocorticoids.

Answer: Androgens, glucocorticoids and mineralocorticoids are all steroid hormones derived from cholesterol. Androgens are responsible for development of sexual characteristics of males and for sperm production. In addition, androgens control libido and aggressiveness. The principal androgen is testosterone, which is produced by interstitial cells of the testis. Glucocorticoids and mineralocorticoids are steroid hormones produced by the adrenal cortex. Glucocorticoids regulate metabolism, specifically of carbohydrate, protein and lipid. The major glucocorticoid is cortisol. It stimulates gluconeogenesis and amino acid uptake by the liver and kidney. In adipocytes (fat cells) it inhibits glucose uptake and stimulates lipolysis. Glucocorticoids also have anti-inflammatory properties. Mineralocorticoids regulate extracellular fluid volume by modulating potassium uptake in the kidney. The principal mineralocorticoid is aldosterone.

14. Understanding structures and properties of terpenes
Look through your refrigerator, your medicine cabinet, and your cleaning solutions shelf or cabinet, and find at least three commercial products that contain fragrant monoterpenes. Identify each one by its scent and then draw its structure.

Answer: Here are a few. Limonene is found in orange oil and lemon oil, phellandrene in spearmint, pinene in pine and eucalyptus, camphene in firs, sweet fennel and nutmeg, myrcene in coriander, ginger, cinnamon and nutmeg.

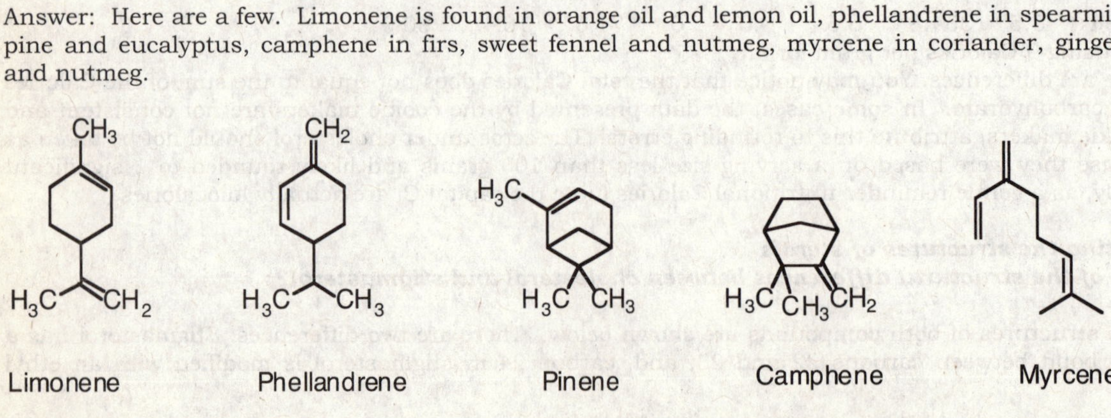

Limonene   Phellandrene   Pinene   Camphene   Myrcene

150

15. *Understanding the chemistry of soaps*
Our ancestors kept clean with homemade soap (page 222), often called "lye soap." Go to http://www.wikihow.com/Make-Your-Own-Soap and read the procedure for making lye soap from vegetable oils and lye (sodium hydroxide). What chemical process occurs in the making of lye soap? Draw reactions to explain. How does this soap work as a cleaner?

Answer: Lye is first dissolved in water and cooled to around 37°C. Next, coconut oil and vegetable shortening are heated to liquefy them and then olive oil is added. This mixture is then brought to 37°C and slowly added to the lye solution.

Lye is sodium hydroxide, which is very soluble and will dissociate into Na⁺ and OH⁻ making a solution with very high pH. The various oils are triacylglycerols and the reaction that ensues is saponification: base-catalyzed hydrolysis of an ester. The reaction is shown below:

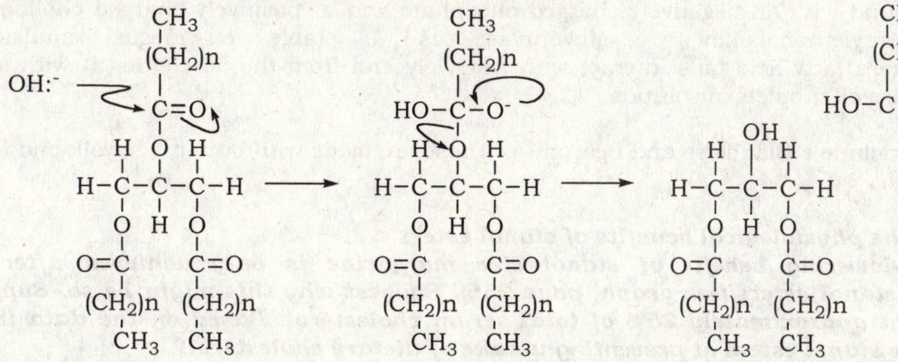

This reaction occurs at all three carbons on the glycerol backbone producing glycerol and three fatty acids that are unprotonated and thus negatively charged. These carboxylates bind sodium to form a salt of a fatty acid that aggregates as soap.

Soaps clean by forming fatty acid micelles that present a hydrophobic interior in which water-insoluble materials can dissolve. These substances are then, in effect, made water soluble by being coated with negative charge from the carboxyl groups on the fatty acids.

The mixture of lipids used for this homemade soap deserves comment. Vegetable shortening is hydrogenated vegetable oil typically made by passing hydrogen gas through oil containing a nickel catalyst. This process saturates some of the double bounds found in vegetable oils that make them liquid at room temperature. Hydrogenation, however, also produces some trans fatty acids. Removal of unsaturated fatty acids allows hydrocarbon chains to interact more effectively via van der Waals force to produce a solid at room temperature. Coconut oil is actually a solid at room temperature because it contains long-chained fatty acids in its triacylglycerols. The use of coconut oil and vegetable oil will make the soap form firm cakes because of the favorable van der Waals interactions long chained fatty acids and trans fatty acids can make. By varying the proportions of triacylglycerol sources one can vary the hardness of the soap cakes.

This is not to be confused with hard soap and soft soap. Hard soap is a consequence of saponification using sodium hydroxide. Soft soaps are made using potassium hydroxide. Sodium salts of fatty acids tend to be less soluble than potassium salts and in general the solubility tends to increase with group I elements. Salts of fatty acids and group II elements, in particular magnesium and calcium, are insoluble. Water with high concentrations of divalent cations is termed "hard water". Soaps work poorly in hard water and leave a "ring around the bathtub" as calcium and magnesium salts of fatty acids precipitate out of solution. Water softening agents are agents like citrate or polyphosphates that chelate divalent cations preventing them from interacting with soap molecules.

16. *Understanding the chemistry of common foods*
Mayonnaise is mostly vegetable oil and vinegar. So what's the essential difference between oil and vinegar salad dressing and mayonnaise? Learn for yourself: Combine a half cup of pure vegetable oil (olive oil will work) with two tablespoons of vinegar in a bottle, cap it securely, and shake the mixture vigorously. What do you see? Now let the mixture sit undisturbed for an hour. What do you see now? Add one egg yolk to the mixture, and shake vigorously again. Let the mixture stand as before. What do you see after an hour? After two hours? Egg yolk is primarily phosphatidylcholine. Explain why the egg yolk caused the effect you observed.

Answer: In the first experiment combining pure vegetable oil and vinegar and shaking vigorously produces an unstable emulsion that eventually separates into an oil layer floating on an aqueous vinegar layer. Vinegar is essentially a dilute solution of acetic acid in water ranging from 3% to 10% acetic acid. An emulsion is a suspension of two immiscible fluids and the emulsion we are attempting to make is an oil-water emulsion.

To make a stable oil-water emulsion we are going to need an emulsifying agent. This is a compound that can interact with both the oil phase and the aqueous phase and so it must be amphiphilic. A detergent would fit the bill nicely and a component of egg yolk serves this purpose.

Dry egg yolk is approximately 60% lipid and 30% protein. Of the lipid 65% is triacylglycerol, 4% cholesterol and most of the rest is phospholipid, principally phosphatidylcholine (26%) and phosphatidylethanolamine (4%). Phosphatidylcholine is often termed lecithin. The head group of lecithin is phosphocholine, a zwitterionic compound with an negatively charged phosphate and a positively charged choline. (Choline is essentially trimethylethanolamine.) Mayonnaise is a stable oil-vinegar emulsion because phosphatidylcholine's fatty acid tails interact with triacylglycerol from the oil coating it with strongly polar groups that stabilize oil droplets in solution.

Other emulsions include Hollandaise and Béarnaise. These are made with butter, egg yolk and lemon juice or vinegar.

17. Evaluating the physiological benefits of stanol esters
The cholesterol-lowering benefit of stanol-ester margarine is only achieved after months of consumption of stanol esters (see graph, page 236). Suggest why this might be so. Suppose dietary sources represent approximately 25% of total serum cholesterol. Based on the data in the graph, how effective are stanol esters at preventing uptake of dietary cholesterol?

Answer: The cholesterol-lowering activity of stanol esters is complex and involves both the secretion of cholesterol in bile and lowering the uptake of cholesterol in the small intestine. Blocking uptake of cholesterol occurs in the small intestine where stanol esters incorporate in micelles formed in the gut from bile salts, phospholipids, free sterols –including cholesterol and phytosterols- and fatty acids. Stanol-esters are more soluble than cholesterol in micelles and so they reduce the levels of cholesterol. Micelles are then taken up by the intestine and repackaged as chylomicrons. Stanol, a phytosterol, is actively excreted back into the lumen of the intestine by sterolin, an integral membrane protein that functions as a heterodimer of ABCG5 and ABCG8 to actively pump phytosterols out of the intestine. Sterolin is also active in the liver where it secretes phytosterols and cholesterol into bile.

If the cholesterol-lowering effect of phytosterols is only due to blockage of uptake then one might expect an immediate lowering of cholesterol levels. That it takes months to achieve lower cholesterol levels may be an indication that gene expression is at work. Higher dietary stanol would lead to higher uptake of stanol that is actively removed from the body by the liver. The liver might be responding to higher stanol levels by expression of ABCG5 and ABCG8, which in turn actively pumps phytosterols and cholesterol into bile secretions.

The graph on page 236 shows approximately a 14% decrease in cholesterol levels over time (240 mg/dl to 210 mg/dl). If the 240 gm/dl level is being maintained by dietary cholesterol, which represents 25% of total cholesterol, then (25% of 240) 60 mg/dl is due to diet alone. Stanol-ester lowers cholesterol by around 30 mg/dl. Thus, stanol-ester treatment appears to be blocking 50% of cholesterol uptake. This assumes that endogenous cholesterol production remains constant. There is evidence that cholesterol production does in fact increase so stanol-esters are likely blocking considerably more than 50% of dietary cholesterol.

18. Comparing therapies for cholesterol lowering
Statins are cholesterol-lowering drugs that block cholesterol synthesis in the human liver (see Chapter 24). Would you expect the beneficial effects of stanol esters and statins to be duplicative or additive? Explain.

Answer: The effects of stanol esters and statins are additive. Statins are drugs that block the enzyme 3-hydroxy-3-methylglutaryl coenzyme A reductase (HMGCoA reductase), an enzyme that catalyzes the rate-limiting step in cholesterol biosynthesis. Stanol esters reduce cholesterol levels by blocking uptake of dietary cholesterol in the intestine and increasing secretion by the liver of cholesterol in bile.

19. Assessing the sterol and stanol content of food products
If most plant-derived food products contain plant sterols and stanols, would it be as effective (for cholesterol-lowering purposes) to simply incorporate plant fats in one's diet as to use a sterol- or stanol-fortified spread like Benecol? Consult a suitable reference (for example, http://lpi.oregonstate.edu/infocenter/phytochemicals/sterols/#sources at the Linus Pauling Institute) to compose your answer.

Answer: According to the Micronutrient Information Center, a single serving (14 g or 1 tablespoon) of Benecol spread has 850 mg of stanol esters (500 mg free stanols). Normal daily dietary intakes of phytosterol range from 150 to 450 mg and only around 10% is stanol. Thus, one or two servings of Benecol provide far more stanol than a typical diet. While dietary phytosterols likely play a role in lowering cholesterol levels there are not many studies showing this. Studies on diets supplemented with phytosterols show that the effective intake levels range from 1.5 to 3 grams per day. Saturated fatty acids have an effect on cholesterol levels and plant triacylglycerols tend to have less saturated fatty acids so increasing one's intake of plant-derived food products would have an indirect effect on cholesterol independent of phytosterols.

20. Evaluating the history and culture of anabolic steroids
Tetrahydrogestrinone is an anabolic steroid. It was banned by the U.S. Food and Drug Administration in 2003, but it has been used illegally since then by athletes to increase muscle mass and strength. Nicknamed "The Clear," it has received considerable attention in high-profile steroid-abuse cases among athletes such as baseball player Barry Bonds and track star Marion Jones. Use your favorite Web search engine to learn more about this illicit drug. How is it synthesized? Who is "the father of prohormones" who first synthesized it? Why did so many prominent athletes use The Clear (and its relative, "The Cream") when less expensive and more commonly available anabolic steroids are in common use? (Hint: There are at least two answers to this last question.)

Answer: Tetrahydrogestrinone is synthesized from gestrinone and as its name implies the two compounds are related by simple oxidation/reduction chemistry: Tetrahydrogestrinone is reduced gestrinone. Tetrahydrogestrinone is synthesized starting with gestrinone and reducing it using hydrogen gas and a suitable catalyst like palladium on carbon. The two compounds are shown below. Reduction of the carbon/carbon triple bond converts gestrinone into tetrahydrogestrinone.

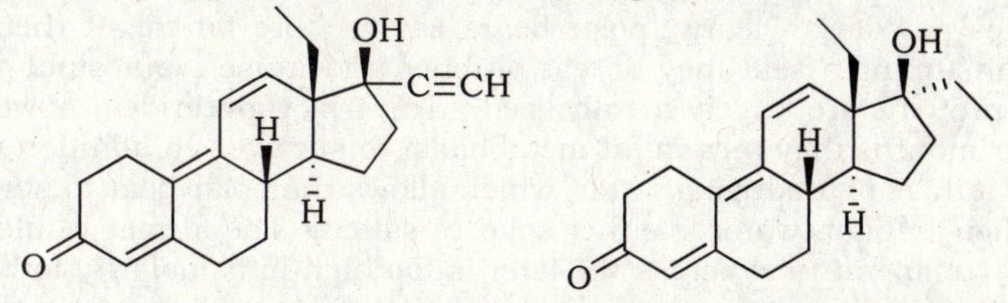

Gestrinone

Tetrahydrogestrinone

The "father of prohormones is Patrick Arnold, a chemist who is credited with first synthesizing tetrahydrogestrinone. (See http://www.steriod.com) This compound became popular because it was undetectable by standard techniques employed to monitor androgen abuse in sports. The World Anti-Doping Agency (WADA), created in 1999, is an international organization that monitors doping in sports. The use of chemicals to enhance sports performance is prohibited by WADA and many other organizations. WADA is concerned for the welfare of athletes and it is committed to insuring that athletes compete in doping-free sports, that the integrity of sports is maintained and that athletes are protected against chemicals that are a health threat. (See http://www.wada-ama.org/en/)

Monitoring of illegal drugs is done by mass spectrometry. Typically, urine samples are analyzed and compared against a library of compounds of known structure to detect particular drugs. The shortcoming of mass spectrometry is that a compound needs to be previously characterized and designer drugs are novel compounds not detected by mass spectrometry screening. Tetrahydrogestrinone was identified as an androgenic steroid, an undetectable "designer drug" given the nickname "The Clear". (See Rapid Commun. Mass Spectrom. 2004; 18, 1245-1249). A spent syringe was sent anonymously to the United States Anti-Doping Agency in June of 2003. Scientists at the UCLA Olympic Analytical Laboratory analyzed the content of the syringe and characterized the compound as tetrahydrogestrinone. This analysis provided a standard against which mass spectrometry is used to detect this androgenic steroid.

Androgens and erythropoietin are the most illicitly used drugs to enhance sport performance. (See Asian J Androl 2008; 10:403-415). Androgenic compounds, as mentioned above, are detected by mass spectrometry.

Mass spectrometry cannot, however, easily distinguish use of a natural androgen like for example testosterone or dihydrotestosterone because signals from these natural androgens are present prior to doping. To combat the problem of circumventing androgen doping by using testosterone or dihydrotestosterone, measurements are made both on the natural androgen and on another natural steroid that is known to be secreted. The other natural androgen is often epitestosterone. Cells produce both epitestosterone and testosterone from pregnenolone so a physiological increase in testosterone should be accompanied by an increase in epitestosterone. The ratio of testosterone to epitestosterone is thus a characteristic of natural steroid metabolism. Doping is easily detected by an increase ratio. "The Cream" is a mixture of testosterone and epitestosterone that maintains a natural ratio of these two androgens and therefore makes doping detection more difficult.

Doping and detection are involved in a virtual arms race. As detection techniques become more sophisticated drug designers produce novel drugs to avoid detection and "The Clear" is an example of this. There are other approaches to detecting drugs. Measurements of stable isotope ratio (C13/C12) are being used because commercial androgen synthesis uses plant sterols as starting material. Plant products have a lower C13/C12 signature as a consequence of isotope discrimination during photosynthesis. Bioassays are also being used to detect any compound that has androgen-like properties. Yeast cells are engineered to contain a human androgen receptor gene and a reporter gene regulated by androgen responsive elements (ARE). Androgens work by binding to androgen receptors. The complex migrates to the nucleus and binds to DNA elements, ARE's, resulting in an increase in expression of particular genes. Androgens are detected looking for novel gene expression in yeast cultured with reporter genes like β-galactosidase, luciferase and green fluorescent protein.

Preparing for the MCAT® Exam

21. Evaluating the biochemistry of polar bears
Make a list of the advantages polar bears enjoy from their nonpolar diet. Why wouldn't juvenile polar bears thrive on an exclusively nonpolar diet?

Answer: Polar bears largely eat seals, which they consume between April and July. It is estimated that they need approximately 2 kg of fat per day to survive. At approximately 9 Calories per gram this amounts to a whopping 18,000 Calories per day! Clearly, polar bears eat to store fat to get them through the summer, oddly enough. When consuming a seal they do eat blubber and muscle but since the body does not store excess amino acids the proteins are largely metabolized. The triacylglycerides, however, are stored for later use. During the summer months they rely on fat metabolism to survive. In addition to being a rich source of calories, this has the advantage of producing water, which allows the polar bear to survive without the need to drink liquid water. In their habitat, water is either solid or salted. The former would require calories simply to melt and bring to body temperature whereas the later is too high in osmolarity to be of use. Juvenile polar bears require, in addition to high calorie diets, diets rich in amino acid because they are growing. (Polar bears need ice from which to hunt seals. So, in the colder months they stock up on seals to get them through the warm months. Global warming may have severe consequences for polar bears because they will have to build up even larger fat stores to survive the longer ice-free summer periods.)

22. Evaluating the biochemistry of snake venom
Snake venom phospholipase A₂ causes death by generating membrane-soluble anionic fragments from glycerophospholipids. Predict the fatal effects of such molecules on membrane proteins and lipids.

Answer: Phospholipase A₂ hydrolyzes the fatty acid located on carbon 2 of phospholipids to produce a free fatty acid and 1-acylglycerophospholipid also known as 2-lysolecithin. Both fatty acids and lysolecithin are detergents capable of dissolving membrane components. Thus, phospholipase A₂ activity is expected to lead to cell lysis. This results in excessive tissue damage.

Questions for Self Study

1. Fill in the blanks. ___ ___ are important biomolecules composed of a long hydrocarbon chain or tail and a carboxyl group. When all of the carbon-carbon bonds are single bonds the compound is said to be ___. This term also indicates that the carbons in the tail are associated with a maximum number of ___ atoms. Compounds of this type with one carbon-carbon double bond are ___ whereas those with multiple carbon-carbon double bonds are ___. Usually there are an ___ number of carbons atoms. These compounds are

components of fats and oils in which they are joined to a ___ backbone in ___ linkage. The hydrolysis of fats or oils with alkali is called ___.

2. True of False
    a. 2-methyl-1,3-butadiene is also known as isoprene. ___
    b. Cholesterol is a phospholipid. ___
    c. The androgens are a class of terpene-based lipids involved in absorption of dietary lipids in the intestine. ___
    d. Vitamins A, E, and K are highly water-soluble vitamins. ___
    e. Cholesterol is a hydrocarbon composed of three six-membered rings and one five-membered ring in addition to a hydrocarbon tail. ___

3. Identify the following from the structures shown below: phosphatidic acid, phosphatidylcholine, phosphatidylserine, phosphatidylinositol, ceramide, phosphatidylethanolamine.

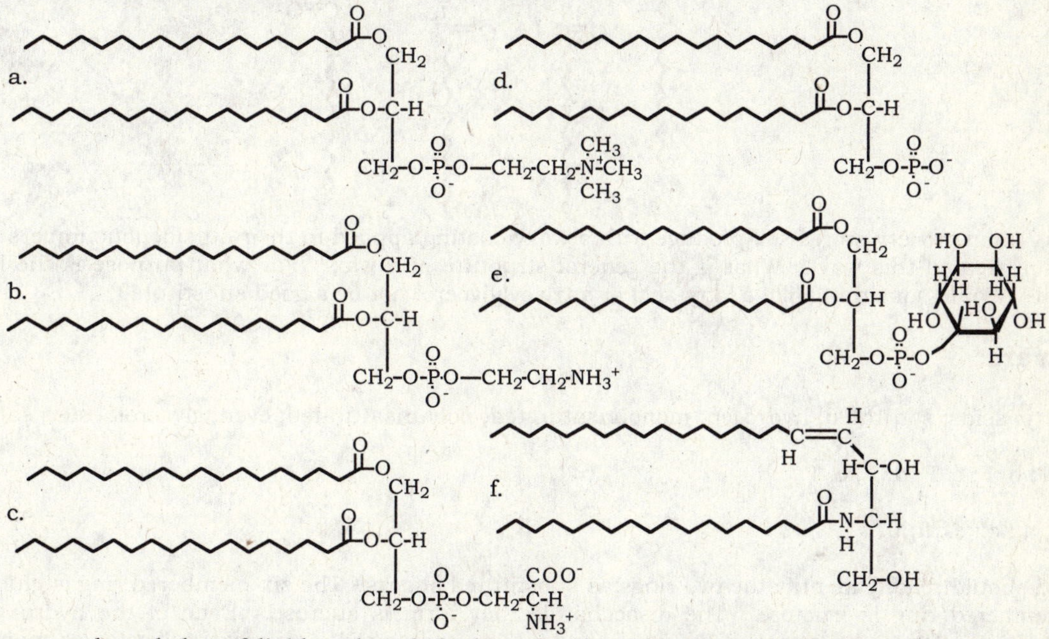

4. Based on your knowledge of lipid and carbohydrate biochemistry identify components of the following compound and state how this compound is chemically similar in structure to triacylglycerols? How does it differ biochemically?

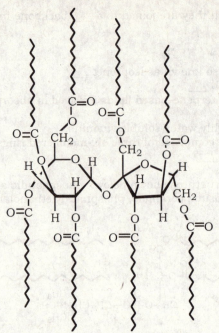

5. Very often grocery stores sell produce with a waxy coating applied to their outside (cucumbers and turnips are often treated this way). What is the general structure of a wax? For what purpose is the layer of wax applied? Would something like a fatty acid or a triacylglycerol not be a good substitute?

Answers

1. Fatty; acids; saturated; hydrogen; monounsaturated; polyunsaturated; even; glycerol; ester; saponification.

2. a. T; b. F; c. F; d. F; e. T.

3. d.; a.; c.; e.; f.; b.

4. You should readily identify the two rings as substituted sugars. The six-membered ring is glucose and the five-membered ring is fructose. The disaccharide they form is sucrose. Each of the hydroxyl groups of sucrose has a fatty acid attached by ester bonds. The compound is sucrose polyester or more commonly known as olestra (Trade name: Olean). Olestra is currently being used as a fat substitute because it has properties identical to fats and oils but is not digested.

Triacylglycerols contain fatty acids esterified to glycerol, a three carbon alcohol. Both triacylglycerol and olestra are amphiphilic molecules with uncharged, weakly polar head groups and hydrocarbon tails.

5. Waxes are composed of a long-chain alcohol and a long-chain fatty acid joined in ester linkage. Waxes are often used to make surfaces water impermeable thus a waxy coating will prevent water loss and prolong shelf life.

A layer of triacylglycerol might accomplish the same results; however, typical fats and oils have lower melting temperatures and would not be expected to form as stable a layer as wax.

Additional Problems

1. More stable mayonnaise can be produced by pre-treating egg yolk with phospholipase A_2. Suggest a rational for this.

2. In regions with mineral-rich water supplies (hard water) it is often difficult to work up a rich lather using hand soaps. Why?

3. At a romantic candle-light dinner, the conversation turns to properties of waxes and what exactly happens when a candle burns. Contribute to the conversation.

4a. Margarine is made from vegetable oil by a process called hydrogenation in which the oil is reacted with hydrogen gas in the presence of a small amount of nickel that functions as a catalyst. Hydrogenation saturates double bonds. Explain why hydrogenated vegetable oil is a solid.
b. Margarines may be purchased in stick-form or in small tubs. What is the important chemical difference between these two kinds of margarines?

5. Venom from honey bees (*Apis mellifera*) contains a phospholipase and several other components including a small amount of a polypeptide that acts as a detergent. Can you suggest a function for the polypeptide in terms of phospholipase activity?

6. Oil-water emulsions like mayonnaise and Hollandaise sauce contain, in addition to egg yolk and a lipid source (like oil or butter), sodium chloride and a source of acid (vinegar or lemon juice). Salt and acid are added for flavor but they also contribute to emulsion formation. They do this by interacting with egg yolk proteins. What might these agents be doing to yolk proteins to help with emulsion formation?

Abbreviated Answers

1. Phospholipase A_2 cleaves phospholipids to produce a free fatty acid and lysophospholipid, both of which are good emulsifying agents.

2. Mineral-rich water contains, among other things, high levels of divalent cations. Divalent cations will interact with and precipitate fatty acids. Water softeners are agents that chelate the offending divalent cations. Water can also be softened by ion exchange with resins that bind divalent cations.

3. Waxes are esters of long-chain alcohols and fatty acids. For example, in beeswax, straight-chain alcohols 24 to 36 carbons in length are esterified to long, straight-chain fatty acids up to 36 carbons in length. The melting temperature of beeswax is around 63°C. When a candle burns, the lit wick produces heat that melts the wax. The liquid wax is drawn up the wick to be consumed in the flame. A good candle will produce very little dripping wax because the wax is all consumed in the flame.
   Wax is a rich source of oxidizable hydrocarbons and serves the same purpose as oil does in a lamp, or gasoline does in an internal combustion engine. However, waxes and oils do not explode because they have very low vapor pressures.

4. a. Margarines are typically made from vegetable oils such as corn oil and soybean oil. They are oils, liquids at room temperature, because their composition includes greater than 50% unsaturated fatty acids. Hydrogenation is the addition of hydrogen to double-bonds producing a saturated hydrocarbon. For a given chain length, saturated hydrocarbons have a higher melting temperature than do unsaturated hydrocarbons. Thus, saturation of the double-bonds in the fatty acids in corn and soy oil reduces the level of unsaturated fatty acids and as a consequence the melting temperature is increased.
   b. The difference between tub-margarine and stick-margarine is the degree of saturation. Tub-margarine is distributed in a container because it is softer than stick-margarine due to a lower degree of hydrogenation.

5. The phospholipase of bee venom requires free lipids as substrates and the purpose of the detergent-like polypeptide may be to dissolve some of the victim's membrane to provide substrates. Once the reaction starts, the products, free fatty acids and lysolecithin, are both detergent-like molecules that will aid in dissolving membranes.

6. The ion components of salts might be interacting with the charged amino acid side chains of yolk proteins to neutralize charge. This might lessen repulsive forces among proteins. In addition, low pH produced by vinegar or lemon juice will alter the charge on yolk proteins.

Summary

Lipids are a large and diverse class of cellular compounds defined by their insolubility in water and solubility in organic solvents. Lipids serve several biological functions. As highly reduced forms of carbon, lipids yield large amounts of energy in the oxidative reactions of metabolism. As hydrophobic molecules, lipids allow membranes to act as effective barriers to polar molecules. The unique bilayer structure of membranes derives mainly from the amphipathic nature of membrane lipids. Certain lipids also play roles as cell-surface components involved in immunity, cell recognition and species specificity. Other lipids act as intracellular messengers and triggers, which regulate a variety of processes.

Most fatty acids found in nature have an even number of carbon atoms. Fatty acids may either be saturated or unsaturated, and double bonds are normally of the *cis* configuration. "Essential" fatty acids, including linoleic and linolenic acids, are not synthesized by mammals, but are required for growth and life. Triacylglycerols consist of a glycerol molecule with three fatty acids esterified. Triacylglycerols in animals are found primarily in adipose tissue, and serve as a major metabolic reserve for the organism.

Glycerophospholipids, a major class of lipids, are composed of an sn-glycerol-3-phosphate with fatty acids esterified at the 1- and 2- positions. Many different "head groups" can be esterified to the phosphate, including choline, ethanolamine, serine, glycerol and inositol. Ether glycerophospholipids are glycerol-based lipids but with a fatty alcohol in ether linkage at carbon 1. Sphingolipids are lipids formed from sphingosine, a long chain fatty alcohol, to which another fatty acid is attached in amide linkage to form a ceramide. Sphingomyelin is a phosphate-containing sphingolipid. Glycosphingolipids consist of a ceramide backbone with one or more sugars. Cerebroside is a glycosphingolipid containing either glucose or galactose. Gangliosides have three or more sugars esterified, one of which must be a sialic acid. Glycosphingolipids are present in only small amounts, but serve numerous important cell functions. Terpenes are a class of lipids derived from isoprene units. The steroids, including cholesterol, are an important class of terpene-based lipids. Other steroids in animals, including the androgens and estrogens (male and female hormones, respectively) and the bile acids (used in digestion) are derived from cholesterol.

Waxes are esters of long-chain fatty acids and long-chain alcohols. Because of their low water solubility and ability to aggregate, waxes are used to form water-impermeable surfaces.

In addition to being components of biological membranes certain lipids serve as substrates to produce chemical signaling molecules. For example, action of phospholipase A_2 on a phosphatidic acid containing arachidonic acid leads to production of free arachidonic acid, which is the precursor of the eicosanoids, and lysophosphatidic acid, an extracellular signaling molecule. Digestion of phosphatidylinositol by phospholipase C releases diacylglycerol, an activator of protein kinase C, and inositol phosphate, which regulates cellular calcium levels.

Chapter 9

Membranes and Membrane Transport

• •

Chapter Outline

❖ Membrane functions:
  - ⅄ Boundary for cell and organelle
  - ⅄ Surface on which reactions can occur
  - ⅄ Regulation of material flux through membrane proteins
  - ⅄ Signal transduction interface
  - ⅄ Specialized properties: Photosynthesis, electron transport, electrical activity
❖ Membrane components
  - ⅄ Lipids: Amphipathic molecules arranged as bilayers with polar groups out and nonpolar groups in
  - ⅄ Proteins: Surface associated or embedded in bilayer
❖ Plasma membrane
  - ⅄ Delimits cell
  - ⅄ Excludes and retains certain ions and molecules
  - ⅄ Major role in energy transduction
  - ⅄ Cell locomotion
  - ⅄ Reproduction
  - ⅄ Signal transduction
  - ⅄ Interactions with other cells or extracellular matrix
❖ Lipid interactions
  - ⅄ Monolayers: Formation of single-molecule-thick layer at air/water interface with polar groups in contact with water
  - ⅄ Micelles: Lipid spheres with polar groups out and hydrophobic tails in the center: Critical micelle concentration is the concentration of amphiphilic compound at which micelles form
  - ⅄ Lipid bilayer: Two lipid monolayers with hydrophobic surfaces face to face
  - ⅄ Liposomes: Vesicles formed by lipid bilayers
❖ Fluid Mosaic Model
  - ⅄ Singer and Nicholson, 1972
  - ⅄ Phospholipid bilayer forming fluid matrix
  - ⅄ Three classes of membrane proteins
    - • Peripheral (extrinsic) proteins
    - • Integral (intrinsic) proteins
    - • Lipid-anchored proteins
❖ Membrane Proteins
  - ⅄ Peripheral membrane proteins (extrinsic): Surface proteins held by variety of weak forces like ionic bonds, hydrogen bonds and hydrophobic interactions
  - ⅄ Integral membrane proteins (intrinsic): Two classes
    - • Single transmembrane segment proteins: Hydrophobic alpha helix that spans the lipid bilayer: Glycophorin: 19-amino acid long alpha helix that spans the membrane with extracellular domain decorated with oligosaccharides that are ABO and MN blood group antigens
    - • Multi-transmembrane segment proteins: Essentially globular proteins embedded in membrane
      - • Bacteriorhodopsin: Seven alpha helical segments embedded in bilayer: Segments organized into a channel

- Porins: Beta sheet motifs
- Wza: α-Helices used to form α-helical barrel

⅄ Hydropathy plot: Tool to identify membrane protein topology
- Calculate average hydrophobicity over window of amino acids
- Move window by one residue and recalculate average hydrophobicity and repeat
  - Single segment proteins: Hydrophobic helix
  - Multi-segmented protein: Multiple hydrophobic helices

⅄ Proline in transmembrane α-helix
- Proline common in interior of membrane α-helix
- Proline distorts α-helix by introduction of kink

⅄ Amino acid location preferences in transmembrane helix
- Hydrophobic amino acids interior
- Charged residues at lipid-water interface
  - Positive inside rule: Lys and Arg often found on cytoplasmic face

❖ Lipid-anchored membrane proteins: Four types
- Amide-linked myristoylated proteins
  - Myristic acid (14:0 fatty acid)
  - Amide linkage to amino group of N-terminal glycine
- Thioester-linked: Fatty acid attached to cysteine as thioester or Ser or Thr as ester
- Thioether-linked prenylated proteins
  - Prenyl: Long-chain isoprene polymers: Farnesyl or geranylgeranyl
  - Attachment as thioether to C-terminal cysteine of CAAX (A= Aliphatic)
  - AXX cleaved after phrenyl addition
- Amide-linked glycosyl phosphotidylinositol (GPI) anchors
  - Lipid: Oligosaccharide-modified phosphoinositol
  - Linkage: Carboxy terminus attached via phosphoethanolamine to mannose residue of oligosaccharide

❖ Membrane asymmetry
⅄ Lateral asymmetry arises from clustering of membrane components within the plane
- Lipid clustering: Phase separation induced by divalent cations and influenced by lipid type
- Protein clustering: Self-associating membrane proteins e.g., bacteriorhodopsin
⅄ Transverse asymmetry
- Lipids: Lipid asymmetry due to two processes
  - Asymmetric synthesis
  - Energy-dependent transport: Flippases and floppases
- Proteins: Asymmetric molecules
- Carbohydrates: Glycoproteins and glycolipids on outer surface

❖ Membrane mobility
⅄ Lipids
- Rapid lateral movement
- Slow transverse movement
  - Flippases: ATP-dependent movement of phosphatidyl serine from outer to inner surface of membrane
  - Floppases: ATP-dependent movement from inner to outer surface of membrane
  - Scramblases: Ca^{2+}-dependent randomize of lipids: Degrades transverse asymmetry
- Lipid bilayer states
  - Gel phase: Solid-ordered state: S_o: Lipid chains extended and packed
  - Liquid crystalline phase: Liquid-disordered state: L_d: Acyl bond rotation and bending
    o Membrane thickness decreases and surface area increases
  - Gel to liquid phase transition: Phase transition
    o Characterize by melting temperature, T_m
  - Liquid-ordered state: L_o:Acyl chain order like S_o and translational disorder of L_d
    o Microdomains: Membrane rafts: Regions in L_o phase
⅄ Protein-dependent modifications of membrane properties

- Hop diffusion: Membrane-skeleton fences restrict lateral diffusion to regions
- Membrane curvature
  - Lipid composition
  - Membrane proteins
  - Scaffolding proteins and amphipathic α-helices: Bind to membrane surface
  - Caveolae: Flask-shaped indentions in plasma membrane
    - Caveolins: Integral membrane proteins: Palmitoyl lipid anchored
- Vesicle formation and membrane fusion
  - SNAREs: Integral membrane proteins with single transmembrane domain and SNARE domain
    - Q_a-SNARE and Q_{bc}-SNARE: Plasma membrane protein
    - R-SNARE: Vesicle membrane
      - SNAREs interact to form complex stabilized by protein complexin
      - Ca^{2+} binding to synaptotagmin displaces complexin and promotes membrane fusion
- ❖ Membrane transport: Three types
  - ⅄ Passive diffusion
    - Entropically driven process: Molecules move down a concentration gradient
      - $\Delta G = RT\ln([C_2]/[C_1])$ for uncharged molecule
      - $\Delta G = RT\ln([C_2]/[C_1]) + Z\mathcal{F}\Delta\Psi$ for charged molecules
      - $R = 8.3145$ J/K·mol, $T = $ K, $Z = $ charge, $\mathcal{F} = 96485$ J/V·mol, $\Delta\Psi = $ electrical potential
    - Rate depends on concentration gradient and lipid solubility
  - ⅄ Facilitated diffusion
    - Entropically driven process as in passive diffusion
    - Involves integral membrane protein
      - Rate depends on concentration but is saturable
      - Specificity and affinity due to protein/transported molecule interaction
    - Membrane channel proteins
      - Channels: Single channels to multi channels
        - Aqueous cavity or funnels lead to selectivity filter
        - Selectivity filters
          - Often bind multiple transported species
          - Gated
    - Examples:
      - Potassium channels
        - Selectivity filter that distinguishes K^+ from Na^+
          - TVGYG in KcsA K^+ channel protein
          - Sequence modified in other ion specific channel proteins
        - Gates: Ligand-gated or voltage-gated
      - Pentameric Mg^{2+} channel (CorA)
        - Transmembrane pore of 5 α-helices
        - Cytosolic domain functions as basic sphincter
  - ⅄ Active transport: Energy driven process
    - Primary active transport: Energy sources
      - ATP hydrolysis (most common)
      - Light energy
      - Redox chains
    - Secondary active transport (Energy is ion gradient formed by some other process)
    - Electrogenic transport: Active transport of ions and net charge both occur
- ❖ Na^+,K^+-ATPase (sodium pump)
  - ⅄ 120 kD α-subunit; 35 kD β-subunit, 6.5 kD, γ-subunit
  - ⅄ 3 Na^+ out, 2 K^+ in per ATP hydrolyzed: Electrogenic
  - ⅄ Ouabain: Cardiac glycoside that inhibits sodium pump

161

❖ Calcium ATPase
  ⅄ 2 Ca^{2+} out of cytoplasm per ATP hydrolyzed
  ⅄ Restores/maintains low cytoplasmic calcium
❖ H$^+$,K$^+$-ATPase
  ⅄ 1 H$^+$ out, 1 K$^+$ in per ATP hydrolyzed
  ⅄ Gastric enzyme: ΔpH largest gradient known
❖ Vacuolar ATPases: Pump H$^+$ in number of vacuoles and cells
❖ ABC Transporters: ATPase that drives import or export
  ⅄ Multidrug resistance efflux pump: Export cellular wastes and toxins
  ⅄ Cancer cells activate these and become drug resistant
❖ Light-energy driven pumps
  ⅄ Bacteriorhodopsin
    • Light-driven H$^+$ pump
  ⅄ Halorhodopsin
    • Light-driven Cl$^-$ pump
❖ Secondary active transport systems
  ⅄ Na$^+$ or H$^+$ coupled movement of amino acids or sugars
  ⅄ Symport: Ion and substance move in same direction
  ⅄ Antiport: Ion and substance move in opposite directions

Chapter Objectives

Lipids associate to form two- and three-dimensional structures. Understand the forces responsible for this behavior, including hydrophobic interactions and van der Waals forces. Know why monolayers of lipids form at an air/water interface, and what a micelle is and how it forms. Lipids are also capable of forming bilayers, an important structural component of biological membranes.

Biological membranes are composed of various lipids arranged in a bilayer and, embedded in the bilayer, integral (or intrinsic) proteins. The fluid mosaic model of membranes suggests that both lipids and proteins are free to move within a bilayer. The two surfaces of bilayers of biological membranes are asymmetric with respect to protein, lipid and carbohydrate composition.

Membrane phase transitions occur when membrane components, in particular lipids, interact in a manner causing loss of fluidity. The temperature of this transition, a transition from solid to liquid, is known as the melting temperature (T_m). What are the effects of degree of saturation, of chain length, of cholesterol on T_m?

Two types of membrane proteins are peripheral and integral proteins. Peripheral proteins interact through electrostatic bonds and hydrogen bonds with the surfaces of bilayers. Integral proteins are strongly associated with the bilayer. There are three kinds of protein motifs responsible for anchoring integral proteins to membranes. Certain integral proteins have a single transmembrane segment, in the form of an α-helix composed of hydrophobic amino acid residues, anchoring the protein to the lipid bilayer. Another structural motif found is the 7-helix, transmembrane segment used by integral proteins involved in transport and signaling activities. Some proteins use β-sheets to form β-barrels. Certain proteins have covalently linked lipid molecules that serve as anchors. You should understand the four kinds of anchors.

Passive Diffusion
Passive diffusion proceeds down a concentration gradient. The driving force is a change in free energy given by $\Delta G = RT\ln([C_2]/[C]_1)$ where $[C_2] < [C_1]$ and the substance moves from side 1 to side 2. For a charged species, the driving force is an electrochemical potential given by $\Delta G = RT\ln([C_2]/[C_1]) + Z\mathcal{F}\Delta\psi$ where Z is the charge, $\mathcal{F}$ is Faraday's constant, and $\Delta\psi$ is the membrane potential.

Facilitated Diffusion
Facilitated diffusion is reminiscent of enzyme kinetics because it is a carrier-mediated process and as such depends on an interaction between a carrier and a transported molecule. The flux is still dependent on a difference in concentration and it occurs from high concentration to low concentration but the dependence is no longer linear. The flux shows saturation at high concentrations and is critically dependent on stereochemistry of the compound. The glucose transporter and the anion transporter (both in erythrocytes) are examples of facilitated diffusion.

Active Transport

Unlike passive and facilitated diffusion, active transport can move a substance against a concentration gradient. However the overall ΔG of the reaction must be favorable and this is achieved by coupling transport to some other energy-yielding process like ATP hydrolysis, capture of light energy, and coupling to other gradients. The sodium pump or Na^+,K^+-ATPase is a well characterized active transporter for movement of 3 Na^+ out of the cell and 2 K^+ into the cell coupled to hydrolysis of ATP. The enzyme is an intrinsic membrane protein that exists in two conformational states that differ in ion- and ATP-binding properties. Understand how transient phosphorylation leads to conformational changes and movement of ions. The cardiac glycosides are important inhibitors of the sodium pump. Understand the consequences of sodium pump inhibition to calcium ion levels in heart muscle. Finally, because of the difference in charge transported (a difference of one positive charge), the sodium pump is electrogenic leading to formation of a membrane potential. The calcium transporter of sarcoplasmic reticulum is also an ATP-dependent transporter, but of calcium. It has a similar mechanism of action to the sodium pump, shuffling between two conformational states with ATP hydrolysis driving calcium uptake. This transporter is a key player in relaxation of muscle and is also electrogenic. The H^+,K^+-ATPase moves protons out of the cell and potassium back into the cell with ATP hydrolysis. This nonelectrogenic pump is capable of producing extremely high concentration gradients of protons.

Light energy-driven active transport systems include bacteriorhodopsin, a H^+-pump, and halorhodopsin, a Cl^--pump. There are many important examples of transport systems driven by ion gradients. Proton gradients, produced by electron-transport driven proton pumping or by proton-ATPases, sodium gradients, produced by the sodium pump, and other cation and anion gradients are used to move a range of molecules, including sugars and amino acids. You should know the terms symport and antiport.

Problems and Solutions

1. Understanding occurrence of natural phospholipids

In Problem 1(b) in Chapter 8 (page 257) you were asked to draw all the phosphatidylserine isomers that can be formed from palmitic and linoleic acids. Which of these PS isomers are not likely to be found in biological membranes?

Answer: Phosphatidylserine with unsaturated lipids at position 1 are very rare. Unsaturated fatty acids are usually found at position 2. Glycerophospholipids with two unsaturated chains, or with a saturated chain at C-1 and an unsaturated chain at C-2, are commonly found in biomembranes.

2. Calculation of phospholipid-to-protein ratios

The purple patches of the Halobacterium halobium membrane, which contain the protein bacteriorhodopsin, are approximately 75% protein and 25% lipid. If the protein molecular weight is 26,000 and an average phospholipid has a molecular weight of 800, calculate the phospholipid to protein mole ratio.

Answer:

Let x = the weight of bacteriorhodopsin-lipid complex.

Weight of lipid in the complex = 0.25x

Weight of protein in the complex = 0.75x

$$\text{Moles lipid} = \frac{0.25x}{800\,\dfrac{g}{mole}} = 3.13 \times 10^{-4}x, \text{ and}$$

$$\text{Moles protein} = \frac{0.75x}{26,000\,\dfrac{g}{mole}} = 2.88 \times 10^{-5}x$$

$$\text{Molar ratio (lipid:protein)} = \frac{3.13 \times 10^{-4}x}{2.88 \times 10^{-5}x} = 10.8$$

3. Understanding densities of membrane components

Sucrose gradients for separation of membrane proteins must be able to separate proteins and protein-lipid complexes having a wide range of densities, typically 1.00 to 1.35 g/mL.

> *a. Consult reference books (such as the CRC Handbook of Biochemistry) and plot the density of sucrose solutions versus percent sucrose by weight (g sucrose per 100 g solution), and versus percent by volume (g sucrose per 100 mL solution). Why is one plot linear and the other plot curved?*
>
> *b. What would be a suitable range of sucrose concentrations for separation of three membrane-derived protein-lipid complexes with densities of 1.03, 1.07, and 1.08 g/mL?*

Answer: The density, at 20° C (ρ, g/mL), of sucrose solutions and their percent by volume (g per 100 mL) are shown in the first two columns. The third column shows the corresponding percent by weight for values shown in the second column.

ρ (g/mL)	% Sucrose (g per 100 mL)	% Sucrose (g per 100 g)*
0.9988	0	0.00
1.0380	10	9.63
1.0806	20	18.51
1.1268	30	26.62
1.1766	40	34.00
1.2299	50	40.65
1.2867	60	46.63
1.3470	70	51.97

*The values in this column are calculated as follows. First, the weight of a weight per volume solution is calculated. For example, a 10% weight per volume solution of sucrose has a density of 1.0380 g/mL. Therefore, 100 mL of this solution has a weight of:

$$100 \text{ mL} \times 1.038 \frac{g}{mL} = 103.8 \text{ g}$$

The percent weight is determined by calculating the amount of sucrose required to make 100 g of solution.

$$\frac{10 \text{ g}}{103.8 \text{ g}} = \frac{x}{100 \text{ g}}$$

$$x = 9.63 \text{ g sucrose in 100 mL.}$$

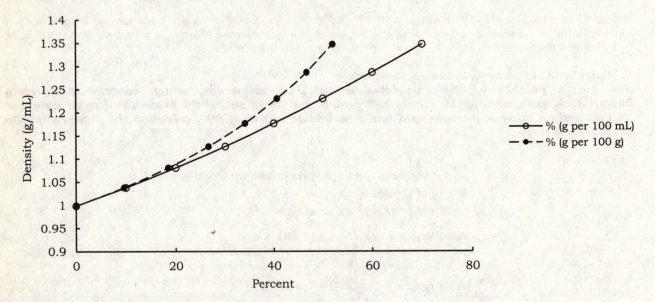

Why is there a difference in the two plots?

For the 10% solution (g per 100 mL), 100 mL of this solution weighs:

$100 \, ml \times 1.0380 \frac{g}{mL} = 103.8$ g of which 10 g was sucrose and

103.8 g-10 g=93.8 g was water.

The volume of 93.8 g of water is $\dfrac{93.8 \, g}{0.9988 \frac{g}{mL}} = 93.91$ mL.

Thus, the 100 mL volume of 10% sucrose is composed of 93.91 mL water and

100 mL-93.91 mL=6.09 mL of sucrose.

Therefore, the 10 g of sucrose occupied 6.09 mL corresponding to a density of

$\dfrac{10g}{6.09mL} = 1.64 \frac{g}{mL}$

To prepare a 10 g per 100 g solution, 90 g of water is mixed with 10 g sucrose.

The final volume is $\dfrac{90 \, g}{0.9988 \frac{g}{mL}} + \dfrac{10 \, g}{1.64 \frac{g}{mL}} = 96.21$ mL

The solution's density is $\dfrac{100 \, g}{96.21 \, mL} = 1.0394 \frac{g}{mL}$

Thus, the 10% (g per 100 g) solution's density ($1.0394 \frac{g}{mL}$) is greater than

the 10% (g per 100mL) solution's density ($1.0380 \frac{g}{mL}$).

4. Understanding diffusion of phospholipids in membranes

Phospholipid lateral motion in membranes is characterized by a diffusion coefficient of about 1 x 10^{-8} cm²/sec. The distance traveled in two dimensions (in the membrane) in a given time is r = (4Dt)$^{1/2}$, where r is the distance traveled in centimeters, D is the diffusion coefficient, and t is the time during which diffusion occurs. Calculate the distance traveled by a phospholipid in a bilayer in 10 msec (milliseconds).

Answer:

$$\text{For D=}1 \times 10^{-8} \frac{cm^2}{sec}, t=10 \text{ msec}=10 \times 10^{-3} \text{ sec}$$

$$r = \sqrt{4 \times D \times t}$$

$$= \sqrt{4 \times (1 \times 10^{-8} \frac{cm^2}{sec}) \times (10 \times 10^{-3} \text{ sec})}$$

$$= 2.0 \times 10^{-5} cm = 2.0 \times 10^{-7} m = 0.2 \ \mu m \text{ or } 200 \text{ nm}$$

5. Understanding diffusion of proteins in membranes

Protein lateral motion is much slower than that of lipids because proteins are larger than lipids. Also, some membrane proteins can diffuse freely through the membrane, whereas others are bound or anchored to other protein structures in the membrane. The diffusion constant for the membrane protein fibronectin is approximately 0.7 x 10 cm^{-12} cm²/sec, whereas that for rhodopsin is about 3 x 10^{-9} cm²/sec.

    a. Calculate the distance traversed by each of these proteins in 10 msec.

    b. What could you surmise about the interactions of these proteins with other membrane components?

Answer: a.

For fibronectin, $D = 0.7 \times 10^{-12} \dfrac{cm^2}{sec}$

For rhodopsin, $D = 3.0 \times 10^{-9} \dfrac{cm^2}{sec}$

$t = 10\ msec = 10 \times 10^{-3}\ sec$

$r = \sqrt{4 \times D \times t}$

For fibronectin $r = \sqrt{4 \times (0.7 \times 10^{-12} \dfrac{cm^2}{sec}) \times (10 \times 10^{-3}\ sec)} = 1.67 \times 10^{-7} cm$

For fibronectin $r = 1.67 \times 10^{-7} cm \times \dfrac{1\ m}{100\ cm} = 1.67 \times 10^{-9} m = 1.67\ nm$

For rhodopsin $r = \sqrt{4 \times (3.0 \times 10^{-9} \dfrac{cm^2}{sec}) \times (10 \times 10^{-3}\ sec)} = 1.10 \times 10^{-5} cm$

For rhodopsin $r = 1.10 \times 10^{-5} cm \times \dfrac{1\ m}{100\ cm} = 1.10 \times 10^{-7} m = 110 \times 10^{-9} m = 110\ nm$

b. The diffusion coefficient is inversely dependent on size and unless we know the size of each protein we can surmise very little. The M_r of rhodopsin and fibronectin are 40,000 and 460,000 respectively. For spherical particles D is roughly proportioned to $[M_r]^{-1/3}$. We might expect the ratio of diffusion coefficients (rhodopsin/fibronectin) to be

$$\dfrac{(40,000)^{-1/3}}{(460,000)^{-1/3}} = 2.3$$

The measured ratio is 4286! Clearly the size difference does not explain this large difference in diffusion coefficients. Fibronectin is a peripheral membrane protein that anchors membrane proteins to the cytoskeleton. Its movement is severely restricted.

6. Understanding the phase transitions of membrane phospholipids
Discuss the effects on the lipid phase transition of pure dimyristoyl phosphatidylcholine vesicles of added (a) divalent cations, (b) cholesterol, (c) distearoyl phosphatidylserine, (d) dioleoyl phosphatidylcholine, and (e) integral membrane proteins.

Answer: Myristic acid is a 14 carbon saturated fatty acid and as a component of dimyristoyl phosphatidylcholine is expected to participate in hydrophobic interactions and van der Waals interactions. At a particular temperature, T_m, these forces are strong enough to produce local order in a bilayer of this phospholipid.

a. Divalent cations (e.g., Mg^{2+}, Ca^{2+}) interact with the negatively charged phosphate group and thus stabilize bilayers and increase the T_m.

b. Cholesterol does not change the T_m; however, it broadens the phase transition. As a lipid, it can participate in hydrophobic and van der Waals interactions. Above the T_m of dimyristoyl phosphatidylcholine, cholesterol stabilizes interactions; however, below the T_m, it interferes with the packing of dimyristoyl phosphatidylcholine.

c. Distearoyl phosphatidylserine contains stearic acid, an 18-carbon, fully saturated fatty acid, which should participate favorably in van der Waals interactions and hydrophobic bonds. Its slightly longer chain length may perturb the geometry of vesicles. Also, the longer chain and negatively-charged head group should raise T_m.

d. Oleic acid is an 18-carbon fatty acid with a single double bond in *cis* configuration between carbons 9 and 10. Although capable of hydrophobic interactions, the unsaturated fatty acids are expected to interfere with van der Waals interactions. The T_m will be decreased.

e. Integral proteins will broaden the phase transition and could either raise or lower the T_m depending on the nature of the protein.

166

7. Determining the free energy of a membrane galactose gradient
Calculate the free energy difference at 25°C due to a galactose gradient across a membrane, if the concentration on side 1 is 2 mM and the concentration on side 2 is 10 mM.

Answer:

$$\Delta G = RT \ln \frac{[C_2]}{[C_1]},$$

where $[C_1]$ and $[C_2]$ are the concentrations of C on opposites of the membrane.

$$\Delta G = 8.314 \times 10^{-3} \frac{kJ}{K \cdot mol} \times 298 \ K \times \ln \frac{10 \ mM}{2 \ mM}$$

$$\Delta G = 4.0 \frac{kJ}{mol}$$

8. Determining the electrochemical potential of a membrane sodium ion gradient
Consider a phospholipid vesicle containing 10 mM Na^+ ions. The vesicle is bathed in a solution that contains 52 mM Na^+ ions, and the electrical potential difference across the vesicle membrane $\Delta\psi = \psi_{outside} - \psi_{inside} = -30$ mV. What is the electrochemical potential at 25°C for Na^+ ions?

Answer: The electrochemical potential is given by the following formula:

$$\Delta G = RT \ln \frac{[C_2]}{[C_1]} + Z\mathcal{F}\Delta\Psi$$

where R is the gas constant, T the temperature in degrees Kelvin, $\mathcal{F}$ is Faraday's constant (96.49 kJ/K·mol) and Z is the charge on the ion: +1 in this case.

$$\Delta G = 8.314 \times 10^{-3} \frac{kJ}{K \cdot mol} \times 298 \ K \times \ln \frac{52 \ mM}{10 \ mM} + (+1) \times 96.49 \frac{kJ}{V \cdot mol} \times (-30 \times 10^{-3} V)$$

$$\Delta G = 1.19 \frac{kJ}{mol}$$

9. Assessing the nature of transmembrane histidine transport
Transport of histidine across a cell membrane was measured at several histidine concentrations:

[Histidine], μM	Transport, μmol/min
2.5	42.5
7	119
16	272
31	527
72	1220

Does this transport operate by passive diffusion or by facilitated diffusion?

Answer: A characteristic of transport by passive diffusion is that the rate of transport is linearly dependent on concentration and so a plot of transport rate versus concentration will be linear. For facilitated diffusion, the transported molecule interacts with a carrier protein in the membrane. The rate of transport will be dependent on concentration but the dependence is not linear. Rather the dependence is reminiscent of Michaelis-Menten enzyme kinetics in that it shows saturation. A plot of rate versus concentration is presented below. The data show a linear relationship indicating that transport is by passive diffusion. However, a facilitated transport system with a high K_m relative to the concentrations of histidine tested here will also be approximately linear. The concentrations tested here are high relative to physiologically reasonable concentrations of histidine and if facilitated diffusion is at work it may not be of physiological importance. One way to confirm that the transport is passive is to retest transport using D-histidine. Using a different stereoisomer of histidine will have no affect on passive diffusion. Facilitated diffusion will show a specificity for one of the stereoisomers.

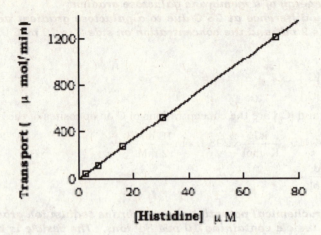

10. Determining the concentration limits of an active transport system
(Integrates with Chapter 3.) Fructose is present outside a cell at 1 μM concentration. An active transport system in the plasma membrane transports fructose into this cell, using the free energy of ATP hydrolysis to drive fructose uptake. What is the highest intracellular concentration of fructose that this transport system can generate? Assume that one fructose is transported per ATP hydrolyzed; that ATP is hydrolyzed on the intracellular surface of the membrane; and that the concentrations of ATP, ADP, and P_i are 3 mM, 1 mM, and 0.5 mM, respectively. T=298 K (Hint: Refer to Chapter 3 to recall the effects of concentration on free energy of ATP hydrolysis.)

Answer: The free energy of hydrolysis of ATP is given by

$$\Delta G = \Delta G^{o\prime} + RT\ln\frac{[ADP][P_i]}{[ATP]}$$

$$\Delta G = -30\frac{kJ}{mol} + 8.314\times10^{-3}\frac{kJ}{K\cdot mol}\times298\ K\times\ln\frac{1\ mM\times0.5\ mM}{3\ mM}$$

$$= -52.05\frac{kJ}{mol}$$

The free energy of a gradient of a substance across a membrane is given by

$$\Delta G = RT\ln\frac{[C_2]}{[C_1]}$$

We can set this equal to the free energy of hydrolysis calculated above but with the opposite sign and solve for C_2 given that C_1 is equal to 1 μM.

$$\Delta G = RT\ln\frac{[C_2]}{[C_1]} = 52.05\frac{kJ}{mol}\ \text{ and}$$

$$[C_2] = [C_1]\times e^{\frac{52.05\frac{kJ}{mol}}{RT}} = 1.0\times10^{-6}M\times e^{\frac{52.05\frac{kJ}{mol}}{8.314\times10^{-3}\frac{kJ}{K\cdot mol}\times298K}}$$

$$[C_2] = 1,330\ M\ !$$

11. Assessing the energy coupling of a transport process
In this chapter we have examined coupled transport systems that rely on ATP hydrolysis, on primary gradients of Na^+ or H^+, and on phosphotransferase systems. Suppose you have just discovered an unusual strain of bacteria that transports rhamnose across its plasma membrane. Suggest experiments that would test whether it was linked to any of these other transport systems.

Answer: If uptake is sensitive to ion gradients, ionophores may be used to destroy the gradients. Uncouplers like dicumarol or dinitrophenol can be used to degrade proton gradients. Ouabain can be used to inhibit the sodium pump. Dependence on ATP hydrolysis may be determined by using nonhydrolyzable ATP analogs. PEP dependent mechanisms similar to PTS in *E. coli* are sensitive to fluoride.

12. Characterization of a myristoyl lipid anchor
Which of the following peptides would be the most likely to acquire a N-terminal myristoyl lipid anchor?
    a. VLIHGLEQN
    b. THISISIT
    c. RIGHTHERE
    d. MEMEME
    e. GETREAL

Answer: Myristoylation occurs on N-terminal glycine residues and the only peptide that qualifies is "e".

13. Characterization of a prenyl lipid anchor
Which of the following peptides would be the most likely to acquire a prenyl anchor?
    a. RIGHTCALL
    b. PICKME
    c. ICANTICANT
    d. AINTMEPICKA
    e. none of the above

Answer: Prenylations typically occur on cysteines in sequences CAAX where the A's' are aliphatic amino acids and X is any residue. The only peptide to fit this description is "a". This peptide has a cysteine followed by alanine and two leucines on its C-terminus.

14. Creating and analyzing a hydropathy plot online
What would the hydropathy plot of a soluble protein look like, compared to those in Figure 9.14? Find out by creating a hydropathy plot at www.expasy.ch. In the search box at the top of the page, type in "bovine pancreatic ribonuclease" and click "Go." The search engine should yield UniProtKB/Swiss-Prot entry P61823. Scroll to the bottom of the page and click "ProtScale" under Sequence Analysis Tools. On the next page, select the radio button for "Hphob. / Kyte and Doolittle," then scroll to the bottom of the page, and click "Submit." On the next page, scroll to the bottom of the page and click "Submit" again. At the bottom of the next page, after a few seconds, you should see a hydropathy plot. How does the plot for ribonuclease compare to those in Figure 9.14? You should see a large positive peak at the left side of the plot. This is the signal sequence portion of the polypeptide.

Answer: To get the search to work I had to change the search site from ExPASy web site to UniProtKB and P61823 was the 5th hit.

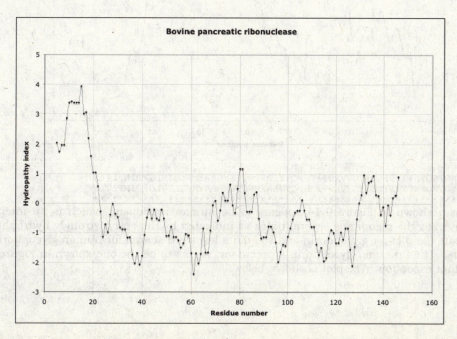

The plot shown above is for bovine pancreatic ribonuclease. I followed directions but recreated the plot in Excel using data from ProtScale found in "Numerical format (verbose)". This option and others are presented at the bottom of the ProtScale results page. Essentially I copied the data into an Excel spread sheet and plotted it.

Hydrophobic regions are positive values and hydrophilic regions are negative values. The plot starts at residue 5 because the default window size of 9 was used and 5 is the middle integer. The values plotted on the y-axis are the average hydrophobicity of 9 amino acid residues. For example, the very first y-value is 2.044. This corresponds to the average hydrophobicity of the first 9 amino acid residues, which are MALKSLVLL. The scale used for the Kyle & Doolittle plot is given near the top of the results page. MALKSLVLL = (1.900 + 1.800 + 3.800 - 3.900 - 0.800 + 3.800 + 4.200 + 3.800 + 3.800)/9 = 2.044.

The large peak near the N-terminus is the signal sequence, which is required for this enzyme to be secreted by the pancreas. The rest of the plot has regions of weak hydrophobicity and regions of hydrophilicity. Ribonuclease has 4 disulfide bonds that stabilize the structure, which is composed of helices and sheets. The protein is soluble in aqueous solution and globular in shape although the globular shape is not a consequence of a large hydrophobic core.

I tried to recreate the plots shown in Figure 9.14 and started with rhodopsin. At first I tried bovine rhodopsin but the very first peak was not correct so I then tried human rhodopsin and got a perfect match. I was curious why bovine rhodopsin didn't match and so I did a sequence alignment of the two proteins and discovered that near the N terminus there are a few substitutions, which accounted for the difference. This is shown below for the first 66 residues along with the sequence alignment over this region.

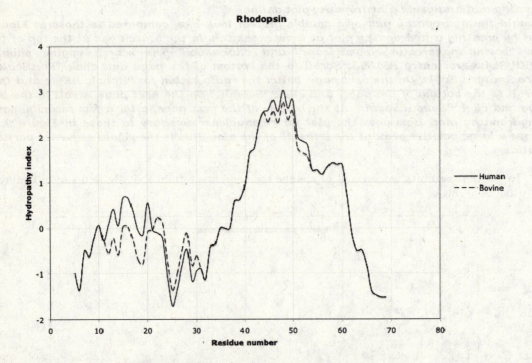

```
OPSD_BOVIN 1 MNGTEGPNFYVPFSNKTGVVRSPFEAPQYYLAEPWQFSMLAAYMFLLIMLGFPINFLTLY
OPSD_HUMAN 1 MNGTEGPNFYVPFSNATGVVRSPFEYPQYYLAEPWQFSMLAAYMFLLIVLGFPINFLTLY
             *************** ******** ************************ **********
```

So, the rhodopsin shown in Figure 9.14 appears to be human rhodopsin, which is an integral membrane protein that traverses the membrane several times as indicated by the seven strongly hydrophobic regions. I wanted to recreate the plot for glycophorin and so I did a key word search for human glycophorin at SwissProt and got four hits. The first one, glycophorin A precursor, seems to be the one shown in Figure 9.14 but with one very important exception. The plot is shown below.

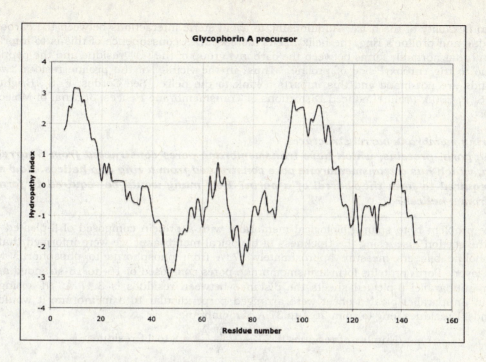

Glycophorin A is an intrinsic red cell membrane protein with an N terminus located on the outside of the membrane and decorated with carbohydrate. To get the N terminus outside the membrane it must have had a signal sequence, which is evidenced by the hydrophobic peak near the N terminus in the above plot. The glycophorin shown in Figure 9.14a starts at about residue number 20. It is the processed protein lacking its signal sequence.

15. Assessing the nature of proline in transmembrane α-helices
Proline residues are almost never found in short α-helices; nearly all transmembrane α-helices that contain proline are long ones (about 20 residues). Suggest a reason for this observation.

Answer: Proline is a helix breaker when located deeper than 3 residues from the N terminus of an α-helix. You might recall that proline is in fact an imido acid, not an amino acid. Proline's nitrogen is connected to both its α-carbon and its δ-carbon to form a ring and so the nitrogen has only one hydrogen covalently bonded to it. When proline's nitrogen participates in a peptide bond it loses its only hydrogen and so the peptide bond lacks a hydrogen bond donor. Any amino acid located within the first three residues of the N terminus of an α-helix does not have its amino hydrogen participating in hydrogen bonds because H bonding in an α-helix is between the carbonyl group of the ith residue and the amino group of the i + 4th residue. Put conversely: the amino group of the ith residue and the carbonyl group of the i - 4th residue. So proline is okay in the first three positions of an α-helix but not any deeper. The stability of a helix depends on hydrogen bonds and long helices can accommodate the disruption in the hydrogen bond that results from an internal proline.

16. Analyzing the structure of proline-containing α-helices
As described in this chapter, proline introduces kinks in transmembrane α-helices. What are the molecular details of the kink, and why does it form? A good reference for this question is von Heijne, G., 1991. Proline kinks in transmembrane α-helices. Journal of Molecular Biology 218:499–503. Another is Barlow, D. J., and Thornton, J. M., 1988. Helix geometry in proteins. Journal of Molecular Biology 201:601–619.

Answer: When positioned within an α-helix, proline causes problems because the peptide plane to which proline donated its nitrogen lacks a hydrogen bond donor. This occurs because proline is an imido acid and thus its nitrogen has only one covalently-bound hydrogen, which is lost in peptide bond formation. Thus, proline is a helix breaker and it is not common to find proline located within short helices of globular proteins. In α-helical transmembrane proteins, however, proline is common but it causes distortion of the helix producing a kink. The kink is a consequence of proline's inability to serve as a hydrogen-bond donor. The carbonyl group of the i-4th residue has no hydrogen bond donor partner. The lack of this hydrogen bond contributes to flexibility of the helix. Additionally, to avoid steric interactions between the carbonyl group of the i-4th residue and proline's ring, the helix rise is increased. A consequence of this is to interfere with the hydrogen bond that normally forms between the carbonyl group of the i-3th residue and the peptide hydrogen of the residue to the carboxyl side of proline. Thus, in the vicinity of the proline residue two stabilizing hydrogen bonds are not made and this imparts a kink to the helix. See Cordes, F. S., Bright, J. N., and Sansom, M. S. P. 2003. Proline-induced Distortions of Transmembrane Helices. Journal of Molecular Biology 323:951-960.

17. Comparing membrane barrel structures
Compare the porin proteins, which have transmembrane pores constructed from β-barrels, with the Wza protein, which has a transmembrane pore constructed from a ring of α-helices. How many amino acids are required to form the β-barrel of a porin? How many would be required to form the same-sized pore from α-helices?

Answer: The problem is to span a biological membrane with a protein composed of β-pleated sheets or α-helices. In the section discussing the thickness of biological membranes we were informed that dipalmitoyl phosphtidylcholine bilayers measure approximately 37 Å from phosphorus to phosphorus with a 26 Å hydrophobic layer. Porin proteins form transmembrane pores composed of 16- to 18-stranded antiparallel β-barrels. For antiparallel β-pleated sheets the distance between residues is 3.47 Å. If a single strand of polypeptide in antiparallel pleated sheet were arranged perpendicular to a membrane it would have to be between 8 and 11 residues long to span 26 Å to 37 Å of membrane.

$$\text{Number of residues} = \frac{26 \text{ Å or } 37 \text{ Å}}{3.47 \dfrac{\text{Å}}{\text{residue}}} = 8 \text{ to } 11 \text{ residues}$$

To make an 18-stranded β-barrel, requires

$$8 \text{ or } 11 \frac{\text{residues}}{\text{strand}} \times 18 \text{ strands} = 144 \text{ to } 198 \text{ residues}$$

To make a pore using α-helices arranged perpendicular to the membrane surface to form a barrel would require α-helices approximately 17 to 25 residues long using 1.5 Å per residue as Δz.

$$\text{Number of residues} = \frac{26 \text{ Å or } 37 \text{ Å}}{1.5 \dfrac{\text{Å}}{\text{residue}}} = 17 \text{ to } 25 \text{ residues}$$

In the Wza protein, a barrel is formed by association of 8 α-helices and this would require

$$17 \text{ or } 25 \frac{\text{residues}}{\text{strand}} \times 8 \text{ strands} = 136 \text{ to } 200 \text{ residues}$$

These calculations, however, are based on incorrect models. For the porins and for the Wza protein the secondary structure elements are not perpendicular to the membrane but tilted. For porins, strands are tilted from 30° to 60°. In OmpF, an *E. coli* porin protein with 16 strands, strands range from 7 residues to 18 residues (See 2OMF) with 16 being the most frequent length. A barrel of 16 strands each 16 residues long contains 256 residues. For OmpF (2OMF) there are 206 residues in β-sheets out of a total of 340 residues. For the Wza protein the C-terminus that forms the pore is organized into a 28-residue irregular helix. So, Wza invests (28 x 8 =) 224 residues to form its pore.

To answer this question I used Swisspdb viewer and studied 2OMF (porin) and 2J58 (Wza) both found at the RSCB Protein Data Bank (http://www.rcsb.org/pdb/home/home.do). In Swisspdb viewer the protein's sequence is displayed in a separate "control panel" that shows the sequence and indicates in which polypeptide chain any residue is located in addition to the residue's involvement in α-helix or β-sheet. It is an easy task to identify, for example, the C terminal helical residues in Wza that contribute to the pore. These residues in the helix can be viewed alone and the length of the helix measured or the number or residues in the helix determined.

18. Assessing the structural consequences of the hop-diffusion model
The hop-diffusion model of Akihiro Kusumi suggests that lipid molecules in natural membranes diffuse within "fenced" areas before hopping the molecular fence to an adjacent area. Study Figure 9.29 and estimate the number of phospholipid molecules that would be found in a typical fenced area of local diffusion. For the purpose of calculations, you can assume that the surface area of a typical phospholipid is about 60 Å².

Answer: In a paper from Kusumi's laboratory (Fujiwara, T et al. 2002 Phospholipids undergo hop diffusion in compartmentalized cell membrane J. Cell Biol. 157:1071-1081) the cell membrane was found to be compartmentalized into 230 nm diameter compartments, which is about the size of the diffusion tracks shown in Figure 9.29. To determine the number of phospholipids one simply divides the surface area by 60 Å^2 per phospholipid.

$$\text{Number of phospholipids} = \frac{\pi \times \left(\dfrac{230 \text{ nm}}{2}\right)^2}{60 \,\dfrac{\text{Å}^2}{\text{phospholipid}} \times \left(\dfrac{\text{nm}}{10 \text{ Å}}\right)^2} = 69{,}245 \text{ molecules}$$

This calculation assumes the surface area to be a circle 230 nm (0.23 μm) in diameter. If one uses instead a square of 0.3 μm the answer is 150,000 molecules.

19. Assessing the energetic consequences of snorkeling in membrane proteins
What are the energetic consequences of snorkeling for a charged amino acid? Consider the lysine residue shown in Figure 9.16. If the lysine side chain was reoriented to extend into the center of the membrane, how far from the center would the positive charge of the lysine be? The total height of the peak for the lysine plot in Figure 9.15 is about 4kT, where k is Boltzmann's constant. If the lysine side chain in Figure 9.16 was reoriented to face the membrane center, how much would its energy increase? How does this value compare with the classical value for the average translational kinetic energy of a molecule in an ideal gas (3/2kT)?

Answer: To answer this question one needs to know the distance from the alpha carbon to the side chain amino group of a lysine. To get an estimate of this distance I used Swisspdb viewer. I restricted viewing to single lysines (in the porin 2OMF) and simply measured the distance between the alpha carbon and side chain amino group nitrogen. The longest distance I found was 6.33 Å. (Variation occurs because of different orientations about the dihedral angles in the side chain.) The thickness of the membrane shown in Figure 9.16 is 30 Å or 15 Å from surface to center. If the alpha carbon is located such that lysine's amino nitrogen is just on the surface of the membrane in its snorkeling position then the alpha carbon is 15 Å – 6.3 Å = 8.7 Å from the center of the membrane. If lysine's side chain now reorients toward the center of the membrane it would be 8.7 Å – 6.3 Å = 2.4 Å from the center of the membrane.

Lysine's total movement is 6.3 Å×2 = 12.6 Å. From Figure 9.15a, the energy required to move 15 Å (to the center of a membrane) is 4 kT. Thus,

$$\frac{4 \text{ kT}}{15 \text{ Å}} = \frac{\text{x}}{12.6 \text{ Å}}$$

Solving for x we find x = 3.36 kT. The ratio of this energy to 3/2 kT is

$$\text{Ratio of energies} = \frac{3.36 \text{ kT}}{\dfrac{3}{2}\text{kT}} = 2.24$$

20. Assessing the dissociation behavior of aspartate residues in a membrane
As described in the text, the pKa values of Asp85 and Asp96 of bacteriorhodopsin are shifted to high values (more than 11) because of the hydrophobic environment surrounding these residues. Why is this so? What would you expect the dissociation behavior of aspartate carboxyl groups to be in a hydrophobic environment?

Answer: The pKa of an aspartic acid in aqueous solution is 3.9 (see Table 4.1) thus at neutral pH an aspartic acid is expected to be unprotonated and negatively charged. When placed in a hydrophobic environment the more stable state of an aspartic acid is its uncharged, protonated form. To maintain its proton, aspartic acid's pKa must be shifted up making it a much weaker acid in a hydrophobic environment.

Another way of looking at this is to consider the force between an unprotonated carboxyl group side chain of an aspartic acid residue and a proton. The force is given by the following equation:

$$F = -\frac{e_1 e_2}{Dr^2}$$

Where D is the dielectric constant, r the distance and e's the charges on each species. The force is a force of attraction that is inversely dependent on the distance between the two charges and inversely dependent on the dielectric constant of the solvent. The dielectric constant of water (see Table 2.1) is 78.5 and that of hexane is 1.9. By placing the two charges in an environment with a low dielectric constant the attractive force is greatly increased relative to that of an aqueous environment, with a correspondingly higher energy required for dissociation.

21. Assessing the dissociation behavior of lysine and arginine residues in a membrane
Extending the discussion from problem 20, how would a hydrophobic environment affect the dissociation behavior of the side chains of lysine and arginine residues in a protein? Why?

Answer: Lysine and arginine have high pKa's for their side chains, 10.5 and 12.5 respectively (see Table 4.1), and when protonated they are positively charged. In a hydrophobic environment the unprotonated and hence uncharged forms are more stable and to achieve this the pKa's have to be lowered.

(In dealing with the carboxyl group of an aspartic acid in problem 20 the pKa is raised to make aspartic acid a weaker acid because its acidic form is protonated and uncharged. In the case of lysine and arginine, the acidic form is protonated and charged and so the pKa's need to be shifted to favor deprotonation at lower pH to achieve the uncharged conjugate base form.)

22. Analyzing the nature of light-driven proton transport
In the description of the mechanism of proton transport by bacteriorhodopsin, we find that light-driven conformation changes promote transmembrane proton transport. Suggest at least one reason for this behavior. In molecular terms, how could a conformation change facilitate proton transport?

Answer: Absorption of light energy must be inducing conformational changes that alter the pKa's of the aspartic acids. The mechanism of proton transport for bacteriorhodopsin is presented in Figure 9.58. Light causes photoisomerization of the all trans retinal and this brings the protonated Schiff base near Asp^{85}. A proton is transferred from the Schiff base to Asp^{85} and this is followed in a series of events by release of a proton to the extracellular membrane surface. The Schiff base is then reprotonated by Asp^{96} but this occurs on the cytoplasmic membrane surface. Light energy drives the configurational change in retinal and this and the deprotonation of the Schiff's base induce a conformational change in the protein that likely accounts for the pKa shifts.

Preparing for the MCAT® Exam

23. Evaluating Singer and Nicolson's fluid mosaic model of membrane structure
Singer and Nicolson's fluid mosaic model of membrane structure presumed all of the following statements to be true EXCEPT:
   a. The phospholipids bilayer is a fluid matrix.
   b. Proteins can be anchored to the membrane by covalently linked lipid chains.
   c. Proteins can move laterally across the membrane.
   d. Membranes should be about 5 nm thick.
   e. Transverse motion of lipid molecules can occur occasionally.

Answer: When Singer and Nicolson proposed the fluid mosaic model membrane proteins anchored by covalently-linked lipids had not been discovered. Thus, "b" is the exception. Lipid-anchored proteins would, however, fit nicely into the model.

Questions for Self Study

1. Match the items in the two columns
   a. Singer and Nicolson
   b. Extrinsic protein
   c. Integral protein
   d. Liposome
   e. Micelle
   f. Flippase
   g. Transition temperature

   1. Peripheral protein.
   2. Lipid bilayer structure.
   3. Lipid transfer from outside to inside.
   4. Phase change.
   5. Intrinsic protein.
   6. Lipid monolayer structure.
   7. Fluid mosaic model.

2. Explain why, for proteins with a single transmembrane segment, the segment is a hydrophobic helix. Why a helix? Why hydrophobic residues?

3. Give three examples of lipid anchoring motifs.

4. Explain the term critical micelle concentration.

5. For each of the statements below state if each applies to one or more of the following: passive diffusion (P), facilitated diffusion (F), and active transport (A).
   a. Can only move down a concentration gradient. ___
   b. Is expected to transport L-amino acid and D-amino acid at the same rate. ___
   c. Can be saturated. ___
   d. Can occur in both directions across a biological membrane. ___
   e. Can be used to concentrate substances. ___
   f. Movement is coupled to exergonic process. ___
   g. Rate is linearly proportional to concentration difference. ___
   h. Movement across biological membrane dependent on lipid solubility. ___
   i. Sodium pump. ___

6. Match the active transport system with an appropriate function.
   a. Na^+,K^+-ATPase   1. Acidifies membrane bound compartments.
   b. Ca^{2+}-ATPase   2. Transports a host of cytotoxic drugs.
   c. H^+,K^+-ATPase   3. Resets levels of important second messangers after stimulation.
   d. Vacuolar ATPase   4. Electrogenic pump inhibited by cardiac glycosides.
   e. MDR ATPase   5. Responsible for production of the largest concentration gradient
                                 known in eukaryotic cells.

7. What is a symport? Antiport? How can they be used to move a substance against its concentration gradient?

8. Bacteriorhodopsin and halorhodopsin are active transport proteins for the movement of protons and chloride ions respectively. What energy source do they use to support ion pumping?

Answers

1. a. 7; b. 1; c. 5; d. 2; e. 6; f. 3; g. 4.

2. The hydrogen bonding groups in the peptide bond are all involved in hydrogen bonds in a helix. Hydrophobic residues can interact with the hydrophobic interior of membranes through hydrophobic interactions.

3. Amide-linked myristoyl anchors, thioester-linked fatty acyl anchors, thioether-linked prenyl anchors, and amide-linked glycosyl phosphatidylinositol anchors.

4. The critical micelle concentration is that concentration of lipid at which micelle formation is supported. Concentrations of lipid below the critical micelle concentration do not form micelles. Lipid solutions whose concentration is greater than the critical micelle concentration contain micelles in equilibrium with free lipid molecules. The concentration of the free lipid is equal to the critical micelle concentration.

5. a. P and F; b. P; c. F and A; d. P and F; e. A; f. A; g. P; h. P; i. A.

6. a. 4; b. 3; c. 5; d. 1; e. 2.

7. A symport is a transport system that couples movement of two substances in the same direction. An antiport couples the movement of two substances in opposite directions. They can be used to move substances against a concentration gradient if transport of the coupled substance is down a concentration gradient. In this case the energy of the concentration gradient of the co-transported substance is used to drive uptake.

8. Light.

Additional Problems

1. The transport properties of two potassium ionophores were being studied in synthetic lipid bilayers with a phase transition temperature of 50°C. Ionophore X transports potassium at a rate proportional to temperature from 20°C to 70°C. In contrast, ionophore Y transports potassium very well above 60°C; however, from 60°C to 40°C the rate of transport falls off precipitously to very low values below 40°C. Based on this information, can you suggest modes of action for these two ionophores?

2. For integral membrane proteins, α-helices play an important role in anchoring the protein to the membrane. Explain why a helix is a thermodynamically stable structural element to embed in a membrane.

3. How might you expect helical wheel plots of α-helical segments from the following proteins to differ: (a) a typical globular protein, (b) an integral membrane protein with a single transmembrane segment, and (c) an integral membrane protein with several α-helices forming a channel through which a water-soluble compound is transported?

4. Explain how soaps and detergents help to remove water insoluble substances.

5. Explain how lipid membrane asymmetry might arise in a natural membrane through action of a flippase that does not couple lipid movement to another thermodynamically favorable process (like ATP hydrolysis).

6. Compare the expected rate of passive diffusion across a cell membrane for the following list of compounds: glyceraldehyde, glyceraldehyde-3-phosphate, erythrose, ribulose, ribose-5-phosphate, D-glucose, L-glucose.

7. Activity of the sodium pump results in the net movement of a positive charge across the membrane. How does this lead to a change in the electrical potential of the membrane?

8. Would you expect proton pumps to be capable of creating large proton gradients if the pumps operated by an electrogenic mechanism? Explain.

9. Construct a helical wheel plot of melittin, whose amino acid sequence is:
Gly-Ile-Gly-Ala-Val-Leu-Lys-Val-Leu-Thr-Thr-Gly-Leu-Pro-Ala-Leu-Ile-Ser-Trp-Ile-Lys-Arg-Lys-Arg-Gln-Gln-Gly. Assume that this peptide forms an α-helix and comment on the structure.

Abbreviated Answers

1. Ionophore X may be a channel-forming ionophore, perhaps like the antibiotic gramicidin. Channel-forming ionophores span the membrane forming a channel through which an ion can diffuse across the membrane. Ionophore Y may be a carrier-ionophore. Carrier-ionophores form a complex with the ion to be transported. This complex diffuses across the membrane and dissociates, releasing the ion on the opposite side of the membrane. Valinomycin is an example of a potassium ionophore of this type.

2. The hydrophobic interior of membranes excludes polar compounds. The peptide bond is a polar bond with both hydrogen bond donor and acceptor groups that would normally be excluded from membranes. However, in a helical conformation the donors and acceptors are involved in intrachain hydrogen bonds making a helix a stable conformation even in a hydrophobic environment.

3. In globular proteins containing α-helices, the helices often contribute to the hydrophobic core of the protein with hydrophobic amino acid residues located along one face of the helix. Integral membrane proteins, with a single stretch of α-helix responsible for anchoring the protein into the membrane, have a helix composed of hydrophobic amino acids. For integral proteins anchored by several helices, the helices are often amphipathic with both a hydrophobic surface and a hydrophilic surface. The hydrophobic surfaces contact the fatty acid side chains of the membrane lipid component, whereas the hydrophilic surfaces face inward and may form a pore though which hydrophilic substances diffuse.

4. Soaps and detergents form micelles in solution. The interior of the micelles is a hydrophobic environment into which nonpolar molecules may dissolve.

5. Energy must be expended to produce an asymmetric distribution of lipids. In the case of a flippase that simply equilibrates lipids in response to a concentration gradient of free lipids, lipid asymmetry might arise if lipids preferentially interact with a membrane protein.

6. For passive diffusion, the flux across the membrane is linearly dependent on the permeability coefficient given by $P = (KD)/x$, where K is the partition coefficient, D the diffusion coefficient, and x the membrane thickness. All of the compounds are water soluble and are expected to have relatively low partition coefficients, decreasing in value in going from glyceraldehyde to glucose. The presence of phosphate groups on two of the compounds will decrease diffusion relative to the unphosphorylated compounds because of an increase in size (leading to a decrease in D) and a decrease in K. A difference in diffusion rates is not expected between the glucose isomers because diffusion does not involve molecular recognition.

7. There are two ways of looking at this. The net movement of a positive charge gives rise to an imbalance of charge across the membrane, resulting in an electrical potential. Alternatively, the movement of charge across the membrane represents a current. Current flowing across the resistance of the membrane will produce a voltage change.

8. For a non-electrogenic proton pump, the captured energy input would be in the form of a proton chemical gradient. For an electrogenic proton pump, the same energy input would be captured in a smaller proton chemical gradient, with the rest of the energy present in an electric potential across the membrane. Accordingly, for the same energy input the proton gradient would be higher with the non-electrogenic pump.

9. The helical wheel plot of melittin in an α-helical conformation is shown below. One face of the helix is lined with hydrophobic amino acids, and a cluster of basic amino acids is found at the C-terminus. A proline residue is positioned approximately in the middle of the helix.

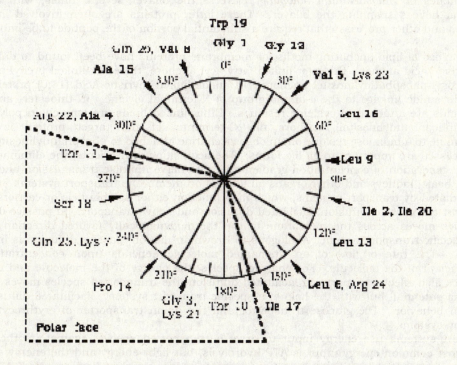

Summary

Lipids form a variety of structures spontaneously in solution, including monolayers, micelles and bilayer structures. Lipid bilayers have a polar surface, composed of charged or neutral lipid head groups, and a nonpolar interior, composed of hydrophobic lipid chains. The fluid mosaic model of membrane structure, proposed by Singer and Nicholson, pictures the lipid bilayer as a fluid, dynamic matrix, with lipids and proteins able to undergo free, rapid lateral motion. A variety of experiments has now confirmed the essential features of the fluid mosaic model.

The fatty acid chains in membrane lipids are oriented roughly perpendicular to the bilayer plane, and this ordering is more pronounced near the bilayer surface. As one proceeds into the bilayer interior, the ordering of the lipid chains decreases, so that the interior is a highly fluid environment. Transverse motion of lipids and proteins is very slow. Different lipid classes show different distributions between the inner and outer monolayers of the membrane bilayer. This lipid asymmetry is a consequence of ATP-dependent flippases that move lipids from the outer leaflet to the inner leaflet of the plasma membrane. Floppases work similarly but in the opposite direction. Finally, scramblases degrade transverse asymmetries by bidirectional movement of lipids in a calcium-dependent process. Proteins are asymmetrically distributed between the two faces of the bilayer, allowing a variety of vectorial (i.e., directionally dependent) functions including transport processes. Lateral asymmetries also exist in membranes, with proteins and lipids able to arrange themselves in clusters or aggregates important to cell function. Lipids in membranes exhibit dramatic, cooperative changes of state at characteristic temperatures. Such phase transitions between the solid, gel-like state at lower temperatures and the fluid, liquid-crystalline state at higher temperatures, are sensitive to the lipid composition and to the presence of proteins, and may be important in a host of biological functions.

Membrane proteins are of three fundamental types. Peripheral proteins form ionic interactions or hydrogen bonds with the surface of the lipid bilayer. Integral proteins intercalate into the lipid bilayer and are strongly associated with the membrane. The lipid anchored proteins attach to membranes via covalently linked lipid moieties. Peripheral proteins can be extracted with high salt, EDTA or urea, while integral proteins can only be removed with organic solvents or detergents. Detergents are amphipathic molecules, with both polar and nonpolar moieties, and function by intercalating into the membrane and solubilizing lipids and proteins. At the critical micelle concentration (CMC), detergents spontaneously form micelles and become much more effective solubilizing agents.

Integral membrane proteins take on a variety of conformations in membranes. Proteins such as glycophorin of the erythrocyte membrane have a single hydrophobic α-helix extending across the bilayer, with hydrophilic segments extending on either side of the lipid bilayer. Other proteins, such as bacteriorhodopsin of the purple patches of *Halobacterium halobium,* traverse the bilayer several times, with six or more hydrophobic alpha helices spanning the bilayer. These latter proteins are often involved in membrane transport activities and other processes that require a substantial portion of the peptide to be imbedded in the membrane.

Four different types of lipid anchoring motifs for membrane proteins have been found to date, including amide-linked myristic acid anchors, thioester-linked fatty acyl anchors, thioether-linked prenyl anchors, and amide-linked glycosyl-phosphatidylinositol anchors. In amide-linked myristic acid (14:0) proteins the fatty acid is attached in amide linkage to the α-amino group of N-terminal glycine. In thioester- and thioether-linked proteins lipids are attached to cysteine residues. Ether-linked lipids are long-chain polyisoprenoids. In GPI (glycosylphosphatidylinositol) anchors the C-terminus of the target protein is linked via phosphoethanolamine to a mannose residue on an oligosaccharide attached to phosphatidylinositol.

Transport processes are important to all life forms. The acquisition of nutrients, the elimination of waste materials and the generation of concentration gradients vital to nerve impulse transmission and the normal function of brain, heart, kidneys and other organs all depend on membrane transport systems. All transport processes are mediated by transport proteins, which may function either as channels or carriers. The three classes of transport are passive diffusion, facilitated diffusion and active transport. In passive diffusion, the transported species moves across the membrane in the thermodynamically favored direction without the assistance of a specific transport system. Analogous to Brownian motion, passive diffusion is in essence an entropic process. The rate of flow of an uncharged molecule depends upon concentration and the permeability coefficient of the molecule. For charged species, the charge of the molecule and the electrical potential difference also affect transport. In facilitated diffusion, the transported species moves according to its thermodynamic potential, but with the help of a specific transport system. Facilitated diffusion systems display saturation behavior. The glucose transporter and the anion transporter of erythrocytes are both facilitated diffusion systems.

Active transport systems use energy input to drive a transported species against its thermodynamic potential. The most common energy input is ATP hydrolysis, but light energy and the energy stored in ion gradients may also be used. All active transport systems are energy-coupling devices. Na,K-ATPase, which transport Na^+ ions out of cells and transports K^+ ions into the cells, is an active transport system. Na,K-ATPase consists of a 120 kD α subunit and a 35 kD β subunit. The enzyme mechanism involves an aspartyl phosphate intermediate. Na,K-ATPase is strongly and specifically inhibited by cardiac glycosides such as ouabain. Calcium transport across the sarcoplasmic reticulum membrane is mediated by Ca-ATPase, an enzyme that is homologous to Na,K-ATPase. The gastric H,K-ATPase likewise transports protons across the membrane of stomach mucosal cells, generating the high concentrations of acid in the stomach that are essential to digestion of food. The H,K-ATPase is homologous to Na,K-ATPase and Ca-ATPase. Proton pumps in osteoclasts enable these cells to degrade the mineral matrix of bone during the remodeling and reconstruction of bone tissue. ATPases also transport peptides and drugs. Another transport protein known as the multidrug resistance (MDR) ATPase actively transports a wide spectrum of drugs out of human cells. This transport system is induced by the chronic administration of drugs (in cancer chemotherapy, for example).

Bacteriorhodopsin (bR) is a light-driven proton transport system from *Halobacterium halobium*. The characteristic purple color of this transport protein arises from a molecule of retinal covalently bound in a Schiff base linkage with the ε-amino group of Lys-216 on the protein. bR is a 26 kD transmembrane protein that packs so densely in the membrane that it naturally forms a two-dimensional crystal. The structure consists of seven transmembrane helical protein segments, with the retinal moiety lying parallel to the membrane plane, about 10 Å below the extracellular surface of the membrane. The mechanism of proton transport involves conversion of the retinal chromophore from the all-*trans* configuration to the 13-*cis* configuration upon light absorption. An analogous transport protein, halorhodopsin, mediates light-driven anion transport across the *H. halobium* membrane.

Secondary active transport systems use the ion and proton gradients established by primary active transport systems to transport amino acids, sugars and other species in certain cells. Most of these operate as symport systems, with the ion or proton and the transported amino acid or sugar moving in the same direction. The lactose permease of *E. coli* is a lactose/H^+ symport system, which actively transports lactose into *E. coli* cells, deriving energy from the proton-motive force across the bacterial membrane.

In addition to the specific systems described above, several rather nonspecific systems also carry out transport processes. For example, Gram-negative bacteria enable the transport of small molecule nutrients and certain other molecules through the outer membrane via porins, which form large, non-specific pores in the outer membrane. Certain porins are specific, such as *LamB* and *Tsx* of *E. coli* and porins P and D1 of *P. aeruginosa*, which possess specific binding sites for maltose and related oligosaccharides, nucleosides, anions and glucose, respectively.

Chapter 10

Nucleotides and Nucleic Acids

• •

Chapter Outline

❖ Nucleic acids: DNA and RNA: Polymers of nucleoside monophosphates: Base + sugar + phosphate
❖ Heterocyclic nitrogenous aromatic bases
  ⅄ Pyrimidines: Six-member heterocyclic aromatic ring with 2 nitrogens
    • Cytosine: DNA and RNA
    • Uracil: RNA
    • Thymine: DNA
  ⅄ Purines: Five-member imidazole ring fused to six-member pyrimidine
    • Adenine: 6-Amino purine: DNA and RNA
    • Guanine: 2-Amino-6-oxy purine: DNA and RNA
  ⅄ H-bond properties
  ⅄ UV-light absorbing properties
  ⅄ Keto to enol tautomerization
❖ Sugars: 5-Carbon pentose: Furanose rings: Numbering system primed
  ⅄ D-ribose: RNA
  ⅄ 2'-deoxy-D-ribose: DNA
❖ Nucleoside: N-glycosidic linkage of base to anomeric carbon of sugar
  ⅄ Anomeric carbon in β configuration
  ⅄ Nomenclature
    • Pyrimidines: + idine: Cytidine, uridine, thymidine
    • Purines: + osine: Adenosine, guanosine
  ⅄ Two conformations
    • Syn: Base over furanose ring: Purines
    • Anti: Base not over furanose ring: Pyrimidines and purines
❖ Nucleotides: Typically 5' nucleoside phosphate
  ⅄ Phosphate esters
    • AMP, GMP, CMP, UMP: 5'-Ribonucleoside monophosphates
    • cAMP, cGMP: 3',5'-Cyclic ribonucleoside monophosphates: Regulation of cellular metabolism
  ⅄ Phosphoanhydrides
    • 5'-Ribonucleoside diphosphates: ADP, GDP, CDP, UDP
      • ATP (energy currency), GTP (protein synthesis and signal transduction), CTP (lipid synthesis), UTP (carbohydrate and polysaccharide synthesis): 5'-Ribonucleoside triphosphates
    • 5'-Deoxyribonucleoside triphosphates : dATP, dGTP, dCTP, TTP (dTTP): DNA synthesis:
❖ Nucleic acids: Nucleoside monophosphates in phosphodiester linkage 3' to 5'
  ⅄ DNA: Genetic material: Typically double stranded
    • dsDNA: Strands antiparallel
    • Interchain H bonds form base pairs
    • Chargaff's rules: A=T, G=C, Purines = Pyrimidines
    • X-ray diffraction of Franklin and Wilkins and model building of Watson and Crick

- ⋏ RNA: Typically single stranded: Produced during transcription
  - • mRNA: Carries information encoded in genes to direct protein synthesis on ribosomes
    - ● Derive from heterogeneous nuclear RNA (hnRNA)
    - ● RNA processed by splicing (removal of introns and joining of exons), capping (5' end), and polyA tail addition (3' end)
  - • rRNA: Components of ribosome: Protein synthesis
    - ● Small subunit of ribosome: Single rRNA
    - ● Large subunit of ribosome: Large subunit rRNA, 5S rRNA, and in eukaryotes 5.8S rRNA
  - • tRNA: Carriers of activated amino acids used by ribosome for protein synthesis
  - • snRNA: Small nuclear RNAs: Mediate splicing
  - • siRNAs: Small interfering RNAs: Degrade mRNAs
  - • miRNAs: Bind to mRNA and block translation
  - • snoRNAs: Required for certain RNA modifications
- ⋏ DNA vs RNA
  - • Composition
    - ● T in DNA not U to distinguish from T formed by deamination of C
    - ● 2' OH in RNA accounts for instability of RNA phosphodiester bond
  - • Hydrolysis
    - ● RNA: Sensitive to base hydrolysis: Resistant to acid hydrolysis
    - ● DNA: Resistant to base hydrolysis: Depurinates by acid hydrolysis (apurinic base)
- ❖ Nucleases: Enzymes that hydrolyze nucleic acids
  - ⋏ Exo-: Attack from ends
  - ⋏ Endo-: Attack internally
  - ⋏ a-Side cleavage: Attacks of 3' side produces 5' phosphorylated product
  - ⋏ b-Side cleavage: Attacks of 5' side produces 3' phosphorylated product
  - ⋏ RNases: RNA specific
  - ⋏ DNases: DNA specific
  - ⋏ Nucleases: Cleave either DNA or RNA
  - ⋏ Restriction endonucleases
    - • Type I and Type III: Cleavage requires ATP
    - • Type II: Cleavage site 4,6,8 base sequence with two-fold axis of symmetry

Chapter Objectives

Understand the difference between a nucleoside and a nucleotide. The nucleotides are phosphorylated derivatives of nucleosides, compounds made up of a sugar and a nitrogenous base. (See Figure 10.11)

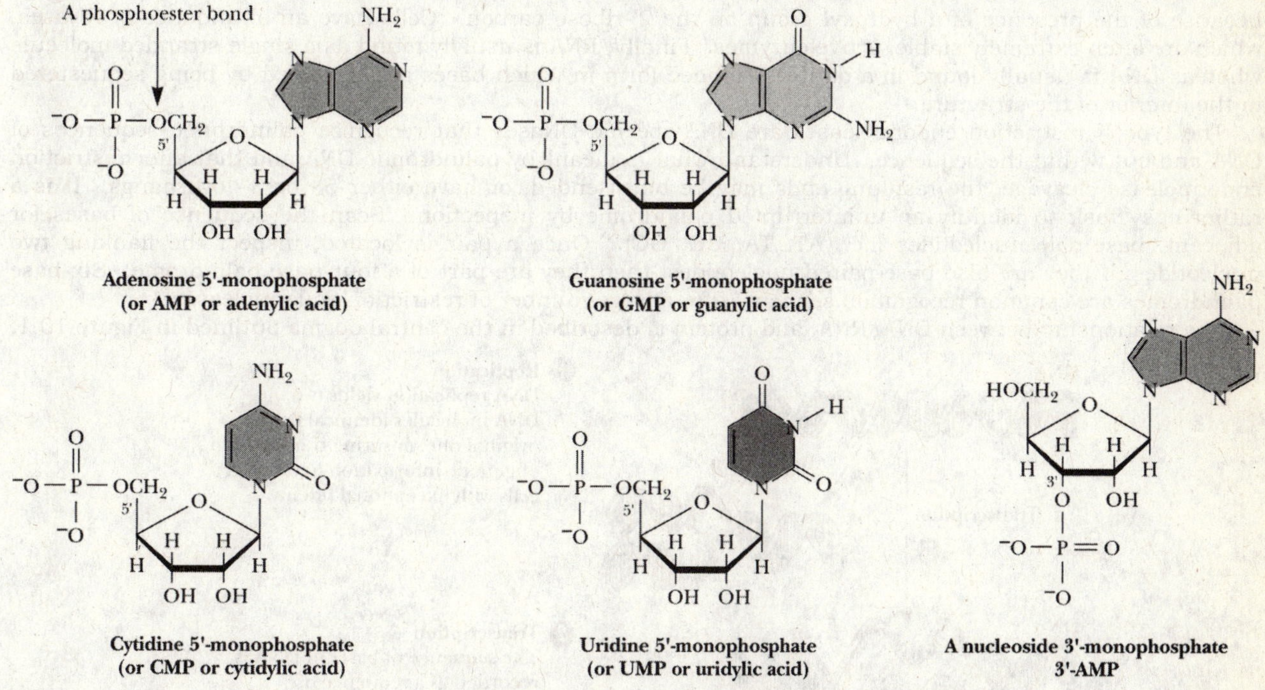

Adenosine 5'-monophosphate
(or AMP or adenylic acid)

Guanosine 5'-monophosphate
(or GMP or guanylic acid)

Cytidine 5'-monophosphate
(or CMP or cytidylic acid)

Uridine 5'-monophosphate
(or UMP or uridylic acid)

A nucleoside 3'-monophosphate
3'-AMP

Figure 10.11 The common ribonucleosides - cytidine, uridine, adenosine, and guanosine. Also, inosine drawn in the anti conformation.

It is important to know the structures of the pyrimidines and purines. Pyrimidines are six-membered heterocyclic, conjugated rings, whereas purines have a five-membered imidazole ring fused to a six-membered ring. (The small pyrimidine has a large name whereas the large purine has a small name.) The commonly occurring pyrimidines have two nitrogens separated by a carbonyl group. In uracil and thymine, an additional carbonyl group is located adjacent to one of the nitrogens and adjacent to it is a carbon with either a hydrogen (in uracil) or a methyl group (in thymine). Cytosine has an amino group replacing one of the carbonyl oxygens of uracil (oxidative deamination of cytosine produces uracil). In the purine double ring, the carbon separating the two nitrogens of the six-membered ring section has a hydrogen in adenine and an amino group in guanine. A carbonyl group is located adjacent to one of the nitrogens of the six-membered ring in guanosine (2-amino-6-hydroxypurine), which is substituted for an amino group in adenine (6-aminopurine).

You should be able to draw the structures of the bases and be able to identify hydrogen-bond donor and acceptor groups. Also, know the groups involved in base pairing in double-stranded DNA.

The compounds like ADP, ATP, GTP, UTP, CTP, NAD+ FAD, Coenzyme A, etc. are all either nucleotides or nucleotide derivatives of the ribonucleotides. The phosphate groups are usually attached to the 5' carbon of the ribose sugar. The dNTPs are used as precursors for DNA.

The convention for writing a nucleic acid sequence is to write, from left to right, the one-letter abbreviations of the bases from the 5'-end to the 3'-end.

The abbreviations most commonly found are A, C, G, T, U, N (any nucleotide, base unspecified), R (purine), and Y (pyrimidine).[1]

Know what phosphodiester bonds are and that they join the 3'-hydroxyl group of a nucleotide to the 5' phosphate of an adjacent nucleotide in DNA and RNA. The single-stranded polymers thus formed have a 5'-end and a 3'-end. DNA is predominantly double-stranded, a fact reflected in Chargaff's rules.

---

[1] There is an extended code, less commonly used, that covers other combinations of bases. B specifies C, G, or T (i.e., anything but A). D indicates A, G, or T (i.e., not C). H is anything but G. V refers to A, C, or G. W signifies either A or T, the weak base-pair formers. S refers to G or C, the strong base-pair formers. A or C is denoted by M, for amino, and G or T is specified by K for keto.

The total DNA content of a haploid cell is its genome size. Diploid cells have two copies of DNA while haploid cells have a single copy of DNA. The discrete molecules of cellular DNA are chromosomes. Cells may have a single chromosome or several chromosomes. Chromosomes are circular DNA molecules in some organisms and linear DNA molecules in others.

RNA molecules are much less stable than DNA molecules. RNA is sensitive to base-catalyzed cleavage because of the presence of a hydroxyl group on the 2'-ribose carbon. Cells have an abundance of RNases, which are often extremely stable, active enzymes. Finally, RNA is usually found as a single-stranded molecule whereas DNA is usually found in a double-stranded form in which bases are protected by being sequestered in the interior of the structure.

The type II restriction endonucleases are DNA-specific DNases that recognize palindromic sequences of DNA and cut within the sequence. Understand what is meant by palindromic DNA and that after restriction endonuclease cleavage, the resulting ends may be blunt-ended, or have either 5'- or 3'- overhangs. It is a rather easy task to identify an uninterrupted palindrome by inspection. Scan the sequence of bases for adjacent, base-pair nucleotides (i.e., AT, TA, CG, GC). Once a pair is located, inspect the flanking two nucleotides; if they are also base-paired nucleotides, then they are part of a four-base palindrome. Six-base palindromes are common recognition sequences for a large number of restriction endonucleases.

The relationship between DNA, RNA, and protein is described in the central dogma outlined in Figure 10.1.

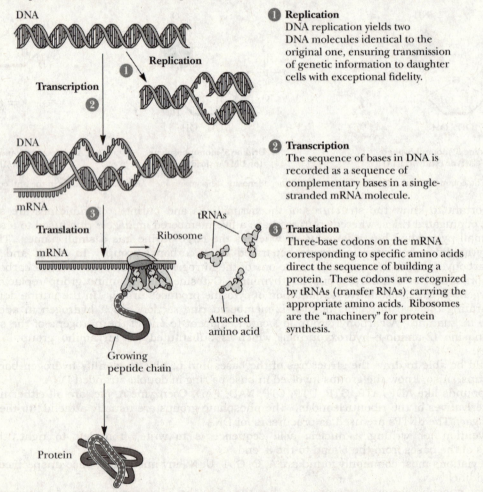

1 Replication
DNA replication yields two DNA molecules identical to the original one, ensuring transmission of genetic information to daughter cells with exceptional fidelity.

2 Transcription
The sequence of bases in DNA is recorded as a sequence of complementary bases in a single-stranded mRNA molecule.

3 Translation
Three-base codons on the mRNA corresponding to specific amino acids direct the sequence of building a protein. These codons are recognized by tRNAs (transfer RNAs) carrying the appropriate amino acids. Ribosomes are the "machinery" for protein synthesis.

Figure 10.1 The fundamental process of information transfer in cells. Information encoded in the nucleotide sequence of DNA is transcribed through synthesis of an RNA molecule whose sequence is dictated by the DNA sequence. As the sequence of this RNA is read (as groups of three consecutive nucleotides) by the protein synthesis machinery, it is translated into the sequence of amino acids in a protein. This information transfer system is encapsulated in the dogma: DNA★RNA★protein.

Problems and Solutions

1. The structure and ionization properties of nucleotides
Draw the principal ionic species for 5'-GMP occurring at pH 2.

Answer:

N-7 pKa 2.4 ⟶   ⟵ N-1 pKa 9.4

pKa's 0.7, 6.1

2. Oligonucleotide structure
Draw the chemical structure of pACG.

3. Chargaff's rules for the base composition of DNA
Chargaff's results (Table 10.1) yielded a molar ratio of 1.29 for A to G in ox DNA, 1.43 for T to C, 1.04 for A to T, and 1.00 for G to C. Given these values, what are the mole fractions of A, C, G, and T in ox DNA?

Answer: For ox DNA: A/G = 1.29; T/C = 1.43; A/T =1.04, G/C = 1.00; (A + G)/(T + C) = 1.1

What are the mole fractions of A, C, G, T? Before starting we should recognize that the data are inconsistent. For example,

$$\frac{T}{C} \times \frac{A}{T} = \frac{A}{C} \text{ and}$$

$$\frac{A}{G} \times \frac{G}{C} = \frac{A}{C} \text{ so}$$

Does $\dfrac{T}{C} \times \dfrac{A}{T} = \dfrac{A}{G} \times \dfrac{G}{C}$?

$$\frac{T}{C} \times \frac{A}{T} = 1.43 \times 1.04 = 1.487$$

$$\frac{A}{G} \times \frac{G}{C} = 1.29 \times 1.00 = 1.290$$

Clearly, there are problems with the data. It is likely that the original data were concentrations of A, G, T and C and if we had these data it would be trivial to calculate the mole fractions of each. Using the data given, we could get several different answers depending on which of the ratios we used. Here is one possible answer. Let,

$$f_A + f_G + f_C + f_T = 1.0$$

where the terms are the mole fraction of each nucleotide.

$$f_A + f_G + f_T + f_C = 1.0$$

But, $f_A = 1.04\, f_T$ and $f_G = 1.0\, f_C$ thus,

$$1.04\, f_T + 1.0\, f_C + f_T + f_C = 1.0, \text{ or}$$

$$2.04\, f_T + 2.0\, f_C = 1.0$$

By substituting $f_T = 1.43\, f_C$ we find that

$$2.04 \times 1.43\, f_C + 2 f_C = 1.0, \text{ and}$$

$$f_C = 0.20$$

And since $\dfrac{G}{C} = 1.0$, $f_G = 0.20$

$$f_A = 1.29 f_G = 1.29 \times 0.20$$

$$f_A = 0.26$$

And, finally since

$$\frac{f_T}{f_C} = 1.43,$$

$$f_T = 1.43 \times f_C = 1.43 \times 0.02 = 0.29$$

The final answer is:

$$f_A = 0.26, f_C = 0.20, f_G = 0.20, f_T = 0.29$$

These are consistent with some of the data but not all of it. Below are other solutions you might come up with that are consistent with some of the data but not all of it.

A	C	G	T
0.30	0.23	0.23	0.33
0.29	0.20	0.20	0.28
0.29	0.20	0.20	0.28

By treating the data as several equations in A, G, T and C one can come up with a solution that best fits all the data. Using A/G = 1.29, here is an example of one such equation:

$$A - 1.29G + 0*C + 0*T = 0$$

The others are (from T/C = 1.43, A/T = 1.04, A/G = 1, A+G/C+T = 1.1 and A + T + C + G = 1,):

$$0*A + 0*G - 1.43C + T = 0$$

$$A + 0*G + 0*C + 1.04T = 0$$

$$A - G + 0*C + 0*T = 0$$

$$A + G - 1.1C - 1.1T = 0$$

$$A + G + C + T = 1.0$$

We have six equations. What we want to do is find the best solutions for A, G, C and T that, when substituted into these equations, make them as consistent as possible. To do this we need to rearrange them such that they all are functions that equal zero. (The first five are already in this form. The fifth would be A + G + C + T −1 = 0.

We then square each, sum them and then find solutions for A, G, C and T that minimize the sum of the squares. (In effect, this is a form of a least squares solution.) (The solution can be obtained using matrix algebra.) The answer is A = .295, G = .223, C = .201 and T = .280.

As an overkill, it must be that since the input data were A, G, T and C, some of the relationships presented in Table 10.1 are not truly independent. For example, the ratio of purines to pyrimidines must have been calculated with A, G, C, and T as input. If we eliminate this relationship from the analysis the answer is A = .289, T = .286, G = .218 and C = .206.

4. Abundance of the different bases in the human genome
Results on the human genome published in Science (Science 291:1304-1350 [2001]) indicate that the haploid human genome consists of 2.91 gigabase pairs (2.91 x 10⁹ base pairs) and that 27% of the bases in human DNA are A. Calculate the number of A, T, G, and C residues in a typical human cell.

Answer: We are dealing with double-stranded DNA so the number of bases is twice the number of base pairs. However, we are asked for the number in a typical human cell. Since the vast majority of human cells are diploid we will be dealing with a cell containing twice the DNA amount listed for the haploid cell. The percentage of A bases is given as 27%. It must be true that there are an equal number of T's. The number of G's and C's must be equal and they are equal to the total number of bases minus the number of A's and T's.

$$A+T+G+C = 2 \times 2 \times 2.91 \times 10^9$$

And, $A = T$

And, $A = 0.27 \times 2 \times 2 \times 2.91 \times 10^9$

And, $A = 3.14 \times 10^9$

$$T = 3.14 \times 10^9$$

Substituting these values into the first equation and setting G = C,

$$3.14 \times 10^9 + 3.14 \times 10^9 + G + G = 2 \times 2 \times 2.91 \times 10^9$$

Solving for for G we get $G = 2.68 \times 10^9$

And since C = G, $C = 2.68 \times 10^9$

5. The base sequence in the two polynucleotide chains of a DNA double helix is complementary
Adhering to the convention of writing nucleotide sequences in the 5'✗3' direction, what is the nucleotide sequence of the DNA strand that is complementary to
d-ATCGCAACTGTCACTA?

Answer:

5'-TAGTGACAGTTGCGAT-3'

Note: The complementary sequence written '5' to 3' is sometimes referred to as the reverse complementary sequence. (As an example, consider the sequence GAGGCTT. Its reverse is TTCGGAG i.e., the sequence literally written in reverse. For the reverse sequence by exchanging each base for its complementary base (i.e., A for T and G for C) we have AAGCCTC, which is the strand that is complementary to GAGGCTT but written 5' to 3'. When the reverse complementary sequence is so defined the complementary sequence is defined literally. Thus, the complementary sequence of GAGGCTT is CTCCGAA.)

6. The relationship between the nucleotide sequence of an mRNA and the DNA strand from which it is transcribed
Messenger RNAs are synthesized by RNA polymerases that read along a DNA template strand in the 3'✗5' direction, polymerizing ribonucleotides in the 5'✗3' direction (see Figure 10.20). Give the nucleotide sequence (5'✗3') of the DNA template strand from which the following mRNA segment was transcribed:
5'-UAGUGACAGUUGCGAU-3'.

Answer: For a mRNA of sequence

5'-UAGUGACAGUUGCGAU-3'

the complementary DNA sequence is

5'-ATCGCAACTGTCACTA-3'

7. The sequence relationship between an antisense RNA strand and its template DNA strand
The DNA strand that is complementary to the template strand copied by RNA polymerase during transcription has a nucleotide sequence identical to that of the RNA being synthesized (except T residues are found in the DNA strand at sites where U residues occur in the RNA). An RNA transcribed from this nontemplate DNA stand would be complementary to the mRNA synthesized by RNA polymerase. Such an RNA is called antisense RNA because its base sequence is complementary to the "sense" mRNA. A promising strategy to thwart the deleterious effects of genes activated in disease states (such as cancer) is to generate antisense RNAs in affected cells. These antisense RNAs would form double-stranded hybrids with mRNAs transcribed from the activated genes and prevent their translation into protein. Suppose transcription of a cancer-activated gene yielded an mRNA whose sequence included the segment 5'-UACGGUCUAAGCUGA. What is the corresponding nucleotide sequence (5'✗3') of the template strand in a DNA duplex that might be introduced into these cells so that an antisense RNA could be transcribed from it?

Answer: For a mRNA 5'-UACGGUCUAAGCUGA-3', what is the corresponding sequence of a template strand in DNA that might make antisense RNA? Antisense RNA has a sequence complementary to the mRNA shown above. It would have to be transcribed from DNA with a sequence complementary to itself. Thus, the DNA must have a template sequence identical (except T for U) to the original mRNA:

5'-TACGGTCTAAGCTGA-3'

In the spring of 1998 Phase III clinical trials for an antisense drug for the treatment of cytomegalovirus retinitis in AIDS patients were completed and a new drug application for this antisense drug was filed with the FDA.

8. Restriction endonuclease mapping of a DNA fragment
A 10-kb DNA fragment digested with restriction endonuclease EcoRI yielded fragments 4 kb and 6 kb in size. When digested with BamHI, fragments 1, 3.5, and 5.5 kb were generated. Concomitant digestion with both EcoRI and BamHI yielded fragments 0.5, 1, 3, and 5.5 kb in size. Give a possible restriction map for the original fragment.

Answer:

EcoRI produces 4 kb and 6 kb fragments giving two possibilities:

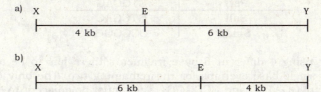

a)

X		E		Y
	4 kb		6 kb	

b)

X		E		Y
	6 kb		4 kb	

BamHI produces 1.0, 3.5, and 5.5 kb fragments giving six possibilities:

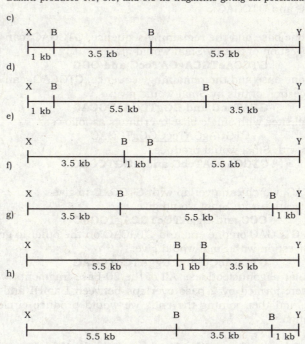

c)

X	B		B		Y
	1 kb	3.5 kb		5.5 kb	

d)

X	B		B		Y
	1 kb	5.5 kb		3.5 kb	

e)

X		B	B		Y
	3.5 kb		1 kb	5.5 kb	

f)

X		B		B	Y
	3.5 kb		5.5 kb		1 kb

g)

X		B	B		Y
	5.5 kb		1 kb	3.5 kb	

h)

X		B		B	Y
	5.5 kb		3.5 kb		1 kb

The double digest produces 0.5, 1.0,. 3.0, and 5.5 kb fragments. To generate a 5.5 kb fragment there must be a BamHI site on the 6 kb EcoRI fragment and within 0.5 kb of either end. The only possibilities are:

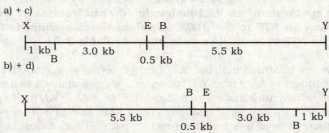

a) + c)

X		E	B		Y
1 kb	3.0 kb		0.5 kb	5.5 kb	
B					

b) + d)

X		B	E		Y
	5.5 kb		3.0 kb		1 kb
		0.5 kb			B

9. Design of DNA sequences with overlapping restriction endonuclease sites
Based on the information in Table 10.2, describe two different 20-base nucleotide sequences that have restriction sites for BamHI, PstI, SalI, and SmaI. Give the sequences of the SmaI cleavage products of each.

Answer: Table 10.5 lists type II restriction endonucleases, their recognition sequence, where in the sequence the enzyme cuts in addition to other information. BamHI PstI, SalI and SmaI all recognize six-base restriction sites. Since we are asked to produce a 20-base sequence we are going to have to have restriction sites overlap. We are not asked to order the sites so this leaves us free to try several combinations.

Enzyme	Recognition Sequence

BamHI	GGATCC
PstI	CTGCAG
SalI	GTCGAC
SmaI	CCCGGG

To make a 20 base fragment using 4 different 6 base fragments there has to be an overlap of two bases between two nucleotides and single-base overlaps for the remaining two. The only 6-base recognition sites that can overlap by two base are GGATCC and CCCGGG. They must be adjacent to each other but arranged in either order. Thus, 1. <u>GGATccCGGG</u> or 2. <u>CCCGggATCC</u>. The overlapping bases are in lower case. The remaining two sites are CTGCAG and GTCGAC.

For 1, CTGCAG can overlap by one base and the remaining sequence, GTCGAC, can overlap it by one, giving GTCGA<u>cTGCAgGATccCGGG</u>. Digestion of this by SmaI would give:

GTCGAcTGCAgGATccC and GGG

For 1, GTCGAC can overlap by one base and the remaining sequence, CTGCAG, can overlap it by one, giving <u>GGATccCGGg</u>TCGAcTGCAG Digestion of this by SmaI would give:

GGATccC and GGgTCGAcTGCAG

For 1, we could add PstI and SalI sites with single base overhangs as follows:

CTGCA<u>gGATccCGGg</u>TCGAC

Digestion of this oligonucleotide with SmaI would produce:

CTGCAgGATccC and GGGTCGAC

For 2, it can overlap with CTGCAG, which can overlap with GTCGAC to give <u>CCCGggATCc</u>TGCAgTCGAC. Digestion with SmaI would give:

CCC and GggATCcTGCAgTCGAC

Alternatively 2 can overlap with GTCGAC on one end and CTGCAG on the other to give GTCGA<u>cCCCGggATCc</u>TGCAG. Digestion with SmaI would give:

GTCGAcCC and GggATCcTGCAG

There are others including circular oligonucleotides. All of the 20-base fragments we explored begin and end in either G's or C's and they were formed by 2-base overlaps between BamHI and SmaI sites. By making these overlap by only one base and then joining the ends we would produce circles 20 base in size. They would linearize with SmaI digestion.

10. Calculate the free energy change for synthesis of a polynucleotide
(Integrates with Chapter 3) The synthesis of RNA can be summarized by the reaction:

$$n\ NTP \rightleftharpoons (NMP)_n + n\ PP_i$$

What is the $\Delta G^{o'}_{overall}$ for synthesis of an RNA molecule 100 nucleotides in length, assuming the $\Delta G^{o'}$ for transfer of an NMP from an NTP to the 3'-OH of polynucleotide chain is the same as the $\Delta G^{o'}$ for transfer of an NMP from an NTP to H_2O? (Use data given in Table 3.3.)

Answer: From Table 3.3 we are informed that the $\Delta G^{o'}$ for hydrolysis of ATP to AMP and PP_i is –32.3 kJ/mol. This reaction is in effect a transfer of AMP from ATP to water. We are allowed to assume that transfer of AMP from ATP to the 3' end of a polynucleotide chain has $\Delta G^{o'}$ of +32.3 kJ/mol. To make a polymer 100 nucleotides long let us start with hydrolysis of a single NTP to NMP and PP with a very favorable $\Delta G^{o'}$ of –32.3 kJ/mol. The NMP would then be used in a transfer reaction with NTP to make the first phosphodiester bond with an unfavorable $\Delta G^{o'}$ of +32.3 kJ/mol. To make 99 phosphodiester bonds, the number in a 100-base-long polynucleotide, the reaction would have to be repeated 98 times. Thus,

$$\Delta G^{o'}_{overall} = -32.3 + 99 \times 32.3\ kJ/mol = 3,165.4\ kJ/mol$$

If we were to have simply made 99 phosphodiester bonds it would require:

$$\Delta G^{o'}_{overall} = 99 \times 32.3\ kJ/mol = 3,198\ kJ/mol$$

However, the RNA would have a triphosphate at its 5' end.

11. Protein-DNA interactions
Gene expression is controlled through the interaction of proteins with specific nucleotide sequences in double-stranded DNA.
a. List the kinds of noncovalent interactions that might take place between a protein and DNA.
b. How do you suppose a particular protein might specifically interact with a particular nucleotide sequence in DNA? That is, how might proteins recognize specific base sequences within the double helix?

Answer: a. The weak forces are hydrophobic interaction, van der Waals interactions, hydrogen bonds and ionic bonds. Of these we might not expect excessive hydrophobic interactions to occur between a protein and dsDNA because dsDNA presents very little in the way of hydrophobic surfaces to the solvent. The bases themselves are hydrophobic but they are stacked in dsDNA. One might expect hydrogen bonding to occur between a protein and bases in a DNA sequence. One might also expect ionic bonding between the negatively-charged sugar/phosphate backbone of DNA and positively-charged amino acid residues from arginine and lysine.

b. Without unwinding dsDNA a protein would have to make base-specific interactions via either the major groove or minor groove. There is more information regarding the identity of base pairs in the major groove. In addition, the dimensions of an α-helix are compatible with the space in the major groove. Thus, one might expect proteins to bind via the major groove.

12. The properties restriction endonucleases must have
Restriction endonucleases also recognize specific base sequences and then act to cleave the double-stranded DNA at a defined site. Speculate on the mechanism by which this sequence recognition and cleavage reaction might occur by listing a set of requirements for the process to take place.

Answer: Restriction enzymes bind double-stranded DNA at sites that have two-fold rotational symmetry and they cleave both strands at identical places on the DNA. For this event to occur efficiently, both strands have to be recognized and cleaved at the same time. An efficient way of doing this is to have the enzyme function as a homodimer binding to the major groove.

One could envision a restriction endonuclease functioning as a monomer, binding to dsDNA via the major groove and hydrolyzing a single strand. To simultaneously cleave the second strand would require a second catalytic site on the enzyme. Alternatively, the enzyme might, upon cleavage of one strand, release from the DNA and rebind to it to cleave the second strand. Nicks in dsDNA can unwind, making it difficult to be recognized by an enzyme that initially bound to dsDNA. One could imagine a DNA binding site that recognizes single-stranded DNA sequences. This, however, would complicate binding to dsDNA. So, one could envision a protein with two catalytic sites or two DNA binding sites, one ssDNA specific and one dsDNA specific. Dimerization, however, seems like a simple way to avoid these complications.

13. The properties that carbohydrates contribute to nucleosides, nucleotides, and nucleic acids
A carbohydrate is an integral part of a nucleoside.
　　a. What advantage does the carbohydrate provide?
　　Polynucleotides are formed through formation of a sugar-phosphate backbone.
　　b. Why might ribose be preferable for this backbone instead of glucose?
　　c. Why might 2-deoxyribose be preferable to ribose in some situations?

Answer: a. A nucleoside is a base in glycosidic linkage to a sugar. Bases are poorly water-soluble and attachment to a sugar will improve their water solubility. Sugars present numerous hydroxyl groups that contribute to water solubility and provide attachment points for other molecules.

b. c. Moving from hexose to ribose to deoxyribose there is a decrease in the number of hydroxyl groups. DNA is much more stable to alkaline hydrolysis than RNA because it lacks a 2' hydroxyl, which in RNA can attack nearby phosphodiester linkages. One could imagine hexoses being even worse in this regard. The use of a hexose in place of a ribose might make packaging of DNA more difficult. The presence of additional hydroxyl groups would require more exposure to water molecules. Further, there would be an increased likelihood that unwanted side reactions involving hydroxyl groups would occur.

14. How does the presence of phosphate affect the properties of nucleotides
Phosphate groups are also integral parts of nucleotides, with the second and third phosphates of a nucleotide linked through phosphoric anhydride bonds, an important distinction in terms of the metabolic role of nucleotides.
　　a. What property does a phosphate group have that a nucleoside lacks?
　　b. How are phosphoric anhydride bonds useful in metabolism?
　　c. How are phosphate anhydride bonds an advantage to energetics of polynucleotide synthesis?

Answer: a. The addition of phosphates will further increase the water solubility of nucleosides by adding negative charge.

b. c. Phosphoanhydride bonds are strongly exergonic and their hydrolysis can be used to drive reactions including polynucleotide synthesis.

15. Calculate the frequency of occurrence of an RNAi target sequence
The RNAs acting in RNAi are about 21 nucleotides long. To judge whether it is possible to uniquely target a particular gene with a RNA of this size, consider the following calculation: What is the expected frequency of occurrence of a specific 21-nt sequence?

Answer: The probability of finding a specific 21-nucleotide sequence is simply $(1/4)^{21}$ or 2.27×10^{-12} assuming that we are working with nucleic acid with roughly equal amounts of A, G, C and T. The frequency of occurrence of this sequence is $1/2.27 \times 10^{-12}$ or one in 4.40×10^{12} nucleotides. The general formula used is $(1/4)^n$ where n is the number of bases and $(1/4)$ is the probability or frequency of a particular base.

16. Calculate the length of a nucleotide sequence whose expected frequency is once per haploid human genome
The haploid human genome consists of 3×10^9 base pairs. Using the logic in problem 15, one can calculate the minimum length of a unique DNA sequence expected to occur by chance just once in the human genome. That is, what is the length of a double-stranded DNA whose expected frequency of occurrence is once in the haploid human genome?

Answer: To answer this we need to solve the following equation for n:

$$\left(\frac{1}{4}\right)^n = \frac{1}{3 \times 10^9}$$

This is done by taking the log. Thus,

$$n \times (-\log 4) = -\log(3 \times 10^9) \text{ or}$$

$$n = \frac{\log(3 \times 10^9)}{\log 4} = 15.7 \approx 16$$

17. Design a sequencing strategy for nucleic acids
Snake venom phosphodiesterase is an a-specific exonuclease (Figure 10.28) that acts equally well on single-stranded RNA or DNA. Design a protocol based on snake venom phosphodiesterase that would allow you to determine the base sequence of an oligonucleotide. Hint: Adapt the strategy for protein sequencing by Edman degradation, as described on pages 10-11 and 114-115.

Answer: One approach might be to simply partially digest the oligonucleotide for varying lengths of time and then analyze the products by mass spectrometry, specifically MALDI-TOF (Matrix assisted laser desorption ionization –time of flight). Complete digestion will only produce individual nucleotides but partial digestion will produce a collection of nucleotides with ragged ends and whose molecular weights differ by loss of a particular nucleotide.

In Edman degradation controlled cleavage of an immobilized peptide releases one amino acid at a time and something similar could be tried with a nucleic acid and an exonuclease, but it could be tricky.

18. Calculate the mass of DNA in a human cell
From the answer to problem 4 and the molecular weights of dAMP (331 D), dCMP (307 D), dGMP (347 D), and dTMP (322 D), calculate the mass (in daltons) of the DNA in a typical human cell.

Answer: In problem 4 we calculated the moles of A, T, C and G in a typical diploid human cell and found the following: A = T = 3.14×10^9 and G = C = 2.68×10^9. The molecular weights given above need to be corrected for loss of water upon phosphodiester bond formation since what gets incorporated into a DNA is a nucleoside monophosphate residue. So, one needs to correct the molecular weights given above by subtracting 18 from each.

Mass of DNA =(331-18) mole of A + (322-18) mole of T +(347-18) mole of G +(307-18) mole of C
Mass of DNA = 3.59×10^{12} D

Preparing for the MCAT® Exam

19. The bases of nucleotides and polynucleotides are "information symbols." Their central role in providing information content to DNA and RNA is clear. What advantages might bases as "information symbols" bring to the roles of nucleotides in metabolism?

Answer: In metabolism, nucleotides are used both to energize compounds and to tag them for specific metabolic fates. For example, glucose destined for storage in glycogen is converted to UDP-glucose. In lipid metabolism, CTP plays a role in activation of phospholipids. Finally, GTP is used in protein synthesis and signal transduction pathways.

20. Structural complementarity is the key to molecular recognition, a lesson learned in Chapter 1. The principle of structural complementarity is relevant to answering problems 5, 6, 7, 11, 12, and 19. The quintessential example of structural complementarity in all of biology is the DNA double helix. What features of the DNA double helix exemplify structural complementarity?

Answer: Structural complementarity of double-stranded DNA is due to complementary base pairs. The fact that bases pair allows for efficient replication and repair of DNA. Even in the case of double-stranded breaks, repair mechanisms based on homologous recombination can rejoin DNA molecules.

Questions for Self Study

1. Fill in the blanks. The two basic kinds of nucleic acids are ___ and ___. They are composed of building blocks termed ___; however, the building blocks are not identical for the two kinds of nucleic acids. One contains the five carbon sugar ___ whereas the other has a modified form of this sugar or ___. The building blocks all contain nitrogenous bases attached to the sugar by ___ bonds. The nitrogenous bases are either derivatives of the 6-membered heterocyclic ring compound ___ or of purines, a compound composed of a 6-membered heterocyclic ring with a 5-membered ___ ring fused to it. The two common purines are ___ and ___. The 6-membered heterocyclic ring compounds include ___, ___, and ___. A ring compound attached to a sugar is termed a ___.

2. Answer True or False
   a. ATP is an example of a deoxynucleoside triphosphate ___.
   b. cAMP is a 3'-5' cyclic form of AMP ___.
   c. The α-phosphate of GTP is the phosphate closest to the sugar moiety ___.
   d. The only biological function of dCTP is as a building block in synthesis of DNA ___.
   e. The only biological function of CTP is as a building block in synthesis of RNA ___.
   f. The most common ribonucleoside triphosphates have phosphate attached to the 5' carbon of the sugar moiety ___.

3. Chargaff's rules provided an important clue to solve the structure of double-stranded DNA. What are Chargaff's rules?

4. Answer True of False
   a. An A/T base pair and a C/G base pair have about the same physical dimensions ___.
   b. If GGGGCCCC represents the sequence of bases in one strand of a double-stranded DNA then the complementary strand must have the sequence CCCCGGGG ___.
   c. mRNA is single-stranded ___.
   d. Heterogeneous nuclear RNA or hnRNA are RNA molecules made in the nucleus and processed into mRNA ___.
   e. rRNA and tRNA are devoid of unmodified nucleosides ___.

5. DNA and RNA react differently to acid and base conditions. Explain.

6. Of the following statements, which are true for type II restriction enzymes?
   a. They are usually exonucleases.
   b. Their recognition sequences are palindromic.
   c. Cleavage is by hydrolysis of both strands.
   d. Cleavage produces 3'-phosphates and 5' hydroxyl groups.
   e. A single restriction enzyme can produce blunt ends or protruding ends depending on the salt conditions.

Answers

1. DNA; RNA; nucleotides (or nucleoside monophosphates); ribose; deoxyribose; glycosidic; pyrimidine; imidazole; adenine; guanine; cytosine; uracil; thymine; nucleoside.

2. a.F; b.T; c.T; d.T; e.F; f.T.

3. [A]= [T]; [G] = [C]; [pyrimidines] = [purines]

4. a.T; b.F (sequences are always written 5' to 3'); c.T; d.T; e.F.

5. RNA is relatively resistant to dilute acid whereas DNA undergoes hydrolysis of glycosidic bonds to purines. DNA is not susceptible to alkaline hydrolysis whereas RNA is readily hydrolyzed to nucleotides in alkaline solution.

6. b and c.

Additional Problems

1. The nucleotides are an important class of biomolecules used as components of the nucleic acids, DNA and RNA. Describe the structure of the nucleoside monophosphates found in DNA and RNA. In your description be sure to describe the three chemical groups that make up a nucleotide and be certain to indicate any difference between deoxyribonucleotides and ribonucleotides.

2. Draw a base pair involving either G and C or A and T.

3. DNA methylation is known to most commonly occur at position 6 in A and position 5 in C. However, the formation of 5-methylcytosine can create so-called mutational hot-spots. Cytosine can undergo oxidative deamination at position 4. In the unmethylated form, this deamination can be corrected, because it is easily recognized. Deamination of 5-methylcytosine, however, can create problems. Explain.

4 In the following sequence of DNA, underline a 4-base palindrome, a 6-base palindrome and a purine-rich sequence.

GAGAAATATAGATCAGAGTTAACTC

5. In the following compounds, the order of solubility is adenine < deoxyadenosine < adenosine < adenosine triphosphate. Explain what each is (i.e. base, nucleoside or nucleotide), how they differ and why they show this relative solubility.

6. When a solution of double-stranded DNA is heated, a sharp transition is observed in the ultraviolet absorption properties of the solution. The solution begins to absorb greater amounts of light at high temperatures, corresponding to the process of denaturation. What physical changes in the double-stranded DNA molecules occur during this process of denaturation?

7. The restriction endonuclease *Bam*HI recognizes the sequence GGATCC and cleaves between the Gs. The enzyme *Bgl* II ("Bagel two") recognizes AGATCT and cuts between AG. What kind of overhangs are generated (i.e., 5' or 3')? Are the overhangs complementary to one another? If the cleavage products are joined together (i.e., a *Bam*HI fragment joined to a *Bgl* II fragment), is the hybrid molecule cleavable by either endonuclease?

Abbreviated Answers

1. The nucleoside monophosphates found in DNA and RNA are composed of a phosphate group, a sugar moiety, and a nitrogenous base. The sugar is a 5-carbon compound, deoxyribose for DNA and ribose for RNA. The phosphate group is attached to the 5'-carbon of the sugar. Attached to the 1'-carbon is a nitrogenous base. Bases are either purines, adenine or guanine, or pyrimidines, cytosine in DNA and RNA, or thymine in DNA only or uracil in RNA only. (Thymine is 5-methyluracil.)

2.

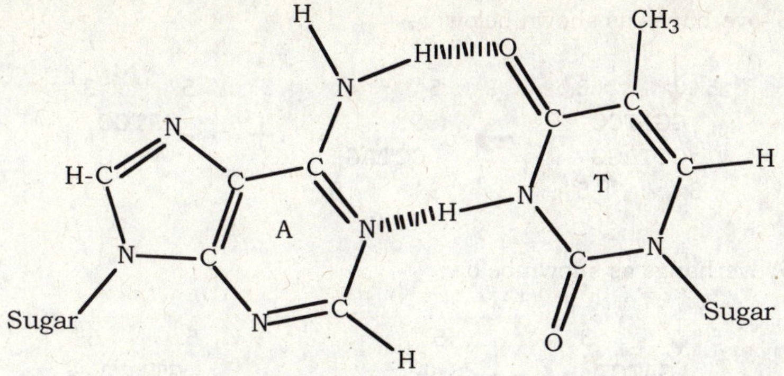

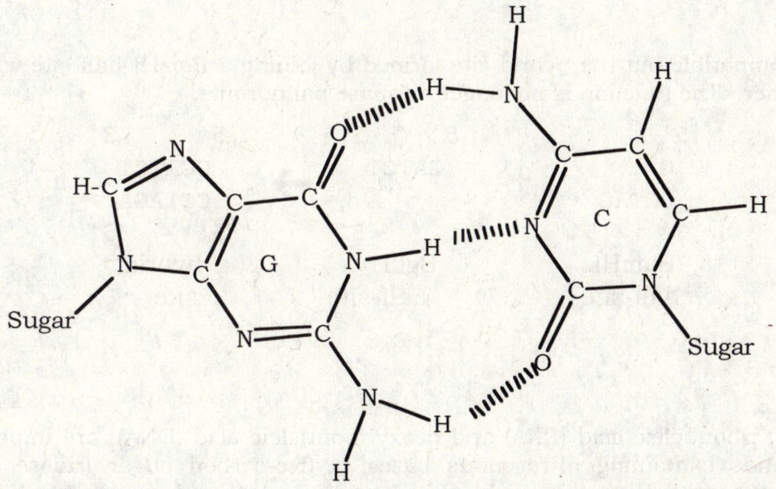

3. Oxidative deamination of cytosine produces uracil, a base not commonly found in DNA. Cells have a repair system that scans DNA for uracils and removes them. The presence of 5-methylcytosine causes problems for this repair process. Oxidative deamination of 5-methylcytosine produces 5-methyluracil also known as thymine. Thymine is a natural component of DNA.

4.

4-base
palindromes
‾‾‾‾ ‾‾‾

GAGAAATATAGATCAGA<u>GTTAAC</u>TC
----
purine rich   6-base
                                   palindrome

5. Adenine is a base. In general, the bases are poorly soluble in aqueous solution. Deoxyadenosine is a nucleoside composed of adenine and the 5-carbon sugar deoxyribose. Adenosine is a ribonucleoside containing the 5-carbon ribose sugar. Both deoxyadenosine and adenosine are more soluble than adenine by virtue of the hydroxyl groups on their sugar moieties. Adenosine triphosphate is ATP, a ribonucleoside triphosphate. The presence of the triphosphate group greatly increases solubility.

6. The forces holding double-stranded DNA together include hydrogen bonds and base stacking. Denaturation of double-stranded DNA results in strand-separation and unstacking of bases. Base stacking occurs by an interaction of π-electrons located above and below the base-planes. One of the consequences of base-stacking is a decrease in the molar extinction coefficient. Stacked bases have a lower extinction coefficient than unstacked bases. Thus, denaturation is accompanied by an increase in ultraviolet light absorption, a phenomenon known as the hyperchromic effect. The ultraviolet absorbance of denatured DNA is approximately 40% higher than that of native DNA.

7. *Bam*HI produces 5'-overhangs as shown below:

```
5' ↓    3'              5'                   5'       3'
   GGATCC        →      G         +          GATCC
   CCTAGG               CCTAG                     G
3'    ↑ 5'           3'      5'               3'
```

Bgl II also produces 5'-overhangs as shown below:

```
5' ↓    3'              5'                   5'       3'
   AGATCT        →      A         +          GTACT
   TCTAGA               TCTAG                    C
3'    ↑ 5'           3'      5'               3'
```

The overhangs are compatible but the hybrid site formed by joining a *Bam*HI half site with a *Bgl*II half site is not recognized by either. The junction is no longer a 6-base palindrome.

```
5'                  5'       3'              5'           3'
 G                   GATCT             →      GGATCT
 CCTAG        +      A                        CCTAGA
3'      5'              5'                  5'           3'
BamHI               BglII                    hybrid
half-site           half-site               site
```

Summary

The nucleic acids, ribonucleic acid (RNA) and deoxyribonucleic acid (DNA), are important biopolymers of nucleotides, compounds containing nitrogenous bases, a five-carbon sugar (ribose or deoxyribose) and phosphate. This chapter describes the basic biochemistry of nucleic acids and nucleotides. The nitrogenous bases come in two types, pyrimidines and purines. Pyrimidines are 6-membered, heterocyclic, aromatic ring structures containing two nitrogens. There are three principal pyrimidines: cytosine, uracil and 5-methyluracil or thymine. Cytosine is found in both DNA and RNA whereas uracil is in RNA and thymine in DNA. The general structure of a purine is a 5-membered imidazole ring fused to a pyrimidine ring. In DNA and RNA, there are two common purines, namely adenine and guanine. Both purines and pyrimidines contain numerous groups that can participate in hydrogen bonds as donors or acceptors or both. In fact, complementary groups exist on adenine and uracil (or thymine) and on guanine and cytosine such that they can form hydrogen-bonded pairs. This base-pairing is the foundation for the structure of double-stranded DNA.

Because the bases have extensive, conjugated double bonds, they absorb light strongly in the UV range. The conjugated bond system allows delocalization of π-electrons, forming electron clouds above and below the base plane. Delocalized π-electrons can interact by base stacking, another force stabilizing double-stranded DNA. Base stacking also affects the efficiency of electronic transitions of π-electrons, such that stacked bases absorb less UV light than unstacked bases. The transition from an ordered nucleic acid structure stabilized by hydrogen bonds and base stacking to an unordered structure is accompanied by an increase in UV absorbance. This transition is known as denaturation and it results in a hyperchromic shift in UV absorption.

Purines and pyrimidines have low water solubility which improves when they are attached to either ribose or deoxyribose sugars through N-glycosidic bonds. The resulting compounds are known as nucleosides. The common nucleosides are cytidine, uridine, thymidine, adenosine and guanosine. Phosphorylated derivatives of nucleosides are known as nucleotides, with phosphoric acid esterified normally to the 5' carbon of the sugar. DNA and RNA are polymers of deoxyribonucleoside monophosphates and ribonucleoside monophosphates, respectively. But, nucleoside monophosphates are not used directly to biosynthesize DNA and RNA. Rather, triphosphate derivatives serve this purpose. The nucleotides used to biosynthesize DNA are: deoxyadenosine 5'-triphosphate (dATP); deoxyguanosine 5'-triphosphate (dGTP); deoxycytosine 5'-triphosphate (dCTP); and, deoxy thymidine 5'-triphosphate (TTP or dTTP). The corresponding nucleoside triphosphates for RNA are: adenosine 5'-triphosphate (ATP); guanosine 5'-triphosphate (GTP); cytosine 5'-triphosphate (CTP); and uridine 5'-triphosphate (UTP).

The deoxyribonucleoside triphosphates are used exclusively as building blocks for DNA. However, the ribonucleoside triphosphates, in addition to serving as the building blocks for RNA, have additional uses. ATP is the major energy currency of the cell, in addition to being a component of coenzymes such as NAD^+ and FAD. GTP is used during protein synthesis and in cell signaling; CTP is involved in phospholipid synthesis; and, UTP is involved in various aspects of carbohydrate metabolism.

Nucleic acids are linear polymers of nucleoside monophosphates linked by phosphodiester bonds between the 3'-hydroxyl of one nucleotide and the 5' phosphate of another. Because of this, nucleic acids are vectorial molecules, with two distinct ends, the 5'-end and the 3'-end. By convention, the sequence of nucleotides in a nucleic acid is represented by the one-letter abbreviations of the bases starting from the 5'-end and continuing toward the 3'-end.

The only biological role of DNA is as genetic material, and the vast majority of organisms use double-stranded DNA for this purpose. In double-stranded DNA (dsDNA), two strands of deoxyribonucleoside monophosphate polymers are joined together in an antiparallel fashion by hydrogen bonds. The hydrogen bonds occur between pairs of complementary bases: A and T; and G and C. Compositional analysis of dsDNA reveals: [A] = [T]; [C] = [G]; and, [purine] = [pyrimidine]. These relationships are known as Chargaff's rules. Depending on the organism, genomic DNA may be a single DNA molecule (or chromosome) or may be divided into several discrete DNA molecules (chromosomes).

RNA serves a number of biological roles including informational, catalytic and structural. As an informational molecule, messenger RNA functions to bring genetic information, encoded in a sequence of bases in DNA, to the ribosome, the site of cellular protein synthesis. This information is used to direct the sequence of amino acids to be joined to form a protein. mRNA production begins with transcription, in which an RNA copy of a sequence of bases along one strand of DNA is made. In eukaryotic organisms, RNA transcription is localized to the nucleus. The primary transcripts, the initial products of transcription also known as heterogeneous nuclear RNA (hnRNA), are processed into mRNA in a series of reactions. Processing includes: (1) Addition of a G residue to the 5'-end of the primary transcript and methylation of this residue and nearby ribose sugars and bases, in a process known as capping; (2) Cleavage of the primary transcripts to produce a shortened 3'-end at which adenylic acid residues are added to form polyA tails; and, (3) Removal of various internal sequences (introns or intervening sequences) by cleavage and subsequent ligation (of exons).

The process of protein synthesis reveals structural and catalytic aspects of RNAs. Ribosomes are ribonucleoprotein complexes of ribosomal proteins and unique RNA molecules known as ribosomal RNA (rRNA). All ribosomes are composed of two parts or subunits, the small subunit and the large subunit. The structure and function of subunits during protein synthesis is determined in large part by rRNAs. The small subunit contains a single rRNA (approximately 1500 bases in prokaryotic rRNA and 1800 bases in eukaryotic rRNA); the large subunit has at least two rRNAs: a 120 bases rRNA known as the 5S rRNA; and a larger rRNA (approximately 3000 bases in prokaryotes and 5000 bases in eukaryotes). (Eukaryotes often have a so-called 5.8S rRNA which corresponds in sequence to the 5'-end of the large subunit rRNA of prokaryotes.)

The amino acids used by the ribosome during protein synthesis are attached to the 3'-end of small RNA molecules known as transfer RNAs (tRNA). Transfer RNAs are used to decode a sequence of three adjacent nucleotides (a codon) in terms of a unique amino acid. In this process, tRNAs form hydrogen bonds between codons on mRNA and a three-base sequence, the anticodon, on tRNA. In addition to tRNAs, cells contain other relatively small, stable RNA molecules. In particular, a class of RNA molecules known as small nuclear RNAs or snRNAs are responsible for mRNA processing in the nucleus of eukaryotic cells.

Apart from the number of strands, there are only minor differences between DNA and RNA: deoxyribose versus ribose sugars, and, thymine versus uracil. The absence of a 2'-OH group in deoxyribose makes DNA stable against base-catalyzed hydrolysis. In fact, DNA is a very stable molecule ideally suited as genetic material. The choice between thymine and uracil arises as a result of the tendency of cytosine to deaminate to uracil. Since thymine is found in DNA, any uracil in DNA must result from deamination of cytosine. Cells have enzymes to remove uracil in DNA, replacing it with thymine and thus preventing mutations.

There is a large number of enzymes, termed nucleases, that catalyze cleavage of phosphodiesterase bonds by hydrolysis. These include DNA-specific enzymes (DNases), RNA-specific enzymes (RNases) or nonspecific nucleases that attack either internal phosphodiester bonds (endonucleases) or phosphodiester bonds on the end of a polymer (5'-exonucleases or 3'-exonucleases). Further, the phosphodiester bond may be attacked on the *a* side producing a 5'-phosphate or on the *b* side producing a 3'-phosphate. Restriction endonucleases are DNA-specific endonucleases. There are three types (I, II, and III) of restriction endonucleases. Type I cleaves DNA without regard to sequence; type II recognizes specific sequences, often palindromic, and cleaves within the sequence; type III recognizes a specific DNA sequence and cleaves nearby. Type II restriction endonucleases are important tools in genetic engineering and molecular biology.

Chapter 11

Structure of Nucleic Acids

•

Chapter Outline

❖ DNA sequencing
  ⅄ Chain termination (dideoxynucleotide): Sanger
    • Enzymatic synthesis of DNA using DNA polymerase
    • Primer extended using dNTPs but terminated using specific ddNTPs (lack 3' hydroxyl)
    • ddNTPs labeled with fluorescent dyes of different color
    • Analyze products by electrophoresis
❖ Double stranded DNA: Strands antiparallel: Base pairs AT and GC
  ⅄ B-form DNA: Right-handed helix
    • Parameters: 0.34 nm/base pair, 10 bp/turn, P = 3.4 nm
    • H-bonded base pairs AT and GC inside
    • Base pairs perpendicular and stacked
    • Negative charged phosphate and sugar outside
    • Major and minor groove
  ⅄ A-form DNA: Right-handed helix
    • Parameter: 0.224 nm/base pair, 11 bp/turn, P = 2.46 nm
    • Base pairs tilted
    • dsRNA and DNA/RNA hybrids
  ⅄ Z-form DNA: Left-handed helix
    • Dinucleotide repeat: Pyrimidine-purine
    • Conformation: Purine -syn, pyrimidine -anti
    • Zigzag pattern of sugar-phosphate backbone
    • 5-methylC and supercoils stabilize
❖ Intercalating agents: Planar, hydrophobic molecules that insert between stacked base pairs: Ethidium bromide, acridine orange, actinomycin D
❖ Alternative hydrogen-bonded structures
  ⅄ Cruciforms: Inverted repeats form stems and loops
  ⅄ Hoogsteen pairs:
    • Alternatives to AT and GC pairing that does not interfere with canonical pairs
    • H-DNA: Triple helix formed by Hoogsteen base pairing
    • Quadruplex DNA: Four-stranded DNA in G-rich dsDNA
❖ DNA denaturation: Strand separation and loss of base stacking
  ⅄ Temperature, pH, ionic strength: Disrupt H bonds
  ⅄ Hyperchromic shift of UV absorbance upon denaturation
  ⅄ T_m: Midpoint of transition, $T_m = 69.3 + 0.42(\%G+C)$ in 0.2 M Na^+
❖ Renaturation
  ⅄ Reannealing of single stranded nucleic acids
  ⅄ For genomic DNA: Rate reflects complexity of DNA
  ⅄ Nucleic acid hybridization
    • Measure evolutionary relationships
    • Identify specific gene
❖ DNA molecules

- ⋏ Linear: Chromosomes of eukaryotes
- ⋏ Circular: Bacterial chromosomes, plasmids, some viruses
- ❖ Supercoils: Release of tortional stress due to underwinding or overwinding of DNA
  - ⋏ Negative supercoils: Underwound DNA
  - ⋏ Positive supercoils: Overwound DNA
  - ⋏ L = T + W
    - • L = linking number
    - • T = twist: Number of helical turns
    - • W = writhe: Number of supercoils (W = 0 for relaxed DNA)
  - ⋏ Superhelical density, $\sigma = (L-Lo)/Lo$
  - ⋏ Natural DNA negatively supercoiled as a result of topoisomerase activity
    - • Type I topoisomerases: Cut one strand
    - • Type II topoisomerases: Cut two strands
- ❖ Eukaryote chromosomes
  - ⋏ Nucleoprotein complexes called chromatin
  - ⋏ Nucleosomes: Histone H2A, H2B, H3, H4 octamers with 146 bp of dsB-DNA wrapped in left-handed coil making about 1.65 turns
  - ⋏ Histones: Basic: Arginine and lysine rich proteins
  - ⋏ Histone H1: Link consecutive nucleosomes together
  - ⋏ Solenoid: 30 nm Filaments: 6 Nucleosomes/turn
  - ⋏ Nonhistone proteins
    - • SMC proteins: ATPases involved in chromosome dynamics
      - • Multiple domains: DNA-binding, ATP-binding, Hinge, dimerization domain
- ❖ Chemical synthesis of DNA
  - ⋏ Solid phase synthesis using "gene machine"
  - ⋏ Phosphoramidite chemistry
  - ⋏ 3' to 5' Direction
- ❖ RNA structure: Typically single-stranded molecules: Intrastrand H bonding produces secondary structures
  - ⋏ Stems: A-form dsRNA ± internal bulges
  - ⋏ Loops
    - • U turns
    - • Tetraloops
  - ⋏ Junctions
    - • Region where several stem-loops meet
  - ⋏ Others
    - • Coaxial stacking of two stems
    - • Pseudoknots
      - • Pairing of bases in loop of stem and loop
      - • Pseudo because pairing is less than a full turn of the helix
    - • Ribose zippers
- ❖ RNA classes
  - ⋏ tRNA: Small (73 to 94 nucleotides) RNA charged with amino acids
    - • Aminoacylated tRNA: Substrate for protein synthesis and decoding function
    - • Cloverleaf secondary structure
      - • Acceptor stem: CCA 3': End at which amino acid is attached
      - • D-loop: Modified base
      - • Anticodon stem and loop: Three-base anticodon: Decodes mRNA
      - • TψG loop: Modified base
  - ⋏ rRNA: Component of ribosome: Protein synthetic machinery
    - • Small subunit: Single rRNA: 16S (1500 nt) or 18S (1800 nt)
    - • Large subunit: Two or three rRNAs
      - • Large subunit rRNA: 23S (2900 nt) or 28S (5000 nt)
      - • 5S rRNA: 120 nt
      - • 5.8S rRNA: 160 nt (eukaryotic cytoplasmic ribosomes)

200

Chapter Objectives

Sequencing Nucleic Acids

It is important to know how DNA is sequenced. F. Sanger developed the chain termination or dideoxy method. DNA is produced enzymatically using a template and a primer that defines the 5'-end of the DNA that is produced. As the primer is being extended, base specific chain-termination events produce a nested set of extension products. The sequence of the synthesized strand is determined by separating the products according to size using electrophoresis. The sequence of the synthesized strand is read from the gel or from an autoradiograph of the gel starting from short products and moving toward long products, reading 5' to 3'.

DNA Secondary Structure

Understand the structure of B-form double-stranded DNA. The two strands are antiparallel and are held together by complementary base-pairs. The strands wrap around each other forming a right-handed helix with 10 base-pairs per helix turn, 3.4 nm per turn, and 0.34 nm between base pairs along the helix axis. The helix axis runs perpendicular to the planar base-pairs. Because the helix axis is centered within the base-pairs, the helix has a major and a minor groove. Compared to B-form DNA, A-form DNA has 11 base-pairs per helical turn and as a consequence the helix is broader, the base-pairs are tilted with respect to the helix axis, and they are closer together. Double-stranded RNA and DNA-RNA hybrids assume the A-form conformation. Z-DNA is a left-handed structure requiring regions of alternating pyrimidine-purine sequence to form.

Solution Properties of DNA

Many techniques in genetic engineering and molecular biology exploit the properties of double-stranded DNA. For example, DNA solutions can be freed of protein by extraction with organic solvents such as liquid phenol or chloroform because DNA is a highly charged, polar molecule. It is not soluble and will therefore not partition into these immiscible organic solvents. To precipitate DNA, an alcohol like ethanol or propanol is added to an aqueous DNA solution. However, a counter ion (Na^+ or NH_4^+) must be present in order for the negatively-charged DNA to form a salt and precipitate. Many techniques require single-stranded DNA. DNA can be denatured by high temperatures or by treatment with urea or formamide to disrupt hydrogen bonds. Additionally, pH values in excess of 11.5 will disrupt hydrogen bonds.

Intracellular DNA

In general, cellular DNAs are extremely large molecules and, in order to fit inside a cell, DNA must be compacted. Prokaryotic organisms accomplish this, in part, by having circular DNAs that are supercoiled. Eukaryotic organisms take a different approach: DNA is associated with specific proteins, termed histones, in complexes known as nucleosomes. Nucleosomes are octamers of two each of histones H2A, H2B, H3, and H4 around which double-stranded DNA is wrapped to give the appearance of "beads-on-a-string". The nucleosomes are further organized into solenoids having six nucleosomes per turn to produce 30-nm filaments. The filaments are further organized to form miniband units of the chromosome. The structures actually seen by light microscopy during mitosis or meiosis are even more highly condensed.

RNA

RNAs are typically single-stranded molecules. However, they may have complementary regions that interact by base-pairing to form short stretches of A-form helix. An example of this is the cloverleaf secondary structure model of tRNAs. Know the general structure of a tRNA and be able to identify important regions such as the D-loop, anticodon, variable loop, TψC loop, and acceptor stem. Like proteins, RNAs have tertiary structure, and in tRNA, the interactions responsible for tertiary structure are made between residues far removed from each other in the sequence. In phylogenetic sequence comparisons, these residues have been identified as highly-conserved positions and are virtually identical in all tRNAs. Phylogenetic comparisons have been made with small subunit-ribosomal RNAs and with large-subunit RNAs. These large, stable RNA molecules play a key role in protein synthesis, a fundamental cellular reaction.

Problems and Solutions

1. Predicting a Sanger sequencing pattern
The oligonucleotide d-AGATGCCTGACT was subjected to sequencing by Sanger's dideoxy method, using fluorescent-tagged dideoxynucleotides and capillary electrophoresis, essentially as shown in Figure 11.3. Draw a diagram of the gel-banding pattern within the capillary.

Answer: In order to sequence the oligonucleotide d-AGATGCCTGACT by the Sanger dideoxy method, a reaction is set up such that the oligonucleotide serves as a template for primer extension using a DNA polymerase. As the primer is extended sequence-specific terminations occur, producing a set of extended primer products that differ in length on their 3' ends. The sequence actually read in Sanger sequencing is the sequence of the extended primer strand (which is complementary to the template strand).

The sequence-specific stops are generated by chain termination during polymerization. The polymerase is supplied with deoxynucleotide triphosphates -dATP, dGTP, dCTP and dTTP- plus small amounts of fluorescently labeled dideoxynucleotides -ddATP, ddGTP, ddCTP and ddTTP. When the polymerase begins a round of elongation it fills its nucleotide elongation site with a nucleotide complementary to the template. The solution, however, is a mixture of regular deoxynucleotide and dideoxynucleotide and so either a normal deoxynucleotide or a dideoxynucleotide is selected. If a normal deoxynucleotide is used the polymerase simply extends the nascent single-stranded DNA molecule by one nucleoside monophosphate. If a dideoxynucleotide is used the nascent chain has this nucleotide attached to its 3' end. Since elongation depends on a 3' hydroxyl group the chain terminates with addition of the dideoxynucleoside. Additionally, the chain is now modified with a fluorescent dye. When the mixture is subject to capillary electrophoresis fragments will separate by size with smaller fragments migrating faster. The electrophoresis unit is equipped with a laser fluorescent spectroscope that scans at a fixed position in the capillary and as DNA molecules migrate past the laser they are excited and fluoresce. The wavelength and magnitude of the fluorescence is recorded to identify the dideoxynucleotide. Figure 11.3 shows a snapshot of capillary electrophoresis with bands at different locations in the capillary system, each showing a color dependent on the dideoxynucleotide on its 3' end.

To get an idea of what the gel might look like let's assume a 20-base sequencing primer was used. The following set of products would be produced with their 3' ends labeled with different fluorescent dyes attached to each of the dideoxynucleotides. In the table below the column headings indicate which dideoxy nucleotide was found and the numbers represent the sizes of the fragments in bases. For example, 21 in column A indicates that an extension fragment 21 bases in size was detected as being labeled with dideoxy A. This product is a consequence of the 3' end of the 20 base primer being extended by an A (which happened to be a dideoxy A). The A is complementary to the T located on the 3' end of the oligonucleotide.

G	A	T	C
22	21	23	24
26	25	30	28
27	29	32	31
36	36	36	36

(Included in each column were limiting values (i.e., 36) to simply show the top of the gel.) Reading from bottom to top gives the sequence of the extended primer 5' to 3'. The sequence is:

5' AGTCAGGCATCT 3'

This corresponds to a template sequence:

5' AGATGCCTGACT 3'

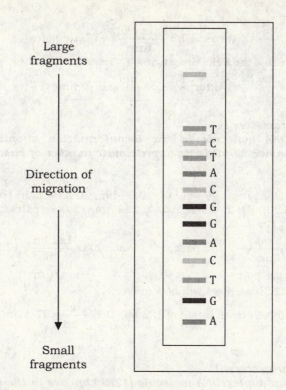

Large
fragments

Direction of
migration

Small
fragments

2. Deducing DNA sequence from Sanger sequencing results

The output of an automated DNA sequence determination by the Sanger dideoxy chain termination method performed as illustrated in Figure 11.3 is displayed at right*. What is the sequence of the original oligonucleotide?

* You will have to refer to the figure in the textbook.

Answer: The color code used in Figure 11.3 is orange for A, green for G, blue for C and red for T. From left to right the peaks represent DNA fragments of increasing size capped on their 3' ends by fluorescently labeled dideoxynucleotides. The data correspond to the following sequence:

5'-AAGGTTGACTTTGCGCTATC-3'

The template sequence is the reverse complement of this sequence.

5'-GATAGCGCAAAGTCAACCTT-3'

3. Helical parameters for a double-stranded DNA

X-ray diffraction studies indicate the existence of a novel double-stranded DNA helical conformation in which ΔZ (the rise per base pair) = 0.32 nm and P (the pitch) = 3.36 nm. What are the other parameters of this novel helix: (a) the number of base pairs per turn, (b) Δ ø (the mean rotation per base pair), and (c) c (the true repeat)?

Answer: For dsDNA with ΔZ = 0.32 nm/base pair, P = 3.36 nm/turn

(a) The number of base pairs per turn is

$$\frac{3.36 \frac{\text{nm}}{\text{turn}}}{0.32 \frac{\text{nm}}{\text{base pair}}} = 10.5 \text{ per turn}$$

(b) The mean rotation per base is

$$\frac{360 \frac{\text{deg}}{\text{turn}}}{10.5 \frac{\text{base pairs}}{\text{turn}}} = 34.3° \text{ per base pair}$$

(c) To determine the true repeat, first find what values of X and Y satisfy the following equation:

$$X \text{ turns} \times 10.5 \frac{\text{base pairs}}{\text{turn}} = Y \text{ base pairs}$$

After 21 base pairs, and two turns, the helix repeats itself. The true repeat is

$$2 \text{ turns} \times 3.36 \frac{\text{nm}}{\text{turn}} = 6.72 \text{ nm}$$

4. B- and Z-DNA helical parameters I
A 41.5 nm-long duplex DNA molecule in the B-conformation adopts the A-conformation upon dehydration How long is it now? What is its approximate number of base pairs?

Answer:

B-DNA, P = 3.40 nm, ΔZ = 0.34 nm, 10 base pairs/turn
A-DNA, P = 2.46 nm, ΔZ = 0.224 nm, 11 base pairs/turn

For B-DNA, 41.5 nm has

$$\frac{41.5 \text{ nm}}{0.34 \dfrac{\text{nm}}{\text{base pair}}} = 122 \text{ base pairs that make } \frac{122 \text{ bp}}{10 \dfrac{\text{bp}}{\text{turn}}} = 12.2 \text{ turns}$$

If converted to A-DNA, these 122 base pairs will now make:

$$\frac{122 \text{ bp}}{11 \dfrac{\text{bp}}{\text{turn}}} = 11.1 \text{ turns with an overall length of } 122 \text{ bp} \times 0.224 \frac{\text{nm}}{\text{bp}} = 27.3 \text{ nm} \ (= 11.1 \text{ turns} \times 2.46 \frac{\text{nm}}{\text{turn}})$$

5. B- and Z-DNA helical parameters II
If 80% of the base pairs in a duplex DNA molecule (12.5 kbp) are in the B-conformation and 20% are in the Z-conformation, what is the long dimension of the molecule?

Answer: Z-DNA contains 12 base pairs per turn, 0.36 to 0.38 nm rise per base pair and a pitch of 4.86 nm. If 20% of a 12.5 kbp DNA molecule is in Z-conformation and the rest is B-conformation this represents 2.5 kbp in Z-DNA and 10.0 kbp in B-DNA. The overall length is:

$$2{,}500 \text{ bp} \times 0.37 \frac{\text{nm}}{\text{bp}} + 10{,}000 \text{ bp} \times 0.34 \frac{\text{nm}}{\text{bp}} = 4{,}325 \text{ nm} = 4.33 \ \mu\text{m}$$

6. DNA supercoiling parameters I
A "relaxed," circular, double-stranded DNA molecule (1600 bp) is in a solution where conditions favor 10 bp per turn. What is the value of L_0 for this DNA molecule? Suppose DNA gyrase introduces 12 negative supercoils into this molecule. What are the values of L, W, and T now? What is the superhelical density, σ?

Answer: A 1600 bp DNA makes 160 turns of B-DNA. The linking number is the number of times a strand intertwines. For a relaxed circle, this is equal to the number of helical turns. L_0 = 160.
In general,

$$L = T + W$$

By introducing 12 negative supercoils, W = -12. If the DNA remains as B-DNA, T = 160. So,

$$L = 160 - 12 = 148$$

The superhelical density, σ, is:

$$\sigma = \frac{\Delta L}{\Delta L_0} = \frac{-12}{160} = -0.075$$

7. DNA supercoiling parameters II
Suppose one double-helical turn of a superhelical DNA molecule changes conformation from B-form to Z-form. What are the changes in L, W, and T? Why do you suppose the transition of DNA from B-form to Z-form is favored by negative supercoiling?

Answer: Z-DNA has 12 bp/turn and B-DNA has 10 bp/turn. For 1 turn of B-form helix:

$$L_B = 1.0 + W_B$$

Now, this turn includes 10 bp which in Z-DNA makes only $(10/12)^{\text{th}}$ of a turn; thus:

$$L_z = \frac{10}{12} + W_z = 0.83 + W_z$$

For a transition from B to Z, strands are not broken, so:

$$L_B = L_z, \text{ or}$$
$$1.0 + W_B = 0.83 + W_z, \text{ or}$$
$$W_z - W_B = 0.167$$

That is, in going from B- to Z-DNA, there is a positive change in superhelicity. Given some number of negative supercoils in B-DNA, this number will be reduced in Z-DNA. Thus, negative supercoils favor Z-DNA.

8. Calculate the number of nucleosomes in a human diploid cell
There is one nucleosome for every 200 bp of eukaryotic DNA. How many nucleosomes are in a diploid human cell? Nucleosomes can be approximated as disks 11 nm in diameter and 6 nm long. If all the DNA molecules in a diploid human cell are in the B-conformation, what is the sum of their lengths? If this DNA is now arrayed on nucleosomes in the "beads-on-a-string" motif, what is its approximate total length?

Answer: Human diploid cells consist of 46 chromosomes and 6×10^9 bp. The number of nucleosomes is given by:

$$\frac{6 \times 10^9 \text{bp}}{200 \dfrac{\text{bp}}{\text{nucleosome}}} = 3 \times 10^7 \text{bp nucleosomes}$$

The length of B-DNA, 6×10^9 bp long, is given by

$$6 \times 10^9 \text{bp} \times 0.34 \frac{\text{nm}}{\text{bp}} = 2.04 \times 10^9 \text{nm} = 2.04 \text{ meters in length!}$$

This DNA as 3×10^7 bp nucleosomes has a length given by:

$$3 \times 10^7 \text{bp nucleosomes} \times 6 \frac{\text{nm}}{\text{nucleosome}} = 18 \times 10^7 \text{nm} = 0.18 \text{ m} = 180 \text{ mm}$$

9. The linear arrangement of complementary sequences along the tRNA strand
The characteristic secondary structures of tRNA and rRNA molecules are achieved through intrastrand hydrogen bonding. Even for the small tRNAs, remote regions of the primary sequence interact via H bonding when the molecule adopts the cloverleaf pattern. Using Figure 11.36 as a guide, draw the primary structure of tRNA, indicating the positions of various self-complementary regions along its length.

Answer:

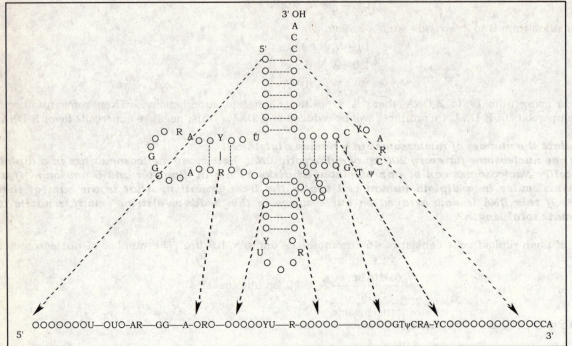

10. Using Chargaff's results to order DNAs according to their T_ms
Using the data in Table 10.1, arrange the DNAs from the following sources in order of increasing T_m:
human, salmon, wheat, yeast, E. coli.

Answer:

The T_m of DNA is proportional to its %(G+C) content which is calculated as:

$$\%(G+C)=\frac{G+C}{G+C+A+T}\times100\% \text{ but since G=C and A=T}$$

$$\%(G+C)=\frac{G}{G+A}\times100\% \text{ or}=\frac{C}{C+T}\times100\%$$

From Table 10.1 we are given various molar ratios.

For dsDNA the Adenine to Guanine ratio should equal the Thymine to Cytosine ratio.

By inspecting Table 10.1 we see that this is not always the case.

Therefore, we will calculate %(G+C) using $\frac{A}{G}$ and $\frac{T}{C}$ ratios separately.

Let $\frac{A}{G}$ =x and $\frac{T}{C}$ =y

$$\%(G+C)=\frac{1}{1+x}\times100\% \text{ or}=\frac{1}{1+y}\times100\%$$

Substituting values of x (first column) or y (second column) we find:

Organism	x	y	%(G+C)
Human	1.56	1.75	39-36
Salmon	1.43	1.43	41
Wheat	1.22	1.18	45-46
Yeast	1.67	1.92	37-34
E. coli	1.05	0.95	49-51

Order of T_m: Yeast < Human < Salmon < Wheat < *E. coli*

11. Calculating T_ms and separating DNA molecules that differ in G:C content
At 0.2 M Na+ the melting temperature of double-stranded DNA is given by the formula $T_m = 69.3 = 0.41(\%G + C)$. The DNAs from mice and rats have (G + C) contents of 44% and 40%, respectively. Calculate the T_ms for these DNAs in 0.2 M NaCl. If samples of these DNAs were inadvertently mixed, how might they be repurified from one another?

Answer: The T_m in 0.2 M Na is given by:

$$T_m = 69.3 + 0.41 \times \%(G+C)$$

$$T_{m\,(mouse)} = 69.3 + 0.41 \times 44 = 87.3°C$$

$$T_{m\,(rat)} = 69.3 + 0.41 \times 40 = 85.7°C$$

The density of DNA is dependent on base composition and so isopycnic centrifugation in CsCl could be attempted. (However, it might prove far easier to just start out with new rats and mice.)

12. Calculating the density of DNAs that differ in G:C content
The buoyant density of DNA is proportional to its (G + C) content. (G:C base pairs have more atoms per volume than A:T base pairs.) Calculate the density (p) of avian tubercle bacillus DNA from the data presented in Table 10.1 and the equation p = 1.660 + 0.098 (GC), where GC is the mole fraction of (G + C) in DNA.

Answer: Avian tubercle bacillus has the following composition:

$$\frac{A}{G} = \frac{T}{C} = 0.4$$

$$(GC) = \frac{G+C}{G+A+C+T} \text{ and since } G = C \text{ and } A = T \text{ we can write}$$

$$(GC) = \frac{G+G}{G+A+G+A} = \frac{2G}{2G+2A} \text{ and from } \frac{A}{G} = 0.4 \text{ we can substitute } A = 0.4 \times G$$

$$(GC) = \frac{G}{G+0.4 \times G} = 0.714$$

Substituting this value in $\rho = 1.660 + 0.098 \times (GC)$ we find

$$\rho = 1.660 + 0.098 \times 0.714 = 1.730 \text{ g/mL}$$

13. Draw a Ψ:G base pair
(Integrates with Chapter 10.) Pseudouridine (Ψ) is an invariant base in the TΨC loop of tRNA; Ψ is also found in strategic places in rRNA. (Figure 10.23 shows the structure of pseudouridine.) Draw the structure of base pair that Ψ might form with G.

Answer: Uridine and pseudouridine are drawn below. In pseudouridine, because the base is attached to a carbon (and not a nitrogen as in uridine), pseudouridine has an additional hydrogen-bond donor. To form a pair with guanine we must match up H-bond donors and acceptors.

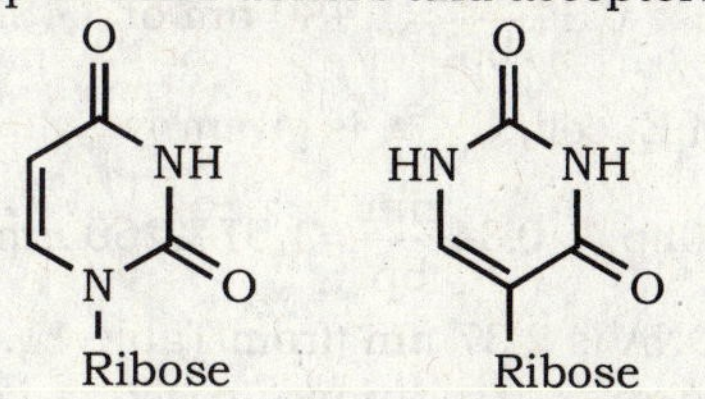

Uridine, U   Pseudouridine, Ψ

In the figure shown on the next page a canonical G:C base pair is drawn along with a G:U pair and two possible pairings of G and Ψ. G pairs with U but the U has to be positioned in a slightly different location relative to the C of a GC pair. The G:U pair is an example of a wobble interaction. One of the two G:Ψ pairs below looks like a G:U pair whereas the other has the ribose in a very different location. The base U has two hydrogen bond acceptors (keto oxygens) and a single hydrogen bond donor. Pseudouridine, however, has two donors and two acceptors.

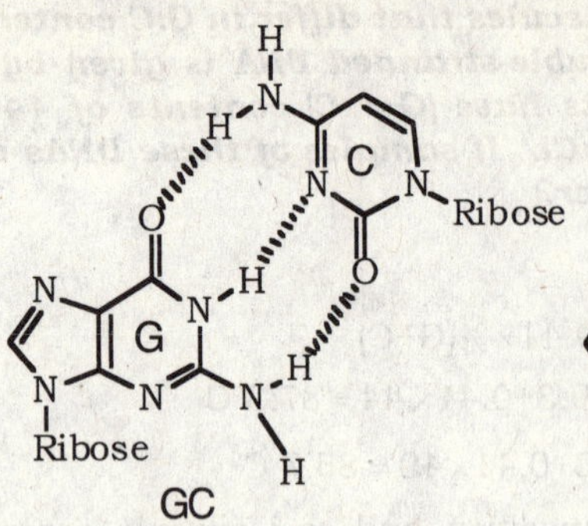

GC

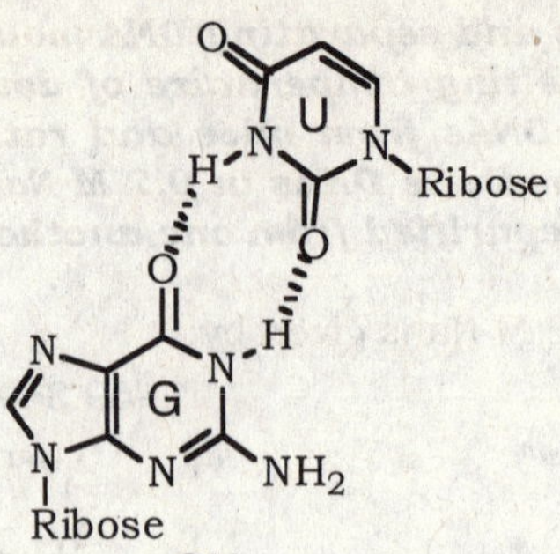

GU

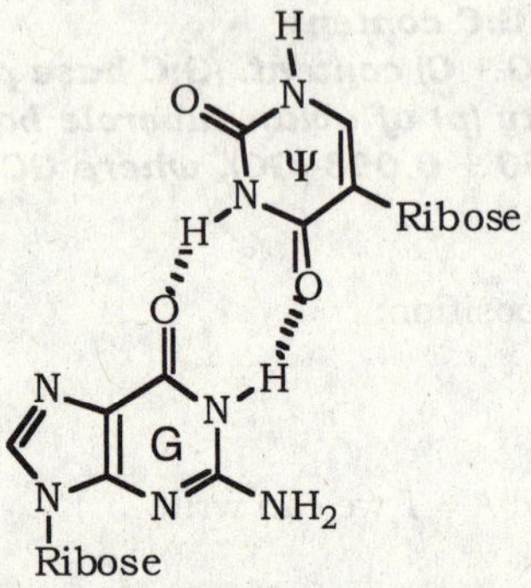

GΨ

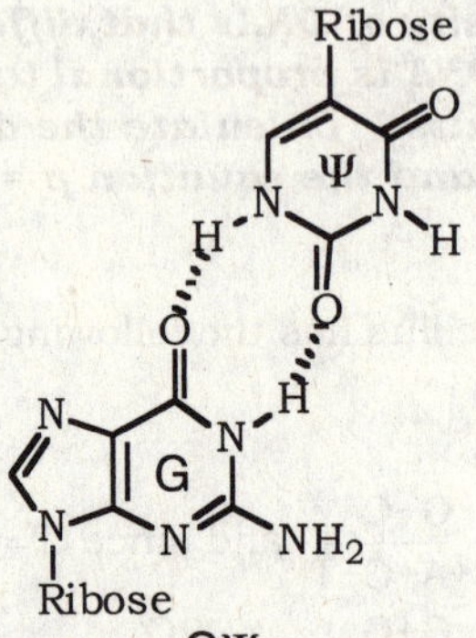

GΨ

14. The dimensions of closed circular DNA molecules

The plasmid pBR322 is a closed circular dsDNA containing 4363 base pairs. What is the length in nm of this DNA (that is, what is its circumference if it were laid out as a perfect circle)? The E. coli K12 chromosome is a closed circular dsDNA of about 4,639,000 base pairs. What would be the circumference of a perfect circle formed from this chromosome? What is the diameter of a dsDNA molecule? Calculate the ratio of the length of the circular plasmid pBR322 to the diameter of the DNA of which it's made. Do the same for the E. coli chromosome.

Answer: From Section 11.2 and Figure 11.7 we know that double-stranded B-form DNA has a Δz (distance along the helical axis between base pairs) of 0.34 nm. (Table 11.1 gives this parameter as 0.332 ± 0.019 nm/bp.)

Circumference of pBR322, C_{pBR322}, is given by

$$C_{pBR322} = 4363bp \times 0.34\frac{nm}{bp} = 1,480 \text{ nm or } 1.48\mu m$$

Circumference of E. coli, $C_{E.coli}$, is given by

$$C_{E.coli} = 4,639,000bp \times 0.34\frac{nm}{bp} = 1,577,260 \text{ nm or } 1.58 \text{ mm}$$

The diameter of double-stranded B-form DNA is 2.37 nm (from Table 11.1).

Length:Width Ratio of pBR322, R_{pBR322}, is given by

$$R_{pBR322} = \frac{1,480 \text{ nm}}{2.37 \text{ nm}} = 624$$

Length:Width Ratio of E. coli, $R_{E.coli}$, is given by

$$R_{E.coli} = \frac{1,577,260 \text{ nm}}{2.37 \text{ nm}} = 665,511$$

15. Identifying DNA structural and functional elements from nucleotide sequence information
Listed below are five DNA sequences. Which one contains a type-II restriction endonuclease ("six-cutter") hexanucleotide site? Which one is likely to form a cruciform structure? Which one is likely to be found in Z-DNA? Which one represents the 5'-end of a tRNA gene? Which one is most likely to be found in a triplex DNA structure?

a. CGCGCGCCGCGCACGCGCTCGCGCGCCGC
b. GAACGTCGTATTCCCGTACGACGTTC
c. CAGGTCTCTCTCTCTCTCTCTC
d. TGGTGCGAATTCTGTGGAT
e. ATCGGAATTCATCG

Answer: The six-cutter is sequence "e". Type II restriction endonucleases typically digest sequences that are palindromic, meaning the sequence of the top strand is the same as the complementary strand when read 5' to 3'. While some type II enzymes recognize interrupted palindromes most have uninterrupted palindromic sequences as targets. An easy way to search by eye for a uninterrupted palindromic sequence is to look for two adjacent bases that are complementary like AT or TA or GC or GC. These are in fact two base uninterrupted palindromes. To see if the two base palindrome happens to be part of a larger palindrome we need to look at the flanking bases to see if they are complementary. If they are then the sequence is a four base palindrome. There is actually a 6-base palindrome in sequence "a" (CGCGCG) right at the 5-end and this sequence repeats itself further down the sequence. There are, however, no enzymes that recognize this sequence. The palindrome GCGCGC, which is recognized by the restriction endonuclease BssHII, is also found twice in sequence "a". But, this sequence also has many CG (pyrimidine-purine) repeats characteristic of Z DNA. Sequence "b" has CGTACG near its 3' end, which is recognized by BswWI. Sequence "c" also has a six-base sequence recognized by BsaI, which recognizes GGTCTC and cuts downstream of this sequence. Sequences "d" and "e" both have GAATTC, which is the recognition site for EcoRI. So, they all have "six-cutter" sites but sequence "e" only matches to first question and not any of the others.

To be honest, after I got done visually inspecting the sequences as explained above I ran the sequences through NEB cutter (http://tools.neb.com/NEBcutter2/index.php). This is a very useful site for analysis of DNA for restriction endonuclease recognition sites. After submitting a sequence I selected "Custom digest" from the "Main options" menu and selected "Enzymes with a particular site length" and restricted this to 6 base recognition sites.

Cruciform structure is b. Cruciform structures require inverted repeats. These are sequences that show twofold symmetry: the sequence 5' to 3' is identical in both strands. Palindromes are inverted repeats but they have problems folding into cruciforms because the terminal loops are too tight. Sequences that form cruciforms are interrupted palindromes. In sequence "a" we decided that there were two six-base recognition sites for the enzyme BssHII and they are separated by several bases and might form a cruciform. Sequence "b" has a 10 base inverted repeat. The first 10 bases on the strand shown are duplicated on the bottom strand and so sequence "b" could form a cruciform with a 10 base pair stem and a 6 base loop. One way to look for palindromes is to do a dot plot of the sequence versus its reverse complement or inverse complement. The reverse complement is simply the complement written 5' to 3'. Generating reverse complements by hand is easy enough for short sequences but the following URL has a program to do this. (http://www.bioinformatics.org/sms/index.html) A simple dot plot program can be found at http://www.vivo.colostate.edu/molkit/dnadot/index.html. This same site also has a sequence manipulation program that will generate inverse complements. Inverted repeats show up along the diagonal in a dot plot and the only sequence to give a strong diagonal is "b".

Z-DNA is formed in regions containing pyrimidine-purine repeats and sequence "a" fits the bill. A way of solving this problem is to submit sequences to ZHunt on line at http://gac-web.cgrb.oregonstate.edu/zDNA/index. The program asks for a FASTA file as input and I created one with all the sequences in the question combined into one sequence. I saved the file as a "text only" file and when I submitted it the program predicted Z-DNA for sequence "a".

The 3'-end of tRNAs end in 5'...CCA-3' and so we need to look for a sequence complementary to this, namely 5' TGG...3' and sequence "d" works.

The only sequence we haven't used is "c", which can form a triplex DNA. Sequence requirements for triple-stranded DNA formation include two pyrimidine-rich strands and a third purine-rich strand. Sequence "c" and its complement are pyrimidine-rich and purine-rich respectively.

16. Draw the cloverleaf structure of a tRNA from RNA T_ms sequence information
The nucleotide sequence of E. coli tRNA^{Gln} is as follows:
UGGGGUAUCG$_{10}$CCAAGCGGU$_{20}$AAGGCACCGG$_{30}$AUUCUGAΨΨC$_{40}$CGGCAUUCCG$_{50}$AGGTΨCGAAU$_{60}$
CCUCGUACCC$_{70}$CAGCCA$_{76}$
From this primary structure information, draw the secondary structure (cloverleaf) of this RNA and identify its anticodon.

Answer: Figure 11.36 will be useful in answering this question because it presents the general structure of a tRNA. All tRNAs end with NCCA-3', where N is any nucleotide, in single strand. Further, all tRNAs, except those used for initiation, have their 5' ends base paired at the beginning of a 7-base stem structure with the base preceding N on the 3' end. The so-called D-loop contains dihydrouridine and is on a 3- or 4-base paired stem. The next secondary structure element is a 5-base pair stem at the end of which is located the anticodon loop. The variable loop is just that, namely variable in size and so we need to turn to the next secondary structure element. The T-pseudouridine-C loop is located on the end of a 5-base pair arm. With this information we should be able to deduce the secondary structure of this tRNA.

The primary structure is shown below with stem regions labeled. The 3' end is single stranded and ends in NCCA. In this case N is G and the 3' unpaired acceptor end is GCCA. Since the 5' end is paired and involved in the 7-base pair acceptor stem it is easy to recognize which bases form this secondary structure. The anticodon is CUG. The two codons for glutamine are CAA and CAG, which require anticodons UUG and CUG and the tRNA anticodon presented here has CUG anticodon. The *E. coli* genome has four tRNA genes for gln. Two have anticodons CUG and the other two UUG. The first position (5') of an anticodon pairs with 3' base of the codon, which is degenerate (i.e., A or G) and U in the first position can pair with A or G (with wobble) but C only pairs with G. Thus, the tRNA we are working with would not suffice to decode all gln codons by itself.

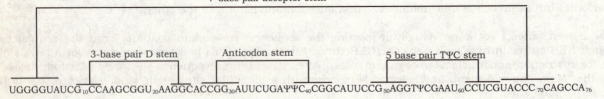

7-base pair acceptor stem

3-base pair D stem   Anticodon stem   5 base pair TΨC stem

UGGGGUAUCG$_{10}$CCAAGCGGU$_{20}$AAGGCACCGG$_{30}$AUUCUGAΨΨC$_{40}$CGGCAUUCCG$_{50}$AGGTΨCGAAU$_{60}$CCUCGUACCC$_{70}$CAGCCA$_{76}$

17. Use the tools of the Protein Data Bank to explore ribosome structure
The Protein Data Bank (PDB) is also a repository for nucleic acid structures. Go to the PDB at www.rcsb.org and enter the PDB ID = 1YI2. 1Y12 is the PDB ID for the structure of the H. marismortui 50S ribosomal subunit with erythromycin bound. Erythromycin is an antibiotic that acts by inhibiting bacterial protein synthesis. In the list of the display options under the image of the 50S subunit, click on one of the viewing options to view the structure. Using the tools of the viewer, zoom in and locate erythromycin within this structure. If the 50S ribosomal subunit can be compared to a mitten, where in the mitten is erythromycin?

Answer: It was a bit hard for me to view the structure clearly using KiNG viewer but I managed. The 50S subunit looks like a catcher's mitt and the erythromycin-binding site is in the pocket (or palm) of the mitt. I had much better luck viewing the molecule with SwissPdb viewer and I am sure RasMol would have worked also.

There are other viewing programs at PDB include Jmol, SimpleViewer, Protein Workshop, and Kiosk. To find these you have to be on the "Asymmetric Unit" view in the image box on the right hand side of the page. The choices are "Biological Assembly" or "Asymmetric Unit". To move between these two you have to activate either the leftward-pointing arrow on the top of the image box or the rightward-pointing arrow. As a word of caution this is a big structure and may take some time to download.

Using Jmol the 50S subunit looks like a catcher's mitt and the erythromycin-binding site is in the pocket (or palm) of the mitt. To simplify the default view use the Display Style pull down menu to view backbone only. If you are having a problem locating erythromycin try this. Right click and activate Select, Hetero, Ligand (hetero and not solvent). This will select 391 items. Next right click again, activate Select and then Display Selected Only. This will display only a small number of ligands of which erythromycin is one in addition to ions like sodium, chloride, and magnesium. With this simplified view you should be able to find erythromycin. Deselecting Display Selected Only should return the full view.

SimpleViewer took a few minutes to open the file and it gave, not surprisingly, a simple view with proteins and RNA shown as ribbons and the heteroatoms in ball and stick. Other viewing options are not possible with SimpleViewer and PDB suggests using Protein Workshop.

Protein Workshop allows a variety of viewing options. To simplify the view select the Visibility tool (item 1. In the right-hand column). From 2 choose "Ribbons: and from 4 click on the whole molecule (1YI2). This should remove protein and RNA leaving heteroatoms including erythromycin. To add back the protein and RNA but only in simple ribbon form from the "Select your tool" section select "Styles". Have styles act on "Ribbons" and in section 3 change the ribbon style to simple line. In item 4 click on 1YI2. To then visualize this change, select in 1 the "Visibility tool" and apply it to 1YI2.

18. Use the human genome database to explore the relationship between genes, proteins, and diseases.
Online resources provide ready access to detailed information about human genome. Go the National Center for Biotechnology Information (NCBI) genome database at http://www.ncbi.nlm.nih.gov/Genomes/index.html and click on Homo sapiens in the Map Viewer genome annotation updates list to access the chromosome map and organization of the human genome. In the "Search For" box, type in the following diseases to discover the chromosomal location of the affected gene and, by exploring links highlighted by the search results, to discover the name of the protein affected by the disease:
  a. Sickle-cell anemia
  b. Tay-Sachs disease
  c. Leprechaunism
  d. Hartnup disorder

Answer: a. A search for "sickle-cell anemia" at Map Viewer returned one hit near the tip of the short arm of chromosome 11. The human genome view on which the hit is highlighted shows the 22 human autosomes, X and Y chromosomes and the mitochondria. The autosomes are arranged (and named) from biggest to smallest with short arm on top. In addition, the search result returned hits to different assemblies. Human sickle-cell anemia is caused by a mutation in the beta-globin gene. Activating the link below chromosome 11 will return a more detailed map. On the left hand side is an Ideogram showing the location of the hit referenced against chromosome banding patterns. You should see that the beta-globin gene is located between band 11p15 and the telomere. You can use the "Maps & Options" button to change maps. Be sure one of the maps is the Morbid map and that it is the master map. You can do a simple "find" command to locate sickle cell anemia. Activating its GeneID number (3043) will return a detailed report and on this page under "Genomic context" is an option to "See HBB in MapViewer". Activating this will return a MapView with the Gene map as the master map and it will have links to OMIM (Online Mendelian Inheritance in Man). Activating this link returned information about sickle cell anemia and other diseases associated with the hemoglobin beta gene.

b. Tay-Sachs disease yielded hits on chromosome 15. You could take the same approach as outlined above for sickle-cell anemia. With the Morbid map as the master map locate Tay-Sachs and activate its GeneID number (3073). You will learn that the gene for hexosaminidase A (alpha polypeptide) is associated with this disease. Activating "See HEXA in MapViewer" will return you to MapViewer with the gene map as the master map and links to OMIM and other databases. By using the link to OMIM we learn that Tay-Sachs disease is an autosomal, progressive neurodegenerative disorder that is fatal at an early age and involves the gene for the protein hexosaminidase A. Using the protein link (pr) (you only see this link and others when the gene map is the master map) we find a summary of hexosaminidase A where we learn that this protein is a subunit of the lysosomal enzyme beta –hexosaminidase. This enzyme is responsible for degradation of ganglioside GM2 and other N-acetyl hexosamines. The protein link returns several hits and the two more useful ones are P06865.2, which is a hit on the SwissProt database, and NP_000511.2, which is the NCBI reference sequence protein database.

c. Leprechaunism returned a hit on the short arm of chromosome 19. From the Morbid map (as the master) activation of GeneID 3643 returned information on the insulin receptor (INSR). Activating "See INSR in MapViewer" returned a MapView with the Gene map as the master map, which contains a link to OMIM. There are various records in OMIM with a variety of information and these are distinguished by symbols whose meanings can be found in the "Symbols" link under FAQ in the left-hand column (on the OMIM homepage). The most useful records for our purpose are labeled with an asterisk (*), number symbol (#) and plus sign (+). Asterisk- and number symbol-labeled records contain information about a gene of known sequence. Number sign records also include a description of the phenotype. Record #246200 is for Donohue syndrome, also known as Leprechaunism, which is caused by a defect in the insulin receptor gene. This is an autosomal recessive disorder that involves mutations in the insulin receptor gene that impair its function. The insulin receptor gene is located on the short arm of chromosome 19 between bands 19p13.3 and 19p13.2. (Bands on the short arm are labeled "p" for petite whereas bands on the long arm are labeled "q".) Linking to the protein database (from OMIM or Map Viewer) should yield record P06213.4, which is the SwissProt record for the human insulin receptor. Under the comment section the function is described. Insulin binding to the receptor stimulates tyrosine protein kinase activity. Also under the comments section there information about diseases caused by defects in this gene and one is Leprechaunism. The disease descriptions are in OMIM (MIM).

d. A search for Hartnup disorder yielded hits on chromosome 5 to a protein involved in transport of neutral amino acids. The gene is located on the tip of the short arm of chromosome 5 in band 5p15. In Map Viewer with the gene map as the master map links to OMIM and protein databases are provided. At OMIM we learn that among other symptoms, " ...a pellagra-like light-sensitive rash..." is common. Pellagra (rough skin in Italian) is caused by niacin deficiency and niacin is produced from tryptophan, which is poorly absorbed as a consequence of a defect in an amino acid transport protein. The link from the OMIM page for Hartnup Disorder to the SLC6A19 gene (608893) explains that the protein belongs to the soluble carrier family 6 neurotransmitter transporter family of proteins. It is a neutral amino acid transporter that is active in absorption of neutral amino acids in the kidney and intestine.

Preparing for the MCAT® Exam

19. (Integrates with Chapter 10.) Erwin Chargaff did not have any DNA samples from thermoacidophilic bacteria such as those that thrive in the geothermal springs of Yellowstone National Park. (Such bacteria had not been isolated by 1951 when Chargaff reported his results.) If he had obtained such a sample, what do you think its relative G:C content might have been? Why?

Answer: Genomes of thermophilic organisms are typically G:C rich to help stabilize dsDNA to high temperatures. Recall that G:C base pairs involve three hydrogen bonds whereas A:T pairs involve only two. Thus, dsDNA rich in G and C will have higher melting temperatures. To get a listing of GC content for prokaryotes visit http://insilico.ehu.es/ and activate the OligoWeb link. On the returned page activate "Show all", which will return a page listing genomes, their sizes, and GC content. I copied the data and pasted it as unformatted text into an MSWord file. I then converted it to a table and transferred the information to an Excel spread sheet and sorted by GC content. The GC content ranged from about 75% to 14%.

20. Think about the structure of DNA in its most common B-form double helical conformation and then list its most important structural features (deciding what is "important" from the biological role of DNA as the material of heredity). Arrange your answer with the most significant features first.

Answer: The information content of DNA is its linear sequence of bases and this is its most important structural feature with regard to information. Some viruses actually use ssDNA to carry their genetic material. Base pairing allows the sequence of one strand to determine the sequence of the other. This is of key importance during replication and so base pairing is also an important informational feature of DNA. The fact that dsDNA is helical with base pairs sequestered in the middle of the helix makes good sense in terms of protecting the base pairs from inadvertent reactions, which could lead to genetic errors. So base stacking and the helical arrangement are important for information stability. The lack of a 2'-hydroxyl group imparts further stability to dsDNA (as compared to dsRNA).

Questions for Self Study

1. For DNA sequencing using the chain termination or dideoxy technique, which of the following are true and which are false?
   a. The 5' ends of synthesized DNAs have dideoxynucleotides on them.
   b. The 5' ends of synthesized DNAs are determined by the sequence of the primer.
   c. The sequence read is the sequence of the template strand.
   d. The DNA polymerase moves along the template strand 3' to 5'.
   e. In automatic sequencing, the 3' ends of products are labeled with fluorescent agents.
   f. The 3' ends of extension products are all the same.
   g. Only double-stranded DNA can be sequenced using this technique.

2. Double stranded DNA is composed of canonical base pairs.
   a. What are they?
   b. What is the orientation between the two strands of dsDNA?
   c. What is the consequence of the two glycosidic bonds holding the bases in each base pair not being directly opposite one another?

3. Answer the following questions with A (A form DNA), B (B form DNA) or Z (Z-DNA).
   a. The double helical form adopted by double stranded RNA ___.
   b. Left-handed helix ___.
   c. Approximately 10 base pairs per turn ___.
   d. Plane of base pair perpendicular to helix axis ___.
   e. Accommodates syn glycosyl bond conformation ___.
   f. Is stabilized by cytosine methylation and by supercoiling ___.
   g. Formation is sequence dependent ___.
   h. Approximately 11 base pairs per turn ___.
   i. Helix axis runs through the base pairs ___.
   j. Helix axis runs through the major groove ___.

4. Fill in the blanks. The transition from dsDNA to single strand DNA is called ___. This process occurs when DNA is subjected to conditions such as ___ or ___ that disrupt ___ bonds. The transition is accompanied by an ___ in the UV absorbance, or a ___ shift, caused by ___ of bases. The temperature at which one-half of the dsDNA has converted to ssDNA is termed the ___ temperature. This temperature is dependent on the solution conditions. For example, as salt concentration increases, the transition temperature ___. High values of pH and high concentrations of urea will ___ the transition temperature. The reverse process, namely the transition from ssDNA to dsDNA is termed ___. The rate of this transition is dependent on the complexity of the DNA such that highly repetitive DNA will produce dsDNA at a ___ rate whereas single-copy, complex DNA will produce dsDNA at a ___ rate.

5. Order the following terms: miniband, chromosome, histone HI, nucleosome, solenoid, base pair, histone octamer.

6. On the diagram of the secondary structure of tRNA shown below indicate the location of the following features:

   a. Anticodon
   b. Acceptor stem
   c. CCA
   d. D loop
   e. Variable loop
   f. Location of highly conserved bases involved in tertiary interactions
   g. Anticodon stem
   h. 5' end
   I. 3' end

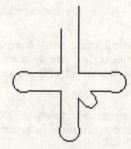

Answers

1. a. F; b. T; c. F; d. T; e. T; f. F; g. F

2. a. AT and GC; b. The two strands are antiparallel.; c. The helix has a major groove and a minor groove.

3. a. A; b. Z; c. B; d. B; e. Z; f. Z; g. Z; h. A; I. B; j. A

4. denaturation (or melting); pH; temperature (or changes in ionic strength, urea concentration, formamide concentration); hydrogen; increase; hyperchromic; unstacking; melting (or transition or T_m); increases; decrease; renaturation (or reannealing); fast; slow.

5. base pair < histone HI < histone octamer < nucleosome < solenoid < miniband < chromosome.

6.

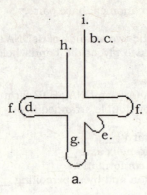

Additional Problems

1. Phylogenetic sequence comparisons have played a key role in determining the secondary structure of ribosomal RNAs. Helical regions are first identified as complementary regions within an RNA sequence from a particular organism. However, the existence of complementary bases does not prove that a helix forms between two regions. Evidence in support of a secondary structure is gained by comparing the same regions from other organisms. The object of this comparison is to find examples of compensatory base-changes for each complementary pair suspected of being part of a helix. Compensatory base-changes identify conserved base-pairs as opposed to conserved sequences that happen to be complementary. The sequences presented below are a small portion of the small-subunit rRNA from three different organisms. Each can form hairpin structures.
a). Identify the complementary regions involved in hairpin formation. (In addition to the canonical base-pairs, G-U pairs are found in RNA helices.)
b). Locate compensatory base-changes within the helical regions of the hairpin.

Organism	Sequence
Halobacterium cutirbrum	UAGCCGUAGGGGAAUCUGCGGCUG
Homo sapiens	UUUCCGUAGGUGAACCUGCGGAAG
Homo sapiens (mitochondria)	UAAGUGUACUGGAAAGUGCACUUG

2. A common technique used to isolate plasmid DNA from bacterial cells is to lyse cells in a solution containing SDS and sodium hydroxide, neutralize the lysate using potassium acetate, and centrifuge the neutralized lysate to remove chromosomal DNA and cellular debris. Explain why this technique separates chromosomal DNA from plasmid DNA.

3. Ethidium bromide is an intercalating agent used to stain DNA. It can be removed from DNA solutions by extraction with butanol. How does ethidium bromide interact with DNA and how does butanol extraction overcome this interaction?

4. There is currently a great deal of interest in repetitive DNA as markers to identify an individual's DNA. Sketch a c_0t curve for a DNA sample composed of 40% repetitive DNA and 60% non-repetitive DNA. Suggest a possible technique for isolating repetitive DNA.

5. During synthesis of an oligonucleotide using phosphoramidite chemistry, a failure occurred at the very last coupling cycle such that only half of the oligonucleotides got extended to full length. The oligonucleotide was to be used for two purposes: as a primer in dideoxynucleotide sequencing, and as a mutagenic primer in site-specific mutagenesis. (In oligonucleotide-directed, site-specific mutagenesis, an oligonucleotide, with a single-base difference from a wild-type sequence, is used as a primer to synthesize a mutant copy of a template DNA.) Would it be wise for the researcher to use the synthesized oligonucleotide without purifying the full-length oligonucleotide product?

Abbreviated Answers

1. The primary and secondary structures are drawn below with compensatory base-changes indicated by *'s.

Primary Structure	Secondary Structure
Halobacterium cutirbrum 5' 3' UAGCCGUAGGGGAAUCUGCGGCUG	5' G UAGCCGUAGG G |||||||||| GUCGGCGUCU A 3' * * A
Homo sapiens 5' 3' UUUCCGUAGGUGAACCUGCGGAAG	5' U UUUCCGUAGG G |||||||||| GAAGGCGUCC A 3'** * A
Homo sapiens (mitochondria) 5' 3' UAAGUGUACUGGAAAGUGCACUUG	5' G UAAGUGUACU G |||||||||| GUUCACGUGA A 3' *** ** A

2. The combination of SDS and sodium hydroxide will lyse cells. SDS is a strong protein denaturant that will remove protein from DNA. Sodium hydroxide will denature DNA. Closed-circular, plasmid DNA is resistant to denaturation because as base pairs are broken and DNA is unwound, supercoils are introduced in plasmid DNA. Supercoiling prevents complete denaturation from occurring. Upon addition of potassium acetate, the basic sodium hydroxide is rapidly neutralized. Plasmid DNA renatures whereas chromosomal DNA forms a tangled mass of partially single-stranded and partially double-stranded DNA. Further, a potassium-SDS salt forms that is insoluble and precipitates. This precipitate together with cellular debris, denatured protein, and tangled, chromosomal DNA are all removed from solution upon centrifugation leaving behind plasmid DNA. (Because RNA is sensitive to base-catalyzed cleavage, some of the cellular RNA is destroyed as well.)

3. The structure of ethidium bromide is shown below.

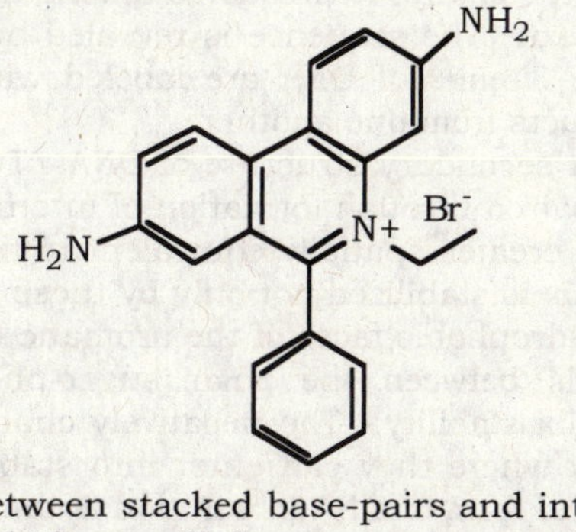

This planar molecule will intercalate between stacked base-pairs and interact via π-bonding. (In this state or when dissolved in organic solvents, ethidium will fluoresce when excited with ultraviolet light. Ethidium fluorescence is commonly used to detect DNA in agarose or acrylamide gels.) The presence of delocalized π-electrons makes ethidium very soluble in organic solvents. During extraction of DNA/ethidium solutions with butanol, the ethidium will partition into the organic phase.

4. It may be possible to isolate repetitive DNA using isopycnic centrifugation provided the density of the repetitive DNA is different than the average density of the non-repetitive DNA.

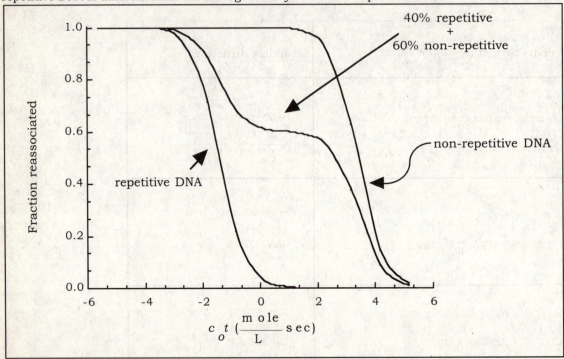

5. Provided a capping step is carried out during each cycle of oligonucleotide synthesis, a failure at any one step will produce an oligonucleotide of length equal to the cycle number at which the failure occurred. Thus, failure at the last step produces a polymer one nucleotide shorter than the full-length oligonucleotide. The full-length product and the failure product have the same 3'-ends but differ by one base on their 5'-ends. Using a mixture of the two as primers for sequencing will produce two sets of sequence products that differ at each band by a single base, giving the sequencing gel the appearance of having "shadow bands".

The mixture of products should pose no problems in oligonucleotide-directed, site-specific mutagenesis.

Summary

In the chain termination method of DNA sequencing, a DNA strand of unknown sequence is enzymatically copied using DNA polymerase, dATP, dGTP, dCTP and dTTP, plus four dideoxynucleoside triphosphate analogs uniquely labeled with fluorescent dyes. Incorporation of a dideoxynucleotide into the growing chain terminates subsequent extension of the chain. Represented among the products is a "ladder" of DNA strands, each one residue longer than the next. The sequence is revealed by considering which dideoxy-NTP analog specifically ended the strand. The 3'-ends of DNA are labeled with dideoxynucleotides carrying different fluorescent dyes to distinguish products from one another.

The double helix is the dominant secondary structure of DNA. Two DNA strands of opposite polarity and complementary base sequence are paired through formation of interchain hydrogen bonds. Exclusive pairing of A only with T and G only with C creates spatially equivalent units, the canonical, or "Watson-Crick", A:T and G:C base pairs. The double helix is stabilized not only by these hydrogen bonds between base pairs, but also by π,π interactions as the flat hydrophobic faces of the aromatic base pairs stack upon one another down the helix center. Hydrogen bonds between the polar sugar-phosphates of the two strands and the surrounding water contribute to helix stability. The negatively charged phosphates along the backbone are localized to the exterior of the helix where they can enter into stabilizing ionic interactions with cations in solution. The prevalent double-helical form of DNA in solution is the B conformation, in which the base pairs are arranged virtually at right angles to the helix axis, and each base pair circumscribes an angle of 36° around the helix. Thus there are about 10 base pairs per helix turn. Each base pair contributes 0.34 nm to the length of the helix, whose pitch then is 3.4 nm. Because the N-glycosidic bonds of the paired bases are not directly across the helix diameter from one another, the sugar-PO₄ backbones are not spaced symmetrically about the helix circumference and the helix has a major and a minor groove.

DNA can adopt double-helical forms other than the B conformation. An alternative right-handed form is A-DNA, a "shorter, squatter" helix formed upon slight dehydration of B-DNA. A-DNA has 11 base pairs per turn. DNA:RNA hybrids and double-stranded regions of RNA chains are typically of the A conformation. Z-DNA is a left-handed double helix. Typically, Z-DNA or regions of DNA in the Z-conformation have an alternating purine-pyrimidine nucleotide sequence, as in dCpGpCpGpCpG. Along either strand, the GpC dinucleotide is conformationally different than the CpG next in line. GpC dinucleotides circumscribe an angle of -45° about the helix axis, but CpGs cover only -15°. Thus, the GpCs are the 'horizontal' zigs and the CpGs are the 'vertical' zags of this "zigzag" or Z-DNA. Z-DNA has 12 base pairs per turn and a single narrow groove.

The double helix in solution is a dynamic, flexible structure, best described in terms of tertiary structure as a semi-rigid random coil. Even its short-range structure is not truly represented as a smooth, featureless barberpole-like molecule. Bases differ sufficiently in structural character that a short sequence of them lends a structural "signature" to its localized segment of DNA, generating the possibility for sequence-specific recognition by DNA-binding proteins.

Bases are capable of pairing in a variety of ways to form novel DNA structures like cruciforms, triplexes and quadruplexes. Cruciform DNA arises from canonical (AT, TA, GC, GC) pairing of inverted repeats within the same strand of DNA. Triplexes and quadruplexes depend on Hoogsteen pairing between T and A and protonated C and G. Hoogsteen pairs do not interfere with canonical base pairing and so both can occur leading to H-DNA, which is triplex DNA. Quadruplex structures are formed with G-rich regions interacting in parallel via Hoogsteen bonding.

Environmental extremes of temperature, pH or ionic strength can disrupt the hydrogen bonds between the base pairs, leading to separation of the DNA strands, so-called "denaturation" or "melting" of DNA. The strands will "re-anneal" if appropriate conditions for H-bond formation are restored. The rate of renaturation depends on the relative concentrations of complementary strands in the DNA solution, and thus is a measure of the sequence complexity of a DNA.

Supercoils represent yet another structural complexity in DNA. They form within dsDNA that in effect has no free ends and arise because of underwinding or overwinding of the DNA helix. Both positive and negative supercoils render the DNA more compact in structure, as compared to "relaxed" DNA, which lacks supercoils. All naturally occurring circular DNA molecules are negatively supercoiled. Negative supercoiling stabilizes Z-DNA.

Chromosomes are nucleoprotein complexes, consisting principally of DNA and a set of basic proteins known as histones. The histones are organized into nucleosomes, octameric units composed of four pairs of the histones H2A, H2B, H3 and H4, around which the DNA is wound. Nucleosomes are then wound into solenoids to form a 30 nm filament, a fundamental structural motif in chromosome structure. This 30 nm filament is believed to arrange into loops of DNA perhaps a million base pairs long. A cytologically distinct structure in chromosomes, called the mini-band, may consist of 18 loops ordered around the circumference of the chromosome axis. Chromosomes contain hundreds of mini-bands.

DNA can be chemically synthesized in the laboratory using phosphoramidite derivatives of the nucleotides and solid-phase techniques of polymerization. Genes greater than 1,000 base pairs in length have been assembled from oligonucleotide precursors made in this fashion, opening possibilities for the controlled manipulation of the nucleotide sequence of a gene and a facile analysis of the effects brought about by such changes.

RNA molecules are usually single-stranded. Nevertheless, RNA species have ornate, highly conserved secondary structures derived from extensive intrachain sequence complementarity in the strands. The single strands of tRNA, upon alignment of intrastrand complementary regions, form a cloverleaf secondary structure consisting of three H-bonded loops and an H-bonded stem. Hydrogen-bonding interactions between the loops then create a "bent" or L-shaped tertiary structure. Ribosomal RNA molecules also possess a high degree of intrastrand sequence complementarity, yielding a complex secondary structure on alignment. The folding patterns observed in the various rRNA species appear to have been highly conserved across the phylogenetic spectrum from archaea and bacteria to higher eukaryotes, suggesting a common design and purpose for these molecules that has persisted over evolutionary time. mRNA displays secondary structural properties as well, but little is known about the details of such structures in these rather labile RNAs.

Chapter 12

Recombinant DNA: Cloning and Creation of Chimeric Genes

• •

Chapter Outline

❖ Genetic engineering: Application of recombinant DNA technology
❖ Cloning vectors: Requirements
  ⋏ Origin of replication
  ⋏ Selectable marker: Typically antibiotic resistance
  ⋏ Cloning site: Polylinker in region not required for vector function
❖ Vector Types: dsDNAs
  ⋏ Plasmids: Circular elements that can accept up to 10 kbp of foreign DNA
  ⋏ Bacteriophage λ: Linear element used to clone 20 kbp inserts
  ⋏ Cosmids: λ cohesive ends separated by 40 kbp insert DNA
  ⋏ Shuttle vector: Origins of replication active in two different organisms
  ⋏ Expression vector: Polylinker downstream of promoter
  ⋏ Yeast artificial chromosomes (YACs): Linear DNA with autonomously replicating sequence (ori), telomeres, and centromere
❖ Chimeric or recombinant plasmids
  ⋏ Linearized vector and compatible foreign insert DNA
  ⋏ DNA ligase joins DNA
  ⋏ Transformation incorporates DNA into host cell
❖ DNA libraries: A collection of recombinant plasmids with insert DNA from particular organism
  ⋏ $N = [\ln(1-P)/\ln(1-f)]$
    • P is probability that library of size N has DNA fragment of interest
    • f = (insert size)/(genome size)
❖ cDNA libraries
  ⋏ cDNA: mRNA directed synthesis of DNA
    • PolyA mRNA isolated using oligo(dT)
    • Reverse transcriptase produces DNA from mRNA
    • Expressed sequence tags: Short cDNA's from mRNA
❖ Library screening
  ⋏ Southern analysis: DNA
    • Replica plate onto nitrocellulose
    • Lyse and denature using alkaline pH
    • Probe with complementary nucleic acid probe
  ⋏ Northern analysis: RNA: Used to identify RNAs
  ⋏ Probes
    • Degenerate oligonucleotides
    • Heterologous probe: DNA fragment from gene in related organism
❖ Microarrays: Gene chips: DNA sequence arrays used to detect large numbers of nucleic acids
❖ Expression vectors
  ⋏ RNA expression: Transcription of cloned gene driven by RNA polymerase promoter
  ⋏ Protein expression of cloned DNA

- Bacterial vectors and host
  - DNA must lack introns
  - Vector must contain bacterial promoter and ribosome binding site
  - Promoters: Strong and regulatable
  - Eukaryotic vector elements
    - Viral promoters
- ❖ Fusion proteins: cDNA cloned in frame with protein coding region on vector
- ❖ Screening of expression libraries: Antibodies to detect specific protein or fusion protein
- ❖ Reporter gene: Used to clone promoters: Easily detected protein product like GFP
- ❖ Yeast two-hybrid system: Identify protein-protein interactions
  - Bait: DNA-binding domain of GAL 4 fused to characterized protein
  - Target (prey): Library of GAL 4 transcriptional activation domain-target protein fusions
- ❖ Immunoprecipitation: Used to detect protein-protein interactions
- ❖ PCR: Polymerase chain reaction: In vitro amplification of DNA
- ❖ RNA interference: dsRNA that causes inactivation of specific mRNA: Gene knockdown
  - dsRNA with one strand complementary to mRNA target expressed in cells
  - Dicer processes dsRNA to produce small interfering RNA (siRNA)
  - siRNA binds to RNA-induced silencing complex
    - siRNA unwound
    - Strand complementary to mRNA paired
    - dsRNA cleaved
- ❖ Site-specific in vitro mutagenesis: Specific alteration of DNA sequence by in vitro technique

Chapter Objectives

Cloning Vectors

Understand the purpose of vectors in recombinant DNA technology. They are the vehicles by which foreign DNA is incorporated into cells in a form that will be amplified by the cell. Some important vectors include the following:

1) Plasmids: small, circular, double-stranded DNA elements containing a selectable marker, typically a gene conferring resistance to an antibiotic.
2) Bacteriophage λ: a linear double-stranded DNA-containing virus used to clone large (up to 16 kbp) DNA fragments.
3) Cosmid: a hybrid cloning vector, containing a plasmid origin of replication and a λ *cos* site, capable of carrying DNA inserts up to 40 kbp.
4) M13: a filamentous, single-stranded, circular DNA-containing virus whose life cycle includes a double-stranded circular DNA that can be manipulated much like a plasmid. Recombinant M13s are used as templates for chain termination sequencing.
5) YACs: yeast artificial chromosomes, the "jumbo jets" of genetic engineering, are used to clone extremely large DNA fragments (100s of kbp).

Formation of Recombinants

Generally, vectors contain multiple restriction endonuclease recognition sites in a single region known as the polylinker. To produce a recombinant or chimeric vector, the vector is digested with a restriction enzyme that recognizes a site in the polylinker, and the DNA to be inserted is digested with an enzyme that will produce ends compatible to the vector-digest. The DNAs are combined using DNA ligase and the recombinants are introduced into cells by transformation. Transformation may be accomplished chemically (a cell capable of being transformed is said to be competent). Alternatively, a brief, intense, electrical shock may be used to transiently produce holes in cells through which DNA can diffuse into the cell. This process is known as electroporation. Vectors that are infectious, such as λ or M13, enter cells by a process known as transfection.

Detecting Recombinants

Producing recombinants is an easy task, but finding the correct recombinant in a library of recombinants is often difficult. Know the techniques available for this purpose including Southern analysis, selecting for expression, screening for expression, and production of hybrid proteins for antibody screening. Reporter gene constructs are used to clone DNA elements involved in regulation of gene expression.

PCR

PCR is a powerful technique for amplifying DNA *in vitro*. Understand the principles behind this technique. You should be able to show that as a function of cycle-number, the full-length product defined by flanking oligonucleotide primers, is amplified exponentially.

Problems and Solutions

1. Construction of a recombinant plasmid
A DNA fragment isolated from an EcoRI digest of genomic DNA was combined with a plasmid vector linearized by EcoRI digestion so sticky ends could anneal. Phage T4 DNA ligase was then added to the mixture. List all possible products of the ligation reaction.

Answer: Both the genomic DNA and the vector were digested with the same restriction enzyme thus all DNA fragments have compatible ends. The ligation products will include linear and circular DNA molecules made with either the genomic fragment or the vector DNA or a combination of both. Because both DNAs have the same "sticky ends", they may join in any orientation. (Depending on the ratio of genomic fragment DNA to vector DNA, recombinants between the two DNAs may be favored.) Only ligation products with vector DNA will successfully transform cells. Further, ligated DNA products with more than one copy of vector DNA are typically unstable in cells.

2. Overlapping restriction sites in a plasmid vector polylinker
The nucleotide sequence of a polylinker in a particular plasmid vector is
-GAATTCCCGGGGATCCTCTAGAGTCGACCTGCAGGCATGC-
This polylinker contains restriction sites for BamHI, EcoRI, PstI, SalI, SmaI, SphI, and XbaI. Indicate the location of each restriction site in this sequence. (See Table 10.2 of restriction enzymes for their cleavage sites.)

Answer: Restriction endonuclease recognition sites are often palindromic sequences and so it may be useful to search for palindromes in the sequence. They are readily identified by scanning the sequence for two adjacent nucleotides that are a complementary pair: AT, TA, GC, CG. These sequences are the middle sequences of all uninterrupted palindromes. Once one of these dinucleotides is found the sequences surrounding the dinucleotide are inspected to see if a palindrome exists. By inspecting the nucleotides on either side of the dinucleotide, a palindrome is indicated by the flanking nucleotides also forming complementary pairs. For example, in the sequence given, an AT dinucleotide is encountered at positions 3 and 4. This dinucleotide is flanked by AT (positions 2 & 5) and G and C (positions 1 & 6) indicating a 6-base palindrome (GAATTC). This sequence is recognized by *Eco*RI. The next dinucleotide found is CG at position 8 & 9. Clearly these two positions form the palindrome CCCGGG recognized by *Sma*I. This *Sma*I site overlaps another palindrome GGATCC (centered on the dinucleotide AT), which is recognized by *Bam*HI. The next dinucleotide is TA in TCTAGA recognized by *Xba*I. The *Sal*I site is next; its sequence is GTCGAC. This is followed by CTGCAG and GCATGC recognized by *Pst*I and *Sph*I. The location of sites is as follows:

```
GAATTCCCGGGGATCCTCTAGAGTCGACCTGCAGGCATGC
GAATTC              TCTAGA              GCATGC
EcoRI               XbaI                SphI
     CCCGGG              GTCGAC
     SmaI                SalI
        GGATCC              CTGCAG
        BamHI               PstI
```

A very useful tool for identifying restriction endonuclease recognition sites is NEBcutter at http://tools.neb.com/NEBcutter2/index.php. This resource is provided by New England BioLabs, a laboratory and supply house for reagents –including restriction endonucleases- for the life sciences. Cut and paste the sequence in the textbox, submit it for digestion and on the returned page under "Main options" do a custom digest restricted for enzymes with six-base recognition sequences. You will discover the same set of enzyme sites but 9 enzymes will be listed because CCCGGG, recognized by SmaI, is also recognized by two other enzymes, TspMI and XmaI that cut at different locations within this sequence. The three enzymes SmaI, XmaI and TspMI are isoschizomers: Enzymes that recognize the same sequence.

3. Design a vector polylinker for directional cloning
A vector has a polylinker containing restriction sites in the following order: HindIII, SacI, XhoI, BglII, XbaI, and ClaI.
a. Give a possible nucleotide sequence for the polylinker.

b. The vector is digested with HindIII and ClaI. A DNA segment contains a HindIII restriction site fragment 650 bases upstream from a ClaI site. This DNA fragment is digested with HindIII and ClaI, and the resulting HindIII-ClaI fragment is directionally cloned into the HindIII-ClaI digested vector. Give the nucleotide sequence at each end of the vector and the insert and show that the insert can be cloned into the vector in only one orientation.

Answer: a. The sequence of a polylinker with sites for *Hind*III, *Sac*I, *Xho*I, *Bgl*II *Xba*I and *Cla*I will have the following six-base recognition sites:

Site	Enzyme
AAGCTT	*Hind*III
GAGCTC	*Sac*I
CTCGAG	*Xho*I
AGATCT	*Bgl*II
TCTAGA	*Xba*I
ATCGAT	*Cla*I

Combining these we find:

```
AAGCTTGAGCTCGAGATCTAGATCGAT
AAGCTT    CTCGAG TCTAGA
HindIII   XhoI   XbaI
          GAGCTC AGATCT  ATCGAT
          SacI   BglII   ClaI
```

Note: The following pairs overlap: *Sac* I and *Xho* I; *Xho* I and *Bgl* II; *Bgl* II and *Xba* I; and, *Xba* I and *Cla* I.

b. *Cla*I recognizes ATCGAT and cuts after the first T. *Hind*III recognizes AAGCTT and cuts after the first A. Assuming the vector is circular the polylinker is shown below with the location of cut sites (in bold) for both enzymes on both strands.

```
-AAGCTTGAGCTCGAGATCTAGATCGAT-
-TTCGAACTCGAGCTCTAGATCTAGCTA-
```

Digestion will produce:

```
-A          AGCTTGAGCTCGAGATCTAGAT         CGAT-
-TTCGA          ACTCGAGCTCTAGATCTAGC          TA-
```

Note: the top and bottom strands are shown and the dashes indicate continuation of the sequence of the vector. Two fragments are produced by digestion of the vector with ClaI and HindIII. The polylinker is cut out of the vector. One end of the polylinker and one end of the vector have a 4-base 5' overhang of AGCT. The other ends of both contain a 2-base 5' overhang of CG. (Keeping track of 5' and 3' ends can be a bit tricky. The reference sequence is the top sequence, which starts AAGCT.... It is written conventionally 5' left to 3' right. Its complementary sequence, the bottom strand, is shown 3' to 5'. Left ends of top strands are 5' and right ends are 3'. The reverse is true for bottom strands.)

We are informed that a DNA fragment that contains a HindIII site 650 bases upstream of a ClaI site was also digested with both enzymes. The original DNA looks like this:

```
-AAGCTT---650---ATCGAT-
-TTCGAA---650---TAGCTA-
```

And, when digested it gives the following products:

```
-A          AGCTT---650---AT         CGAT-
-TTCGA          A---650---TAGC          TA-
```

The 650 base pair fragment has ends identical to the digested polylinker shown in the first digest and so it could replace it in a ligation reaction. It could not, however, be cloned in the reverse orientation because the ends would no longer be compatible. (The 2-base 5' overhang is not compatible with the 4-base 5' overhang.) So, the fragment can clone in only one direction.

4. The number of clones needed to screen a yeast genomic library at 99% confidence
Yeast (Saccharomyces cerevisiae) has a genome size of 1.21 x 10^7 bp. If a genomic library of yeast DNA was constructed in a bacteriophage λ vector capable of carrying 16 kbp inserts, how many individual clones would have to be screened to have a 99% probability of finding a particular fragment?

222

Answer: For a genome 1.21 x 10^7 bp, the number of 16 kbp cloned inserts that would have to be screened to have a 99% probability of finding a particular fragment is given by:

$$N = \frac{\ln(1-P)}{\ln(1-f)} \text{ where P=0.99 and } f = \frac{16 \text{ kbp}}{1.21 \times 10^7} = \frac{16 \times 10^3}{1.21 \times 10^7} = 1.32 \times 10^{-3}$$

$$N = \frac{\ln(1-0.99)}{\ln(1-1.32 \times 10^{-3})} = 3,480$$

5. The number of clones needed to screen a very large genomic library at 99% confidence

The South American lungfish has a genome size of 10.2 x 10^{10} bp. If a genomic library of lungfish DNA was constructed in a cosmid vector capable of carrying inserts averaging 45 kbp in size, how many individual clones would have to be screened to have a 99% probability of finding a particular DNA fragment?

Answer:

$$N = \frac{\ln(1-P)}{\ln(1-f)} \text{ where P=0.99 and } f = \frac{45 \text{kbp}}{1.02 \times 10^{11}} = \frac{45 \times 10^3}{1.02 \times 10^{11}} = 4.41 \times 10^{-7}$$

$$N = \frac{\ln(1-0.99)}{\ln(1-4.41 \times 10^{-7})} = 1.04 \times 10^7 = 10.4 \text{ million}$$

6. Designing primers for PCR amplification of a DNA sequence

Given the following short DNA duplex of sequence (5'✱3')

ATGCCGTAGTCGATCATTACGATAGCATAGCACAGGGATCACACATGCACACACATGACATAGGACAGATAGCAT

what oligonucleotide primers (17-mers) would be required for PCR amplification of this duplex?

Answer: Amplification by PCR requires two oligonucleotides (17-mers). The oligonucleotide used to amplify from the 5'-side has a sequence identical to the first 17 nucleotides of the fragment:

5'-ATGCCGTAGTCGATCAT-3'

The oligonucleotide for the 3'-end is a 17-mer with a sequence complementary to the last 17 bases:

5'-ATGCTATCTGTCCTATG-3'

(In selecting oligonucleotides for PCR, care must be taken to insure that: the oligonucleotide does not bind at alternative sites in the template DNA; the oligonucleotide does not have large regions of self-complementarity; and, that the 3'-ends are not in A/T-rich regions.)

7. A polylinker for expression of a β-galactosidase fusion protein

Figure 12.3 shows a polylinker that falls within the β-galactosidase coding region of the lacZ gene. This polylinker serves as a cloning site in a fusion protein expression vector where the cloned insert is expressed as a β-galactosidase fusion protein. Assume the vector polylinker was cleaved with BamHI and then ligated with an insert whose sequence reads

GATCCATTTATCCACCGGAGAGCTGGTATCCCCAAAAGACGGCC...

What is the amino acid sequence of the fusion protein? Where is the junction between β-galactosidase and the sequence encoded by the insert? (Consult the genetic code table on the inside front cover to decipher the amino acid sequence.)

Answer: *Bam*HI recognizes GGATCC and cuts after the first G giving

```
G          GATCC
CCTAG          G
```

Once ligated with insert, the sequence becomes

GGATCCATTTATCCACCGGAGAGCTGGTATCCCCAAAAGACGGCC....

What is the amino acid sequence of the fusion protein? To answer this we must know the reading frame at the *Bam*HI site. This is shown in Figure 12.3. The reading frame is such that GAT in the *Bam*HI site is in frame. This gives:

```
G|GAT|CCA|TTT|ATC|CAC|CGG|AGA|GCT|GGT|ATC|CCC|AAA|AGA|CGG|CC...
  Asp Pro Phe Ile His Arg Arg Ala Gly Ile Pro Lys Arg Arg Pro
```

8. *Using PCR for the in vitro mutagenesis of a protein coding sequence*
The amino acid sequence across a region of interest in a protein is

Asn-Ser-Gly-Met-His-Pro-Gly-Lys-Leu-Ala-Ser-Trp-Phe-Val-Gly-Asn-Ser

The nucleotide sequence encoding this region begins and ends with an EcoRI site, making it easy to clone out the sequence and amplify it by polymerase chain reaction (PCR). Give the nucleotide sequence of this region. Suppose you wished to change the middle Ser residue to a Cys to study the effects of this change on the protein's activity. What would be the sequence of the mutant oligonucleotide you would use for PCR amplification?

Answer: We are asked for the nucleotide sequence of a gene that codes for the polypeptide NSGMHPGKLASWFVGNS. Because the sequence begins and ends with an *Eco*RI site, GAATTC must appear at both ends. The codon for Asn (N), AAY, is part of an *Eco*RI site as AAT. Thus, the 5'-end of the mRNA must begin with (G)AAU. (In the DNA fragment G would have been separated from the rest of the sequence by *Eco*RI digestion because the enzyme cleaves between G and A.) The 3' end codes for Ser which has six codons, UCN and AGY. The codon set UCN must appear on the 3' end with the UC forming the last two bases of the *Eco*RI site.

We are asked to convert Ser to Cys by site-directed mutagenesis. The codons of serine are UCN and AGY; Cys codons are UGY. Let us assume that the internal serine is AGY, one base different from the cysteine codon. The mRNA and DNA sequences are:

```
mRNA
                           CUN
5'(G)AAU-UCN-GGN-AUG-CAY-CCN-GGN-AAR-UUY-GCN-AGY-UGG-UUY-GUN-GGG-AAU-UCN 3'
     Asn-Ser-Gly-Met-His-Pro-Gly-Lys-Leu-Ala-Ser-Trp-Phe-Val-Gly-Asn-Ser

DNA
                           CTN
5'(G)AAT-TCN-GGN-ATG-CAY-CCN-GGN-AAR-TTY-GCN-AGY-TGG-TTY-GTN-GGG-AAT-TCN 3'

Primer Sequence
                    3' CCN-TTY-AAR-CGN-ACR-ACC-AAR-CAN-CCC-TTA-AGN-NNN 5'
                                        Cys
```

Notice that the primer sequence is written 3' to 5' (but it is labeled this way.) When it binds to the *Eco*RI fragment it will produce a single-base mismatch that will convert the codon for Ser (AGY) to the codon for Cys (UGY). Since the Ser residue is encoded nearer the 3'-end of the *Eco*RI fragment, the mutant oligonucleotide for PCR amplification should encompass this end. It includes several extra bases at the 5'-end (NNN) to place the *Eco*RI site internal and 12 bases at the 3'-end beyond the A-site directed mutation. These additional bases are included to insure that the primer's 3' end binds in a G/C-rich region.

To produce a mutation by PCR, we can use this primer and a second primer with a sequence identical to a sequence upstream of the *Eco*RI site on the 5'-end of the fragment. Using this primer and the mutagenic primer, a PCR fragment can be amplified that has *Eco*RI sites on each end and contains the desired mutation.

9. *Combinatorial libraries: calculating the number of sequence possibilities for oligonucleotides and peptides*
Combinatorial chemistry can be used to synthesize polymers such as oligopeptides or oligonucleotides. The number of sequence possibilities for a polymer is given by x^y, where x is the number of different monomer types (for example, 20 different amino acids in a protein or 4 different nucleotides in a nucleic acid) and y is the number of monomers in the oligomers.
   a. Calculate the number of sequence possibilities for RNA oligomers 15 nucleotides long.
   b. Calculate the number of amino acid sequence possibilities for pentapeptides.

Answer: For RNA oligomers, the total number of sequences is given by 4^{15} = 1,073,741,824 or about a billion different 15-mers. For a pentapeptide, the total number of sequences is given by 20^5 = 3,200,000 or about 3 million.

10. *Using the yeast two-hybrid system to discover protein-protein interactions*
Imagine that you are interested in a protein that interacts with proteins of the cytoskeleton in human epithelial cells. Describe an experimental protocol based on the yeast two-hybrid system that would allow you to identify proteins that might interact with your protein of interest.

Answer: To run a yeast two-hybrid experiment, we first construct a fusion protein between the DNA binding domain of yeast GAL4 gene and the cytoskeletal protein we are interested in studying (e.g., actin, tubulin or another cytoskeletal protein). A GAL4⁻ yeast strain with our reporter gene regulated by the GAL4 promoter is transformed with this construct containing the "bait" fused to a domain that will bind to the GAL4 promoter. It will not, however, activate the promoter because it lacks a transcriptional activation domain. Next, a library of fusion proteins is made from human epithelial cell mRNA cloned to the transcriptional activator domain of the yeast GAL4 protein. The GAL4⁻ yeast strain previously transformed with the "bait" construct is now transformed with the library of "prey" and screened for expression of the reporter gene. The reporter gene will be expressed if the DNA binding domain is joined to the transcriptional activator domain. One way this can happen is if the bait protein interacts with a prey.

11. Preparing cDNA libraries from different cells
Describe an experimental protocol for the preparation of two cDNA libraries, one from anaerobically grown yeast cells and the second from aerobically grown yeast cells.

Answer: mRNA is isolated from total RNA of both aerobically and anaerobically grown yeast. The mRNA is used as template with reverse transcriptase and oligo dT to produce ssDNA, which is then converted to dsDNA and cloned into a vector. To insure that full length mRNA is being used to produce the cDNA library there are protocols that exploit the 5' cap of full-length mRNA's.

12. Using gene chips to study differential gene expression
Describe an experimental protocol based on DNA microarrays (gene chips) that would allow you to compare gene expression in anaerobically grown yeast versus aerobically grown yeast.

Answer: Yeast has approximately 6,000 open reading frames and there are DNA microarrays (gene chips) with all of this information on them. mRNA isolated from both aerobically and anaerobically grown yeast are used to produce ssDNA, which are end-labeled with different fluorescent dyes, one for aerobic and one for anaerobic conditions. The ssDNA's are hybridized to the DNA microarray and the location of fluorescent ssDNA (derived from mRNA) is determined. Regions of the microarray with both fluorescent dyes represent genes actively transcribed under both conditions. Genes with only one of the dyes represent genes active in one condition and not the other.

13. Using antibodies to screen cDNA libraries
You have an antibody against yeast hexokinase A (hexokinase is the first enzyme in the glycolytic pathway). Describe an experimental protocol using the cDNA libraries prepared in problem 11 that would allow you to identify and isolate cDNA for hexokinase. Consulting Chapter 5 for protein analysis protocols, describe an experimental protocol to verify that the protein you have identified is hexokinase A.

Answer: The cDNA library would have to be cloned into an expression vector. That is, the cDNA library would have to be cloned behind a promoter so that the sequence information is used to produce a protein. Libraries of clones would then be screened for production of an antigen recognized by the hexokinase A antibody. The antibody could be coupled to a matrix to be used to isolate the gene product. The gene product could then be subjected to trypsin digestion and mass spectrometry to identify the protein. Alternatively, the cloned gene could be sequenced.

14. Design of a reporter gene construct to study promoter function
In your experiment in problem 12, you discover a gene that is strongly expressed in anaerobically grown yeast but turned off in aerobically grown yeast. You name this gene nox (for "no oxygen"). You have the "bright idea" that you can engineer a yeast strain that senses O₂ levels if you can isolate the nox promoter. Describe how you might make a reporter gene construct using the nox promoter and how the yeast strain bearing this reporter gene construct might be an effective oxygen sensor.

Answer: The *Saccharomyces cerevisiae* genome was sequenced in 1996 and it is well characterized. The hits you got in your microarray experiment done for question 12 will tell you which genes are activated. From the genomic sequence you should probably be able to recognize a promoter and its control elements. If not, you will at least know the open reading frame (ORF) that was expressed differentially. You could then use PCR primers to amplify a region starting upstream of the ORF and ending in the 5' end of the ORF. This PCR fragment would then be cloned into a yeast vector as a fusion with a convenient reporter gene like the coding region of green fluorescent protein (GTP). The vector would then be transformed into yeast and tested aerobically and an anaerobically for appropriate expression.

Biochemistry on the Web

15. Search the National Center for Biotechnology Information (NCBI) website at http://www.ncbi.nlm.nih.gov/sites/entrez?db=Genome to discover the number of organisms whose genome sequences have been completed. Explore the rich depository of sequence information available here by selecting one organism from the list and browsing through the contents available.

Answer: An alternative entrance to NCBI is the following URL: http://www.ncbi.nlm.nih.gov/. From this location activate the link to Genome.

To find information on completed genomes use the links under the column titled "Browse by organism group". Under "Browse by organism groups", activate "Eukaryota" and on the returned page use the "Sequence Status" pull down menu to limit information to completed projects (40 complete genomes as of June 2011). These genomes and others are also presented in Map Viewer. Map Viewer is particularly useful for locating genes within a genome. For microbial genomes including archaea and bacteria use the "Prokaryota" link under "Browse by organism group". Use the pull down menu to restrict records to "All Archaea" (110 complete genomes as of June 2011) or "All Bacteria" (1517 complete genomes as of June 2011). (To get a sense of how fast these data grow, I rechecked this information about a week later and found 111 and 1523 for all archaea and all bacteria respectively.)

Preparing for the MCAT® Exam

16. Figure 12.1 shows restriction endonuclease sites for the plasmid pBR322. You want to clone a DNA fragment and select for it in transformed bacteria by using resistance to tetracycline and sensitivity to ampicillin as a way of identifying the recombinant plasmid. What restriction endonucleases might be useful for this purpose?

Answer: We need to insert DNA into the ampicillin resistance gene and so restriction endonucleases that cut only in this region of pBR322 will be useful for this purpose. Referring to the restriction map shown in Figure 12.1 the following restriction sites are identified: SspI, ScaI, PvuI, PstI, and PpaI.

A more general way of answering this question is to use the NEBcutter tool at New England Biolab's restriction enzyme database, Rebase, at http://rebase.neb.com/rebase/rebase.html. NEBcutter is located at http://tools.neb.com/NEBcutter2/index.php. pBR322 is included as one of the standard sequences in a pull-down menu. Select it, indicate that the sequence is circular and submit the sequence for analysis. (The default for restriction enzymes is NEB enzymes). The results page gives a crude map of pBR322 with ampicillin resistance and tetracycline resistance genes highlighted. (The Amp^r gene is bla, which stands for beta lactamase. Ampicillin is a beta lactam antibiotic and the enzyme, beta-lactamase hydrolyzes it.) While several of the enzymes are the same as shown in Figure 12.1, there are a few others. NEBcutter lists AdhI, BsaI and AseI as having sites within the ampicillin resistance gene. AseI recognizes ATTAAT and cuts between the T's giving 5' TA overhangs. AhdI recognizes GACNNNNNGTC and cuts between the 3rd and 4th N. The N signifies that the enzyme recognizes any base at the positions labeled with N. Thus, fragments generated with AhdI may not necessarily be useful because they could produce incompatible ends. Finally, BsaI is listed and PpaI is not. It turns out that these enzymes are isoschizomers. They recognize GGTCTC but cut downstream of the sequence. (There recognition sequence is listed as GGTCTC (1/5), which means that they cut one base further downstream from GGTCTC on the top strand and 5 bases further down from GGTCTC on the bottom strand.) This enzyme is very likely to produce DNA fragments with incompatible ends so it may not be useful in a cloning experiment.

The restriction sites shown on the map are those for enzymes that cut only once in pBR322: digestion would linearize the plasmid. We could use an enzyme that cuts more than once provided all the cut sites are located within the ampicillin resistance gene. Changing the display to "2 cutters" shows location of restriction sites cut twice. There are two enzymes BtsI and BsrDI that cut within the ampicillin gene but they have cut sites outside their recognition sequences and so may produce DNA fragments with incompatible ends.

17. Suppose in the Figure 12.8 known amino acid sequence, tryptophan was replaced by cysteine. How would that affect the possible mRNA sequence? (Consult the inside front cover of this textbook for amino acid codons.) How many nucleotide changes are necessary in replacing Trp with Cys in this coding sequence? What is the total number of possible oligonucleotide sequences for the mRNA if Cys replaces Trp?

Answer: The sequence presented in Figure 12.8 is FMEWHKN (PheMetGluTrpHisLysAsn). We are asked to change W to C and then state how many different mRNA sequences could code for this peptide. Tryptophan

(W) is coded by UGG whereas cysteine is UGY (Y = pyrimidine = U or C) so we expect there to be only one change necessary in converting a tryptophan codon into a cysteine codon. Possible mRNA's are shown below first using the extended one letter code in which Y is U or C and R (purine) is A or G. With the exception of Met, all the other amino acids are coded for by two codons. Thus, six of the amino acids have two possibilities giving $2^6 = 64$ different mRNA's.

Phe	Met	Glu	Cys	His	Lys	Asn
UUY	AUG	GAR	UGY	CAY	AAR	AAY
UUU	AUG	GAA	UGU	CAU	AAA	AAA
UUU	AUG	GAA	UGU	CAU	AAA	AAG
UUU	AUG	GAA	UGU	CAU	AAG	AAA
UUU	AUG	GAA	UGU	CAU	AAG	AAG
UUU	AUG	GAA	UGU	CAC	AAA	AAA
UUU	AUG	GAA	UGU	CAC	AAA	AAG
UUU	AUG	GAA	UGU	CAC	AAG	AAA
UUU	AUG	GAA	UGU	CAC	AAG	AAG
UUU	AUG	GAA	UGC	CAU	AAA	AAA
UUU	AUG	GAA	UGC	CAU	AAA	AAG
UUU	AUG	GAA	UGC	CAU	AAG	AAA
UUU	AUG	GAA	UGC	CAU	AAG	AAG
UUU	AUG	GAA	UGC	CAC	AAA	AAA
UUU	AUG	GAA	UGC	CAC	AAA	AAG
UUU	AUG	GAA	UGC	CAC	AAG	AAA
UUU	AUG	GAA	UGC	CAC	AAG	AAG
UUU	AUG	GAG	UGU	CAU	AAA	AAA
UUU	AUG	GAG	UGU	CAU	AAA	AAG
UUU	AUG	GAG	UGU	CAU	AAG	AAA
UUU	AUG	GAG	UGU	CAU	AAG	AAG
UUU	AUG	GAG	UGU	CAC	AAA	AAA
UUU	AUG	GAG	UGU	CAC	AAA	AAG
UUU	AUG	GAG	UGU	CAC	AAG	AAA
UUU	AUG	GAG	UGU	CAC	AAG	AAG
UUU	AUG	GAG	UGC	CAU	AAA	AAA
UUU	AUG	GAG	UGC	CAU	AAA	AAG
UUU	AUG	GAG	UGC	CAU	AAG	AAA
UUU	AUG	GAG	UGC	CAU	AAG	AAG
UUU	AUG	GAG	UGC	CAC	AAA	AAA
UUU	AUG	GAG	UGC	CAC	AAA	AAG
UUU	AUG	GAG	UGC	CAC	AAG	AAA
UUU	AUG	GAG	UGC	CAC	AAG	AAG
UUC	AUG	GAA	UGU	CAU	AAA	AAA
UUC	AUG	GAA	UGU	CAU	AAA	AAG
UUC	AUG	GAA	UGU	CAU	AAG	AAA
UUC	AUG	GAA	UGU	CAU	AAG	AAG
UUC	AUG	GAA	UGU	CAC	AAA	AAA
UUC	AUG	GAA	UGU	CAC	AAA	AAG
UUC	AUG	GAA	UGU	CAC	AAG	AAA
UUC	AUG	GAA	UGU	CAC	AAG	AAG
UUC	AUG	GAA	UGC	CAU	AAA	AAA
UUC	AUG	GAA	UGC	CAU	AAA	AAG
UUC	AUG	GAA	UGC	CAU	AAG	AAA
UUC	AUG	GAA	UGC	CAU	AAG	AAG
UUC	AUG	GAA	UGC	CAC	AAA	AAA
UUC	AUG	GAA	UGC	CAC	AAA	AAG
UUC	AUG	GAA	UGC	CAC	AAG	AAA
UUC	AUG	GAA	UGC	CAC	AAG	AAG
UUC	AUG	GAG	UGU	CAU	AAA	AAA
UUC	AUG	GAG	UGU	CAU	AAA	AAG
UUC	AUG	GAG	UGU	CAU	AAG	AAA
UUC	AUG	GAG	UGU	CAU	AAG	AAG
UUC	AUG	GAG	UGU	CAC	AAA	AAA
UUC	AUG	GAG	UGU	CAC	AAA	AAG
UUC	AUG	GAG	UGU	CAC	AAG	AAA
UUC	AUG	GAG	UGU	CAC	AAG	AAG
UUC	AUG	GAG	UGC	CAU	AAA	AAA
UUC	AUG	GAG	UGC	CAU	AAA	AAG
UUC	AUG	GAG	UGC	CAU	AAG	AAA
UUC	AUG	GAG	UGC	CAU	AAG	AAG
UUC	AUG	GAG	UGC	CAC	AAA	AAA

UUC	AUG	GAG	UGC	CAC	AAA	AAG
UUC	AUG	GAG	UGC	CAC	AAG	AAA
UUC	AUG	GAG	UGC	CAC	AAG	AAG

Questions for Self Study

1. What three common features are shared by all cloning vectors?

2. How is the enzyme T4 DNA ligase used to produce chimeric plasmids?

3. Match the terms in the columns below:
   - a. Bacteriophage λ
   - b. Cosmid
   - c. M13
   - d. Shuttle Vector
   - e. Transformation
   - f. YAC
   - g. Electroporation

   1. A plasmid capable of propagation in two different organisms.
   2. Cellular uptake of exogenous DNA.
   3. Vector used to clone extremely large DNA molecules.
   4. A filamentous, ssDNA phage.
   5. A transformation technique.
   6. A bacterial virus used as a cloning vector.
   7. A hybrid vector containing *cos* sites.

4. Which of the following apply to Southern analysis?
   - a. Can be used to detect specific recombinants.
   - b. Is useful in chromosome walking.
   - c. Uses antibodies to detect gene expression.
   - d. Used labeled RNA or DNA as a probe.
   - e. Is used to identify RNA.

5. What is the difference between fusion protein expression and reporter gene expression in genetic engineering?

6. Polymerase chain reaction or PCR is a *in vitro* technique used to amplify specific sequences of DNA. It consists of repeated cycles of a three-step process. What are the three steps that constitute a cycle and what happens during each step?

Answers

1. An origin or replication, a selectable marker, a cloning site.

2. T4 DNA ligase is used to covalently join foreign DNA into a cloning site on a plasmid.

3. a. 6; b. 7; c. 4; d. 1; e. 2; f. 3; g. 5

4. a., b., d.

5. In fusion gene expression a protein coding region under study is cloned into an expression vector to produce a recombinant sequence that is expressed as a hybrid or fusion protein. In this case the experimenter is interested in studying the expressed protein. Reporter genes are used to identify transcriptional regulatory elements like promoters and enhancers.

6. Denaturation: dsDNA template is converted to ssDNA by high temperature. Annealing: Temperature is lowered to allow DNA primers to bind to template DNA. Polymerization (extension): Primer extension occurs at the temperature optimum for thermostable DNA polymerase.

Additional Problems

1. An experimenter decided to use PCR to isolate a particular gene from an exotic organism. The gene under study had previously been characterized in several different organisms. The experimenter was able to locate two conserved regions within the gene, to make oligonucleotides complementary to these two regions, and to set up a PCR amplification. The experimenter programmed the thermocycler and left for the weekend. After two cycles, a power failure occurred, erasing the memory in the thermocycler. A colleague, realizing the

importance of the experiment, decided to restart the thermocycler. However, the colleague knew nothing about PCR, and so the person asked you what to do.

a. Describe a typical PCR cycle and explain the purpose of each step in the cycle.

b. The person takes your advice and is in fact so excited by your description of PCR that the person runs a gel. No products were found. Desperate for results, the person decides to change one of the temperatures in the cycle and to perform a second PCR run. Which temperature, which direction, and why?

c. The adjustment results in production of only one faint band, not an impressive yield for such a marvelous procedure. The person decides to make one more attempt and triples the total number of cycles. This time several bands are produced. Can the results be trusted? Explain.

2. One of the constraints of standard PCR is having to know the sequence of two separate regions of a gene. Can you suggest a method that eliminates the necessity of having to know two separated sequences within a gene?

3. In problem 5 above, you were asked to calculate the genome library size of lungfish DNA necessary to have a 99% probability of finding a particular DNA fragment. The answer came out to 10 million. If the library is plated onto Petri dishes, with approximately 1000 colonies per plate, calculate the number of plates required, the total surface area, and the height if the plates were stacked on top of each other. (A typical Petri dish is 100 x 15 mm.)

4a. A biochemist is attempting to clone a eukaryotic gene and, using Southern analysis of an *Eco*RI digest of total genomic DNA, learns that the gene is located on a 2.5 kbp *Eco*RI fragment. Further, it was discovered that the *Eco*RI fragment lacks *Bam*HI restriction sites. Suggest an appropriate vector to use in this case.

b. The strategy the biochemist decides to take is to digest total genomic DNA with *Eco*RI, separate the fragments by agarose gel electrophoresis, cut out of the gel the 2.5 kbp fragment, elute the DNA from the gel slice, and ligate the DNA into a suitable vector. "Piece of cake!" the biochemist exclaims, "This will be a one-step cloning and we'll have our gene without having to screen our recombinants by Southern analysis." This is an optimistic statement. Can you explain why?

5. Vectors based on λ are available to simplify chromosome walking. They contain two different bacteriophage RNA polymerase promoters (from SP6, T3 or T7 phage) flanking a cloning site. Further, the RNA polymerase promoters are on opposite strands and they are oriented such that expression from these promoters produces RNAs complementary to the insert ends. Explain how these vectors facilitate chromosome walking.

6. In one version of oligonucleotide-directed, site-specific mutagenesis, a recombinant M13 is used as a template for second-strand synthesis using a mutagenic oligonucleotide. The mutagenic oligonucleotide is complementary to several bases on both sides of the target base but at the target base it produces a single-base mismatch. The oligonucleotide is used *in vitro* to synthesize a complementary strand whose ends are ligated together by DNA ligase. Thus, a single-stranded M13 is converted to a double-stranded, closed, circular form (often referred to as the RF or replicative form). Suggest a simple assay to check for second-strand synthesis and for DNA ligase activity.

Abbreviated Answers

1a. In a typical PCR reaction, double-stranded template DNA and two primers, complementary to two sequences flanking the region to be amplified, are mixed with a thermostable DNA polymerase and dNTPs. The mixture is first heated to 90-95°C to denature the template DNA. Then, the temperature is lowered to between 40°C and 60°C to allow primer binding to the template. Finally, the temperature is adjusted to the temperature optimum for the DNA polymerase to allow primer extension to occur (72-75°C). The process of heating to denature, cooling to allow primer binding, and heating to extend primer represents a cycle that is repeated a number of times, the result of which is an exponential amplification of the DNA between the two primers.

b. Failure to amplify may be a result of inefficient primer binding to the template because the temperature of the second stage of the cycle, namely primer binding, exceeds the T_m of the primer. One possibility is to lower the primer binding temperature.

c. The production of only a faint band may be parent nature's way of telling you that something is not quite right with the reaction. It may be the case that the primers are inappropriate for the template DNA. Typically, PCR reactions are set up to produce a generous amplification after a specified number of cycles. Increasing substantially the cycle number may allow for amplification of DNA contaminating the template

DNA. In addition, increasing the cycle-number provides additional opportunities for random mispriming to occur and to be subsequently amplified.

2. In a process known as inverse-PCR, DNA can be amplified from a region of known sequence outward in both directions. Template DNA is digested with a restriction enzyme known to cut outside of the region of interest. (Southern analysis can be used to screen for an appropriate enzyme.) The fragmented DNA is circularized by ligation and then used as template with primers that recognized the known sequence. Linear fragments, produced by primer extension using the circular DNA as template, will subsequently act as template during later rounds of PCR. A diagram of the process is shown below.

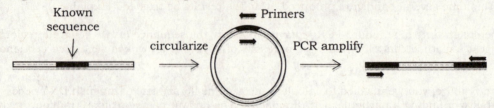

In anchored-PCR, template DNA is fragmented by restriction digestion and the ends ligated to a primer whose sequence is known. Amplification occurs between this primer and an internal, previously characterized sequence within the gene. This is shown below.

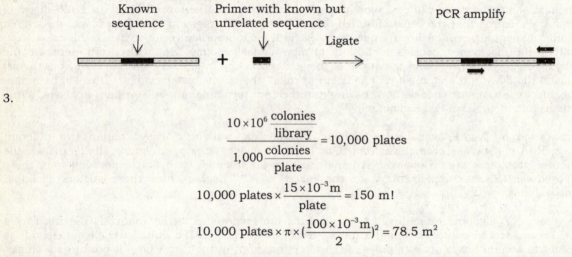

3.

$$\frac{10 \times 10^6 \dfrac{\text{colonies}}{\text{library}}}{1,000 \dfrac{\text{colonies}}{\text{plate}}} = 10,000 \text{ plates}$$

$$10,000 \text{ plates} \times \frac{15 \times 10^{-3}\text{m}}{\text{plate}} = 150 \text{ m!}$$

$$10,000 \text{ plates} \times \pi \times (\frac{100 \times 10^{-3}\text{m}}{2})^2 = 78.5 \text{ m}^2$$

4a. The 2.5 kbp fragment is a small fragment that could be cloned in a plasmid. Cosmids, λ, and YACs are all more appropriate for much larger fragments.

b. The cloning strategy is sound; however, the biochemist may be a bit optimistic in thinking that the 2.5 kbp fragment identified by Southern analysis is a unique fragment. Eukaryotic genomes are large, greater than 10×10^6 bp for yeast, a simple eukaryote, and a digest with a restriction enzyme with a six-base recognition sequence is expected to generate a very large number of fragments. For example, the probability of finding a particular 6-base sequence is $(1/4)^6 = (1/4,096)$. For a genome of 10×10^6 bp one might expect to have around $10 \times 10^6 \times (1/4,096) = 2,441$ sites. Therefore, an enzyme with a six-base recognition sequence is expected to generate thousands of bands, which will form essentially a continuum upon electrophoresis. The fact that Southern analysis highlights a single band only means that, of all of the bands in the continuum at that particular size, there is one (or possibly more) with the sequence of interest. The biochemist will have to screen the products of the cloning procedure.

5. In chromosome walking, the ends of a cloned fragment are used as probes to search for other clones in the library containing overlapping segments of the DNA. The vectors described are ideally suited for the purpose of producing RNAs from DNA at the ends of the cloned fragment. The RNA can be labeled and used in Southern analysis to search for overlapping clones.

6. One easy assay is to examine the products by agarose gel electrophoresis. Single-stranded template M13 has a faster electrophoretic mobility than double-stranded M13. Therefore, second-strand synthesis will be accompanied by a gel shift of the M13 template to a slower migrating form. If second-strand synthesis is

successful, a double-stranded, nicked circle is produced. Successful ligation will convert the nicked circle to a covalently-closed double-stranded molecule. Electrophoresis in the presence of ethidium bromide is capable of distinguishing between these two. Ethidium bromide binding to covalently-closed, double-stranded DNA will induce positive supercoils. Thus, the mobility of the ligated product will be slightly faster than relaxed circles in the presence of ethidium.

Summary

Plasmids are circular, extrachromosomal DNA molecules. Plasmids have the ability to self-replicate and often carry genes for novel metabolic capacities (such as antibiotic resistance) that allow host cells to grow under otherwise adverse conditions. These facts make plasmids ideal vehicles for the amplification of foreign DNA segments. The discovery of restriction endonucleases provided an essential tool for generating big DNA fragments with predictable features, such as defined ends. In particular, restriction endonucleases that create 5'- and 3'- overhangs, "sticky ends", ease the task of assembling chimeric plasmids *in vitro*. The circular plasmid is opened, or "linearized", by restriction endonuclease digestion, a foreign DNA segment is annealed to it via hybridization of its ends with the complementary "sticky ends" of the plasmid, and the two DNA sequences are covalently ligated using DNA ligase, a process that recloses the plasmid circle. Using these techniques and related developments, essentially any DNA sequence can be cloned.

Since the size of foreign DNA inserts in plasmids is inherently limited to 10 kbp, other vector systems for the propagation of larger DNA segments have emerged, such as bacteriophage λ-based cosmid cloning systems and even the creation of artificial yeast chromosomes (YACs) for cloning purposes.

The overriding aim of most cloning work is the isolation and characterization of specific genes. Since any given gene represents only a tiny fraction of an organism's genome or total DNA content, the challenge is to find the particular DNA sequence that represents the gene among a large population of extraneous DNA segments. Genomic libraries are prepared by partially digesting the organism's DNA and cloning the fragments into a suitable vector. The library is then screened for the gene of interest, using probes designed to uniquely identify the gene. Typically, a probe is a single-stranded DNA molecule whose nucleotide sequence is complementary to some region of the target gene. Genes can also be cloned by identifying a clone derived from a gene on the same chromosome and using this clone as a probe to find clones of neighboring DNA segments, then repeating the process, progressively moving along the chromosome. This tactic is called "chromosome walking". Libraries can also be constructed by copying the mRNA population isolated from a selected cell type into DNA using reverse transcriptase. The product, cDNA, is converted to duplex DNA and cloned into an appropriate vector. cDNA libraries cloned into expression vectors where the DNA inserts are both transcribed and translated in recipient cells can be screened for protein products of the gene using immunological techniques and other methods. The properties of gene regulatory sequences can be investigated by cloning them into sites adjacent to genes encoding easily assayed protein products, so-called reporter genes. Expression of the reporter gene serves as an index of the transcriptional efficiency of the cloned regulatory sequence.

The major problem with cloning is identifying a particular clone of interest and this is often accomplished by nucleic acid hybridization. Complementary single-stranded DNA or RNA can be used to identify cloned DNA. In Southern hybridization, denatured DNA is transferred to a solid filter and the filter then probed with DNA with a sequence of interest to the experimenter. The transferred DNA could be from a plate of bacterial colonies each harboring recombinant plasmids or DNA could be from an agarose or polyacrylamide gel. DNA microarrays rely on hybridization to detect expression of many genes under a variety of conditions. The "gene chips" are solid supports decorated with many single strand DNA probes derived from genomic information. One can use gene chips to detect differential expression of genes under particular environmental conditions. This is accomplished by isolating mRNA under particular conditions, converting it to cDNA and then, in effect, determining how efficiently cDNA hybridizes to the gene chips.

Polymerase chain reaction (PCR) is a technique for greatly amplifying the amount of any DNA segment in a genome, provided sequence information is available about regions flanking the desired segment. This information is used in synthesizing pairs of complementary oligonucleotides that anneal next to the segment (one oligonucleotide at each 3'-end). These oligonucleotides serve to prime synthesis of the region by DNA polymerase. The resultant duplex DNA is melted by heating to 95°C, and the solution is cooled to initiate another cycle of synthesis. Repeating this thermal cycle and synthesis regime for 25 cycles leads to a million-fold amplification in the concentration of the original DNA sequence. PCR has been automated through the invention of thermal cyclers that carry out the entire process in less than 4 hours. PCR is thus an automated, cell-free amplification procedure for generation of specific DNA sequences.

With cloned genes in hand, alteration of the nucleotide sequences of these genes in order to observe the consequences of these "mutations" on the biological function of the gene or gene product became feasible. *In vitro* mutagenesis has been used to change the nucleotide sequence in a systematic way so that the effects of

Chapter 13

Enzyme—Kinetics and Specificity

$$\bullet \ \bullet \ \bullet \ \bullet \ \bullet \ \bullet \ \bullet \ \bullet \ \bullet \ \bullet \ \bullet \ \bullet \ \bullet \ \bullet \ \bullet \ \bullet \ \bullet \ \bullet \ \bullet \ \bullet$$

Chapter Outline

- ❖ Enzymes
  - ᐱ Biological catalysts that function in dilute, aqueous solutions under mild conditions (e.g., pH) and low temperatures to increase reaction rate
  - ᐱ Catalytic power: Ratio of catalyzed rate to uncatalyzed rate
  - ᐱ Specificity
- ❖ Nomenclature
  - ᐱ Common names
  - ᐱ International Commission on Enzymes (EC numbers): Six classes
- ❖ Terms
  - ᐱ Holoenzyme: Protein and prosthetic group
  - ᐱ Apoenzyme: Protein lacking prosthetic group
  - ᐱ Prosthetic group: Firmly bound coenzyme
  - ᐱ Coenzyme: Organic compound cofactor
- ❖ Chemical kinetics
  - ᐱ Rate law: $v = -dA/dt = k[A]^n$
    - • k = rate constant
    - • n = order of reaction,
      - • n=1, first order $[A]$
      - • n=2, second order $[A]^2$ or $[A][B]$
  - ᐱ Molecularity: Number of molecules that must simultaneously react
- ❖ Reaction rates
  - ᐱ Limited by activation barrier: Free energy needed to reach transition state
  - ᐱ Arrhenius equation $k = Ae^{-\Delta G^*/RT}$: ΔG^* = Free energy of activation
  - ᐱ Influenced by
    - • Temperature
    - • Catalysts
- ❖ Enzyme kinetics
  - ᐱ Saturatable: Zero-order kinetics at high $[S]$
  - ᐱ Michaelis-Menten: $v_0 = (V_{max}[S])/(K_m+[S])$
    - • $K_m = (k_{-1}+k_2)/ k_1$
    - • $[S] = K_m$ when $v_0 = V_{max}/2$
- ❖ Enzyme quantities
  - ᐱ One international unit: Amount to catalyze formation of one micromole of product in one minute
  - ᐱ Turnover number = $k_{cat} = k_2 = V_{max}/[E_T]$
  - ᐱ k_1 sets upper limit on catalytic efficiency: Reaction can go no faster than rate at which E and S form ES
  - ᐱ k_{cat}/K_m = catalytic efficiency
- ❖ Plots
  - ᐱ Direct plot: v versus $[S]$: $v_0 = (V_{max}[S])/(K_m+[S])$
    - • Rectangular hyperbola

233

- Asymptotically approaches V_{max} when [S] high
- $[S] = K_m$ when $v_0 = V_{max}/2$
- ⋏ Lineweaver-Burk (double-reciprocal): $1/v_0$ versus $1/[S]$: $1/v_0 = (K_m/V_{max})(1/[S]) + 1/V_{max}$
  - Linear
  - Slope $= K_m/V_{max}$
  - y-intercept $= 1/V_{max}$
  - x-intercept $= -1/K_m$
- ⋏ Hanes-Woolf: $[S]/v_0$ versus [S]: $[S]/v_0 = (1/V_{max})[S] + (K_m/V_{max})$
  - Linear
  - Slope $= 1/V_{max}$
  - y-intercept $= K_m/V_{max}$
  - x-intercept $= -K_m$
- ❖ pH dependence: Enzymes generally active over pH range
- ❖ Temperature dependence of enzyme-catalyzed reactions
  - ⋏ Below 50°C: Q_{10}: Ratio of activities at two temperatures 10° apart: For typical enzyme $Q_{10} = 2$
  - ⋏ Above 50°C: Typically enzyme denatures
- ❖ Inhibition
  - ⋏ Reversible inhibition: Noncovalent
    - Competitive inhibition: Inhibitor and substrate compete for same binding site
      - V_{max} unchanged
      - $K_m = K_m(1 + [I]/K_I)$
    - Noncompetitive inhibition
      - Pure noncompetitive inhibition: S and I bind at different sites
        - ○ K_m unchanged
        - ○ $V_{max. app} = V_{max}/(1 + [I]/K_I)$
      - Mixed noncompetitive inhibition: I binding influences S binding
        - ○ $K_I < K_{I'}$: $V_{max,app}$ decreased: $K_{m,app}$ increased
        - ○ $K_{I'} < K_I$: $V_{max,app}$ decreased: $K_{m,app}$ decreased
      - Uncompetitive: I bonds to ES complex
        - ○ $V_{max,app}$ decreased: $K_{m,app}$ decreased
  - ⋏ Irreversible inhibition: Covalent
    - Suicide substrate or Trojan Horse substrate
    - Site-specific affinity label
- ❖ Bisubstrate reactions: Two substrates converted to two products
  - ⋏ Sequential or single displacement
    - Random: Creatine kinase: Either substrate binds to enzyme, either product is released
    - Ordered: NAD^+-dependent dehydrogenases: Leading substrate binds first
  - ⋏ Ping-Pong or double-displacement: Aminotransferase: Leading substrate binds: Enzyme modified: Product released: Second substrate binds: Enzyme unmodified: Second product released
- ❖ Enzyme specificity
  - ⋏ Induced fit: Conformational change with substrate binding
  - ⋏ Active site binds transition state molecule
  - ⋏ Conformational changes by 1st substrate binding may be required for 2nd substrate binding
- ❖ Catalytic biomolecules
  - ⋏ Enzymes: Proteins
  - ⋏ Ribozymes: RNA
    - RNase P RNA component
    - Self-splicing RNAs
    - Ribosomal RNA: Peptidyl transferase property of rRNA
  - ⋏ Abzymes: Antibodies

Chapter Objectives

Michaelis-Menten enzyme kinetics

One of the keys to understanding Michaelis-Menten enzyme kinetics is to remember the model and the assumptions. The enzyme-substrate complex is in rapid equilibrium with free substrate. Product formation involves a fast catalytic step followed by a slow release of product. The velocity, v, is the initial velocity of the reaction measured immediately upon addition of substrate when product concentration is very small. The Michaelis-Menten equation is not a complicated equation and is easy to understand.

$$v = \frac{V_{max}[S]}{K_m + [S]} = \frac{k_2[E_T][S]}{\frac{k_{-1} + k_2}{k_1} + [S]}$$

It is easy to remember that K_m has units of concentration because the denominator of the Michaelis-Menten equation is the term $K_m + [S]$. V_{max} is equal to the product of k_2 and $[E_T]$. V_{max} is reached when substrate concentration is large. How large? Large relative to K_m. The equation is an example of a rectangular hyperbola. In a plot of v vs $[S]$, V_{max} is approached asymptotically as $[S]$ increases and K_m is the substrate concentration that supports $v = V_{max}/2$.

Graphical representation

The Lineweaver-Burk double-reciprocal plot uses $1/v$ and $1/[S]$ as variables because they are linearly related and give rise to straight lines. A minor disadvantage to double-reciprocal plots is that they use inverse space: as $[S]$ increases, $1/[S]$ decreases. Values of v at low $[S]$ tend to dominate the plot. With this in mind it is easy to remember that the $1/v$ intercept is equal to $1/V_{max}$, because the intercept occurs at the smallest real value of $1/v$. Smaller values of $1/v$ occur only to the left of the $1/v$-axis, where $1/[S]$ is negative. (Negative concentrations do not exist in the real world.) The $1/[S]$ intercept is equal to $-1/K_m$. Alternative variables related linearly are $[S]/v$ vs. $[S]$, used in Hanes-Woolf plots. Concentrate on direct plots (i.e., v vs. $[S]$) and on double-reciprocal plots. Examples of a double-reciprocal plot and a Hanes-Woolf plot are presented in Figures 13.9 and 13.10 below.

$$\frac{1}{v} = \frac{K_m}{V_{max}}\left(\frac{1}{[S]}\right) + \frac{1}{V_{max}}$$

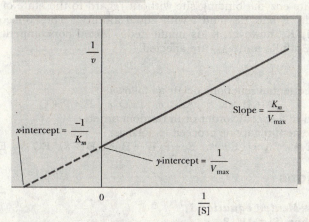

Figure 13.9 The Lineweaver-Burk double reciprocal plot, depicting extrapolations that allow the determination of the x - and y -intercepts and slope.

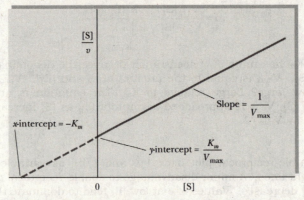

$$\frac{[S]}{v} = \left(\frac{1}{V_{max}}\right)[S] + \frac{K_m}{V_{max}}$$

Figure 13.10 A Hanes-Woolf plot of $[S]/v$ versus $[S]$, another straight-line rearrangement of the Michaelis-Menten equation.

Inhibitors

Competitive inhibition occurs whenever an inhibitor competes with a substrate for the substrate binding site in a reversible manner. How will the presence of a competitive inhibitor affect enzyme kinetics? Clearly, at high $[S]$, S will out compete the inhibitor. Thus, V_{max} is unaffected. The apparent K_m is increased to $K_m(1 + [I]/K_I)$. (The term $[I]/K_I$ is the inhibitor concentration normalized to the inhibitor dissociation constant, K_I.)

In noncompetitive inhibition, the inhibitor binds reversibly to a site different than the substrate binding site but the inhibitor-bound enzyme is inactive. The easiest case to remember is pure noncompetitive inhibition. Here, I binds to its enzyme binding site without regard to the state of occupancy of the substrate binding site. In other words, E and ES bind I with equal affinity. The consequence is a decrease in the apparent V_{max} to $V_{max}/(1 + [I]/K_I)$; however, K_m is unaffected. Mixed noncompetitive inhibition occurs when I binding to E and ES differ. Both K_m and V_{max} are affected.

Bisubstrate reactions

Single-displacement or sequential reactions occur as follows:
$$A + B + E \star \ AEB \star \ PEQ \star \ E + P + Q$$
The two substrates may bind in a specific order or in random order.
Ping-Pong or double-displacement reactions proceed as follows:
$$A + E \star \ EA \star \ E'P \star \ E' + P; E' + B \star \ E'B \star \ EQ \star \ E + Q$$

Problems and Solutions

1. Exploring the Michaelis-Menten equation I
What is the v/V_{max} ratio when $[S] = 4K_m$?

Answer:

$$v = \frac{V_{max}[S]}{K_m + [S]}$$

When $[S] = 4K_m$

$$v = \frac{V_{max} \times 4K_m}{K_m + 4K_m} = \frac{4V_{max}}{5}, \text{ or}$$

$$\frac{v}{V_{max}} = 0.8$$

2. Exploring the Michaelis-Menten equation II
If V_{max} = 100 µmol/mL·sec and K_m = 2 mM, what is the velocity of the reaction when [S] = 20 mM?

Answer: For V_{max} = 100 µmol/sec and K_m = 2 mM, [S] = 20 mM

$$\upsilon = \frac{V_{max}[S]}{K_m + [S]}$$

$$= \frac{100\dfrac{\mu mol}{mL \cdot sec} \times 20\ mM}{2\ mM + 20\ mM}$$

$$= \frac{100\dfrac{\mu mol}{mL \cdot sec} \times 20\ mM}{22\ mM}$$

$$= 91\frac{\mu mol}{mL \cdot sec}$$

3. Exploring the Michaelis-Menten equation III
For a Michaelis-Menten reaction, k_1 = 7 x 10^7/ M·sec, k_{-1} = 1 x 10^3/sec and k_2 = 2 x 10^4/sec. What are the values of K_s and K_m? Does substrate binding approach equilibrium or does it behave more like a steady-state system?

Answer: When k_1 = 7 x 10^7/M·sec, k_{-1} = 1 x 10^3/sec, k_2 = 2 x 10^4/sec

$$K_s = \frac{k_{-1}}{k_1}$$

$$= \frac{\dfrac{1 \times 10^3}{sec}}{\dfrac{7 \times 10^7}{M \cdot sec}}$$

$$K_s = 1.43 \times 10^{-5}M$$

$$K_m = \text{Michaelis-Menten constant} = \frac{k_{-1} + k_2}{k_1}$$

$$= \frac{\dfrac{1 \times 10^3}{sec} + \dfrac{2 \times 10^4}{sec}}{\dfrac{7 \times 10^7}{M \cdot sec}}$$

$$K_m = 3.0 \times 10^{-4}M$$

Once the ES complex forms the substrate may be either converted to product or released as S. Formation of product is governed by a first order rate constant, k_2, that is 20 times larger than the first order rate constant for dissociation of ES (i.e., k_{-1}). Thus, the enzyme behaves like a steady state system. Only in the case where k_2 is rate-limiting will substrate binding approach equilibrium.

4. Graphing the results from kinetics experiments with enzyme inhibitors
The following kinetic data were obtained for an enzyme in the absence of inhibitor (1), and in the presence of two different inhibitors (2) and (3) at 5 mM concentration. Assume $[E_T]$ is the same in each experiment.

[S] (mM)	(1) υ (µmol/mL·sec)	(2) υ (µmol/mL·sec)	(3) υ (µmol/mL·sec)
1	12	4.3	5.5
2	20	8	9
4	29	14	13
8	35	21	16
12	40	26	18

Plot these data as double-reciprocal Lineweaver-Burk plots and use your graph to answers a. and b.

a. Determine V_{max} and K_m for the enzyme.

b. *Determine the type of inhibition and the K_I for each inhibitor.*

Answer: The data may be analyzed using double-reciprocal variables. For each [S] and corresponding υ, we will calculate $1/[S]$ and $1/\upsilon$.

[S] (mM)	1/[S] M^{-1}	υ (1) μmol/mL·sec	1/ υ (1) mL·sec/μmol	υ (2) μmol/mL·sec	1/ υ (2) mL·sec/μmol	υ (3) μmol/mL·sec	1/ υ (3) mL·sec/μmol
1	1000	12	8.33×10^4	4.3	2.33×10^5	5.5	1.82×10^5
2	500	20	5.00×10^4	8	1.25×10^5	9	1.11×10^5
4	250	29	3.45×10^4	14	7.14×10^4	13	7.69×10^4
8	125	35	2.86×10^4	21	4.76×10^4	16	6.25×10^4
12	83.3	40	2.50×10^4	26	3.85×10^4	18	5.56×10^4

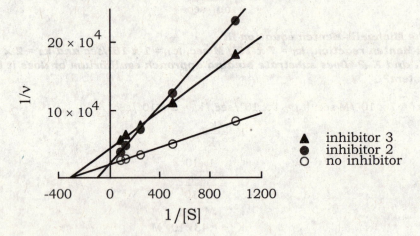

Plots of $1/\upsilon$ vs. $1/[S]$ indicate straight lines given by

(1) $1/\upsilon = \ \ 63.2(1/[S]) + 1.95 \times 10^4$
(2) $1/\upsilon = \ 211.8(1/[S]) + 2.00 \times 10^4$
(3) $1/\upsilon = \ 137.2(1/[S]) \ + 4.38 \times 10^4$

In general,

$$\frac{1}{\upsilon} = \frac{K_m}{V_{max}} \times \frac{1}{[S]} + \frac{1}{V_{max}}$$

Thus, the y-intercept is equal to $1/V_{max}$ and the x-intercept is equal to $-1/K_m$.

(a)

Condition	V_{max} (μmol/mL·sec)	K_m (mM)
No inhibitor	51.2	3.2
	$V_{max,app}$	$K_{m,app}$
5 mM inhibitor (2)	49.9	10.6
5 mM inhibitor (3)	22.8	3.1

(b) Inhibitor (2) increases the apparent K_m of the enzyme without affecting V_{max}. This is characteristic of a competitive inhibitor. In this case K_I is calculated as follows

$$K_{m,app} = K_m \left(1 + \frac{[I]}{K_I}\right) \text{ or}$$

$$10.6 \text{ mM} = 3.2 \text{ mM}\left(1 + \frac{5 \text{ mM}}{K_I}\right)$$

Solving for K_I we find

$$K_I = \frac{5 \text{ mM}}{\left(\frac{10.6 \text{ mM}}{3.2 \text{ mM}} - 1\right)} = 2.16 \text{ mM}$$

Inhibitor (3) decreases V_{max} but leaves K_m unchanged, an example of noncompetitive inhibition. In this case K_I is calculated as follows

$$\frac{1}{V_{max,app}} = \frac{1}{V_{max}}\left(1 + \frac{[I]}{K_I}\right) \text{ or}$$

$$\frac{1}{22.8} = \frac{1}{51.2}\left(1 + \frac{[I]}{K_I}\right)$$

Solving for K_I we find

$$K_I = \frac{5 \text{ mM}}{\left(\frac{51.2}{22.8} - 1\right)} = 4.01 \text{ mM}$$

5. How varying the amount of enzyme or the addition of inhibitors affects v versus [S] plots
Using Figure 13.7 as a model, draw curves that would be obtained from v versus [S] plots when
a. twice as much enzyme is used.
b. half as much enzyme is used.
c. a competitive inhibitor is added.
d. a pure noncompetitive inhibitor is added.
e. an uncompetitive inhibitor is added.
For each example, indicate how V_{max} and K_m change.

Answer: Figure 13.7 is a direct plot of initial velocity versus substrate.
a. With twice as much enzyme the plot should asymptotically approach a value of V_{max} that is twice the value shown in the figure. But, since the K_m is unchanged it will reach 1/2 its V_{max} at the same value of [S] independent of enzyme concentration.
b. Half as much enzyme will reach a V_{max} value half that shown in Figure 13.7 but it will have the same K_m.
c. A competitive inhibitor will not influence V_{max}, so the plot will asymptotically approach V_{max} but the K_m will be larger by a factor of $(1 + [I]/K_I)$ where [I] is the inhibitor concentration and K_I is the dissociation constant for inhibitor binding to the enzyme.
d. For a pure noncompetitive inhibitor, K_m will be unchanged but V_{max} will be lower by a factor of $1/(1 + [I]/K_I)$.
e. For an uncompetitive inhibitor K_m will be decreased by the factor $1/(1 + [I]/K_{I'})$ and V_{max} will be decreased by the same factor where $K_{I'}$ is the inhibitor dissociation constant for binding to the enzyme/substrate complex.

Plots for a. and b. are shown below along with a control plot (in the middle). It is clear that while each reaches a different V_{max}, they all show the same K_m.

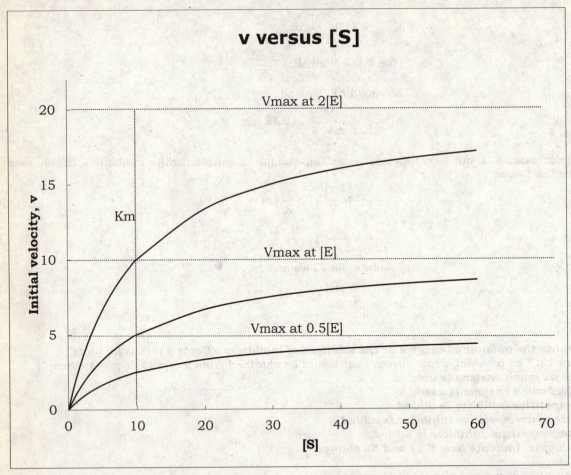

The next graph shows the effects of inhibitors. The competitive inhibitor Increases the K_m but does not affect the V_{max}.   For pure noncompetitive inhibition, V_{max} is depressed whereas K_m is unchanged.   Finally, for uncompetitive inhibition both K_m and V_{max} are effected but in interesting ways. V_{max} is decreased but so is K_m. If you think of K_m apparent as the concentration of substrate at which the enzyme is at one-half V_{max}, it would appear that a lower $K_{m, apparent}$ suggests better functioning of the enzyme!

As a cautionary note: The value of [S] at which the enzyme is at one-half V_{max} is the K_m when the enzyme is tested without inhibitors. When inhibitors are tested, it is more appropriate to refer to this value as the apparent K_m. The relationships between K_m and apparent K_m's are given in Table 13.6 of the textbook.

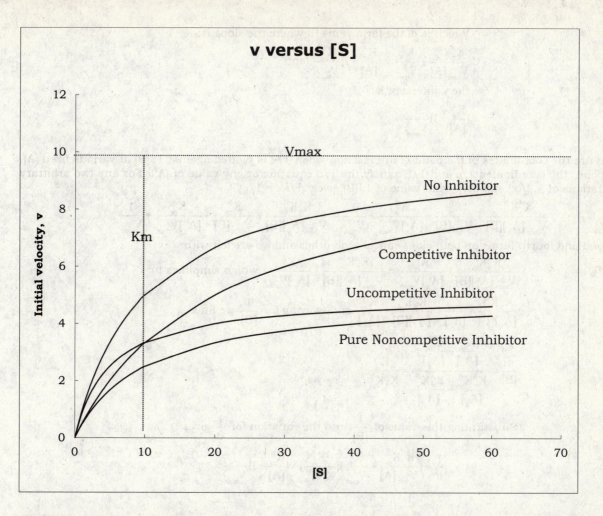

v versus [S]

6. Using graphical methods to derive the kinetic constants for an ordered, single-displacement reaction

The general rate equation for an ordered, single-displacement reaction where A is the leading substrate is

$$\upsilon = \frac{V_{max}[A][B]}{K_S^A K_m^B + K_m^A[B] + K_m^B[A] + [A][B]}$$

Write the Lineweaver-Burk (double-reciprocal) equivalent of this equation and calculate algebraic expressions for (a) the slope; (b) the y-intercepts; and (c) the horizontal and vertical coordinates of the point of intersection, when 1/υ is plotted versus 1/[B] at various fixed concentrations of A.

Answer: (a, b) To write the Lineweaver-Burk equation we take the inverse of both sides of the equation:

$$\frac{1}{\upsilon} = \frac{K_S^A K_m^B + K_m^A[B] + K_m^B[A] + [A][B]}{V_{max}[A][B]} = \frac{K_S^A K_m^B}{V_{max}[A][B]} + \frac{K_m^A}{V_{max}[A]} + \frac{K_m^B}{V_{max}[B]} + \frac{1}{V_{max}}$$

Rearranging the expression for $\dfrac{1}{\upsilon}$ as a function of $\dfrac{1}{[B]}$ gives:

$$\frac{1}{\upsilon} = \left(\frac{K_S^A K_m^B}{V_{max}[A]} + \frac{K_m^B}{V_{max}}\right)\frac{1}{[B]} + \left(\frac{K_m^A}{[A]} + 1\right)\frac{1}{V_{max}}$$

Which is of the form y=mx+b where the slope is:

$$\frac{K_S^A K_m^B}{V_{max}[A]} + \frac{K_m^B}{V_{max}} = (\frac{K_S^A}{[A]} + 1)\frac{K_m^B}{V_{max}}, \text{ and}$$

the y-intercept is:

$$(\frac{K_m^A}{[A]} + 1)\frac{1}{V_{max}}$$

(c) What are the coordinates of the point of intersection when $1/\upsilon$ is plotted against $1/[B]$ at various fixed [A]. At this point, the coordinates $1/\upsilon$ and $1/B$ satisfy the $1/\upsilon$ equation at any value of [A]. For any two arbitrary concentrations of A, $[A_x]$ and $[A_y]$, what value of $1/[B]$ gives $1/\upsilon x = 1/\upsilon y$?

$$\frac{K_S^A K_m^B}{V_{max}[A_x][B]} + \frac{K_m^B}{V_{max}[B]} + \frac{K_m^A}{[A_x]V_{max}} + \frac{1}{V_{max}} = \frac{K_S^A K_m^B}{V_{max}[A_y][B]} + \frac{K_m^B}{V_{max}[B]} + \frac{K_m^A}{[A_y]V_{max}} + \frac{1}{V_{max}}$$

The second and fourth terms on both sides cancel each other and we are left with:

$$\frac{K_S^A K_m^B}{V_{max}[A_x][B]} + \frac{K_m^A}{[A_x]V_{max}} = \frac{K_S^A K_m^B}{V_{max}[A_y][B]} + \frac{K_m^A}{[A_y]V_{max}} \text{ which simplifies to:}$$

$$\frac{K_S^A K_m^B}{[A_x][B]} + \frac{K_m^A}{[A_x]} = \frac{K_S^A K_m^B}{[A_y][B]} + \frac{K_m^A}{[A_y]} \text{ and, solving for } \frac{1}{[B]} \text{ we find:}$$

$$\frac{1}{[B]} = \frac{\dfrac{K_m^A}{[A_y]} - \dfrac{K_m^A}{[A_x]}}{\dfrac{K_S^A K_m^B}{[A_x]} - \dfrac{K_S^A K_m^B}{[A_y]}} = \frac{-K_m^A}{K_S^A K_m^B}$$

Substituting this value of $\dfrac{1}{[B]}$ into the equation for $\dfrac{1}{\upsilon}$ gives:

$$\frac{1}{\upsilon} = (\frac{K_S^A K_m^B}{V_{max}[A]} + \frac{K_m^B}{V_{max}})(\frac{-K_m^A}{K_S^A K_m^B}) + (\frac{K_m^A}{[A]} + 1)\frac{1}{V_{max}}$$

$$\frac{1}{\upsilon} = \frac{1}{V_{max}}(1 - \frac{K_m^A}{K_S^A})$$

7. Interpreting kinetics experiments from graphical patterns
The following graphical patterns obtained from kinetic experiments have several possible interpretations depending on the nature of the experiment and the variables being plotted. Give at least two possibilities for each.

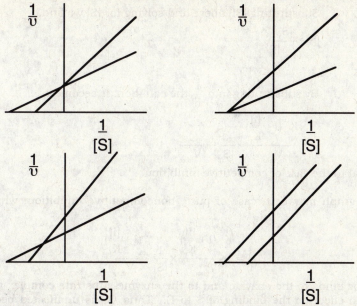

Answer: In the top left graph, the results may be due to a competitive inhibitor. The line with the steeper slope is from enzyme kinetic measurements in the presence of an inhibitor whereas the other line is enzyme kinetic data without inhibitor.   In competitive inhibition, V_{max}, the reciprocal of the $1/\upsilon$ intercept, is independent of the presence of inhibitor. However, $K_{m,app}$ increases, as reflected in the decrease (in absolute magnitude) of the $1/[S]$ intercept which equals $1/K_{m,app}$. In competitive inhibition, I and S compete for the same enzyme-binding site.   The apparent K_m increases because, in the presence of inhibitor, higher concentrations of substrate are required to half-saturate the enzyme. The lines are described by the following equation:

$$\frac{1}{\upsilon} = \frac{K_m(1+\frac{[I]}{K_I})}{V_{max}} \times \frac{1}{[S]} + \frac{1}{V_{max}}$$

An alternative interpretation of the data is that the inhibitor binds to and forms a complex with the substrate. In this case the equation

$$\frac{1}{\upsilon} = \frac{K_m}{V_{max}} \times \frac{1}{[S]} + \frac{1}{V_{max}}$$

must be modified to take this inhibitor-substrate interaction into account. In this equation, [S] is the initial concentration of substrate available to the enzyme. The concentration of substrate used to construct the plots is the total substrate concentration, $[S_T]$. Under normal conditions, $[S_T] = [S]$, but in this case $[S_T] = [S] +$ [SI] because some of the substrate is complexed to I. We can derive an expression for [S] as follows:

From $[S_T] = [S] + [SI]$, we have

$[S] = [S_T] - [SI]$, and using

$K_I = \frac{[I][S]}{[SI]}$, we see that

$[SI] = \frac{[I][S]}{K_I}$

Substituting [SI] above and solving for [S] we find,

$$[S] = \frac{[S_T]}{1 + \dfrac{[I]}{K_I}}$$

By substituting into $\dfrac{1}{\upsilon}$ the equation, it becomes:

$$\frac{1}{\upsilon} = \frac{K_m \left(1 + \dfrac{[I]}{K_I}\right)}{V_{max}} \times \frac{1}{[S_T]} + \frac{1}{V_{max}}$$

This equation is identical to the case of competitive inhibition.

The top right hand graph may be a case of pure, noncompetitive inhibition, which is described by the following equation

$$\frac{1}{\upsilon} = \frac{K_m \left(1 + \dfrac{[I]}{K_I}\right)}{V_{max}} \times \frac{1}{[S]} + \frac{\left(1 + \dfrac{[I]}{K_I}\right)}{V_{max}}$$

In this case the inhibitor binds to the enzyme and to the enzyme-substrate complex with equal affinity. The binding of I to E has no effect on the binding of S to E. Thus, K_m is unaffected because the inhibitor-free enzyme is capable of normal catalysis. V_{max} is decreased to $V_{max}/(1 + [I]/K_I)$. In effect, the inhibitor lowers the active enzyme concentration. The data can also be explained as the result of experiments conducted at two different enzyme concentrations. (In this case no inhibition whatsoever is involved.) Another possibility, involving an inhibitor, is that the inhibitor binds irreversibly to E and inactivates it. This is the same as lowering the total enzyme concentration. Another possibility is that we may be looking at a case of random, single-displacement bisubstrate reaction. Look at Figure 13.20 and compare it to this plot. In this plot the top line would be from data generated at a fixed concentration of the second substrate whereas the second (lower) line from conditions in which the second substrate concentration is higher. Binding of substrate A is not affected by binding of B and vice versa. Finally, an inhibitor that simply inactivated the enzyme irreversibly would give similar data. The inhibitor, in effect, lowers the concentration of active enzyme, which would give the same K_m but lower V_{max}.

The bottom, left graph may be an example of mixed noncompetitive inhibition in which the inhibitor I binds to both E and ES but unequally. In this case, I binds to ES with higher affinity (i.e., lower K'_I) than to S. V_{max} is decreased and $K_{m,app}$ is increased. Alternatively, this may be an example of a single-displacement bisubstrate mechanism characterized by the following equation

$$\frac{1}{\upsilon} = \frac{1}{V_{max}}\left(K_m^A + \frac{K_S^A K_m^B}{[B]}\right) \times \frac{1}{[A]} + \frac{1}{V_{max}}\left(1 + \frac{K_m^B}{[B]}\right)$$

Compare this to Figure 13.19. The line with the shallower slope would be data collected at higher second substrate concentration.

The bottom right graph is characteristic of uncompetitive inhibition. Uncompetitive inhibitors bind only to the enzyme-substrate complex, ES. As a consequence V_{max} is decreased by the factor $(1 + [I]/K_I)$ but so is K_m, apparent. The enzyme appears to function better in terms of substrate concentration (it half-saturates at lower concentrations) but its V_{max} is lower. This happens because I binding to ES shifts the equilibrium to the right making it easier to half-saturate with substrate. Alternatively, the graph is typical of double-displacement (ping-pong) bisubstrate mechanisms as shown in Figure 3.22

8. Using the equations of enzyme kinetics to treat methanol intoxication
Liver alcohol dehydrogenase (ADH) is relatively nonspecific and will oxidize ethanol or other alcohols, including methanol. Methanol oxidation yields formaldehyde, which is quite toxic, causing, among other things, blindness. Mistaking it for the cheap wine he usually prefers, my dog Clancy ingested about 50 mL of windshield washer fluid (a solution 50% in methanol). Knowing that methanol would be excreted eventually by Clancy's kidneys if its oxidation could be blocked, and realizing that, in terms of methanol oxidation by ADH, ethanol would act as a competitive inhibitor, I decided to offer Clancy some wine. How much of Clancy's favorite vintage (12% ethanol) must he consume in order to lower the activity of his ADH on methanol to 5% of its normal value if the K_m values of canine ADH for ethanol and methanol are 1 millimolar and 10 millimolar, respectively? (The K_I for ethanol in its role as competitive inhibitor of methanol oxidation by ADH is the same as

its K_m). Both the methanol and ethanol will quickly distribute throughout Clancy's body fluids, which amount to about 15 L. Assume the densities of 50% methanol and the wine are both 0.9 g/mL.

Answer: The K_m values of alcohol dehydrogenase for ethanol and methanol are l mM and 10 mM. How many moles of methanol are in 50 mL of a 50% solution (v/v)? The solution was made by adding 25 mL methanol and adjusting the volume to 50 mL with water. This amount of methanol (i.e., 25 mL) weighs 25 mL x 0.9 g/mL = 22.5 g. The molecular composition of methanol is CH_4O with a weight of 32 g/mol. Thus, Clancy consumed 22.5 g ÷ 32 g/mol) = 0.7 moles of methanol. Fortunately for him, he diluted it into 15 L of body fluid giving a final concentration of 0.7 moles ÷15 L = 46.9 mM. This concentration is well above the K_m of alcohol dehydrogenase for methanol at 10 mM. We expect the enzyme to function at

$$\upsilon = \frac{V_{max} \times 46.9 \text{ mM}}{10 \text{ mM} + 46.9 \text{ mM}}$$
$$= 0.82\, V_{max}$$

Now, how much ethanol must Clancy consume to lower υ to 5% of 0.82 V_{max} or to about .041 V_{max}?

$$\upsilon = \frac{V_{max}[S]}{K_m(1 + \frac{[I]}{K_I}) + [S]}, \text{ or}$$

$$0.041\, V_{max} = \frac{V_{max} \times 46.9 \text{ mM}}{10 \text{ mM}(1 + \frac{[\text{Ethanol}]}{1.0 \text{ mM}}) + 46.9 \text{ mM}}$$

Solving for [Ethanol], we find [Ethanol] = 109 mM

To raise his alcohol concentration to 109 mM, he must drink:
$$109 \text{ mM} \times 15 \text{ L} = 1.63 \text{ moles of ethanol.}$$

If Clancy gets his alcohol (ethanol) from wine, how much wine must he drink? The Mr of ethanol (C_2H_6O) is 46, so 1.63 moles represents:

$$1.63 \text{ mol} \times 46\frac{g}{mol} = 75g \text{ ethanol.}$$

At a density of 0.9 g/mL, this represents

$$\frac{75g}{0.9\frac{g}{mL}} = 83 \text{ mL of pure ethanol, or}$$

$$\frac{83 \text{ mL}}{V_T} = 0.12$$

Solving for V_T,
V_T= 694 mL of 12% ethanol.   Clancy needs about one 750-mL bottle of wine.

9. Quantitative relationships between rate constants to calculate K_m, kinetic efficiency (k_{cat}/K_m) and V_{max} –I
Measurement of the rate constants for a simple enzymatic reaction obeying Michaelis-Menten kinetics gave the following results:
$k_1 = 2 \times 10^8$ $M^{-1}sec^{-1}$
$k_{-1} = 1 \times 10^3$ sec^{-1}
$k_2 = 5 \times 10^3$ sec^{-1}
   a. What is K_s, the dissociation constant for the enzyme-substrate complex?
   b. What is K_m, the Michaelis constant for this enzyme?
   c. What is k_{cat} (the turnover number) for this enzyme?
   d. What is the catalytic efficiency (k_{cat}/K_m) for this enzyme?
   e.   Does this enzyme approach "kinetic perfection"? (That is, does k_{cat}/K_m approach the diffusion-controlled rate of enzyme association with substrate?)
   f. If a kinetic measurement was made using 2 nanomoles of enzyme per mL and saturating amounts of substrate, what would V_{max} equal?
   g.   Again, using 2 nanomoles of enzyme per mL of reaction mixture, what concentration of substrate would give $\upsilon = 0.75$ V_{max}?

h. If a kinetic measurement was made using 4 nanomoles of enzyme per mL and saturating amounts of substrate, what would V_{max} equal? What would K_m equal under these conditions?

Answer: a. The dissociation constant K_s is simple the ratio of k_{-1} to k_1, the forward and reverse rate constants for formation of the ES complex. An easy way to remember this is that the units of dissociation constants are concentration. If fact, a functional definition of the dissociation constant is the concentration of free ligand that half-saturates the binding site. Given units for the rate constants it is clear that $K_s = k_{-1}/k_1$.

$$K_s = \frac{k_{-1}}{k_1} = \frac{1 \times 10^3 sec^{-1}}{2 \times 10^8 M^{-1} sec^{-1}}$$

$$K_s = 0.5 \times 10^{-5} M \text{ or } 5 \text{ }\mu M$$

b. The Michaelis constant, K_m, is $(k_{-1}+k_2)/k_1$. Again, an easy way to remember this is to recall that K_m is the substrate concentration at which the enzyme is functioning at one-half its maximum velocity. Thus, K_m must have units of concentration.

$$K_m = \frac{k_{-1} + k_2}{k_1} = \frac{1 \times 10^3 sec^{-1} + 5 \times 10^3 sec^{-1}}{2 \times 10^8 M^{-1} sec^{-1}}$$

$$K_m = 3 \times 10^{-5} M \text{ or } 30 \text{ }\mu M$$

c. The turnover number, k_{cat}, is given by $V_{max}/[E_T]$, which is simply equal to k_2. Thus, k_{cat} is 5×10^3 sec^{-1}.

d. The catalytic efficiency is the turnover number divided by K_m, both of which we have determined above.

$$\frac{k_{cat}}{K_m} = \frac{5 \times 10^3 sec^{-1}}{3 \times 10^{-5} M}$$

$$\frac{k_{cat}}{K_m} = 1.67 \times 10^8 M^{-1} sec^{-1}$$

e. The very fastest we could expect the enzyme to function is given by the rate constant for diffusion of a small molecule, which is in the range 10^8 to 10^9 M^{-1} sec^{-1}. The catalytic efficiency of the enzyme is within this range so, yes, this enzyme does approach kinetic perfection.

f. With saturating amounts of substrate, the enzyme functions at V_{max}. Given k_2 and the enzyme concentration, V_{max} is calculated as follows:

$$V_{max} = k_2 \times [E_{total}] = 5 \times 10^3 sec^{-1} \times \frac{2 \times 10^{-9} mol}{mL}$$

$$V_{max} = 1 \times 10^{-5} mol / mL \cdot sec$$

g. We must solve the Michaelis-Menten equation for [S], given $v = 0.75$ V_{max}.

$$v = \frac{V_{max} \times [S]}{K_m + [S]}$$

$$v(K_m + [S]) = V_{max} \times [S]$$

$$[S](V_{max} - v) = vK_m$$

$$[S] = \frac{vK_m}{(V_{max} - v)}$$

Given $v = 0.75$ V_{max} and $K_m = 30$ μM

$$[S] = \frac{0.75 \, V_{max} \times 30 \, \mu M}{(V_{max} - 0.75 \, V_{max})}$$

$$[S] = 90 \mu M$$

h. We are asked here to determine V_{max} given twice the enzyme amount used in "f". The V_{max} is simply proportional to the enzyme concentration and, so, is twice the value determined in f.

$$V_{max} = k_2 \times [E_{total}] = 5 \times 10^3 sec^{-1} \times \frac{4 \times 10^{-9} mol}{mL}$$

$$V_{max} = 2 \times 10^{-5} mol / mL \cdot sec$$

The K_m is, however, unchanged at 30 μM.

246

10. Quantitative relationships between rate constants to calculate K_m, kinetic efficiency (k_{cat}/K_m) and V_{max} –II

Triose phosphate isomerase catalyzes the conversion of glyceraldehyde-3-phosphate to dihydroxyacetone phosphate.

$$Glyceraldehyde\text{-}3\text{-}P \rightleftharpoons dihydroxyaceone\text{-}P$$

The K_m of this enzyme for its substrate glyceraldehyde-3-phosphate is 1.8 x 10⁻⁵ M.

When [glyceraldehyde-3-phosphate] = 30 μM the rate of the reaction, v, was 82.5 μmole mL⁻¹ sec⁻¹.

    a. What is V_{max} for this enzyme?

    b. Assuming 3 nanomoles per mL of enzyme was used in this experiment ([E_{total}] = 3 nanomol/mL), what is k_{cat} for this enzyme?

    c. What is the catalytic efficiency (k_{cat}/K_m) for triose phosphate isomerase?

    d. Does the value of k_{cat}/K_m reveal whether triose phosphate isomerase approaches "catalytic perfection"?

    e. What determines the ultimate speed limit of an enzyme-catalyzed reaction? That is, what is it that imposes the physical limit on kinetic perfection?

Answer:

a. We are given K_m, v and [S] and are asked to determine V_{max}. Starting with the Michaelis-Menten equation, we solve it for V_{max} as follows.

$$v = \frac{V_{max} \times [S]}{K_m + [S]}$$

$$V_{max} \times [S] = v \times (K_m + [S])$$

$$V_{max} = \frac{v \times (K_m + [S])}{[S]}$$

Given $v = 82.5$ μmol mL⁻¹sec⁻¹, $K_m = 1.8 \times 10^{-5}$ M = 18 μM, [S] = 30 μM

$$V_{max} = \frac{82.5 \text{ μmol mL}^{-1}\text{sec}^{-1} \times (18 + 30)\text{μM}}{30 \text{ μM}}$$

$$V_{max} = 132 \text{ μmol mL}^{-1}\text{sec}^{-1}$$

b. The turnover number, k_{cat}, is given by $V_{max}/[E_T]$. Thus,

$$k_2 = \frac{V_{max}}{E_{total}}$$

Given $V_{max} = 132$ μmol mL⁻¹sec⁻¹ and $E_{total} = 3$ nanomol/mL

$$k_2 = \frac{132 \times 10^{-6}\text{mol mL}^{-1}\text{sec}^{-1}}{3 \times 10^{-9} \text{ mol/mL}}$$

$$k_2 = 44,000 \text{ sec}^{-1}$$

c. The catalytic efficiency is the turnover number divided by K_m, both of which we know.

$$\frac{k_{cat}}{K_m} = \frac{44,000 \text{ sec}^{-1}}{1.8 \times 10^{-5}\text{M}}$$

$$\frac{k_{cat}}{K_m} = 2.44 \times 10^9 \text{M}^{-1}\text{sec}^{-1}$$

d. The very fastest we could expect the enzyme to function is given by the rate constant for diffusion of a small molecule, which is in the range 10^8 to 10^9 M⁻¹ sec⁻¹. The catalytic efficiency of the enzyme actually exceeds this range.

e. The ultimate limit is how fast ES forms. More precisely, it is the rate at which E encounters S. Given that enzymes are typically large molecules relative to their substrates, the limit of the reaction is determined by the rate of diffusion of S.

11. Quantitative relationships between rate constants to calculate K_m, kinetic efficiency (k_{cat}/K_m) and V_{max} –III

The citric acid cycle enzyme fumarase catalyzes the conversion of fumarate to form malate.

Fumarate + H₂O ⇌ malate

The turnover number, kcat, for fumarase is 800/sec. The Km of fumarase for its substrate fumarate is 5 μM.

 a. In an experiment using 2 nanomole/mL of fumarase, what is Vmax?
 b. The cellular concentration of fumarate is 47.5 μM. What is v when [fumarate] = 47.5 μM?
 c. What is the catalytic efficiency of fumarase?
 d. Does fumarase approach "catalytic perfection"?

Answer:

a. We can determine V_{max}, given k_{cat} (i.e., k_2) and the concentration of fumarase as follows:

$$V_{max} = k_2 \times [E_{total}] = 800 \text{ sec}^{-1} \times \frac{2 \times 10^{-9} \text{mol}}{\text{mL}}$$

$$V_{max} = 1.6 \times 10^{-6} \text{mol} / \text{mL} \cdot \text{sec}$$

$$V_{max} = 1.6 \text{ μmol} / \text{mL} \cdot \text{sec}$$

b. To determine v given the substrate constant, K_m and V_{max} (just calculated) we can use the Michaelis-Menten equation as follows:

$$v = \frac{V_{max} \times [S]}{K_m + [S]}$$

Given $V_{max} = 1.6 \text{ μmol} / \text{mL} \cdot \text{sec}$ (from a.) , $K_m = 5 \text{ μM} = 18 \text{ μM}$, $[S] = 47.5 \text{ μM}$

$$v = \frac{1.6 \text{ μmol} / \text{mL} \cdot \text{sec} \times 47.5 \text{ μM}}{5 \text{ μM} + 47.5 \text{ μM}}$$

$$v = 1.45 \text{ μmol mL}^{-1}\text{sec}^{-1}$$

c. The catalytic efficiency is the turnover number divided by K_m, both of which we know.

$$\frac{k_{cat}}{K_m} = \frac{800 \text{sec}^{-1}}{5 \text{ μM}}$$

$$\frac{k_{cat}}{K_m} = 1.60 \times 10^8 \text{M}^{-1}\text{sec}^{-1}$$

d. The very fastest we could expect the enzyme to function is given by the rate constant for diffusion of a small molecule, which is in the range 10^8 to 10^9 M⁻¹ sec⁻¹. The catalytic efficiency of this enzyme is in this range so, yes, fumarase approaches perfection.

12. Quantitative relationships between rate constants to calculate Km, kinetic efficiency (kcat/Km) and Vmax –IV
Carbonic anhydrase catalyzes the hydration of CO₂.

CO₂ + H₂O ⇌ H₂CO₃

The Km of carbonic anhydrase for CO₂ is 12 mM. Carbonic anhydrase gave an initial velocity v₀ = 4.5 μmols H₂CO₃ formed/mL sec when [CO₂] = 36 mM.

 a. What is the Vmax for this enzyme?
 b. Assuming 5 pmol/mL (5 x 10⁻¹² moles/mL) of enzyme were used in this experiment, what is kcat for this enzyme?
 c. What is the catalytic efficiency of this enzyme?
 d. Does carbonic anhydrase approach "catalytic perfection"?

Answer:

a. We are given K_m and the initial velocity of a reaction at a given substrate concentration. We can use this information and the Michaelis-Menten equation to calculate V_{max} as follows:

$$v = \frac{V_{max} \times [S]}{K_m + [S]}$$

$$V_{max} \times [S] = v \times (K_m + [S])$$

$$V_{max} = \frac{v \times (K_m + [S])}{[S]}$$

Given $v = 4.5\ \mu mol\ mL^{-1}sec^{-1}$, $K_m = 12\ mM$, $[S] = 36\ mM$

$$V_{max} = \frac{4.5\ \mu mol\ mL^{-1}sec^{-1} \times (12 + 36)\ mM}{36\ mM}$$

$$V_{max} = 6\ \mu mol\ mL^{-1}sec^{-1}$$

b. To calculate k_{cat}, given the enzyme concentration and V_{max}, recall that $k_{cat} = k_2$, which is determined as follows.

$$k_2 = \frac{V_{max}}{E_{total}}$$

Given $V_{max} = 6\ \mu mol\ mL^{-1}sec^{-1}$ and $E_{total} = 5\ picomol/mL$

$$k_2 = \frac{6 \times 10^{-6} mol\ mL^{-1}sec^{-1}}{5 \times 10^{-12}\ mol/mL}$$

$$k_2 = k_{cat} = 1.2 \times 10^6\ sec^{-1}$$

c. The catalytic efficiency is the turnover number divided by K_m, both of which we know.

$$\frac{k_{cat}}{K_m} = \frac{1.2 \times 10^6 sec^{-1}}{12\ mM}$$

$$\frac{k_{cat}}{K_m} = 1.00 \times 10^8 M^{-1}sec^{-1}$$

d. The very fastest we could expect the enzyme to function is given by the rate constant for diffusion of a small molecule, which is in the range 10^8 to $10^9\ M^{-1}\ sec^{-1}$. The catalytic efficiency of this enzyme is in this range so, yes, carbonic anhydrase approaches perfection.

13. *Quantitative relationships between rate constants to calculate K_m, kinetic efficiency (k_{cat}/K_m) and V_{max} –V*
Acetylcholinesterase catalyzes the hydrolysis of the neurotransmitter acetylcholine:
Acetylcholine + H₂O → acetate + choline
The K_m of acetylcholinesterase for its substrate acetylcholine is $9 \times 10^{-5}\ M$. In a reaction mixture containing 5 nanomoles/mL of acetylcholinesterase and 150 μM acetylcholine, a velocity
$v_0 = 40\ \mu mol/mL \cdot sec$ was observed for the acetylcholinesterase reaction.
　　a. Calculate the V_{max} for this amount of enzyme.
　　b. Calculate k_{cat} for acetylcholinesterase.
　　c. Calculate the catalytic efficiency (k_{cat}/K_m) for acetylcholinesterase.
　　d. Does acetylcholinesterase approach "catalytic perfection"?

Answer:
a. Again, we are given K_m, the initial velocity or a reaction at a given substrate concentration. We can use this information and the Michaelis-Menten equation to calculate V_{max} as follows:

$$v = \frac{V_{max} \times [S]}{K_m + [S]}$$

$$V_{max} \times [S] = v \times (K_m + [S])$$

$$V_{max} = \frac{v \times (K_m + [S])}{[S]}$$

Given $v = 40$ μmol mL^{-1}sec^{-1}, $K_m = 9 \times 10^{-5}$M $= 90$ μM, $[S] = 150$ μM

$$V_{max} = \frac{40 \ \mu mol \ mL^{-1}sec^{-1} \times (90 + 150) \ \mu M}{150 \ \mu M}$$

$$V_{max} = 64 \ \mu mol \ mL^{-1}sec^{-1}$$

b. To calculate k_{cat}, given the enzyme concentration and V_{max}, recall that $k_{cat} = k_2$, which is determined as follows.

$$k_2 = \frac{V_{max}}{E_{total}}$$

Given $V_{max} = 64$ μmol mL^{-1}sec^{-1} and $E_{total} = 5$ nanomol/mL

$$k_2 = \frac{64 \times 10^{-6} mol \ mL^{-1}sec^{-1}}{5 \times 10^{-9} \ mol/mL}$$

$$k_2 = k_{cat} = 1.28 \times 10^4 \ sec^{-1}$$

c. The catalytic efficiency is the turnover number divided by K_m, both of which we know.

$$\frac{k_{cat}}{K_m} = \frac{1.28 \times 10^4 \ sec^{-1}}{9 \times 10^{-5}M}$$

$$\frac{k_{cat}}{K_m} = 1.42 \times 10^8 \ M^{-1}sec^{-1}$$

d. The very fastest we could expect the enzyme to function is given by the rate constant for diffusion of a small molecule, which is in the range 10^8 to 10^9 M^{-1} sec^{-1}. The catalytic efficiency of this enzyme is in this range so, yes, acetylcholinesterase approaches perfection.

14. Quantitative relationships between rate constants to calculate K_m, kinetic efficiency (k_{cat}/K_m) and V_{max} –VI
The enzyme catalase catalyzes the decomposition of hydrogen peroxide:
$$2 \ H_2O_2 \leftrightharpoons 2 \ H_2O + O_2$$
The turnover number (k_{cat}) for catalase is 40,000,000 sec^{-1}. The K_m of catalase for its substrate H_2O_2 is 0.11 M.
  a. In an experiment using 3 nanomole/mL of catalase, what is V_{max}?
  b. What is v when $[H_2O_2]$ is 0.75 M?
  c. What is the catalytic efficiency of catalase?
  d. Does catalase approach "catalytic perfection"?

Answer:
a. Given K_m for catalase as 0.11 M and k_{cat} (i.e., k_2) as 40,000,000 sec^{-1}, V_{max} is calculated as follows.

$$V_{max} = k_2[E_{total}]$$

Given $k_2 = 40,000,000$ sec^{-1}, $[E_{total}] = 3$ nanomol/mL

$$V_{max} = 40,000,000 \ sec^{-1} \times 3 \times 10^{-9} \ mol / mL$$

$$V_{max} = 0.12 \ mol \ mL^{-1}sec^{-1} \ or \ 120 \ mmol \ mL^{-1}sec^{-1}$$

b. To determine v given the substrate constant, K_m and V_{max} (just calculated) we can use the Michaelis-Menten equation as follows:

$$v = \frac{V_{max} \times [S]}{K_m + [S]}$$

Given $V_{max} = 0.12$ mol mL^{-1}sec^{-1} (from a.) , $K_m = 0.11$ M, $[S] = 0.75$ M

$$v = \frac{0.12 \text{ mol mL}^{-1}\text{sec}^{-1} \times 0.75 \text{ M}}{0.11 \text{ M} + 0.75 \text{ M}}$$

$$v = 0.1047 \text{ mol mL}^{-1}\text{sec}^{-1}$$

$$v = 104.7 \text{ mmol mL}^{-1}\text{sec}^{-1}$$

c. The catalytic efficiency is the turnover number divided by K_m, both of which we know.

$$\frac{k_{cat}}{K_m} = \frac{40{,}000{,}000 \text{ sec}^{-1}}{0.11 \text{M}}$$

$$\frac{k_{cat}}{K_m} = 3.64 \times 10^8 \text{ M}^{-1}\text{sec}^{-1}$$

d. The very fastest we could expect the enzyme to function is given by the rate constant for diffusion of a small molecule, which is in the range 10^8 to 10^9 M^{-1} sec^{-1}. The catalytic efficiency of this enzyme is in this range so, yes, catalase approaches perfection.

15. *The effect of the accumulation of P, the reaction product, on the Michaelis-Menten equation Equation 13.9 presents the simple Michaelis-Menten situation where the reaction is considered to be irreversible ([P] is negligible). Many enzymatic reactions are reversible, and P does accumulate.*

a. Derive an equation for v, the rate of the enzyme-catalyzed reaction S→P in terms of a modified Michaelis-Menten model that incorporates the reverse reaction that will occur in the presence of product, P.

b. Solve this modified Michaelis-Menten equation for the special situation when v = 0 (that is, S⇄P is at equilibrium, or in other words, $K_{eq} = [P]/[S]$).

(J. B. S. Haldane first described this reversible Michaelis-Menten modification, and his expression for K_{eq} in terms of the modified M-M equation is known as the Haldane relationship.)

Answer: a. Equation 13.9 is simply modified to include conversion of P back to S. The model we need to work with is:

$$E + S \rightleftarrows ES \rightleftarrows E + P$$

And, we must consider rate constants k_1, k_{-1}, k_2 and now k_{-2} for formation of ES by combination of E + P. The rate of change of the concentration of ES is given by the following:

$$\frac{d[ES]}{dt} = k_1[E][S] + k_{-2}[E][P] - k_{-1}[ES] - k_2[ES]$$

If we assume [ES] reaches steady state then $\frac{d[ES]}{dt} = 0$

This allows us to solve for [ES] in terms of the rate constants, [S] and [P]

$$[ES] = [E]\frac{k_1[S] + k_{-2}[P]}{k_{-1} + k_2} \quad (1)$$

The total amount of enzyme is given by:

$[E_{total}] = [E] + [ES]$ (2)

Substituting (1) into (2) and solving for [E]:

$$[E] = \frac{[E_{total}]}{1 + \frac{k_1[S] + k_{-2}[P]}{k_{-1} + k_2}} \quad (3)$$

The velocity of the S to P reaction is given by:

$$v = k_2[ES] \text{ or } k_2([E_{total}] - [E])$$

Substituting for $[E]$ using (3) we get:

$$v = k_2 \left([E_{total}] - \frac{[E_{total}]}{1 + \dfrac{k_1[S] + k_{-2}[P]}{k_{-1} + k_2}} \right)$$

Rearranging this equation we get:

$$v = k_2[E_{total}] \left(\frac{1 + \dfrac{k_1[S] + k_{-2}[P]}{k_{-1} + k_2} - 1}{1 + \dfrac{k_1[S] + k_{-2}[P]}{k_{-1} + k_2}} \right)$$

$$v = k_2[E_{total}] \left(\frac{\dfrac{k_1[S] + k_{-2}[P]}{k_{-1} + k_2}}{1 + \dfrac{k_1[S] + k_{-2}[P]}{k_{-1} + k_2}} \right)$$

$$v = \left(\frac{\dfrac{k_1 k_2[E_{total}][S]}{k_{-1} + k_2} + \dfrac{+k_2[E_{total}]k_{-2}[P]}{k_{-1} + k_2}}{1 + \dfrac{k_1[S] + k_{-2}[P]}{k_{-1} + k_2}} \right)$$

We define $k_2[E_{total}] = V_{max}^f$, $-k_{-2}[E_{total}] = V_{max}^r$

(Note the minus sign in the last identity.)

$$\frac{k_1}{k_{-1} + k_2} = \frac{1}{K_M^S} \text{ and } \frac{k2}{k_{-1} + k_2} = \frac{1}{K_M^P}$$

The equation simplifies to:

$$v = \frac{\dfrac{V_{max}^f[S]}{K_M^S} - \dfrac{V_{max}^r[P]}{K_M^P}}{1 + \dfrac{[S]}{K_M^S} + \dfrac{[P]}{K_M^P}}$$

b. When $v = 0$, the above equation should give the equilibrium constant for the reaction.

$$0 = \frac{\dfrac{V_{max}^f[S]}{K_M^S} - \dfrac{V_{max}^r[P]}{K_M^P}}{1 + \dfrac{[S]}{K_M^S} + \dfrac{[P]}{K_M^P}}$$

$$\frac{V_{max}^f[S]}{K_M^S} - \frac{V_{max}^r[P]}{K_M^P} = 0, \text{ or}$$

$$\frac{V_{max}^f[S]}{K_M^S} = \frac{V_{max}^r[P]}{K_M^P}, \text{ or}$$

$$K_{eq} = \frac{[P]}{[S]} = \frac{V_{max}^f K_M^P}{V_{max}^r K_M^S}$$

16. Enzyme A follows simple Michaelis-Menten kinetics.
   a. The K_m of enzyme A for its substrate S is K_m^S = 1 mM. Enzyme A also acts on substrate T and its K_M^T = 10 mM. Is S or T the preferred substrate for enzyme A?
   b. The rate constant k_2 with substrate S is 2×10^4 sec^{-1}, with substrate T, $k_2 = 4 \times 10^5$ sec^{-1}. Does enzyme A use substrate S or substrate T with greater catalytic efficiency?

Answer:

a. It would appear as if S is the preferred substrate because the enzyme half-saturates at 1mM for [S] but requires 10 mM for [T].

b. The catalytic efficiency is k_{cat}/K_m (k_2/K_m). For S and T, catalytic efficiencies are:

$$\frac{k_2}{K_M} = \frac{2 \times 10^4 sec^{-1}}{1 \ mM} \text{ for S}$$

$$\frac{k_2}{K_M} = \frac{2 \times 10^4 sec^{-1}}{1 \times 10^{-3} M} = 2 \times 10^7 sec^{-1} M^{-1}$$

$$\frac{k_2}{K_M} = \frac{4 \times 10^5 sec^{-1}}{10 \ mM} \text{ for T}$$

$$\frac{k_2}{K_M} = \frac{4 \times 10^5 sec^{-1}}{10 \times 10^{-3} M} = 4 \times 10^7 sec^{-1} M^{-1}$$

The enzyme is more efficient using T as substrate.

17. Use Figure 13.12 to answer the following questions.
   a. Is the enzyme whose temperature versus activity profile is shown in Figure 13.12 likely to be from an animal or a plant? Why?
   b. What do you think the temperature versus activity profile for an enzyme from a thermophilic bacterium growing in a 80°C pool of water would resemble?

Answer:

a. The enzyme whose activity versus temperature is shown in Figure 13.12 has a temperature maximum at 37°C. In addition, the profile is quite sharp suggesting that the enzyme is maximized for function at or near this temperature. Many animals are endothermic, maintaining a constant body temperature and this enzyme's activity profile would be consistent with an enzyme from an animal of this kind. Plants, on the other hand, typically experience a broader range of temperatures.

b. For an enzyme from a thermophile, one might expect a similar activity profile but one whose maximum is at higher temperatures.

Questions for Self Study

1. Enzymes have three distinctive features that distinguish them from chemical catalysts. What are they?

2. Chose from the list following each enzyme systematic name the term that best fits the enzyme class.
   a. Oxidoreductase
      1. Cleavage using water. 2. *Cis-trans* conversion. 3. Uses NAD or NADP.
   b. Transferase
      1. C-C bond formation. 2. Methylation reaction. 3. Hydrolysis.
   c. Hydrolase
      1. O_2 functions as acceptor. 2. Acetyl-group transfer. 3. Cleavage of peptide bond.
   d. Isomerase
      1. Methyltransferase. 2. Aldose to ketose conversion. 3. Peptide bond hydrolysis

3. Match the terms in the first column with the descriptions in the second column.
   a. A $\star$ P
   b. υ
   c. $d[P]/dt = d[S]/dt$
   d. k in $\upsilon = k[A][B]$
   e. k_{-1}/k_1
   f. $(k_{-1} + k_2)/k_1$
   g. υ at $[S] = \infty$
   h. $[S]$ when $\upsilon = V_{max}/2$
   i. k_{cat}
   j. $1/\upsilon$ when $1/[S] = 0$

   1. K_m
   2. V_{max}
   3. First-order reaction
   4. Michaelis constant
   5. V_{max}/E_{Total}
   6. The velocity or rate of a reaction.
   7. Enzyme:substrate dissociation constant
   8. $1/V_{max}$
   9. Equilibrium
   10. Second-order rate constant

4. Fill in the blanks. The energy barrier preventing thermodynamically favorable reactions from occurring is known as the ____. In general, catalysts ____ this energy barrier by providing an alternate reaction pathway leading from substrate to product. The intermediate along this reaction pathway that has the highest free energy is known as the ___.

5. Answer True or False
   a. For competitive inhibition the apparent K_m is larger than K_m by the factor $(1 + [I]/K_I)$. ___
   b. For noncompetitive inhibition the apparent V_{max} is larger than V_{max} by the factor $(1 + [I]/K_I)$. ___
   c. A plot of V_{max} versus temperature is linear for a typical enzyme. ___
   d. Ordered, single-displacement reactions refer to bisubstrate reactions in which two substrates bind to the enzyme before catalysis occurs. ___
   e. The reaction catalyzed by the enzyme glutamate:aspartate aminotransferase is a good example of a double-displacement or Ping-Pong reaction. ___

6. What is a ribozyme? What is an abzyme?

Answers

1. Enormous catalytic power, specificity, and the ability to be regulated.

2. a. 3; b. 2; c. 3; d. 2.

3. a. 3; b. 6; c. 9; d. 10; e. 7; f. 1 or 4; g. 2; h. 1 or 4; i. 5; j. 8.

4. Energy of activation; lower; transition state.

5. a. T; b. F; c. F; d. T; e. T

6. A ribozyme is a catalytic RNA. An abzyme is a catalytic antibody.

Additional Problems

1. A researcher was studying the kinetic properties of β-galactosidase using an assay in which o-nitrophenol-β-galactoside (ONPG), a colorless substrate, is converted to galactose and o-nitrophenolate, a brightly colored, yellow compound. Upon addition of 0.25 mM substrate to a fixed amount of enzyme, o-nitrophenolate (ONP) production was monitored as a function of time by spectrophotometry at $\lambda = 410$ nm. The following data were collected:

Time (sec)	$A_{410\,nm}$
0	0.000
15	0.158
30	0.273
45	0.360
60	0.429
75	0.484
90	0.529
150	0.652
210	0.724
270	0.771
330	0.805
390	0.830
450	0.849
510	0.864

   a. Convert A_{410} nm to concentration of o-nitrophenolate, [ONP], using $\varepsilon = 3.76$ mM^{-1}cm^{-1} as the extinction coefficient.
   b. Plot [ONP] versus time.
   c. Determine the initial velocity of the reaction.
   d. Explain why the curve is nearly linear initially and later approaches a plateau.
   e. Describe how K_m and V_{max} for β-galactosidase can be determined with additional experimentation.

2. Under what conditions can an enzyme assay be used to determine the relative amounts of an enzyme present?

3. For many reactions, it is often difficult to measure product formation, so coupled assays are often used. In a coupled enzyme assay, the activity of an enzyme is determined by measuring the activity of a second enzyme that uses as substrate the product of a reaction catalyzed by the first enzyme. The utility of coupled assays is that the product of the second enzyme is easy or convenient to measure. You are asked to design a coupled assay to measure the relative amounts of a particular enzyme in several samples. In qualitative terms with respect to K_m and V_{max} for both enzymes, explain how an appropriate coupled assay is set up.

4. In the case of a competitive inhibitor, explain how the inhibitor dissociation constant, K_I, can be determined from enzyme kinetic data.

5. Is K_m a good measure of the dissociation constant for substrate binding to enzyme? Explain.

6. Derive an expression for υ as a function of [S] for mixed, noncompetitive inhibition.

7. In Figure 13.16a, the presence of inhibitor decreases V_{max}; however, it also decreases $K_{m,app}$. This seems to suggest that in the presence of an inhibitor the remaining active enzyme's performance is improved since it saturates at a lower substrate concentration. Can you explain this paradox?

Abbreviated Answers

1a. The concentration of ONP, [ONP], is calculated using Beer's Law, $A = \varepsilon \cdot C \cdot l$, where ε is the extinction coefficient, C is the concentration, and l is the path length (1 cm in this case). $[ONP] = A_{410} \, nm/(\varepsilon \cdot l \, cm)$.

Time (sec)	$A_{410 \, nm}$	[ONP] (mM)
0	0.00	0.00
15	0.158	0.042
30	0.273	0.073
45	0.36	0.096
60	0.429	0.114
75	0.484	0.129
90	0.529	0.141
150	0.652	0.173
210	0.724	0.193
270	0.771	0.205
330	0.805	0.214
390	0.83	0.221
450	0.849	0.226
510	0.864	0.230

b. The plot of [ONP] versus time is shown below

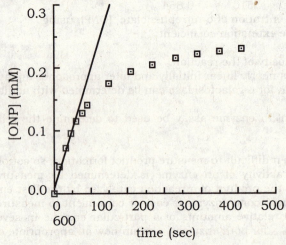

c. The initial velocity is equal to the slope of a line drawn through the initial points as shown above and has a value of 0.00188 mM/sec.

d. The enzyme is converting substrate to product and the curve is the time course of this process. The initial points are linear because the enzyme is obeying Michaelis-Menten kinetics. At later times, two things happen to cause the rate of product formation, given by the slope of a tangent drawn along the graph, to decrease: [S] decreases; and, [P] increases. The curve is approaching 2.5 mM, the initial value of substrate, indicating that nearly all of the initial substrate has been converted to product.

e. By repeating the experiment at different initial substrate concentrations, the initial velocities at varying substrate concentrations can be measured. These data form a υ versus [S] data set, which can be used to determine K_m and V_{max}.

2. At substrate concentrations high relative to K_m, an enzyme functions at or near V_{max}. V_{max}, given by $V_{max} = k_2 E_T$, is proportional to the total enzyme concentration.

3. The following reaction sequence represents a coupled enzyme assay.

$$
\begin{array}{cc} E_1 & E_2 \\ S \rightarrow P_1 \rightarrow P_2 \end{array}
$$

We are interested in determining the relative amount of E_2 by measuring the rate of production of P_2. The concentration of the substrate for E_1, namely S, must be large relative to K_{m1} in order for E_1 to function at

256

V_{max}. Also, the rate at which E_2 produces P_2 must be constant. This requires that $[P_1]$ be at a steady-state concentration. The rate of change of $[P_1]$ is given by

$$\frac{d[P_1]}{dt} = V_{max,E_1} - \frac{V_{max,E_2}[P_1]}{K_{m2} + [P_1]}$$

For $[P_1]$ to be at steady state, $\frac{d[P_1]}{dt} = 0$, and solving for $[P_1]$ we find:

$$[P_1]_{steady\ state} = \frac{K_{m2} \times V_{max,E_1}}{V_{max,E_2} - V_{max,E_1}}$$ which is positive only when

$$V_{max,E_2} > V_{max,E_1}$$

Thus, [S] must be high enough to saturate E1 and $k_2 \cdot [E_{2,T}]$ must be greater than $V_{max,E1}$.

4. For competitive inhibition, $K_{m,app} = K_m(1 + [I]/K_I)$. By measuring $K_{m,app}$ at several different inhibitor concentrations and plotting $K_{m,app}$ against [I], the slope of the resulting straight line divided by the y-intercept (equal to K_m) is $1/K_I$.

5. K_m and K_D are defined as follows:

$$K_m = \frac{k_{-1} + k_2}{k_1}, \text{ and } K_D = \frac{k_{-1}}{k_1}$$

When k_{-1} is large relative to k_2, $K_m \approx K_D$

6. The following reactions occur for mixed, noncompetitive inhibition:
$$E + S \leftrightarrows ES \rightarrow P; E + I \leftrightarrows EI; ES + I \leftrightarrows IES$$
Binding of I to E and to ES are governed by:

$$K_I = \frac{[E][I]}{[EI]}; \ K_I' = \frac{[ES][I]}{[IES]}$$

These equations can rearranged to give:

$$[EI] = \frac{[E][I]}{K_I}; \ [IES] = \frac{[ES][I]}{K_I'}$$

The rate of change of [ES] is given by:

$$\frac{d[ES]}{dt} = -k_2[ES] - k_{-1}[ES] + k_1[E][S]$$

Using this equation and making the steady-state assumption we find that

$$[ES] = \frac{k_1[E][S]}{k_2 + k_{-1}} = \frac{[E][S]}{K_m}$$

The total concentration of enzyme, E_T, is given by:

$$[E_T] = [E] + [ES] + [EI] + [IES]$$

Substituting the expressions for [ES], [EI], and [IES] above we find that:

$$[E_T] = [E] + \frac{[E][S]}{K_m} + \frac{[E][I]}{K_I} + \frac{[E][S][I]}{K_m K_I'}, \text{ and solving for [E] we find:}$$

$$[E] = \frac{[E_T] \times K_m}{\left(1 + \dfrac{[I]}{K_I'}\right)} \Bigg/ \left(K_m \frac{\left(1 + \dfrac{[I]}{K_I}\right)}{\left(1 + \dfrac{[I]}{K_I'}\right)} + [S] \right)$$

The rate of product formation is given by:

$$\upsilon = k_2[ES] = k_2 \frac{[E][S]}{K_m}$$

Finally, substituting the expression for [E] above we find:

$$v = \frac{\dfrac{k_2[E_T]}{(1+\dfrac{[I]}{K_I'})} \times [S]}{K_m \times \dfrac{(1+\dfrac{[I]}{K_I})}{(1+\dfrac{[I]}{K_I'})} + [S]} = \frac{V_{max,app}[S]}{K_{m,app}+[S]}$$

Clearly, $V_{max,app}$ is always less than V_{max}. $K_{m,app}$ may be less than, equal to, or greater than K_m depending on the magnitude of K_I relative to $K_{I'}$.

7. In Figure 13.16a, I binds to ES with greater affinity (lower dissociation constant) than to E. Therefore, the addition of I is expected to affect the $E + S \leftrightarrows ES$ equilibrium by shifting it to the right allowing more ES to form at lower [S].

Summary

Enzymes are the catalysts of metabolism, allowing living systems to achieve kinetic control over the thermodynamic potential within organic reactions. Enzymes have three distinctive attributes that are essential to their biological purpose: enormous catalytic power, great selectivity in catalyzing only very specific reactions, and the ability to be regulated so that their activity is compatible with the momentary needs of the cell. Formally, enzymes are classified by a system of nomenclature based on the particular reaction they catalyze, although certain trivial names for enzymes often enjoy common usage. Enzymes may be simple proteins or they may be proteins complexed with nonprotein components called cofactors. Cofactors include metal ions and organic molecules known as coenzymes.

An enzyme accelerates the rate of a process by lowering the thermodynamic barrier to reaction, known as the free energy of activation. The reactant in an enzyme-catalyzed reaction is called the substrate. The fact that graphs of enzyme activity as a function of substrate concentration show a limiting plateau or saturation level was the important clue that an enzyme (E) actually binds its substrate (S) to form an enzyme:substrate (ES) complex that can react to yield product (P). Enzyme kinetic analysis is based on the notion put forth by Michaelis and Menten that the E and S reversibly interact, and ES can react to form P. Briggs and Haldane extended this model by assuming that the system $E + S \leftrightarrows ES \rightarrow P$ quickly reaches a steady state condition where $d[ES]/dt = 0$. The standard Michaelis-Menten equation for an enzymatic reaction is then derived:

$$v = \frac{V_{max}[S]}{K_m+[S]}$$

V_{max} is the maximal velocity attained when [S] is so great as to not be rate limiting; K_m is a constant reflecting the relative affinity and reactivity of the enzyme-substrate interaction. Because graphs of v versus [S] are not linear, the Michaelis-Menten expression is often rearranged to give equations that yield linear relationships between rate and substrate concentration.

Because their activity is highly dependent on maintenance of their native protein structure, enzymes are very sensitive to pH and temperature. Enzyme inhibition has been an important means for investigating enzymes and their mode of action. Such studies have also contributed to human health through the science of pharmacology, since many drugs exert their action as inhibitors of enzymes. Some inhibitors act by competing with the substrate for binding to the active site of the enzyme, altering the apparent K_m. Other inhibitors uniquely affect the apparent V_{max} while still others affect both $K_{m,app}$ and $V_{max,app}$. Inhibitors, which react covalently with the enzyme, can lead to irreversible inhibition of its activity.

Most enzymatic reactions involve more than one substrate. These multi-substrate reactions can be analyzed kinetically to determine whether two (or more) substrates are bound before reaction occurs (so-called sequential or single-displacement reactions), or whether one substrate binds and reacts with the enzyme before the other substrate is bound (Ping-Pong or double-displacement reactions). Exchange reactions are another useful tool in enzyme kinetic analysis.

Recent discoveries have broadened our view of biocatalysis. Certain RNA molecules have the essential features of enzymes: rate enhancement, catalytic turnover and specificity of reaction. This finding has important implications to our view of molecular or pre-biotic evolution. Further, biocatalysts in the form of antibodies have been "tailor-made" by using transition state analogs of a reaction as antigens. The antibodies elicited in response to these analogs act as biocatalysts by facilitating entry of the reactants into a reactive transition-state intermediate. The creation of "designer enzymes" has become a realistic possibility.

Chapter 14

Mechanisms of Enzyme Action

• •

Chapter Outline

- ❖ Transition state intermediate
    - ⅄ Energy difference between ES and $EX^{\ddagger}$ is smaller than that between S and $X^{\ddagger}$
    - ⅄ Enzyme stabilizes transition state more than substrate
- ❖ Catalytic mechanisms or factors
    - ⅄ Destabilization of ES complex
        - • Entropy loss upon formation of ES complex
        - • Destabilization of S because E designed to bind transition state
            - Structural strain: E binds transition state better than S
            - Desolvation: Desolvated ionic groups show increased reactivity
            - Electrostatic effects: Charge proximity
    - ⅄ Stabilization of $EX^{\ddagger}$ complex: Transition state analogs are potent inhibitors
    - ⅄ Covalent catalysis
        - • Nucleophilic catalysis
            - Nucleophilic centers on enzymes: Amines, carboxylates, hydroxyls, imidazole, thiol
            - Electrophilic targets on substrates: Phosphoryl, acyl, glycosyl groups
        - • Electrophilic catalysis: Coenzymes
    - ⅄ General acid-base catalysis
        - • Specific acid-base catalysis: H^+ or OH^- involved: No dependence on buffer concentration
        - • General acid-base catalysis
            - AH (protonated acid) donates
            - B^- (unprotonated base) accepts protons
        - • Low-barrier hydrogen bond: H bond between groups with matching pK_as
            - LBHB formation stabilizes transition state
            - LBHB may facilitate hydrogen tunneling
    - ⅄ Metal ion catalysis
        - • Metalloenzymes: Metal ion tightly bound
        - • Metal activated enzymes: Metal ion loosely bound
        - • Metal functions
            - Stabilize formation of negative charge
            - Powerful nucleophile at neutral pH
    - ⅄ Proximity and orientation
- ❖ Serine proteases
    - ⅄ Enzyme diversity
        - • Trypsin, chymotrypsin, elastase: Digestive enzymes
        - • Thrombin: Blood clotting
        - • Subtilisin: Bacterial enzyme
        - • Plasmin, TPA: Blood clot breakdown
        - • Acetylcholinesterase (not a protease): Degrades neurotransmitter acetylcholine

259

⤊ Digestive enzyme specificity: Cleavage (hydrolysis) of peptide bond on carbonyl side
- Chymotrypsin: Phe, Trp, Tyr: Hydrophobic pocket near catalytic triad
- Trypsin: Lys, Arg: Aspartic acid at bottom of hydrophobic pocket
- Elastase: Small neutral amino acids: Mouth of pocket closed by threonine and valine

⤊ Burst kinetics: Rapid release of amino end: Carboxyl end attached to enzyme and slowly hydrolyzed

⤊ Catalytic triad: His[57], Asp[102], Ser[195]
- Ser[195]: Acyl-enzyme intermediate formed by hydroxylate attack on carbonyl carbon of peptide bond
- His[57]
  - General base: Deprotonates Ser[195]: Deprotonates water
  - General acid: Protonates carbonyl oxygen: Protonates peptide nitrogen
- Asp[102]: Orients His[57]

❖ Aspartic proteases
⤊ Enzyme diversity
- Pepsin, chymosin: Digestive enzymes
- Cathepsin D and cathepsin E: Protein degradation in lysosomes
- Renin: Blood pressure regulation
- AIDS HIV-1 protease: Viral life cycle

⤊ Catalytic dyad: Two aspartic acids
- Low-barrier hydrogen bond between two aspartic acids and catalytic water
- Substrate participates in flow of electrons to and from substrate to aspartic acids

❖ Chorismate Mutate: Chorismate to prephenate
⤊ Claisen rearrangement
⤊ Enzymatic activity: N terminus of P protein
⤊ Dimerization via antiparallel coiled-coil: 7-residue repeat of leucines
⤊ Transition state analog: Consistent with chair transition state
- Near attack conformation promoted by electrostatic and hydrophobic interactions of active site residues

Chapter Objectives

General Considerations

To appreciate why reactions can be slow, you must understand that even the most favorable path from reactant to product passes through some transition state intermediate. Formation of the transition state intermediate is an energy-requiring process, and the energy necessary for its formation, the activation energy, is an energy barrier that must be overcome for the reaction to occur. Enzyme-catalyzed reactions also have an activation energy but it is not the energy difference between the substrate and the transition state that matters; rather it is the energy difference between the enzyme-substrate complex and the enzyme-transition state complex. Enzymes increase reaction rates because the energy barrier between ES and EX‡ is less than the barrier between S and X‡ (See Figure 14.1).

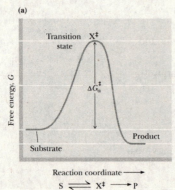
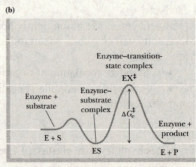

Figure 14.1 Enzymes catalyze reactions by lowering the activation energy. Here the free energy of activation for (a) the enzyme-catalyzed reaction, $\Delta G_e^{\ddagger}$, is smaller than that for (b) the noncatalyzed reaction, $\Delta G_u^{\ddagger}$.

Understand the catalytic mechanisms used by enzymes to achieve enormous rate accelerations, which include: 1. Entropy loss in ES formation; 2. Destabilization of ES due to strain, desolvation or electrostatic effects; 3. Near-attack conformations; 3. Covalent catalysis; 4. General acid or base catalysis; 5. Low-barrier hydrogen bonds; 6. Metal ion catalysis; and 7. Proximity and orientation.

Serine Proteases - Covalent Catalysis

The serine proteases and related enzymes are arguably the best-characterized group of enzymes. The key to catalysis is a catalytic triad of His[57], Asp[102], and Ser[195] (in chymotrypsin). Understand the role of this serine in catalysis, the fact that it is transiently, covalently modified during catalysis accounting for the burst kinetics, a rapid release of one product followed by a slow release of the second product. The mechanism based on this catalytic triad is found in a number of enzymes, including trypsin, chymotrypsin, elastase, thrombin, subtilisin, plasmin, tissue plasminogen activator, and acetylcholinesterase, to name a few. We have already encountered a few of the digestive serine proteases in Chapter 4 as tools used in primary sequence determination of proteins. Substrate specificity is determined by a substrate-binding pocket that is distinct from the catalytic triad.

Aspartic Proteases – Low-barrier hydrogen bonds

Pepsin, chymosin, cathepsin D and E, rennin, and HIV-1 protease are examples of aspartic proteases. The mechanism of action was initially thought to be acid/base catalysis with two aspartic acid residues functioning as a general acid and a general base. The pH profile of enzymes with this type of catalysis is bell-shaped, indicating the involvement of a proton donor and acceptor. Aspartic acid residues participate in low-barrier hydrogen bonds with a water molecule to form a cyclic structure. Substrate joins the structure and a flow of electrons first to the substrate to form a tetrahedral intermediate and then from the substrate to cleave the peptide bond releases the carboxyl group. Flow of electrons from the amino group of the cleaved bond results in a second tetrahedral intermediate formed with the enzyme, which is resolved by a water molecule.

Chorismate Mutase: Stabilization of transition state

This enzyme participates in the pathway leading to biosynthesis of phenylalanine by converting chorismate to prephenate. The reaction mechanism involves stabilization of a transition state: The chair conformation of the substrate, which is stabilized by electrostatic and hydrophobic interactions from active-site amino acid residues. The transition state is a near-attack conformation that positions reactive groups nearby.

Problems and Solutions

1. Characterizing a covalent enzyme inhibitor
Tosyl-L-phenylalanine chloromethylketone (TPCK) specifically inhibits chymotrypsin by covalently labeling His[57].
   a. Propose a mechanism for the inactivation reaction, indicating the structure of the product(s).
   b. State why this inhibitor is specific for chymotrypsin.
   c. Propose a reagent based on the structure of TPCK that might be an effective inhibitor of trypsin.

Answer: The structure of tosyl-L-phenylalanine chloromethylketone (TPCK) and the reaction mechanism leading to covalent labeling of His[57] are shown below:

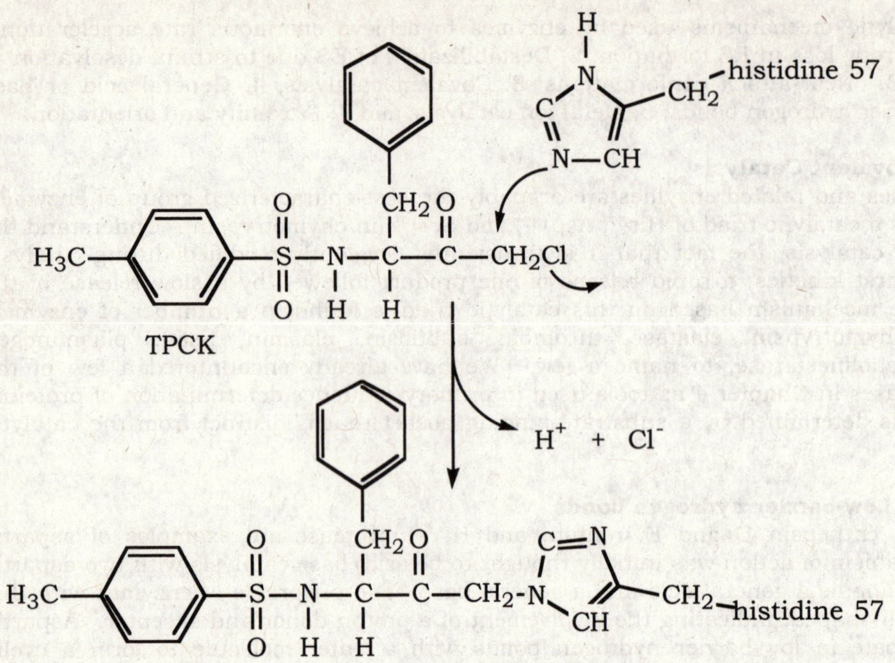

H

histidine 57

CH$_2$

C

N—CH

H$_3$C—

O

CH$_2$ O

S—N—C—C—CH$_2$Cl

O H H

TPCK

H$^+$ + Cl$^-$

H$_3$C—

O

CH$_2$ O

S—N—C—C—CH$_2$—N

O H H

C=N

CH$_2$—histidine 57

CH

b. The phenyl group of TPCK binds to the specificity pocket of chymotrypsin by hydrophobic interactions. This brings the chloromethylketo group into the active site.

c. Trypsin shows specificity for peptide bonds in which the carbonyl carbon is from a positively charged amino acid such as arginine or lysine. Reagents specific for trypsin may be produced by replacing the phenylalanine residue of TPCK with arginine or lysine.

2. Using site-directed mutants to understand an enzyme mechanism
In this chapter, the experiment in which Craik and Rutter replaced Asp^{102} with Asn in trypsin (reducing activity 10,000-fold) was discussed.
 a. On the basis of your knowledge of the catalytic triad structure in trypsin, suggest a structure for the "uncatalytic triad" of Asn-His-Ser in this mutant enzyme.
 b. Explain why the structure you have proposed explains the reduced activity of the mutant trypsin.
 c. See the original journal articles (Sprang, et al., 1987. Science 237:905-909 and Craik, et al., 1987. Science 237:909-13) to see what Craik and Rutter's answer to this question was.

Answer: a. Catalysis involves ionization of Ser^{195}, which is accomplished in the case of wild-type trypsin by loss of a proton to His^{57} whose pK$_a$ is influenced by Asp^{102} as shown below. In the case of Asn^{102} substitution, the amino group of Asn^{102} functions as a hydrogen-bond donor thus preventing His^{57} from accepting a proton from Ser^{195}.

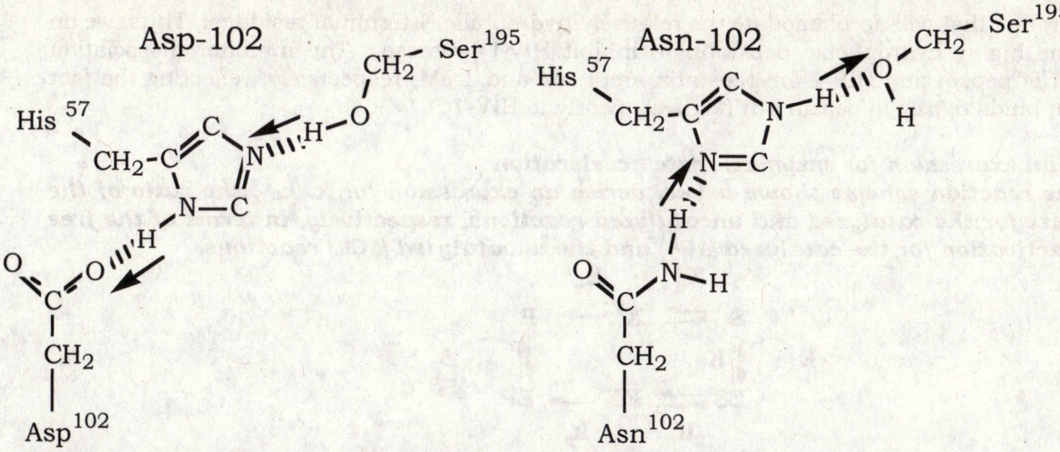

Asp-102   CH₂ Ser¹⁹⁵   Asn-102   CH₂ Ser¹⁹⁵

His⁵⁷ ... CH₂-C ... His⁵⁷ ... CH₂-C ...

Asp¹⁰²   Asn¹⁰²

b. In the wild-type enzyme, Asp¹⁰² functions as a hydrogen-bond donor and acceptor. In the mutant enzyme, Asn¹⁰² predominantly functions as a hydrogen-bond donor as shown above.

3. Assessing the action of an aspartic protease inhibitor
Pepstatin (below) is an extremely potent inhibitor of the monomeric aspartic proteases, with K_i values of less than 1 nM.

    a. Based on the structure of pepstatin, suggest an explanation for the strongly inhibitory properties of this peptide.

    b. Would pepstatin be expected to also inhibit the HIV-1 protease? Explain your answer.

Answer: The structure of pepstatin is:

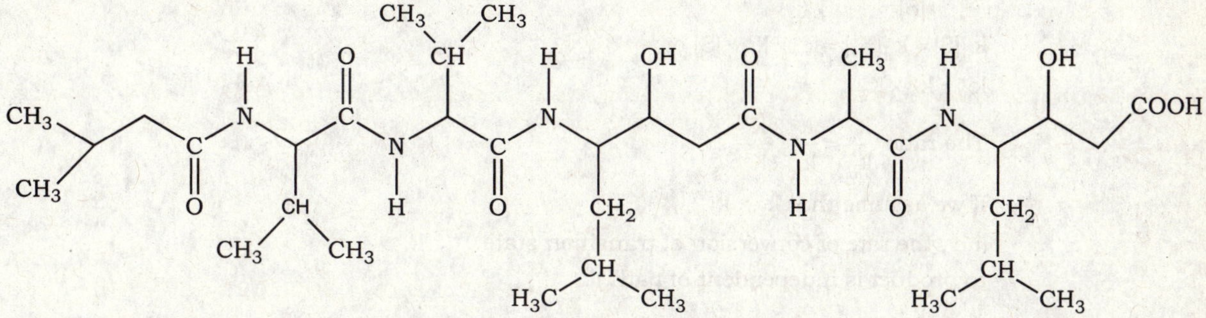

It contains 5 peptide bonds joining together the following groups: methylbutanoic acid; valine; valine; 4-amino-3-hydroxy methylheptanoic acid (statine); alanine; and, 4-amino-3-hydroxymethyl heptanoic acid. Thus, pepstatin has a structure very similar to a 6 residue polypeptide chain of hydrophobic amino acids. The active site of pepsin can accommodate a chain of up to 7 residues in length. Further, pepsin cleaves peptide bonds between hydrophobic amino acids. Pepstatin binds tightly to the active site because of the nature of its hydrophobic side chains. Assuming that the catalytic site is in the middle of the hydrophobic binding pocket, pepstatin will be positioned with a carbon-carbon single bond at the active site.

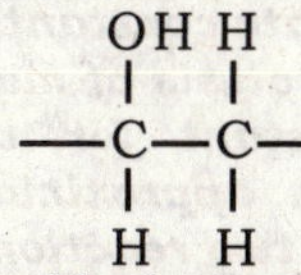

OH H
| |
—C—C—
| |
H H

Thus, a hydroxyl group replaces a carbonyl. The high pK_a of the hydroxyl will resist catalysis.

The HIV-1 protease cleaves between Tyr and Pro in the sequence Ser-Gln-Asn-Tyr-Pro-Ile-Val. The structure of HIV-1 protease is similar to that of mammalian aspartic proteases. The active site is a cleft between two identical subunits with two aspartic acids contributed from both subunits at the heart of the catalytic mechanism. The mammalian aspartic protease preferentially cleaves between hydrophobic amino acids residues. The sequence shown above has only two hydrophobic residues on the C terminus. We might expect that the catalytic pocket of HIV-1 protease to be lined with hydrophilic groups that will accommodate the relatively hydrophilic N-terminal residues. Thus, we do not expect the highly hydrophobic pepstatin to inhibit HIV-I protease. The inhibitor dissociation constants (K_I) of pepsin and HIV-1 for pepstatin are 1 nM and 1 μM, respectively, reflecting the fact that pepstatin binds tightly to pepsin but relatively poorly to HIV-1

4. Deriving an expression for enzymatic rate acceleration
Based on the reaction scheme shown below, derive an expression for k_e/k_u , the ratio of the rate constants for the catalyzed and uncatalyzed reactions, respectively, in terms of the free energies of activation for the catalyzed ($\Delta G^{\ddagger}_e$) and the uncatalyzed ($\Delta G^{\ddagger}_u$) reactions.

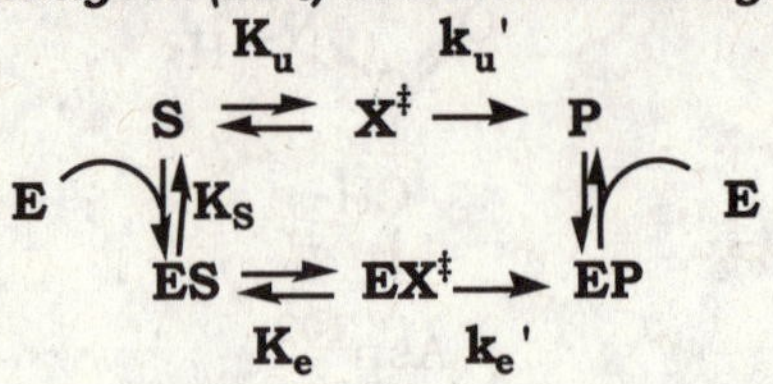

$$S \underset{}{\overset{K_u}{\rightleftharpoons}} X^{\ddagger} \overset{k_u{}'}{\longrightarrow} P$$

$$ES \underset{}{\overset{K_e}{\rightleftharpoons}} EX^{\ddagger} \overset{k_e{}'}{\longrightarrow} EP$$

with E, K_S linking the two rows.

Answer:

The enzyme-catalyzed rate is given by $v_e = k_e[ES] = k_e^{'}[EX^{\ddagger}]$, but

$K_e^{\ddagger} = \dfrac{[EX^{\ddagger}]}{[ES]}$ and $[EX^{\ddagger}] = K_e^{\ddagger} \times [ES]$ thus

$k_e[ES] = k_e^{'}[EX^{\ddagger}] = k_e^{'} \times K_e^{\ddagger} \times [ES]$ or

$k_e = k_e^{'} \times K_e^{\ddagger}$

The uncatalyzed rate is given by $v_u = k_u[S] = k_u^{'}[X^{\ddagger}]$, but

$K_u^{\ddagger} = \dfrac{[X^{\ddagger}]}{[S]}$ and $[X^{\ddagger}] = K_u^{\ddagger} \times [S]$ so

$k_u[S] = k_u^{'}[X^{\ddagger}] = k_u^{'} \times K_u^{\ddagger} \times [S]$ or

$k_u = k_u^{'} \times K_u^{\ddagger}$

The ratio $\dfrac{k_e}{k_u} = \dfrac{k_e^{'} \times K_e^{\ddagger}}{k_u^{'} \times K_u^{\ddagger}}$

If we assume that $k_e^{'} \approx k_u^{'}$

(i.e., the rate of conversion of transition state

to product is independent of path.)

$\dfrac{k_e}{k_u} = \dfrac{K_e^{\ddagger}}{K_u^{\ddagger}}$

In general, $\Delta G = -RT\ln K$, or $K = e^{\frac{-\Delta G}{RT}}$, thus

$\dfrac{k_e}{k_u} = \dfrac{e^{\frac{-\Delta G_e^{\ddagger}}{RT}}}{e^{\frac{-\Delta G_u^{\ddagger}}{RT}}} = e^{\frac{\Delta G_u^{\ddagger} - \Delta G_e^{\ddagger}}{RT}}$

5. Comparison of enzymatic and nonenzymatic rate constants
The k_{cat} for alkaline phosphatase-catalyzed hydrolysis of methylphosphate is approximately 14/sec at pH 8 and 25°C. The rate constant for the uncatalyzed hydrolysis of methylphosphate under the same conditions is approximately 1×10^{-15}/sec. What is the difference in the activation free energies of these two reactions?

Answer: Using equation 14.3, we can calculate the difference in activation energies between the uncatalyzed and catalyzed reaction:

$$\frac{k_e}{k_u} = \frac{K_S}{K_T}$$

Where K_S is the dissociation constants for enzyme-substrate complex, and

K_T is the dissociation constants for enzyme-transition state complex.

In general, $K = e^{\frac{-\Delta G}{RT}}$, and

$$\frac{k_e}{k_u} = \frac{K_S}{K_T} = \frac{e^{\frac{-\Delta G_S}{RT}}}{e^{\frac{-\Delta G_T}{RT}}} = e^{\frac{(\Delta G_T - \Delta G_S)}{RT}} \text{ or}$$

$$\Delta G_T - \Delta G_S = RT \ln \frac{k_e}{k_u} = RT \ln \frac{\frac{14}{\text{sec}}}{\frac{1 \times 10^{-15}}{\text{sec}}}$$

$$\Delta G_T - \Delta G_S = 8.314 \times 298 \times 37.18$$

$$\Delta G_T - \Delta G_S = 92 \frac{kJ}{mol}$$

6. Understanding the actions of proteolytic enzymes
Active α-chymotrypsin is produced from chymotrypsinogen, an inactive precursor. The first intermediate –π-chymotrypsin– displays chymotrypsin activity. Suggest proteolytic enzymes that might carry out these cleavage reactions effectively.

Answer: Conversion of chymotrypsinogen to π-chymotrypsin involves a single peptide bond cleavage at arginine 15 of chymotrypsinogen. Trypsin specificity for arginine and lysine can produce this cleavage. Chymotrypsin itself catalyzes the conversion of π-chymotrypsin to α-chymotrypsin.

7. Understanding the mechanism of carboxypeptidase A
Consult a classic paper by William Lipscomb (1982. Accounts of Chemical Research 15:232-238), and on the basis of this article write a simple mechanism for the enzyme carboxypeptidase A.

Answer: Lipscomb's paper begins with a discussion of factors that influence enzyme catalysis and discusses entropy changes, transition-state binding, chemical catalysis and other effects. He then goes on to look for these in the mechanism of carboxypeptidase A. It is clear that entropy changes are involved in substrate binding because the enzyme restricts rotational and translational degrees of freedom of the substrate upon binding. In the active site, regions around the Zn^{2+} and several charged amino acids compensate for loss of substrate entropy by releasing waters of solvation. Transition state binding was difficult to access but the active site contains a number of reactive groups including Glu-270, Tyr-248, and Zn^{2+} that contribute to catalysis. The pH optimum of the enzyme, 7.5, suggested involvement of either a water molecule and or Glu-270, whose pK_a would have to be shifted to around neutrality from a typical value for a side-chain carboxylate of around 4.5. The enzyme hydrolyzes peptide bonds starting at the C-terminus of its substrate. The mechanism proposed by Lipscomb involves binding of the carboxy terminus to Arg145 and further restriction of the scissile peptide bone with a hydrogen bond to Tyr248. The pathway for catalysis involves general acid/base chemistry in which glu270 promotes the attack of water on the carbonyl carbon of the scissile peptide bond. Zinc is thought to either stabilize a negatively-charged oxyanionic intermediate or help activate a water molecule. The Figure below shows these points.

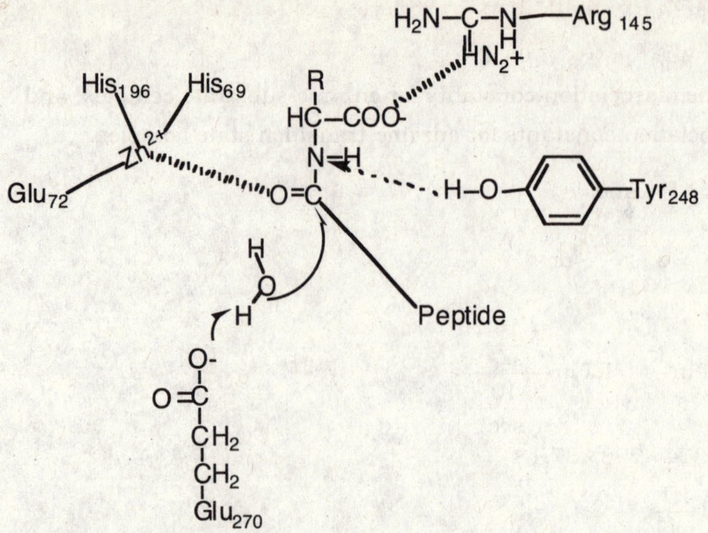

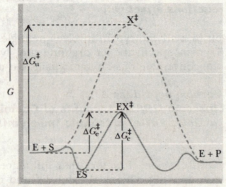

8. Calculation of rate enhancement from energies of activation
The relationships between the free energy terms defined in the solution to Problem 4 above are shown in the following figure:

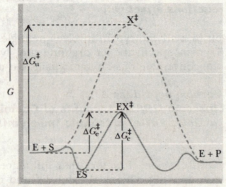

Reaction coordinate

If the energy of the ES complex is 10 kJ/mol lower than the energy of the E + S, the value of $\Delta G_e{}^‡$ is 20 kJ/mol, and the value of $\Delta G_u{}^‡$ is 90 kJ/mol. What is the rate enhancement achieved by an enzyme in this case?

Answer: In question 4 above we determined the ratio of catalyzed rate constant to uncatalyzed rate constant as:

$$\frac{k_e}{k_u} = \frac{e^{\frac{-\Delta G_e^‡}{RT}}}{e^{\frac{-\Delta G_u^‡}{RT}}} = e^{\frac{(\Delta G_u^‡ - \Delta G_e^‡)}{RT}}$$

Further, we are given of $\Delta G_u{}^‡$ as 90 kJ/mol $\Delta G_e{}^‡$ as 20 kJ/mol. At room temperature, (273 + 25 K) and using for R 8.31451 x 10^{-3}, the ratio is:

$$\frac{k_e}{k_u} = e^{\frac{(\Delta G_u^‡ - \Delta G_e^‡)}{RT}} = e^{\frac{(90-20)}{298 \times 8.3145 \times 10^{-3}}}$$

$$\frac{k_e}{k_u} = 1.86 \times 10^{12}$$

9. Understanding the implications of transition state stabilization
As noted on page 451, a true transition state can bind to an enzyme active site with a K_T as low as 7 x10^{-26} M. This is a remarkable number, with interesting consequences. Consider a hypothetical solution of an enzyme in equilibrium with a ligand that binds with a K_D of 1 x 10^{-27} M. If the

concentration of free enzyme, [E], is equal to the concentration of the enzyme–ligand complex, [EL], what would [L], the concentration of free ligand, be? Calculate the volume of solution that would hold one molecule of free ligand at this concentration.

Answer: For the reaction:

$$EL \rightleftharpoons E + L$$

The equilibrium dissociation constant is:

$$K_D = \frac{[E][L]}{[EL]}$$

If the concentration of enzyme-ligand complex, [EL] is equal to the free enzyme concentration, [E], we see that

$$K_D = [L]$$

That is, the equilibrium dissociation constant is equal to the free ligand concentration that half-saturates the ligand-binding molecule. (At half saturation [EL] = [E].) Thus, $[L] = K_D = 1 \times 10^{-27}$ M.

The volume that would contain one molecule is:

$$1 \times 10^{-27} \frac{mol}{L} \times \frac{6.02 \times 10^{23} \, molecules}{mol} \times x = 1 \, molecule$$

Solving for x we find:

$$x = \frac{1 \, molecule}{1 \times 10^{-27} \dfrac{mol}{L} \times \dfrac{6.02 \times 10^{23} \, molecules}{mol}} = 1,661 \, L$$

10. Understanding the very tight binding of transition states
Another consequence of tight binding (problem 9) is the free energy change for the binding process. Calculate $\Delta G^{o\prime}$ for an equilibrium with a K_D of 1×10^{-27} M. Compare this value to the free energies of the noncovalent and covalent bonds with which you are familiar. What are the implications of this number, in terms of the nature of the binding of a transition state to an enzyme active site?

Answer: To calculate $\Delta G^{o\prime}$ we use the following equation:
$$\Delta G^{o\prime} = -RT \ln K_D$$
$$\Delta G^{o\prime} = -8.31 \, J/K \cdot mol \times 273 + 25) \, K \times \ln(1 \times 10^{-27})$$
$$\Delta G^{o\prime} = -154 \, kJ/mol$$
Covalent bond energies are in the range of around 300 (carbon-nitrogen single bond) to around 600 (carbon-carbon double bond) whereas hydrogen bonds range from 10 to 30 kJ/mol. The ligand binding energy in this case is equivalent to several strong hydrogen bonds or one weak covalent bond. (For example, the covalent bond in dioxgen is around 150 kJ/mol.)

11. Assessing the metabolic consequences of life without enzymes
The incredible catalytic power of enzymes can perhaps best be appreciated by imagining how challenging life would be without just one of the thousands of enzymes in the human body. For example, consider life without fructose-1,6-bisphosphatase, an enzyme in the gluconeogenesis pathway in liver and kidneys (see Chapter 22), which helps produce new glucose from the food we eat:

Fructose-1,6-bisphosphate + H_2O → Fructose-6-P + P_i
The human brain requires glucose as its only energy source, and the typical brain consumes about 120 g (or 480 calories) of glucose daily. Ordinarily, two pieces of sausage pizza could provide more than enough potential glucose to feed the brain for a day. According to a national fast-food chain, two pieces of sausage pizza provide 1340 calories, 48% of which is from fat. Fats cannot be converted to glucose in gluconeogenesis, so that leaves 697 calories potentially available for glucose synthesis. The first-order rate constant for the hydrolysis of fructose-1,6-bisphosphate in the absence of enzyme is 2×10^{-20}/sec. Calculate how long it would take to provide enough glucose for one day of brain activity from two pieces of sausage pizza without the enzyme.

Answer: If a typical brain consumes 120 g of glucose this corresponds to

$$\frac{120\ g}{180\ \dfrac{g}{mol}} = 0.667\ \text{mol glucose}$$

We will assume that all this glucose comes from gluconeogenesis and hence all of it must pass through fructose-1,6-bisphosphatase. Further, we will assume that it all derives from the liver (gluconeogenesis is most active in the liver) and an average volume of a human liver is around 1500 mL. (See *Alcohol and Alcoholism* 35:531-532, (2000)). Next we need to know the concentration of fructose-1,6-bisphosphate in liver. This is difficult to determine because the concentration might be expected to vary depending on physiological conditions. We could get an estimate by looking for the K_m of the enzyme. This can be found using SwissProt, looking for the human enzyme and then following links to NiceZyme view of the protein linked via the enzyme's E.C. number and then to BRENDA, a comprehensive enzyme information system. There were several listings with an average of around 2 μM. It is typically the case that the K_m for enzymes is within the physiological concentration range of its substrates and so we will assume a concentration of 2 μM for fructose-1,6-bisphosphate.

Using the first-order rate constant for non-enzymatic hydrolysis of fructose-1,6-bisphosphate we find that fructose-6-phosphate is being supplied at the following rate:

$$\text{rate} = \frac{2\times10^{-20}}{sec}\times 2\times10^{-6}M = 2\times10^{-26}\ \frac{M}{sec}$$

Converting this to moles per second we must multiply by the volume (1500 mL or 1.5 L) which gives:

$$2\times10^{-26}\ \frac{M}{sec}\times 1.5\ L = 3\times10^{-26}\ \frac{moles}{sec}$$

To produce 0.667 moles would take:

$$\frac{0.667\ mol}{3\times10^{-26}\ \dfrac{moles}{sec}} = 2.2\times10^{25}\ sec\ \times\frac{min}{60\ sec}\times\frac{hr}{60\ min}\times\frac{day}{24\ hr}\times\frac{yr}{356\ days}$$

$$= 7\times10^{17}\ \text{yrs!}$$

Preparing for the MCAT® Exam

The following graphs show the temperature and pH dependencies of four enzymes, A, B, X, and Y. Problems 12 through 18 refer to these graphs.

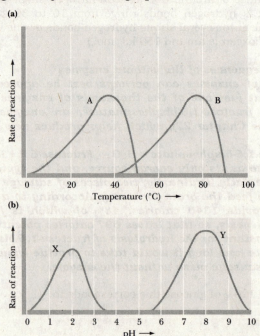

(a)

(b)

12. Assessing the localization of proteases
Enzymes X and Y in the figure are both protein-digesting enzymes found in humans. Where would they most likely be at work?
    a. X is found in the mouth, Y in the intestine.
    b. X in the small intestine, Y in the mouth.
    c. X in the stomach, Y in the small intestine.
    d. X in the small intestine, Y in the stomach.

Answer: The correct answer is "c". The pH optima for X and Y are 2 and 8 respectively. The acidic pH of 2 suggests to us that enzyme X functions in the stomach. Enzyme Y likely functions in the small intestine, which is maintained at a slightly alkaline pH. (When chyme passes from the stomach into the small intestine it is neutralized by bicarbonate making the pH slightly alkaline.)

13. Understanding enzymatic reaction parameters
Which statement is true concerning enzymes X and Y?
    a. They could not possibly be at work in the same part of the body at the same time.
    b. They have different temperature ranges at which they work best.
    c. At a pH of 4.5, enzyme X works slower than enzyme Y.
    d. At their appropriate pH ranges, both enzymes work equally fast.

Answer: The correct answer is "a". Referring to the pH plot, it is clear that the enzymes have different pH ranges. So, it seems unlikely they function in the same environment. Answer "b" is not correct as it refers to enzymes A and B and not X and Y. Answers "c" and "d" are incorrect.

14. Understanding enzymatic reaction parameters
What conclusion may be drawn concerning enzymes A and B?
    a. Neither enzyme is likely to be a human enzyme.
    b. Enzyme A is more likely to be a human enzyme.
    c. Enzyme B is more likely to be a human enzyme.
    d. Both enzymes are likely to be human enzymes.

Answer: The correct answer is "b". We are given temperature range data for A and B and it is clear that A's temperature range is centered around 37°C whereas B's temperature range is centered at around 70°C. Normal human body temperature is 37°C.

15. Understanding enzymatic reaction parameters
At which temperatures might A and B both work?
    a. Above 40°C.
    b. Below 50°C.
    c. Above 50°C and below 40°C.
    d. Between 40°C and 50°C.

Answer: The correct answer is "d". The activity curves for both enzymes overlap in the range 40°C to about 50°C.

16. Understanding the response of enzymes to environmental conditions
An enzyme-substrate complex can form when the substrate(s) bind(s) to the active site of the enzyme. Which environmental condition might alter the conformation of an enzyme in the figure to the extent that its substrate is unable to bind?
    a. Enzyme A at 40°C.
    b. Enzyme B at pH 2.
    c. Enzyme X at pH 4.
    d. Enzyme Y at 37°C.

Answer: The correct answer is "c". Enzyme A at 40°C is active. We have no idea how enzyme B functions at different pH values nor do we know now enzyme Y functions at 37°C.

17. Understanding the response of enzymes to environmental conditions
At 35°C, the rate of the reaction catalyzed by enzyme A begins to level off. Which hypothesis best explains this observation?
    a. The temperature is too far below optimum.

> **b. The enzyme has become saturated with substrate.**
> **c. Both a and b.**
> **d. Neither a nor b.**

Answer: The correct answer is "b". Clearly "a" is not true for enzyme A as its temperature optimum is around 37°C. This also eliminates "c". If the "level off" in the question refers to leveling off of activity when substrate concentration increases then "b" is the only correct answer.

18. Understanding the response of enzymes to environmental conditions
In which of the following environmental conditions would digestive enzyme Y be unable to bring its substrate(s) to the transition state?
> **a. At any temperature below optimum.**
> **b. At any pH where the rate of reaction is not maximum.**
> **c. At any pH lower than 5.5.**
> **d. At any temperature higher than 37°C.**

Answer: The correct answer is "c". Below pH 5.5 enzyme Y is inactive. Answer "a" is incorrect because there are many temperatures below the optimum temperature at which the enzyme is still active and presumable still binding the transition state. The same is true for pH values so "b" is incorrect. We have no information at all about the temperature profile of enzyme Y.

19. Comparing Covalent and General-Acid Base Reaction Mechanisms
Review the mechanisms of the serine and aspartic proteases, and compare these two mechanisms carefully. Are there steps in the mechanisms that are similar? How are they similar? How are they different? Suggest experiments that could support or refute your hypotheses.

Answer: While serine proteases exhibit covalent catalysis the mechanism also involves general acid-base catalysis. In the catalytic triad of aspartic acid, histidine, and serine, serine is activated by deprotonation by histidine. Histidine also functions as a general acid by protonating the leaving amino group. In addition a low-barrier hydrogen bond between aspartic acid (102) and histidine (57) functions during catalysis by helping to form a tetrahedral intermediate between the substrate and serine. Experimental proof for these aspects of the catalytic mechanism come from a variety of studies. For example, using model substrates like esters it is possible to stabilize the covalent adduct on the active-site serine. The dependence on the acid-base properties of histidine is reflected in the pH dependence of the reaction, which is consistent with involvement of a single group with a pK_a of a histidine. (See: Catalytic mechanism of serine proteases: Reexamination of the dependence of the histidyl $^1J_{13C2-H}$ coupling constant in the catalytic triad of α-lytic protease by Bachovchin, W. W. et al. (1981) PNAS 78:7323-7326). X-ray crystallography and NMR were used in a study by Kidd and co-workers (See: Breaking the low barrier hydrogen bond in a serine protease by Kidd, R. E., et al (1999) Protein Science 8:410-417) to show the importance of a low barrier hydrogen bond between histidine and aspartic acid that was important for serine protease activity.

The mechanism of aspartic proteases clearly does not involve covalent modification but is dependent on low barrier hydrogen bonds and to some extend general acid base catalysis. Evidence supporting general acid-general base catalysis is the pH dependence of the reaction with a single peak consistent with a pK_a for aspartic acid. However, structural studies indicate that the two active-site aspartic acid residues form a hydrogen bonded network with a water molecule. The hydrogen bond network functions as a proton and electron sink.

Questions for Self Study

1. List the six catalytic mechanisms or factors that contribute to the performance of enzymes.

2. What is a transition-state analog?

3. Match an item in the first column to one in the second column.
   a. Covalent catalysis   1. Sensitive to changes in [metal ion].
   b. Specific acid-base catalysis   2. Formation of modified amino acid residue.
   c. General acid-base catalysis   3. Tightly bound Zn^{2+}.

    d. Metalloenzyme   4. Catalysis involves specific H^+ donor.
    e. Metal activated enzyme   5. Rate is pH dependent but [Buffer] independent.

4. What three amino acids form the catalytic triad of the serine proteases?

5. The aspartic proteases show a bell-shaped curve of enzyme activity versus pH centered around pH = 3. How is this information consistent with the involvement of two aspartic acid residues in catalysis?

6. The active sites of many enzymes are organized to stabilize the conformation of substrates that bring the reacting atoms of the substrates in van der Waals contact and at bond angles resembling the transition state. What are these conformations called and how might they aide in catalysis?

7. What conditions are necessary for two groups to form low-barrier hydrogen bonds?

Answers

1. Entropy loss in ES formation; destabilization of ES due to strain, desolvation or electrostatic effects; covalent catalysis; general acid or base catalysis; metal ion catalysis; proximity and orientation.

2. A transition-state analog is a molecule that is chemically and structurally similar to the transition state.

3. a. 2; b. 5; c. 4; d. 3; e. 1.

4. Aspartic acid, histidine, and serine.

5. The fact that the center of the curve is around pH = 3.0 indicates that an acidic amino acid such as aspartic acid or glutamic acid is involved. The bell-shape is consistent with the involvement of both a general acid and a general base i.e., two distinct groups.

6. The conformation of the substrate is termed a near-attack conformation and it is along a reaction path leading to the transition state.

7. The hydrogen bond distances must be relatively short and the hydrogen bonded pairs must have pK_a values that are close to each other.

Additional Problems

1. The pH profile of pepsin is bell-shaped, indicating the involvement of a general acid and a general base in catalysis with pK_as around 4.3 and 1.4, respectively. The two candidates for these roles are two aspartic acid residues, Asp^{32} and Asp^{215}. However, the pK_as of aspartic acid residues in polypeptides range from 4.2 to 4.6. How might the pK_a of an aspartic acid residue be influenced by protein structure?

2. State the specificity exhibited by the serine proteases trypsin, chymotrypsin, and elastase, and explain how this specificity is determined by the protein.

3. a. *p*-Nitrophenyl acetate is slowly hydrolyzed by chymotrypsin to *p*-nitrophenol and acetic acid. One nice property of this compound is that the product, *p*-nitrophenol, when ionized, is brightly yellow-colored, making it convenient to follow enzyme activity spectrophotometrically. Draw [*p*-nitrophenylate] produced as a function of time.
b. On the same graph, draw [acetate] production as a function of time.
c. Why are the two plots different?

4. Discuss the importance of histidine in enzyme catalysis.

5. a. Papain is a sulfhydryl protease from papaya that shows broad specificity. (It is used as a

component of meat tenderizers.) During catalysis, an acyl-enzyme intermediate is formed as an acylthioester. The active residue is Cys[25], which is readily modified by iodoacetate, a potent inhibitor of papain. How might the role of Cys[25] in papain be similar to Ser[95] in chymotrypsin?

b. Suggest a mechanism by which iodoacetate inhibits papain.

Abbreviated Answers

1. The pK_a of the side-chain carboxyl group of aspartic acid is around 4.3. Thus, the general acid in pepsin has a pK_a in the normal range of pK_as exhibited by aspartic acids. For an aspartic acid to function as a general base with $pK_a = 1.4$, its pK_a must be lowered considerably. How might this be achieved? In a protein, the polypeptide chain may fold in a manner to place an aspartic acid in an environment that influences its pK_a. One way to lower the pK_a is to have positive-charge residues nearby stabilizing the negatively charged carboxylate form of the aspartic acid side chain.

2. Chymotrypsin cleaves peptide bonds in which the carbonyl group is donated by amino acids with bulky, hydrophobic groups such as Phe, Trp, and Tyr. Trypsin's specificity is for Arg and Lys, large, positively charged amino acids. Elastase hydrolyzes after Ala and Gly. The substrate-binding pocket in chymotrypsin is a deep, hydrophobic pocket adjacent to the catalytic site. A similar pocket is found in trypsin but with an aspartic acid residue located deep in the pocket (in a position occupied by a serine residue in chymotrypsin). In elastase, the substrate-binding pocket is filled by a valine residue and a threonine residue. Analogous positions in chymotrypsin and trypsin are occupied by glycine residues.

3. a and b

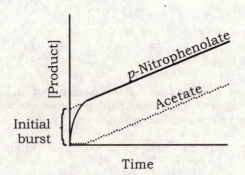

c. Initially, rapid production of *p*-nitrophenolate occurs without accompanying acetate production. Only at later times are the rates of production of *p*-nitrophenolate and acetate identical. This "burst" of activity produces a molar equivalent of *p*-nitrophenolate. Hydrolysis of *p*-nitrophenylacetate is a two-step process. First, the acetate group is transferred from the substrate to Ser[195] on the enzyme and *p*-nitrophenolate is released. Subsequently, acetate is hydrolyzed from Ser[195]. The second step is the rate-limiting step. The acetate production curve exhibits an initial lag corresponding to formation of acyl-Ser[195].

4. Histidine is the only amino acid with a pK_a around neutral pH. It can function as a general acid or general base, as seen for His[57] in the serine proteases.

5a In chymotrypsin catalysis involves a nucleophilic attack of the carbonyl carbon of the peptide bond by Ser[195] forming an acylester intermediate. This was facilitated by Ser[195] losing a proton to nearby His[57]. In papain, the active-site residue is a cysteine. Cysteine has a much lower pK_a than serine and will more readily function as a nucleophile. An acylthioester is formed as an intermediate in catalysis in papain.

b.

Summary

Enzymes accelerate reaction rates by reducing the activation energies for biochemical reactions. To do so, enzymes must stabilize the transition state complex EX‡, more than they stabilize the enzyme-substrate complex, ES. Put another way, enzymes are designed by nature to bind the transition state more tightly than the substrate or product. Enzyme-catalyzed reactions are typically 10^7 to 10^{14} times faster than their uncatalyzed counterparts. The catalytic mechanisms which contribute to the prowess of enzymes include a) entropy loss in ES formation, b) destabilization of ES due to strain, desolvation or electrostatic effects, c) covalent catalysis, general acid-base catalysis, metal ion catalysis, and proximity and orientation effects. When a substrate binds to an enzyme, the (usually large) intrinsic binding energy is compensated by entropy loss and destabilization in the ES complex, but not in the EX‡ complex. The transition state for any reaction is a 'moving target' - a transient species that exists for only 10^{-14} - 10^{-13} s. The short lifetime of the transition state prevents quantitative studies of this complex, but the dissociation constant for the EX‡ complex can be estimated to be on the order of 10^{-15} M. Transition state analogs are stable molecules that resemble the transition state for a particular reaction, and that therefore bind very tightly to their respective enzymes. Since they are only approximations of the transition state structure, transition state analogs cannot, however, bind as tightly as the transition state itself.

Some enzyme reactions derive rate acceleration from the formation of a covalent bond with the substrate. Many of the side chains of amino acids in proteins offer suitable nucleophilic centers for catalysis, including amines, carboxylates, hydroxyls, imidazole and thiol groups. Electrophilic catalysis may also occur, but usually involves coenzyme adducts that possess or generate electrophilic centers. General acid/base catalysis, in which a proton is transferred from a suitable donor/acceptor to or from the substrate in the transition state, may also provide significant rate accelerations. General acid/base catalysis is most effective when the pK_a of the donating or accepting group is near the ambient pH. Metal ions may also exert catalytic power in enzyme reactions. Metals may act as electrophilic catalysts, stabilizing the increased electron density or negative charge, which may develop in a reaction, or may act to coordinate, and thus increase the acidity of nucleophilic groups at the active site. Enzymes also provide significant rate acceleration via proximity and orientation effects, i.e. by providing an active site environment in which all the reacting groups are proximal and oriented optimally for the desired reaction.

The serine proteases, including trypsin, chymotrypsin, elastase, thrombin, subtilisin, plasmin, tissue plasminogen activator and many similar enzymes, cleave peptide chains in a mechanism involving a covalent intermediate and general acid/base catalysis. The active sites of serine proteases contain a highly conserved catalytic triad consisting of Asp[102], His[57] and Ser[195]. Asp[102] functions to immobilize and orient His[57], which accepts a proton from Ser[195], facilitating attack by Ser[195] on the acyl carbon of the scissile bond. The mechanisms of the serine proteases were elucidated in studies of hydrolysis of model organic esters. The observation of burst kinetics, the rapid release of one of the reaction products in an amount equal to the enzyme concentration, followed by a slower, steady state product release, provided evidence for a multi-step reaction sequence and the eventual identification of covalent acyl enzyme intermediates.

Aspartic proteases are active at acidic pH and their catalytic mechanism involves two aspartic acid residues. A number of important enzymes are aspartic proteases including pepsin and chymosin (digestive enzymes), cathepsin D and E (lysosomal proteins), renin (blood physiology and the HIV-1 protease (processing of HIV polyprotein). Catalysis involves formation of a low-barrier hydrogen bond with the two aspartic acid residues, water and the peptide bond substrate. The low-barrier hydrogen bond forms a network that facilitates movement of electrons to and from the substrate and water. The HIV-1 protease, which cleaves the polyproteins produced by the AIDS virus, is an aspartic protease and is a dimer of identical subunits, which mimics the two-lobed monomeric structure of pepsin and other aspartic proteases.

Chorismate mutase catalyzes the conversion of chorismate to prephenate, an intermediate in biosynthesis of phenylalanine. Its chemistry is Claisen rearrangement, which is catalyzed by stabilization of a near-attack conformation in which the substrate is in the chair conformation, stabilized by numerous hydrogen bonds and ionic interactions. A near-attack conformation is a conformation reached along a reaction pathway in which reacting atoms are at van der Waals contact and at an angle that represents the transition state. The uncatalyzed reaction in water occurs via the same pathway but the presence of the enzyme makes formation of the near-attack conformation much more likely to occur.

Chapter 15

Enzyme Regulation

● ●

Chapter Outline

❖ Enzyme regulation
  ⋏ Approach to equilibrium
  ⋏ K_m of enzymes in the range of in vivo substrate concentrations
  ⋏ Genetic controls
  ⋏ Covalent modification
  ⋏ Allosteric regulation
  ⋏ Others
    Zymogens: Proenzymes or zymogens: Activated by proteolysis
    • Proinsulin: Insulin
    • Chymotrypsinogen: Chymotrypsin
    • Blood clotting factors: Serine protease cascade converting fibrinogen to fibrin
  ⋏ Isozymes: Lactate dehydrogenase: A_4, A_3B_2, A_2B_2, A_1B_3, B_4
❖ General properties of regulatory proteins
  ⋏ Kinetic properties
    • Do not follow simple Michaelis-Menten kinetics
    • Activity sigmoidal: Higher order dependence on substrate concentration
    • Cooperativity
  ⋏ Allosteric inhibition
  ⋏ Often regulated by activation
  ⋏ Oligomeric organization
  ⋏ Effectors alter distribution of conformational isomers
❖ Cooperativity models
  ⋏ Monod, Wyman, Changeux (1965): Symmetry model
    • Two conformations: T (tense or taut) and R (relaxed)
    • R state high affinity, T state lower affinity
    • Positive homotropic effectors: Substrate binding shifts equilibrium to R
    • Heterotropic effectors
      • Positive effector: Binds to R State
      • Negative effector: Binds to T state
  ⋏ Koshland, Nemethy, Filmer (1966) Sequential model: Induced fit: S-binding induces conformational change
❖ Covalent modification through phosphorylation
  ⋏ Protein kinases: Converter enzymes that add phosphate from ATP
    • Target amino acids: Ser, Thr or Tyr
    • Target amino acids within a consensus sequence
      • Intrasteric control: Pseudosubstrate sequence (lacking phosphate target amino acid) binds to kinase and inactivates it
  ⋏ Protein phosphatase: Converter enzymes that hydrolyze phosphate

❖ Other covalent modifications: Adenylylation, uridylylation, ADP-ribosylation, methylation, disulfide bond formation and others
❖ Glycogen Phosphorylase: Cooperative enzyme that removes glucose residues from glycogen by phosphorolysis
⅄ Structure
⋅ Homodimer: 842-amino acid long subunits
⋅ Glycogen binding sites
• One is a substrate site
• The other localizes the enzyme to glycogen
⋅ Phosphate binding site: Phosphate is positive homotropic effector
⋅ Serine[14]: Site of covalent phosphorylation of enzyme: Phosphoprotein active and insensitive to allosteric regulation
⋅ ATP/AMP binding site: Heterotropic effectors: ATP inhibits/AMP activates
⋅ Glucose-6-phosphate binding site: Negative heterotropic regulator
⋅ Caffeine: Allosteric inhibitor
⋅ Pyridoxal phosphate bound to Lys[680]
❖ Regulation of glycogen phosphorylase
⅄ Enzyme conforms to MCW two state model
⋅ R state: Promoted by AMP
⋅ T state: Promoted by ATP, glucose-6-P, caffeine
⋅ T to R conversion: Asp displaced and replaced by Arg: Negatively charged Asp blocks phosphate binding whereas positively charged Arg facilitates phosphate
⅄ Covalent modification of Ser 14 by phosphorylation produces active form independent of AMP, ATP, glucose-6-P and caffeine
⋅ Kinase is glycogen phosphorylase kinase
⋅ Phosphate removed by phosphoprotein phosphatase I
❖ Enzyme cascade: Regulates activation of glycogen phosphorylase
⅄ Hormone binds to cell surface receptor releasing GDP from heterotrimeric G-protein
⅄ GTP binds to G_α subunit causing it to dissociate as G_α:GTP complex
⅄ G_α:GTP complex stimulates adenylyl cyclase
⋅ GTPase activity produces inactive G_α:GDP complex
⅄ Adenylyl cyclase produces cAMP
⅄ cAMP binds to regulatory subunits of R_2C_2 cAMP-dependent protein kinase
⋅ R subunit dissociates
⋅ C subunit becomes active
⅄ Protein kinase phosphorylates glycogen phosphorylase kinase
⅄ Glycogen phosphorylase kinase phosphorylates serine 14 of glycogen phosphorylase
❖ Oxygen binding proteins hemoglobin and myoglobin
⅄ Hemoglobin: Transports oxygen from lung to tissue: Sigmoidal oxygen binding
⅄ Myoglobin: Oxygen storage protein: Simple oxygen binding curve
❖ Oxygen binding group
⅄ Ferrous iron
⅄ Protoporphyrin ring
⅄ Histidine F8 5th iron coordinate
⅄ 6th Iron coordinate
⋅ In oxy-protein: Oxygen
⋅ In deoxy-protein: Nothing
⋅ In met-protein: Water: Iron is ferric (met): Does not bind oxygen
❖ Myoglobin: Monomeric protein
❖ Hemoglobin: Dimer of dimers: $(\alpha\beta)_2$
⅄ Sigmoidal oxygen binding: Cooperativity: Oxygen is positive homotropic regulator
⅄ Negative heterotropic effectors: Protons, carbon dioxide, chloride, 2,3-BPG
⋅ Protons lower oxygen affinity: Bohr effect: Oxygen unloaded at low pH
⋅ Carbon dioxide
• Covalently modifies α-amino groups as carbamate: Carbamate forms salt link to stabilize T state
• Carbon dioxide also converted to carbonic acid: Source of protons for Bohr effect

- 2,3-Bisphosphoglycerate: Binds to Hb: Stabilizes T state and thus lowers oxygen affinity
  - Physiology
    - Fetal hemoglobin: Binds 2,3-BPG poorly: High affinity oxygen binding allows fetal Hb to get oxygen from adult Hb
    - Sickle cell anemia: Amino acid substitution responsible for hemoglobin polymerization
- ❖ Oxygen binding in Hb accompanied by conformational changes
  - Oxygen binding to heme iron moves iron into porphyrin plane
  - Histidine F8 accompanies this movement by moving helix F, EF corner and FG corner
  - Deoxy Hb is T state and oxy Hb is R state conformations: MWC model
  - Oxygen binding to α-chain indices conformational change that is transmitted to β-chain: Sequential model

Chapter Objectives

Enzymes are proteins that act as biological catalysts but unlike inorganic catalysts, enzymes exhibit incredible specificity and can be regulated in numerous ways. Know the factors controlling enzymatic activity including: product build-up and approach to equilibrium, availability of substrates and cofactors, regulation of enzyme production and degradation, covalent modification, allosteric regulation, biosynthesis of enzymes as zymogens, isozymes, and regulation by modulator proteins. Keep in mind that enzymes are proteins and as such may be capable of assuming different conformations and that the substrate-binding site may undergo local conformational change upon substrate binding.

Allosteric Regulation

In order to understand allosteric regulation, you must first understand standard Michaelis-Menten kinetics. Plots of υ versus [S] are hyperbolic, approaching V_{max} as [S] becomes large and having K_m defined as the substrate concentration where $\upsilon = 0.5 \times V_{max}$. Allosteric proteins exhibit sigmoidal plots of υ versus [S]. The shape of the sigmoidal curve may be influenced by inhibitors or activators that bind at sites distinct from the substrate-binding site yet still influence catalysis.

Myoglobin and Hemoglobin

These were the first proteins whose structures were solved by x-ray crystallography and they serve as a paradigm for understanding allosteric regulation. Be sure to understand the physiological function of these two proteins: myoglobin is an oxygen-storage protein, whereas hemoglobin is an oxygen-transporting protein. Comparison of the structures of myoglobin and hemoglobin is a classic study in the four levels of protein structure: amino acid sequence, secondary structure, three-dimensional structure, and, for hemoglobin, quaternary structure. To a first approximation, hemoglobin appears to be simply a tetramer of myoglobin-like subunits but the subunits are in fact not identical and hemoglobin binds O_2 cooperatively. Oxygen saturation curves for myoglobin and hemoglobin are presented below in Figure 15.21. Be sure to understand cooperativity and the affects of H^+, CO_2 and 2,3-bisphosphoglycerate on O_2-binding curves for hemoglobin.

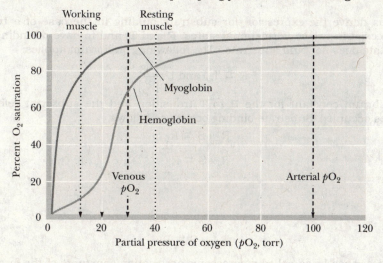

Figure 15.21 O_2-binding curves for hemoglobin and myoglobin.

Allosteric Models

Be familiar with the two models for allosteric behavior: the symmetry model of Monod, Wyman, and Changeux; and, the sequential model of Koshland, Nemethy, and Filmer. The symmetry model postulates the existence of two conformational states, **R** (relaxed) and **T** (taut) that differ in their affinity for substrate with **R** having a higher affinity than **T**. Cooperativity arises because substrate-binding shifts the **R/T** equilibrium towards the **R** state. Allosteric activators and inhibitors bind at specific sites distinct from the substrate-binding site but influence substrate binding by influencing the **R/T** equilibrium. Inhibitor binding favors the **T** state whereas activator binding favors the **R** state. In the sequential model, ligand-binding sites are in communication with each other such that the state of occupancy of a ligand-binding site influences the state of occupancy of other sites.

Problems and Solutions

1. General controls over enzyme activity
List six general ways in which enzyme activity is controlled.

Answer: Enzyme activity may be controlled by:
 a. Accumulation of product as the reaction approaches equilibrium. As substrate is converted to product, [P] increases.
 b. Availability of substrates and cofactors.
 c. Regulation of the amounts of enzyme synthesized or degraded by cells.
 d. Covalent modifications catalyzed by modifying enzymes or converter enzymes may activate or inhibit an enzyme or alter its kinetic properties.
 e. Allosteric regulation leading to either inhibition or activation.
 f. Regulation by interaction with modulator proteins.
 g. Activation or inactivation by proteolysis.
 h. Isozymes

2. Why zymogens are an advantage
Why do you suppose proteolytic enzymes are often synthesized as inactive zymogens?

Answer: Proteolytic enzymes hydrolyze peptide bonds of proteins, the most diverse and abundant class of biopolymers. Clearly, cells must take care in the biosynthesis of proteases to guard against inadvertent digestion of cellular proteins. Thus, proteolytic enzymes are produced as inactive zymogens that are activated only at the site at which their action is required.

3. Graphical analysis of MWC allosteric enzyme kinetics
(Integrates with Chapter 13.) Draw both Lineweaver-Burk plots and Hanes-Woolf plots for an MWC allosteric enzyme system, showing separate curves for the kinetic response in (a) the absence of any effectors; (b) the presence of allosteric activator A; and (c) the presence of allosteric inhibitor I.

Answer: First, let us derive the expression for substrate binding in the case of a two-state model. The protein is assumed to exist in two conformational states, **R** and **T**, and to have n binding sites. Further, it is assumed that the **T** state does not bind substrate. The following equilibrium applies:

$$R_0 \leftrightharpoons T_0, \text{ and } L = \frac{[T_0]}{[R_0]}$$

Where L is the equilibrium constant for the **R** to **T** transition and the subscript refers to the number of substrate binding-sites occupied. Substrate binding occurs as follows:

$$R_0 + S \leftrightharpoons R_1$$
$$R_1 + S \leftrightharpoons R_2$$

$$R_{n-1} + S \leftrightharpoons R_n$$

Substrate binding is described by so-called macroscopic association constants of the form:

$$K_i = \frac{[R_i]}{[R_{i-1}][S]}$$

Substrate binding, normalized to total protein concentration is given by:

$$\upsilon = \frac{[R_1] + 2[R_2] + \cdots + n[R_n]}{[T_0] + [R_0] + [R_1] + [R_2] + \cdots + [R_n]}$$

The macroscopic association constants can be rearranged to give each $[R_i]$ as a function of K_i, $[R_0]$, and $[S]$ as follows:

$$[R_i] = K_i \times [R_{i-1}][S] = K_i K_{i-1} \times [R_{i-2}][S]^2 = \cdots = [S]^i [R_0] \prod_i K_i$$

$$\text{Where } \prod_i K_i = K_1 K_2 \cdots K_i$$

Substituting the expressions for $[R_i]$s into the equation for υ we have:

$$\upsilon = \frac{\displaystyle\sum_{i=1}^{n} i \times [S]^i [R_0] \prod_i K_i}{[T_0] + [R_0] + \displaystyle\sum_{i=1}^{n} [S]^i [R_0] \prod_i K_i}$$

Now, the macroscopic association constants refer to substrate binding to any site as opposed to a specific site. The association constant for binding to a specific site is the microscopic association constant, K^*. The two are related as follows:

$$K_i = \frac{(n+1-i)}{i} K^*$$

Substituting for K_i and using $[T_0] = L[R_0]$, the expression for υ becomes:

$$\upsilon = \frac{\displaystyle\sum_{i=1}^{n} i \times [S]^i \prod_i \frac{(n+1-i)}{i} K^*}{1 + L + \displaystyle\sum_{i=1}^{n} [S]^i \prod_i \frac{(n+1-i)}{i} K^*}$$

The numerator and the denominator are equal to:

$$n[S]K^*(1 + [S]K^*)^{n-1} \text{ and } L + (1 + [S]K^*)^n$$

Upon substituting these terms above, υ simplifies to:

$$\upsilon = \frac{n[S] \times K^*(1 + [S]K^*)^{n-1}}{L + (1 + [S]K^*)^n}, \text{ or}$$

$$\upsilon = \frac{n\alpha(1 + \alpha)^{n-1}}{L + (1 + \alpha)^n}$$

Where $\alpha = K^*[S]$. (Because $K^*_a = 1/K^*_d$, the term α is a normalized concentration; it is simply $[S]$ divided by the dissociation constant K^*_d.

This equation can be generalized to take into account heterotropic interactions. Heterotropic interactions refer to binding of molecules at sites other than substrate-binding sites that influence the **R** to **T** transition. For simplicity let us assume that an inhibitor I binds to the **T** state at a single site whereas an activator A binds to the **R** state at a single site. If we define $\beta = [I]/K_I$ and $\gamma = [A]/K_A$, where $[I]$ and $[A]$ refer to inhibitor and activator concentrations, and K_I and K_A are inhibitor and activator dissociation binding constants, the expression for υ becomes:

$$\upsilon = \frac{n\alpha(1 + \alpha)^{n-1}}{L \dfrac{(1+\beta)^n}{(1+\gamma)^n} + (1 + \alpha)^n}$$

Lineweaver-Burk plots and Hanes-Woolf plots are shown below for a two-site protein ($n = 2$) with: $L = 0$ (no cooperativity); $L = 1000$ ($\beta = 0, \gamma = 0$); $L = 1000$, $\beta = 1$ ($\gamma = 1$); and, $L = 1000$, $\gamma = 1$ ($\beta = 0$).

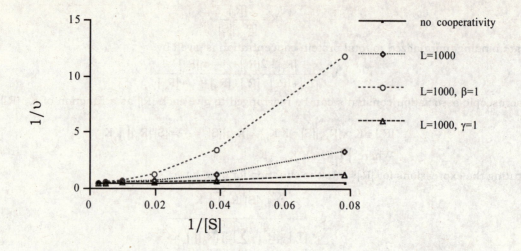

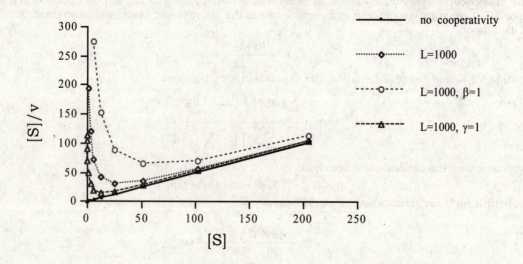

4. Graphic analysis of negative cooperativity in KNF allosteric enzyme kinetics
The KNF model for allosteric transitions includes the possibility of negative cooperativity. Draw Lineweaver-Burk and Hanes-Woolf plots for the case of negative cooperativity in substrate binding. (As a point of reference, include a line showing the classic Michaelis-Menten response of v to [S].)

Answer: For the case of negative cooperativity, substrate binding lowers the affinity of subsequent substrate binding. This cannot be accommodated by the Monod-Changeux model. The Koshland, Nemethy and Filmer model, however, does work for negative cooperativity. Negative cooperativity is characterized in a Hill plot (see next question) by n < 1.0. That is, the dependence of velocity on substrate concentration is to power n, where n < 1.0. The following curves were generated in Excel using the following equation:

$$v = \frac{V_{max}\left[S\right]^{n}}{K_m + \left[S\right]^{n}}$$

Values for [S] ranged from $0.1K_m$ to $10K_m$. For the plots shown below V_{max} = 10 and n = 0.5. Negative cooperativity is shown in open circles. The ×'s indicate a classic Michaelis-Menten enzyme.

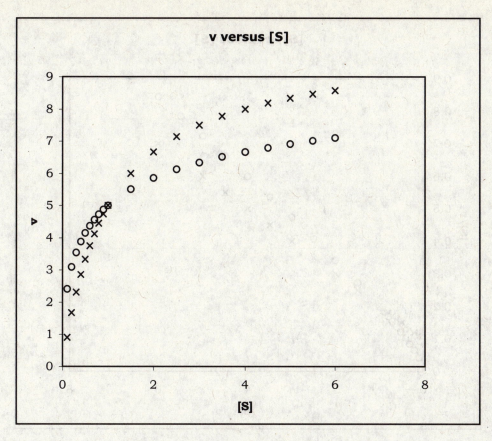

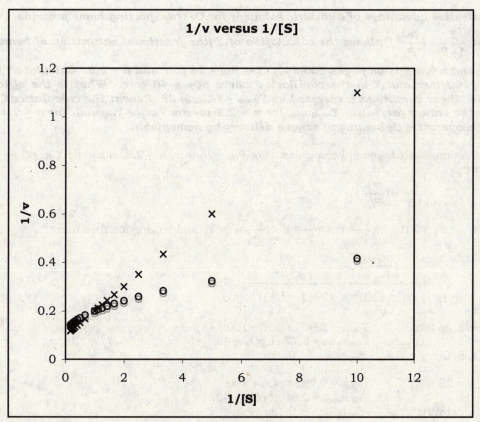

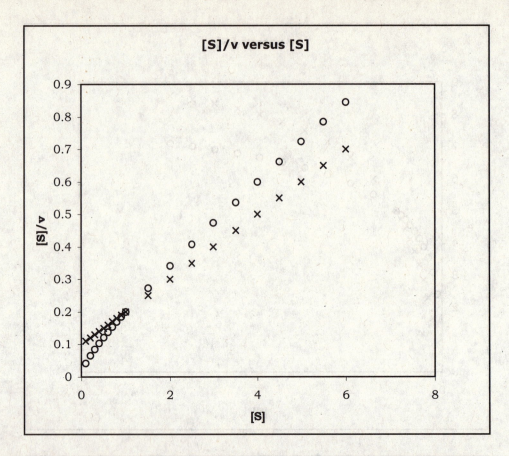

[S]/v versus [S]

5. The quantitative advantage of allosteric behavior for O₂-transporting heme proteins

The equation $\frac{Y}{1-Y} = (\frac{pO_2}{P_{50}})^n$ allows the calculation of Y (the fractional saturation of hemoglobin with

O₂) given P_{50} and n (see box on pages 500-501). Let P_{50} = 26 torr and n = 2.8. Calculate Y in the lungs where pO_2 = 100 torr and Y in the capillaries where pO_2 = 40 torr. What is the efficiency of O₂ delivery under these conditions (expressed as Y_{lungs} - $Y_{capillaries}$)? Repeat the calculations, but for n = 1. Compare the values for Y_{lungs} - $Y_{capillaries}$ for n = 2.8 versus Y_{lungs} - $Y_{capillaries}$ for n = 1 to determine the effect of cooperative O₂-binding of oxygen delivery by hemoglobin.

Answer: Y = fractional saturation of hemoglobin. For P_{50} = 26 torr, n = 2.8, calculate Y at pO_2 = 100 torr and at pO_2 = 40 torr.

$$\frac{Y}{1-Y} = (\frac{pO_2}{P_{50}})^n$$

$$Y = (1-Y)(\frac{pO_2}{P_{50}})^n = (\frac{pO_2}{P_{50}})^n - (\frac{pO_2}{P_{50}})^n \times Y, \text{ and solving for Y we find}:$$

$$Y = \frac{(\frac{pO_2}{P_{50}})^n}{1+(\frac{pO_2}{P_{50}})^n} = \frac{(pO_2)^n}{(pO_2)^n + (P_{50})^n}$$

For n = 2.8, P_{50} = 26 torr: Y_{lungs} = 0.98 at pO_2 = 100.
                                          $Y_{capillaries}$ = 0.77 at pO_2 = 40.
Efficiency of O₂ delivery = Y_{lungs} - $Y_{capillaries}$ = 0.21

For n = 1.0, P_{50} = 26 torr: Y_{lungs} = 0.79 at pO_2 = 100.
                                          $Y_{capillaries}$ = 0.61 at pO_2 = 40.
Efficiency of O₂ delivery = Y_{lungs} - $Y_{capillaries}$ = 0.18

At an oxygen tension of 100 torr, hemoglobin with n = 2.8 is nearly completely saturated. Under the same condition hemoglobin with n = 1.0 is only 79% saturated. Thus, for the case of n = 2.8, hemoglobin can carry a maximum load of oxygen away from the lung. In the tissues at a pO_2 = 40 torr, hemoglobin with n = 2.8 versus n = 1.0 can change its fractional saturation by 0.21 and 0.18, respectively. When expressed as a percentage of the initial load ([efficiency ÷ Y_{lung}] *x* 100%), O_2 delivery by hemoglobin with n = 2.8 and n = 1.0 is 21% and 23%, a modest difference.

6. Predict the effect of caffeine consumption on glycogen phosphorylase activity
The cAMP formed by adenylyl cyclase (Figure 15.19) does not persist because 5'-phosphodiesterase activity prevalent in cells hydrolyzes cAMP to give 5'AMP. Caffeine inhibits 5'-phosphodiesterase activity. Describe the effects on glycogen phosphorylase activity that arise as a consequence of drinking lots of caffeinated coffee.

Answer: By inhibiting phosphodiesterase activity cAMP levels will slowly rise or will remain elevated for longer periods of time. This will prolong stimulation of cAMP-dependent protein kinase leading to persistent phosphorylation of glycogen phosphorylase kinase and in turn glycogen phosphorylase. However, caffeine is an allosteric inhibitor of glycogen phosphorylase a and b. Caffeine stabilizes the less-active T conformation.

(Caffeine can influence three physiological conditions: Functioning of adenosine receptors, intracellular calcium levels, and cyclic nucleotide levels. Caffeine is a competitive inhibitor of adenosine receptors and is effective at low levels. In contrast, intracellular calcium levels and cyclic nucleotide levels are increased by caffeine but only at high levels of caffeine.)

7. Predict the O_2-binding properties of stored blood whose hemoglobin is BPG-depleted
If no precautions are taken, blood that has been stored for some time becomes depleted in 2,3-BPG. What will happen if such blood is used in a transfusion?

Answer: The compound 2,3-bisphosphoglycerate (2,3-BPG) binds to a single site on the hemoglobin tetramer. The binding site is located on the two-fold symmetry axis and is made up of 8 basic groups, 4 each from the two β chains. 2,3-BPG binds with ionic interactions and stabilizes hemoglobin in the T state. Because 2,3-BPG is a side product of glycolysis, it can be metabolized along this pathway when glycolytic intermediates are low. Blood stored for long periods may be low in 2,3-BPG for this reason. In red cells depleted of 2,3-BPG, hemoglobin will bind O_2 avidly and not release it to the tissues. Transfusions with such blood could actually lead to suffocation.

8. Estimate the quantitative effects of allosteric regulators on glycogen phosphorylase activity.
Enzymes have evolved such that their K_m values (or $K_{0.5}$ values) for substrate(s) are roughly equal to the in vivo concentration(s) of the substrate(s). Assume that glycogen phosphorylase is assayed at [P_i] ≈ $K_{0.5}$ in the absence and presence of AMP or ATP. Estimate from Figure 15.15 the relative glycogen phosphorylase activity when (a) neither AMP or ATP is present, (b) AMP is present, and (c) ATP is present. (Hint: Use a ruler to get relative values for the velocity v at the appropriate midpoints of the saturation curves.)

Answer: $K_{0.5}$ is defined as the concentration of substrate at which the enzyme functions at $0.5 \times V_{max}$. From Figure 15.15b we can estimate the location of $K_{0.5}$ on the x-axis by measuring the distance to V_{max} along the υ-axis, drawing a line parallel to the x-axis at a value of υ equal to $0.5 \times V_{max}$, and, where this line intersects the ATP curve (upper curve), drawing a perpendicular. The point of intersection of the perpendicular on the x-axis is equal to $K_{0.5}$. The intersection of this perpendicular on the +ATP curve is υ for = $K_{0.5}$ in the presence of ATP.

    a. υ = $0.5 \times V_{max}$ in the absence of ATP and AMP (by definition).
    b. υ ≈ $0.85 \times V_{max}$ in the presence of AMP.
    c. υ ≈ $0.12 \times V_{max}$ in the presence of ATP.

9. Describe the effects on cAMP and glycogen levels in cells exposed to cholera toxin
Cholera toxin is an enzyme that covalently modifies the G_α-subunit of G proteins. (Cholera toxin catalyzes the transfer of ADP-ribose from NAD^+ to an arginine residue in G_α, an ADP-ribosylation reaction.) Covalent modification of G_α inactivates GTPase activity. Predict the consequences of cholera toxin on cellular cAMP and glycogen levels.

Answer: Adenylyl cyclase, the enzyme that produces cyclic AMP, is stimulated by G_α-GTP. This stimulation is turned off by GTPase activity of G_α. By inhibiting this GTPase activity we expect adenylyl cyclase to be continuously stimulated leading to an increase in cellular cAMP levels, which, in turn, should lead to a decrease in glycogen levels. (A similar mechanism is exploited by cancer cells involving Ras proteins. These proteins regulate cellular responses to extracellular signals. Ras proteins are activated and released from membrane localization by GTP exchange for GDP. GTP-Ras is active but its intrinsic GTPase activity inactivates it. In many human carcinomas, Ras proteins are resistant to hydrolysis of GTP, which leads to constant signaling by GTP-Ras.)

10. One way negative cooperativity might make metabolic sense
Allosteric enzymes that sit at branch points leading to several essential products sometimes display negative cooperativity for feedback inhibition (allosteric inhibition) by one of the products. What might be the advantage of negative cooperativity instead of positive cooperativity in feedback inhibitor binding to such enzymes?

Answer: The enzyme in question produces an intermediate that may be metabolized along several different pathways leading to several different products. The products, in turn, regulate this enzyme by negative cooperativity. This means that each of the products binds to the enzyme and inhibits it, but the inhibition is cooperative. Enzyme activity will never be completely shut off, only minimized and only when there is a build-up of all of the products. Thus, the intermediate is made independent of any one final product. This prevents any one product from completely shutting down production of the other products.

The use of positive cooperativity in this case seems inappropriate. Positive cooperativity would have to occur in response to substrate, which is being supplied from the wrong end of the sequence.

11. Protein kinase specificity via consensus target sequences and intrasteric control
Consult Table 15.2 and
> **a. Suggest a consensus amino acid sequence within phosphorylase kinase that makes it a target of protein kinase A (the cAMP-dependent protein kinase).**
> **b. Suggest an effective amino acid sequence for a regulatory domain pseudosubstrate sequence that would exert intrasteric control on phosphorylase kinase by blocking its active site.**

Answer: cAMP-dependent protein kinase A phosphorylates serine or threonine in a sequence Arg-(Arg or Lys)-any amino acid-(Ser or Thr) or [-R(R/K)X(S/T)]. We must search for a region within the sequence of phosphorylase kinase to see if one is present. To get the sequence do a full text search at SwissProt (http://us.expasy.org/cgi-bin/sprot-search-ful). From the results page activate KPB1_HUMAN (P46020), which will bring you an information page for human glycogen phosphorylase kinase. The features section actually tells you where in the sequence protein kinase A phosphorylates. But, let's see if we can find the sequence on our own.

After looking over all this wonderful information about this enzyme scroll down to the Sequences section of this report and activate the FASTA format link (P46020 in FASTA format). Copy the all the data (except any line beginning with a ">" as these are comments lines" and paste it into a word document and edit it (using the Replace command in the Edit menu) by replacing paragraph marks (under "special") with "deletes". This action should convert the sequence into a single word as shown below.

```
MRSRSNSGVRLDGYARLVQQTILCHQNPVTGLLPASYDQKDAWVRDNVYSILAVWGLGLAYRKNADRDEDKAKAYELEQSVVKLMRGLL
HCMIRQVDKVESFKYSQSTKDSLHAKYNTKTCATVVGDDQWGHLQLDATSVYLLFLAQMTASGLHIIHSLDEVNFIQNLVFYIEAAYKT
ADFGIWERGDKTNQGISELNASSVGMAKAALEALDELDLFGVKGGPQSVIHVLADEVQHCQSILNSLLPRASTSKEVDASLLSVVSFPA
FAVEDSQLVELTKQEIIITKLQGRYGCCRFLRDGYKTPKEDPNRLYYEPAELKLFENIECEWPLFWTYFILDGVFSGNAEQVQEYKEALE
AVLIKGKNGVPLLPELYSVPPDRVDEEYQNPHTVDRVPMGKLPHMWGQSLYILGSLMAEGFLAPGEIDPLNRRFSTVPKPDVVVQVSIL
AETEEIKTILKDKGIYVETIAEVYPIRVQPARILSHIYSSLGCNNRMKLSGRPYRHMGVLGTSKLYDIRKTIFTFTPQFIDQQQFYLAL
DNKMIVEMLRTDLSYLCSRWRMTGQPTITFPISYSMLDEDGTSLNSSILAALRKMQDGYFGGARVQTGKLSEFLTTSCCTHLSFMDPGP
EGKLYSEDYDDNYDYLESGNWMNDYDSTSHARCGDEVARYLDHLLAHTAPHPKLAPTSQKGGLDRFQAAVQTTCDLMSLVTKAKELHVQ
NVHMYLPTKLFQASRPSFNLLDSPHPRQENQVPSVRVEIHLPRDQSGEVDFKALVLQLKETSSLQEQADILYMLYTMKGPDWNTELYNE
RSATVRELLTELYGKVGEIRHWGLIRYISGILRKKVEALDEACTDLLSHQKHLTVGLPPEPREKTISAPLPYEALTQLIDEASEGDMSI
SILTQEIMVYLAMYMRTQPGLFAEMFRLRIGLIIQVMATELAHSLRCSAEEATEGLMNLSPSAMKNLLHHILSGKEFGVERSVRPTDSN
VSPAISIHEIGAVGATKTERTGIMQLKSEIKQVEFRRLSISAESQSPGTSMTPSSGSFPSAYDQQSSKDSRQGQWQRRRRLDGALNRVP
VGFYQKVWKVLQKCHGLSVEGFVLPSSTTREMTPGEIKFSVHVESVLNRVPQPEYRQLLVEAILVLTMLADIEIHSIGSIIAVEKIVHI
ANDLFLQEQKTLGADDTMLAKDPASGICTLLYDSAPSGRFGTMTYLSKAAATYVQEFLPHSICAMQ
```

My single "word" has 1223 character. Next, using MS Word search for sequences like RR?S, RR?T, RK?S, RK?T. (The ? is a wild card meaning any character will be accepted.) RRFS and RRLS were the only two found.

284

```
MRSRSNSGVRLDGYARLVQQTILCHQNPVTGLLPASYDQKDAWVRDNVYSILAVWGLGLAYRKNADRDEDKAKAYELEQSVVKLMRGLL
HCMIRQVDKVESFKYSQSTKDSLHAKYNTKTCATVVGDDQWGHLQLDATSVYLLFLAQMTASGLHIIHSLDEVNFIQNLVFYIEAAYKT
ADFGIWERGDKTNQGISELNASSVGMAKAALEALDELDLFGVKGGPQSVIHVLADEVQHCQSILNSLLPRASTSKEVDASLLSVVSFPA
FAVEDSQLVELTKQEIITKLQGRYGCCRFLRDGYKTPKEDPNRLYYEPAELKLFENIECEWPLFWTYFILDGVFSGNAEQVQEYKEALE
AVLIKGKNGVPLLPELYSVPPDRVDEEYQNPHTVDRVPMGKLPHMWGQSLYILGSLMAEGFLAPGEIDPLN**RRFS**TVPKPDVVVQVSIL
AETEEIKTILKDKGIYVETIAEVYPIRVQPARILSHIYSSLGCNNRMKLSGRPYRHMGVLGTSKLYDIRKTIFTFTPQFIDQQQFYLAL
DNKMIVEMLRTDLSYLCSRWRMTGQPTITFPISYSMLDEDGTSLNSSILAALRKMQDGYFGGARVQTGKLSEFLTTSCCTHLSFMDPGP
EGKLYSEDYDDNYDYLESGNWMNDYDSTSHARCGDEVARYLDHLLAHTAPHPKLAPTSQKGGLDRFQAAVQTTCDLMSLVTKAKELHVQ
NVHMYLPTKLFQASRPSFNLLDSPHPRQENQVPSVRVEIHLPRDQSGEVDFKALVLQLKETSSLQEQADILYMLYTMKGPDWNTELYNE
RSATVRELLTELYGKVGEIRHWGLIRYISGILRKKVEALDEACTDLLSHQKHLTVGLPPEPREKTISAPLPYEALTQLIDEASEGDMSI
SILTQEIMVYLAMYMRTQPGLFAEMFRLRIGLIIQVMATELAHSLRCSAEEATEGLMNLSPSAMKNLLHHILSGKEFGVERSVRPTDSN
VSPAISIHEIGAVGATKTERTGIMQLKSEIKQVEF**RRLS**ISAESQSPGTSMTPSSGSFPSAYDQQSSKDSRQGWQRRRRLDGALNRVP
VGFYQKVWKVLQKCHGLSVEGFVLPSSTTREMTPGEIKFSVHVESVLNRVPQPEYRQLLVEAILVLTMLADIEIHSIGSIIAVEKIVHI
ANDLFLQEQKTLGADDTMLAKDPASGICTLLYDSAPSGRFGTMTYLSKAAATYVQEFLPHSICAMQ
```

The correct one happens to be RRLS with the serine located at 1018.

b. To regulate phosphorylase kinase with a pseudosubstrate sequence we must first look at Table 15.2 for its consensus target sequence: -KRKQIS*VRGL. The serine in this sequence is phosphorylated by the kinase reaction. A suitable regulatory domain pseudosubstrate sequence would mimic the target sequence but have an amino acid substitution at serine. A convenient substitution is A (alanine) for S (serine) since both amino acids are about the same size. Thus, a potential pseudosubstrate sequence might be KRKQIAVRGL.

12. Allosteric regulation versus covalent modification
What are the relative advantages (and disadvantages) of allosteric regulation versus covalent modification?

Answer: Allosteric regulation involves substrates or metabolites that are linked metabolically in some fashion. Thus, the regulatory signal is part of a metabolic process. And, the process is regulated by its immediate concentration of allosteric regulator. There is no possibility for anticipating metabolic needs. This is an advantage and a disadvantage. By being responsive to local metabolic signals, an allosterically regulated system will be very efficient in using resources and reliance on reversible binding will allow rapid response to changes in local metabolic signals. The major disadvantage, however, is an inability to be overridden by signals not directly involved in the pathway.

Covalent modification allows for inputs not directly in the metabolic pathway. In addition, since the modifications are covalent they can have long lasting effects. The disadvantage to covalent modification is the necessity of having to have ways of stopping the modifying enzymes and removing the covalent modification. This often involves a cascade of events.

13. Potential treatment of sickle-cell anemia by drugs targeted to HbS
You land a post as scientific investigator with a pharmaceutical company that would like to develop drugs to treat people with sickle-cell anemia. They want ideas from you! What molecular properties of Hb S might you suggest as potential targets of drug therapy?

Answer: Hemoglobin S differs from normal hemoglobin by a single amino acid change in the β chain: a Glu to Val at position 6. This change replaces an acidic, and therefore negatively charged amino acid, for a hydrophobic amino acid. The problem with sickle cell hemoglobin is that at low oxygen tension it polymerizes into long polymers that cause red cells to assume an elongated, inflexible shape, which causes them to move poorly through small-diameter capillaries. You might suggest developing a drug that binds to this region of hemoglobin and lowers its hydrophobicity. You might think about some N-terminal chemistry that alters the characteristics of the N-terminus, which is located only six amino acid residues away.

One of the few drugs currently being used to treat sickle cell anemia is hydroxyurea. However, hydroxyurea does not address the defective hemoglobin. It induces formation of fetal hemoglobin and the presence of fetal hemoglobin lessens sickling.

14. Nitric oxide is a metabolic regulator, but what kind?

Under appropriate conditions, nitric oxide (NO·) combines with Cys 93β in hemoglobin and influences its interaction with O₂. Is this interaction an example of allosteric regulation or covalent modification?

Answer: The reaction between nitric oxide and Cys 93 produces S-nitrosocysteine, which is shown below. Clearly it is a covalent modification. The reaction, however, is reversible and does so especially in deoxy-hemoglobin. The binding and release of nitric oxide is a property of hemoglobin that appears to be modulated by reversible binding of oxygen.

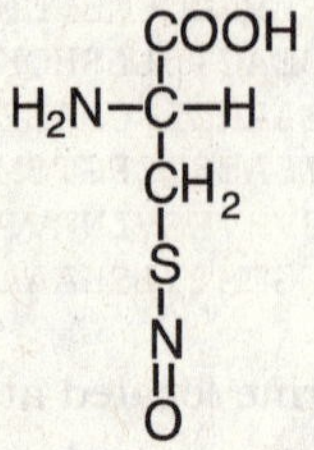

$$\begin{array}{c} COOH \\ | \\ H_2N-C-H \\ | \\ CH_2 \\ | \\ S \\ | \\ N \\ \| \\ O \end{array}$$

15. Is lactate an allosteric effector of Mb?

Lactate, a metabolite produced under anaerobic conditions in muscle, lowers the affinity of myoglobin for O₂. This effect is beneficial, because O₂ dissociation from Mb under anaerobic conditions will provide the muscle with oxygen. Lactate binds to Mb at a site distinct from the O₂-binding site at the heme. In light of this observation, discuss whether myoglobin should be considered an allosteric protein.

Answer: Lactate production is a consequence of anaerobic metabolism of pyruvate (fermentation). Pyruvate's aerobic fate is to be fed into the citric acid cycle, a process that requires oxygen. Thus, buildup of lactate is a signal that pyruvate is not being metabolized sufficiently fast. Since myoglobin is a store of oxygen, regulation of oxygen affinity by lactate makes physiological sense.

Is myoglobin an allosteric protein given this observation? Allostery refers to regulation of enzymatic activity or ligand binding by another molecule at another binding site on the protein. A consequence of allostery is modulation of protein structure either by shifting the equilibrium between two conformational states (T and R) or modulating the properties of neighboring binding sites (induced fit). That lactate influences oxygen affinity of myoglobin by binding to a site distinct from the oxygen binding site is consistent with allostery. Additionally, oxymyoglobin and deoxymyoglobin exhibit different affinities for lactate. These would indicate that myoglobin is an allosteric protein. The title of a paper by Giardia et al (Functional modulation by lactate of myoglobin A monomeric allosteric hemoprotein (1996) J. Biol. Chem. 29:16999-17001) clearly indicates how they view the question.

Allostery has traditionally been thought of as a property of multimeric proteins, like hemoglobin.

16. Morpheeins – a model for allosteric regulation based on the reversible association-dissociation of a monomeric-oligomeric protein

An allosteric model based on multiple oligomeric states of a protein has been proposed by E. K. Jaffe (2005. Morpheeins: A new structural paradigm for allosteric regulation. Trends in Biochemical Sciences 30:490–497). This model coins the term morpheeins to describe the different forms of a protein that can assume more than one conformation, where each distinct conformation assembles into an oligomeric structure with a fixed number of subunits. For example, conformation A of the protein monomer forms trimers, whereas conformation B of the monomer forms tetramers. If trimers and tetramers have different kinetic properties (Kₘ and kcat values), as in low-activity trimers and high-activity tetramers, then the morpheein ensemble behaves like an allosterically regulated enzyme. Drawing on the traditional MWC model as an analogy, diagram a simple morpheein model in which wedge-shaped protein monomers assemble into trimers but the alternative conformation for the monomer (a square shape) forms tetramers. Further, the substrate, S, or allosteric regulator, A, binds "only" to the square conformation, and its binding prevents the square from adopting the wedge conformation. Describe how your diagram yields allosteric behavior.

Answer: In the diagram below the monomer can exist in two conformational states indicated by the third-of-a-circle wedge and the rectangle with a round indentation in one side. The indentation is a substrate binding site. The wedge can associate into a trimer that does not bind ligand or that exhibits low catalytic properties.

The rectangle can associate into a tetramer with functional ligand-binding sites. A positive allosteric activator molecule, shown as a small, shaded rectangle, binds to the rectangular conformation only. In the presence the activator the equilibrium is shifted to the rectangular conformation and then to the tetrameric conformation. In this state, ligand binding is tightest.

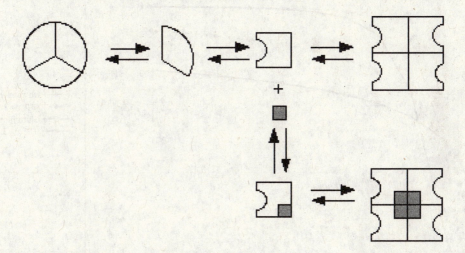

17. Draw substrate saturation curves for the substrates of CTP synthetase which show both positive and negative cooperativity
CTP synthetase catalyzes the synthesis of CTP from UTP:
$$UTP + ATP + glutamine \rightleftharpoons CTP + glutamate + ADP + P_i$$
The substrates UTP and ATP show positive cooperativity in their binding to the enzyme, which is an α_4-type homotetramer. However, the other substrate, glutamine, shows negative cooperativity. Draw substrate saturation curves of the form v versus [S]/K$_{0.5}$ for each of these three substrates that illustrate these effects.

Answer: If the enzyme exhibits positive cooperativity for UTP and ATP then the Hill coefficient must be greater than one but likely less than four. (Since the enzyme is a homotetramer it would only have a Hill coefficient of 4 if it were perfectly cooperative.) The $K_{0.5}$ values for the three substrates are expected to be different but values for UTP and ATP might be similar (but not necessarily equal). The Hill coefficients for UTP and ATP need not be identical. Finally, since the enzyme exhibits negative cooperativity for glutamine the Hill coefficient in this case should be less than 1. The question asks for plots of v versus [S] normalized to $K_{0.5}$ values and this will allow all three plots to be presented on the same graph. The graph below includes the case of no cooperativity.

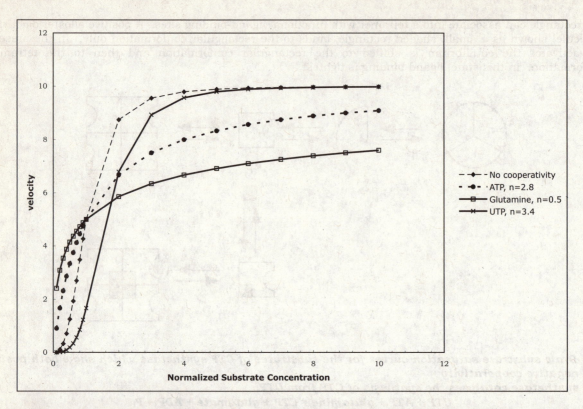

18. Draw a model for the conformational states of glyceraldeyde-3-P dehydrogenase, a tetrameric enzyme displaying negative cooperativity

Glyceraldehyde-3-phosphate dehydrogenase catalyzes the synthesis of 1,3-bisphosphoglycerate:

$$Glyceraldehyde\text{-}3\text{-}P + P_i + NAD^+ \rightleftharpoons 1,3\text{-}BPG + NADH + H^+$$

The enzyme is a tetramer. NAD⁺ binding shows negative cooperativity. Draw a diagram of possible conformational states for this tetrameric enzyme and its response to NAD⁺ binding that illustrates negative cooperativity.

Answer: Negative cooperativity requires use of the induced fit model. The enzyme has binding sites for glyceraldehyde-3-phosphate, inorganic phosphate and NAD⁺ and NAD⁺-binding exhibits negative cooperativity. The enzyme functions as a tetramer. For negative cooperativity to come into play binding of NAD⁺ to one subunit in the tetramer must induce a conformational change that is transmitted to the other subunits resulting in lower-affinity NAD⁺-binding sites. The binding sites for NAD⁺ are initially shown as small open circles. When the first NAD⁺ binds (depicted as a small, filled circle) it induces a conformational change in the subunit that is transmitted to neighboring subunits. The conformational change in the neighboring subunits changes the affinity of the NAD⁺-binding site and this is shown as a transition from circles to squares with rounded corners to squares.

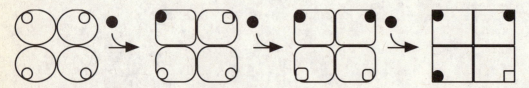

19. Proteomics as a tool to study metabolic regulation
Proteomics studies have revealed that protein acetylation is an important mechanism for regulation of metabolic enzymes. Describe how proteomics might have led to this discovery.

Answer: The first acetylated proteins to be discovered were histones that were acetylated on the side chain nitrogen of specific lysine residues. The acetyl group derives from acetyl-CoA and the enzymes responsible for acetylation are histone acetyl transferases (HATs). Histone deacetylases (HDACs) are responsible for removal of acetyl groups and thus acetylation is a reversible process. Side-chain lysine acetylation is known as Nᵉacetylation and it modifies the protein's behavior. N^α-acetylation is also known to occur with the N-terminal nitrogen of a protein being acetylated. This acetylation is involved in protein stability.

The consequence N-acetylation is to lose a positive charge. This happens because the lone-pair on the nitrogen is delocalized by the carbonyl group of the acetyl moiety thus preventing it from being protonated. Modification of histones by acetylation was at first thought to cause a weakening of protein/DNA interactions due to loss of a positive charge on a histone side chain. It is now known that acetylated lysines are recognized by so-called bromodomains and so lysine acetylation regulates protein-protein interactions.

HATs and HDACs are specific examples of KATs and KDACs, lysine (one-letter code is K) acetyltransferases and lysine deacetylases, enzymes that act on lysine residues in specific proteins. It is now known that many proteins are subject to acetylation/deacetylation regulation. A good review on the subject is: *Comprehensive lysine acetylomes emerging from bacteria to humans* by Go-Woon Kim and Xian-Jiao Yang in Trends in Biochemical Sciences (2011) 36:211-220. The review discusses the history of discovery of acetylated lysine and discusses the importance of the use of antibodies against modified lysines and mass spectrometry to identify and characterize acetylated lysines.

Preparing for the MCAT® Exam

20. On the basis of the graphs shown in Figures 15.29 and 15.30 and the relationship between blood pH and respiration (Chapter 2), predict the effect of hyperventilation on Hb:O_2 affinity.

Answer: Hyperventilation will increase arterial pO_2 and decrease pCO_2. The decrease in pCO_2 causes a change in blood pH by the following mechanism catalyzed by carbonic anhydrase.

$$H^+ + HCO_3^- \rightleftharpoons H_2CO_3 \rightleftharpoons H_2O + CO_2$$

Lower pCO_2 pulls the equation to the right resulting in a decrease in $[H^+]$ and an increase in pH. Looking at Figure 15.29 this will result in an increase in affinity of Hb for oxygen and a modest increase in the percent saturation of hemoglobin. Figure 15.30 indicates that CO_2 lowers hemoglobin's affinity for oxygen. The mechanism is carbamate formation on alpha amino groups. This negatively-charged adduct will form a salt link that stabilizes the deoxy or T state. Lowering of pCO_2 will remove carbamate and also favor the oxy state. Thus, hyperventilation is expected to increase oxygen affinity.

21. Figure 15.18 traces the activation of glycogen phosphorylase from hormone to phosphorylation of the b form of glycogen phosphorylase to the a form. These effects are reversible when hormone disappears. Suggest reactions by which such reversibility is achieved.

Answer: The whole process is activated when hormone binds to a cell-surface receptor causing the receptor's intercellular domain to exchange GTP for GDP on the $G\alpha$ subunit of a $G_\alpha G_\beta G_\gamma$ protein heterotrimer. G_α-GTP dissociates from $G_\beta G_\gamma$, binds to adenylyl cyclase, which, in turn, produces cAMP. cAMP activates cAMP-dependent protein kinase, which phosphorylates glycogen phosphorylase kinase, which phosphorylates glycogen phosphorylase thus activating it. To shut off this process hormone release must stop. G_α-GTP has GTPase activity that converts GTP to GDP, which results in inactivation of G_α. For G_α to become active again it must pick up GTP, which occurs only when hormone is bound to the receptor allowing $G_\beta G_\gamma$ to function as a guanine nucleotide exchange factor. Cyclic AMP levels are regulated by adenylyl cyclase activity and phosphodiesterase. When hormone is no longer present G_α-GTP is no longer available to stimulate adenylyl cyclase. Phosphodiesterase continues to hydrolyze cAMP to AMP thus shutting off this signal. Without cAMP, cAMP-dependent protein kinase is no longer stimulated and thus proteins are not actively phosphorylated. Protein phosphatase I, however, continues to remove phosphate groups from proteins by hydrolysis and as a consequence glycogen phosphorylase kinase and glycogen phosphorylase are inactivated.

Questions for Self Study

1. What is the essential difference between the "lock and key" hypothesis of enzyme substrate interaction and the "induced fit" hypothesis?

2. List six factors that affect or modulate the activity of enzymes.

3. How are the proteins trypsinogen, chymotrypsinogen, pepsinogen, procarboxypeptidase, proelastase, related to active enzymes?

289

4. Lactate dehydrogenase is a tetramer that catalyzes the interconversion of lactate and pyruvate. There are two different subunits, A and B, that can interact to form either homotetramers of heterotetramers with different kinetic properties. Write the five possible tetrameric combinations. What kind of enzyme regulation does lactate dehydrogenase typify?

5. Fill in the blanks. Oxygen binding to myoglobin is ___ by pH whereas decreasing pH _____ the affinity of hemoglobin for oxygen. This is known as the ____ effect. One physiological source of protons that will decrease pH in tissue is produced by the enzyme carbonic anhydrase, which catalyzes the following reaction: ____. One of the substrates of this reaction, namely ____, by itself lowers the oxygen affinity of hemoglobin by reacting with free α-amino groups to produce a ___, a negatively charged species that can form salt bridges and stabilize the ____ form of hemoglobin. Another compound that will stabilize this form of hemoglobin is ___, which binds at a single site on hemoglobin.

6. Fetal hemoglobin has (higher or lower) affinity for oxygen than does adult hemoglobin because it has (higher or lower) affinity for 2,3-bisphosphoglycerate.

7. The enzyme aspartate transcarbamolyase catalyzes the condensation of carbamoyl phosphate and aspartic acid to form N-carbamoylasparate. A plot of initial velocity versus aspartate concentration is not a rectangular hyperbola as found for enzymes that obey Michaelis-Menten kinetics. Rather, the plot is sigmoidal. Further, CTP shifts the curve to the right along the substrate axis and makes the sigmoidal shape more pronounced. ATP shifts the curve to the left and makes the sigmoidal shape less pronounced. What kind of enzyme is aspartate transcarbamoylase and what roles do CTP and ATP play in enzyme activity?

8. Match the items in the left hand column with items in the right hand column.
   a. Symmetry Model
   b. Monod, Wyman, and Changeux
   c. K system
   d. V system
   e. Koshland, Nemethy, and Filmer
   f. Induced fit
   g. Hill coefficient
   h. Increases [R conformation]
   i. Decreases [R conformation]
   j. Homotropic effector

   1. Second noncatalytic substrate binding site.
   2. Substrate binding concentration dependence.
   3. Negative effector.
   4. Conformation [substrate] dependent.
   5. $K_{0.5}$ changes in response to effectors.
   6. V_{max} regulated.
   7. Sequential model.
   8. $R_0 \star T_0$
   9. Two conformational states.
   10. Positive effector.

Answers

1. The "lock and key" hypothesis depicts the enzyme as a rigid template into which the substrate fits. In the "induced fit" hypothesis enzymes are highly flexible, dynamic molecules and the shape of the enzyme's active site is modified by substrate binding.

2. Product inhibition, substrate availability, genetic controls such as induction and repression, covalent modifications, allosteric regulation by effector molecules, zymogens, isozymes, and modulator proteins.

3. These proteins are precursors of the enzymes trypsin, chymotrypsin, pepsin, carboxypeptidase, and elastase. They are converted to active enzymes by hydrolysis of peptide bonds.

4. A_4, A_3B, A_2B_2, AB_3, B_4; isozymes.

5. unaffected; lowers; Bohr; $CO_2 + H_2O \star H_2CO_3 \star H^+ + HCO_3^-$; carbon dioxide; carbamate; T (or tense or taut); 2,3-bisphosphoglycerate.

6. Higher; lower

7. Aspartate transcarbamoylase is a regulatory enzyme. CTP is a negative heterotropic effector and ATP is a positive heterotropic effector.

8. a. 8 or 9; b. 8 or 9; c. 5; d. 6; e. 7; f. 4; g. 2; h. 10; i. 3; j. 1.

Additional Problems

1. Enzyme activity may be controlled in several ways (see answer to Questions for Self Study 2 above). It is clear that some of these regulatory mechanisms are effective nearly instantly whereas others are slowly activated. With respect to duration, certain controls are short-lived whereas others are long-lasting. Compare and contrast enzyme control mechanisms with respect to these parameters.

2. The heme-binding site of both myoglobin and hemoglobin is a pocket in the hydrophobic core of these proteins. This pocket is lined with hydrophobic amino acid side-chains with the exception of histidine F8 (the 8th residue of helix F). Generally, hydrophilic residues such as histidine are not located in hydrophobic interiors. What are the important functions of histidine F8 that require it to be buried in a hydrophobic environment?

3. The histidine discussed above is the so-called distal histidine. A second histidine, the proximal histidine, is located on the opposite side of the heme ring. How does the proximal histidine influence ligand specificity?

4. Free heme is an effective oxygen-binding molecule but it has two shortcomings: its iron is readily oxidized, and it forms a heme:O_2:heme complex that has low solubility. How has incorporation of heme into a protein corrected these heme flaws?

5. Discuss the role of amino acid substitution in adapting fetal hemoglobin and adult hemoglobin to different physiological functions.

6. It is rather easy to determine if an accidental cut involves a vein or an artery. Arterial cuts produce crimson-colored blood whereas venous cuts produce darker, bluish-tinted blood. Explain.

7a. A lecturer wants to demonstrate the spectral changes upon oxygenation and deoxygenation of hemoglobin. So, pure hemoglobin, obtained as a lyophilized (freeze-dried) powder from a biochemical supply company, is resuspended in a buffered salt solution. The resulting solution is brownish-red in color indicating a mixture of oxyhemoglobin and met-hemoglobin. What is met-hemoglobin and how can it be converted to oxyhemoglobin?
b. The conversion is made to oxyhemoglobin and the lecturer now decides to produce deoxyhemoglobin by lowering the O_2-tension by placing the solution under partial vacuum. This is only moderately successful. Can you suggest modifications to the procedure that will favor deoxyhemoglobin formation?

8. The expression $\upsilon = \dfrac{n\alpha(1+\alpha)^{n-1}}{L+(1+\alpha)^n}$ derived in the answer to problem 3 above assumes that substrate binds to the **R** state but not the **T** state. If this assumption is changed allowing substrate binding to both states, the equation becomes:

$$\upsilon = \frac{\alpha n(1+\alpha)^{n-1}+Lc\alpha n(1+c\alpha)^{n-1}}{L(1+\alpha)^n+(1+\alpha)^n}$$

where $c = K_R/K_T$, the ratio of dissociation constants for substrate binding to the **R** and **T** states. (a) Under what conditions is substrate binding to protein dominated by T-state binding. (b) Plot υ versus α for varying L and fixed c. (c) Plot υ versus α for varying L and fixed c.

9. Using the Hill plot show below, estimate K_1 and K_4 for hemoglobin and K for myoglobin. Estimate the Hill coefficient for myoglobin and for hemoglobin at low, intermediate, and high values of pO_2.

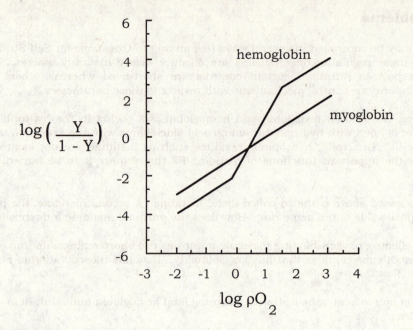

Abbreviated Answers

1. Variations in [product], [S], and [cofactors] have immediate consequences for enzyme activity. The same is true for allosteric regulation. Further, these controls last as long as the signals are present. Genetic controls, such as induction and repression leading to an increase or decrease in enzyme synthesis, are fast, although not immediate, and can be long lasting. Clearly, degradation is long-lasting. However, a delicate balance between synthesis and degradation can lead to a very responsive system of control. Covalent modification may lead to rapid and long lasting changes in enzyme activity. Balancing the activity of modifying enzymes against enzymes that can reverse these modifications creates a sensitive system of enzyme control. Finally, zymogens and isozymes are examples of enzyme control mechanisms that can be extremely long lasting.

2. This histidine, known as the distal histidine because it is on the opposite side of the oxygen-binding site where the so-called proximal histidine is located (see following question), is one of the iron coordinates in both myoglobin and hemoglobin. In hemoglobin, movement of the distal histidine in a single subunit of Hb can influence the accessibility of the oxygen-binding site in the other subunits.

3. The proximal histidine is located close enough to the heme iron so as to prevent carbon monoxide from binding to hemoglobin with optimum geometry.

4. By forming a protein:heme complex, the apparent solubility of heme is greatly increased. Thus, higher concentrations can be maintained without the problem of limited solubility, especially in the presence of oxygen. Because the heme is partially buried in a hydrophobic pocket, it is protected from oxidation. Oxidation produces met-Hb which does not transport oxygen.

5. Oxygen is supplied to the fetal blood system via the placenta where gas exchange occurs between fetal and maternal circulatory systems. Thus, the fetus must successfully compete with adult, maternal hemoglobin for oxygen. This is possible because fetal hemoglobin has a higher affinity for oxygen than adult hemoglobin. The β-chains of adult-Hb are replaced by γ-chains in fetal-HB. A critical function of β-chains is formation of the 2,3-BPG binding-site at the β/β interface of the (αβ) dimer junction. The binding site is composed of eight positive charges, four from each β chain of which three are from side chains and the fourth is from the N-terminus. In fetal-HB, the 2,3-BPG binding-site is considerably weaker due to an amino acid substitution, Ser-143 for His-143, at the 2,3-BPG binding site. As a consequence fetal-HB binds 2,3-BPG less avidly and therefore has a high oxygen affinity.

6. In arterial blood, hemoglobin is oxygenated whereas in venous blood, deoxy-hemoglobin predominates. The change in color upon change in oxygenation state is due to an alteration in the iron coordination. In

oxyHb, iron is six-fold coordinated with the sixth position occupied by oxygen. In deoxyHb, iron is five-fold coordinated.

7a. Met-hemoglobin is hemoglobin with its iron atom in the +3 oxidation state as ferric (Fe^{3+}) iron. It can be converted to oxyhemoglobin by treatment with a reducing agent such as sodium hydrosulfite (sodium dithionite).

b. A couple of things come to mind. Since it is pure hemoglobin, it may lack 2,3-bisphosphoglycerate (2,3-BPG), accounting for the tenacity of oxygen binding. Alternatively, the pH of the solution could be lowered and CO_2 added to decrease oxygen affinity.

8a. When the product Lc is much greater than 1, T-state binding dominates. Typically, $K_R > K_L$ and so c > 1.
b. and c. The upper four curves shown below are for c = 0.1 and L = 1, 10, 100, and 1,000. The lower three curves are for L = 10,000 and c = 0.1, 0.03, and 0.01.

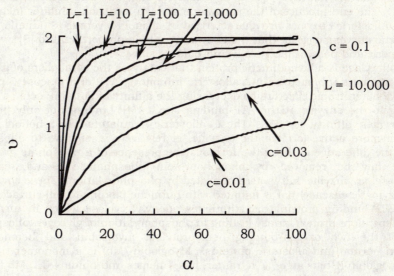

9. From the Hill equation, $\log(\frac{Y}{1-Y}) = n \times \log(pO_2) - K$, it can be seen that when Y is equal to 1/2, K = x-intercept = $n(\log pO_2)$. The x-intercept is determined graphically as shown below.

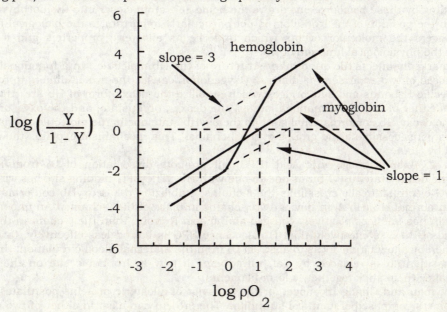

Summary

Enzymes display an extraordinary specificity with regard to the substrates they act upon and the reactions they catalyze. Molecular recognition through structural complementarity is the basis of this specificity. The active sites of enzymes are pockets or clefts in the protein structure whose shape and charge distribution are complementary to the substrate's own properties. In turn, substrate binding induces changes in protein conformation so that the substrate is embraced by the enzyme and catalytic groups are brought to bear. In effect, the enzyme:substrate complex mimics the transition state intermediate of the reaction.

Regulation of enzyme activity can be achieved at several levels: enzyme synthesis or degradation, reversible covalent modifications catalyzed by converter enzymes, noncovalent interactions with cellular metabolites (allosteric regulation), zymogen activation, isozymes or modular proteins.

Allosteric regulation occurs when regulatory metabolites modulate the activity of allosteric enzymes by binding at sites distinct from the active site. Such effectors may activate (positively regulate) or inhibit (negatively regulate) enzyme activity. Often, an allosteric enzyme catalyzes the first step in an essential metabolic pathway. The end product of the pathway may be a feedback inhibitor (negative regulator) of this allosteric enzyme. Allosteric enzymes are characterized by sigmoid υ versus [S] kinetic patterns. The sigmoid shape of such kinetic curves suggests that the enzymatic rate is proportional to $[S]^n$, where $n > 1$. Allosteric enzymes are oligomeric and possess multiple ligand-binding sites.

Glycogen phosphorylase is a homodimeric protein that catalyzes the catabolism of glycogen into glucose-1-phosphate. The enzyme is inhibited by two allosteric inhibitors, ATP and glucose-6-phosphate, which both function as negative heterotropic effectors by decreasing the affinity of P_i. AMP acts as a positive heterotropic effector by binding to the enzyme at the ATP-binding site. AMP binding not only blocks ATP binding but increases the enzymes affinity for P_i. These allosteric regulators all function on phosphorylase b. Phosphorylase a, a more active form of the enzyme, is relatively insensitive to allosteric regulation. The physical basis for the difference in these two forms is the presence of a phosphate group on serine 14. The phosphate group may be removed by phosphoprotein phosphatase I and reattached by glycogen phosphorylase kinase, an enzyme that is itself activated by phosphorylation. These covalent modifications are the result of an enzymatic cascade that is initiated with hormone binding to cell surface receptors, transduced into release of a GTP-binding protein that stimulates adenylyl cyclase which in turn produces cAMP that stimulates cAMP-dependent protein kinase leading to phosphorylation of glycogen phosphorylase kinase.

Comparison of the two oxygen-binding hemoproteins, myoglobin and hemoglobin, illustrates the differences between normal and allosteric proteins. Myoglobin (Mb) is a monomer that has one heme and binds one O_2. Hemoglobin (Hb) is an $\alpha_2\beta_2$ tetramer, has 4 hemes and binds 4 O_2. Mb functions as an oxygen storage protein in muscle, while Hb serves as the vertebrate oxygen-transport protein. The oxygen-binding curve for Mb resembles the substrate saturation curves seen for normal, "Michaelis-Menten" enzymes, but the Hb O_2-binding curve is sigmoid, indicating that the binding of O_2 to Hb is cooperative. Cooperative binding means that the binding of one O_2 to an Hb molecule enhances the binding of additional O_2 molecules by the same Hb. O_2-binding by these hemoproteins draws the heme Fe^{2+} atom closer into the porphyrin ring system. This small shift in the position of the iron atom is not particularly important to the biological function of Mb, but in Hb, it triggers the molecular events which underlie its allosteric properties and it has profound consequences for the physiological function of Hb.

Oxygenation markedly affects the quaternary structure of the Hb molecule. In a functional sense, the Hb molecule is composed of two equivalent $\alpha\beta$-dimers. Oxygenation causes the two $\alpha\beta$-dimers to rotate relative to one another. Specific H-bonds and salt bridges which stabilize the deoxy form of Hb are ruptured when O_2 binds. Hb adopts either of two conformational states, the deoxy-, or **T**, state, and the oxy-, or **R**, state. The transition from the deoxy Hb conformational state to the oxyHb state opens the molecule so that O_2 now gains access to previously inaccessible heme groups of the β-chains. This mechanism explains the cooperative O_2-binding exhibited by Hb.

H^+, CO_2, and 2,3-bisphosphoglyceric acid (BPG) all promote dissociation of O_2 from Hb by binding preferentially to the deoxy state. BPG binds one-to-one with Hb. The BPG-binding site lies within the central cavity of Hb. BPG electrostatically crosslinks the β-chains, stabilizing the deoxyHb conformation. Fetal Hb (Hb *F*) differs from normal Hb (Hb *A*) in having two γ-chains in place of the β-chains. An important difference is that each γ-chain has one less positively charged amino acid R-group than the β-chain in the BPG-binding site, so Hb *F* binds BPG less effectively than Hb *A*. Since BPG is bound less effectively, O_2 is bound more avidly by Hb *F*, thereby allowing the fetus to abstract O_2 from the maternal blood circulation. In sickle-cell Hb (Hb *S*), Glu at position $\beta6$ is replaced by Val, introducing a new hydrophobic site on the surface of Hb molecules that makes them stick together to form filaments.

The Monod, Wyman and Changeux model for the behavior of allosteric proteins postulates that allosteric proteins are oligomers composed of identical subunits. The protein can exist in either of two conformational states, the **T** or "taut" state, and the **R** or "relaxed" state. The distribution of the protein between the two conformational states is characterized by the equilibrium constant, L: L = **T₀/R₀**. The different states have different affinities for the various ligands. Usually, it is assumed that the substrate S binds "only" to the **R** state; the positive allosteric effector A binds "only" to the **R** state; and the negative allosteric effector I binds "only" to the **T** state. Given these parameters, the sigmoid rate response of allosteric enzymes to substrate concentration and the effects exerted by allosteric effectors are understood. The system described above is called a "K" system, because the substrate concentration giving half-maximal velocity, defined as $K_{0.5}$, varies, while V_{max} remains constant. In the "V" system of allosteric control, **R** and **T** states have the same affinity for S, but the **R** state is catalytically active, while the **T** state is not. A and I still differ in their relative affinities for **R** and **T**. Both the K and the V systems assume that the oligomeric allosteric protein undergoes a concerted conformational transition so that all subunits have identical conformations at the same time. The Koshland, Nemethey and Filmer model rests on the "induced fit" hypothesis. That is, via tightly linked subunit interactions, the binding of ligand to one subunit can influence the ligand-binding affinity of neighboring subunits. Therefore, an oligomeric protein could pass through a sequential series of conformations with differing ligand affinities, and different subunits may not necessarily have the same conformation at the same time.

Chapter 16

Molecular Motors

• •

Chapter Outline

❖ Molecular motors: Energy transducing molecules that convert chemical energy into mechanical energy
❖ Muscle contraction: Actin/myosin-based system
   ⅄ Muscle cells: Multinucleated long cells known as muscle fibers
   ⅄ Myofibrils: Linear arrays of muscle proteins within muscle fibers
   ⅄ Sarcomere: Repeat unit along myofibril
   ⅄ Sarcolemma: Specialized plasma membrane surrounding muscle fibers
      • Transverse (or t) tubules: Extensions of plasma membrane (sarcolemma) that contact myofibrils
      • Sarcoplasmic reticulum: Specialized endoplasmic reticulum membrane that surrounds myofibrils
❖ Sarcomeres: Basic structural units of contraction: Repeat regions along myofibrils
   ⅄ Z line to Z line: Attachment site of f-actin
   ⅄ I band: Centered around Z lines: F actin of opposite polarity on each side, tropomyosin and troponin
   ⅄ A band: Centered within sarcomere: Composed of myosin
   ⅄ H zone: Centered on M disc within A bond: Opposite myosin polarity on neither side of M disc
      • Region of A band outside H zone: Actin and myosin overlap
❖ Thin filaments
   ⅄ Filamentous (F) actin: Polymers of globular (G) actin
      • Polar filaments
      • Stimulates ATPase activity of myosin
   ⅄ Tropomyosin: Heterodimeric protein: Coiled coils of alpha helix
      • Dimers form head to tail polymers
      • Polymers bind to F actin
      • Stoichiometry: 7 Actin per tropomyosin heterodimer
   ⅄ Troponin
      • Troponin T: Binds to tropomyosin
      • Troponin I: Binds to tropomyosin and actin
      • Troponin C: Binds to Troponin I
❖ Thick filaments: Bipolar myosin filaments
   ⅄ Myosin composition
      • Two heavy chains
      • Two LC1 (light chains): Essential light chains
      • Two LC2: Regulatory light chains
   ⅄ Myosin structure
      • Heavy chain tails associate to form filaments
        • 7-residue repeats form coiled-coil
      • Heads: Heavy and light chains: ATPase activity
❖ Muscle contraction: Sliding filament model

- Muscle tension (when accompanied by muscle shortening i.e., nonisometric contractions) is a result of thin filaments sliding relative to thick filaments
  - Sarcomere length decreases
  - I band and H zone decrease
- Sliding depends on ATP and calcium
  - ATP: Myosin heads are actin-activated ATPase
    - ATP dissociates myosin from actin
    - Myosin hydrolyzes ATP but requires actin to release products
    - Release of products is accompanied by conformational change
    - ATP binding reverses conformational change
- ❖ Cytoskeleton
  - Functions
    - Cell shape, mechanical strength, movement, transport of organelles, separation of chromosomes
  - Components
    - Microtubule, microfilaments, intermediate filaments
- ❖ Microtubule-based systems
  - Microtubules
    - Hollow tubes composed of the heterodimeric protein tubulin: αβ–tubulin dimer
    - Tubulin subunits add onto the (+)-end
- ❖ Motor proteins: Myosins, kinesins, dyneins
  - Kinesin 1: Tetramer of two heavy and two light chains
    - Heavy chain N-terminal motor domain (binds to microtubules), coiled-coil stalk, C-terminal light chain binding sites and cargo binding site
    - Light chains
  - Dyneins: Microtubule motors: Axonemal and cytoplasmic varieties
    - Cytoplasmic dynein: Dimer of heavy chains and several light chains
      - Heavy chain: Motor domain and microtubule-binding stalk domain
  - Myosin V: Multimer of 16 chains
    - Heavy chains form dimers: Domains include motor head, calmodulin binding neck and long tail with light chains
  - Organelle movement along microtubules: microtubules point + end towards cell periphery
    - Cytosolic dyneins move vesicles from + end to – end (back towards cell center)
    - Kinesis move from – end towards + end
  - Motor characteristics
    - Dyneins, myosins and kinesins are processive: Move along track without releasing
    - Movement governed by ATP binding and hydrolysis
      - Kinesins move hand-over-hand
- ❖ Motor molecules that move along DNA and RNA: Translocases
  - Translocase that unwind DNA are helicases
  - Translocases related to RecA protein
  - Translocase movement is processive and driven by ATP hydrolysis
    - Processivity by two different mechanisms
      - Hexameric helicases form rings around DNA
      - Rep helicase move hand-over-hand
        - Basis is negative cooperativity between DNA binding sites
- ❖ Rotary motors in bacteria
  - Extracellular protein filaments: Polymers of flagellin: Function as propellers
  - Membrane localized motors
    - Attached to flagellin
    - Use proton gradient for energy source
    - Reversal of motor causes change in direction of swimming

298

Chapter Objectives

Microtubules

Microtubules are hollow cylindrical structures composed of the heterodimeric protein tubulin. Know the basic structure of the microtubule, the fact that it is composed of 13 protofilaments made up of a head-to-tail arrangement of tubulin dimers and that the structure has two distinct ends, the plus end and the minus end. These designations come from the fact that subunit addition to a microtubule occurs predominantly on the plus end. Microtubules function as cytoskeletal elements and as the principle polymeric structure in cilia and eukaryotic flagella. In cilia and flagella, microtubules are arranged in a characteristic 9 + 2 array known as the axoneme. A key, non-tubulin component of the axoneme is the microtubule-dependent ATPase dynein. The axoneme is a mechanochemical energy transducer converting the chemical energy released upon hydrolysis of ATP by dynein into the energy of axonemal movement. Cytoplasmic microtubules are involved in a number of cell functions including movement of particles within cells. This is best typified by axonal transport in neurons. Axonal microtubules like cytoplasmic microtubules are arranged with their plus ends towards the periphery of the cell or, in this case, towards the axon ends, whereas their minus ends are towards the cell body. Movement of organelles and vesicles from the plus end of a microtubule to the minus end is mediated by cytosolic dyneins. Movement in the opposite direction is dependent on kinesins, ATPases similar to cytosolic dyneins.

Muscle Cell Anatomy

Skeletal muscle cells contain fiber bundles termed myofibrils composed predominantly of actin and myosin filaments. The myofibrils are divided into functional units called sarcomeres. Know how the basic anatomy of a sarcomere is composed of actin filaments and myosin filaments and how the two interdigitate. A sarcomere is a banded structure bordered by two Z lines. The A band marks the location of myosin filaments and is centered in the sarcomere. The bipolar myosin filaments contain a central portion devoid of myosin heads. The actin filaments are located in the I band and in part of the A band as they interdigitate with the myosin fibers. The region of the A band not containing interdigitated actin filaments is termed the H zone. The central bare region of the myosin bipolar filaments is centered in the H zone. During contraction, the I band shortens because actin and myosin filaments increase their regions of overlap. As a consequence, the H band shortens. The t-tubules are extensions of the sarcolemma that reach into the muscle cell interior and terminate close to the sarcoplasmic reticulum. Calcium ion release from the sarcoplasmic reticulum is the key event in initiation of muscle contraction and calcium release is set off by electrical signals propagated from the cell surface to the sarcoplasmic reticulum via the t-tubules.

Thin Filaments

The thin filaments are filaments of globular actin studded at regular intervals with a troponin complex of TnT, TnI, and TnC and lined with tropomyosin. Myosin binding sites on F-actin can be covered and exposed by movement of tropomyosin in a calcium-dependent manner mediated by TnC. The thin filaments attach to the Z lines by association with α-actinin, a protein homologous to spectrin the major structural protein of the red blood cell cytoskeleton. The ends of actin filaments are capped with β-actinin and two other components, γ-actinin, and paratropomyosin, are also found associated with the thin filament.

Thick Filaments

Myosin is the principal component of the thick filaments. A myosin molecule is composed of two heavy chains and two pairs of light chains, LC1 and LC2. Myosin is an asymmetric molecule with two heads (on one end) and a tail. The thick filaments are bundles of myosins, interacting in a tail-to-tail fashion to give rise to a "dumbbell" like filament with a bare zone in the middle (tails only) and the heads sticking out around each end. The heads are the business end of the molecule; they have ATPase activity and are responsible for muscle contraction. The thick filaments contain additional proteins including creatine kinase and several proteins found in the M disk, a structure within the H zone important for organizing myosin filaments. The thick filaments are linked to the Z lines by β-connectin.

Sliding Filaments and Myosin ATPase Cycle

Myosin is an actin-activated ATPase that cycles between two conformational states. You should be familiar with the ATPase cycle of myosin and how it leads to force generation. In the resting state, the nucleotide binding site of myosin is filled with ADP and P_i. Calcium, released upon stimulation, binds to TnC and induces a conformational change that causes movement of tropomyosin resulting in exposure of myosin binding sites on actin. Myosin binds to actin, releases P_i and subsequently undergoes a conformational change and release of ADP. The conformational change results in movement of the actin filament relative to myosin in a power stroke. ATP will now bind to the nucleotide binding site causing myosin to dissociate from actin. Myosin now hydrolyzes ATP as it returns to the resting conformation.

Calcium Ion Regulation
   Calcium release is mediated through a series of calcium channels. The t-tubules contain dihydropyridine-sensitive calcium channels whereas calcium channels in the terminal cisternae of the sarcoplasmic reticulum are sensitive to ryanodine. Uptake of calcium leading to muscle relaxation is mediated by calcium pumps.

Problems and Solutions

1. Assessing the biochemical consequences of animal anatomy
The cheetah is generally regarded as nature's fastest mammal, but another amazing athlete in the animal kingdom (and almost as fast as the cheetah) is the pronghorn antelope, which roams the plains of Wyoming. Whereas the cheetah can maintain its top speed of 70 mph for only a few seconds, the pronghorn antelope can run at 60 mph for about an hour! (It is thought to have evolved to do so in order to elude now-extinct ancestral cheetahs that lived in North America.) What differences would you expect in the muscle structure and anatomy of pronghorn antelopes that could account for their remarkable speed and endurance?

Answer: Clearly the pronghorn antelope must be a master of oxidative phosphorylation and because of this must be capable of supplying muscle with oxygen-rich blood. The organs involved in oxygen homeostasis, namely the circulatory system, including the heart, blood, and blood vessels, are optimized for this task. The animal has a large, powerful heart, blood with a high hematocrit and red cells with high amounts of hemoglobin, and well vascularized muscle and lung tissue. The animal is light and compact and of efficient design. In addition, muscle cells contain large numbers of mitochondria and numerous short muscle fibers.

2. Analyzing the action of a muscle contraction inhibitor
An ATP analog, β,γ-methylene-ATP, in which a -CH₂- group replaces the oxygen atom between the β- and γ phosphorus atoms, is a potent inhibitor of muscle contraction. At which step in the contraction cycle would you expect β,γ-methylene-ATP to block contraction?

Answer: β,γ–methylene-ATP is a nonhydrolyzable analog of ATP that binds to ATP-binding sites and is capable of inducing conformational changes associated with ATP binding but is not destroyed by hydrolysis. During muscle contraction, myosin heads interact with actin in a cyclic manner that results in muscle contraction. Myosin in the resting state, not in contact with actin, has bound ADP and P_i. Binding of actin stimulates release of P_i, which is followed by a conformational change of the myosin heads and ADP release. The conformational change, called the power stroke, is responsible for force generation. Binding of ATP or the ATP analog will subsequently cause myosin to dissociate from actin. To reestablish the myosin conformation prior to the power stroke, ATP hydrolysis is required. The nonhydrolyzable ATP analog will cause myosin dissociation but without subsequent hydrolysis no conformational change occurs to reestablish the myosin head in its original conformation.

3. Understanding the role of phosphocreatine
ATP stores in muscle are augmented or supplemented by stores of phosphocreatine. During periods of contraction, phosphocreatine is hydrolyzed to drive the synthesis of needed ATP in the creatine kinase reaction:
$$Phosphocreatine + ADP \rightleftharpoons creatine + ATP$$
Muscle cells contain two different isozymes of creatine kinase, one in the mitochondria and one in the sarcoplasm. Explain.

Answer: Creatine is phosphorylated at the expense of ATP in the mitochondria and then exported to the sarcoplasm. In the sarcoplasm, it functions as a store of high-energy phosphate bonds to be used during intense muscle activity to replenish ATP supplies. Creatine phosphorylation in the mitochondria is accomplished by a mitochondrial creatine kinase. Phosphorylation of ADP by phosphocreatine in the sarcoplasm is accomplished by a sarcoplasmic creatine kinase. Free creatine and phosphocreatine thus form a circuit with phosphocreatine leaving the mitochondria and creatine entering.

4. Understanding the biochemistry of rigor mortis
Rigor is a muscle condition in which muscle fibers, depleted of ATP and phosphocreatine, develop a state of extreme rigidity and cannot be easily extended. (In death, this state is called rigor mortis, the rigor of death.) From what you have learned about muscle contraction, explain the state of rigor

in molecular terms.

Answer: Rigor occurs when the myosin head is in contact with actin in the absence of ATP. When ATP levels are low, sarcoplasmic levels of calcium ion increase thereby causing a conformational change in troponin that leads to tropomyosin movement and uncovering of myosin-binding sites on actin. Myosin binds, releases P_i, and undergoes a conformational change leading to force generation and release of ADP. However, in the absence of ATP, myosin remains attached to actin in the contracted state leading to stiff, contracted muscles typical of rigor mortis.

5. Assessing the contraction potential of human muscle
Skeletal muscle can generate approximately 3 to 4 kg of tension or force per square centimeter of cross-sectional area. This number is roughly the same for all mammals. Since many human muscles have large cross-sectional areas, the force that these muscles can (and must) generate is prodigious. The gluteus maximus (on which you are probably sitting as you read this) can generate a tension of 1200 kg! Estimate the cross-sectional areas of all of the muscles in your body and the total force that your skeletal muscles could generate if they all contracted at once.

Answer: A person might have a maximum of 50% of her or his total body mass as muscle. For a 70 kg person the muscle would account for approximately 35 kg. Assuming a density of 1 g/ml (muscle is slightly more dense than water but not by much) the 35 kg is approximately 35,000 mL or 35,000 cm³. If this total mass of muscle were in the shape of a cylinder 1 cm thick the cylinder's cross sectional area would be 35,000 cm².

We are informed that the gluteus maximus can generate 1200 kg of force and that the average force is 3 to 4 kg per cm².   If muscle generates 3 to 4 kg of tension per cm² then 35,000 cm² can generate between 1.05×10^5 and 1.40×10^5 kg of force. This corresponds to 154.3 tons. (One kg is approximately 0.0011 ton.)

6. Assessing structure of tubulin
Calculate a diameter for a tubulin monomer, assuming that the monomer M_r is 55,000, that the monomer is spherical, and that the density of the protein monomer is 1.3 g/ml. How does the number that you calculate compare to the dimension portrayed in Figure 16.2?

Answer: A single tubulin monomer weighs:

$$\frac{55,000 \frac{g}{mol}}{6.023 \times 10^{23} \frac{atom}{mol}} = 9.13 \times 10^{-20} g$$

The volume occupied by this amount of protein is determined as follows:

$$\frac{9.13 \times 10^{-20} g}{1.3 \frac{g}{mL}} = 7.03 \times 10^{-20} mL$$

Assuming this volume is as a sphere, the diameter is calculated as follows:

$$V = \frac{4}{3} \pi \frac{D^3}{8} \quad \text{(which is simply } \frac{4}{3} \pi r^3 \text{expressed as a function of diameter)}$$

$$\frac{4}{3} \pi \frac{D^3}{8} = 7.02 \times 10^{-20} mL \times \frac{cm^3}{mL} \times \left(\frac{10^{-2}}{cm}\right)^3$$

Solving for D we find that:

$$D = \sqrt[3]{\frac{7.02 \times 10^{-20} ml \times \frac{cm^3}{ml} \times \left(\frac{10^{-2}}{cm}\right)^3}{\frac{4}{3} \times 3.14159 \times \frac{1}{8}}}$$

$$D = 5.12 \times 10^{-9} m = 5.12 \text{ nm}$$

To determine the diameter of a tubulin monomer from Figure 16.2, I got out my ruler and measured the diameter of a monomer and compared it to the scale shown in the figure (for 24 nm). I came up with a measurement of 5.14 nm!

7. Assessing the tubulin content of a liver cell
Use the number you obtained in problem 6 to calculate how many tubulin monomers would be found in a microtubule that stretched across the length of a liver cell. (See Table 1.2 for the diameter of a liver cell.)

Answer: The length of a liver cell is 20,000 nm. So, a microtubule 20,000 nm long would have:

$$\frac{20,000 \text{ nm}}{5.12 \dfrac{\text{nm}}{\text{monomer}}} = 3,906 \text{ monomers}$$

But, there are 13 protofilaments per cross section. Thus, there are

$$3,906 \text{ monomers} \times 13 = 50,781 \text{ monomers}$$

8. Understanding axonal vesicle transport
The giant axon of the squid may be up to 4 inches in length. Use the value cited in this chapter for the rate of movement of vesicles and organelles across axons to determine the time required for a vesicle to traverse the length of this axon.

Answer: Vesicles move 2 to 5 μm/sec. How long would it take a vesicle moving at this speed to travel 4 inches?

$$\frac{4 \text{ inches} \times 2.54 \dfrac{\text{cm}}{\text{inch}} \times \dfrac{10^{-2}\text{cm}}{\text{m}}}{2 \text{ to } 5 \dfrac{\text{um}}{\text{sec}} \times \dfrac{10^{-6}\text{m}}{\text{um}}} = 20,000 \text{ to } 50,800 \text{ sec}$$

$$\text{or } \frac{(20,000 \text{ to } 50,800 \text{ sec})}{60 \dfrac{\text{sec}}{\text{min}} \times 60 \dfrac{\text{min}}{\text{hr}}} = 5.6 \text{ to } 14 \text{ hours}$$

9. Understanding the molecular structure of myosin
As noted in this chapter, the myosin molecules in thick filaments of muscle are offset by approximately 14 nm. To how many residues of a coiled coil structure does this correspond?

Answer: A coiled coil has a pitch of 0.51 nm and 3.5 residues per turn (slightly different from the 0.54 pitch of an alpha helix with 3.6 residues per turn). The number of residues of coiled coil represented by 14 nm is given by:

$$\frac{14 \text{ nm}}{0.51 \dfrac{\text{nm}}{\text{turn}}} = 27.5 \text{ turns} \times \frac{3.5 \text{ residues}}{\text{turn}} = 96 \text{ residues}$$

10. Assessing the energetics of sarcoplasmic reticulum calcium transport
Use the equations of Chapter 9 to determine the free energy difference represented by a Ca^{2+} gradient across the sarcoplasmic reticulum membrane if the luminal (inside) concentration of Ca^{2+} is 1 mM and the concentration of Ca^{2+} in the solution bathing the muscle fibers is 1 μM.

Answer:

$$\Delta G = RT \times \ln \frac{1 \text{ mM}}{1 \text{ μM}} = 8.31541 \frac{\text{J}}{\text{mol K}} \times 298 \text{ K} \times \ln \frac{1 \times 10^{-3}}{1 \times 10^{-6}}$$

$$\Delta G = 17,115 \frac{\text{J}}{\text{mol}}$$

11. Understanding the energetic driving force of the Ca-ATPase
Use the equations of Chapter 3 to determine the free energy of hydrolysis of ATP by the sarcoplasmic reticulum Ca-ATPase if the concentration of ATP is 3 mM, the concentration of ADP is 1 mM, and the concentration of P_i is 2 mM

Answer:

For hydrolysis of ATP

$$\Delta G = \Delta G_o{}' + RT \times \ln \frac{[ADP][P_i]}{[ATP]} = -30,500 \text{ J} + 8.31541 \frac{J}{mol \text{ } K} \times 298 \text{ K} \times \ln \frac{\left(1 \times 10^{-3}\right) \times \left(2 \times 10^{-3}\right)}{3 \times 10^{-3}}$$

$$\Delta G = -30,500 \text{ J} + 8.31541 \frac{J}{mol \text{ } K} \times 298 \text{ K} \times \ln \frac{\left(1 \times 10^{-3}\right) \times \left(2 \times 10^{-3}\right)}{3 \times 10^{-3}}$$

$$\Delta G = -48,620 \frac{J}{mol}$$

12. Understanding the stoichiometry of sarcoplasmic reticulum calcium transport
Under the conditions described in problems 10 and 11, what is the maximum number of Ca^{2+} ions that could be transported per ATP hydrolyzed by the Ca-ATPase?

Answer: Movement of calcium ions up this gradient would require 17,100 J/mol. One mole would require 17,100 Joules, 2 moles would require 34,200 Joules and 3 moles would require 51,300 Joules. Hydrolysis of a mole of ATP releases –48,620 Joules, which is sufficient to move 2 calcium ions but not quite enough to move 3.

13. Understanding the energetics of motor proteins
For each of the motor proteins in Table 16.2, calculate the force exerted over the step size given, assuming that the free energy of hydrolysis of ATP under cellular conditions is -50 kJ/mol.

Answer: Hydrolysis of ATP releases -50 kJ of energy per mol of ATP and accounts for movement of the various motor molecules by one step. This movement was accomplished by a force acting through a step distance.
$$\text{Energy} = \text{Force} \times \text{distance}$$
A Joule of energy is a Newton of force acting through 1 meter of distance. Thus, 50 kJ is equivalent to 50,000 newton·meter.

$$\text{Force} = \frac{\text{Energy}}{\text{distance}}$$

$$\text{Force}_{\text{Kinesin 1}} = \frac{50 \text{ kJ}}{8 \text{ nm}} = \frac{50,000 \text{ newton} \cdot \text{m}}{8 \times 10^{-9} \text{ m}} = 6.25 \times 10^{12} \text{ newtons}$$

$$\text{Force}_{\text{Myosin V}} = \frac{50 \text{ kJ}}{8 \text{ nm}} = \frac{50,000 \text{ newton} \cdot \text{m}}{36 \times 10^{-9} \text{ m}} = 1.39 \times 10^{12} \text{ newtons}$$

$$\text{Force}_{\text{Kinesin 1}} = \frac{50 \text{ kJ}}{8 \text{ nm}} = \frac{50,000 \text{ newton} \cdot \text{m}}{28 \times 10^{-9} \text{ m}} = 1.79 \times 10^{12} \text{ newtons}$$

14. Relating myosin function and stoichiometry to weight lifting
When you go to the gym to work out, you not only exercise many muscles but also involve many myosins (and actins) in any given exercise activity. Suppose you lift a 10-kg weight a total distance of 0.4 m. Using the data in Table 16.2 for myosin, calculate the minimum number of myosin heads required to lift this weight and the number of sliding steps these myosins must make along their associated actin filaments.

Answer: First let's calculate the energy expended in lifting the 10-kg weight 0.4 m. The 10-kg is the object's mass and to convert it to weight we must multiply by 9.8 m/sec². Then we will determine how many ATP molecules would have to be hydrolyzed to supply this energy.
$$\text{Energy} = \text{Weight} \times \text{distance}$$

$$\text{Energy} = 10 \text{ kg} \times 9.8 \frac{m}{sec^2} \times 0.4 \text{ m} = 39.2 \text{ kg} \frac{m^2}{sec^2} = 39.2 \text{ J}$$

ATP hydrolysis accounts for 50 kJ per mol ATP.

$$\frac{39.2 \text{ J}}{50 \times 10^3 \dfrac{\text{J}}{\text{mol}}} = 7.84 \times 10^{-4} \text{mol}$$

The number of molecules is:

$$7.84 \times 10^{-4} \text{mol} \times \frac{6.02 \times 10^{23} \text{molecules}}{\text{mol}} = 4.72 \times 10^{20} \text{molecules}$$

In the "Critical Developments in Biochemistry" box on page 526 we are informed that a single cycle of a single myosin molecule involves movement of between 4 nm and 11 nm, generates a force of 1.7 to 4 pN and this corresponds to energy consumption (force times distance) of 0.68×10^{-20}J to 4.4×10^{-20}J per cycle. Using the larger value we see that

$$\frac{39.2 \text{ J}}{4.4 \times 10^{-20} \dfrac{\text{J}}{\text{cycle}}} = 8.9 \times 10^{20} \text{cycles}$$

For a single myosin to move 0.4 m, it would have to make multiple steps. Using 11 nm as the single myosin step size (see above), to move 0.4 m requires the following number of steps:

$$\frac{0.4 \text{ m}}{11 \dfrac{\text{nm}}{\text{step}}} = \frac{0.4 \text{ m}}{11 \times 10^{-9} \dfrac{\text{m}}{\text{step}}} = 3.64 \times 10^7 \text{steps}$$

Each step is accompanied by hydrolysis of ATP by a myosin molecule. Thus, the number of steps we calculated above is equivalent to steps per myosin molecule. If we divide the number of ATP molecules by the number of steps per myosin this will give us the number of myosins involved.

$$\frac{4.72 \times 10^{20} \text{molecules ATP}}{3.64 \times 10^7 \dfrac{\text{steps}}{\text{myosin}}} = 1.30 \times 10^{13} \text{myosins}$$

As an alternative, we calculated above that it would take 8.9×10^{20}cycles to generate 39.2 J of energy. If movement by a single myosin requires 3.64×10^7steps and each step is a cycle then the number of myosins required is

$$\frac{8.9 \times 10^{20} \text{cycles}}{3.64 \times 10^7 \dfrac{\text{steps or cycles}}{\text{myosin}}} = 2.45 \times 10^{13} \text{myosins}$$

15. Understanding human smooth muscle
In which of the following tissues would you expect to find smooth muscle?
   a. Arteries
   b. Stomach
   c. Urinary bladder
   d. Diaphragm
   e. Uterus
   f. The gums in your mouth

Answer: If you take a deep breath before answering this one you might realize that you have voluntary control over your diaphragm suggesting that it is not composed of smooth muscle. So, pick any of the others.

16. Analyzing the mechanism of muscle contraction
When an action potential (nerve impulse) arrives at a muscle membrane (sarcolemma), in what order do the following events occur?
   a. Release of Ca^{2+} ions from the sarcoplasmic reticulum
   b. Hydrolysis of ATP, with release of energy
   c. Detachment of myosin from actin
   d. Sliding of myosin along actin filament
   e. Opening of switch 1 and switch 2 on myosin head

Answer: The action potential must first release calcium from the sarcoplasmic reticulum so "a" is the first step. Calcium binds to troponin C and conformational change makes actin available for myosin binding. When myosin binds switch 1 and switch 2 open (see Figure 16.9). So, "e" is the next step in the sequence. Sliding of myosin along the actin filament occurs "d" and then the myosin must detach from actin "c" (by ATP binding to myosin) before ATP hydrolysis can occur "b". The order is a-e-d-c-b.

17. Essay question on the functional relevance of molecular motors
(Essay question.) You are invited by the National Science Foundation to attend a scientific meeting to set the agenda for funding of basic research related to molecular motors for the next 10 years. Only basic research will be funded, ruling out studies on human subjects. You are asked to suggest the research area most worthy of scientific research. Your presentation must include (1) a brief background on what we currently know about the subject; (2) identification of a key research topic about which more needs to be known; and (3) a justification of why additional knowledge in this area is critical for advancing the field (that is, why investigations in this area are especially important). You are not being asked to provide the methods or experiments that might be used to address the problem—only the concept. Base your presentation on what you have learned in this chapter (you may consult and include references from the Further Reading section), and limit your presentation to 300 words

Answer: Start out by reading a few brief reviews on the subject of molecular motors. It would make a great deal of sense to focus on one area like myosins or RNA helicases to make this a manageable assignment. Good luck.

Preparing for the MCAT® Exam

18. Analyzing the stepping behavior of the kinesin motor
Consult Figure 16.17 and use the data in problem 8 to determine how may steps a kinesin motor must take to traverse the length of the squid giant axon.

Answer: In Figure 16.9 we are told that the kinesin motor moves 80 Å per step. In question 8 we are told that a squid giant axon is 4 inches. The number of steps is calculated as follows:

$$\frac{4 \text{ inches} \times 2.54 \frac{\text{cm}}{\text{inch}} \times \frac{\text{m}}{100 \text{ cm}}}{80 \text{ Å} \times \frac{\text{m}}{10^{10} \text{Å}}} = 1.27 \times 10^7 = 12.7 \text{ million steps}$$

19. Understanding the biochemical basis of muscle cramps
When athletes overexert themselves on hot days, they often suffer immobility from painful muscle cramps. Which of the following is a reasonable hypothesis to explain such cramps?
  a. Muscle cells do not have enough ATP for normal muscle relaxation.
  b. Excessive sweating has affected the salt balance within the muscles.
  c. Prolonged contractions have temporarily interrupted blood flow to parts of the muscle.
  d. All of the above.

Answer: The correct answer is "d". However, in the specific situation "c" has its virtues. The truth is that muscle cramps occur for lots or reasons or more correctly, the exact causes of muscle cramps are unknown. Electrolyte imbalance of sodium, calcium, potassium and magnesium contributes as does dehydration. Low ATP levels would prevent muscle from relaxing. Temporary interruption of blood flow would block the muscle cell's ability to replenish ATP stores.

20. Understanding the genetics of Duchenne muscular dystrophy
Duchenne muscular dystrophy is a sex-linked recessive disorder associated with severe deterioration of muscle tissue. The gene for the disease:
  a. is inherited by males from their mothers.
  b. should be more common in females than in males.
  c. both a and b.
  d. neither a nor b.

Answer: The correct answer is "a". Duchenne muscular dystrophy is a disease caused by defects in the gene coding for the protein dystrophin. Dystrophin functions in muscle cells as a structural support anchoring the muscle cytoskeleton to the muscle cell membrane. Symptoms of Duchenne muscular dystrophy include progressive muscle weakness and loss of muscle mass. The inability to produce dystrophin causes cell membranes to become permeable leading to cell lysis, which may cause an immune response that adds to muscle tissue damage.

A key word search for Duchenne muscular dystrophy in SwissProt returned a "NiceProt" results page with lots of information about this protein. The location of the gene can be determined by activating the GeneCards link, which shows a map of the X-chromosome with the dystrophin gene highlighted near p21.1 ("p" stands for petite and refers to the short arm of a chromosome). The gene is located on the X chromosome and is recessive. Duchenne muscular dystrophy largely occurs in males. Males are more susceptible because they have only one copy of the X gene (males are XY where as females are XX). A male has a 50% of getting the disease if his mother is a carrier. Some carriers show mild symptoms such as muscle weakness.

Questions for Self Study

1. Fill in the blanks. Cytoskeletal elements composed of hollow, cylindrical structures 24 nm in diameter are known as ___. These polymers are made up of the protein ___, a dimeric protein composed of ___ and ___. These cytoskeletal elements are the major structural elements of cilia and flagella. An ATPase, _____, is responsible for production of a sliding motion between polymers that is converted to a bending motion typical of ciliary or flagellar motion. A cytosolic variety of this ATPase is found in axons where it mediates the movement of organelles and vesicles from the plus end of a microtubule to the minus end. (The plus end is the end at which dimer subunits are ___ to the microtubule.) Another ATPase, ___, mediates movement of organelles and vesicles in the opposite direction.

2. What are the four kinds of muscle? Which is multinucleated?

3. On the microscopic level, skeletal and cardiac muscles display light and dark bands. What is the general name given to muscle with banding patterns?

4. Fill in the blanks. The basic structural units of skeletal muscle contraction are ___. They are arranged as linear arrays into small fibers called ___. The fibers are covered by a specialized endoplasmic reticulum called ___. The ends of the fibers are bounded by extension of the muscle cell plasma membrane or ___. The extensions form ___ that connect the specialized sarcoplasmic reticulum with the plasma membrane.

5. In the diagram presented below label the following: I band, Z line, A band, H zone, location of thin filaments only, location of thick filaments only, overlap of thin and thick filaments.

6. Sort the terms below into thick filament or thin filament components.
a. G-actin
b. Heavy chains
c. Troponin I
d. Regulatory light chain
e. Tropomyosin
f. S1 fragments
g. HMM
h. Creatine kinase
i. β-Actinin
j. Troponin C
k. Light meromyosin
l. Troponin T
m. Essential light chain
n. Myosin

7. In skeletal muscle, cycles of contraction and relaxation are a result of a calcium cycle. Describe this calcium cycle.

Answers

1. Microtubules; tubulin; α-tubulin; β-tubulin; dynein; added; kinesin.

2. Skeletal (multinucleated), cardiac, smooth, and myoepithelial.

3. Striated.

4. Sarcomeres; myofibrils; sarcoplasmic reticulum; sarcolemma; transverse (or T) tubules.

z.

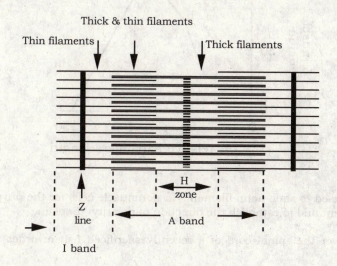

6.

Thick filament terms	Thin filament terms
b. Heavy chains	a. G-actin
d. Regulatory light chain	c. Troponin I
f. S1 fragments	e. Tropomyosin
g. HMM	i. β-Actinin
h. Creatine kinase	j. Troponin C
k. Light meromyosin	l. Troponin T
m. Essential light chain	
n. Myosin	

7. Calcium is released from the sarcoplasmic reticulum in response to electrical stimulation. This results in an increase in the concentration of sarcoplasmic calcium which leads to calcium binding to troponin C and movement of troponin I and tropomyosin. The result of tropomyosin movement is to expose myosin-binding sites on actin. Relaxation occurs when the sarcoplasmic calcium ion concentration is lowered as a result of active transport of calcium back into the sarcoplasmic reticulum by calcium pumps.

Additional Problems

1. Design a simple experiment to determine the association constant for assembly of tubulin dimers to the ends of a microtubule.

2. The concentration of tubulin dimers determined above is the critical dimer concentration for assembly of microtubules. At equilibrium, dimers are being added to the polymer but if dimer assembly causes the free tubulin dimer concentration to fall below the critical concentration, the microtubule will disassemble, thereby releasing free dimers to increase the concentration of dimers. The plus end and the minus end of microtubules have different critical dimer concentrations with the minus end being higher than the plus end. How does this lead to treadmilling?

3. The tenderloin is a muscle that runs along the back of an animal and is the source of filet mignon, a tender cut of meat, so tender in fact that when cooked, it can be cut with a fork. Can you suggest a reason why this muscle is so tender?

4. The graph below shows tension developed by a sarcomere held at various fixed lengths where 100 refers to the natural resting length of the sarcomere. Based on your knowledge of sarcomere microanatomy, explain the line segments "ab", "bc", "cd", and "de".

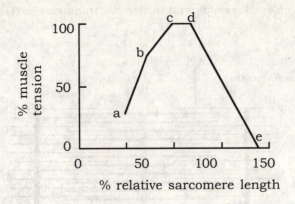

5. Myosin heads can be used to stain actin filaments in nonmuscle cells for the purpose of identification of a filament as an actin filament and to establish the direction of polarity. Explain.

6. It is customary to sever the spinal cord of a recently sacrificed fish in order to prevent the fish from becoming stiff. Why?

Abbreviated Answers

1. If we assume that the thermodynamics of elongation is more favorable than initiation, then, in conditions that favor microtubule formation, microtubules will be in equilibrium with free tubulin dimers or:

$$MT_{n-1} + T \leftrightarrow MT_n \text{ where}$$

$$K = \frac{[MT_n]}{[MT_{n-1}][T]}$$

Recognizing that $[MT_{n-1}] = [MT_n]$ we see that $K_A = [T]^{-1}$. If we could measure the concentration of free tubulin dimer in equilibrium with microtubules we could determine K. This can be accomplished by simply centrifuging the sample at a speed that will sediment microtubules but not tubulin dimers. By measuring the concentration of tubulin in the supernatant, we can determine K.

2. Consider a solution of tubulin polymerizing into microtubules. As the concentration of free dimers decreases to the critical concentration of minus end assembly, addition of dimers to the minus end stops. However, since the critical concentration of plus end assembly is lower than minus end assembly, dimers will continue to add to the plus end. This depletes the concentration of free subunits below the critical concentration of minus end assembly, causing subunits to dissociate from the minus end. The system reaches a steady state at which minus end disassembly is balanced by plus end assembly.

3. There are a couple of reasons. First, the muscle is composed of a relatively small number of very large muscle fibers with little connective tissue. Second, the muscle is not a extremely active muscle like the muscles of the shoulder. Finally, the filet is a cross sectional cut out of the muscle across the long axis of the muscle fibers. The meat can be easily cut into smaller pieces by simply separating muscle fibers with a fork.

4. At position "e", the sarcomere has been stretched so that thick and thin filaments no longer overlap and no tension is possible. Moving from "e" to "d", the length corresponds to an increase in overlap between thick and thin filaments and therefore a decrease in sarcomere length. At "d", actin filaments maximally overlap the region of thick filaments containing heads. At point "c", the thin filaments from opposite sides of the sarcomere come in contact. Moving from "d" to "c" no change in the number of cross bridges overlapping with thin filaments occurs and the sarcomere can generate maximum tension. Sarcomere length between "c" and "b" result in overlap of thin filaments from opposite sides of the sarcomere. The drop in tension in moving from "b" to "a" results when actin filaments from opposite sides of the sarcomere begin to overlap with myosin heads on the opposite side of the thick filament. The following diagrams illustrate these points.

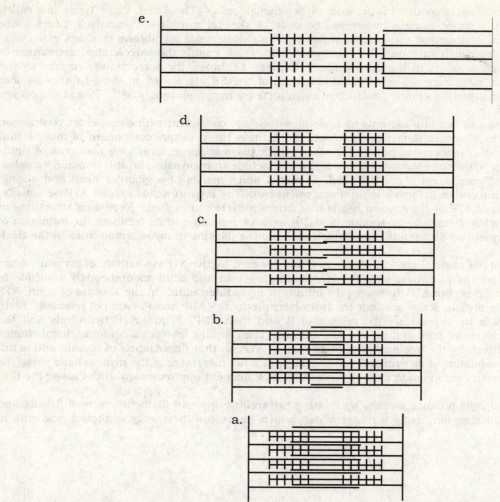

5. The S1 fragment of myosin will bind to actin fibers and decorate them in a characteristic arrowhead pattern. The tips of the arrowheads point away from the Z line in a sarcomere.

6. Fish become stiff because of rigor mortis as a result of a release of calcium and depletion of ATP stores. After death, the nerves quickly deteriorate and begin to randomly fire leading to muscle stimulation and rapid depletion of ATP stores. By severing the spinal cord and icing the fish the depletion of ATP stores can be delayed.

Summary

All organisms (at least at some stage in life) are capable of movement or motion. The energy for these processes is chemical energy that is transduced into mechanical energy by motor proteins. Motor proteins found in microtubule-based motile systems and those found in actin-dependent systems utilize hydrolysis of ATP whereas the rotary motor found in bacterial flagella uses the energy of a proton gradient.

Microtubules are self-assembling structures that are fundamental components of cilia and eukaryotic flagella. Microtubules are formed by polymerization of tubulin, a dimeric protein comprised of similar 55 kD subunits. By the GTP-dependent process of treadmilling, tubulin units are added at the plus end of the microtubule and removed from the minus end. With their plus ends oriented toward the cell periphery and the minus ends toward the center of the cell, microtubules form a significant part of the cytoskeleton. The motion of cilia and flagella results from the ATP-driven sliding or walking of dyneins along one microtubule while remaining firmly attached to an adjacent microtubule. Within eukaryotic cells, cellular dyneins and kinesin proteins are responsible for microtubule-based transport of organelles and vesicles.

Actin and myosin are two proteins responsible for various kinds of motion in a number of cell types. The best characterized system is muscle. Muscle contraction is essential to higher organisms for limb movement, locomotion, beating of the heart and other functions. Four different kinds of muscle are found in higher

309

organisms: skeletal, cardiac, smooth muscle, and myoepithelial cells. The latter three types are called myocytes and contain a single nucleus, whereas the cells of skeletal muscle - the muscle fibers - are multinucleate. Skeletal and cardiac muscle are striated. The sarcolemmal membrane contains extensions called transverse tubules which surround the sarcomeres and which enable the sarcolemmal membrane to contact the ends of each myofibril in the muscle fiber. Between t-tubules, the sarcomere is covered by the sarcoplasmic reticulum membrane. The terminal cisternae of the SR are joined to the t-tubules by foot structures. The trigger signal for muscle contraction is an increase in cytoplasmic $[Ca^{2+}]$. Relaxation occurs when $[Ca^{2+}]$ is reduced.

Electron micrographs of banding patterns in skeletal muscle are consistent with a model for contraction in which thick and thin filaments slide along each other. Actin is the principal component of muscle thin filaments. It consists of a monomeric protein, G actin, which polymerizes to form long filaments of right-handed helical F-actin. Other components of thin filaments include tropomyosin and the troponin complex. Myosin is the principal component of muscle thick filaments and consists of a globular head and a long, helical tail. The tails intertwine to form a left-handed coiled coil. The myosin heads exhibit ATPase activity, and it is the hydrolysis of ATP by the myosin heads that drives muscle contraction. Repeating structures in the myosin tails, including 7-residue, 28-residue, and 196-residue repeating units facilitate the formation of myosin-solvent and myosin-myosin interactions which stabilize the packing of myosin molecules in the thick filaments.

The molecular events of skeletal muscle contraction are powered by the ATPase activity of myosin. Actin increases the ATPase activity of myosin by 200-fold. Further, myosin and actin spontaneously associate to form an actomyosin complex, but ATP decreases the affinity of myosin for actin. In the absence of actin, ATP is rapidly cleaved at the myosin active site, but the hydrolysis products - ADP and P_i - are not released. Actin activates myosin ATPase by stimulating the release of P_i and then ADP. Thus, ATP hydrolysis and the association and dissociation of actin and myosin are coupled. The coupling involves a conformational change in the myosin head, driven by the free energy of hydrolysis of ATP, so that dissociation of myosin and actin, hydrolysis of ATP and rebinding of myosin and actin occur with a net movement of the myosin head along the actin filament. Each such conformation change moves the thick filament approximately 100 Å along the thin filament.

Flagellated bacterial cells produce motility by rotating extracellular protein filaments, termed flagella and composed of the protein flagellin, using a rotary motor, which harnesses the energy of proton gradients to produce movement.

Chapter 17

Metabolism: An Overview

• •

Chapter Outline

❖ Metabolic diversity
  ⅄ Autotrophs: Utilize carbon dioxide as carbon source
  ⅄ Heterotrophs: Require carbon in organic molecules
  ⅄ Phototrophs: Use light as source of energy
  ⅄ Chemotrophs: Use chemical reactions as source of energy
❖ Role of oxygen
  ⅄ Obligate aerobes: Require oxygen
  ⅄ Obligate anaerobes: No need for oxygen: Oxygen poisonous
  ⅄ Facultative anaerobes: Adapt to anaerobic conditions
❖ Metabolic processes: Two types
  ⅄ Catabolism
    • Exergonic: Energy transduced into ATP
    • Oxidative degradation of complex molecules into simpler compounds
    • Electrons captured by NAD^+
  ⅄ Anabolism
    • Complex biomolecules produced from simpler precursors
    • Synthesis driven by energy, typically ATP hydrolysis
    • NADPH serves as source of electrons
  ⅄ Convergence of pathways
    • Stage 1: Nutrient macromolecules broken down into building block monomers
    • Stage 2: Building block monomers degraded to limited set of simpler intermediates
    • Stage 3: Combustion of intermediates to CO_2 and H_2O
  ⅄ ATP is energy connection between catabolism and anabolism
  ⅄ NAD^+ and NADPH are oxidizing and reducing agents in metabolism
  ⅄ Regulation
    • Anabolism and catabolism separately regulated
    • Competing pathways often localized in different cellular compartments
    • Anabolic and catabolic pathways (involving same end points) differ in at least one or more steps
      • Allows pathways to be regulated separately
      • Avoids simultaneous functioning of both pathways
❖ Enzymes organization
  ⅄ Separate, soluble entities: Intermediates supplied by diffusion
  ⅄ Multienzyme complexes: Intermediates passed between components of complex
  ⅄ Membrane-bound enzymes or complexes
❖ Experimental methods
  ⅄ Metabolic inhibitors
  ⅄ Genetic mutations
  ⅄ Isotopic traces
    • Radioactive isotopes: Detected when isotope decays by release of energy

- Stable, heavy isotopes: Alter density of product
  - ↲ NMR: Signal sensitive to electronic environment of atomic nucleus
- ❖ Nutrition: Use of foods by organisms: Five components
  - ↲ Proteins: Source of amino acids
    - Essential amino acids
    - Degradation of amino acids
      - Glucogenic: Some amino acids can be used to produce glucose
      - Ketogenic: Some amino acids can be used to produce fatty acids
  - ↲ Carbohydrates: Metabolic energy derived from catabolism of carbohydrates
  - ↲ Lipids
    - Metabolic energy sources
    - Components of membranes
    - Sources of essential fatty acids: Linoleic and linolenic acids
  - ↲ Fiber: Cellulose, hemicellulose, lignins
  - ↲ Vitamins: Two classes
    - Water soluble: Used as coenzymes or coenzyme precursors (except vitamin C)
    - Fat soluble

Chapter Objectives

Understand the terms metabolism, catabolism, and anabolism. Know the classification of organisms, with respect to carbon and energy requirements, into photoautotrophs, photoheterotrophs, chemoautotrophs, and chemoheterotrophs. (What are you? Your houseplants?) Understand the terms obligate aerobes, facultative anaerobes, and obligate anaerobes.

The remainder of the text will concentrate on details of metabolism and take a step-by-step tour of individual pathways. This chapter gives the big picture. For example, oxygen moves in a cycle from photoautotrophic cells as O_2 to heterotrophic cells, from heterotrophic cells to heterotrophic cells as oxy-compounds, and from heterotrophic cells back to photoautotrophic cells as CO_2 and H_2O. Carbon cycles between photoautotrophs and heterotrophs as CO_2 and organic compounds.

Understand the role of ATP in metabolism. ATP is an energy-rich compound used to couple exergonic to endergonic processes. In general, catabolic pathways are exergonic and produce ATP whereas anabolic pathways are endergonic and consume ATP. The cofactors NAD^+ and NADPH play key roles in oxidation-reduction reactions. Catabolic pathways are often oxidative in nature with NAD^+ serving as the oxidant. In contrast, anabolic pathways are reductive with NADPH serving as a reductant.

Intermediary metabolism can be overwhelming at first glance but appreciate the fact that the pathways converge to or diverge from a small number of key intermediates. In catabolism, large macromolecules such as proteins, polysaccharides, and lipids, are broken down into their component building blocks, which are catabolized into a small set of common metabolic intermediates that may be degraded to simple end products (e.g., CO_2, NH_3, H_2O). Alternatively, these intermediates can be used in anabolism to resynthesize macromolecules. Resynthesis pathways may share a number of intermediates with degradive pathways, but at least one step is different. This allows for independent regulation of the competing actions. How is regulation achieved? Clearly reactions must be sensitive to concentrations of substrates and products but more sophisticated controls are often imposed at key reactions. These controls include allosteric regulation, covalent modification, and alterations of the levels or types of enzymes used.

Problems and Solutions

1. Global carbon dioxide cycling expressed as human equivalents
If 3×10^{14} kg of CO_2 are cycled through the biosphere annually, how many human equivalents (70-kg persons composed of 18% carbon by weight) could be produced each year from this amount of CO_2?

Answer: The amount of carbon in a 70 kg person is:

$$0.18 \times 70 \text{ kg} = 12.6 \text{ kg}$$

$$\text{Carbon dioxide is } \frac{12}{12+16+16} \times 100\% = 27.3\% \text{ by weight in carbon}$$

So, 3×10^{14} kg of CO_2 represents:

$$3 \times 10^{14} \text{kg} \times 0.273 = 8.18 \times 10^{13} \text{kg}$$

This represents

$$\frac{8.18 \times 10^{13} \text{kg}}{12.6 \dfrac{\text{kg}}{\text{person}}} = 6.5 \times 10^{12} \text{people}$$

That is, 6,500,000,000,000 or 6.5 trillion people.

2. The different modes of carbon and energy metabolism
Define the difference in carbon and energy metabolism between photoautotrophs and photoheterotrophs, and between chemoautotrophs and chemoheterotrophs.

Answer: Autotrophs can utilize CO_2 to synthesize biomolecules, whereas heterotrophs require more complex compounds. For example, an autotroph utilizes CO_2 as a carbon source to produce reduced carbon compounds like monosaccharides. Heterotrophs require compounds already containing reduced carbons to synthesize biomolecules.

In terms of energy, phototrophs can transduce light energy into chemical energy. Chemotrophs rely on chemical reactions, principally oxidation/reduction reactions during which a compound is oxidized in a series of steps. During this process energy is transduced into other chemical forms.

Photoautotrophs are at the base of the food pyramid and capture the energy of sunlight to reduce CO_2 into carbohydrates. Photoheterotrophs utilize sunlight but are incapable of fixing CO_2. Thus, they require a source of reduced carbon. Chemoautotrophs can utilize CO_2 and do so using a chemical reaction, such as oxidation of an inorganic compound, as an energy source. Chemoheterotrophs require carbon in reduced form for both synthesis of biomolecules and for energy transduction.

3. Where do the O atoms in organisms come from?
Name the three principal inorganic sources of oxygen atoms that are commonly available in the inanimate environment and readily accessible to the biosphere?

Answer: Most photosynthetic organisms use H_2O as a source of electrons to reduce CO_2 during carbon fixation. Thus, water is an abundant source of oxygen that is readily available. During the course of this process, oxygen gas (O_2) is liberated into the atmosphere. Aerobic organisms utilize O_2 as an electron acceptor during oxidation of reduced carbon compounds; CO_2 is released. This CO_2 is used by photosynthetic autotrophs to produce biomolecules. Oxygen thus cycles between H_2O, O_2, and CO_2.

4. How do catabolism and anabolism differ?
What are the features that generally distinguish pathways of catabolism from pathways of anabolism?

Answer: Catabolic pathways are oxidative and result in the conversion of complex molecules into simple compounds. The reactions are overall thermodynamically spontaneous (negative ΔG). Catabolic pathways can result in production of a high-energy compound, such as ATP from ADP and P_i, and/or support reduction of NAD^+ to NADH.

Anabolic pathways are generally reductive in nature and must be driven thermodynamically, often by coupling key steps in a pathway to hydrolysis of ATP. The electrons for reduction often are supplied by NADPH, a reduced, phosphorylated form of NAD^+. Anabolic pathways convert simple compounds into more complex biomolecules.

5. How are the enzymes of metabolic pathways organized?
Name the four principal modes of enzyme organization in metabolic pathways.

Answer: A metabolic pathway consists of all of the individual reactions involved in transforming a substrate into an end product. The reactions or steps are almost always catalyzed by enzymes. The enzymes may be separate proteins localized within a cell or cellular compartment but otherwise soluble. The product of one reaction is made available to the next reaction by diffusion. Alternatively, two or more enzymes may be organized into a multienzyme complex. Here, the enzymes catalyze sequential steps in the pathway. In this organization, the product of one enzyme is released directly to the next enzyme in the pathway. Finally, enzymes and enzyme complexes may be confined to a particular membrane within the cell. The purpose may be to provide a bridge between compartments, or to localize an enzymatic activity in a two-dimensional surface. Substrates confined to move in or along the membrane are only required to do a two-dimensional search to encounter enzyme.

6. Why do anabolic and catabolic pathways differ?
Why is the pathway for the biosynthesis of a biomolecule at least partially different than the pathway for its catabolism? Why is the pathway for the biosynthesis of a biomolecule inherently more complex than the pathway for its degradation?

Answer: Catabolic pathways are exergonic with one or more steps having large negative ΔG values. These steps represent thermodynamic barriers to an anabolic sequence functioning by reversal of the catabolic pathway. Thus, separate reactions are required to bypass them. For example, in one step in glycolysis, phosphofructokinase catalyzes the transfer of a phosphate group from ATP to fructose-6-phosphate to produce ADP and fructose-1,6-bisphosphate. In gluconeogenesis, the antithesis of glycolysis, the conversion of fructose-1,6-bisphosphate to fructose-6-phosphate is accomplished by a simple hydrolysis reaction catalyzed by the enzyme fructose-1,6-bisphosphatase. Under physiological concentrations of ATP, ADP and P_i, the kinase reaction in glycolysis favors production of fructose-1,6-bisphosphate whereas the phosphatase reaction favors production of fructose-6-phosphate. Parallel catabolic and anabolic pathways must include at least one different step to allow for independent regulation of each pathway.

Biosynthesis of complex biomolecules from simple compounds is an endergonic process. Thus, for anabolic pathways to function they must be coupled to highly exergonic reactions such as hydrolysis of ATP. These coupling reactions increase the complexity of anabolic pathways. For example, conversion of phosphoenolpyruvate (PEP) to pyruvate, with phosphorylation of ADP to ATP, is a single reaction in glycolysis. The reversal of this step in gluconeogenesis proceeds in two stages: Pyruvate is first carboxylated to oxaloacetate, with hydrolysis of ATP driving the reaction; second, oxaloacetate is converted to PEP with GTP serving as a phosphoryl group donor.

7 What are primary metabolic roles of the three principal nucleotides?
(Integrates with Chapters 1 and 3.) What are the metabolic roles of ATP, NAD+, and NADPH?

Answer: The living state requires a constant source of energy for biosynthesis, for movement, for replication, in short, for all of the activities characterizing the living state. The immediate source of energy depends on the cell type (see problem 2) and includes light energy, chemical energy, or both. However, independent of the source of energy, all cells transduce energy into the chemical energy of the phosphoric anhydride bonds of ATP. ATP is the energy currency of cells. Hydrolysis of the high-energy bonds of ATP may be used to drive biosynthetic pathways and to support all of the other activities of the living state. Thus, ATP serves to couple exergonic processes to endergonic processes. For example, ATP synthesized during catabolic, exergonic pathways, is used in hydrolysis reactions to drive anabolic, endergonic pathways.

The major catabolic pathways common to virtually all cells are based on oxidation of carbon compounds. For example, the conversion of glucose to C_2O and H_2O includes three oxidative sequences. In glycolysis, glucose is converted to pyruvate. Next, pyruvate is oxidized to acetyl-coenzyme A. Finally, acetyl-coenzyme A is metabolized to CO_2 and H_2O in the citric acid cycle. These three oxidative sequences of necessity are accompanied by electron transfers and NAD+ serves as the primary electron acceptor.

In contrast, anabolic pathways are often reductive; the immediate reductant or electron donor is NADPH. Thus, this reduced, phosphorylated form of NAD+ plays a role in metabolism exactly opposite to that of NAD+.

8. How is metabolism regulated?
(Integrates with Chapter 15.) Metabolic regulation is achieved via regulating enzyme activity in three prominent ways: allosteric regulation, covalent modification, and enzyme synthesis and degradation. Which of these three modes of regulation is likely to be the quickest; which the slowest? For each of these general enzyme regulatory mechanisms, cite conditions in which cells might employ that mode in preference over either of the other two.

Answer: Allosteric regulation of enzyme activity is the quickest regulatory mechanism. In the course of our studies in biochemistry, we will encounter many examples of allosteric enzymes. We will find allosteric sites responsive to a host of compounds including: end-products of a metabolic sequence in which the enzyme functions; key substrates or substrate derivatives of the reaction; and, important biochemical intermediates such as ATP and its derivatives, ADP, AMP, and cyclic AMP, and, NAD+ and NADH. In short, allosteric regulation is responsive to local metabolic signals.

Allosteric regulatory mechanisms probably evolved out of a necessity for cells to be responsive to local metabolic signals independently of the concentrations of the particular substrates and products of an allosterically regulated enzyme. For example, many allosteric enzymes are regulated by an end-product of a

314

metabolic pathway in which the enzyme catalyzes one step. In this case, enzyme activity is responsive not only to its immediate substrates and products but to the concentration of some intermediate into which its product is eventually converted.

Covalent modification leading to enzyme regulation is used by cells under conditions in which the immediate metabolic state of the cell is to be ignored. For example, in muscle cells, glycogen metabolism is regulated, in part, by hormones. Adrenaline stimulates glycogen breakdown into glucose by activating glycogen phosphorylase. This activation is caused by a specific, covalent modification of glycogen phosphorylase, a phosphorylation of a particular serine residue on the enzyme. Phosphorylated glycogen phosphorylase is active, even under conditions of high concentrations of glucose-6-phosphate, an inhibitor of unphosphorylated glycogen phosphorylase.

The slowest mode for regulating enzyme activity is by protein synthesis and degradation. In general, enzyme synthesis may involve all aspects of gene expression including: transcription of DNA into RNA; RNA processing into mRNA; translation of mRNA into a protein; and, post-translational modification of the protein to the final enzymatic form.

Metabolic regulation via synthesis and degradation of enzymes is often used when cells experience major changes in their environment. Wild type *E. coli* are capable of growth on a defined medium containing salts and minerals and a reduced carbon source. This is possible because *E. coli* is capable of synthesizing important building-block molecules, as for example, the amino acids. However, the bacteria stop synthesizing many of the enzymes involved in amino acid biosynthesis when grown in a medium containing amino acids; the enzymes are resynthesized when amino acids are no longer in the medium.

9. What are the advantages of metabolic compartmentalization?
What are the advantages of compartmentalizing particular metabolic pathways within specific organelles?

Answer: Compartmentalization provides an easy solution to the problem of regulating two opposing pathways. For example, the catabolic pathway for fatty acid metabolism, known as β–oxidation, occurs in the mitochondria. In contrast, fatty acid biosynthesis largely occurs in the cytosol. Compartments may also provide a particular set of conditions favorable to only certain reactions. There is a group of enzymes known as acid hydrolases that are localized in lysosomes. These enzymes include proteases, nucleases, lipases and glycosidases and they function at acidic pH. The enzymes are confined by a membrane barrier, and their activity is enhanced by the acidic pH maintained within lysosomes. If inadvertently released into the cytosol, these enzymes fail to cause damage to cellular components because the cytosol is maintained at a higher pH.

10. Why are metabolomic analyses challenging?
Name and discuss four challenges associated with metabolomic measurements in biological systems.

Answer: Metabolomics is the systematic identification and quantitation of low molecular weight metabolites in living systems. A glance at a metabolic map will convince you that metabolism is complicated. It involves a large number of different molecules. Metabolites can range from compounds with only minor or subtle differences to molecules that are very different from each other. Metabolomic measurements have to detect these differences. A further complexity arises due to differences in biological half-lives of intermediates and their concentrations. Some compounds are stable and in high concentrations whereas others are unstable and present at very low concentrations. Stability can also be greatly influenced by conditions such as temperature, pH, salt concentration, etc., and thus in vitro and in vivo measurements can differ greatly. The complexity of living organisms adds to the difficulty of metabolomic measurements. It is often the case that a starting point for metabolomics is a model organism or an easily collected biological fluid (e.g., blood, urine). These might be a good starting point but they may not reflect accurately the physiology of the organism or the physiology in more complex organisms.

11. Which is 'better': NMR or MS?
Compare and contrast mass spectrometry and NMR in terms of their potential advantages and disadvantages for metabolomic analysis.

Answer: Mass spectrometry is a very sensitive detection technique that requires small amounts of a compound. It can be used to identify unknown compounds, quantify known compounds and determine the structure and chemical properties of compounds. It is used in studies on compounds ranging from low molecular weight metabolites to proteins. Mass spectrometry is often combined with a variety of separation techniques like gas chromatography, liquid chromatography and electrophoresis. It is not an in vivo technique nor does it give 3-dimensional information.

NMR has a variety of uses ranging from in vivo to in vitro. It is a noninvasive technique that can be applied to organisms, cells, biological fluids or pure or partially purified samples. It can be used to monitor chemical reactions and dynamic processes. It can also provide structural information.

12. How do vitamin-derived coenzymes aid metabolism?
What chemical functionality is provided to enzyme reactions by pyridoxal phosphate (see Chapter 13)? By coenzyme A (see Chapter 19)? By vitamin B$_{12}$? By thiamine pyrophosphate (see Chapter 19)?

Answer: Pyridoxal-5-phosphate (PLP) is the biologically active form of vitamin B$_6$. It participates in a number of reactions involving amino acids including: transamination, α-decarboxylation, β-decarboxylation, β-elimination, γ-elimination, racemization and aldol reactions. The key to catalysis is the ability of pyridoxal-5-phosphate to form a stable Schiff base adduct with amino acids. The structure of PLP attached to a lysine residue is shown below.

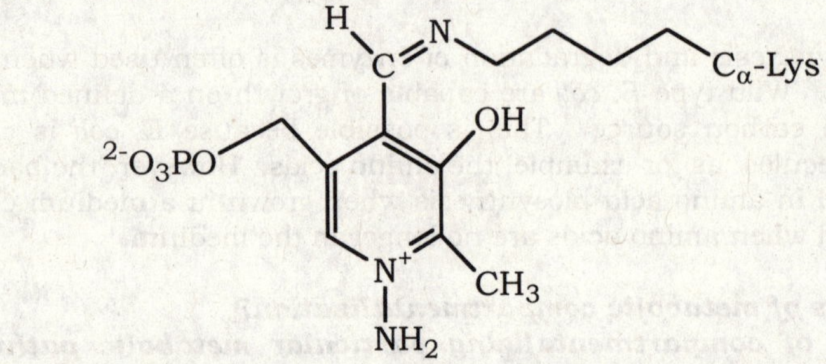

Vitamin B$_{12}$ or cyanocobalamin participates in three types of reactions: intramolecular rearrangements, reductions of ribonucleotides to deoxyribonucleotides, and methyl group transfers.

Coenzyme A functions as an acyl group carrier. Its structure is shown below.

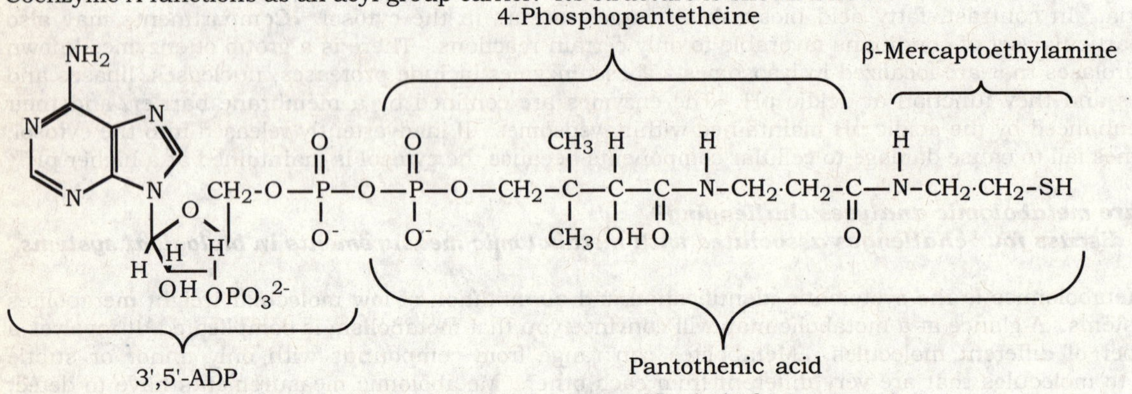

The acyl group is attached in thiolester linkage to the sulfhydryl of coenzyme-A.

The structure of thiamine pyrophosphate is shown below.

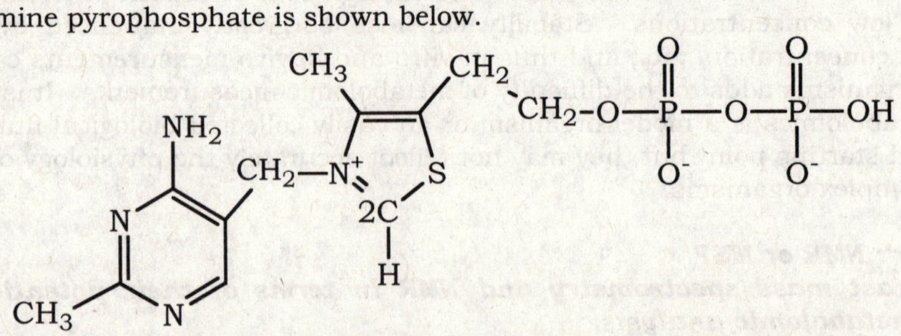

The carbon (C-2) in the five-membered ring between N and S is readily ionized to a carbanion that can function as a nucleophile. Thiamine pyrophosphate participates in decarboxylations of α-keto acids and the formation and cleavage of α-hydroxyketones; examples include yeast pyruvate decarboxylase (metabolizes pyruvate to acetaldehyde which in yeast can be converted to ethanol) and transketolase reactions (which will be encountered in the pentose phosphate shunt pathway and in the Calvin cycle). The key to catalysis is nucleophilic attack by C-2 on a carbonyl carbon.

13. What are the features of the series of –omes?
Define the following terms:
   a. Genome
   b. Transcriptome
   c. Proteome
   d. Metabolome
   e. Fluxome

Answer: The genome is the total complement of genetic material of an organism. In the case of eukaryotes is refers to the total sequence of bases in DNA of chromosomes in a haploid cell. Eukaryotic cells often contain additional genetic material such as mitochondrial and/or chloroplast DNA. For all organisms the genome refers to double stranded DNA. Virus genomes comprise a variety of nucleic acids such as single stranded or double stranded DNA or RNA.

The transcriptome is the full complement of mRNAs from a cell or particular tissue at a particular time or under a particular set of conditions. It represents expressed genes.

The proteome is the collections of all the proteins present in a cell, organism or tissue at a particular time or under a particular set of conditions.

The metabolome is the set of small molecule compounds found in a cell, organism or a tissue at a particular time and under a particular set of conditions.

The metabolome is clearly dependent on the proteome but compounds in the cell's environment influence it. The proteome derives from translated mRNAs but protein half-lives can have a dramatic influence on it. The transcriptome derives from genes found in the genome but not all genes are expressed at the same time and to the same level. Further, mRNAs exhibit a range of half-lives and hence biological stability.

The fluxome is the study of the movement of molecules through metabolic pathways. A cell's response to some particular condition may be an alteration of fluxes through integrated biochemical networks. The consequence is an alteration in the set of small molecule compounds that constitute the metabolome. The response is clearly dependent on a cell's genome, its transcriptome, and its proteome and involves a change in RNA and/or protein levels in addition to changes in protein activity. The consequence is to shift or alter the movement of molecules through metabolic pathways.

14. What is alcohol dehydrogenase enzyme mechanism?
The alcohol dehydrogenase reaction, described in Figure 17.11, interconverts ethyl alcohol and acetaldehyde and involves hydride transfer to and from NAD⁺ and NADH, respectively. Write a reasonable mechanism for the conversion of ethanol to acetaldehyde by alcohol dehydrogenase.

Answer: The conversion of ethanol to acetaldehyde is an oxidation and so an oxidant must be involved. Alcohol dehydrogenase uses NAD^+ for this purpose. A simple mechanism would involve polarization of the hydroxyl group by a base on the enzyme followed by reduction of NAD^+. Alcohol dehydrogenase, however, is a zinc-requiring enzyme. Zinc, as Zn^{2+}, in effect lowers the hydroxyl group's pK_a, which allows the reduction of NAD^+ to proceed as follows:

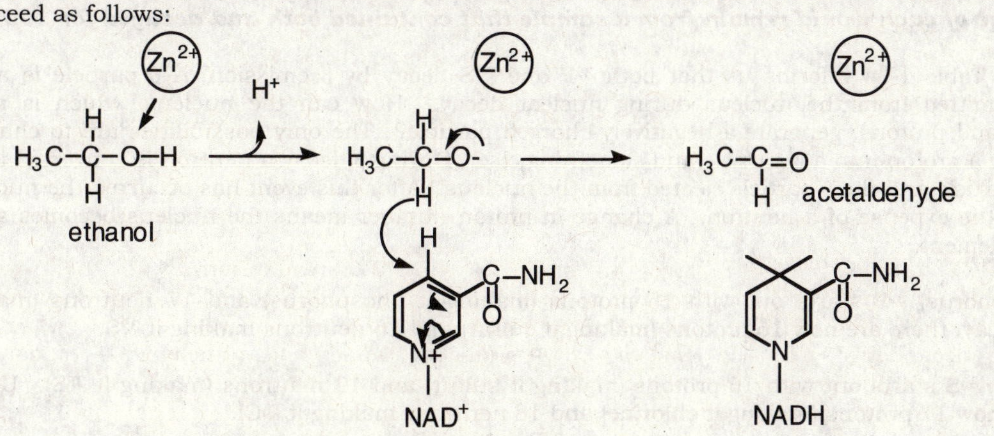

15. Where are various metabolic pathways localized and what are their inputs?
For each of the following metabolic pathways, describe where in the cell it occurs and identify the
starting material and end product(s):
    a. Citric acid cycle
    b. Glycolysis
    c. Oxidative phosphorylation
    d. Fatty acid synthesis

Answer: The citric acid cycle occurs in the mitochondria in eukaryotic cells and in the cytoplasm of prokaryotic cells. It is a cycle that has as input two-carbon acetyl groups attached to coenzyme A (acetyl-CoA). The acetyl group is attached to oxaloacetate, a 4-carbon compound, to produce citrate, which is metabolized back to oxaloacetate in the cycle. Two carbons enter the cycle and two carbons exit it as CO_2. The cycle is oxidative and electrons from it are used to reduce NAD^+ and FAD (to NADH and $FADH_2$).

Glycolysis occurs in the cytosol where glucose is converted to two molecules of pyruvate, a 3-carbon compound.

Oxidative phosphorylation is the formation of ATP (from ADP and P_i) that depends on a variety of membrane complexes to move electrons from NADH and $FADH_2$ to some terminal electron acceptor like O_2 in the case of aerobes. The process occurs on the inner mitochondrial membrane in eukaryotes and on the plasma membrane of prokaryotes.

Fatty acid synthesis is a process that occurs in the cytosol. It uses two-carbon acetyl units in an activated 3-carbon form –malonly CoA- to synthesize fatty acids.

16. A possible solution to global warming
Many solutions to the problem of global warming have been proposed. One of these involves
strategies for carbon sequestration -the removal of CO_2 from the earth's atmosphere by various
means. From your reading of this chapter, suggest and evaluate a strategy for carbon sequestration
in the ocean.

Answer: The U.S. Department of Energy has a website that discusses carbon sequestration at http://csite.esd.ornl.gov/. One approach is to sequester carbon in underground repositories. This involves capturing CO_2 at the point of production and taking it out of the active carbon cycle. Plants capture carbon dioxide and it was through plants that fossil fuels were produced. Another approach is to enhance carbon fixation by terrestrial plants and then remove the fixed carbon from the biosphere. The ocean represents a huge reservoir of carbon-based organisms including phytoplankton that can fix and reduce carbon dioxide, and organisms like corals and mollusks that use carbon dioxide to produce calcium carbonate. Finally, a better understanding of the biology of microorganisms that produce methane or hydrogen gas might present novel solutions to the problem.

17. Radioactive isotopes as metabolic tracers
Consult Table 17.4, and consider the information for ^{32}P and ^{33}S. Write reactions for the decay
events for these two isotopes, indicating clearly the products of the decays, and calculate what
percentage of each would remain from a sample that contained both and decayed for 100 days.

Answer: Table 17.4 informs us that both ^{32}P and ^{33}S decay by β emission. A β particle is a high-speed electron emitted from the nucleus during nuclear decay. How can the nucleus, which is made up on neutrons and protons, generate a negatively-charged particle? The only possibilities are to change either a neutron or a proton into a β particle and something else. During β decay a neutron is converted into a proton and a β particle and the β particle ejected from the nucleus. After this event has occurred the nucleus gains a proton at the expense of a neutron. A change in proton number means the nucleus becomes an atom of a different element.

For phosphorus, ^{32}P starts out with 15 protons (making it phosphorus) and 17 neutrons (making it ^{32}P). Upon β decay, there are now 16 protons (making it sulfur) and 16 neutrons making it ^{32}S.

$$^{33}P \rightarrow \beta^- + {}^{32}S$$

For sulfur, ^{35}S starts out with 16 protons (making it sulfur) and 19 neutrons (making it ^{35}S). Upon β decay, there are now 17 protons (making it chlorine) and 18 neutrons making it ^{35}Cl.

$$^{33}P \rightarrow \beta^- + {}^{32}S$$

The amount of isotope remaining after 100 days is calculated using the following equation:

318

$$\frac{A}{A_0} = e^{\frac{-0.693t}{t_{1/2}}}$$

The term A_0 is the amount of isotope at time zero, A the amount of isotope after time, t, and $t_{1/2}$ is the half-life of the isotope. The value 0.693 is simply the natural log of 2. The equation can be easily derived recognizing that radioactive decay is first order in A, with some rate constant k. Thus,

$$\frac{dA}{dt} = -kA$$

Rearranging

$$\frac{dA}{A} = -kdt$$

Now, integrating between A0 at t = 0 and A at t we get:

$$\ln\frac{A}{A_0} = -kt \text{ or}$$

$$\frac{A}{A_0} = e^{-kt} \quad (1)$$

We can express k as a function of $t_{1/2}$ by recognizing that $A = A_0/2$ when $t = t_{1/2}$. Thus,

$$\ln\frac{\frac{A_0}{2}}{A_0} = -kt_{1/2}$$

$$k = \frac{\ln2}{t_{1/2}} = \frac{\ln2}{t_{1/2}} = \frac{0.693}{t_{1/2}}$$

Substituting this into (1) we get:

$$\frac{A}{A_0} = e^{\frac{0.693t}{t_{1/2}}}$$

For ^{32}P with a half-life of 14.3 days after 100 days we have:

$$\frac{A}{A_0} = e^{\frac{0.693 \times 100 \text{ days}}{14.3 \text{ days}}} = 7.86 \times 10^{-3}$$

The percent remaining is 0.786 %

For ^{35}S with a half-life of 87.1 days after 100 days we have:

$$\frac{A}{A_0} = e^{\frac{0.693 \times 100 \text{ days}}{87.1 \text{ days}}} = 0.451$$

The percent remaining is 45.1 %

Preparing for the MCAT® Exam

18. **Which statement is most likely to be true concerning obligate anaerobes?**
    a. **These organisms can use oxygen if it is present in their environment.**
    b. **These organisms cannot use oxygen as their final electron acceptor.**
    c. **These organisms carry out fermentation for at least 50% of their ATP production.**
    d. **Most of these organisms are vegetative fungi.**

Answer: The correct answer is "b". Obligate anaerobes cannot use oxygen. In addition they cannot tolerate oxygen, and so oxygen in their environment is detrimental. Facultative anaerobes can produce ATP in the absence of oxygen via fermentation but not obligate anaerobes, which eliminates "c". Finally, "d" is incorrect because fungi are aerobes.

19. Which of the following experimental approaches to metabolic analysis cannot be used directly on living cells?
    a. NMR spectroscopy
    b. Metabolic inhibitors
    c. Mass spectroscopy
    d. Isotopic tracers

Answer: The answer is "Mass spectroscopy" (c). It is the only destructive technique. NMR is actually used in hospital settings to image patients but it is called MRI (magnetic resonance imaging). NMR uses high magnetic fields, which are apparently harmless to living organisms. Metabolic inhibitors could be thought of as pharmaceuticals. Almost any chemical compound is potentially harmful at high enough concentrations but at low doses and infrequent use they can have minimal long-term effects. Short-lived radioactive isotopes are used as isotopic tracers.

Questions for Self Study

1. The entries in the last four columns of Table 17.1 have been rearranged in an incorrect order. Correct them.

Classification	Carbon Source	Energy Source	Electron Donors	Examples
Phototrophs	Organic compounds	Oxidation-reduction reactions	Organic compounds	Nonsulfur purple bacteria
Photoheterotrophs	CO_2	Oxidation-reduction reactions	H_2O, H_2S, S, other inorganic compounds	All animals, most microorganism, nonphotosynthetic plant tissue such as roots, photosynthetic cells in the dark
Chemotrophs	Organic compounds	Light	Organic compounds e.g., glucose	Green plants, algae, cyanobacteria, photosynthetic bacteria
Chemoheterotrophs	CO_2	Light	Inorganic compounds: H_2, H_2S, NH_4^+, NO_2^-, Fe^{2+}, Mn^{2+}	Nitrifying bacteria; hydrogen, sulfur, and iron bacteria

2. Match terms in the two columns
    a. Obligate aerobes
    b. Facultative anaerobes
    c. Obligate anaerobes
    d. Aerobes
    e. Anaerobes

    1. Use O_2 as electron acceptor.
    2. Cannot use O_2 as electron acceptor.
    3. O_2 required for life.
    4. Use O_2 but can adapt to lack of O_2.
    5. Cannot use or tolerate O_2.

3. For each of the items listed below are they indicative of an anabolic (A) pathway or a catabolic (C) pathway?
    a. Consumes ATP.
    b. Involves the cofactor NAD^+.
    c. Results in reduction of compounds using NADPH.
    d. Energy-yielding.
    e. Production of glucose by photoautotroph.
    f. Results in decrease in molecular complexity.

4. Metabolic regulation of anabolic and catabolic pathways is accomplished by regulating the activities of enzymes. What three levels of enzymatic control are most common?

5. Anabolic pathways are endergonic whereas catabolic pathways are exergonic. What role does ATP play in coupling the two kinds of pathways?

6. How are radioactive isotopes and stable, heavy isotopes used in biochemical research?

Answers

1.

Classification	Carbon Source	Energy Source	Electron Donors	Examples
Phototrophs	CO_2	Light	H_2O, H_2S, S, other inorganic compounds	Green plants, algae, cyanobacteria, photosynthetic bacteria
Photoheterotrophs	Organic compounds	Light	Organic compounds	Nonsulfur purple bacteria
Chemotrophs	CO_2	Oxidation-reduction reactions	Inorganic compounds: H_2, H_2S, NH_4^+, NO_2^-, Fe^{2+}, Mn^{2+}	Nitrifying bacteria; hydrogen, sulfur, and iron bacteria
Chemoheterotrophs	Organic compounds	Oxidation-reduction reactions	Organic compounds e.g., glucose	All animals, most microorganism, nonphotosynthetic plant tissue such as roots, photosynthetic cells in the dark

2. a. 3; b. 4; c. 5; d. 1; e. 2.

3. a. A; b. C; c. A; d. C; e. A; f. C.

4. Allosteric regulation, covalent modification, and enzyme synthesis and degradation.

5. Steps in anabolic pathways are often driven by hydrolysis of ATP. One result of catabolism is the phosphorylation of ADP to ATP.

6. Both types of isotopes are used to label compounds. Radioactive isotopes are unstable and release energy upon decay. The presence of a radioactively labeled compound can be detected by the energy emitted upon radioactive decay. Stable, heavy isotopes increase the density or mass of a compound into which they are incorporated.

Additional Problems

1. Table 17.4 lists the properties of various radioisotopes. For ^{14}C the following information is given: Type - radioactive, Radiation Type -β^-, Half-life- 5700 yr. Explain this information.

2. Calculate the percent density difference between double-stranded DNA containing ^{14}N (the normal isotope) and ^{15}N (the heavy isotope). Assume that the average molecular weight of a base-pair in dsDNA is 625 and that the DNA is 50% A-T rich.

3. A gasoline-powered, internal-combustion engine and a polar bear metabolizing blubber run very similar reactions overall to produce energy; however, the details are quite different. Explain.

4. NMR spectroscopy has been used to measure cellular pH by determining the chemical shift of cellular P_i. Phosphorus NMR is particularly convenient because ^{31}P has a large magnetic moment and is the naturally occurring isotope of phosphorus. But, why is inorganic phosphate studied? Why not one of the phosphates of ATP? Typically [ATP] is greater than [P_i].

Abbreviated Answers

1. The designation "radioactive" indicates that ^{14}C is unstable and will be converted into another isotope or element. Radiation type β^- informs us that the conversion occurs by beta emission; an electron is emitted from the nucleus. The atomic mass remains constant (the atomic mass is the sum of neutrons and protons); however, the atomic number increases by 1 to 7 and ^{14}C is converted to ^{14}N, a stable isotope. The emitted electron may be detected by liquid scintillation counting or by exposure to film. In liquid scintillation counting, the electron excites fluorescent molecules that emit visible light upon returning to their ground state. The amount of light is measured using a photomultiplier. In film exposure, the electron reduces silver grains in the film. Under certain conditions, the number of grains reduced is proportional to the number of

decay events. The half-life refers to the time it takes for one-half of the radioisotope to decay. So, after 5700 years only half of the starting material remains.

2. Our first task is to determine the number of nitrogens in an "average base-pair". A, G, C, and T have 5, 5, 3, and 2 nitrogens, respectively. Therefore A:T and G:C base-pairs have 7 and 8 nitrogens and an average base-pair in a 50% A-T rich dsDNA has 7.5 nitrogens. The molecular weight of a ^{14}N-containing base-pair is 625. For a ^{15}N-containing base-pair, the molecular weight is

$$625 - (14 \times 7.5) + (15 \times 7.5) = 632.5$$

The percent density difference is given by:

$$\% \text{ Difference} = \frac{632.5 - 625}{625} \times 100\% = 1.2\%$$

3. The engine and the polar bear are both relying on carbon oxidation reactions for energy and the general equation that describes both reactions is:

$$(-CH_2-)_n + O_2 \rightarrow CO_2 + H_2O$$

In an engine, gasoline vapor mixed with oxygen is ignited by a spark to initiate carbon oxidation. (The reaction is often not complete and CO or carbon monoxide and other partially oxidized carbon compounds are produced.) The polar bear takes a different approach. Fat, with an average carbon not unlike a typical hydrocarbon found in gasoline, is metabolized via β-oxidation, the citric acid cycle, and electron transport to produce CO_2; O_2 is consumed during electron transport to produce H_2O.

4. The chemical shift of phosphorus depends on, among other things, the local magnetic environment in which the phosphorus is located. Protonation of phosphate groups will result in a chemical shift. The advantage to using inorganic phosphate to monitor intracellular pH is that phosphorus has a pK_a around neutral pH. The pK_as of phosphate in nucleotides are approximately one pH unit lower.

Summary

From a chemical perspective, living organisms are open thermodynamic systems far displaced from equilibrium. Further, cellular constituents are characteristically in a chemically more reduced state than the inanimate matter that surrounds them. Nevertheless, cells must somehow extract energy and reducing power from the environment and use this energy and reducing power to maintain the complex chemical activities that collectively constitute the living state. The sum of all these processes is metabolism.

Maps of intermediary metabolism portray hundreds of enzymatic reactions, but these represent only the principal metabolic pathways. It is estimated that the cells of higher organisms carry out literally thousands of metabolic reactions, each catalyzed by a specific enzyme. Despite the great variations and adaptations found in organisms, virtually all have the same basic set of major metabolic pathways, a fact providing strong evidence that they evolved from a common ancestor. Still, considerable metabolic diversity exists between organisms in terms of the sources of carbon and energy that they exploit. Photoautotrophs require only light as a source of energy and CO_2 as a source of carbon; photoheterotrophs use light as a sole source of energy but must have an organic supply of carbon in order to synthesize their complement of biomolecules. Chemoautotrophs can live on CO_2 as their sole source of carbon provided an oxidizable inorganic substrate, such as Fe^{2+}, NO_2^- or NH_4^+, is available to supply energy. Chemoheterotrophs are organisms needing an organic carbon source for both energy and biomolecular synthesis. The flows of energy, carbon and oxygen within the biosphere are intimately related. Phototrophs use solar energy to fix CO_2 into more reduced organic compounds; the electrons for this reduction come from water and O_2 is evolved. Heterotrophs feed on the organic products of photosynthesis, harvesting the energy in these compounds in oxidative reactions that release carbon dioxide, consume O_2, and form H_2O.

Metabolism consists of two major realms: the degradative pathways of catabolism and the biosynthetic pathways of anabolism. A metabolic pathway is composed of a sequence of enzymatic reactions, transforming some precursor to a specific end product via a consecutive series of intermediates. The enzymes of a particular pathway may exist either as physically separate entities, as a stable multi-enzyme complex, or as a membrane-associated system. Metabolic pathways often contain many individual steps for energetic reasons: energy transactions typically fall in the range of 0-50 kJ/mol, to fit within the prevailing free energy change for ADP phosphorylation/ATP hydrolysis in cells. Catabolic pathways exist to degrade a wide variety of fuel molecules - fats, carbohydrates, proteins - but these pathways converge to a few end products, namely acetyl-CoA and certain citric acid cycle intermediates. Water, ammonium and carbon dioxide are the ultimate end products of catabolism. In contrast, anabolic pathways diverge from a limited set of building blocks (amino acids, sugars, nucleotides and acetyl-CoA) to the end results of their synthetic activity, the astounding variety of complex biomolecules. Corresponding pathways of catabolism and anabolism differ for essential reasons. First, thermodynamics dictates that more energy is required for the synthesis of a biomolecule than can be realized from its degradation. Second, in order to have independent regulation of opposing metabolic sequences, the sequences must consist, at least in part, of unique reactions mediated by separate enzymes that serve as distinct sites for regulation.

ATP is the energy currency of the cell. ATP is a major product of catabolism; in turn, ATP is a principal substrate for anabolism, providing the chemical energy that serves as the driving force for biosynthesis. Electrons released in oxidative reactions of catabolism are collected in the pyridine nucleotide coenzymes, NADH and NADPH. The electrons carried in these two coenzymes have very different fates. Those in NADH are passed to O_2 to form H_2O in an energetic process known as oxidative phosphorylation, which couples NADH oxidation to ATP synthesis. In contrast, the electrons in NADPH provide the reducing power necessary to drive reductive anabolic reactions.

Metabolism is a tightly regulated, integrated process. Metabolic regulation is achieved by controlling the activity of key enzymes on one of a number of levels: allosteric regulation, covalent modification and enzyme synthesis/degradation. The reactions comprising various metabolic pathways have been revealed through a growing arsenal of experimental methods. Early approaches relied on cell-free assays of metabolic intermediates and the enzymatic reactions transforming them. Enzyme inhibitors were a useful tool in these studies. Genetic mutations result in specific metabolic blocks that are very informative about metabolic reactions. Isotopic tracers are particularly good probes of metabolism since isotopically labeled compounds have essentially the same chemical behavior as their unlabeled counterparts. Isotopes can be traced through their radioactivity or their difference in mass (or density). NMR spectroscopy has the power to reveal detailed information about the chemical fate of substances in a noninvasive, nondestructive way. Evidence regarding the compartmentation of metabolic processes within specific organelles has established that the flow of intermediates in metabolism is channeled spatially as well as chemically.

Nutrition is the use of foods by organisms. Nutrients include proteins, carbohydrates, lipids, fibers, and vitamins. Proteins serve as a source of essential amino acids. Excess amino acids may be converted to glucose, fatty acids, or completely degraded. Catabolism of carbohydrates represent an important source of metabolic energy. Lipid degradation is also an important source of metabolic energy; however, lipids also provide essential fatty acids. Fibers, including cellulose, hemicellulose, and lignin, play important roles in digestion and absorption. Vitamins are nutrients, which are required in the diet, usually in trace amounts, if they cannot be synthesized by the organism itself. Except for vitamin C, the water soluble vitamins are all components or precursors of coenzymes, low molecular weight species that bring unique functionalities to certain enzymes. The fat-soluble vitamins are not related to coenzymes, but still have essential roles in various biological processes.

Chapter 18

Glycolysis

• •

Chapter Outline

- ❖ Glycolysis: Embden-Meyerhof pathway: Glucose to pyruvate
- ❖ Energy consuming phase: Five reactions: Two ATP consumed
    - ⅄ Production of glucose-6-P: ATP consumed: 1st Priming step
        - Hexokinase
            - Low K_m for glucose
            - Reacts with many different 6-carbon sugars
            - Glucose-6-P allosteric inhibitor
        - Glucokinase: Liver enzyme
            - High K_m for glucose: Functions only when glucose abundant
            - No allosteric inhibition
            - Enzyme levels regulated by insulin
        - Advantages to phosphorylated sugar
            - Impermeable: Remains inside cell
            - Glucose concentration gradient maintained
            - Glucose carbon skeleton committed to metabolism
    - ⅄ Fructose-6-P formation: Aldose-ketose isomerization
        - Phosphoglucoisomerase
        - Prepares C-1 for second phosphorylation
        - Prepares C-3 for eventual cleavage
    - ⅄ Fructose-1,6-bisphosphate formation: ATP consumed: 2nd Priming step
        - Phosphofructokinase
        - Commits carbon skeleton to glycolysis
        - Allosteric regulation
            - Inhibitors
                - ○ ATP: Increases K_m for Fructose-6-P
                - ○ Citrate
            - Activators
                - ○ AMP: Reverses inhibition of ATP: Formed by adenylate kinase from ADP
                - ○ Fructose-2,6-bisphosphate: Increase affinity for F-6-P: Decreases ATP inhibition
    - ⅄ Formation of DHAP and glyceraldehyde-3-P
        - Fructose bisphosphate aldolase
        - Reverse of aldol condensation
        - Two classes of aldolases
            - Class I: Lysine forms Schiff base with carbonyl carbon: $NaBH_4$ inhibits
            - Class II: Zinc-containing enzyme: EDTA inhibits
    - ⅄ Trios phosphate isomerase
        - DHAP/glyceraldehyde-3-P interconversion
        - Ketose to aldose isomerization
- ❖ Energy yielding phase: Five reactions: Four ATP produced per glucose

- ↗ Glyceraldehyde-3-phosphate dehydrogenase: Product 1,3-Bisphosphoglycerate
  - • Oxidation/reduction
    - • NAD^+ reduced to NADH
    - • Glyceraldehyde oxidized to glycerate
  - • Carboxylic acid/phosphoric acid mixed anhydride formed
  - • Site of arsenate poisoning
- ↗ Phosphoglycerate kinase: ATP produced
  - • Substrate level phosphorylation
  - • 1,3-Bisphosphoglycerate also used to form 2,3-BPG by mutase
- ↗ 3-Phosphoglycerate to 2-phosphoglycerate conversion
  - • Phosphoglycerate mutase
  - • 2,3-BPG cofactor
- ↗ Enolase reaction
  - • 2-Phosphoglycerate to PEP
  - • Dehydration reaction
- ↗ Pyruvate kinase reaction
  - • PEP to pyruvate
  - • ADP to ATP
  - • Regulation
    - • Activated by AMP, F-1,6-BP
    - • Inhibited by ATP, acetyl CoA, alanine
    - • Covalent phosphorylation inhibits pyruvate kinase
- ❖ Fate of NADH and pyruvate
  - ↗ Aerobic respiration
    - • Pyruvate to acetyl-Coenzyme A to citric acid cycle to carbon dioxide
    - • NADH to electron transport
  - ↗ Anaerobic metabolism: Fermentation
    - • Lactic acid fermentation: Pyruvate to lactic acid regenerates NAD^+
    - • Ethanol fermentation
      - • Pyruvate to acetaldehyde
      - • Acetaldehyde to ethanol regenerates NAD^+
- ❖ Metabolism of other sugars
  - ↗ Fructose
    - • Fructokinase at C-1: Aldolase forms DHAP and glyceraldehyde: Triose kinase to produce glyceraldehyde-3-P
    - • Hexokinase at C:6 to produce fructose-6-P
  - ↗ Mannose: Hexokinase: Mannose-6-P isomerase to fructose-6-P
  - ↗ Galactose: Leloir pathway
    - • Galactokinase at C-1 to produce galactose-1-P
    - • Galactose-1-P + UDP-glucose to glucose-1-P and UDP-galactose
    - • UDP-galactose epimerase: UDP-glucose
  - ↗ Glycerol
    - • Glycerol kinase
    - • Glycerol phosphate dehydrogenase to DHAP
  - ↗ Hypoxic Stress: Response to oxygen limitation
    - • Increase in glycolytic enzymes
    - • Regulated by hypoxia inducible factor (HIF): Transcription factor active in low oxygen
      - • DNA-binding protein: Binds to hypoxia responsive elements and activates
      - • Hydroxylated by prolyl hydroxylase when oxygen plentiful: Hydroxylated HIF is ubiquitinated and degraded
      - • Hydroxylated by factor-inhibiting HIF (FIH) blocks association with coactivator p300

Chapter Objectives

Glycolysis

Glycolysis is the metabolic pathway leading from glucose to two molecules of pyruvate. Before jumping into the details it may be instructive to look at some basic chemistry. Glucose and pyruvic acid are shown below.

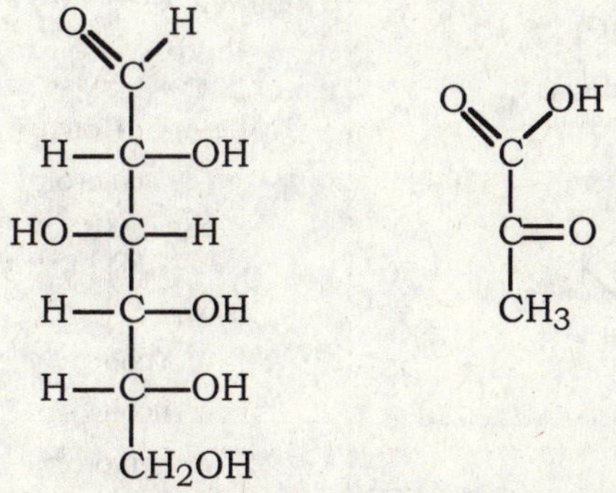

The carboxylate carbon and the keto carbon of pyruvate are at a more oxidized state whereas the methyl carbon is at a more reduced state relative to an "average" carbon of glucose. Therefore, this conversion, from glucose to pyruvate, must be accompanied by an additional reduction reaction. (In effect, the production of one of the oxidized carbons of pyruvate is balanced by production of the reduced carbon.) What gets reduced is NAD^+. When pyruvate is further metabolized to lactic acid or ethanol and CO_2, NADH is used to reduce the carbonyl carbon to an alcoholic carbon. The net result is regeneration of NAD^+, which allows glycolysis to continue. Glycolysis produces ATP and the energy for the reaction is the result of an internal oxidation/reduction.

Understand the basic organization of glycolysis into an energy-consuming phase and an energy-producing phase (see Figure 18.1). The two kinases, hexokinase (or glucokinase) and phosphofructokinase are key enzymes in phase I. Know how each is regulated but focus on phosphofructokinase. Phase II is the pay-back portion of glycolysis. Key enzymes in this phase include: glyceraldehyde-3-phosphate dehydrogenase, responsible for oxidation of the carbon skeleton; phosphoglycerate kinase, which catalyzes a substrate-level phosphorylation of ADP; and pyruvate kinase, responsible for the second substrate-level phosphorylation.

The Fate of Pyruvate

The ability to produce pyruvate from glucose depends on availability of NAD^+. But, this coenzyme is in limiting amounts in cells and, thus, must be recycled for glycolysis to continue. In fermentation, the end product of glycolysis, namely pyruvate, is used as a substrate to recycle NAD^+. Understand alcoholic fermentation and lactic acid fermentation. An advantage to relying on these processes is the ability to produce ATP rapidly. The disadvantages include the production of problematic waste products containing reduced carbons.

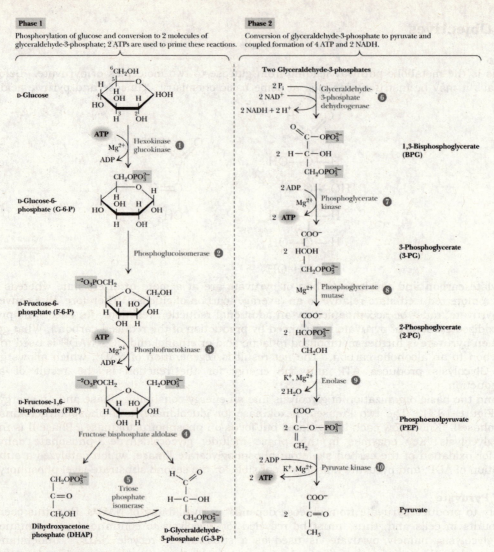

Phase 1

Phosphorylation of glucose and conversion to 2 molecules of glyceraldehyde-3-phosphate; 2 ATPs are used to prime these reactions.

Phase 2

Conversion of glyceraldehyde-3-phosphate to pyruvate and coupled formation of 4 ATP and 2 NADH.

Figure 18.1 The glycolytic pathway

Problems and Solutions

1. Characterizing glycolysis

List the reactions of glycolysis that:

a. are energy-consuming (under standard state conditions).

b. are energy-yielding (under standard state conditions).

c. consume ATP.

d. yield ATP.

e. are strongly influenced by changes in concentration of substrate and product because of their molecularity.

f. are at or near equilibrium in the erythrocyte (see Table 18.2).

Answer: Energy-consuming reactions have positive values of ΔG whereas energy-yielding reactions have negative values of ΔG. Under standard conditions the same rules apply but to values of $\Delta G°'$. Using Table 18.1, we can classify reactions into these two types as follows:

a.

Energy-consuming reaction	Enzyme	$\Delta G^{\circ\prime}$ kJ/mol
Glucose-6-phosphate → fructose-6-phosphate	phosphoglucoisomerase	+1.67
Fructose-1,6-bisphosphate → dihydroxyacetone-P + glyceraldehyde-P	fructose biphosphate aldolase	+23.9
Dihydroxyacetone-P → glyceraldehyde-P	triose phosphate isomerase	+7.56
Glyceraldehyde-P + P_i + NAD^+ → 1,3-bisphosphoglycerate+ NADH + H^+	glyceraldehyde-3-P dehydrogenase	+6.30
3-Phosphoglycerate → 2-phosphoglycerate	phosphoglycerate mutase	+4.4
2-Phosphoglycerate → phosphoenolpyruvate + H_2O	enolase	+1.8

b.

Energy-yielding reaction	Enzyme	$\Delta G^{\circ\prime}$ kJ/mol
α-D-glucose + ATP^{4-} → glucose-6-phosphate + ADP^{3-}	hexokinase, glucokinase	-16.7
Fructose-6-phosphate + ATP^{4-} → fructose-1,6-bisphosphate + ADP^{3-} + H^+	phosphofructokinase	-14.2
1,3-Bisphosphoglycerate + ADP^{3-} → 3-P-glycerate + ATP^{4-}	phosphoglycerate kinase	-18.9
Phosphoenolpyruvate + ADP^{3-} + H^+ → 1,3-bisphosphoglycerate+ NADH + H^+	pyruvate kinase	-31.7
Pyruvate + NADH + H^+ → lactate + NAD^+	lactate dehydrogenase	-25.2

c.

ATP-utilizing reaction	Enzyme	$\Delta G^{\circ\prime}$ kJ/mol
α-D-glucose + ATP^{4-} → glucose-6-phosphate + ADP^{3-}	hexokinase, glucokinase	-16.7
Fructose-6-phosphate + ATP^{4-} → fructose-1,6-bisphosphate + ADP^{3-} + H^+	phosphofructokinase	-14.2

d.

ATP-producing reaction	Enzyme	$\Delta G^{\circ\prime}$ kJ/mol
1,3-Bisphosphoglycerate + ADP^{3-} → 3-P-glycerate + ATP^{4-}	phosphoglycerate kinase	-18.9
Phosphoenolpyruvate + ADP^{3-} + H^+ → 1,3-bisphosphoglycerate+ NADH + H^+	pyruvate kinase	-31.7

e. What reactions are strongly influenced by concentration? In general,

$$\Delta G = \Delta G^{\circ\prime} + RT \times \ln \frac{\prod_i [P_i]}{\prod_i [R_i]}$$

Reactions that generate an increase or decrease in the number of products over reactants are expected to be sensitive to concentration changes. This is the case for reactions catalyzed by two enzymes: fructose bisphosphate aldolase and glyceraldehyde-3-phosphate dehydrogenase.

f. Reactions that are at or near equilibrium have values of ΔG close to zero. Using this criterion, the reactions catalyzed by the following enzymes must be close to equilibrium: phosphoglucoisomerase, phosphofructokinase, fructose bisphosphate aldolase, triose phosphate isomerase, phosphoglycerate kinase, phosphoglycerate mutase, and enolase.

2. Radiotracer labeling of pyruvate from glucose

Determine the anticipated location in pyruvate of labeled carbons if glucose molecules labeled (in separate experiments) with ^{14}C at each position of the carbon skeleton proceed through the glycolytic pathway.

Answer: Starting with glucose labeled at a particular carbon, the label will remain at that carbon until the fructose bisphosphate aldolase reaction. This reaction produces glyceraldehyde-3-phosphate and dihydroxyacetone phosphate. Carbons 1 (the aldehyde), 2 and 3 of glyceraldehyde derive from carbons 4, 5 and 6 of glucose. The phosphorylated carbon of dihydroxyacetone phosphate derives from carbon 1 of glucose and the ketone carbon from carbon 2. However, the next two reactions are an isomerization of the two triose phosphates into glyceraldehyde-3-phosphate and conversion to 1,3-bisphosphoglycerate. The isomerization interconverts the triose phosphates. Thus, carbon 3 of glyceraldehyde-3-phosphate derives from both carbons 1 and 6 of glucose; carbon 2 is from carbons 2 and 5 of glucose; and, carbon 1 is from carbons 3 and 4 of glucose. The steps from glyceraldehyde to pyruvate convert carbon 1 of glyceraldehyde-3-phosphate into the carboxyl carbon of pyruvate. So, the carboxyl carbon of pyruvate will acquire label from carbons 3 and 4 of glucose. The keto carbon of pyruvate is from carbon 2 of glyceraldehyde-3-phosphate, which derives from carbons 2 and 5 of glucose. Finally, the methyl carbon of pyruvate derives from carbons 1 and 6 of glucose.

3. Effects of concentration changes in glycolysis

In an erythrocyte undergoing glycolysis, what would be the effect of a sudden increase in the concentration of (a) ATP? (b) AMP? (c) fructose-1,6-bisphosphate? (d) fructose-2,6-bisphosphate? (e) citrate? (f) glucose-6-phosphate?

Answer: Erythrocytes lack mitochondria and derive energy largely from glycolysis. Glycolysis is regulated at the two steps catalyzed by phosphofructokinase and pyruvate kinase. Phosphofructokinase is a key enzyme that serves as a "valve" regulating the movement of glucose units into glycolysis. Prior to formation of fructose-1,6-bisphosphate, there are other metabolic options (e.g., pentose phosphate pathway, glycogen synthesis). The phosphofructokinase reaction commits glucose to glycolysis. This enzyme is subject to a number of regulatory mechanisms. At low ATP levels, the enzyme has a low K_m for fructose-6-phosphate; however, at high ATP levels, the enzyme behaves cooperatively and has an increased K_m for fructose-6-phosphate and a sigmoidal dependence of activity on fructose-6-phosphate concentration. This behavior is a result of allosteric regulation by ATP. Phosphofructokinase has an additional ATP binding site distinct from the substrate site. ATP binding to the allosteric site causes activity to be less sensitive to fructose-6-phosphate. This allosteric inhibition by ATP is reversed by competitive binding of AMP. The inhibitory effect of ATP is enhanced by citrate. This citric acid cycle intermediate is produced in the mitochondria, and, when mitochondrial concentrations are high, citrate may be exported to the cytosol to be used in fatty acid biosynthesis. Cytosolic citrate also functions as an inhibitor of glycolysis by binding to phosphofructokinase and enhancing the allosteric effects of ATP. Since erythrocytes lack mitochondria, citrate is not expected to play an important role in regulation of glycolysis. Phosphofructokinase is also allosterically regulated by fructose-2,6-bisphosphate whose concentration is regulated by the enzymatic activities: phosphofructokinase 2 and fructose-2,6-bisphosphatase. High concentrations of fructose-2,6-bisphosphate is a signal to increase glycolysis by increasing the affinity of phosphofructokinase for its substrate, fructose-6-phosphate, and by decreasing the inhibitory effects of ATP.

The second key enzyme in regulation of glycolysis is pyruvate kinase. This enzyme has allosteric binding sites for AMP and fructose-2,6-bisphosphate leading to activation of the enzyme. Further, the enzyme is allosterically inhibited by ATP (and by alanine, and acetyl-coenzyme A).

330

With these points in mind we predict that increased concentrations of ATP will lead to inhibition of glycolysis, and increased concentrations of AMP will stimulate glycolysis. An increase in fructose-1,6-bisphosphate will stimulate glycolysis. Fructose-2,6-bisphosphate is a signal to increase the rate of glycolysis. Citrate will inhibit glycolysis. Finally, increased concentrations of glucose-6-phosphate may stimulate glycolysis. Glucose-6-phosphate is isomerized to fructose-6-phosphate, a substrate of phosphofructokinase. Depending on the ATP, AMP, citrate and fructose-2,6-bisphosphate levels, phosphofructokinase may be quite responsive to increased levels of fructose-6-phosphate. (The first reaction of glycolysis, phosphorylation of glucose to glucose-6-phosphate may be sensitive to glucose-6-phosphate levels as well. In muscle, the enzyme catalyzing this reaction is inhibited by glucose-6-phosphate, whereas in liver the enzyme is not.)

4. Understanding NADH cycling in glycolysis
Discuss the cycling of NADH and NAD⁺ in glycolysis and the related fermentation reactions.

Answer: Nicotinamide adenine dinucleotide plays a role in metabolism as an enzyme cofactor in oxidation-reduction reactions. It is found in limited concentrations and cycles between reduced (NADH) and oxidized (NAD^+) forms. Conversion of glucose to two molecules of pyruvate results in the net production of two molecules of ATP and the reduction of two molecules of NAD^+ to NADH. To continue the process of glycolysis, NAD^+ must be regenerated. Under anaerobic conditions, this is accomplished by using pyruvate or a derivative of pyruvate (acetaldehyde) to oxidize NADH back to NAD^+. What is the net effect? Glucose is converted to two molecules of lactic acid (or two molecules of CO_2 and ethanol). The reaction driving the synthesis of ATP, namely glucose to lactate, is an internal oxidation-reduction reaction. This is easily seen by comparing the carbons of glucose to the carbons of pyruvate and lactate:

$$
\begin{array}{ccc}
\text{CHO} & & \\
\text{H---C---OH} & \text{COOH} & \text{COOH} \\
\text{HO---C---H} & \text{C=O} & \text{HO---C---H} \\
\text{H---C---OH} & \text{CH}_3 & \text{CH}_3 \\
\text{H---C---OH} & & \\
\text{CH}_2\text{OH} & & \\
\text{D-glucose} & \text{pyruvic acid} & \text{lactic acid}
\end{array}
$$

The carboxyl carbons of pyruvic and lactic acids are oxidized relative to the carbons in glucose, whereas the methyl carbons are reduced. For lactic acid, carbon 2 is at the same oxidation state of an average carbon atom in glucose. Thus, conversion of glucose to lactic acid is in effect an internal oxidation-reduction. The same is not true for pyruvic acid because the keto carbon is oxidized relative to the average carbon in glucose. During pyruvic acid production NAD^+ was reduced. Conversion of pyruvic acid to lactic acid, a reduction, is accompanied by oxidation of NADH to NAD^+. Cycling of nicotinamide adenine dinucleotide between its oxidized and reduced forms, in effect, allows an internal oxidation-reduction reaction to occur within glucose half molecules.

5. Understanding the reaction mechanisms of glycolysis

For each of the following reactions, name the enzyme that carries out this reaction in glycolysis and write a suitable mechanism for the reaction.

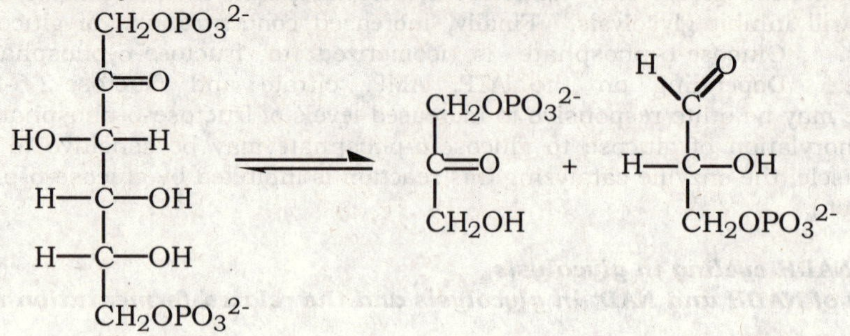

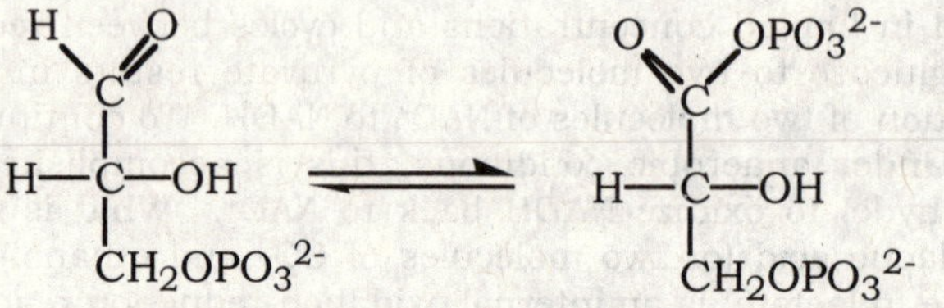

$$CH_2OPO_3^{2-} \quad C=O \quad HO-C-H \quad H-C-OH \quad H-C-OH \quad CH_2OPO_3^{2-} \quad \rightleftharpoons \quad CH_2OPO_3^{2-} \quad C=O \quad CH_2OH \quad + \quad H-C=O \quad H-C-OH \quad CH_2OPO_3^{2-}$$

$$H-C=O \quad H-C-OH \quad CH_2OPO_3^{2-} \quad \rightleftharpoons \quad O=C-OPO_3^{2-} \quad H-C-OH \quad CH_2OPO_3^{2-}$$

Answer: The first reaction is reaction 4 of glycolysis catalyzed by fructose bisphosphate aldolase. There are two classes of aldolase enzymes found in nature: Class I aldolase and Class II aldolase. Class I aldolase catalysis involves the formation of a Schiff base intermediate and because of this NaBH₄ is a potent inhibitor of this enzyme. Class II aldolase has an active-site Zn^{2+} that is required for catalysis. Divalent cation chelating agents, like EDTA, will inhibit Class II aldolases.

Catalysis for a Class I aldolase begins with Schiff base formation between an active-site lysine and the keto carbon of fructose-1,6-bisphosphate with elimination of water. The hydroxyl group of carbon 4 loses a proton to a nearby cysteine followed by elimination of the enolate anion. The resulting enamine is reprotonated, and DHAP is released by hydrolysis as shown below.

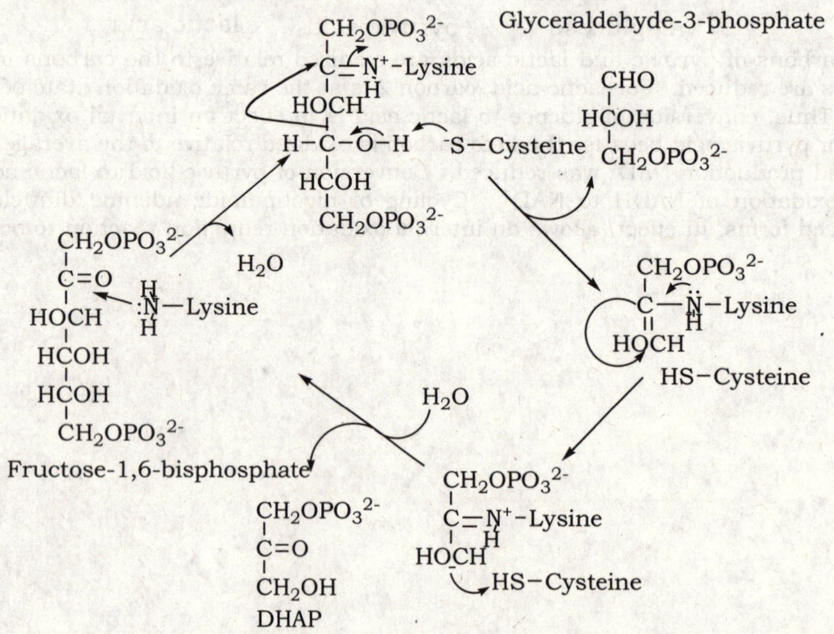

For Class II aldolases, an active-site Zn^{2+} acts as an electrophile to polarize the carbonyl group of the substrate and stabilizes the enolate intermediate as shown below.

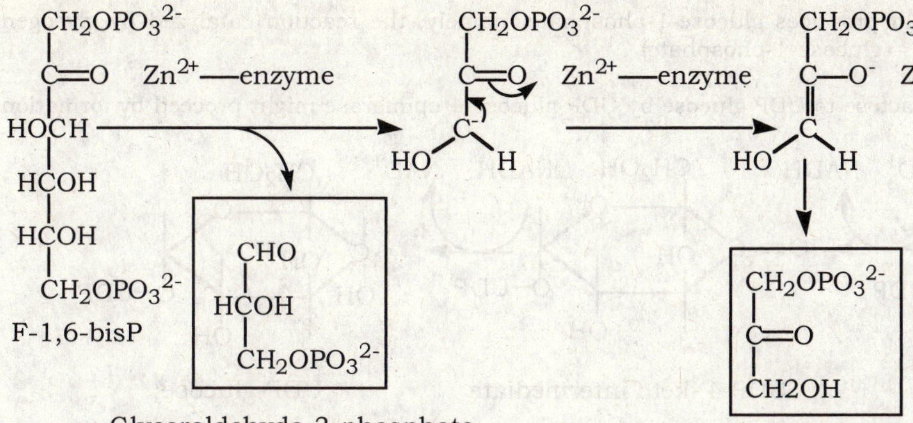

$$CH_2OPO_3^{2-} \quad Zn^{2+}\text{---enzyme}$$

C=O

HOCH

HCOH

HCOH

$$CH_2OPO_3^{2-}$$

F-1,6-bisP

CHO

HCOH

$$CH_2OPO_3^{2-}$$

Glyceraldehyde-3-phosphate

$$CH_2OPO_3^{2-} \quad Zn^{2+}\text{---enzyme}$$

C=O

HO—C—H

$$CH_2OPO_3^{2-} \quad Zn^{2+}\text{---enzyme}$$

C—O⁻

HO—C—H

$$CH_2OPO_3^{2-}$$

C=O

CH2OH

DHAP

The second reaction is catalyzed by glyceraldehyde-3-phosphate dehydrogenase. The complete reaction is:

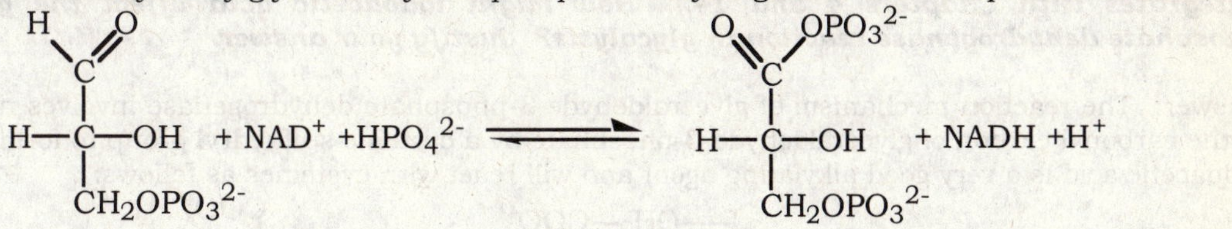

H—C=O

H—C—OH + NAD⁺ +HPO₄²⁻

$$CH_2OPO_3^{2-}$$

⇌

O=C—OPO₃²⁻

H—C—OH + NADH +H⁺

$$CH_2OPO_3^{2-}$$

Glyceraldehyde-3-phosphate   1,3-bisphosphoglycerate

The reaction mechanism involves nucleophilic attack of the carbonyl carbon by a cysteine to form a hemithioacetal, which transfers a hydride to NAD⁺ to form a thioester. The thioester is subsequently attacked by phosphate to produce 1,3-bisphosphoglycerate. This sequence of reactions is shown below.

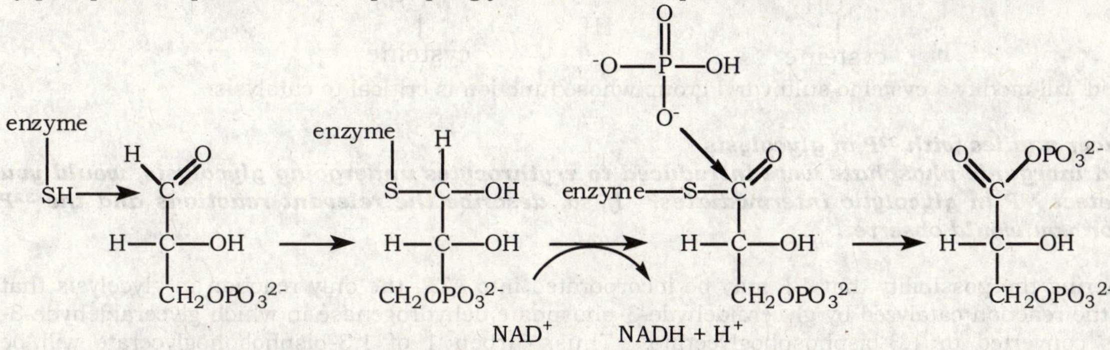

enzyme

SH →

H—C=O

H—C—OH

$$CH_2OPO_3^{2-}$$

enzyme

S—C—OH (H)

H—C—OH

$$CH_2OPO_3^{2-}$$

NAD⁺   NADH + H⁺

O=P—OH

O⁻   O⁻

enzyme—S—C=O

H—C—OH

$$CH_2OPO_3^{2-}$$

O=C—OPO₃²⁻

H—C—OH

$$CH_2OPO_3^{2-}$$

6. The reactions and mechanisms of the Leloir pathway
Write the reactions that permit galactose to be utilized in glycolysis. Write a suitable mechanism for one of these reactions.

Answer: Galactose is first phosphorylated at carbon 1 by galactokinase in the following reaction:

D-Galactose + ATP⁴⁻ → D-galactose-1-phosphate²⁻ + ADP³⁻ + H⁺

The enzyme galactose-1-phosphate uridylyl transferase reacts galactose-1-phosphate with UDP-glucose to produce UDP-galactose and glucose-1-phosphate:

Galactose-1-phosphate + UDP-glucose → UDP-galactose + glucose-1-phosphate

Next, UDP-galactose is converted to UDP-glucose by UDP-glucose-4-epimerase:

UDP-galactose → UDP-glucose

The glucose moiety of UDP-glucose may be released by the second reaction as glucose-1-phosphate. Alternatively, it may be incorporated into glycogen and subsequently released as glucose-1-phosphate. Finally, glucose-1-phosphate is converted to glucose-6-phosphate by phosphoglucomutase:

Glucose-1-phosphate → glucose-6-phosphate.

The UDP-glucose used in the second reaction is produced by UDP-glucose pyrophosphorylase in the following reaction:

Glucose-1-phosphate + UTP → UDP-glucose + PPᵢ

(There is another reaction that produces glucose-1-phosphate, namely, the reaction catalyzed by glycogen phosphorylase: glycogen + P$_i$ → glucose-1-phosphate)

The conversion of UDP-galactose to UDP-glucose by UDP-glucose-4-epimerase might proceed by formation of a 4-keto intermediate.

UDP-Galactose   UDP-4'-keto intermediate   UDP-glucose

7. The effect of iodoacetic acid on the glyceraldehyde-3-P dehydrogenase reaction
(Integrates with Chapters 4 and 14.) How might iodoacetic acid affect the glyceraldehyde-3-phosphate dehydrogenase reaction in glycolysis? Justify your answer.

Answer: The reaction mechanism of glyceraldehyde-3-phosphate dehydrogenase involves nucleophilic attack of the carbonyl carbon of glyceraldehyde-3-phosphate by a cysteine sulfhydryl group to form a hemithioacetyl. Iodoacetic acid is a very good alkylating agent and will react with cysteines as follows:

Iodoacetic acid will modify a cysteine sulfhydryl group whose function is critical to catalysis.

8. Radiotracer studies with ^{32}P in glycolysis
If ^{32}P-labeled inorganic phosphate were introduced to erythrocytes undergoing glycolysis, would you expect to detect ^{32}P in glycolytic intermediates? If so, describe the relevant reactions and the ^{32}P incorporation you would observe.

Answer: Ignoring the possibility that ^{32}P may be incorporated into ATP, the only reaction in glycolysis that utilizes P$_i$ is the reaction catalyzed by glyceraldehyde-3-phosphate dehydrogenase in which glyceraldehyde-3-phosphate is converted to 1,3-bisphosphoglycerate. Thus, carbon 1 of 1,3-bisphosphoglycerate will be labeled. The label is lost to ATP after 1,3-bisphosphate is converted to 3-phosphoglycerate by phosphoglycerate kinase. So, we expect no other glycolytic intermediates to be directly labeled. (Once the label is incorporated into ATP it will show up at carbons 1 and 6 of glucose.)

9. Comparing glycolysis entry points for sucrose
Sucrose can enter glycolysis by either of two routes:
Sucrose phosphorylase
Sucrose + P$_i$ ⇌ fructose + glucose-1-phosphate
Invertase
Sucrose + H$_2$O ⇌ fructose + glucose
Would either of these reactions offer an advantage over the other in the preparation of sucrose for entry into glycolysis?

Answer: The reaction catalyzed by sucrose phosphorylase produces glucose-1-phosphate, which enters glycolysis as glucose-6-phosphate, by the action of phosphoglucomutase. Therefore, glucose-6-phosphate produced by this route, bypasses the hexokinase reaction, normally employed, in which ATP hydrolysis occurs.

10. Assessing the role of Mg²⁺ in glycolysis
What would be the consequences of a Mg²⁺ deficiency for the reactions of glycolysis?

Answer: Kinases actually employ $Mg^{2+}-ATP^{4-}$, and not ATP^{4-}, as substrate. Therefore, a magnesium ion deficiency will affect the following enzymes: hexokinase (glucokinase), phosphofructokinase, phosphoglycerate kinase, and pyruvate kinase.

11. Energetics of the triose phosphate isomerase reaction
(Integrates with Chapter 3.) Triose phosphate isomerase catalyzes the conversion of dihydroxyacetone-P to glyceraldehyde-3-P. The standard free energy change, $\Delta G^{\circ\prime}$, for this reaction is +7.6 kJ/mol. However, the observed free energy change (ΔG) for this reaction in erythrocytes is +2.4 kJ/mol.
  a. Calculate the ratio of [dihydroxyacetone-P]/[glyceraldehyde-3-P] in erythrocytes from ΔG.
  b. If [dihydroxyacetone-P] = 0.2 mM, what is [glyceraldehyde-3-P]?

Answer: The reaction being examined is:
$$\text{Dihydroxyacetone-P} \rightarrow \text{glyceraldehyde-3-P}$$
The observed free energy change (ΔG) is calculated from the standard free energy change (ΔG°) using the following equation:

$$\Delta G = \Delta G^{\circ\prime} + RT \times \ln \frac{[\text{Products}]}{[\text{Reactants}]}$$

In this case, we have

$$\Delta G = \Delta G^{\circ\prime} + RT \times \ln \frac{[\text{glyceraldehyde} - 3 - \text{phosphate}]}{[\text{dihydroxyacetone phosphate}]}$$

Since the question asks us for the ratio of [dihydroxyacetone-P]/glyceraldehyde-3-P], rearrange, using the property $\log (x) = -\log (1/x)$ to give

$$\Delta G = \Delta G^{\circ\prime} - RT \times \ln \frac{[\text{dihydroxyacetone phosphate}]}{[\text{glyceraldehyde} - 3 - \text{phosphate}]}$$

Solving for the desired ratio, we have

$$\frac{[\text{dihydroxyacetone phosphate}]}{[\text{glyceraldehyde} - 3 - \text{phosphate}]} = e^{\frac{(\Delta G^{\circ\prime} - \Delta G)}{RT}}$$

Substituting the numbers, and using 37°C as the temperature at which erythrocytes find themselves, we have

$$\frac{[\text{dihydroxyacetone phosphate}]}{[\text{glyceraldehyde} - 3 - \text{phosphate}]} = e^{\frac{(7.6-2.4)}{8.314 \times 10^{-3} \times 310}} = e^{2.018}$$

$$\frac{[\text{dihydroxyacetone phosphate}]}{[\text{glyceraldehyde} - 3 - \text{phosphate}]} = 7.52$$

This answer tells us that the chemical reaction as written above lies to the left, which we already knew from the positive value for its ΔG.

b. Given that the dihydroxyacetone-P concentration in erythrocytes is 0.2 mM, we can calculate the concentration of glyceraldehyde-3-P from the answer for part a above.

$$[\text{glyceraldehyde} - 3 - \text{phosphate}] = \frac{[\text{dihydroxyacetone phosphate}]}{7.52} = \frac{0.2 \text{ mM}}{7.52}$$

$$[\text{glyceraldehyde} - 3 - \text{phosphate}] = 0.027 \text{ mM} = 27 \ \mu\text{M}$$

12. Energetics of the enolase reaction
(Integrates with Chapter 3.) Enolase catalyzes the conversion of 2-phosphoglycerate to phosphoenolpyruvate + H_2O. The standard free energy change, $\Delta G^{\circ\prime}$, for this reaction is +1.8 kJ/mol. If the concentration of 2-phosphoglycerate is 0.045 mM and the concentration of phosphoenolpyruvate is 0.034 mM, what is ΔG, the free energy change for the enolase reaction, under these conditions?

Answer: The reaction being examined is:

335

$$2\text{-Phosphoglycerate} \rightarrow \text{phosphoenolpyruvate} + H_2O$$

The observed free energy change (ΔG) is calculated from the standard free energy change ($\Delta G^{\circ\prime}$) using the following equation:

$$\Delta G = \Delta G^{\circ\prime} + RT \times \ln\frac{[\text{Products}]}{[\text{Reactants}]}, \text{ or in this case}$$

$$\Delta G = \Delta G^{\circ\prime} + RT \times \ln\frac{[\text{phosphoenoypyruvate}]}{[2-\text{phosphoglycerate}]}$$

Substituting the given values for the concentrations and $\Delta G^{\circ\prime}$, and using 25°C for the temperature, we have:

$$\Delta G = 1.8\,\frac{kJ}{mol} + 8.314 \times 10^{-3}\,\frac{kJ}{mol \cdot K} \times 298K \times \ln\frac{0.034\text{ mM}}{0.045\text{ mM}}$$

$$\Delta G = 1.8\,\frac{kJ}{mol} - 0.69\,\frac{kJ}{mol} = +1.11\,\frac{kJ}{mol}$$

At 37°C the value is 1.08 kJ/mol.

13. Energetics of PEP hydrolysis

(Integrates with Chapter 3.) The standard free energy change (ΔG°) for hydrolysis of phosphoenolpyruvate (PEP) is -62.2 kJ/mol. The standard free energy change (ΔG°) for ATP hydrolysis is -30.5 kJ/mol.

a. What is the standard free energy change for the pyruvate kinase reaction:
$$ADP + phosphoenolpyruvate \rightarrow ATP + pyruvate$$
b. What is the equilibrium constant for this reaction?
c. Assuming the intracellular concentrations of [ATP] and [ADP] remain fixed at 8 mM and 1 mM, respectively, what will be the ratio of [pyruvate]/[phosphoenolpyruvate] when the pyruvate kinase reaction reaches equilibrium?

Answer: a. The reaction
$$ADP + \text{phosphoenolpyruvate} \rightarrow ATP + \text{pyruvate}$$
is the sum of two coupled reactions

Reaction 1:   Phosphoenolpyruvate + $H_2O \rightarrow$ pyruvate + P_i;
$$\Delta G^{\circ\prime}_1 = -62.2 \text{ kJ/mol}$$

and

Reaction 2:   ADP + $P_i \rightarrow$ ATP + H_2O;
$$\Delta G^{\circ\prime}_2 = +30.5 \text{ kJ/mol}$$

The value of $\Delta G^{\circ\prime}$ for the latter reaction is positive (+) because we have written it in the direction of ATP synthesis, rather than in the direction of hydrolysis, for which the value of $\Delta G^{\circ\prime}$ is given in the problem.

The total change in the standard free energy of coupled reactions is given by the sum of the free energy changes of the individual reactions. Accordingly, we have for the reaction in question,

$$\Delta G^{\circ\prime}_{Total} = \Delta G^{\circ\prime}_1 + \Delta G^{\circ\prime}_2 = -62.2\,\frac{kJ}{mol} + 30.5\,\frac{kJ}{mol} = -31.7\,\frac{kJ}{mol}$$

b. The equilibrium constant, K'_{eq}, is derived from the standard free energy change using the following equation:

$$\Delta G^{\circ\prime} = -RT \times \ln K'_{eq}$$

which upon rearranging yields

$$K'_{eq} = e^{-\frac{\Delta G^{\circ\prime}}{RT}}$$

Substituting the appropriate numerical values, with the product RT = 2.48 at 25°C, we have

$$K'_{eq} = e^{\frac{(-31.7)}{2.48}} = 3.56 \times 10^5$$

The equilibrium position of this reaction obviously lies far to the right.

c. The equilibrium constant is the ratio of products to reactants at equilibrium. For this reaction,

$$K'_{eq} = \frac{[ATP][pruvate]}{[ADP][PEP]} = \frac{[ATP]}{[ADP]} \times \frac{[pruvate]}{[PEP]} = \frac{8 \text{ mM}}{1 \text{ mM}} \times \frac{[pruvate]}{[PEP]}, \text{ or}$$

$$\frac{[pruvate]}{[PEP]} = \frac{K'_{eq}}{8} = \frac{3.56 \times 10^5}{8} = 4.45 \times 10^4$$

14. Energetics of fructose-1,6-bisP hydrolysis
(Integrates with Chapter 3.) The standard free energy change (ΔG°') for hydrolysis of fructose-l,6-bisphosphate (FBP) to fructose-6-phosphate (F-6-P) and P_i is -16.7 kJ/mol:
$$FBP + H_2O \rightarrow \text{fructose-6-P} + P_i$$
The standard free energy change (ΔG°') for ATP hydrolysis is -30.5 kJ/mol:
$$ATP + H, \rightarrow ADP + P_i$$
a. What is the standard free energy change for the phosphofructokinase reaction:
$$ATP + \text{fructose-6-phosphate} \rightarrow ADP + FBP$$
b. What is the equilibrium constant for this reaction?
c. Assuming the intracellular concentrations of [ATP] and [ADP] are maintained constant at 4 mM and 1.6 mM, respectively, in a rat liver cell, what will be the ratio of [FBP]/[fructose-6-P] when the phosphofructokinase reaction reaches equilibrium?

Answer: a. The reaction
$$ATP + \text{fructose-6-P} \rightarrow ADP + FBP$$
is derived as the sum of two coupled reactions

Reaction 1:   fructose-6-P + P_i → FBP + H_2O;
$$\Delta G_1^{o'} = + 16.7$$
and
Reaction 2:   ATP + H_2O → ADP + P_i;
$$\Delta G_2^{o'} = - 30.5 \text{ kJ/mol}$$

The value of ΔG°' for the former reaction is positive (+) because we have written it in the direction of P_i removal, rather than in the direction of P_i addition, for which the value of ΔG°' is given in the problem.

The total change in the standard free energy of coupled reactions is given by the sum of the free energy changes of the individual reactions. Accordingly, we have for the reaction in question,
$$\Delta G^{o'}_{Total} = \Delta G^{o'}_1 + \Delta G^{o'}_2 = +16.7 \frac{kJ}{mol} - 30.5 \frac{kJ}{mol} = -13.8 \frac{kJ}{mol}$$

b. The equilibrium constant, K'_{eq}, is derived from the standard free energy change using the following equation:
$$\Delta G^{o'} = -RT \times \ln K'_{eq}$$
which upon rearranging yields
$$K'_{eq} = e^{-\frac{\Delta G^{o'}}{RT}}$$

Substituting the appropriate numerical values, with the product RT = 2.48 at 25°C, we have
$$K'_{eq} = e^{\frac{(-13.8)}{2.48}} = 262$$

The equilibrium position of this reaction lies far to the right. (At 37°C the value is 211.)

The equilibrium constant is the ratio of products to reactants at equilibrium. For this reaction,
$$K'_{eq} = \frac{[ADP][FBP]}{[ATP][\text{fructose}-6-P]} = \frac{[ADP]}{[ATP]} \times \frac{[FBP]}{[\text{fructose}-6-P]}$$
$$K'_{eq} = \frac{1.6 \text{ mM}}{4 \text{ mM}} \times \frac{[FBP]}{[\text{fructose}-6-P]} = 0.4 \times \frac{[FBP]}{[\text{fructose}-6-P]}, \text{ or}$$
$$\frac{[FBP]}{[\text{fructose}-6-P]} = \frac{K'_{eq}}{0.4} = \frac{261}{0.4} = 652$$

(At 37°C the value is 528.)

337

15. Energetics of 1,3-bisphophoglycerate hydrolysis
(Integrates with Chapter 3.) The standard free energy change ($\Delta G°'$) for hydrolysis of 1,3-bisphosphoglycerate (1,3-BPG) to 3-phosphoglycerate (3-PG) and P_i is -49.6 kJ/mol:
$$1,3\text{-BPG} + H_2O \rightarrow 3\text{-PG} + P_i$$
The standard free energy change ($\Delta G°'$) for ATP hydrolysis is -30.5 kJ/mol:
$$ATP + H_2O \rightarrow ADP + P_i$$
    a. What is the standard free energy change for the phosphoglycerate kinase reaction:
$$ADP + 1,3\text{-BPG} \rightarrow ATP + 3\text{-PG}$$
    b. What is the equilibrium constant for this reaction?
    c. If the steady-state concentrations of [1,3-BPG] and [3-PG] in an erythrocyte are 1 μM and 129 μM, respectively, what will be the ratio of [ATP]/[ADP], assuming the phosphoglycerate kinase reaction is at equilibrium?

Answer: a. The reaction
$$ADP + 1,3\text{-BPG} \rightarrow ATP + 3\text{-PG}$$
is derived as the sum of two coupled reactions

    Reaction 1: $1,3\text{-BPG} + H_2O \rightarrow 3\text{-PG} + P_i$;
$$\Delta G_1^{°'} = -49.6 \text{ kJ/mol}$$

and

    Reaction 2: $ADP + P_i \rightarrow ATP + H_2O$;
$$\Delta G_2^{°'} = +30.5 \text{ kJ/mol}$$

The value of $\Delta G°'$ for the latter reaction is positive (+) because we have written it in the direction of ATP synthesis, rather than in the direction of hydrolysis, for which the value of $\Delta G°'$ is given in the problem.

    The total change in the standard free energy of coupled reactions is given by the sum of the free energy changes of the individual reactions. Accordingly, we have for the reaction in question,

$$\Delta G°'_{Total} = \Delta G°'_1 + \Delta G°'_2 = -49.6 \frac{kJ}{mol} + 30.5 \frac{kJ}{mol} = -19.1 \frac{kJ}{mol}$$

b. The equilibrium constant, K_{eq}', is derived from the standard free energy change using the following equation:

$$\Delta G°' = -RT \times \ln K'_{eq}$$

which upon rearranging yields

$$K'_{eq} = e^{-\frac{\Delta G°'}{RT}}$$

Substituting the appropriate numerical values, with the product RT = 2.48 at 25°C, we have

$$K'_{eq} = e^{\frac{(-19.1)}{2.48}} = 2227$$

(At 37°C the value is 1651).

c. The equilibrium constant is the ratio of products to reactants at equilibrium. For this reaction,

$$K'_{eq} = \frac{[ATP][3-PG]}{[ADP][1,3-BPG]} = \frac{[ATP]}{[ADP]} \times \frac{[3-PG]}{[1,3-BPG]}$$

$$K'_{eq} = \frac{[ATP]}{[ADP]} \times \frac{120 \ \mu M}{1 \ \mu M}, \text{ or}$$

$$\frac{[ATP]}{[ADP]} = \frac{K'_{eq}}{120} = \frac{2227}{120} = \frac{2227}{120} = 18.6$$

(At 37°C the value is 13.8).

16. Energetics of the hexokinase reaction
The standard-state free energy change, $\Delta G°'$, for the hexokinase reaction is -16.7 kJ/mol. Use the values in Table 18.2 to calculate the value of ΔG for this reaction in the erythrocyte at 37°C.

Answer: Hexokinase phosphorylates various six-carbon sugars, including glucose, using ATP as the phosphate donor. For glucose, the reaction is:

Glucose + ATP ⇌ glucose-6-phosphate and ADP

Using Table 18.2 we find the following values for concentrations: [glucose] = 5.0 mM, [ATP] = 1.85 mM, [glucose-6-P] = 0.083 mM and [ADP] = 0.14 mM and using them we can first calculate the equilibrium constant. Thus,

$$\Delta G = \Delta G^{\circ\prime} + RT \ln \frac{[glucose-6-P][ADP]}{[glucose][ATP]}$$

$$\Delta G = -16.7 \text{ kJ/mol} + (8.3154 \text{ J/K mol}) (273+37 \text{ K}) \times \ln \frac{0.083 \text{ mM} \times 0.14 \text{ mM}}{5 \text{ mM} \times 1.85 \text{ mM}}$$

$$\Delta G = -16.7 \text{ kJ/mol} - 17.2 \text{ kJ/mol}$$

$$\Delta G = -33.9 \text{ kJ/mol}$$

17. Analyzing the concentration dependence of the adenylate kinase reaction

Taking into consideration the equilibrium constant for the adenylate kinase reaction (page 585), calculate the change in concentration in AMP that would occur if 8% of the ATP in an erythrocyte (red blood cell) were suddenly hydrolyzed to ADP. In addition to the concentration values in Table 18.2, it may be useful to assume that the initial concentration of AMP in erythrocytes is 5 µM.

Answer: The values given for ADP and ATP from Table 18.2 are 0.14 mM and 1.85 mM respectively and the equilibrium constant for the adenylate kinase is 0.44. Using this information and equation 18.7 (see below) we can calculate the [AMP].

$$K_{eq} = \frac{[ATP][AMP]}{[ADP]^2}$$

$$[AMP] = \frac{K_{eq} \times [ADP]^2}{[ATP]} = \frac{0.44 \times \left(0.14 \times 10^{-3}\right)^2}{1.85 \times 10^{-3}} = 4.7 \times 10^{-6} \approx 5 \text{ µM}$$

The same equation is used to calculate the new concentration of AMP after 8% of the initial ATP is hydrolyzed to ADP. This will lower ATP by (0.08×1.85 mM =) 0.148 mM and increase ADP by the same amount, more than doubling the ADP concentration.

$$K_{eq} = \frac{[ATP][AMP]}{[ADP]^2}$$

$$[AMP] = \frac{K_{eq} \times [ADP]^2}{[ATP]} = \frac{0.44 \times \left(0.14 \times 10^{-3} + 0.148 \times 10^{-3}\right)^2}{1.85 \times 10^{-3} - 0.148 \times 10^{-3}} = 21.4 \times 10^{-6} \approx 20 \text{ µM}$$

A small amount of ATP hydrolysis nearly doubles [ADP] but it increases [AMP] four-fold.

18. Distinguishing the mechanisms of Class I and Class II aldolases

Fructose bisphosphate aldolase in animal muscle is a class I aldolase, which forms a Schiff base intermediate between substrate (for example, fructose-1,6-bisphosphate or dihydroxyacetone phosphate) and a lysine at the active site (see Figure 18.12). The chemical evidence for this intermediate comes from studies with aldolase and the reducing agent sodium borohydride, NaBH₄. Incubation of the enzyme with dihydroxyacetone phosphate and NaBH₄ inactivates the enzyme. Interestingly, no inactivation is observed if NaBH₄ is added to the enzyme in the absence of substrate. Write a mechanism that explains these observations and provides evidence for the formation of a Schiff base intermediate in the aldolase reaction.

Answer: Formation of the Schiff base intermediate involves attack by the amino group of a side chain lysine residue on the carbonyl carbon of a keto group. A carbon-nitrogen covalent double bond is formed upon elimination of water. This double bond is the target of sodium borohydride, a potent reducing agent. Sodium borohydride will reduce the double bond forming a stable adduct on the active-site lysine. In the absence of substrate the active site lysine is in its amino form and hence not a target for reduction by sodium borohydride. This is shown below for reaction of the enzyme with dihydroxyacetone phosphate.

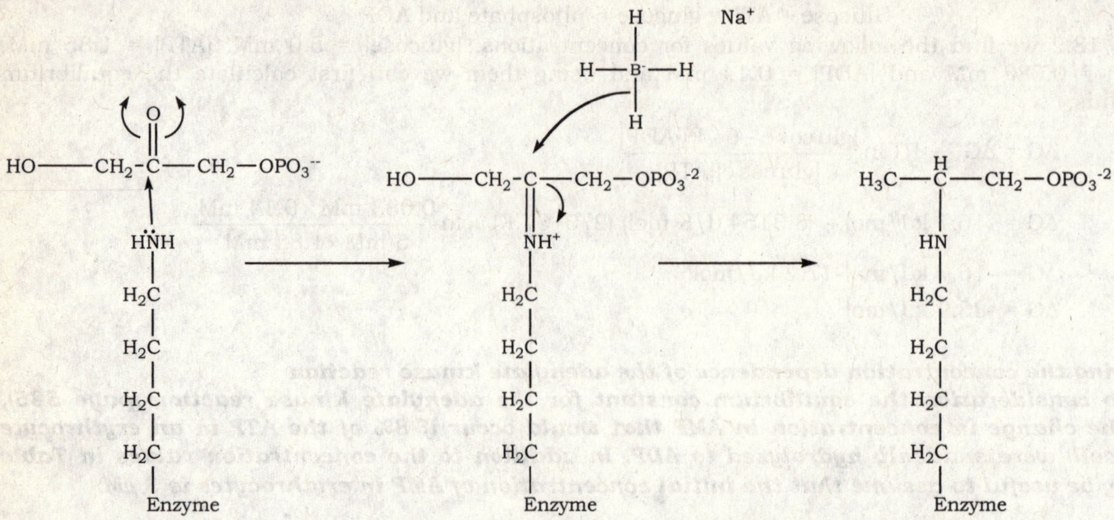

19. Understanding the Ping-Pong mechanism of Gal-1-P uridylyltransferase

As noted on page 600, the galactose-1-phosphate uridylyltransferase reaction proceeds via a ping-pong mechanism. Consult Chapter 13 page 430 to refresh your knowledge of ping-pong mechanisms, and draw a diagram to show how a ping-pong mechanism would proceed for the uridylyltransferase.

Answer: Galactose-1-phosphate uridylyltransferase catalyzes the following reaction:

$$\text{UDP-glucose} + \alpha\text{-D-Galactose-1-P} \rightleftharpoons \alpha\text{-D-Glucose-1-P} + \text{UDP-Galactose}$$

UDP-glucose binds to the active site and UMP is transferred to an active site histidine to make a covalent adduct. Glucose-1-P is subsequently released. Galactose-1-P then binds and UMP is transferred to it to produce UDP-galactose. The reaction is shown below.

Step 1

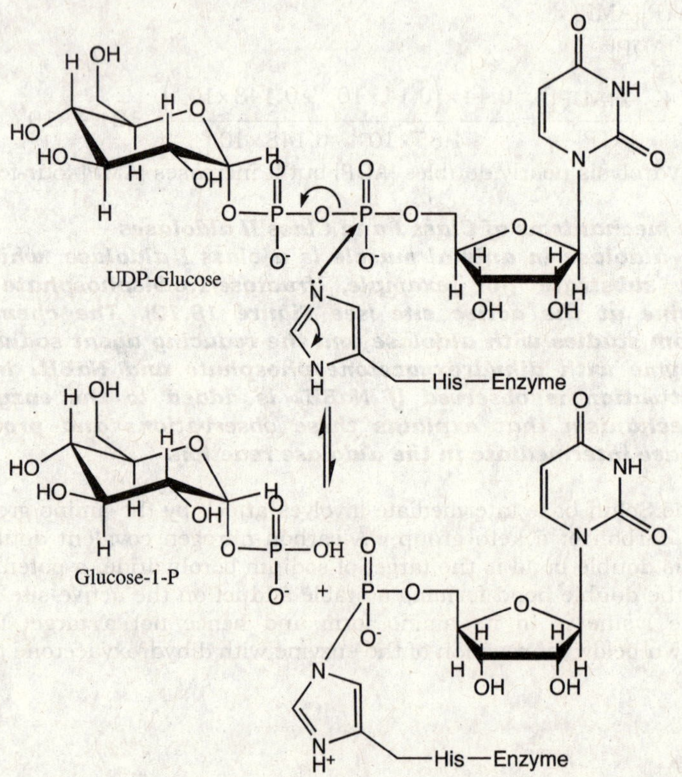

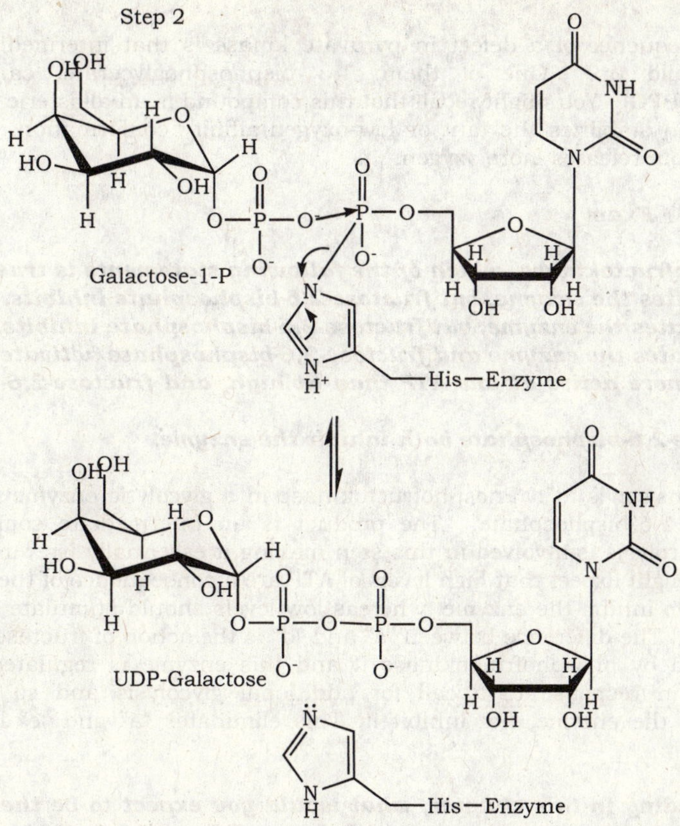

Step 2

Galactose-1-P

His—Enzyme

UDP-Galactose

His—Enzyme

20. Understanding the mechanism of hemolytic anemia
Genetic defects in glycolytic enzymes can have serious consequences for humans. For example, defects in the gene for pyruvate kinase can result in a condition known as hemolytic anemia. Consult a reference to learn about hemolytic anemia, and discuss why such genetic defects lead to this condition.

Answer: One place to start your search is at NCBI (http://www.ncbi.nlm.nih.gov/).   Do a search for "Hemolytic AND anemia AND pyruvate AND kinase" in all databases and activate the hit to OMIM. There are two records related to pyruvate kinase, 609712 and 266200, and activating the later you will learn that deficiencies in pyruvate kinase in red blood cells are the most common cause of hereditary nonspherocytic hemolytic anemia and that the most common abnormality in the glycolytic pathway is due to mutations in one of the pyruvate kinase genes.   Activating 609712 gives information about pyruvate kinase genes and mutations within these genes.   Nonspherocytic hemolytic anemia is a term that describes a number of blood disorders that are transmitted genetically.   The disorders lead to lysis of red blood cells, which exhibit irregular, non-spherical shapes and hemolyze causing the anemia.

You might recall that red cells lack mitochondria and so their sole source of metabolic energy is glycolysis. Defects in pyruvate kinase, the enzyme that converts phosphoenolpyruvate to pyruvate with accompanying phosphorylation of ADP to ATP, will back up glycolytic intermediates and lead to a decrease in ATP levels. Among the many uses of ATP in red cells, maintenance of electrolyte gradients by the sodium-potassium ATPase (sodium pump) is critical to red cell physiology. (See van Wijk, R. and van Solinge, W. W., 2005. The energy-less red blood cell is lost: erythrocyte abnormalities in glycolysis.   Blood 106, 4034-4042) Decreasing levels of ATP lead to alterations in sodium and potassium levels in red cells and one consequence is dehydration leading to changes in red blood cell morphology.   Abnormal red blood cells can lyse or are removed from circulation by the spleen.

The condition is autosomal recessive and so anemia is evident in homozygotes.   Some heterozygotes, however, may actually benefit from a single copy of an abnormal pyruvate kinase gene.   Some studies have shown that certain mutations actually protect against malaria.   The agent that causes malaria, *Plasmodium falciparum*, invades red blood cells and the invasion in cells with lower levels of active pyruvate kinase may stress them so they exhibit abnormal shapes causing them to be removed.

341

Another interesting consequence of a defect in pyruvate kinase is that intermediates in glycolysis before phosphoenolpyruvate build up. One of them, 1,3 bisphosphoglycerate, can be converted to 2,3 bisphosphoglycerate (2,3 BPG). You might recall that this compound is an allosteric affector of hemoglobin. It binds at a single site and stabilizes the taut or low-oxygen-affinity conformation. This actually helps the anemia because hemoglobin releases more oxygen.

Preparing for the MCAT® Exam

21. Regarding phosphofructokinase, which of the following statements is true:
   a. Low ATP stimulates the enzyme, but fructose-2,6-bisphosphate inhibits.
   b. High ATP stimulates the enzyme, but fructose-2,6-bisphosphate inhibits.
   c. High ATP stimulates the enzyme and fructose-2,6-bisphosphate activates.
   d. The enzyme is more active at low ATP than at high, and fructose-2,6-bisphosphate activates the enzyme.
   e. ATP and fructose-2,6-bisphosphate both inhibit the enzyme.

Answer: The correct answer is "d". Phosphofructokinase in a glycolytic enzyme that converts fructose-6-phosphate into fructose-1,6-bisphosphate. The product is an intermediate committed to the glycolytic pathway. Since ATP hydrolysis is involved in this step making it essentially irreversible, this step has to be carefully regulated. We might expect that high levels of ATP are a consequence of the cell not being active and so we might expect ATP to inhibit the enzyme, whereas low levels should stimulate it. With this in mind we can eliminate "b" and "c". The difference between "a" and "d" is the action of fructose-2,6-bisphosphate. This compound is synthesized by phosphofructokinase II and this enzyme is regulated in a number of ways. Activation, however, is in response to a call for additional glycolysis and so we expect fructose-2,6-bisphosphate to activate the enzyme, not inhibit it. This eliminates "a" and "e" leaving us with the best answer, "d".

22. Based on your reading in this chapter, what would you expect to be the most immediate effect on glycolysis if the steady-steady concentration of glucose-6-P were 8.3 mM instead of 0.083 mM?

Answer: The answer depends on where this occurs. For cells using hexokinase, which is inhibited by glucose-6-phosphate, an increase in glucose would lead to an increase in glucose-6-phosphate, which, in turn, would inhibit hexokinase. For cells using glucokinase (liver for example) no inhibition would occur. The conversion to glucose-6-phosphate would continue.

Questions for Self Study

1. Answer True or False to the following statements.
   a. Another name for glycolysis is the Embden-Meyerhof pathway. ___
   b. Glycolysis is confined to the mitochondrial matrix. ___
   c. In the first phase of glycolysis, glucose is converted to two molecules of glyceraldehyde-3-phosphate and NAD⁺ is reduced. ___
   d. The only difference between hexokinase and glucokinase is sugar specificity. ___
   e. Phosphofructokinase hydrolyzes ATP. ___
   f. Conversion of glucose-6-P to fructose-6-P and conversion of glyceraldehyde-3-P to dihydroxyacetone-P are aldose to ketose conversions. ___

2. Phosphofructokinase is the key enzyme regulating glycolysis. For the following compounds indicate their effects on phosphofructokinase activity.
   a. ATP
   b. AMP
   c. Citrate
   d. Fructose-2,6-bisphosphate
   e. Fructose-6-phosphate

3. The ability of a cell to continue glycolysis in the presence of adequate supplies of glucose and a need for ATP is critically dependent on the reaction catalyzed by glyceraldehyde-3-phosphate dehydrogenase. Why? What is the reaction catalyzed by this enzyme?

4. Which two reactions in the second phase of glycolysis are examples of substrate level phosphorylations?

5. a. The production of lactate or ethanol and carbon dioxide from glucose are examples of what process?
b. What is the purpose of this process?

6. Fill in the blanks. The _____ pathway is followed by galactose to enter glycolysis. Galactose is first converted to galactose-1-phosphate by _____ in an ATP-dependent reaction. Galactose-1-phosphate is exchanged for the glucose-1-phosphate moiety of a uridine nucleotide derivative producing _____. This compound is acted on by a glucose-4-epimerase converting it to ___.

7. Conversion of 1,3-bisphosphoglycerate to 2,3-bisphosphoglycerate (a allosteric effector of hemoglobin) and the conversion of 3-phosphoglycerate to 2-phosphoglycerate are examples of reactions catalyzed by what kind of enzymes?

Answers

1. a. T; b. F; c. F; d. F; e. F; f. T.

2. a. inhibits; b. blocks ATP inhibition; c. inhibits; d. stimulates; e. stimulates.

3. The enzyme converts glyceraldehyde-3-phosphate and phosphate to 1,3-bisphosphoglycerate with accompanying reduction of NAD^+ to NADH. The reaction is critical because the levels of NAD^+ are limiting.

4. The reactions catalyzed by phosphoglycerate kinase and pyruvate kinase.

5. a. Fermentation; b. Fermentation provides a pathway for regeneration of NAD^+ from NADH that does not involve oxygen or another external electron acceptor.

6. Leloir; galactokinase; UDP-galactose; UDP-glucose.

7. Mutases

Additional Problems

1. Why is it advantageous to phosphorylate glucose immediately after being transported into cells?

2. Describe the basic chemistry of interconversion of glucose-6-phosphate to fructose-6-phosphate and of dihydroxyacetone phosphate to glyceraldehyde-3-phosphate.

3. What reaction is catalyzed by the enzyme fructose bisphosphate aldolase? Class I aldolases are irreversibly inactivated by NaBH₄ but inactivation requires the presence of either fructose 1,6-bisphosphate or dihydroxyacetone phosphate. The enzyme can be treated with NaBH₄ in the presence of glyceraldehyde-3-phosphate and upon removal of NaBH₄ the enzyme is active. Explain.

4. Describe the reaction catalyzed by phosphoglycerate mutase. What other molecules are required for activity?

5. What key differences exist between hexokinase and glucokinase to insure that glucose is properly apportioned between muscle and liver?

6. Draw the structures of glycerol and of glyceraldehyde. Anaerobic organisms, grown in the absence of oxygen, can use glyceraldehyde as a source of metabolic energy but not glycerol. However, metabolism of glycerol starts with glycerol kinase converting glycerol to 3-phosphoglycerol, which is subsequently metabolized to glyceraldehyde-3-phosphate. Why is glycerol catabolism a problem for anaerobic organisms yet conversion of glyceraldehyde to pyruvate possible?

7. Mannose is a 2' epimer of glucose. However, only one additional enzyme is required to get mannose into glycolysis. Outline a reasonable pathway by which mannose enters glycolysis.

Abbreviated Answers

1. Phosphorylation of glucose makes it essentially impermeable to the cell membrane and traps glucose inside the cell. Phosphorylation also keeps the [glucose] low inside cells allowing uptake to occur down a glucose concentration gradient. Finally, phosphorylation commits glucose to some form of metabolism.

2. Both reactions are isomerizations of aldoses to ketoses.

3. Fructose bisphosphate aldolase interconverts fructose-1,6-bisphosphate with dihydroxyacetone phosphate and glyceraldehyde-3-phosphate. Class I aldolases form a Schiff base during catalysis. (See answer to problem 5, above.) The Schiff base forms at the keto carbon of either fructose-1,6-bisphosphate or dihydroxyacetone phosphate and $NaBH_4$ will reduce the Schiff base to a stable adduct. Once modified in this fashion, the enzyme is irreversibly inactivated. Treating the enzyme with either $NaBH_4$ alone or with $NaBH_4$ and glyceraldehyde-3-phosphate has no effect because a Schiff base is not formed.

4. Phosphoglycerate mutase moves the phosphate group from C-3 of 3-phosphoglycerate to C-2 to produce 2-phosphoglycerate. The yeast and rabbit muscle enzymes use 2,3-bisphosphoglycerate as a cofactor. The cofactor transfers a phosphate group to an active-site histidine to form a phosphohistidine intermediate. The phosphate group is then transferred to C-2 of 3-phosphoglycerate at the active site to form enzyme-bound 2,3-bisphosphoglycerate. The product, 2-phosphoglycerate, is formed when phosphate is subsequently transferred from C-3 to an enzyme histidine. The cofactor 2,3-bis-phosphoglycerate is also used as an allosteric inhibitor of hemoglobin. It is formed from 1,3-bisphosphoglycerate by bisphosphoglycerate mutase.

5. Hexokinase has a low K_m for glucose, approximately 0.1 mM, and is inhibited by glucose-6-phosphate. Thus, cells containing hexokinase, such as muscle, will continue to phosphorylate glucose even when blood-glucose levels are low. However, if glucose is not metabolized immediately and glucose-6-phosphate levels build up, hexokinase is inhibited. Glucokinase, found in the liver, has a much higher K_m for glucose, approximately 10 mM, insuring that liver cells phosphorylate glucose only when blood-glucose levels are high. In addition, glucokinase is not inhibited by glucose-6-phosphate. This allows liver cells to continue to accumulate glucose. In a later chapter we will see that excess glucose is stored as glycogen.

6. The structures of glycerol and glyceraldehyde are shown below.

$$\begin{array}{ccc}
 & & \begin{array}{c} H \quad O \\ \backslash \ \parallel \\ C \end{array} \\
CH_2OH & & | \\
HO-C-H & & H-C-OH \\
| & & | \\
CH_2OH & & CH_2OH \\
\text{glycerol} & & \text{glyceraldehyde}
\end{array}$$

Glycerol is metabolized by first being phosphorylated by glycerol kinase to *sn*-glycerol-3-phosphate, which is converted to dihydroxyacetone phosphate (DHAP) by glycerol phosphate dehydrogenase. DHAP is subsequently isomerized to glyceraldehyde. Since glycerol is converted into glyceraldehyde it is curious that anaerobic organisms can utilize one of these compounds but not the other. The problem lies with the enzyme glycerol phosphate dehydrogenase. This enzyme uses NAD^+ as coenzyme. Thus, glyceraldehyde production by this pathway results in NAD^+ reduction. A second NAD^+ is reduced when glyceraldehyde is metabolized via glycolysis. One NAD^+ is regenerated when pyruvate is converted to lactate or ethanol and CO_2 but since two were reduced, one NADH remains. In the absence of oxygen or some other suitable electron acceptor, NAD^+ levels drop as all of the coenzyme is trapped in the reduced form NADH.

7. Mannose is phosphorylated by hexokinase to produce mannose-6-phosphate, which is isomerized by phosphomannose isomerase to fructose-6-phosphate, a glycolytic intermediate.

Summary

Glycolysis - the degradation of glucose and other simple sugars - is a catabolic process carried out by nearly all cells. An anaerobic process, it provides precursor molecules for aerobic catabolic pathways, such as the critic acid cycle, and serves as an emergency energy source when oxygen is limiting. Glycolysis consists of two phases. The first series of five reactions breaks glucose down to two molecules of glyceraldehyde-3-phosphate. The second phase converts these two molecules of glyceraldehyde-3-P into two molecules of pyruvate. The overall pathway produces two molecules of ATP per glucose consumed. In aerobic organisms, including humans, pyruvate is oxidatively decarboxylated to produce a molecule of acetyl coenzyme A, which fuels the citric acid cycle (and subsequent electron transport). Under anaerobic conditions, the pyruvate formed in glycolysis can be reduced to lactate (in oxygen-starved muscles or in microorganisms) or ethanol (in yeast) in fermentation processes. During glycolysis, a portion of the metabolic energy of the glucose molecule is converted into ATP. The subsequent entry of pyruvate into the TCA cycle provides more energy in the form of ATP and reduced coenzymes.

The initial reaction of glycolysis involves phosphorylation of glucose at C-6, a reaction that consumes an ATP molecule in order to make more ATP later. Hexokinase is the primary catalyst for this reaction, but, when glucose levels rise in the liver, glucokinase becomes active and shares the responsibility for phosphorylation of glucose. The second reaction of glycolysis is the isomerization of glucose-6-phosphate to fructose-6-phosphate, carried out by phosphoglucoisomerase in a reaction involving the transient formation of an ene-diol intermediate. The new primary alcohol group created at C-1 on the carbon skeleton is the site of another phosphorylation in the next reaction, catalyzed by phosphofructokinase. Like the hexokinase/glucokinase reaction, this reaction also consumes a molecule of ATP. The phosphofructokinase reaction is the most elaborately regulated reaction in the glycolytic pathway. ATP is an allosteric inhibitor of phosphofructokinase, whereas AMP can reverse the inhibition due to ATP. Although ATP levels in cells rarely change by more than 10%, the action of adenylate kinase facilitates relatively large changes in AMP levels in response to hydrolysis of ATP. The phosphofructokinase reaction is also inhibited by citrate, providing a critical link between regulation of glycolysis and the citric acid cycle. Fructose-2,6-bisphosphate is a potent allosteric activator that increases the affinity of the enzyme for the substrate fructose-6-phosphate.

The 6-carbon substrate skeleton is cleaved to two 3-carbon units in the fructose bisphosphate aldolase reaction. The products are dihydroxyacetone phosphate and glyceraldehyde-3-phosphate. Two types of fructose bisphosphate aldolase enzymes exist in nature. Animal tissues produce a Class I aldolase that forms a covalent Schiff base adduct with substrate and is thus inhibited by $NaBH_4$. Bacteria and fungi produce a Class II aldolase that does not form Schiff base adducts with substrate, but which contains an active site Zn^{2+} ion which acts as an electrophile to polarize the carbonyl group of the substrate and stabilize the enolate intermediate in the aldolase reaction. The dihydroxyacetone phosphate produced by the aldolase reaction is converted to another molecule of glyceraldehyde-3-phosphate by the fifth reaction of the pathway, triose phosphate isomerase. Although these last two reactions are energetically unfavorable, the earlier priming reactions bring the equilibrium constant for the first five reactions of glycolysis close to 1.

The second phase of glycolysis begins with the oxidation of glyceraldehyde-3-phosphate to produce a high energy phosphate metabolite, 1,3-bisphosphoglycerate and a reduced coenzyme, NADH, in the glyceraldehyde-3-phosphate dehydrogenase reaction. The reaction mechanism involves nucleophilic attack by a cysteine-SH group at the active site on the carbonyl carbon of glyceraldehyde-3-phosphate to form a hemithioacetal intermediate, followed by hydride transfer to NAD^+ and nucleophilic attack by phosphate to displace the product from the enzyme. In the seventh step of the pathway, phosphoglycerate kinase, transfers a phosphoryl group from 1,3-bisphosphoglycerate to ADP to form an ATP, an example of substrate level phosphorylation. Since each glucose molecule sends two molecules of 1,3-bisphosphoglycerate into the phosphoglycerate kinase reaction, this step pays off the ATP debt created by the two priming reactions of glycolysis. The phosphoglycerate mutase reaction converts 3-phosphoglycerate to 2-phosphoglycerate and the enolase reaction then converts 2-phosphoglycerate to phosphoenolpyruvate, a high energy phosphate which can drive the synthesis of another ATP molecule in the pyruvate kinase reaction, the final reaction of the glycolytic pathway. Pyruvate kinase is subject to regulation and possesses allosteric sites for activation by AMP and fructose-2,6-bisphosphate and inhibition by ATP, acetyl-CoA and alanine. Phosphorylation of pyruvate kinase by cAMP-dependent protein kinase makes pyruvate kinase more sensitive to inhibition by ATP and alanine and raises the K_m for PEP. Other sugars, including fructose, mannose and galactose, can enter the glycolytic pathway if they can be converted by appropriate enzymes into glycolysis intermediates.

Chapter 19

The Tricarboxylic Acid Cycle

• •

Chapter Outline

❖ TCA cycle (tricarboxylic acid): Citric acid cycle: Krebs cycle
- Input: Acetyl units
- Output: Carbon dioxide, ATP (or GTP), reduced cofactors
- Related activities
  · Electron transport: Recycles cofactors
  · Oxidative phosphorylation: Transduces energy into ATP
❖ Pyruvate dehydrogenase complex: Major source of acetyl units from glucose
- Reaction
  · Pyruvate to acetyl CoA and carbon dioxide
  · NAD^+ reduced
- Enzyme complex: Three enzyme activities
  · Pyruvate dehydrogenase
  · Dihydrolipoyl transacetylase
  · Dihydrolipoyl dehydrogenase
- Coenzymes
  · Thiamine pyrophosphate
  · Coenzyme A
  · Lipoic acid
  · NAD^+
  · FAD: Protein bound
- Regulation
  · Acetyl-CoA, NADH inhibitors
  · AMP activates: GTP inhibits
  · Covalent modification (mammalian enzyme): Phosphoenzyme inhibited
    • Pyruvate dehydrogenase kinase: Activate by acetyl-CoA and NADH
    • Pyruvate dehydrogenase phosphatase: Calcium activated
❖ Citrate synthase
- Perkin condensation: Acetyl CoA + oxaloacetate to citrate
- Regulation
  · NADH: Allosteric inhibitor
  · Succinyl CoA: Allosteric inhibitor: Intracycle inhibitor
❖ Aconitase
- Citrate to aconitate to isocitrate: Dehydration-rehydration
- Iron-sulfur cluster
- Site of action of fluoroacetate: Trojan horse inhibitor: Requires conversion to fluorocitrate
❖ Isocitrate dehydrogenase
- Oxidative decarboxylation of isocitrate: Forms α-ketoglutarate
- NAD^+ reduced
- Regulation
  · NADH, ATP allosteric inhibitors

- • ADP allosteric activator
❖ α-Ketoglutarate dehydrogenase: Multienzyme complex: Similar to pyruvate dehydrogenase complex
  - ⋏ Oxidative decarboxylation of α-ketoglutarate: Produces succinyl-CoA
  - ⋏ NAD$^+$ reduced
  - ⋏ Regulation
    - • NADH inhibits
❖ Succinyl-CoA synthetase: Succinate thiokinase
  - ⋏ Substrate-level phosphorylation
    - • GTP in mammals: Nucleoside diphosphate kinase moves P to ATP
    - • ATP in plants and bacteria
❖ Succinate dehydrogenase
  - ⋏ FAD reduced
  - ⋏ Membrane-bound enzyme: Covalently bound FAD
  - ⋏ Electron transport: Electrons moved to coenzyme Q
❖ Fumarase: Hydration of fumarate: Produces malate
❖ Malate dehydrogenase
  - ⋏ NAD$^+$ reduced
  - ⋏ Oxaloacetate produced
❖ TCA intermediates connected to biosynthetic pathways
  - ⋏ α-Ketoglutarate: Precursor of Glu
    - • Glu: Precursor of
      - • Pro, Arg, Gln
  - ⋏ Succinyl CoA: Porphyrin ring synthesis
  - ⋏ Fumarate: Precursor of Asp
  - ⋏ Oxaloacetate: Precursor of
    - • Asp: Precursor of
      - • Pyrimidine nucleotides, Asn, Met, Lys, Thr, Ile
    - • PEP: Precursor of
      - • Phe, Tyr, Trp (plants and microorganism)
      - • 3-phosphoglycerate: Precursor of
        - ○ Ser, Gly, Cys
      - • Glucose
❖ Anaplerotic reactions: Produce TCA cycle intermediates
  - ⋏ Pyruvate carboxylase
  - ⋏ PEP carboxykinase
  - ⋏ Malic enzyme
❖ Glyoxylate cycle: Plants and bacteria: Uses acetyl units to produce oxaloacetate: Oxaloacetate to pyruvate: Pyruvate to glucose
  - ⋏ Citric acid cycle to isocitrate
  - ⋏ Isocitrate lyase produces glyoxylate and succinate
  - ⋏ Succinate to oxaloacetate via citric acid cycle
  - ⋏ Malate synthase: Glyoxylate + acetyl CoA to malate

Chapter Objectives

Citric Acid Cycle

The citric acid cycle is a central metabolic cycle to which catabolism of monosaccharides, amino acids, and fatty acids converge. We see that pyruvate, the end product of glycolysis, is introduced as an acetyl unit attached to coenzyme A catalyzed by pyruvate dehydrogenase. (A carbon is lost as CO_2.) The pyruvate dehydrogenase reaction is worth understanding, because we encounter almost identical biochemistry in α-ketoglutarate dehydrogenase catalysis of α-ketoglutarate to succinyl-CoA. Pyruvate is decomposed to CO_2 and a two-carbon acetyl group, which is attached to thiamine pyrophosphate. The acetyl group is transferred to coenzyme A via lipoic acid and in the process, lipoic acid is reduced. To reoxidize lipoic acid, electrons are transferred to protein-bound FAD and then to NAD$^+$.

The citric acid cycle begins with condensation of acetyl-CoA and oxaloacetate to produce citric acid, a tricarboxylic acid (see Figure 19.2). The cycle then converts citrate to oxaloacetate and two molecules of CO_2.

You should understand that the basic chemistry is oxidation. (Input is an acetyl group, outputs are carbon dioxide.) Oxidations are always accompanied by reductions and the reductions that take place include three steps at which NAD^+ is reduced to NADH: isocitrate to α-ketoglutarate conversion catalyzed by isocitrate dehydrogenase, α-ketoglutarate to succinyl-CoA conversion catalyzed by α-ketoglutarate dehydrogenase (similar in reaction mechanism to pyruvate dehydrogenase), and malate to oxaloacetate conversion catalyzed by malate dehydrogenase. The first two reactions are responsible for the two CO_2 molecules produced by the cycle. An additional reduction occurs during the conversion of succinate to fumarate catalyzed by succinate dehydrogenase. This enzyme plays two metabolic roles, one in the TCA cycle and the other as a component of the electron transport chain.

There are a number of important points to remember about the citric acid cycle that will help you with other aspects of biochemistry.

1. When amino acid metabolism is covered we will see that a few of the citric acid cycle intermediates derive from and can be converted into amino acids. Examples include pyruvic acid/alanine, α-ketoglutarate/glutamic acid, and oxaloacetate/aspartic acid.

2. The citric acid cycle has a substrate-level phosphorylation catalyzed by succinyl-CoA synthetase. Synthetases are enzymes that drive product formation with NTP hydrolysis. So, succinyl-CoA synthetase can produce succinyl-CoA using hydrolysis of either GTP or ATP depending on the source of the enzyme. (Citric acid synthase synthesizes citric acid but it does not utilize hydrolysis of ATP to drive the reaction.)

3. The steps from succinate to fumarate to malate to oxaloacetate are almost identical in chemistry to β-oxidation, the principal pathway of fatty acid catabolism.

Regulation of the Cycle

The citric acid cycle is regulated at pyruvate dehydrogenase by acetyl-CoA, CoA, NAD^+, NADH, and ATP; at citrate synthase by NADH, ATP, and succinyl CoA; at isocitrate dehydrogenase by NAD^+, ADP, and ATP; and, at α–ketoglutarate dehydrogenase by NADH, AMP, and succinyl-CoA. The cycle is sensitive to NAD^+ levels, a measure of electron transport activity, to adenine nucleotides, a measure of energy charge, and to immediate substrates like acetyl-CoA or succinyl-CoA.

Citric Acid Cycle Outputs and Inputs

The whole point of the cycle is to consume acetyl units for energy with CO_2 as a byproduct. However, other pathways can produce TCA cycle intermediates and TCA intermediates can feed into other pathways. For example, acetyl units can be derived from catabolism of fatty acids and certain amino acids, whereas citrate is used to move carbons out of the citric acid cycle to synthesize fatty acids. Amino acids and citric acid cycle intermediates are interconvertible. Reactions leading to citric cycle intermediates are known as anaplerotic reactions.

Glyoxylate Cycle

The glyoxylate cycle allows plants and bacteria to convert acetyl units into sugar carbons. Parts of the cycle are identical to the TCA cycle. However, the key differences are conversion of isocitrate to succinate and glyoxylate by isocitrate lyase, and production of malate from glyoxylate and acetyl-CoA by malate synthase. This allows for the production of four-carbon units from two-carbon acetyl units. Later we will study gluconeogenesis, an anabolic pathway leading to synthesis of glucose from certain three- or four-carbon intermediates.

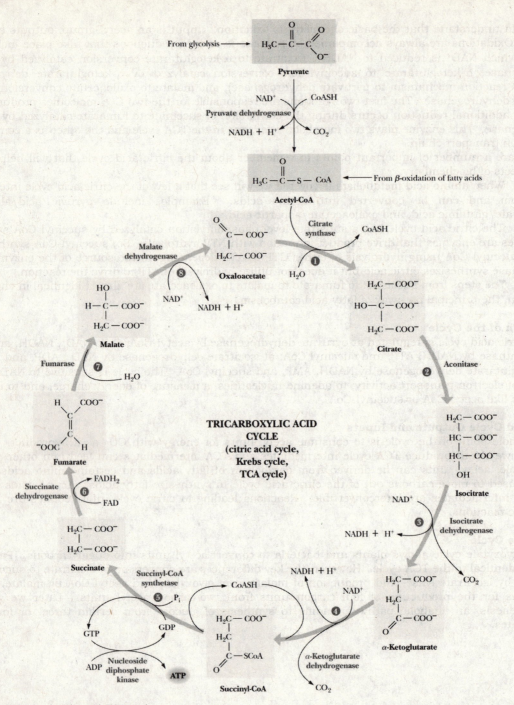

Figure 19.2 The tricarboxylic acid (TCA) cycle.

Problems and Solutions

1. Radiolabelling with ¹⁴C-glutamate
Describe the labeling pattern that would result from the introduction into the TCA cycle of glutamate labeled at the Cγ with ¹⁴C.

Answer: Glutamate may be converted to α-ketoglutarate either by transamination or by the action of glutamate dehydrogenase. In either case, glutamate with label at the Cγ will produce α-ketoglutarate labeled at Cγ as shown below.

$$
\begin{array}{ccc}
\text{COO}^- & & \text{COO}^- \\
^{14}\text{CH}_2 & & ^{14}\text{CH}_2 \\
| & \longrightarrow & | \\
\text{CH}_2 & & \text{CH}_2 \\
| & & | \\
+\text{H}_3\text{N}-\text{C}-\text{COO}^- & & \text{C}=\text{O} \\
| & & | \\
\text{H} & & \text{COO}^-
\end{array}
$$

The label is not lost in the first cycle and shows up in oxaloacetate at either of two positions.

$$
\begin{array}{c}
\text{COO}^- \\
^{14}\text{CH}_2 \\
| \\
^{14}\text{C}=\text{O} \\
| \\
\text{COO}^-
\end{array}
$$

No label is lost in the second round as well; however, in each subsequent cycle, 50% of the label is lost. This occurs because at the end of the second round, label is equally distributed among all the carbons of oxaloacetate. Since each cycle releases carbons from oxaloacetate, one-half of the label is lost. The remaining label is redistributed equally among all the carbons and again one-half is lost with subsequent cycles.

2. Assessing the effects of substrate concentrations on the TCA cycle
Describe the effect on the TCA cycle of (a) increasing the concentration of NAD⁺,(b) reducing the concentration of ATP, and (c) increasing the concentration of isocitrate.

Answer: The citric acid cycle is tightly coupled to electron transport by the reactions catalyzed by the following enzymes: isocitrate dehydrogenase, α-ketoglutarate dehydrogenase, succinate dehydrogenase and malate dehydrogenase. The *in vivo* reactions catalyzed by isocitrate dehydrogenase and α-ketoglutarate are far from equilibrium, with large negative ΔG' values (See Table 19.1), and are important control points for the citric acid cycle.

Isocitrate dehydrogenase is allosterically activated by ADP and NAD⁺ and inhibited by ATP and NADH. Therefore, this enzyme is sensitive to the NAD⁺:NADH and ADP:ATP ratios. An increase in NAD⁺ or a reduction in the concentration of ATP will stimulate this enzyme and the citric acid cycle. Further, increasing the concentration of isocitrate will increase the citric acid cycle. This occurs not because isocitrate is a regulator of the cycle but rather because it is a substrate of isocitrate dehydrogenase, an enzyme catalyzing a reaction far from equilibrium and therefore sensitive to changes in concentration of substrate.

α–Ketoglutarate dehydrogenase is inhibited by NADH and an increase in concentration of NAD⁺ resulting from a decrease in NADH will relieve this inhibition and stimulate the cycle. Finally, the activity of pyruvate dehydrogenase is sensitive to both NAD⁺ and ATP; high NAD⁺ and low ATP lead to stimulation of pyruvate dehydrogenase leading to an increased rate of production of acetyl-CoA if pyruvate is available.

3. Assessing the effect of active-site phosphorylation on enzyme activity
(Integrates with Chapter 15.) The serine residue of isocitrate dehydrogenase that is phosphorylated by protein kinase lies within the active site of the enzyme. This situation contrasts with most other examples of covalent modification by protein phosphorylation, where the phosphorylation occurs at a site remote from the active site. What direct effect do you think such active-site phosphorylation might have on the catalytic activity of isocitrate dehydrogenase? (See Barford, D., 1991. Molecular mechanisms for the control of enzymic activity by protein phosphorylation. Biochimica et Biophysica Acta 1133:55-62.).

Answer: Isocitrate dehydrogenase catalyzes the oxidative decarboxylation of isocitrate to α-ketoglutarate with reduction of NAD⁺ to NADH. Phosphorylation of Ser[113] leads to enzyme inactivation because isocitrate no longer binds to the enzyme. Ser[113] normally forms a hydrogen bond with the γ-carboxyl group of isocitrate. Phosphorylated Ser[113] interferes with isocitrate binding by electrostatic repulsion of the carboxyl group and by steric hindrance.

4. Understanding the mechanism of the α-ketoglutarate dehydrogenase reaction
The first step of the α-ketoglutarate dehydrogenase reaction involves decarboxylation of the substrate and leaves a covalent TPP intermediate. Write a reasonable mechanism for this reaction.

Answer:

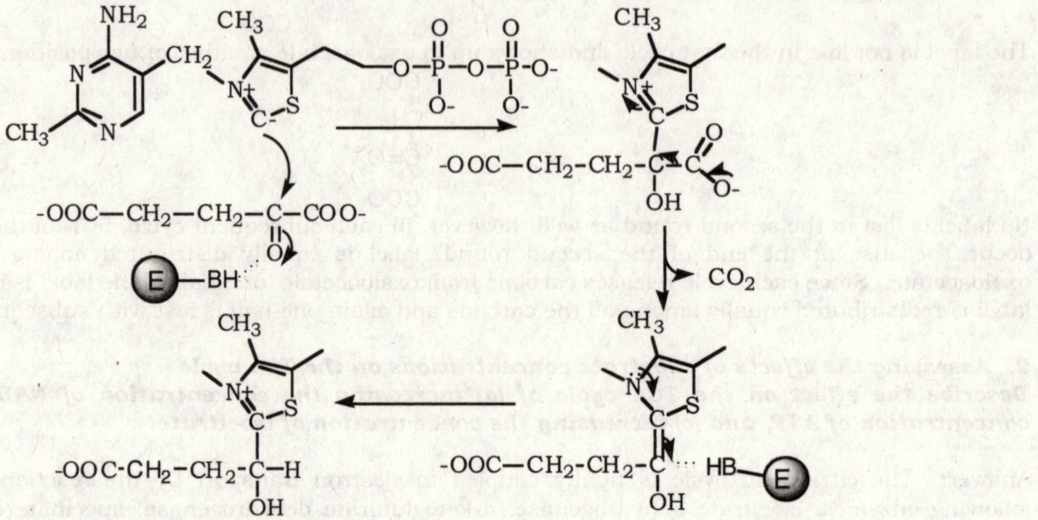

5. Understanding the action of fluoroacetate on the TCA cycle
In a tissue where the TCA cycle has been inhibited by fluoroacetate, what difference in the concentration of each TCA cycle metabolite would you expect, compared with a normal, uninhibited tissue?

Answer: Fluoroacetate is an example of a suicide inhibitor or a Trojan horse inhibitor. When tested *in vitro* against isolated enzymes of the TCA cycle, fluoroacetate has no effect. However, when used on isolated mitochondria or *in vivo*, fluoroacetate is a potent inhibitor of aconitase. Fluoroacetate is converted to fluoroacetyl-coenzyme A by acetyl-CoA synthetase, which is condensed with oxaloacetate by citrate synthase to produce 2R,3S-fluorocitrate. This isomer specifically inhibits aconitase (see problem 19). With aconitase inhibited, we might expect all TCA cycle intermediates to decrease in concentration. Citrate concentrations, however, will increase in concentration but isocitrate and all subsequent metabolites will decrease in concentration. Fluoroacetate poisoning leads to a host of physiological changes including alterations in carbohydrate metabolism. Citrate accumulation leads to inhibition of phosphofructokinase. Low levels of oxaloacetate prevent cells from consuming acetyl units and these eventually lead to ketosis and lactidosis.

6. Designing an assay for succinate dehydrogenase
Based on the description of the physical properties of FAD and FADH₂, suggest a method for the measurement of the enzyme activity of succinate dehydrogenase.

Answer: Succinate dehydrogenase catalyzes the oxidation of succinate to fumarate. This enzyme serves a dual purpose: It is a TCA cycle enzyme and it is a component of the electron transport system. Succinate dehydrogenase is an inner mitochondrial membrane-bound enzyme containing a covalently bound FAD that cycles between oxidized and reduced ($FADH_2$) forms as it passes electrons from succinate to coenzyme Q. A convenient way to measure the activity of this enzyme is to take advantage of the fact that the reduced and oxidized forms of FAD exhibit different absorption spectra. In particular, FAD is yellow in color and has a maximal absorbance at 450 nm whereas $FADH_2$ is colorless. By monitoring the absorbance at 450 nm, the activity of succinate dehydrogenase can be measured.

7. Understanding the oxidative processes in the TCA cycle
Starting with citrate, isocitrate, α-ketoglutarate, and succinate, state which of the individual carbons of the molecule undergo oxidation in the next step of the TCA cycle. Which of the molecules undergo a net oxidation?

Answer:

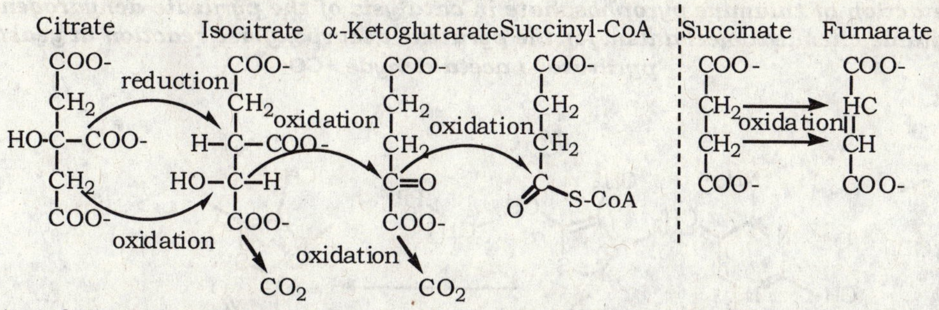

Citrate   Isocitrate   α-Ketoglutarate   Succinyl-CoA   Succinate   Fumarate

$$
\begin{array}{ccccc}
\text{COO}^- & \text{COO}^- & \text{COO}^- & \text{COO}^- & \text{COO}^- \quad \text{COO}^- \\
| & | & | & | & | \quad\quad | \\
\text{CH}_2 & \text{CH}_2 & \text{CH}_2 & \text{CH}_2 & \text{CH}_2 \quad\quad \text{HC} \\
| & | & | & | & | \quad\quad \| \\
\text{HO-C-COO}^- & \text{H-C-COO}^- & \text{CH}_2 & \text{CH}_2 & \text{CH}_2 \quad\quad \text{CH} \\
| & | & | & | & | \quad\quad | \\
\text{CH}_2 & \text{HO-C-H} & \text{C=O} & \text{C} & \text{COO}^- \quad \text{COO}^- \\
| & | & | & \diagdown & \\
\text{COO}^- & \text{COO}^- & \text{COO}^- & \text{O} \quad \text{S-CoA} &
\end{array}
$$

reduction ... oxidation ... oxidation ... oxidation ... oxidation

CO_2 CO_2

In the conversion of citrate to isocitrate, one carbon is reduced and one carbon is oxidized. Hence, there is no net change in the oxidation state in going from citrate to isocitrate. Production of α-ketoglutarate from isocitrate is the result of an oxidation. The enzyme that catalyzes this reaction, isocitrate dehydrogenase, uses NAD^+ as a cofactor, which is reduced to NADH during the reaction. Likewise, conversion of α–ketoglutarate to succinate is an oxidation. In this case, the conversion is a two-step process. In the first step, α–ketoglutarate is converted to succinyl-CoA by α–ketoglutarate dehydrogenase. This is an oxidation (the carbon in a thiol ester bond is at the same oxidation state as a carbon in an ester bond) and so it must be accompanied by a reduction, in this case NAD^+ is reduced to NADH. Succinyl-CoA to succinate is a hydrolysis reaction that is coupled to production of a high-energy phosphate bond (as GTP or ATP). Finally, succinate to fumarate is catalyzed by succinate dehydrogenase. This enzyme has an enzyme-bound FAD coenzyme that is reduced when succinate is oxidized to fumarate.

8. Designing substrate analog inhibitors for the TCA cycle
In addition to fluoroacetate, consider whether other analogs of TCA cycle metabolites or intermediates might be introduced to inhibit other, specific reactions of the cycle. Explain your reasoning.

Answer: Fluoroacetate blocks the citric acid cycle by reacting with oxaloacetate to form fluorocitrate, which is a powerful inhibitor of aconitase. In the aconitase reaction, the elements of water are abstracted from citrate and added back to form isocitrate. Addition of water also occurs in the conversion of fumarate to malate catalyzed by fumarase and we might expect fluorinated derivatives of fumarate or malate to inhibit this enzyme. Malonate, a succinate analog, is a competitive inhibitor of fumarase. 3-Nitro-2-S-hydroxylpropionate, when deprotonated, is a transition state analog of fumarase. These compounds are shown below.

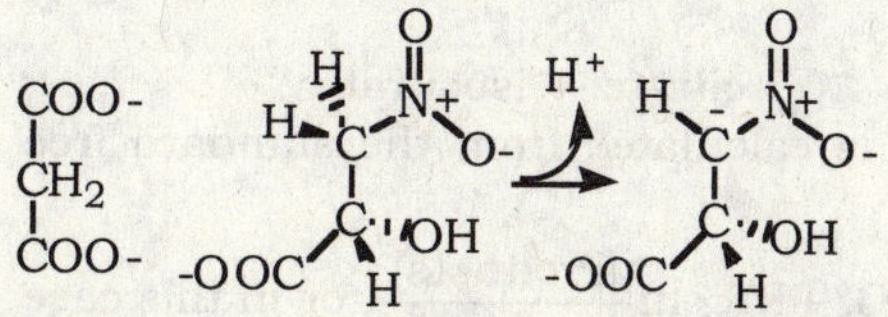

Malonate   S-Nitro-2-S-hydroxypropionate

9. Understanding the mechanism of the pyruvate dehydrogenase reaction
Based on the action of thiamine pyrophosphate in catalysis of the pyruvate dehydrogenase reaction, suggest a suitable chemical mechanism for the pyruvate decarboxylase reaction in yeast:

$$pyruvate \rightarrow acetaldehyde + CO_2$$

Answer

10. Understanding the energetics of the aconitase reaction
(Integrates with Chapter 3.) Aconitase catalyzes the citric acid cycle reaction:

$$citrate \leftrightarrows isocitrate$$

The standard free energy change, $\Delta G°$, for this reaction is +6.7 kJ/mol. However, the observed free energy change (ΔG) for this reaction in pig heart mitochondria is +0.8 kJ/mol. What is the ratio of [isocitrate]/[citrate] in these mitochondria? If [isocitrate] = 0.03 mM, what is [citrate]?

Answer: The reaction being examined is

$$citrate \rightarrow isocitrate$$

The observed free energy change (ΔG) is calculated from the standard free energy change ($\Delta G°$) using the following equation:

$$\Delta G = \Delta G°' + RT \times \ln \frac{[\text{Products}]}{[\text{Reactants}]}, \text{ or in this case}$$

$$\Delta G = \Delta G°' + RT \times \ln \frac{[\text{isocitrate}]}{[\text{citrate}]}$$

This equation may be rearranged to give:

$$\frac{[\text{isocitrate}]}{[\text{citrate}]} = e^{\frac{\Delta G - \Delta G°'}{RT}}$$

Substituting the appropriate numerical values, with the product RT = 2.48 at 25°C, we have:

$$\frac{[\text{isocitrate}]}{[\text{citrate}]} = e^{\frac{+0.8\frac{kJ}{mol} - 6.7\frac{kJ}{mol}}{2.48\frac{kJ}{mol}}} = e^{\frac{-5.9}{2.48}} = 0.093$$

The concentration of citrate is calculated as follows:

$$\frac{[\text{isocitrate}]}{[\text{citrate}]} = 0.093, \text{ or}$$

$$[\text{citrate}] = \frac{[\text{isocitrate}]}{0.093} = \frac{0.03 \text{ mM}}{0.093} = 0.324 \text{ mM}$$

11. Radiolabelling with $^{14}CO_2$ in the pyruvate carboxylase reaction
Describe the labeling pattern that would result if $^{14}CO_2$ were incorporated into the TCA cycle via the pyruvate carboxylase reaction.

Answer: The pyruvate carboxylase reaction converts pyruvate and CO_2 to oxaloacetate, with label appearing on the β-carboxyl group of oxaloacetate. When this is used to synthesize citrate, the labeled carbon shows up as a carboxyl group up on the middle carbon, which is eliminated by the α-ketoglutarate dehydrogenase step as shown below.

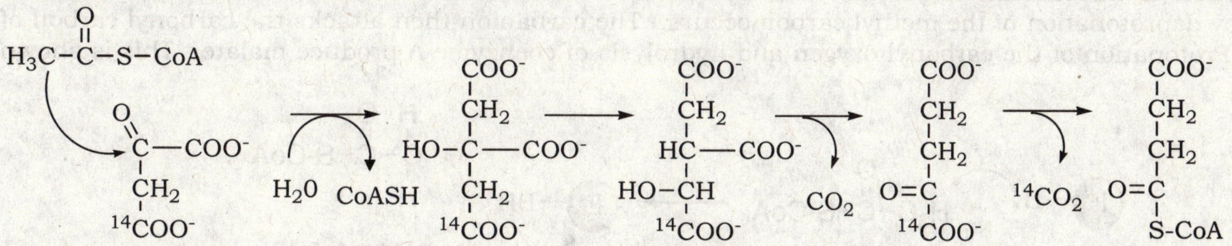

$$H_3C-\overset{\displaystyle O}{\overset{\|}{C}}-S-CoA$$

$$\overset{\displaystyle O}{\overset{\|}{C}}-COO^- \quad \xrightarrow[CoASH]{H_2O} \quad HO-C-COO^- \quad \longrightarrow \quad HC-COO^- \quad \xrightarrow{CO_2} \quad O=C \quad \xrightarrow{^{14}CO_2} \quad O=C$$

(Scheme: oxaloacetate with $^{14}COO^-$ on the β-carboxyl → citrate $HO-C-COO^-$ with CH_2/CH_2 and $^{14}COO^-$ → isocitrate $HC-COO^-$, $HO-CH$, $^{14}COO^-$ → α-ketoglutarate $O=C$, $^{14}COO^-$ losing CO_2 → succinyl-CoA $O=C-S-CoA$)

12. Consequences of radiolabelling with $^{14}CO_2$ in the reductive TCA cycle
Describe the labeling pattern that would result if the reductive, reverse TCA cycle (see A Deeper Look on page 631) operated with $^{14}CO_2$.

Answer: In the reverse TCA cycle, CO_2 is incorporated in going from succinyl-CoA to α-ketoglutarate and again from isocitrate to citrate. In the first cycle, oxaloacetate would be labeled at both of its carboxyl carbons as shown below.

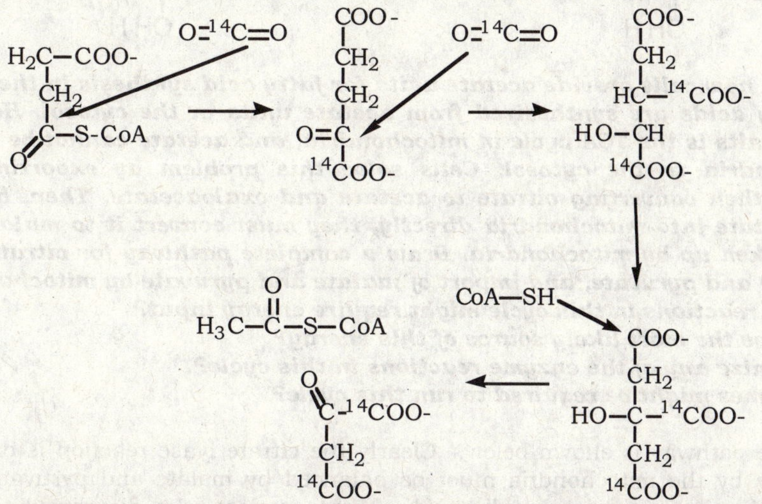

(Scheme showing succinyl-CoA H_2C-COO^-, CH_2, $C-S-CoA$ plus $O=^{14}C=O$ → α-ketoglutarate COO^-, CH_2, CH_2, $O=C$, $^{14}COO^-$ plus $O=^{14}C=O$ → isocitrate COO^-, CH_2, $HC-^{14}COO^-$, $HO-CH$, $^{14}COO^-$ → citrate COO^-, CH_2, $HO-C-^{14}COO^-$, CH_2, $^{14}COO^-$ → via $CoA-SH$ to acetyl-CoA $H_3C-C(=O)-S-CoA$ and oxaloacetate $C-^{14}COO^-$, CH_2, $^{14}COO^-$)

In the second pass through the cycle one of the carbons would be eliminated as the carboxyl group of acetyl-CoA. The remaining labeled carbon would eventually be eliminated as a methyl carbon of acetyl-CoA. Using the double-labeled oxaloacetate and converting it to succinyl-CoA, it is easy to see that a labeled carboxyl group is lost as acetyl-CoA. The remaining labeled carbon is one of the methylene carbons of citrate, which are lost as the methyl carbon of acetyl-CoA.

13. Consequences of radiolabelling with $^{14}CH_3$–acetyl-CoA in the glyoxylate cycle
Describe the labeling pattern that would result in the glyoxylate cycle if a plant were fed acetyl-CoA labeled at the $-CH_3$ carbon.

Answer: The glyoxylate cycle begins in the glyoxysome with synthesis of citrate and conversion to isocitrate. Next, isocitrate is converted to succinate and glyoxylate. The methyl carbon of acetyl-CoA ends up as a methylene carbon of succinate. This would be transported to the mitochondria where it would be converted to oxaloacetate. Since succinate has no chiral carbons, the labeled carbon appears as either the carbonyl carbon or the methylene carbon of oxaloacetate. This, however, occurs in the TCA cycle in the mitochondria. So, the labeling pattern will be as shown in Figure 19.15b.

14. Understanding the mechanism of the malate synthase reaction
The malate synthase reaction, which produces malate from acetyl-CoA and glyoxylate in the glyoxylate pathway, involves chemistry similar to the citrate synthase reaction. Write a mechanism for the malate synthase reaction and explain the role of CoA in this reaction.

Answer: In the citrate synthase reaction, the methyl carbon of an acetyl group attached to coenzyme A is activated by deprotonation using a strong base on the enzyme. The carbanion that is produced attacks the carbonyl carbon of oxaloacetate to produce citrate. In the malate synthase reaction the same activation of acetyl CoA by deprotonation of the methyl carbon occurs. The carbanion then attacks the carbonyl carbon of glyoxylate. Protonation of the carbonyl oxygen and hydrolysis of coenzyme A produce malate. This is shown below.

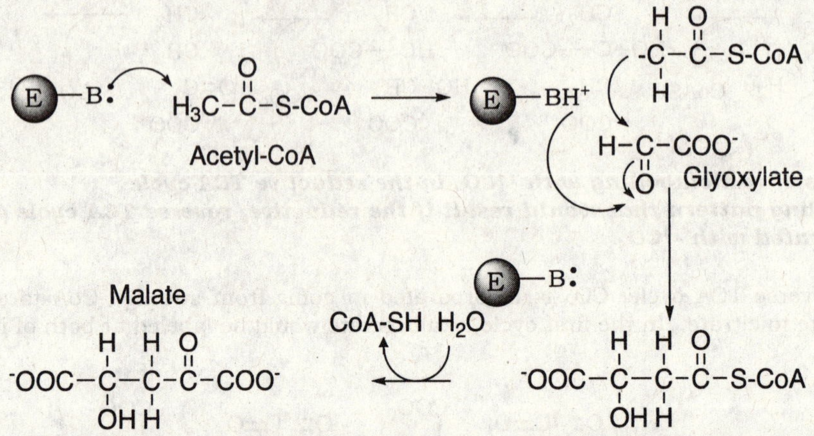

15. Understanding how cells provide acetate units for fatty acid synthesis in the cytosol
In most cells, fatty acids are synthesized from acetate units in the cytosol. However, the primary source of acetate units is the TCA cycle in mitochondria, and acetate cannot be transported directly from the mitochondria to the cytosol. Cells solve this problem by exporting citrate from the mitochondria and then converting citrate to acetate and oxaloacetate. Then, because cells cannot transport oxaloacetate into mitochondria directly, they must convert it to malate or pyruvate, both of which can be taken up by mitochondria. Draw a complete pathway for citrate export, conversion of citrate to malate and pyruvate, and import of malate and pyruvate by mitochondria.
  a. Which of the reactions in this cycle might require energy input?
  b. What would be the most likely source of this energy?
  c. Do you recognize any of the enzyme reactions in this cycle?
  d. What coenzymes might be required to run this cycle?

Answer: The uptake pathway is shown below. Clearly the citrate lyase reaction is driven by hydrolysis of ATP. Citrate release by the mitochondria must be balanced by malate and pyruvate uptake to maintain charge neutrality. (Citrate is a tricarboxylic acid whereas malate is a dicarboxylic acid and pyruvate a monocarboxylic acid.) Conversion of oxaloacetate to malate to pyruvate in the cytosol moves electrons from NADH to NAPDH and the later is used in anabolic pathways. Malate to oxaloacetate to citrate is part of the citric acid cycle and pyruvate dehydrogenase feeds two-carbon acetyl units into the citric acid cycle. Citrate lyase in the cytosol provides acetyl units for fatty acid biosynthesis, which also requires reducing equivalents in the form of NADPH. NADPH is supplied by malic enzyme from electrons originating on NADH. Clearly, NAD$^+$ plays a key role in this process

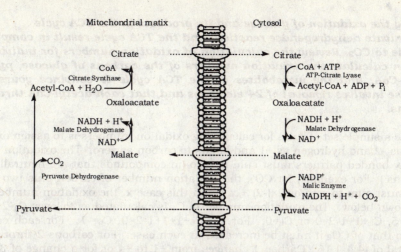

16. Assessing the equilibrium concentrations in the malate dehydrogenase reaction
A typical intramitochondrial concentration of malate is 0.22 mM. If the ratio of NAD+ to NADH in mitochondria is 20, and if the malate dehydrogenase reaction is at equilibrium, calculate the concentration of oxaloacetate in the mitochondrion at 20°C. A typical mitochondrion can be thought of as a cylinder 1 μm in diameter and 2 μm in length. Calculate the number of molecules of oxaloacetate in a mitochondrion. In analogy with pH (the negative logarithm of [H+]), what is pOAA?

Answer: Malate dehydrogenase interconverts malate and oxaloacetate in the following reaction:

$$\text{Malate} + \text{NAD}^+ \rightleftharpoons \text{Oxaloacetate} + \text{NADH} + \text{H}^+$$

In Table 19.1 we are given $\Delta G^{\circ\prime}$ for malate dehydrogenase as 29.7 kJ/mol and we are informed that the reaction is at equilibrium.

$$\Delta G = \Delta G^{\circ\prime} + RT\ln\frac{[\text{oxaloacetate}][\text{NADH}]}{[\text{malate}][\text{NAD}^+]}$$

At equilibrium, $\Delta G = 0$,

$$\Delta G^{\circ\prime} = -RT\ln\frac{[\text{oxaloacetate}][\text{NADH}]}{[\text{malate}][\text{NAD}^+]}$$

$$\frac{[\text{oxaloacetate}][\text{NADH}]}{[\text{malate}][\text{NAD}^+]} = e^{\frac{-\Delta G^{\circ\prime}}{RT}}$$

$$[\text{oxaloacetate}] = [\text{malate}]\frac{[\text{NAD}^+]}{[\text{NADH}]}e^{\frac{-\Delta G^{\circ\prime}}{RT}}$$

$$[\text{oxaloacetate}] = 0.22 \times 10^{-3} \times 20 \times e^{\frac{-29,700}{8.314 \times (273+25)}}$$

$$[\text{oxaloacetate}] = 2.73 \times 10^{-8}\text{M or } 27.3 \text{ nM or } 0.027 \text{ μM}$$

The volume of a mitochondrion is calculated as follows:

$$V = \pi\left(\frac{d}{2}\right)^2 \times \ell$$

$$V = 3.14\left(\frac{1 \times 10^{-6}}{2}\right)^2 \times 2 \times 10^{-6}$$

$$V = 1.57 \times 10^{-18}\text{m}^3 \times \left(\frac{100 \text{ cm}}{1 \text{ m}}\right)^3 \times \left(\frac{1 \text{ L}}{1000 \text{ cm}^3}\right)$$

$$V = 1.57 \times 10^{-15}\text{L}$$

The p[OAA] is the negative logarithm of the concentration of oxaloacetate (base 10), which is 7.56. Finally, the number of molecules is the concentration times volume times Avogadro's number.

$$\text{Number of oxaloacetate molecules} = 2.73 \times 10^{-8}\frac{\text{mol}}{\text{L}} \times 1.57 \times 10^{-15}\text{L} \times 6.02 \times 10^{23}\frac{\text{molecules}}{\text{mol}}$$

Number of oxaloacetate molecules = 25.8 ≈ 25 or 26 molecules

17. Understanding the oxidation of glucose and its products in the TCA cycle
Glycolysis, the pyruvate dehydrogenase reaction, and the TCA cycle result in complete oxidation of a molecule of glucose to CO_2. Review the calculation of oxidation numbers for individual atoms in any molecule, and then calculate the oxidation numbers of the carbons of glucose, pyruvate, the acetyl carbons of acetyl-CoA, and the metabolites of the TCA cycle to convince yourself that complete oxidation of glucose involves removal of 24 electrons and that each acetyl-CoA through the TCA cycle gives up 8 electrons.

Answer: In this case a simple set of rules for calculating oxidation numbers is to assign oxidation numbers to bonds with oxygen as -2 and hydrogen as +1 and bonds to carbon as zero. The oxidation number of a carbon and all its covalently bonded partners must sum to zero (in compounds that are neutrally charged and have no lone pair electrons.) For example, for CO_2, the oxidation number of carbon plus two times the oxidation number of oxygen must sum to zero or $2\times(-2) + x = 0$..In this case x, the oxidation number of carbon is +4 in CO_2. Glucose is shown below. The oxidation number of C_1 is calculated as follows: $C_1-2+1=0$ or $C_1=+1$. For carbons 2 through 5, $C_n+1+1-2=0$ or $C_n=0$. For C_6, $C_6+1+1+1-2=0$, $C_6=-1$. For each carbon to change its oxidation number to that of CO_2 it must be increased in each case. For carbons 2 through 5 the increase is from 0 to 4 for a total of $(4_5 4=)$ 16. Carbon 1 changes from +1 to +4 or for a change of 3. Carbon 6 changes from -1 to +4 for a change of 5. The sum is $(16+3+5 =)$ 24.

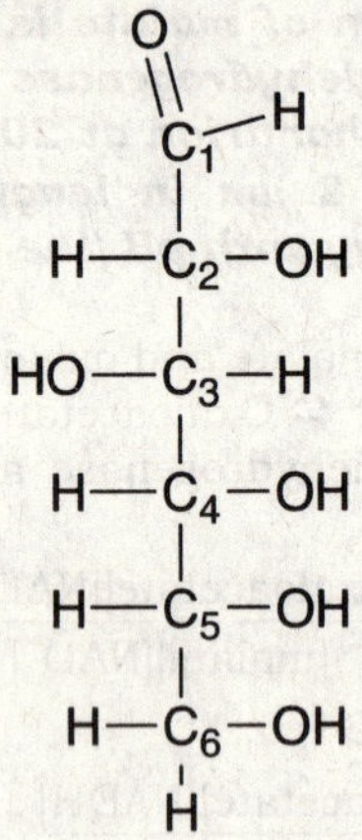

For the acetyl group in acetyl CoA, the methyl carbon's oxidation number is -3 and changing it to the oxidation number of carbon in carbon dioxide (+4) is a change of +7. The carbonyl carbon's oxidation number is +4 (-2 for oxygen and -2 for the sulfur in coenzyme A). However, in order to metabolize the acetyl group, the coenzyme adduct must be removed by hydrolysis. This will convert the carbonyl carbon to a carboxyl carbon with an oxidation number of +3 (-2 for two oxygens and +1 for a proton). Converting it to carbon dioxide is a change of +1 in oxidation number.

18. A simple way to understand and remember the reactions of the TCA cycle
Recalling all reactions of the TCA cycle can be a challenging proposition. One way to remember these is to begin with the simplest molecule— succinate, which is a symmetric four-carbon molecule. Begin with succinate, and draw the eight reactions of the TCA cycle. Remember that succinate → oxaloacetate is accomplished by a special trio of reactions: oxidation of a single bond to a double bond, hydration across the double bond, and oxidation of an alcohol to a ketone. From there, a molecule of acetyl-CoA is added. If you remember the special function of acetyl-CoA (see A Deeper Look, page 616), this is an easy reaction to draw. From there, you need only isomerize, carry out the two oxidative decarboxylations, and remove the CoA molecule to return to succinate.

Answer: Succinate (fumarate, malate, oxaloacetate) is a dicarboxylic acid and since it is a symmetric, four-carbon compound it is easy to remember its structure. The steps from succinate to fumarate to malate to oxaloacetate involve production of a trans double bond, hydration of the double bond and subsequent reduction of the hydroxyl group to a keto group. The trans double bond in fumarate is produced by removal of two hydrogens. Malate is formed by addition of hydrogen to one carbon and a hydroxyl group to the other. Finally, oxidation of malate's hydroxyl group produces oxaloacetate's keto group. Citrate (and isocitrate) is a tricarboxylic acid (TCA cycle, citric acid cycle) formed from oxaloacetate (dicarboxylic acid) and acetyl coenzyme A. The third carboxyl group in citrate must come from the carbonyl carbon in acetyl coenzyme A and thus the methyl carbon of the acetyl group of acetyl coenzyme A must be attached to oxaloacetate. The most likely place for this to occur is on the keto carbon of oxaloacetate. Thus citrate's central carbon has a hydroxyl group and a carboxyl group. Citrate is otherwise a symmetric molecule. Conversion of citrate to isocitrate involves rearranging the elements of water and so the hydroxyl group is moved to an adjacent carbon. We are almost through the cycle and to complete it we need to figure out the structure of the remaining two compounds succinyl-CoA and α-ketoglutarate. If you understand succinate, it is not difficult to understand succinyl-CoA. Succinyl-CoA is succinate in thioester linkage with coenzyme A. Coenzyme A derivatives are produced by rather energetic reactions and the one that produced succinyl-CoA was driven release of CO_2 that derived from a carboxyl group on α-ketoglutarate. So, from succinyl-CoA back to α-ketoglutarate the coenzyme A adduct is replaced by a carboxyl group to produce the 5-carbon dicarboxylic acid intermediate. The gap we need to fill in is between α-ketoglutarate and isocitrate, between a dicarboxylic acid and a tricarboxylic acid. A bit of math will indicate that we need a carboxyl group that must have been attached to isocitrate's central carbon.

19. Understanding the mechanism-based inactivation of aconitase by fluoroacetate
Aconitase is rapidly inactivated by 2R,3R-fluorocitrate, which is produced from fluoroacetate in the citrate synthase reaction. Interestingly, inactivation by fluorocitrate is accompanied by stoichiometric release of fluoride ion (i.e., one F-ion is lost per aconitase active site). This observation is consistent with "mechanism-based inactivation" of aconitase by fluorocitrate. Suggest a mechanism for this inactivation, based on formation of 4-hydroxy-trans-aconitate, which remains tightly bound at the active site. To assess your answer, consult this reference: Lauble, H., Kennedy, M., et al., 1996. The reaction of fluorocitrate with aconitase and the crystal structure of the enzyme-inhibitor complex. Proceedings of the National Academy of Sciences 93:13699–13703.

Answer: Aconitase interconverts citrate and isocitrate and an intermediate in the reaction is aconitate. It does this by abstracting water from citrate (or isocitrate) to produce aconitase, whose double bond is then rehydrated. (Aconitase hydrates the carbon-carbon double bond in aconitate and depending on the orientation of aconitate in the active site either citrate or isocitrate is produced.) We are told that fluorocitrate is converted to 4-hydroxy-trans-aconitate. Clearly the hydroxyl group was moved and it seems reasonable to think that this occurred by dehydration and rehydration. This would move the hydroxyl group from C3 to C4 but it would eliminate the double bond. The product, however, still has a double bond. We are also told that a fluoride ion is also released. If addition of OH- occurs first and it is accompanied by release of F- this would account for movement of the hydroxyl group and loss of fluoride. Figure 4 of the Lauble et al. paper outlines the details. Water is removed from fluorocitrate, the double-bond intermediate flips in the active site and a hydroxide is added to carbon 4. But, this causes the double bond to move resulting in loss of fluoride and production of 4-hydroxy-trans-aconitate. This compound binds to the enzyme noncovalently but with very high affinity.

Preparing for the MCAT® Exam

20. Complete oxidation of a 16-carbon fatty acid can yield 129 molecules of ATP. Study Figure 19.2 and determine how many ATP molecules would be generated if a 16-carbon fatty acid were metabolized solely by the TCA cycle, in the form of 8 acetyl-CoA molecules.

Answer: If one enters the TCA cycle with acetyl-CoA, a high-energy phosphoanydride bond is generated at the succinyl-synthetase step as GTP, which can be used by nucleoside diphosphate kinase to phosphorylate ADP to ATP. Doing this with 8 acetyl-CoA molecules would generate 8 ATP. This, however, ignores completely the potential of $FADH_2$ and NADH, which, when reoxidized by the electron transport chain, lead to much more ATP.

21. Study Figure 19.18 and decide which of the following statements is false.
   a. Pyruvate dehydrogenase is inhibited by NADH.
   b. Pyruvate dehydrogenase is inhibited by ATP.
   c. Citrate synthase is inhibited by NADH.
   d. Succinyl-CoA activates citrate synthase.
   e. Acetyl-CoA activates pyruvate carboxylase.

Answer: Choice "d" is false. Succinyl-CoA inhibits citrate synthase. The citric acid cycle is largely regulated by local metabolic signals like TCA cycle intermediates, ATP and NADH. If succinyl-CoA is building up this must mean that the TCA cycle is not moving because of high ATP and/or high NADH. Succinyl-CoA inhibits citrate synthase thereby stopping production of more citric acid, which would lead to increase in TCA cycle intermediates.

Pyruvate dehydrogenase is inhibited by NADH and ATP. This enzyme, if active, would commit more carbon from pyruvate to the citric acid cycle (or fatty acid synthesis). That it is inhibited by NADH and ATP makes good metabolic sense. Finally, pyruvate carboxylase is an anaplerotic reaction that produces more oxaloacetate, a TCA cycle intermediate. When acetyl-CoA levels increase, again it makes good metabolic sense to increase oxaloacetate levels allowing for an increase in the TCA cycle, which should lead to decrease in acetyl-CoA.

Questions for Self Study

1. a. Complete the following reaction and name the enzyme complex that catalyzes it.
    Pyruvate + CoA + ___ ⋆ acetyl-CoA + __ + __ + CO_2.
    b. This enzyme complex uses five coenzymes. Name the coenzymes.

2. The citric acid cycle is also known as the Krebs cycle and the TCA (tricarboxylic acid) cycle. Identify the tricarboxylic acids in the cycle.

3. Fill in the blanks. The citric acid cycle converts two-carbon ___ units into ___. In the process the carbons are ____ and coenzymes are ____. The coenzymes serve as mobile electron carriers connecting the citric acid cycle to ___.

4. α-Ketoglutarate can be produced by oxidative decarboxylation of a citric acid cycle intermediate or by deamination of an amino acid. Identify these two sources of α-ketoglutarate.

5. The α-ketoglutarate dehydrogenase complex used five coenzymes during catalysis. One of them, lipoic acid, a carboxylic acid, is transiently covalently modified during catalysis. In particular a disulfide bond is broken and reformed. Briefly outline this process.

6. In an earlier chapter we learned that thiol ester bonds are high-energy bonds. In the citric acid cycle there is one substrate-level phosphorylation (of GDP) driven by cleavage of a thiol ester bond. Identify the reaction and the enzyme. (Hint: The enzyme is named for the reverse reaction.)

7. Succinate dehydrogenase converts succinate to fumarate. This enzyme is a member of which two important metabolic sequences?

8. The citric acid cycle results in net oxidation of carbons. The production of succinate leaves a four-carbon unit in the cycle that is converted to oxaloacetate. Order the compounds shown below from least oxidized to most oxidized and identify them.

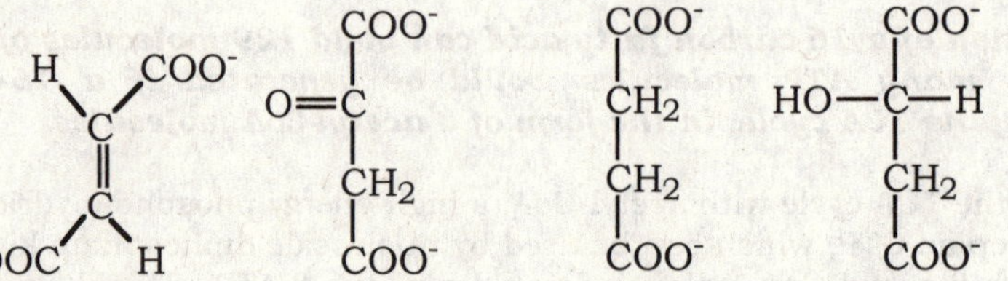

9. What is an anaplerotic reaction?

10. The citric acid cycle is regulated by metabolic signals at pyruvate dehydrogenase leading into the cycle and at three places in the cycle, citrate synthase, isocitrate dehydrogenase, and α-ketoglutarate dehydrogenase. List the metabolic signals and state how they affect the cycle.

11. Fill in the blanks. The ___ cycle is a metabolic sequence active in plants, particularly seedlings, in which acetyl units are converted to ____. In this cycle citrate is produced from oxaloacetate and acetyl-CoA and citrate is converted to isocitrate. These two steps are identical to two steps in the ___ cycle. However, the fate of isocitrate is different in the two cycles. In plants isocitrate is converted to succinate and the two-carbon compound ____. Succinate is subsequently metabolized to oxaloacetate. The two-carbon compound and a second acetyl group from acetyl-CoA are condensed to form a four-carbon compound that is metabolized to ___. Thus, the inputs to the cycle are ___, which are converted into a four-carbon compound.

Answers

1. a. Pyruvate + CoA + NAD^+ ⋆ acetyl-CoA + NADH + H^+ + CO_2; pyruvate dehydrogenase;
   b. Thiamine pyrophosphate, coenzyme A, lipoic acid, NAD^+, and FAD.

2. Citrate and isocitrate.

3. Acetyl; carbon dioxide; oxidized; reduced; electron transport chain.

4. Isocitrate and glutamic acid.

5. During catalysis the disulfide bond is broken by transfer of an acetyl unit to lipoic acid in thiol ester linkage. The acetyl unit is then transferred to CoA leaving the disulfide bond as two sulfhydryl groups. An oxidation must occur for the disulfide to be reformed and this occurs at the expense of FAD reduction. (Subsequently, electrons are passed to NAD^+.)

6. The reaction is conversion of succinyl CoA to succinate and CoA with phosphorylation of GDP to GTP. The enzyme is succinyl-CoA synthetase. (Why is it a synthetase and not a synthase?)

7. Citric acid cycle and electron transport.

8. The compounds are succinate, fumarate, malate, and oxaloacetate. Fumarate and malate (boxed) are at the same oxidation state.

```
COO⁻          ┌─────────────────────────────┐      COO⁻
|             │     H      COO⁻    COO⁻      │      |
CH₂           │      \\    /       |         │      O=C
|             │       C           HO—C—H     │      |
CH₂           │       ‖            |         │      CH₂
|             │       C           CH₂        │      |
COO⁻          │  ⁻OOC/   \\H        COO⁻      │      COO⁻
              └─────────────────────────────┘
```

9. It is a reaction that replenishes citric acid cycle intermediates.

10. NADH inhibits; NAD^+ stimulates; ATP inhibits; ADP and AMP stimulate; Succinyl-CoA inhibits; acetyl-CoA inhibits, CoA stimulates.

11. Glyoxylate; oxaloacetate; TCA (or citric acid); glyoxylate; malate; two acetyl units.

Additional Problems

1a. Identify the following compounds.

```
CH₃          COO⁻         COO⁻
|            |            |
C=O          CH₂          CH₂
|            |            |
COO⁻         C=O          CH₂
             |            |
             COO⁻         C=O
                          |
                          COO⁻
```

  b. If a Schiff base were produced between a side chain lysine and each of these compounds and then subsequently displaced by ammonia, to what would each of these compounds be converted?

2. The glyoxylate cycle is divided between specialized organelles known as glyoxysomes and the mitochondria. Describe the interplay between these two organelles that is necessary for the glyoxylate cycle to function.

3. The four steps in the citric acid cycle from succinate to oxaloacetate are similar in chemistry to the fatty acid catabolic pathway known as β-oxidation. In this pathway, a carbon in a fatty acid chain two positions

away from the carboxylic acid group (the first position is α, next β, the last ω or omega) is oxidized to a keto group. Show that this also occurs when succinate is converted to oxaloacetate.

4. In glycolysis, phosphofructokinase is allosterically inhibited by citrate. What is a high concentration of citrate signaling and why is it important to turn off glycolysis under these conditions?

5. How is succinate dehydrogenase different than the other TCA cycle enzymes?

6. Succinyl-CoA synthetase is responsible for the only substrate-level phosphorylation found in the citric acid cycle. The mammalian enzyme uses GDP as a nucleotide and produces GTP. How is the high-energy phosphate bond of GTP utilized by cells?

7. If a labeled carbon atom of citric acid survives one turn of the cycle it shows up in two positions in oxaloacetate. Which steps are responsible for this?

8. Pyruvate carboxylase is a biotin-containing enzyme that reacts pyruvate with CO_2 to produce oxaloacetate. In general terms, describe this reaction. What is the role of biotin in this reaction? Of what general type of reaction is this an example?

Abbreviated Answers

1a. Pyruvate, oxaloacetate, and α-ketoglutarate.
 b. alanine, aspartic acid, glutamic acid.

2. The glyoxylate cycle begins with an acetyl unit from acetyl-CoA condensing onto oxaloacetate to produce citrate, which is converted to isocitrate. Although identical to steps in the citric acid cycle, which is carried out in mitochondria, these steps are carried out in glyoxysomes. Isocitrate is metabolized into glyoxylate and succinate by isocitrate lyase and a second acetyl unit from acetyl-CoA is condensed onto glyoxylate to produce malate, which can be used to produce monosaccharides. Glyoxysomes are unable to generate oxaloacetate from succinate produced by isocitrate lyase and rely on the mitochondria. Succinate is exported to the mitochondria where it is metabolized via the citric acid cycle to oxaloacetate. Oxaloacetate is converted to aspartate by transamination with glutamate as an amino group donor. The products of transamination, namely aspartic acid and α-ketoglutarate, are transported back to the glyoxysome where the reverse reaction is carried out to produce oxaloacetate and glutamate. Glutamate is exported back to the mitochondria and oxaloacetate is now available for the glyoxylate cycle.

3.

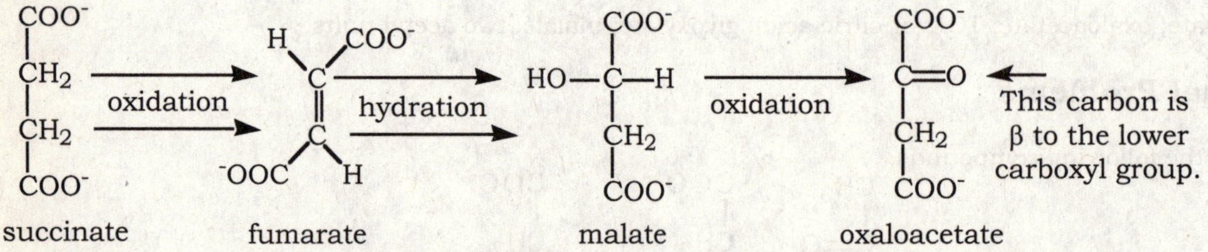

4. High concentrations of citrate are an indication that the citric acid cycle is backing up. This can occur because ATP levels are satisfactory, NAD^+ is unavailable, or an alternate source of acetyl-units, perhaps from fatty acid catabolism, is available. There is no need to continue glycolysis and citrate inhibition at phosphofructokinase insures that glucose units are conserved.

5. Succinate dehydrogenase is responsible for the conversion of succinate to fumarate in the citric acid cycle. This enzyme is also part of the electron transport chain and moves electrons to coenzyme Q and is localized in the inner mitochondrial membrane. The other TCA cycle enzymes are located in the matrix.

6. Signal transduction and protein synthesis both use hydrolysis of GTP. In addition, GTP can be used to phosphorylate ADP by nucleoside diphosphate kinase, which catalyzes the following reaction:
$$GTP + ADP \rightarrow ATP + GDP.$$

7. Both succinate and fumarate are symmetrical molecules with two sets of equivalent carbons. Label at succinate (or fumarate) thus shows up in two positions in later stages.

8. Pyruvate carboxylase catalyzes a two-step reaction leading to oxaloacetate formation. In the first step, bicarbonate is attached to biotin in an ATP-dependent reaction. In the second step, the activated carboxyl group is transferred to pyruvate to produce oxaloacetate. Biotin functions as a carrier of activated carboxyl groups. The reaction is an example of an anaplerotic reaction because it leads to synthesis of a citric acid cycle intermediate.

Summary

Pyruvate produced in glycolysis can be oxidatively decarboxylated to acetyl-CoA and then oxidized in the tricarboxylic acid cycle (also known as the citric acid cycle or Krebs cycle). Electrons produced in this oxidative process are collected in reduced coenzymes (NADH and $FADH_2$) then passed through the electron transport pathway to oxygen (O_2), the final electron acceptor. Discovery of the cycle dates to Hans Krebs' studies of the oxidation of small organic acids by kidney and liver tissue, Szent-Gyorgyi's observation that addition of certain dicarboxylic acids increased the oxygen consumption of muscle tissue and the demonstration by Martius and Knoop that citrate could be converted to isocitrate and α-ketoglutarate.

The oxidative decarboxylation of pyruvate is carried out by pyruvate dehydrogenase, a multienzyme complex composed of three types of protein subunits and five different coenzymes. Acetyl-CoA produced in this reaction enters the TCA cycle, combining with oxaloacetate in the citrate synthase reaction. This irreversible step commits acetate units to oxidation and to energy production. Citrate synthase is allosterically inhibited by NADH and by succinyl-CoA. Citrate is then isomerized to isocitrate by aconitase in a two-step process involving aconitate as an intermediate. This reaction converts a tertiary alcohol - a poor substrate for further oxidation - to a secondary alcohol, which can be readily oxidized. Fluoroacetate blocks the TCA cycle *in vivo*, although it has no apparent effect on any single reaction of the TCA cycle. Citrate synthase converts fluoroacetate to fluorocitrate, which strongly inhibits aconitase.

Isocitrate is oxidatively decarboxylated by isocitrate dehydrogenase in a two-step reaction involving oxidation of the C-2 alcohol of isocitrate to form oxalosuccinate, followed by a β-decarboxylation reaction which expels the central carboxyl group as CO_2, leaving the product α-ketoglutarate. NADH produced in this reaction links the TCA cycle and the electron transport chain. Isocitrate dehydrogenase is allosterically inhibited by NADH and ATP and allosterically activated by ADP. A second oxidative decarboxylation occurs in the α-ketoglutarate dehydrogenase reaction, which involves a multienzyme complex similar in many respects to pyruvate dehydrogenase. The reaction produces NADH and succinyl-CoA - a thiol ester product. Succinyl-CoA is used in the next step to drive the synthesis of GTP (in animals) or ATP (in plants and bacteria). This reaction is catalyzed by succinyl CoA synthetase and is an example of substrate level phosphorylation, the only such reaction in the TCA cycle. The mechanism is postulated to involve formation of succinyl phosphate at the active site, followed by transfer of the phosphoryl group to an active site histidine with release of succinate. Phosphate is then transferred to GDP to form GTP.

The TCA cycle is completed by converting succinate to oxaloacetate. The process involves an oxidation, a hydration reaction and a second oxidation. Succinate dehydrogenase, a membrane-bound enzyme and a part of the electron transport chain, catalyzes the oxidation of succinate to fumarate, linking this reaction to reduction of FAD. Fumarase then catalyzes the trans-hydration of fumarate to produce malate and malate dehydrogenase completes the cycle by oxidizing malate to oxaloacetate. The latter reaction is strongly endergonic, but the reaction is pulled forward by the favorable citrate synthase reaction. The net reaction accomplished by the TCA cycle produces two molecules of CO_2, one ATP and four reduced coenzymes per acetate group oxidized. Carbon introduced to the cycle in any given citrate synthase reaction is not expelled as CO_2 in that same turn of the cycle. The carbonyl carbon of any given acetyl-CoA survives the first turn intact, but is completely lost as CO_2 in the second turn. The methyl carbon survives two full turns, and then undergoes 50% loss in the third turn, 25% loss in the fourth turn, etc. The TCA cycle provides a variety of intermediates for biosynthetic pathways, including citrate, α-ketoglutarate, succinyl-CoA, fumarate and oxaloacetate. On the other hand, the cell also feeds some of these same molecules back into the TCA cycle from other reactions, replenishing TCA cycle intermediates in so-called anaplerotic reactions.

The main sites of regulation in the TCA cycle are pyruvate dehydrogenase, citrate synthase, isocitrate dehydrogenase and α-ketoglutarate dehydrogenase. All these enzymes are inhibited by NADH, and ATP inhibits both pyruvate dehydrogenase and isocitrate dehydrogenase. The cycle is turned on when ADP/ATP and NAD^+/NADH ratios are high. Acetyl-CoA activates pyruvate carboxylase and succinyl CoA inhibits both citrate synthase and α-ketoglutarate dehydrogenase.

Animals use the TCA cycle primarily for energy production and they essentially waste carbon units by giving off CO_2. As a result, they cannot effect net synthesis of carbohydrates from acetyl-CoA. Plants and bacteria, on the other hand, employ a modification of the TCA cycle - the glyoxylate cycle - to conserve carbon units and facilitate eventual biosynthesis of carbohydrates from two-carbon acetate units.   In plants, isocitrate lyase and malate synthase - the enzymes of the glyoxylate cycle - are contained in specialized organelles called glyoxysomes. The glyoxylate cycle permits seeds to grow underground where photosynthesis is impossible. When the plant begins photosynthesis and can fix CO_2 to produce sugars, glyoxysomes disappear.

Chapter 20

Electron Transport and Oxidative Phosphorylation

. . . .

Chapter Outline

- ❖ Oxidative phosphorylation driven by electron transport
  - ⅄ Membrane associated processes
    - • Plasma membrane of bacteria
    - • Mitochondrial membrane of eukaryotes
      - • Outer mitochondrial membrane: Permeable to $M_r < 10,000$ due to protein porin
      - • Intermembrane space: Creatine kinase, adenylate kinase, cytochrome c
      - • Inner mitochondrial membrane: Highly impermeable
        - o High protein content
        - o High content of unsaturated fatty acids
        - o Cardiolipin and diphosphatidylglycerol: No cholesterol
        - o Cristae: Folds that increase surface area
      - • Matrix
        - o TCA cycle enzymes (except succinate dehydrogenase)
        - o Enzymes for catabolism of fatty acids
        - o Circular DNA
        - o Ribosomes, tRNAs
- ❖ Electron transport chain
  - ⅄ Electron mediators
    - • Flavoproteins: Tightly bound FMN or FAD
    - • Coenzyme Q (ubiquinone): 1 or 2 Electron transfers: Mobile within membrane
    - • Cytochromes: Fe^{2+}/Fe^{3+}
      - • Cytochrome a's: Isoprenoid (15-C) on modified vinyl and formyl in place of methyl
      - • Cytochrome b's: Iron-protoporphyrin IX
      - • Cytochrome c's: Iron-protoporphyrin IX linked to cysteine
    - • Iron-sulfur proteins: Fe^{2+}/Fe^{3+}: Several types
    - • Protein-bound copper: Cu^+/Cu^{2+}
  - ⅄ Electron transport complexes: Four
    - • Complex I: NADH-coenzyme Q reductase (NADH reductase)
      - • Electron movement
        - o [FMN] accepts electron pair from NADH
        - o [FMNH₂] donates electrons to Fe-S
        - o Fe-S donates electrons to coenzyme Q
      - • Protons pumped from matrix to cytosol
      - • Supports 3 ATP
    - • Complex II: Succinate-coenzyme Q reductase (succinate dehydrogenase)
      - • Components
        - o FAD
        - o Fe-S centers

- No protons pumped
- Supports 2 ATP
- Similar complexes that reduce coenzyme Q
    - Glycerolphosphate dehydrogenase: No protons pumped
    - Fatty acyl-CoA dehydrogenase: No protons pumped
- Complex III: Coenzyme Q-cytochrome c reductase
    - Components
        - Cytochromes b_L and b_H
        - Rieske protein: Fe-S protein
    - Q-cycle
    - Protons pumped
- Complex IV: Cytochrome c oxidase
    - Components
        - Cytochrome a and Cu_A
        - Cytochrome a_3 and Cu_B
    - Binuclear center: O_2 consumption and H_2O production
    - Protons pumped
- Mitchell's chemiosmotic hypothesis: Proton gradient used to drive ATP synthesis
    - Protons per electron pair
        - From succinate 6
        - From NADH 10
    - Four protons per ATP
        - ADP uptake: 1 proton
        - ATP synthesis: 3 protons
    - One oxygen consumed per electron pair
    - P/O
        - From NADH: 2.5
        - From succinate: 1.5
- ATP synthase: $F_1F_oATPase$
    - F_1: ATP synthesis: Spherical particles on inner membrane
    - F_o: Proton channel in inner membrane
- Inhibitors of oxidative phosphorylation
    - Complex I: Rotenone, ptericidin, amytal, mercurial compounds
    - Complex II: 2-Thenoyltrifluoroacetone, carboxin
    - Complex III: Antimycin, myxothiazol
    - Complex IV: Cyanide, azide, carbon monoxide
    - ATP synthase: Oligomycin, DCCD
- Uncouplers: Stimulate electron transport: Short circuit proton gradient: Block ATP production
    - Proton ionophores: Lipid soluble substance with dissociable proton
    - Thermogenin: Uncoupler protein: Generates heat using proton gradient
- Mitochondrial exchange and uptake
    - ATP/ADP translocate
        - ADP in: ATP out: 1 Proton in
    - Glycerolphosphate shuttle
        - Cytosolic glycerolphosphate dehydrogenase: NADH-dependent
        - Mitochondrial glycerolphosphate dehydrogenase: FAD-dependent
        - DHAP and glycerolphosphate exchanged
    - Malate-aspartate shuttle
        - Cytosolic and mitochondrial malate dehydrogenases both use NADH
        - Aspartate exchanged for glutamate
        - Malate exchanged for α-ketoglutarate
- Cytochrome c and apoptosis
    - Apoptosis: Programmed cell death
        - Set off in response to various signals: Calcium, ROS, lipids, kinases
        - Cytochrome c released by mitochondria

- ATP-ADP translocase (inner membrane) and voltage-dependent anion channel (outer membrane) form pores that pass cytochrome c
- Cytochrome c in cytosol
  - Activates apoptosome
  - Apoptosome activates caspases that destroy cell components

Chapter Objectives

Oxidation/Reduction Reactions

Electron transport involves sequential oxidation/reduction reactions. For any electron carrier, we can think of the carrier as participating in a reaction in which electrons are either produced or consumed. If electrons are consumed, the carrier is reduced and if electrons are produced, the carrier is oxidized. But in reality, free electrons are not just present in solution waiting to react (electrons are not like a typical chemical substrate); rather electrons are exchanged between pairs of reacting molecules. The pairs are an oxidant or oxidizing agent and a reductant or reducing agent. An oxidant accepts electrons, is itself reduced, but oxidizes the reductant, the agent from which the electrons originated. A reductant donates electrons, is itself oxidized, but reduces an oxidant.

The tendency to donate or accept electrons is measured by standard reduction (redox) potentials. It should be clear how these measurements are made. A reference half-cell is connected to a sample half-cell by an agar salt bridge and a low-resistance pathway. (The agar salt bridge simply functions to complete the circuit between the half-cells connected by the low-resistance pathway.) The reference half-cell is H^+/H_2 at 1 M and 1 atmosphere. The reduction potential of this half-cell is 0 V by definition. (Clearly, no potential exists when the reference half-cell is connected to another reference half-cell.) If the reaction in the sample cell consumes electrons, electrons will flow from the reference cell to the sample cell unless a voltage is applied to prevent this flow. In this case a positive voltage is required. Thus, a positive standard redox potential indicates that the sample is an oxidant relative to the reference cell and a negative standard redox potential indicates that the sample is a reductant relative to the reference cell. The two key formulas to remember are:
$$\Delta G^{\circ\prime} = -n\mathcal{F}\Delta\mathcal{E}_o{}' \text{ and } \mathcal{E} = \mathcal{E}_o{}' + (RT/n\mathcal{F})\ln([\text{oxidant}]/[\text{reductant}]).$$

Electron Transport Chain

A simplified way of thinking about the electron transport chain is to divide the chain into two types of components, mobile electron carriers and membrane-bound protein complexes. The mobile electron carriers include: $NAD^+/NADH$, coenzyme Q or ubiquinone, cytochrome c, and oxygen. Substrates, such as malate (for malate dehydrogenase) or α-ketoglutarate (for α-ketoglutarate dehydrogenase) or succinate (for succinate dehydrogenase) can be thought of as mobile electron carriers as well; however, they are peripheral to the electron transport chain. There are four membrane-bound complexes that simply move electrons from one mobile electron carrier to another. The first complex is NADH-coenzyme Q reductase or NADH reductase, which moves electrons from NADH to CoQ. CoQ can also be reduced by succinate-coenzyme Q reductase also known as succinate dehydrogenase, which uses succinate as a source of electrons. Coenzyme Q-cytochrome c reductase moves electrons from $CoQH_2$ to cytochrome c. Finally, cytochrome oxidase moves electrons from reduced cytochrome c to molecular oxygen.

Proton Gradient Formation

Electron transport accomplishes two things: it regenerates reduced cofactors such as NAD^+ and [FAD], and it produces ATP. Understand how the energy of electron transport is used to form a proton gradient, which is used in turn to phosphorylate ADP to ATP. Protons are pumped out of the mitochondria by NADH-coenzyme Q reductase. Protons are also expelled in the Q cycle involving coenzyme Q-cytochrome c reductase. The essential point is that coenzyme Q carries electrons and protons whereas cytochrome c carries only electrons. Cytochrome c oxidase also moves protons but the details of how this is achieved are unknown.

ATP synthase

ADP phosphorylation and proton gradient dissipation are coupled by the ATP synthase or F_1F_o-ATPase. Understand how this protein complex is organized and situated in the inner mitochondrial membrane. The F_o portion (o is for oligomycin) is a proton pore and the F_1 is an ATPase that functions in the reverse direction to produce ATP.

Inhibitors and Uncouplers

Appreciate the difference between an inhibitor and an uncoupler. Inhibitors block the action of some component of electron transport or ATP synthase. Uncouplers do not interfere with electron transport and in fact stimulate it. However, they provide an alternative pathway to dissipate the energy of electron transport.

Problems and Solutions

1. Understanding redox couples
For the following reaction,

$$FAD + 2 \text{ cyt } c \ (Fe^{2+}) \rightarrow FADH_2 + 2 \text{ cyt } c \ (Fe^{3+})$$

determine which of the redox couples is the electron acceptor and which is the electron donor under standard-state conditions, calculate the value of $\Delta \mathcal{E}_0'$, and determine the free energy change for the reaction.

Answer: The reduction half-reactions and their standard reduction potentials for the reaction are (from Table 3.5):

$$FAD + 2H^+ + 2e^- \rightarrow FADH_2, \qquad\qquad \mathcal{E}_0' = -0.219 \text{ V}^*$$

and,

$$\text{cytochrome } c, Fe^{3+} + e^- \rightarrow \text{cytochrome } c, Fe^{2+}, \qquad \mathcal{E}_0' = 0.254 \text{ V}.$$

The standard reduction potential is the voltage that is generated between a sample half-cell and reference half-cell (H^+/H_2). In effect, it is the voltage that must be applied to a circuit connecting a sample half-cell and the reference half-cell to prevent current from flowing. Using this convention we see that if the FAD half-cell is connected to the reference half-cell, a slightly negative voltage of -0.219 V must be applied to prevent electrons from flowing from the reference cell into the sample cell. Conversely, +0.254 V must be applied when the cytochrome c half-cell is connected to the reference cell. Thus, electrons have a greater tendency to flow from the reference cell to cytochrome c than to FAD. Therefore, electrons must move from $FADH_2$ to cytochrome c. Thus, $FADH_2$ is the electron donor and cytochrome c, Fe^{3+} is the electron acceptor.

$$\Delta \mathcal{E}_0' = \mathcal{E}_0'(\text{acceptor}) - \mathcal{E}_0'(\text{donor}) = \mathcal{E}_0'(\text{cyto c}) - \mathcal{E}_0'(FAD)$$

$$\Delta \mathcal{E}_0' = +0.254 - (-0.219) = +0.473 \text{ V}$$

The free energy change for the reaction is given by:

$\Delta G^{\circ\prime} = -n \mathcal{F} \Delta \mathcal{E}_0'$, where n is the number of electrons, and

$$\mathcal{F} = \text{Faraday's constant} = 96.485 \frac{\text{kJ}}{\text{V} \cdot \text{mol}}$$

$$\Delta G^{\circ\prime} = -2 \times 96.485 \frac{\text{kJ}}{\text{V} \cdot \text{mol}} \times 0.473 \text{ V} = -91.3 \frac{\text{kJ}}{\text{mol}}$$

* The value of -0.219 given for FAD in Table 3.5 is for free FAD. Protein-bound FAD has a standard reduction potential in the range of from 0.003 to -0.091 with 0.02 V being a typical value. Using 0.02V, $\Delta G^{\circ\prime}$ = -45.2 kJ/mol.

2. Determining the redox potential of the glyceraldehyde-3-phosphate dehydrogenase reaction
Calculate $\Delta \mathcal{E}_0'$ for the glyceraldehyde-3-phosphate dehydrogenase reaction, and calculate the free energy change for the reaction under standard-state conditions.

Answer: Glyceraldehyde-3-phosphate dehydrogenase catalyzes the following reaction:

$$\text{Glyceraldehyde-3-phosphate} + P_i + NAD^+ \rightarrow 1,3\text{-bisphosphoglycerate} + NADH + H^+$$

The relevant half reactions are (from Table 3.5):

$$NAD^+ + 2H^+ + 2e^- \rightarrow NADH + H^+, \qquad \mathcal{E}_0' = -0.320$$

and,

$$\text{Glycerate-1,3-bisphosphate} + 2 H^+ + 2 e^- \rightarrow \text{glyceraldehyde-3-phosphate} + P_i,$$
$$\mathcal{E}_0' = -0.290$$

Thus, NAD^+ is the electron acceptor and glyceraldehyde-3-phosphate is the electron donor.

$$\Delta \mathcal{E}_0' = \mathcal{E}_0'(\text{acceptor}) - \mathcal{E}_0'(\text{donor}) = \mathcal{E}_0'(NAD^+) - \mathcal{E}_0'(G3P)$$

$$\Delta \mathcal{E}_0' = -0.320 - (-0.290) = -0.030 \text{ V}$$

The free energy change for the reaction is given by:

$\Delta G^{\circ\prime} = -n \mathcal{F} \Delta \mathcal{E}_0'$, where n is the number of electrons, and

$$\mathcal{F} = \text{Faraday's constant} = 96.485 \frac{\text{kJ}}{\text{V} \cdot \text{mol}}$$

$$\Delta G^{\circ\prime} = -2 \times 96.485 \frac{\text{kJ}}{\text{V} \cdot \text{mol}} \times (-0.030 \text{ V}) = 5.79 \frac{\text{kJ}}{\text{mol}}$$

3. Understanding the pH dependence of NAD⁺ reduction
For the following redox reaction,

$$NAD^+ + 2 H^+ + 2 e^- \rightleftharpoons NADH + H^+$$

suggest an equation (analogous to Equation 20.13) that predicts the pH dependence of this reaction, and calculate the reduction potential for this reaction at pH 8.

Answer: The NAD⁺ reduction shown above may be rewritten as:

$$NAD^+ + H^+ + 2 e^- \rightarrow NADH$$

However, a source of electrons is needed for the reaction to occur. Therefore, let us assume that the reaction is in aqueous solution with a general reductant of the form:

$$Reductant \rightleftharpoons Oxidant + 2e^-$$

The overall reaction is:

$$NAD^+ + H^+ + Reductant \rightleftharpoons NADH + Oxidant$$

$$\Delta G = \Delta G^{o\prime} + RT\ln\frac{[Oxidant][NAD^+]}{[Reductant][NADH][H^+]}$$

$$= \Delta G^{o\prime} + RT\ln\frac{[Oxidant][NAD^+]}{[Reductant][NADH]} + RT\ln\frac{1}{[H^+]}$$

$$= \Delta G^{o\prime} + RT\ln\frac{[Oxidant][NAD^+]}{[Reductant][NADH]} - RT\ln[H^+]$$

But, $\ln[H^+]=2.303\log_{10}[H^+]$ and $pH \equiv -\log_{10}[H^+]$, so

$$-\ln[H^+]=-2.303\log_{10}[H^+] = 2.303 \times pH, \text{ thus}$$

$$\Delta G = \Delta G^{o\prime} + RT\ln\frac{[Oxidant][NAD^+]}{[Reductant][NADH]} + 2.303 \times RT \times pH$$

In general, $\Delta G = -n\mathcal{F}\Delta\mathcal{E}$ or $\Delta\mathcal{E} = -\dfrac{\Delta G}{n\mathcal{F}}$ where n is the number of electrons, and

$$F = \text{Faraday's constant} = 96.485 \frac{kJ}{V \cdot mol}$$

$$\Delta\mathcal{E} = \Delta\mathcal{E}_o{}' - \frac{RT}{n\mathcal{F}}\ln\frac{[Oxidant][NAD^+]}{[Reductant][NADH]} - 2.303 \times \frac{RT}{n\mathcal{F}} \times pH$$

To calculate the reduction potential we assume that the reaction is being carried out under standard conditions. Thus, all reactants and products are at 1 M concentration. The above equation simplifies to:

$$\Delta\mathcal{E} = \Delta\mathcal{E}_0{}' - 2.303 \times \frac{RT}{n\mathcal{F}} \times pH$$

4. Understanding an antidote for cyanide poisoning
Sodium nitrite (NaNO₂) is used by emergency medical personnel as an antidote for cyanide poisoning (for this purpose, it must be administered immediately). Based on the discussion of cyanide poisoning in Section 20.10, suggest a mechanism for the life-saving effect of sodium nitrite.

Answer: Cytochrome c oxidase is the principle target of cyanide (CN⁻) poisoning. Cyanide binds to the ferric (Fe^{3+}) or oxidized form of cytochrome a₃. Sodium nitrite may be administered intravenously in an attempt to combat cyanide poisoning. The intention of sodium nitrite treatment is to produce an alternate target for cyanide. Nitrite will oxidize the very abundant hemoglobin to methemoglobin (ferric hemoglobin) that will react with cyanide.

5. Assessing the mechanism of uncoupler action

A wealthy investor has come to you for advice. She has been approached by a biochemist who seeks financial backing for a company that would market dinitrophenol and dicumarol as weight-loss medications. The biochemist has explained to her that these agents are uncouplers and that they would dissipate metabolic energy as heat. The investor wants to know if you think she should invest in the biochemist's company. How do you respond?

Answer: The structures of dicumarol and dinitrophenol are:

Dicumarol   2,4-Dinitrophenol

The biochemistry of the suggestion is sound. (Beware: This is not an endorsement of the idea. Please read on.) Both compounds are uncoupling agents and act by dissipating the proton gradient across the inner mitochondrial membrane. Instead of being used to synthesize ATP, the energy of the proton gradient is dissipated as heat. As ATP levels decrease, electron transport will increase, and, glucose or fatty acids will be metabolized in an attempt to meet the false metabolic demand.

The compounds are both lipophilic molecules and are capable of dissolving in the inner mitochondrial membrane. Their hydroxyl groups have low pK_as because they are attached to conjugated ring systems. On the cytosolic (and higher pH) surface of the inner mitochondrial membrane the compounds will be protonated whereas on the matrix side they will be unprotonated.

So much for the theoretical biochemistry. In working with unfamiliar compounds, it is imperative to consult references to find out what is known about them. One good place to start is the MSDS (Material Safety Data Sheet). The MSDS is a summary of potential hazards of a compound provided by the manufacturer. Another good reference, and one found in virtually every biochemist's laboratory, is the Merck Index (Published by Merck & Co., Rahway, N.J.). The Merck Index informs us, in the section on human toxicity under 2,4-dinitrophenol, that this compound is highly toxic, produces an increase in metabolism (good for a diet), increased temperature, nausea, vomiting, collapse, and, death (a drastic weight loss indeed). Under dicumarol, human toxicity is not mentioned, however, under the subsection, therapeutic category, we are informed that the compound is used as an anticoagulant in humans. Dicumarol was first discovered as the agent responsible for hemorrhagic sweet clover disease in cattle. It is produced by microorganisms in spoiled silage and causes death by bleeding. It is used therapeutically as an anticoagulant but the therapy must be carefully monitored.

6. Assessing the [ATP]/[ADP][Pᵢ] ratio and ATP synthesis

Assuming that 3 H^+ are transported per ATP synthesized in the mitochondrial matrix, the membrane potential difference is 0.18 V (negative inside), and the pH difference is 1 unit (acid outside, basic inside), calculate the largest ratio of [ATP]/[ADP][Pᵢ] under which synthesis of ATP can occur.

Answer: The free energy difference per mole for protons across the inner mitochondrial membrane is given by:

$$\Delta G = -2.303 \times RT \times (pH_{out} - pH_{in}) + \mathcal{F}\Delta\Psi$$

Where, pH_{out} is the pH outside the mitochondria; pH_{in} is the matrix pH; $\Delta\Psi$ is the potential difference across the inner mitochondrial membrane, $V_{in} - V_{out}$; R is the gas constant, 8.314×10^{-3} kJ/mol K; $\mathcal{F}$ is Faraday's constant, 96.485 kJ/V mol; and T is temperature in K (°C + 273).

$$\Delta G = -2.303 \times (8.314 \times 10^{-3})(298) \times (1) + 96.485 \times (-0.18)$$

$$\Delta G = -23.07 \frac{kJ}{mol}$$

For movement of 3 protons we have

$$\Delta G = 3 \times (-23.07 \frac{kJ}{mol}) = -69.2 \text{ kJ}$$

What is the largest value of $\dfrac{[ATP]}{[ADP][P_i]}$ under which synthesis of ATP can occur?

For the reaction, ADP + P_i → ATP + H_2O, ΔG is given by:

$$\Delta G = \Delta G^{o'} + RT \ln \frac{[ATP]}{[ADP][P_i]}$$

$$\Delta G = +30.5 \frac{kJ}{mol} + (8.314 \times 10^{-3})(298) \ln \frac{[ATP]}{[ADP][P_i]}$$

$$\Delta G = (+30.5 + 2.48 \ln \frac{[ATP]}{[ADP][P_i]})kJ$$

For translocation of 3 protons coupled to synthesis of 1 ATP to be thermodynamically favorable the overall ΔG must be negative. Therefore,

$$\Delta G_{3H^+} + \Delta G_{ATP} < 0$$

$$-69.2 + 30.5 + 2.48 \ln \frac{[ATP]}{[ADP][P_i]} < 0$$

Solving for $\frac{[ATP]}{[ADP][P_i]}$ we find:

$$\frac{[ATP]}{[ADP][P_i]} < e^{\frac{69.2-30.5}{2.48}} = e^{15.6} = 6 \times 10^6 M^{-1}$$

At 37°C the answers are 69.9 kJ and $4.36 \times 10^6 M^{-1}$.

7. Analyzing the succinate dehydrogenase reaction
Of the dehydrogenase reactions in glycolysis and the TCA cycle, all but one use NAD+ as the electron acceptor. The lone exception is the succinate dehydrogenase reaction, which uses FAD, covalently bound to a flavoprotein, as the electron acceptor. The standard reduction potential for this bound FAD is in the range of 0.003 to 0.091 V (Table 3.5). Compared to the other dehydrogenase reactions of glycolysis and the TCA cycle, what is unique about succinate dehydrogenase? Why is bound FAD a more suitable electron acceptor in this case?

Answer: Succinate dehydrogenase converts succinate to fumarate, an oxidation of an alkane to an alkene. The other oxidation reactions in the TCA cycle and in glycolysis either convert alcohols to ketones or ketones to carboxyl groups. These oxidations are sufficiently energetic to reduce NAD+. The standard redox potential (from Table 3.5) for reduction of fumarate to succinate is 0.031 V. In contrast, the redox potential for NAD+ is -0.320 V. Thus, under standard conditions, if NAD+ participated in the succinate/fumarate reaction, it would do so as a reductant (i.e., as NADH) passing electrons to fumarate to produce succinate, the exact opposite of what is accomplished in the TCA cycle. To remove electrons from succinate, an oxidant with a higher (more positive) redox potential is required. The covalently bound FAD of succinate dehydrogenase meets this requirement with a standard redox potential in the range of 0.003 to 0.091 V. (We will encounter another example of conversion of an alkane to an alkene with reduction of FAD in fatty acid metabolism or β-oxidation.)

8. Analysis of the NACH-CoQ reductase reaction
a. What is the standard free energy change ($\Delta G^{o'}$), for the reduction of coenzyme Q by NADH as carried out by complex I (NADH-coenzyme Q reductase) of the electron transport pathway if $E_o'(NAD^+/NADH+H^+) = -0.320$ V and $E_o'(CoQ/CoQH_2) = +0.060$ V.
b. What is the equilibrium constant (K_{eq}) for this reaction?
c. Assume that (1) the actual free energy release accompanying the NADH-coenzyme Q reductase reaction is equal to the amount released under standard conditions (as calculated above), (2) this energy can be converted into the synthesis of ATP with an efficiency = 0.75 (that is, 75% of the energy released upon NADH oxidation is captured in ATP synthesis), and (3) the oxidation of 1 equivalent of NADH by coenzyme Q leads to the phosphorylation of 1 equivalent of ATP.
Under these conditions, what is the maximum ratio of [ATP]/[ADP] attainable for oxidative phosphorylation when [P_i] = 1 mM? (Assume $\Delta G^{o'}$ for ATP synthesis = +30.5 kJ/mol.)

Answer: a. The relevant half reactions are (from Table 3.5):
         NAD+ + 2H+ + 2e- → NADH + H+, E_o' = -0.320
and,
         CoQ (UQ) + 2 H+ + 2 e- → CoQH2 (UQH2), E_o' = +0.060
Thus, CoQ is the electron acceptor and NADH is the electron donor.

$$\Delta \mathcal{E}_0' = \mathcal{E}_0'(\text{acceptor}) - \mathcal{E}_0'(\text{donor}) = \mathcal{E}_0'(\text{CoQ}) - \mathcal{E}_0'(\text{NAD}^+)$$

$$\Delta \mathcal{E}_0' = +0.060 - (-0.320) = +0.38 \text{ V}$$

The free energy change for the reaction is given by:

$\Delta G^{\circ\prime} = -n\mathcal{F}\Delta \mathcal{E}_0'$, where n is the number of electrons, and

$$\mathcal{F} = \text{Faraday's constant} = 96.485 \frac{\text{kJ}}{\text{V} \cdot \text{mol}}$$

$$\Delta G^{\circ\prime} = -2 \times 96.485 \frac{\text{kJ}}{\text{V} \cdot \text{mol}} \times (0.38 \text{ V}) = -73.3 \frac{\text{kJ}}{\text{mol}}$$

b. To calculate the equilibrium constant:

From $\Delta G^{\circ\prime} = -RT \ln K_{eq}$ we see that

$$K_{eq} = e^{\frac{-\Delta G^{\circ\prime}}{RT}}$$

$$K_{eq} = e^{\frac{-\Delta G^{\circ\prime}}{RT}} = e^{\frac{-(-73.3)}{(8.314 \times 10^{-3})(298)}} = 7.1 \times 10^{12}$$

c. Assuming that $\Delta G = -73.3$ kJ/mol, the amount of energy used to synthesize ATP is:

$$-73.3 \frac{\text{kJ}}{\text{mol}} \times (0.75) = -55 \frac{\text{kJ}}{\text{mol}}$$

Since $\Delta G = \Delta G^{\circ\prime} + RT \ln \dfrac{[\text{ATP}]}{[\text{ADP}][P_i]}$

$$\frac{[\text{ATP}]}{[\text{ADP}]} = [P_i] \times e^{\frac{\Delta G - \Delta G^{\circ\prime}}{RT}} = (1 \text{ mM}) \times e^{\frac{55-30.5}{(8.314 \times 10^{-3})(298)}} = (1 \times 10^{-3}) \times e^{9.88}$$

$$\frac{[\text{ATP}]}{[\text{ADP}]} = 19.7$$

9. *Determining the impact of the succinate dehydrogenase reaction on oxidative phosphorylation*
Consider the oxidation of succinate by molecular oxygen as carried out via the electron transport pathway.

succinate + ¹/₂ O₂ → fumarate + H₂O

a. What is the standard free energy change ($\Delta G^{\circ\prime}$), for this reaction if $\mathcal{E}_0'$(fumarate/succinate) = +0.031 V and $\mathcal{E}_0'$ (¹/₂O₂/H₂O) = +0.816 V.
b. What is the equilibrium constant (K_{eq}) for this reaction?
c. Assume that (1) the actual free energy release accompanying succinate oxidation by the electron transport pathway is equal to the amount released under standard conditions (as calculated above), (2) this energy can be converted into the synthesis of ATP with an efficiency = 0.70 (that is, 70% of the energy released upon succinate oxidation is captured in ATP synthesis), and (3) the oxidation of 1 succinate leads to the phosphorylation of 2 equivalents of ATP.
Under these conditions, what is the maximum ratio of [ATP]/[ADP] attainable for oxidative phosphorylation when [Pᵢ] = 1 mM? (Assume $\Delta G^{\circ\prime}$ for ATP synthesis = +30.5 kJ/mol.)

Answer: a. The relevant half reactions are (from Table 3.5):

Fumarate + 2H⁺ + 2e⁻ → succinate,	$\mathcal{E}_0'$	= +0.031

and,

¹/₂ O₂ + 2 H⁺ + 2 e⁻ → H₂O,	$\mathcal{E}_0'$	= +0.816

Thus, O₂ is the electron acceptor and succinate is the electron donor.

$$\Delta\mathcal{E}_o'=\mathcal{E}_o'(acceptor)-\mathcal{E}_o'(donor)=\mathcal{E}_o'(O_2)-\mathcal{E}_o'(succinate)$$

$$\Delta\mathcal{E}_o'=+0.816-(0.031)=+0.785\ V$$

The free energy change for the reaction is given by:

$\Delta G^{\circ\prime}=-n\mathcal{F}\Delta\mathcal{E}_o'$, where n is the number of electrons, and

$$\mathcal{F}=Faraday's\ constant=96.485\frac{kJ}{V\cdot mol}$$

$$\Delta G^{\circ\prime}=-2\times96.485\frac{kJ}{V\cdot mol}\times(0.785\ V)=-151.5\frac{kJ}{mol}$$

b. To calculate the equilibrium constant:

From $\Delta G^{\circ\prime}=-RT\ln K_{eq}$ we see that

$$K_{eq}=e^{\frac{-\Delta G^{\circ\prime}}{RT}}$$

$$K_{eq}=e^{\frac{-\Delta G^{\circ\prime}}{RT}}=e^{\frac{-(-151.5)}{(8.314\times10^{-3})(298)}}=3.60\times10^{26}$$

c. Assuming that $\Delta G = -151.3$ kJ/mol, the amount of energy used to synthesize ATP is:

$$-151.5\frac{kJ}{mol}\times(0.70)=-106.1\frac{kJ}{mol}\ \text{for 2 ATP synthesized}$$

Per ATP we have $53.05\frac{kJ}{mol}$

Since $\Delta G=\Delta G^{\circ\prime}+RT\ln\frac{[ATP]}{[ADP][P_i]}$

$$\frac{[ATP]}{[ADP]}=[P_i]\times e^{\frac{\Delta G-\Delta G^{\circ\prime}}{RT}}=(1\ mM)\times e^{\frac{53.05-30.5}{(8.314\times10^{-3})(298)}}$$

$$\frac{[ATP]}{[ADP]}=8.89$$

10. Determining the effects of NADH oxidation on oxidative phosphorylation
Consider the oxidation of NADH by molecular oxygen as carried out via the electron transport pathway

$$NADH + H^+ + \tfrac{1}{2}\ O_2 \rightarrow NAD^+ + H_2O$$

a. What is the standard free energy change ($\Delta G^{\circ\prime}$), for this reaction if $\mathcal{E}_o'(NAD^+/NADH+H^+)$ = -0.320 V and $\mathcal{E}_o'$ ($\tfrac{1}{2}$ O_2/H_2O) = +0.816 V.
b. What is the equilibrium constant (K_{eq}) for this reaction?
c. Assume that (1) the actual free energy release accompanying NADH oxidation by the electron transport pathway is equal to the amount released under standard conditions (as calculated above), (2) this energy can be converted into the synthesis of ATP with an efficiency = 0.75 (that is, 75% of the energy released upon NADH oxidation is captured in ATP synthesis), and (3) the oxidation of 1 NADH leads to the phosphorylation of 3 equivalents of ATP.
Under these conditions, what is the maximum ratio of [ATP]/[ADP] attainable for oxidative phosphorylation when [Pᵢ] = 1 mM? (Assume $\Delta G^{\circ\prime}$ for ATP synthesis = +30.5 kJ/mol.)

Answer
a. The relevant half reactions are (from Table 3.5):

$\quad\quad NAD^+ + 2H^+ + 2e^- \rightarrow NADH + H^+,$ $\quad\quad\quad\quad \mathcal{E}_o'$ = -0.320

and,

$\quad\quad \tfrac{1}{2}\ O_2 + 2H^+ + 2e^- \rightarrow H_2O$ $\quad\quad\quad\quad\quad\quad \mathcal{E}_o'$ = +0.816

Thus, oxygen is the electron acceptor and NADH is the electron donor.

$$\Delta\mathcal{E}_0'=\mathcal{E}_0'(\text{acceptor})-\mathcal{E}_0'(\text{donor})=\mathcal{E}_0'(\text{NAD}^+)-\mathcal{E}_0'(\text{G3P})$$

$$\mathcal{E}_0'=+0.816-(-0.320)=+1.136 \text{ V}$$

The free energy change for the reaction is given by:

$\Delta G^{\circ'}=-n\mathcal{F}\Delta\mathcal{E}_0'$, where n is the number of electrons, and

$$\mathcal{F}=\text{Faraday's constant}=96.485\frac{\text{kJ}}{\text{V}\cdot\text{mol}}$$

$$\Delta G^{\circ'}=-2\times96.485\frac{\text{kJ}}{\text{V}\cdot\text{mol}}\times(1.136\text{ V})=-219.2\frac{\text{kJ}}{\text{mol}}$$

b. To calculate the equilibrium constant:

From $\Delta G^{\circ'} = -RT\ln K_{eq}$ we see that

$$K_{eq} = e^{\frac{-\Delta G^{\circ'}}{RT}}$$

$$K_{eq} = e^{\frac{-\Delta G^{\circ'}}{RT}} = e^{\frac{-(-219.2)}{(8.314\times10^{-3})(298)}} = 2.65\times10^{38}$$

c. Assuming that $\Delta G = -219.2$ kJ/mol, the amount of energy used to synthesize ATP is:

$$-219.2\frac{\text{kJ}}{\text{mol}}\times(0.75)=-164.4\frac{\text{kJ}}{\text{mol}} \text{ for 3 ATP synthesized}$$

Per ATP we have $54.8\dfrac{\text{kJ}}{\text{mol}}$

Since $\Delta G=\Delta G^{\circ'}+RT\ln\dfrac{[\text{ATP}]}{[\text{ADP}][\text{P}_i]}$

$$\frac{[\text{ATP}]}{[\text{ADP}]}=[\text{P}_i]\times e^{\frac{\Delta G-\Delta G^{\circ'}}{RT}}=(2\text{ mM})\times e^{\frac{54.8-30.5}{(8.314\times10^{-3})(298)}}$$

$$\frac{[\text{ATP}]}{[\text{ADP}]}=36.4$$

11. Determining the effect of the cytochrome oxidase reaction on oxidative phosphorylation
Write a balanced equation for the reduction of molecular oxygen by reduced cytochrome c as carried out by complex IV (cytochrome oxidase) of the electron transport pathway.
a. What is the standard free energy change ($\Delta G^{\circ'}$), for this reaction if $\mathcal{E}_0'(cytc(Fe^{3+})/cytc(Fe^{2+}))$ = +0.254 V and $\mathcal{E}_0'$ ($^1/_2$ O_2/H_2O) = +0.816 V.
b. What is the equilibrium constant (K_{eq}) for this reaction?
c. Assume that (1) the actual free energy release accompanying cytochrome c oxidation by the electron transport pathway is equal to the amount released under standard conditions (as calculated above), (2) this energy can be converted into the synthesis of ATP with an efficiency = 0.60 (that is, 60% of the energy released upon cytochrome c oxidation is captured in ATP synthesis), and (3) the reduction of 1 molecule of O_2 by reduced cytochrome c leads to the phosphorylation of 2 equivalents of ATP.
Under these conditions, what is the maximum ratio of [ATP]/[ADP] attainable for oxidative phosphorylation when [P_i] = 3 mM? (Assume $\Delta G^{\circ'}$ for ATP synthesis = +30.5 kJ/mol.)

Answer: The balanced equation for transfer of electrons from cytochrome c to oxygen is:

$$4 \text{ Cytochrome } c(\text{Fe}^{2+}) + O_2 + 4\text{ H}^+ \rightarrow 4 \text{ Cytochrome } c(\text{Fe}^{3+}) + 2H_2O$$

a. The relevant half reactions are (from Table 3.5):

$$\text{Cytochrome } c(\text{Fe}^{3+}) + e^- \rightarrow \text{Cytochrome } c(\text{Fe}^{2+}), \qquad \mathcal{E}_0' = +0.254$$

and,

$$^1/_2\ O_2 + 2\text{ H}^+ + 2\text{ e}^- \rightarrow H_2O \qquad \mathcal{E}_0' = +0.816$$

Thus, oxygen is the electron acceptor and cytochrome c is the electron donor.

$$\Delta \mathcal{E}_0'=\mathcal{E}_0'(\text{acceptor}) - \mathcal{E}_0'(\text{donor}) = \mathcal{E}_0'(O_2) - \mathcal{E}_0'(\text{cytochrome c})$$

$$\Delta \mathcal{E}_0'=+0.816-(+0.254)=+0.562 \text{ V}$$

The free energy change for the reaction is given by:

$\Delta G^{\circ\prime}=-n\mathcal{F}\Delta\mathcal{E}_0'$, where n is the number of electrons, and

$$\mathcal{F}=\text{Faraday's constant}=96.485 \frac{\text{kJ}}{\text{V}\cdot\text{mol}}$$

$$\Delta G^{\circ\prime}=-4 \times 96.485 \frac{\text{kJ}}{\text{V}\cdot\text{mol}} \times (0.562 \text{ V})=-217 \frac{\text{kJ}}{\text{mol}}$$

b. To calculate the equilibrium constant:

From $\Delta G^{\circ\prime} = -RT\ln K_{eq}$ we see that

$$K_{eq} = e^{\frac{-\Delta G^{\circ\prime}}{RT}}$$

The value of RT at 25°C is $(8.314 \times 10^{-3})(298) \frac{\text{kJ}}{\text{mol}} = 2.48 \frac{\text{kJ}}{\text{mol}}$

Thus,

$$K_{eq} = e^{\frac{-\Delta G^{\circ\prime}}{RT}} = e^{\frac{-(-217)}{2.48}} = 1.08 \times 10^{38}$$

c. Assuming that $\Delta G = -217$ kJ/mol, the amount of energy used to synthesize ATP is:

$$-217 \frac{\text{kJ}}{\text{mol}} \times (0.60) = -130.2 \frac{\text{kJ}}{\text{mol}} \text{ for 2 ATP synthesized}$$

Per ATP we have $65.1 \frac{\text{kJ}}{\text{mol}}$

Since $\Delta G = \Delta G^{\circ} + RT\ln \frac{[\text{ATP}]}{[\text{ADP}][P_i]}$

$$\frac{[\text{ATP}]}{[\text{ADP}]} = [P_i] \times e^{\frac{\Delta G - \Delta G^{\circ}}{RT}} = (3 \text{ mM}) \times e^{\frac{65.1-30.5}{2.48}} = (3 \times 10^{-3}) \times e^{\frac{65.1-30.5}{2.48}}$$

$$\frac{[\text{ATP}]}{[\text{ADP}]} = 3437$$

12. Understanding the energetics of the lactate dehydrogenase reaction
The standard reduction potential for (NAD+/NADH+H+) is -0.320 V, and the standard reduction potential for (pyruvate/lactate) is -0.185 V.
a. What is the standard free energy change, ΔG°', for the lactate dehydrogenase reaction:
               NADH + H+ + pyruvate → lactate + NAD+
b. What is the equilibrium constant (Keq) for this reaction?
c. If [pyruvate] = 0.05 mM and [lactate] = 2.9 mM and ΔG for the lactate dehydrogenase reaction = -15 kJ/mol in erythrocytes, what is the (NAD+/NADH) ratio under these conditions?

Answer
a. The relevant half reactions are (from Table 3.5):

    $NAD^+ + 2H^+ + 2e^- \star NADH + H^+$, $\mathcal{E}_0' = -0.320$
and,
    pyruvate $+ 2H^+ + 2e^- \star$ lactate, $\mathcal{E}_0' = -0.185$

Thus, pyruvate is the electron acceptor and NADH is the electron donor.

$$\Delta \mathcal{E}_0'=\mathcal{E}_0'(\text{acceptor}) - \mathcal{E}_0'(\text{donor}) = \mathcal{E}_0'(\text{pyruvate}) - \mathcal{E}_0'(NAD^+)$$

$$\Delta \mathcal{E}_0'=-0.185-(-0.320)=+0.135 \text{ V}$$

The free energy change for the reaction is given by:

$\Delta G^{\circ\prime}=-n\mathcal{F}\Delta\mathcal{E}_0'$, where n is the number of electrons, and

$$\mathcal{F}=\text{Faraday's constant}=96.485 \frac{\text{kJ}}{\text{V}\cdot\text{mol}}$$

$$\Delta G^{\circ\prime}=-2 \times 96.485 \frac{\text{kJ}}{\text{V}\cdot\text{mol}} \times (0.135 \text{ V})=-26.05 \frac{\text{kJ}}{\text{mol}}$$

b. To calculate the equilibrium constant:

From $\Delta G^{\circ\prime} = -RT \ln K_{eq}$ we see that

$$K_{eq} = e^{\frac{-\Delta G^{\circ\prime}}{RT}}$$

The value of RT at 25°C is $(8.314 \times 10^{-3})(298) \frac{kJ}{mol} = 2.48 \frac{kJ}{mol}$

Thus,

$$K_{eq} = e^{\frac{-\Delta G^{\circ\prime}}{RT}} = e^{\frac{-(-26.05)}{2.48}} = e^{10.5} = 3.65 \times 10^4$$

c. The ratio of oxidized to reduced cofactor is calculated as follows:

$$\Delta G = \Delta G^{\circ\prime} + RT \ln \frac{[lactate][NAD^+]}{[pyruvate][NADH]}, \text{ or}$$

$$\frac{[lactate][NAD^+]}{[pyruvate][NADH]} = e^{\frac{\Delta G - \Delta G^{\circ\prime}}{RT}}, \text{ where } \Delta G = -15 \frac{kJ}{mol}, \ \Delta G^{\circ\prime} = -26.05 \frac{kJ}{mol}, \ RT = 2.48 \frac{kJ}{mol}$$

$$\frac{[NAD^+]}{[NADH]} = \frac{[pyruvate]}{[lactate]} e^{\frac{-15-(-26.05)}{2.48}} = \frac{0.05mM}{2.9mM} e^{4.46} = 1.48$$

13. Assessing proton transport and free energy change in bacterial ATP synthesis
Assume that the free energy change, ΔG, associated with the movement of one mole of protons from the outside to the inside of a bacterial cell is -23 kJ/mol and 3 H⁺ must cross the bacterial plasma membrane per ATP formed by the F₁F₀ATP synthase. ATP synthesis thus takes place by the coupled process:

$$3 \ H^+_{out} + ADP + P_i \leftrightharpoons 3 \ H^+_{in} + ATP + H_2O$$

a. If the overall free energy change (ΔG_overall) associated with ATP synthesis in these cells by the coupled process is -21 kJ/mol, what is the equilibrium constant, K_eq, for the process?
b. What is ΔG_synthesis, the free energy change for ATP synthesis, in these bacteria under these conditions?
c. The standard free energy change for ATP hydrolysis is ΔG°'_hydrolysis is -30.5 kJ/mol. If [Pi] = 2 mM in these bacterial cells, what is the [ATP]/[ADP] ratio in these cells?

Answer
a.

$$\Delta G^{\circ\prime} = -RT \ln K_{eq}, \text{ or}$$

$$K_{eq} = e^{\frac{-\Delta G^{\circ\prime}}{RT}} = e^{\frac{-(-21)}{(8.314 \times 10^{-3}) \times 298}} = e^{8.48} = 4.80 \times 10^3 M^{-1}$$

b. The overall free energy change for ATP synthesis accounts for proton movement and ATP synthesis. Because ΔG is a state property we can write:

$$\Delta G_{overall} = \Delta G_{synthesis} + \Delta G_{proton \ movement} \text{ or,}$$

$$\Delta G_{synthesis} = \Delta G_{overall} - \Delta G_{proton \ movement}$$

$$\Delta G_{synthesis} = -21 \ kJ - 3 \ mol \ protons \times (-23 \frac{kJ}{mol \ protons})$$

$$\Delta G_{synthesis} = -21 \ kJ + 69 \ kJ = 48 \ kJ$$

c.

$$\Delta G_{hydrolysis} = \Delta G^{\circ\prime}_{hydrolysis} + RT \ln \frac{[ADP][P_i]}{[ATP]} \text{ or}$$

$$\ln \frac{[ADP][P_i]}{[ATP]} = \frac{\Delta G_{hydrolysis} - \Delta G^{\circ\prime}_{hydrolysis}}{RT}$$

$$\ln\frac{[\text{ADP}][\text{P}_i]}{[\text{ATP}]} = \frac{(-48)-(-30.5)}{(8.314\times10^{-3})(298)} = -7.06$$

$$\frac{[\text{ADP}][\text{P}_i]}{[\text{ATP}]} = e^{-7.06} = 8.56\times10^{-4}$$

$$\frac{[\text{ATP}]}{[\text{ADP}][\text{P}_i]} = \frac{1}{8.56\times10^{-4}}$$

$$\frac{[\text{ATP}]}{[\text{ADP}]} = \frac{[\text{P}_i]}{8.56\times10^{-4}} = \frac{2\times10^{-3}}{8.56\times10^{-4}} = 2.34$$

14. Describing the path of electrons through the Q cycle of complex III
Describe in your own words the path of electrons through the Q cycle of Complex III.

Answer: Complex III moves electrons from coenzyme Q to cytochrome c. A little reflection should tell us that something special has to happen. Cytochrome c is a one-electron carrier whereas CoQ can carry both electrons and protons. Further, it can carry one or two electrons. In the dihydroquinone form, it carries two protons and two electrons. However, in the semiquinone form, it is anionic carrying a single electron (and no protons).

The movement of electrons through Complex III follows the so-called Q-cycle. The point of the cycle is that when an electron is transferred from $CoQH_2$ to cytochrome c through the Rieske protein and cytochrome c_1, two protons are released leaving anionic semiquinone. Semiquinone then passes one electron to cytochrome b_L, which passes it on to cytochrome b_H. The electron is used to convert ubiquinone to semiquinone. This semiquinone remains on the matrix side of the inner mitochondrial membrane until a second cytochrome c is reduced. A second electron via cytochrome b_L and cytochrome b_H reduces semiquinone to dihydroquinone, which protonates with protons from the matrix.

15. Describing the path of electrons through complex IV
Describe in your own words the path of electrons through the copper and iron centers of Complex IV.

Answer: In Complex IV, four electrons and two protons react with O_2 to produce 2 H_2O. This four-electron reduction is accomplished by moving electrons, one at a time from cytochrome c to Cu_A to cytochrome a and then to the binuclear center composed of Cu_B and cytochrome a_3. The Cu^{2+}_B of binuclear center is reduced with the first electron from cytochrome a. Then, the iron center of cytochrome a_3 is reduced by a second electron. With both elements of the binuclear center reduced, oxygen binds and electrons are transferred to both atoms of O_2. The now activated oxygen is cleaved with addition of another electron and two protons to produce a water molecule bound to Cu_B and an anionic oxygen bound to the iron center of cytochrome a_3. This oxygen atom is released as a water molecule upon reduction by a single electron and addition of two protons. In the diagram below, lone pairs are shown as paired black dots. When the two electrons are transferred to O_2, the two oxygens are now joined by a single bond.

$$Fe^{3+} \qquad Cu^{2+}$$
$$\downarrow e^-$$
$$Fe^{3+} \qquad Cu^{2+}$$
$$\downarrow e^-$$
$$O_2 \rightarrow Fe^{2+} \ \ddot{:}O{=}\ddot{O}{:} \ \ Cu^{2+}$$

$$Fe^{3+}-:\ddot{O}-\ddot{O}:- \ Cu^+$$
$$\downarrow e^- \ 2\,H^+$$
$$Fe^{3+} \ -:\ddot{O}\cdot H\pm\ddot{O}\pm H\,Cu^+$$
$$\downarrow$$
$$Fe^{4+} \ -:\ddot{O}:- \quad H_2O \ Cu^+$$
$$\downarrow e^- \ 2\,H^+$$
$$Fe^{3+} \quad H_2O$$

16. Understanding oxidation of unsaturated lipids

In the course of events triggering apoptosis, a fatty acid chain of cardiolipin undergoes peroxidation to release the associated cytochrome c. Lipid peroxidation occurs at a double-bond. Suggest a mechanism for the reaction of hydrogen peroxide with an unsaturation in a lipid chain, and identify a likely product of the reaction.

Answer: Cytochrome c has roles in both electron transport and apoptosis. In electron transport its heme iron interconverts between ferrous and ferric iron (Fe^{2+} to Fe^{3+}) as it moves electrons between complex III and complex IV of the electron transport chain. Its role is apoptosis is rather complex. When released from mitochondria it initiates apoptosome formation leading to activation of caspase and apoptosis. Cytochrome c also has a role in sensing reactive oxygen species and initiating a series of events leading to cytochrome c release from mitochondria. Cytochrome c exists in two forms, soluble and membrane-bound. In its soluble form it participates in single-electron oxidation/reduction reaction through its heme iron. The iron in soluble cytochrome c has bonds to a histidine on one side of the protoporphyrin ring and a methionine on the other. The iron is thus protected from interaction with molecules like oxygen or hydrogen peroxide.

The membrane-bound form of cytochrome c is thought to be associated with cardiolipin in the inner mitochondrial membrane. The fatty acids in cardiolipin (diphosphatidylglycerol) are unsaturated or polyunsaturated and thus susceptible to oxidation. Membrane-bound cytochrome c is conformationally different than soluble cytochrome c. The environment around the heme iron is more open allowing it to interact with hydrogen peroxide. Cardiolipin oxidation is initiated by membrane-bound cytochrome c when hydrogen peroxide binds to the heme. Peroxidation at a carbon-carbon double bond on cardiolipin produces oxidized cardiolipin. Oxidized cardiolipin is thought to migrate to the outer mitochondrial membrane to aid in forming a mitochondrial permeability transition pore that allows soluble cytochrome c to escape. See Kagan, V. E. et al. 2004 Oxidative lipidomics of Apoptosis: Redox catalytic interactions of cytochrome c with cardiolipins and phosphatidylserine *Free Radical Biology and Medicine* 37: 1963-1985.

17. Determining the number of protons in a typical mitochondrion

In problem 16 at the end of Chapter 19, you might have calculated the number of molecules of oxaloacetate in a typical mitochondrion. What about protons? A typical mitochondrion can be thought of as a cylinder 1 μm in diameter and 2 μm in length. If the pH in the matrix is 7.8, how many protons are contained in the mitochondrial matrix?

Answer: The volume of a mitochondrion is calculated as follows

$$V = \pi \left(\frac{d}{2}\right)^2 \times \text{length}$$

$$V = 3.14 \left(\frac{1 \times 10^{-6}}{2}\right)^2 \times 2 \times 10^{-6}$$

$$V = 1.57 \times 10^{-18} \text{m}^3 \times \left(\frac{100 \text{ cm}}{1 \text{ m}}\right)^3 \times \left(\frac{1 \text{ L}}{1000 \text{ cm}^3}\right)$$

$$V = 1.57 \times 10^{-15} \text{L}$$

The hydrogen ion concentration is

$$\text{pH} = -\log[\text{H}^+] \text{ or}$$
$$[\text{H}^+] = 10^{-\text{pH}} = 10^{-7.8}$$
$$[\text{H}^+] = 1.58 \times 10^{-8} \text{M}$$

The number of molecules is the volume times concentration times Avogadro's number or

Number of molecules of $\text{H}^+ = 1.57 \times 10^{-15} \text{L} \times 1.58 \times 10^{-8} \text{M} \times 6.02 \times 10^{23}$

Number of molecules of $\text{H}^+ = 15$

18. Understanding the unique requirement for FAD in the succinate dehydrogenase reaction **Considering that all other dehydrogenases of glycolysis and the TCA cycle use NADH as the electron donor, why does succinate dehydrogenase, a component of the TCA cycle and the electron transfer chain, use FAD as the electron acceptor from succinate, rather than NAD$^+$? Note that there are two justifications for the choice of FAD here—one based on energetics and one based on the mechanism of electron transfer for FAD versus NAD$^+$.**

Answer: Succinate dehydrogenase converts succinate to fumarate. The standard reduction potential for this reaction (actually its reverse) is 0.031 V. The standard reduction potential for NAD$^+$ is -0.320 V and for protein-bound FAD the redox potential ranges from 0.003 to 0.091. (See Table 3.5.) For electrons to be moved from succinate and to a cofactor, the cofactor must have a smaller redox potential. This can be seen from equation 20.2.

$$\Delta G^{o\prime} = -nF \Delta \mathcal{E}_0\prime \text{ or}$$
$$\Delta G^{o\prime} = -n\mathcal{F}(\mathcal{E}_{of}\prime - \mathcal{E}_{oi}\prime)$$

Where $\mathcal{E}_{of}\prime$ and $\mathcal{E}_{oi}\prime$ are the standard reduction potentials of the electron acceptor and the electron donor. For $\Delta G^{o\prime}$ to be negative, $\mathcal{E}_{of}\prime$ must be greater than $\mathcal{E}_{oi}\prime$. To move electrons from succinate it will be the electron donor and it must pass electrons to an acceptor whose standard reduction potential is larger, which is possible for protein-bound FAD but not for NAD$^+$. In other words the oxidation of succinate to fumarate (formation of a carbon-carbon double bond) is not sufficiently energetic enough to support reduction of NAD$^+$ but it will reduce FAD.

Succinate dehydrogenase moves electrons from succinate to coenzyme Q through a number of iron-sulfur centers. In the FeS centers iron participates in single-electron oxidation/reduction reactions and so for electrons to be passed from FADH$_2$ to an FeS center, FAD must be capable of participating in single-electron transfers, which it is.

Preparing for the MCAT® Exam

19. Based on your reading on the F$_1$F$_o$-ATPase, what would you conclude about the mechanism of ATP synthesis:
  a. The reaction proceeds by nucleophilic substitution via the S$_N$2 mechanism.
  b. The reaction proceeds by nucleophilic substitution via the S$_N$1 mechanism.
  c. The reaction proceeds by electrophilic substitution via the E1 mechanism.
  d. The reaction proceeds by electrophilic substitution via the E2 mechanism.

Answer: The correct answer is "a". The reaction involving ADP and P$_i$ can't be an elimination mechanism so the only good choices are substitution mechanisms. Boyer's ^{18}O exchange experiment shows that during hydrolysis of ATP label in incorporated into P$_i$. Water in this case must be serving as a nucleophile, attacking the γ-phosphorus with release of ADP. This is an S$_N$2 reaction.

20. *Imagine that you are working with isolated mitochondria and you manage to double the ratio of protons outside to protons inside. In order to maintain the overall ΔG at its original value (whatever it is), how would you have to change the mitochondria membrane potential?*

Answer: The ΔG for the initial situation is given by:

$$\Delta G_1 = RT\ln\frac{[C2]}{[C1]} + z\mathcal{F}\Delta\Psi_1$$

We are asked to change one side to twice the original value.

Let's change side 2 to $2 \times C2$.

But, we are asked to adjust $\Delta\Psi$ such that ΔG is unchanged.

Let the new potential be $\Delta\Psi_1 + \partial$

$$RT\ln\frac{[C2]}{[C1]} + z\mathcal{F}\Delta\Psi_1 = RT\ln\frac{[2\times C2]}{[C1]} + z\mathcal{F}\left(\Delta\Psi_1 + \partial\right)$$

We now have to solve this equation for ∂.

$$RT\ln\frac{[C2]}{[C1]} + z\mathcal{F}\Delta\Psi_1 = RT\ln 2 + RT\ln\frac{[C2]}{[C1]} + z\mathcal{F}\left(\Delta\Psi_1 + \partial\right)$$

$$\partial = \frac{RT\ln 2}{z\mathcal{F}}$$

Substituting in values for R,T and Faraday constant ($z = +1$)

$$\partial = \frac{8.31451 \times 298 \times \ln 2}{\left(+1\right) \times 96,488}$$

$$\partial = 1.78 \times 10^{-2}$$

$$\partial = 17.8 \text{ mV}$$

Doubling of the proton gradient would change ΔG by $RT\ln 2$ or 1717 J/mol. To compensate for this the electrical potential would have to be changed by the amount indicated. Note: The sign of ΔG would be negative or positive depending on which side, i.e., 1 or 2, we changed in the above equation. Whichever we picked, this would have a tendency to drive protons from the side with the higher concentration. To prevent this we have to change the membrane potential such that that side becomes less positive (or more negative) by 17.8 mV.

Questions for Self Study

1. How does the formula $\Delta G^{o'} = n\mathcal{F}\Delta\mathcal{E}_o'$ predict the direction of electron flow between two reduction half-reactions under standard conditions?

2. The electron transport chain is composed of four complexes. For the statements below indicate which complex or complexes fit the description. The complexes are: NADH-coenzyme Q reductase (complex I); succinate-coenzyme Q reductase (complex II); coenzyme Q-cytochrome c reductase (complex III); cytochrome c oxidase (complex IV).
   a. Contains two copper sites.
   b. Contains protein-bound FAD.
   c. Produces water upon electron transport.
   d. Is involved in the Q cycle.
   e. Moves electrons from substrate directly into the electron transport chain.
   f. Reduces Q.
   g. Is a citric acid cycle enzyme.
   h. Reduces a small protein localized on the outer surface of the inner mitochondrial membrane.
   i. Moves protons across the inner mitochondrial membrane.
   j. Moves electrons from a lipid-soluble mobile electron carrier to a water-soluble mobile electron carrier.
   k. Contains tightly bound FMN.
   l. Transfers electrons to oxygen.
   m. Moves electrons from NADH to CoQ.
   n. Moves electrons from cytochrome c to oxygen.

o. Contains cytochromes.

3. The energy difference for protons across the inner mitochondrial membrane is given by:

$$\Delta G = 2.303RT \times \log \frac{H^+_{out}}{H^+_{in}} + Z\mathcal{F}\Delta\psi$$

How would this equation be modified by the following conditions?
   a. If protons were uncharged species.
   b. If the inner mitochondrial membrane was freely permeable to ions other than protons.
   c. If the equation was written as a function of pH.

4. ATP synthase or F_1F_o-ATPase is composed of two complexes: a peripheral membrane complex and an integral membrane complex. What roles do the complexes play in oxidative phosphorylation?

5. What is the key difference between an electron transport inhibitor and an uncoupler? What are the consequences of each to electron transport, oxygen consumption, and ATP production?

Answers

1. In general, a reaction is spontaneous when $\Delta G < 0$. When the reaction is under standard conditions $\Delta G^{\circ\prime} < 0$. Thus, $\Delta \mathcal{E}_o\prime > 0$, $\mathcal{E}_o\prime_{final}- \mathcal{E}_o\prime_{initial} > 0$ or $\mathcal{E}_o\prime_{final} > \mathcal{E}_o\prime_{initial}$. Electrons will flow from the reduction half-reaction with the smaller standard reduction potential to the reduction half-reaction with the larger standard reduction potential.

2. a. IV; b. II; c. IV; d. III; e. II; f. I and II; g. II; h. III; i. I, III, and IV; j. III; k. I; l. IV; m. I; n. IV; o. III and IV.

3. a. $\Delta G = 2.303RT \times \log \frac{H^+_{out}}{H^+_{in}}$; b. $\Delta G = 2.303RT \times \log \frac{H^+_{out}}{H^+_{in}}$; c. $\Delta G = -2.303RT(pH_{out} - pH_{in}) + Z\mathcal{F}\Delta\psi$

4. The peripheral membrane complex or F_1 unit is the site at which ADP and P_i are joined to form ATP. The integral membrane complex, F_o, is a proton channel.

5. An electron transport inhibitor blocks the movement of electrons through a component of the transport chain. Inhibitors will block electron transport for electrons entering the chain above the blockage point but not for those entering below the blockage point. Oxygen consumption and ATP production will, depending on the source of electrons, be diminished or stopped. Uncoupling agents provide an alternate pathway through which protons can reenter the mitochondrial matrix. Uncouplers stimulate electron transport and oxygen consumption but block ATP synthesis.

Additional Problems

1. Isolated, inner mitochondrial membranes do not transport electrons from NADH to O_2 unless cytochrome c is added. Why?

2. Describe one site in the electron transport chain responsible for pumping hydrogen ions out of the mitochondria during electron transport.

3. Nigericin is an ionophore whose structure is shown below.

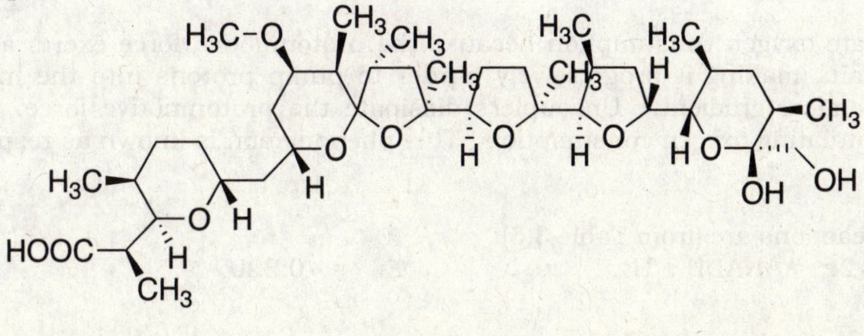

381

It is soluble in the inner mitochondrial membrane in either its acidic form with the carboxyl group protonated or as a potassium salt. Valinomycin is a cyclic compound composed of various L- and D- amino acids and lactic acid and functions as a potassium ionophore. The presence of nigericin, valinomycin, and potassium will effectively destroy the electrochemical potential of mitochondria. Why?

4. An uncoupling agent like dinitrophenol actually stimulates O_2 consumption. Why?

5. In the citric acid cycle, malate dehydrogenase catalyzes the following reaction:

$$\text{malate} + NAD^+ \;\star\; \text{oxaloacetate} + NADH$$

Given the following standard reduction potentials, calculate $\Delta G^{\circ\prime}$

Reduction Half-Reaction	$\mathcal{E}_o'$ (V)
Oxaloacetate $+ 2H^+ + 2\,e^-$ $\star$ malate	-0.166
$NAD^+ + H^+ + 2\,e^-$ $\star$ NADH	-0.320

In which direction will the reaction proceed under standard conditions. Which component functions as the oxidant? Reductant? Does this differ from the physiological direction?

Abbreviated Answers

1. Inner mitochondrial membranes contain all of the components necessary for electron transport except cytochrome c. Cytochrome c is a relatively small protein, M_r = 1,800, localized on the outer surface of the inner mitochondrial membrane and it is readily lost during purification procedures. Cytochrome c moves electrons between coenzyme Q-cytochrome c reductase and cytochrome c oxidase and must be present for electrons to flow from NADH to O_2.

2. There are three sites at which protons are pumped out of the mitochondria in response to electron transport: NADH dehydrogenase, coenzyme Q-cytochrome c reductase, and cytochrome c oxidase. The details of how protons are pumped by cytochrome c oxidase are unknown. Coenzyme Q-cytochrome c reductase is involved in the Q cycle. NADH dehydrogenase moves electrons from NADH to coenzyme Q using a flavoprotein and several Fe-S proteins to move electrons. In NADH dehydrogenase, the flavoprotein accepts electrons and protons from NADH on the matrix side of the inner mitochondrial membrane but donates electrons to a series of Fe-S proteins. CoQ is thought to play two roles in the movement of electrons through NADH dehydrogenase. It is the terminal electron acceptor for the complex. Additionally, CoQ is thought to participate at an intermediate stage of electron flow by accepting electrons from one Fe-S component and donating electrons to another Fe-S component. In the process, CoQ binds protons from the matrix side of the membrane and releases protons into the intermembrane space

3. Since valinomycin is specific for potassium, it will not affect the proton gradient by itself. In the presence of potassium, valinomycin will move potassium into the mitochondrial matrix until potassium is actually concentrated inside the mitochondria. This occurs because initially potassium moves down a potassium chemical gradient (assuming that the mitochondrial matrix has a low potassium concentration) until the potassium concentration is equal on both sides of the membrane. But, potassium continues to accumulate, being driven inside the cell by the electrical potential of the proton gradient. Net movement of potassium stops when the potassium concentration difference across the membrane is sufficient to produce a potassium electrical potential equal in magnitude and opposite in sign from the electrical potential of the proton gradient. However, the proton gradient still exists.

    Nigericin alone will simply replace the proton gradient with a potassium gradient. The chemical potential of the proton gradient is effectively destroyed but an electrical potential, now due to potassium, is still present.

    The combination of nigericin, valinomycin, and potassium will dissipate the proton gradient without creating a potassium gradient.

4. Uncouplers stimulate oxygen consumption because the protonmotive force exerts a backpressure on the electron transport chain, making it progressively harder to pump protons into the inner membrane space against the electrochemical gradient. Uncouplers dissipate the protonmotive force, thereby relieving this backpressure and stimulating oxygen consumption. This phenomenon is known as respiratory control.

5. The relevant half reactions are (from Table 3.5):

$$NAD^+ + 2H^+ + 2e^- \;\star\; NADH + H^+, \qquad \mathcal{E}_o' = -0.320$$

and,

oxaloacetate + 2H$^+$ + 2e^- $\rightleftharpoons$ malate,　　　　$\mathcal{E}_o'$ = -0.166

Thus, oxaloacetate is the electron acceptor and NADH is the electron donor.

$$\Delta\mathcal{E}_o'=\mathcal{E}_o'(\text{acceptor}) - \mathcal{E}_o'(\text{donor}) = \mathcal{E}_o'(\text{NAD}^+) - \mathcal{E}_o'(\text{oxaloacetate})$$

$$\Delta\mathcal{E}_o'=-0.320-(-0.166)=-0.154 \text{ V}$$

The free energy change for the reaction is given by:

$\Delta G^{\circ\prime}=-n\mathcal{F}\Delta\mathcal{E}_o'$, where n is the number of electrons, and

$$\mathcal{F}=\text{Faraday's constant}=96.485\frac{\text{kJ}}{\text{V}\cdot\text{mol}}$$

$$\Delta G^{\circ\prime}=-2\times 96.485\frac{\text{kJ}}{\text{V}\cdot\text{mol}}\times(-0.154 \text{ V})=+29.7\frac{\text{kJ}}{\text{mol}}$$

A positive $\Delta G^{\circ\prime}$ indicates that the reaction, under standard conditions, is not favorable and will proceed in the reverse direction. Electrons will flow from NADH to oxaloacetate. Thus, NADH serves as a reductant and oxaloacetate serves as an oxidant. In the citric acid cycle, removal of oxaloacetate by citrate synthase lowers the concentration of oxaloacetate and drives the reaction toward NADH production.

Summary

Most of the metabolic energy obtained from sugars and similar metabolites entering glycolysis and the TCA cycle is funneled into NADH and [FADH$_2$] via oxidation/reduction reactions. Electrons stored in the form of reduced coenzymes and flavoproteins are then passed through an elaborate and highly organized chain of proteins and coenzymes, the so-called electron transport chain, finally reaching O$_2$, molecular oxygen, the terminal electron acceptor. In the course of electron transport, energy is stored in the form of a proton gradient across the inner mitochondrial membrane. This proton gradient then provides the energy to drive ATP synthesis in oxidative phosphorylation. Electron transport and oxidative phosphorylation are membrane-associated processes. Bacteria carry out these processes at (and across) the plasma membrane, and in eukaryotic cells they are carried out and localized in mitochondria. Mitochondria are surrounded by a simple outer membrane and a more complex inner membrane. The proteins of the electron transport chain and oxidative phosphorylation are associated with the inner mitochondrial membrane.

The tendency of the components of the electron transport chain to transfer electrons to other species is characterized by an electron transfer potential or standard reduction potential, $\mathcal{E}_o'$. These potentials are determined by measuring the voltages generated in reaction half cells. Standard reduction potentials are measured relative to a reference half cell. The H$^+$/H$_2$ half cell normally serves as the reference half cell and is assigned a standard reduction potential of 0.0 V. The sign of $\mathcal{E}_o'$ for a redox couple can be used to indicate whether reduction or oxidation occurs for that couple in any real reaction (involving two redox couples). The redox couple with the more negative $\mathcal{E}_o'$ will act as an electron donor, and the redox couple with the more positive $\mathcal{E}_o'$ will act as an electron acceptor. For redox couples with a large positive standard reduction potential, the oxidized form of the couple has a strong tendency to be reduced.

Fractionation of the membranes containing the electron transport chain results in the isolation of four distinct complexes, including I) NADH-coenzyme-Q reductase, II) succinate-coenzyme-Q reductase, III) coenzyme-Q-cytochrome c reductase, and IV) cytochrome c oxidase. Oxidation of NADH and reduction of coenzyme-Q by complex I is accompanied by transport of protons from the matrix to the cytoplasmic side of the inner mitochondrial membrane. Complex II, succinate-coenzyme-Q reductase or succinate dehydrogenase, is an integral membrane protein, a TCA cycle enzyme and a flavoprotein. The net reaction carried out by this complex is oxidation of succinate and reduction of coenzyme-Q, with intermediate oxidation/reduction of [FADH$_2$]. In complex III, reduced coenzyme-Q passes its electrons to cytochrome c via a complex redox cycle involving three cytochromes, several FeS centers and proton transport across the inner mitochondrial membrane. Cytochrome c, a mobile electron carrier, passes electrons from complex III to complex IV, cytochrome c oxidase. Electron transfer in complex IV involves two hemes and two copper sites in cytochrome a and a$_3$ and results in proton transport across the inner mitochondrial membrane. In spite of their spatial proximity, the four major complexes of the electron transport chain operate and migrate independently in the inner mitochondrial membrane.

The elucidation of the mechanism which couples electron transport and ATP synthesis was one of the great challenges (and triumphs) of modern biochemistry. Peter Mitchell's chemiosmotic hypothesis postulates that a proton gradient generated by the electron transport chain is utilized to drive ATP synthesis. The mitochondrial complex that carries out ATP synthesis is called ATP synthase or F$_1$F$_0$-ATPase. The F$_1$ subunit of this complex is spherical in shape, consists of five major polypeptide chains and catalyzes ATP synthesis. The F$_0$ subunit forms the transmembrane pore or channel through which protons move to drive ATP synthesis. The essence of the Mitchell hypothesis was confirmed in a reconstitution experiment by E. Racker and W. Stoeckenius, using bacteriorhodopsin to generate a proton gradient and the ATP synthase to generate ATP. As shown by Paul Boyer, the energy provided by electron transport drives enzyme conformation changes which regulate the binding and release of substrates on ATP synthase. The mechanism involves catalytic cooperativity between three interacting sites. Many of the details of electron transport and oxidative phosphorylation have been gained from studying the effects of specific inhibitors, including rotenone, which inhibits NADH-UQ reductase, and cyanide, which blocks complex IV, cytochrome c oxidase. Uncouplers dissipate the proton gradient across the inner mitochondrial membrane which was created by electron transport. Endogenous uncouplers including thermogenin provide heat for the organism by uncoupling these processes. Molecular shuttle systems exist to carry the electrons of NADH into the mitochondrial matrix, because NADH and $FADH_2$ cannot be transported across the inner mitochondrial membrane. The two systems that operate in this manner are the glycerol phosphate shuttle and the malate-aspartate shuttle. ATP produced by oxidative phosphorylation is carried out of the mitochondria by an ATP/ADP translocase, which couples the exit of ATP with the entry of ADP.

The net yield of energy from glucose oxidation - either ~30 or ~32 ATP synthesized per glucose - depends upon the shuttle system used. The overall efficiency of metabolism (from glucose to the TCA cycle to electron transport and oxidative phosphorylation) is about 54%.

Chapter 21

Photosynthesis

• •

Chapter Outline

❖ Photosynthesis: Two aspects
  ⋏ Light reactions: NADPH and ATP produced, oxygen evolved, water is electron donor
  ⋏ Dark reactions: Carbon dioxide fixed
❖ Chloroplasts: Plastids
  ⋏ Outer membrane: Porous
  ⋏ Intermembrane space
  ⋏ Inner membrane
  ⋏ Stroma: Volume within inner membrane but outside thylakoids
  ⋏ Thylakoid: Paired folds
    • Lamella
    • Thylakoid vesicles
    • Grana: Stacks of thylakoid vesicles
    • Thylakoid space (lumen)
❖ Photosynthetic pigments
  ⋏ Chlorophyll: Porphyrin-like molecule with Mg^{2+}
  ⋏ Secondary light-harvesting pigments: Broaden wavelength range
    • Carotenoids
    • Phycobilins
❖ Light energy
  ⋏ Absorption of energy
    • $E = h\upsilon$: h = Planck's constant, υ = frequency of light
    • $\upsilon = c/\lambda$: c = speed of light in a vacuum, λ = wavelength
    • Light photons: Packet of energy or quantum: $E = h\upsilon$
  ⋏ Energy excites π electrons
  ⋏ Release of energy
    • Heat
    • Fluorescence: Light but at longer wavelength: Lower frequency: Less energetic
    • Resonance energy transfer: Near neighbors exchange exciton: Förster resonance energy transfer
    • Energy transduction: Molecule with excited electron becomes better reductant
❖ Characteristics of photosynthesis
  ⋏ Membrane-localized
  ⋏ Pigments organized into units
    • Antennae molecules harvest light
    • Reaction center: Energy sink of photosynthetic unit
  ⋏ Eukaryotes: Two photosystems
    • PSI: 700 nm: Excited electrons reduce $NADP^+$
    • PSII: 680 nm: Splits water: Generates oxygen: Pumps protons: Reduces PSI
  ⋏ Prokaryotes: PSII-like but without oxygen evolution capacity
❖ Protein complexes: Two light harvesting complexes

⋏ PSI complex: Provides reducing power for production of NADPH (Used in Calvin cycle to fix carbon dioxide

⋏ PSII complex: Splits water producing oxygen (byproduct) and electrons: Electrons moved to PSI via electron transport chain

⋏ Cytochrome b_6/cytochrome f complex

• Moves electrons from PSII to PSI: Protons pumped (ATP produced): Noncyclic flow of electrons

• Moves electrons from PSI to PSI: Protons pumped (ATP produced): Cyclic flow of electrons

❖ Z Scheme: Movement of electrons from water via PSII through cyto b_6/cyto f complex through PSI to $NADP^+$

❖ PSII activities: H_2O/plastiquinone oxidoreductase

⋏ P680 reaction center complex cycles through S_0, S_1, S_2, S_3, S_4

• Each step removes one electron from water

• With S_4 oxygen is released

⋏ P680 to P680* to $P680^+$ by exciton absorption and electron release

⋏ Electron from P680 passes to pheophytin: Chlorophyll with 2 H^+ in place of Mg^{2+}

⋏ Pheophytin passes electron to plastoquinone

⋏ Plastoquinone (lipid mobile) passes electrons to cyto b_6/cyto f

⋏ $P680^+$ to P680: Electron from water through Mn complex and D (tyrosyl residue)

❖ Cyto b_6/cyto f activities: Plastoquinol/plastocyanin oxidoreductase

⋏ Cytochrome f: Moves electrons from plastoquinone to plastocyanin

• Plastocyanin: Copper atom involved in single-electron transfers

• Diffuses along inside of thylakoid along membrane

⋏ Cytochrome b_6 involved in cyclic electron transport: Cyclic photophosphorylation

❖ PSI activities: Plastocyanin/ferredoxin oxidoreductase

⋏ P700 to P700* to $P700^+$ by exciton absorption and electron release

⋏ Electron from P700 passes to quinone (A1)

⋏ Quinone (A1) passes electron to membrane-bound ferredoxins

⋏ Membrane-bound ferredoxins pass electrons to soluble ferredoxin

⋏ Soluble ferredoxin passes electrons to $NADP^+$: Ferredoxin/$NADP^+$ reductase

❖ Quantum yields

⋏ Eight photons per O_2 evolved: Four per PSI and four for PSII

⋏ One proton per photon: Eight protons pumped

⋏ Three protons per ATP

⋏ Two NADPH produced

❖ Photophosphorylation

⋏ Proton translocations

• Water to PSII

• Q-cycle

⋏ $\Delta p = \Delta \Psi - 2.3RT \times \Delta pH/(\mathcal{F})$

⋏ CF_1CF_oATP synthase couples proton release to ATP production

⋏ Noncyclic photophosphorylation: Electrons from water to $NADP^+$

⋏ Cyclic photophosphorylations: Electrons from PSI through cyto b_6 back to PSI

❖ Calvin-Benson cycle

⋏ Only known pathway of net CO_2 fixation

⋏ Immediate product 3-phosphoglycerate: Reduced to glyceraldehyde-3-P using NADPH

⋏ G-3-P used to produce 4-, 5-, 6-, and 7-C compounds

⋏ Hexose principle product

❖ Rubisco: CO_2 + ribulose-1,5-BP → two 3-PG

⋏ Fixes carbon dioxide to carbon skeleton

⋏ Most abundant enzyme in biosphere

❖ Regulation of CO_2 fixation

⋏ Changes in stromal pH

• Light pumps protons into thylakoid

● Stromal pH increases to 8

o Rubisco carbamylation promoted

o Fructose-1,6-BPase activated

- o Ribulose-5-P kinase activated
- o G-3-P dehydrogenase activated
- • Generation of reducing power
  - • F-1,6-BPase activated by reduction
  - • Sedoheptulose-1,7-BPase activated by reduction
  - • Ribulose-5-P kinase activated by reduction
- • Mg^{2+} efflux upon proton influx (charge balance)
  - • Rubisco activated
  - • F-1,6-BPase activated
- ❖ Photorespiration: Rubisco: Ribulose bisphosphate carboxylase/oxygenase
  - ⅄ Oxygenase activity produces 3-PG + phosphoglycolate: O_2 consumed
  - ⅄ Phosphoglycolate converted to glyoxylate: O_2 consumed
  - ⅄ Glyoxylate to glycine to serine to 3-PB
  - ⅄ Glycine to serine: CO_2 produced
- ❖ Hatch-Slack pathway: C-4 plants: CO_2 delivery system
  - ⅄ PEP + CO_2 to oxaloacetate to malate: Mesophyll cells
  - ⅄ Malate to pyruvate + CO_2: Bundle sheath cells
  - ⅄ Pyruvate and malate exchanged
- ❖ Crassulacean acid metabolism
  - ⅄ Stomata closed during the day: Malate decarboxylated to produce CO_2
  - ⅄ Stomata opened during the night: CO_2 + PEP to oxaloacetate and malate

Chapter Objectives

Photosynthesis

It is convenient to divide photosynthesis into light and dark reactions. Keep in mind that the light reactions harvest light energy and transduce it into the chemical energy of ATP and NADPH. ATP serves its usual role as energy currency. NADPH is a cellular reducing agent. In the dark reactions, CO_2 is fixed into organic compounds and reduced to a hydrated carbon. The objective is to produce carbohydrates from CO_2, the chemistry is carbon reduction, and the biochemistry is the Calvin cycle.

Light Reactions

Light energy is used to split water in a reaction called photolysis producing O_2, H^+, and electrons. The electrons move along an electron transport chain leading to $NADP^+$. Analogous to the electron transport chain in mitochondria, electron flow in chloroplasts (and other photosynthetic systems) is accompanied by H^+ gradient formation, which is used, in turn, to synthesize ATP. Get the big picture down and then fill in the detail.

The abstraction of electrons from water is an energy-requiring process and in photosynthesis, the energy derives from sunlight. How is this energy harvested? Chlorophylls and accessory light-harvesting pigments are responsible. Production of O_2 from 2 H_2O generates 8 electrons. In the "Z" system, generation of each electron requires two quanta of light (only one quantum per electron is needed for the cyclic pathway of PSI). This explains why the pigment molecules are organized into photosynthetic units. The pigment molecules in a unit cooperate to collect light quanta and funnel the energy to a reaction center. Photosynthetic eukaryotes have two separate but interacting photosystems, PS-I and PS-II. PS-II is responsible for O_2 production and has a reaction center P680. PS-1 reduces $NADP^+$ if coupled to PSII or by itself can move electrons in a cycle beginning and ending at its reaction center P700. The movement of electrons from water to $NADP^+$ requires excitation of PS-II, electron transport from PS-II to PS-I, excitation of PS-I, and electron transport from PS-I to $NADP^+$. Electron transport from PS-II to PS-I is accompanied by proton-gradient formation. This gradient is subsequently used to produce ATP. When we get to the Calvin cycle we will see that there is a greater requirement for ATP than NADPH and the additional ATP can be generated by cyclic photophosphorylation. This process involves only PS-I. An excited electron is fed from PS-I into the electron transport chain leading from PS-II to PS-I and thus moves in a circle. However, cyclic electron flow is accompanied by proton-gradient formation and subsequent ATP production.

Dark Reactions

Rubisco is the key player in CO_2 fixation. Understand the carboxylase reaction leading to two phosphoglycerates. Rubisco also catalyzes an oxygenase reaction, which leads to photorespiration. The carboxylase reaction adds carbon at the oxidation state of a carboxylate group to ribulose-1,5-bisphosphate to produce phosphoglycerate. The carboxyl groups now must be reduced to aldehydes for carbohydrate synthesis to occur. The initial steps include phosphorylation to 1,3-BPG and reduction to glyceraldehyde-3-phosphate and P_i. (We saw these steps in glycolysis and they will appear again when we take up gluconeogenesis in the next chapter.) The remainder of the Calvin cycle (Figure 21.25) converts glyceraldehyde to glucose, not an onerous task given our knowledge of gluconeogenesis, and regenerates ribulose-1,5-bisphosphate. This aspect of the cycle is reminiscent of the nonoxidative phase of the pentose phosphate pathway and uses isomerases, aldolases, transketolases, phosphatases, and a kinase to convert three-carbon sugars into five-carbon sugars.

Regulation

Calvin cycle activity is coordinated with the light reactions by three light-induced effects: light-induced pH changes, generation of reducing power, and Mg^{2+} efflux. Know the targets for these controls. Rubisco regulation by Mg^{2+} is dependent on CO_2 availability.

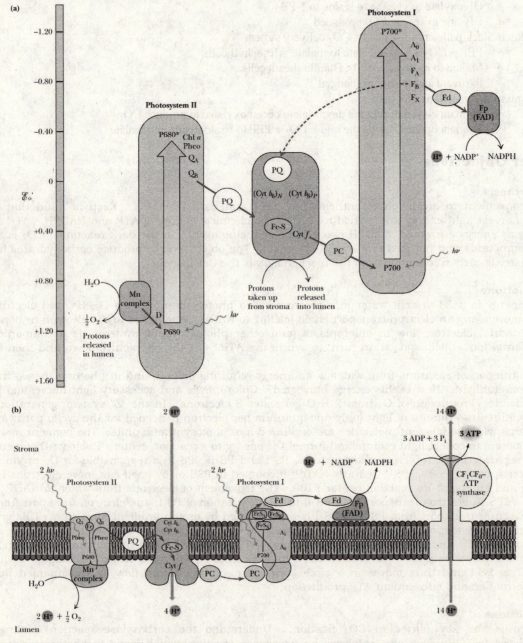

Figure 21.11 The Z-scheme of photosynthesis. The Z-scheme is a diagrammatic representation of photosynthetic electron flow from H_2O to $NADP^+$.

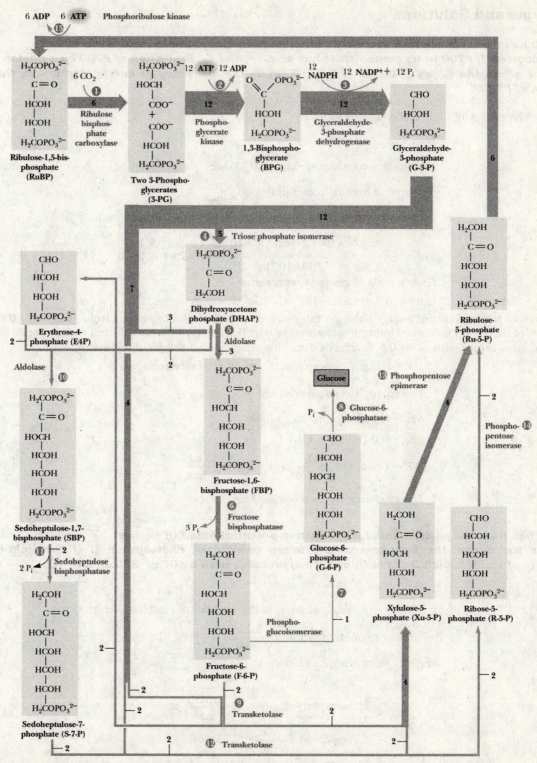

Figure 21.25 The Calvin-Benson cycle of reactions. The number of arrows at each step indicates the number of molecules reacting in a turn of the cycle that produces one molecule of glucose. Reactions are numbered as in Table 21.1 in the Garrett and Grisham text.

Problems and Solutions

1. P700 has the most negative standard reduction potential found in Nature
In Photosystem I, P700 in its ground state has an $\mathcal{E}_o' = +0.4$ V. Excitation of P700 by a photon of 700-nm light alters the $\mathcal{E}_0'$ of P700 to -0.6 V. What is the efficiency of energy capture in this light reaction of P700?*

Answer: The energy of a photon of light is given by:

$$E = \frac{hc}{\lambda},$$

h= Planck's constant=6.626×10^{-34} J · sec

c= speed of light in a vacuum=$3 \times 10^8 \frac{m}{sec}$

λ = wavelength of light. In this case λ=700 nm=700×10^{-9}m

Thus, $E=\dfrac{(6.626 \times 10^{-34} \text{ J} \cdot \text{sec}) \times (3 \times 10^8 \frac{m}{sec})}{700 \times 10^{-9} \text{m}} = 2.84 \times 10^{-19}$ J

And, a mole of photons represents

$(2.84 \times 10^{-19} \text{ J}) \times 6.02 \times 10^{23} = 171$ kJ of energy

Absorption of this amount of energy changes the redox potential of photosystem I from +0.4 V to -0.6 V. We can think of this as an oxidation-reduction reaction in which an electron is transferred from a donor with $\mathcal{E}_o'$ = +0.4 V to an acceptor with $\mathcal{E}_o'$ = -0.6 V. The $\Delta G^{\circ\prime}$ for this reaction can be calculated using:

$\Delta G^{\circ\prime}$=-n$\mathcal{F}\Delta\mathcal{E}_O'$, where n is the number of electrons, and

$\mathcal{F}$=Faraday's constant=$96.485 \frac{kJ}{V \cdot mol}$

$\Delta\mathcal{E}_O'=\Delta\mathcal{E}_0'$(acceptor) - $\Delta\mathcal{E}_0'$(donor)

$\Delta\mathcal{E}_O'$=-0.6-(+0.4)=-1.0 V

$\Delta G^{\circ\prime}$=-1 × 96.485$\frac{kJ}{V \cdot mol}$ × (-1.0 V)=96.5$\frac{kJ}{mol}$

The efficiency =$\dfrac{96.5}{171} \times 100\% = 56.4\%$

2. PSII has the most positive standard reduction potential found in Nature
What is the $\mathcal{E}_o'$ for the light-generated primary oxidant of Photosystem II if the light-induced oxidation of water (which leads to O_2 evolution) proceeds with a $\Delta G^{\circ\prime}$ of -25 kJ/mol?

Answer:

From $\Delta G^{\circ\prime}$=-n$\mathcal{F}$ $\Delta\mathcal{E}_O'$, where n is the number of electrons, and

$\mathcal{F}$=Faraday's constant=$96.485 \frac{kJ}{V \cdot mol}$

$\Delta\mathcal{E}_O'=\mathcal{E}_0'$(acceptor)-$\mathcal{E}_0'$(donor)

We see that:

$\mathcal{E}_0'$(acceptor) - $\mathcal{E}_0'$(donor) = $\dfrac{-\Delta G^{\circ\prime}}{n\mathcal{F}}$, or

$\mathcal{E}_0'$(acceptor) = $-\dfrac{\Delta G^{\circ\prime}}{n\mathcal{F}}+\mathcal{E}_0'$(donor)

$\mathcal{E}_0'$(acceptor)=$-\dfrac{-25\frac{kJ}{mol}}{4 \times 96.485\frac{kJ}{V \cdot mol}}+0.816$

$\mathcal{E}_0'$(acceptor) = 0.881 V

Generation of O_2 is a four-electron transfer, so n = 4.

3. Equating the energy of an oxidation-reduction potential difference and the energy for ATP synthesis
(Integrates with Chapters 3 and 20.) Assuming that the concentrations of ATP, ADP, and P_i in chloroplasts are 3 mM, 0.1 mM, and 10 mM, respectively, what is the ΔG for ATP synthesis under these conditions? Photosynthetic electron transport establishes the proton-motive force driving photophosphorylation. What redox potential difference is necessary to achieve ATP synthesis under the foregoing conditions, assuming 1.3 ATP equivalents are synthesized for each electron pair transferred.

Answer:

$$\Delta G = \Delta G°' + RT \times \ln\frac{[ATP]}{[ADP][P_i]}$$

At T=25°C (298 K),

$$\Delta G = 30.5\frac{kJ}{mol} + (8.314\times10^{-3}\frac{kJ}{mol\cdot K})(298 \text{ K})\times\ln\frac{3\times10^{-3}}{(0.1\times10^{-3})(10\times10^{-3})}$$

$$\Delta G = 30.5\frac{kJ}{mol} + 19.8\frac{kJ}{mol}$$

$$\Delta G = 50.3\frac{kJ}{mol}$$

From $\Delta G = -n\mathcal{F}\Delta\mathcal{E}$, where n is the number of electrons, and

$$\mathcal{F} = \text{Faraday's constant} = 96.485\frac{kJ}{V\cdot mol}$$

$$\Delta\mathcal{E} = -\frac{\Delta G}{n\mathcal{F}} = -\frac{1.3\times50.3\frac{kJ}{mol}}{2\times96.485\frac{kJ}{V\cdot mol}}$$

$$\Delta\mathcal{E} = -0.34 \text{ V}$$

4. The Q cycle in photosynthesis
(Integrates with Chapters 20.) Write a balanced equation for the Q cycle as catalyzed by the cytochrome b_6f complex of chloroplasts.

Answer: In the Q cycle, reduced plastoquinone, PQH_2, passes one electron to cytochrome b_6 and the other to cytochrome f and releases two protons into the thylakoid lumen. Cytochrome f passes electrons to plastocyanin (PC), which carries electrons to photosystem I. The electron on cytochrome b_6 is passed to quinone, Q, to make semiquinone, Q^-. Cytochrome b_6/f is reduced a second time by QH_2, which releases two protons into the thylakoid lumen. This results in reduction of a second PC through cytochrome f. And, reduction of semiquinone to fully reduced quinone, which picks up protons from the stroma to become QH_2. The balanced equation is:

$$PQH_2 + 2H^+_{Stroma} + 2\text{ PC}(Cu^{2+}) \rightarrow PQ + 2\text{ PC}(Cu^+) + 4\text{ } H^+_{Thylakoid}$$

5. The relative efficiency of ATP synthesis in noncyclic versus cyclic photophosphorylation
If noncyclic photosynthetic electron transport leads to the translocation of 3 H^+/e^- and cyclic photosynthetic electron transport leads to the translocation of 2 H^+/e^-, what is the relative photosynthetic efficiency of ATP synthesis (expressed as the number of photons absorbed per ATP synthesized) for noncyclic versus cyclic photophosphorylation? (Assume that the CF_1CF_0 ATP synthase yields 3 ATP/14 H^+).

Answer: In noncyclic photosynthetic electron transport, photosystems I and II are excited by absorption of light quanta. Thus, two quanta are needed to excite one electron, which translocates 3 protons.

$$\frac{3\ H^+}{e^-} \times \frac{e^-}{2\ \text{quanta}} = \frac{3\ H^+}{2\ \text{quanta}}$$

But, ATP is produced by

$$\frac{3\ \text{ATP}}{14\ H^+} \times \frac{3\ H^+}{2\ \text{quanta}} = \frac{9\ \text{ATP}}{28\ \text{quanta}} = \frac{0.32\ \text{ATP}}{\text{quanta}}$$

Quanta (number of photons) per ATP is the inverse of this

$$\frac{3.11\ \text{quanta}}{\text{ATP}}$$

In cyclic photosynthetic electron transport, photosystems I is excited by absorption of light quanta. Thus, one quantum is needed to excite one electron, which translocates 2 protons.

$$\frac{2\ H^+}{e^-} \times \frac{e^-}{1\ \text{quanta}} = \frac{2\ H^+}{\text{quanta}}$$

But, ATP is produced by

$$\frac{3\ \text{ATP}}{14\ H^+} \times \frac{2\ H^+}{\text{quanta}} = \frac{6\ \text{ATP}}{14\ \text{quanta}} = \frac{0.43\ \text{ATP}}{\text{quanta}}$$

Quanta (number of photons) per ATP is the inverse of this

$$\frac{2.33\ \text{quanta}}{\text{ATP}}$$

6. Delta pH and delta psi in the chloroplast proton-motive force
(Integrates with Chapters 20.) In mitochondria, the membrane potential ($\Delta\psi$) contributes relatively more to $\Delta\rho$ (proton-motive force) than does the pH gradient (ΔpH). The reverse is true in chloroplasts. Why do you suppose that the proton-motive force in chloroplasts can depend more on ΔpH than mitochondria can? Why is ($\Delta\psi$) less in chloroplasts than in mitochondria?

Answer: Both processes involve translocation of protons in response to electron transport. Since protons are charged species, net movement should produce both a proton gradient and a charge separation, leading to an electrical potential. In chloroplasts, however, movement of protons into the thylakoid lumen is countered by movement of Mg^{2+} out of the lumen. The chloroplast is, in effect, leaky to charge and so the electrical potential does not build up to the extent it does in mitochondria. The electrochemical potential across the thylakoid membrane is dominated by the proton gradient. In mitochondria, both a proton gradient and a membrane potential occur.

The pH change that occurs in mitochondria happens in the matrix of the mitochondria and in the cytoplasm of the cell (because the outer mitochondrial membrane is permeable to small molecules). In chloroplasts, however, the change in pH is restricted to the stroma and the lumen of the thylakoids. Mitochondrial pH changes may be restricted in magnitude because of pH-induced activity changes that might accompany large pH changes. The cytoplasm of plant cells does not experience the pH change due to chloroplast activity and so a bigger pH change may be tolerated.

7. The role of Try161 in PSII
Predict the consequences of a Y161F mutation in the amino acid sequence of the D1 and the D2 subunits of PSII.

Answer: The notation Y161F indicates a mutation changing the amino acid located at position 161 from tyrosine (Y) to phenylalanine (F). (The codons for tyrosine are UAU an UAC and those for phenylalanine are UUU and UUC. So, a simple one-base transversion of A to T would cause this change.) In photosynthesis, movement of electrons from H_2O to P680 involves manganese and tyrosine 161, which becomes a free radical in its oxidized form. Tyrosine's hydroxyl group participates in the mechanism of electron transfer to a nearby manganese ion. Replacing tyrosine with phenylalanine would clearly prevent this interaction from occurring. This would prevent P680, once activated by light to P680* and then to P680+ by electron transfer, from returning back to the ground state.

8. H⁺/ATP ratio in photosynthetic bacteria

(Integrates with Chapters 20.) **Calculate (in Einsteins and in kJ/mol) how many photons would be required by the Rhodopseudomonas viridis *photophosphorylation system to synthesize 3 ATPs? (Assume that the R. viridis F_1F_0-ATP synthase c-subunit rotor contains 12 c-subunits and that the R. viridis cytochrome b/c_1 complex translocates 2 H⁺/e⁻.)***

Answer: *R. viridis* is a photosynthetic prokaryote. It has only a single photosytem, thus activation of one electron requires a single photon of light. If we assume that 12 c-subunits corresponds to movement of 12 H⁺ per production of 3 ATP (i.e., one turn of the c-subunit rotor to turn the trimeric F_1 complex once, producing 3 ATP).

$$\frac{3\ \text{ATP}}{12\ \text{H}^+} \times \frac{2\ \text{H}^+}{\text{e}^-} \times \frac{1\ \text{e}^-}{\text{quantum}} = \frac{0.5\ \text{ATP}}{\text{quantum}} = \frac{1\ \text{ATP}}{2\ \text{quanta}}$$

The energy of a quantum of light is given by:

$$E = \frac{hc}{\lambda},$$

$$h = \text{Planck's constant} = 6.626 \times 10^{-34}\,\text{J}\cdot\text{sec}$$

$$c = \text{speed of light in a vacuum} = 3 \times 10^{8}\,\frac{\text{m}}{\text{sec}}$$

$$\lambda = \text{wavelength of light.}$$

If we assume $\lambda = 680$ nm, the longest wavelength of light capable of exciting PSII.

$$\lambda = 680\ \text{nm} = 680 \times 10^{-9}\text{m}$$

Thus, $E = \dfrac{(6.626 \times 10^{-34}\,\text{J}\cdot\text{sec}) \times (3 \times 10^{8}\,\frac{\text{m}}{\text{sec}})}{680 \times 10^{-9}\text{m}} = 2.92 \times 10^{-19}\text{J}$

This is the amount of energy in one photon.

One ATP requires 2 photons or $5.84 \times 10^{-19}\text{J}$

An Einstein is the amount of energy in an Avogadro's number of photons. The amount of energy in an Einstein of 680 nm light is:

$$5.84 \times 10^{-19}\text{J} \times 6.023 \times 10^{23} = 351\ \text{kJ}$$

9. The standard-state free energy change for NADP-glyceraldehyde-3-P dehydrogenase

(Integrates with Chapters 3 and 20.) **Calculate $\Delta G^{\circ\prime}$ for the NADP⁺-specific glyceraldehyde-3-P dehydrogenase reaction of the Calvin-Benson cycle.**

Answer: In the Calvin cycle, glyceraldehyde-3-P dehydrogenase moves electrons from NADPH to 1,3-bisphosphoglycerate producing glyceraldehyde-3-phosphate and NADP⁺. The $\Delta G_0{}^\prime$ is calculated as follows:

$$\Delta G^{\circ\prime} = -n\mathcal{F}\Delta\mathcal{E}_0{}^\prime, \text{ where n is the number of electrons, and}$$

$$\mathcal{F} = \text{Faraday's constant} = 96.485\,\frac{\text{kJ}}{\text{V}\cdot\text{mol}}$$

$$\Delta\mathcal{E}_0{}^\prime = \mathcal{E}_0{}^\prime(\text{acceptor}) - \mathcal{E}_0{}^\prime(\text{donor})$$

$$\Delta\mathcal{E}_0{}^\prime = \mathcal{E}_0{}^\prime(1,3\text{BPG}) - \mathcal{E}_0{}^\prime(\text{NADPH})$$

$$\Delta\mathcal{E}_0{}^\prime = -0.290 - (-0.320) = 0.030\ \text{V}$$

$$\Delta G^{\circ\prime} = -2 \times 96.485\,\frac{\text{kJ}}{\text{V}\cdot\text{mol}} \times (0.030) = -5.79\,\frac{\text{kJ}}{\text{mol}}$$

10. Photosynthesis of glucose

Write a balanced equation for the synthesis of a glucose molecule from ribulose-1,5-bisphosphate and CO_2 that involves the first three reactions of the Calvin cycle and subsequent conversion of the two glyceraldehyde-3-P molecules into glucose by a reversal of glycolysis.

Answer:

$$\mathbf{CO_2 + H_2O + RuBP \rightarrow 2\ 3\text{-PG}}$$
$$\mathbf{2\ 3\text{-PG} + 2\ ATP \rightarrow 2\ 1,3\text{-BPG} + 2\ ADP}$$
$$\mathbf{2\ 1,3\text{-BPG} + 2\ NADPH + 2\ H^+ \rightarrow 2\ NADP^+ + 2\ P_i + 2\ G\text{-3-P}}$$
$$\mathbf{G\text{-3-P} \rightarrow DHAP}$$

$$G\text{-}3\text{-}P + DHAP \rightarrow F\text{-}1,6\text{-}BP$$
$$F\text{-}1,6\text{-}BP + H_2O \rightarrow F\text{-}6\text{-}P + P_i$$
$$F\text{-}6\text{-}P \rightarrow G\text{-}6\text{-}P$$
$$G\text{-}6\text{-}P + H_2O \rightarrow glucose + P_i$$

Net: $CO_2 + RuBP + 3\ H_2O + 2\ ATP + 2\ NADPH + 2\ H^+ \rightarrow glucose + 2\ ADP + 2\ NADP^+ + 4\ P_i$

11. Tracing the fate of CO_2 during photosynthesis

^{14}C-labeled carbon dioxide is administered to a green plant, and shortly thereafter the following compounds are isolated from the plant: 3-phosphoglycerate, glucose, erythrose-4-phosphate, sedoheptulose-1,7-bisphosphate, ribose-5-phosphate. In which carbon atoms will radioactivity be found?

Answer: CO_2 fixation is catalyzed by ribulose bisphosphate carboxylase (rubisco) in the following reaction:

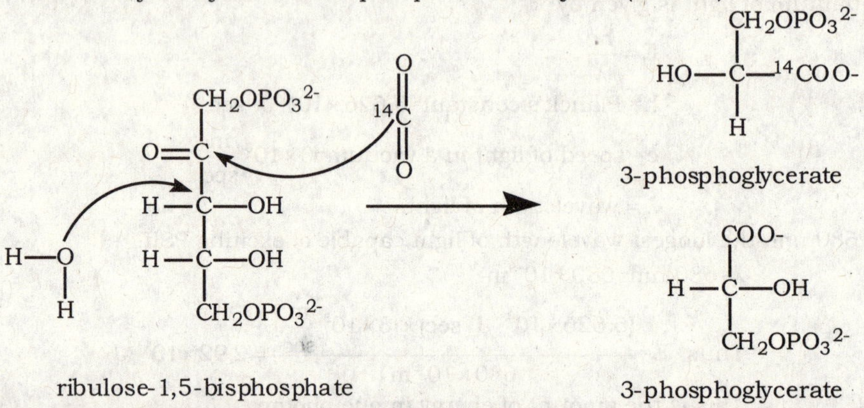

ribulose-1,5-bisphosphate

3-phosphoglycerate

Thus, the first compound labeled is carbon 1 of 3-phosphoglycerate. In the Calvin cycle, this compound is used, ultimately, to produce glucose. Synthesis starts with the conversion of 3-phosphoglycerate to 1,3-bisphosphoglycerate by phosphoglycerate kinase. 1,3-Bisphosphoglycerate is then converted to glyceraldehyde-3-phosphate by glyceraldehyde-3-phosphate dehydrogenase. Starting with carbon 1 of 3-phosphoglycerate labeled, this carbon is carbon 1 of 1,3-bisphosphoglycerate, and carbon 1 of both glyceraldehyde-3-phosphate and dihydroxyacetone phosphate (produced from glyceraldehyde-3-phosphate by triose phosphate isomerase action). Both glyceraldehyde-3-phosphate and dihydroxyacetone phosphate are used to produce glucose. The action of aldolase will combine these two three-carbon compounds to produce fructose-1,6-bisphosphate. This aldol condensation reaction joins carbon 1 of dihydroxyacetone phosphate to carbon 1 of glyceraldehyde-3-phosphate. Thus, carbons 3 and 4 of fructose-1-6-bisphosphate acquire label, which will show up in the same carbons in glucose.

In the Calvin cycle, transketolase is responsible for production of erythrose-4-phosphate. The four carbons of erythrose derive from carbons 3, 4, 5 and 6 of fructose-6-phosphate. Thus, we expect erythrose-4-phosphate to be labeled at carbons 1 and 2.

The transketolase reaction moves carbon 1 and the keto carbon of fructose to glyceraldehyde-3-phosphate to produce xylulose-5-phosphate, which may be converted to ribulose-5-phosphate by phosphopentose epimerase. By this path, label is expected at carbon 3 of ribulose-5-phosphate. (Ribulose is also labeled at carbons 1, 2 and 3. See below.)

The fate of sedoheptulose-1,7-bisphosphate in the Calvin cycle is as follows. This compound is synthesized from erythrose-4-phosphate and dihydroxyacetone phosphate by aldolase. Thus, we expect label at carbons 3, 4, and 5 from carbon 1 of dihydroxyacetone phosphate and carbons 1 and 2 of erythrose-4-phosphate, respectively. Transketolase can convert sedoheptulose to ribose-5-phosphate by transferring carbons 1 and 2 (the keto carbon) to glyceraldehyde-3-P. The ribose formed by this path derives from carbons 3, 4, and 5 of sedoheptulose and will be labeled at C-1, 2, and 3.

12. Regulation of CO_2 fixation during photosynthesis

The photosynthetic CO_2 fixation pathway is regulated in response to specific effects induced in chloroplasts by light. What is the nature of these effects, and how do they regulate this metabolic pathway?

Answer: Ribulose-1,5-bisphosphate carboxylase (rubisco), the enzyme responsible for the initial fixation of carbon dioxide, is activated indirectly by light. Rubisco requires a specific post-translational modification, a carbamylation of lysine-201, and subsequent magnesium ion binding, which is dependent on carbamylation,

394

for activity. One of the substrates of rubisco, namely ribulose-1,5-bisphosphate, is a potent inhibitor of carbamylation. It binds to the active site and prevents carbamylation from occurring. This inhibition is relieved by rubisco activase, a regulatory protein that is activated by light, binds to rubisco, and promotes release of ribulose-1,5-bisphosphate in an ATP-dependent manner.

Light indirectly activates rubisco and several other enzymes in another manner. Rubisco, fructose-1,6-bisphosphatase, ribulose-5-phosphate kinase, and glyceraldehyde-3-phosphate dehydrogenase have alkaline pH optima. The light-dependent proton pumping into the thylakoid space raises the pH of the stroma leading to activation of these enzymes.

Light-dependent proton pumping is accompanied by an efflux of Mg^{2+} ions, which counteracts the charge imbalance caused by H^+. Ribulose bisphosphate carboxylase and fructose-1,6-bisphosphatase are both magnesium-activated enzymes.

Several enzymes including fructose-1,6-bisphosphatase, sedoheptulose-1,7-bisphosphatase, and ribulose-5-phosphate kinase are activated upon reduction of critical disulfide bonds, as mediated by reduced thioredoxin. Thioredoxin is reduced by NADPH, one of the products of noncyclic light-driven photosynthetic electron transport.

13. The formation of phosphoglycerate during photorespiration
Write a balanced equation for the conversion of phosphoglycolate to glycerate-3-P by the reactions of photorespiration. Does this balanced equation demonstrate that photorespiration is a wasteful process?

Answer:

$$2\ (\textbf{Phosphoglycolate} + \textbf{H}_2\textbf{O} \rightarrow \text{Glycolate} + \textbf{P}_i)$$
$$2\ (\text{Glycolate} + \textbf{O}_2 \rightarrow \text{Glyoxylate} + \textbf{H}_2\textbf{O}_2)$$
$$2\ (\text{Glyoxylate} + \textbf{Serine} \rightarrow \text{Glycine} + \text{Hydroxypyruvate})$$
$$2\ \text{Glycine} \rightarrow \textbf{Serine} + \textbf{CO}_2 + \textbf{NH}_3$$
$$2\ (\text{Hydroxypyruvate} + \textbf{NADH} + \textbf{H}^+ \rightarrow \text{Glycerate} + \textbf{NAD}^+)$$
$$2\ (\text{Glycerate} + \textbf{ATP} \rightarrow \textbf{3-Phosphoglycerate} + \textbf{ADP})$$

Net: 2 Phosphoglycolate + 2 H_2O + 2 O_2 + 2 ATP + 2 NADH + 2 H^+ + Serine →3-Phosphoglycerate + CO_2 + ~~NH_3~~ + 2 NAD^+ + 2 ADP + 2 H_2O_2 + 2 P_i

To clean up this reaction, we could use the glutamate dehydrogenase reaction to convert ammonium and α-ketoglutarate into glutamate.

$$NH_3 + \alpha\text{-Ketoglutarate} + \textbf{NADH} + \textbf{H}^+ \rightarrow \text{Glutamate} + \text{H}_2\text{O} + \textbf{NAD}^+$$

Net: 2 Phosphoglycolate + ~~α-Ketoglutarate~~ + H_2O + 2 O_2 + 2 ATP + 3 NADH + 3 H^+ + ~~Serine~~ → ~~Glutamate~~ + 3-Phosphoglycerate + CO_2 + 3 NAD^+ + 2 ADP + 2 H_2O_2 + 2 P_i

Using an aminotransferase we could convert glutamate into serine and α-ketoglutarate with hydroxypyruvate as keto acceptor.

$$\text{glutamate} + \text{hydroxypyruvate} \rightarrow \alpha\text{-Ketoglutarate} + \text{Serine}$$

Net: 2 Phosphoglycolate + ~~hydroxypyruvate~~ + H_2O + 2 O_2 + 2 ATP + 3 NADH + 3 H^+ → 3-Phosphoglycerate + CO_2 + 3 NAD^+ + 2 ADP + 2 H_2O_2 + 2 P_i

Finally, we could convert one phosphoglycerate into glycerate and then into hydroxypyruvate as follows: glutamate into serine and α-ketoglutarate with hydroxypyruvate as keto acceptor.

$$3\text{-Phosphoglycerate} + \text{ADP} \rightarrow \text{Glycerate} + \text{ATP}$$
$$\text{Glycerate} + \text{NAD}^+ \rightarrow \text{Hydroxypyruvate} + \text{NADH} + \text{H}^+$$

Net: 2 Phosphoglycolate + H_2O + 2 O_2 + ATP + 2 NADH + 2 H^+ → 3-Phosphoglycerate + CO_2 + 2 NAD^+ + ADP + 2 H_2O_2 + 2 P_i. According to this balanced equation production of 3-phosphoglycerate requires hydrolysis of ATP and oxidation of 2 NADH, two energy-rich molecules.

14. The source of the oxygen atoms in photosynthetic O_2 evolution
The overall equation for photosynthetic CO_2 fixation is
$$6\ CO_2 + 6\ H_2O \star C_6H_{12}O_6 + 6\ O_2$$
All the O atoms evolved as O_2 come from water; none comes from carbon dioxide. But 12 O atoms are evolved as 6 O_2, and only 6 O atoms as 6 H_2O in the equation. Also, 6 CO_2 have 12 O atoms, yet there are only 6 O atoms in $C_6H_{12}O_6$. How can you account for these discrepancies? (Hint: Consider the partial reactions of photosynthesis: ATP synthesis, $NADP^+$ reduction, photolysis of water, and the overall reaction for hexose synthesis in the Calvin-Benson cycle.)

Answer: The net reaction for the Calvin Cycle given in Table 3.5 is:
$$6\ CO_2 + 18\ ATP + 12\ NADPH + 12\ H^+ + 12\ H_2O \star\ glucose + 18\ ADP + 18\ P_i + 12\ NADP^+$$
But, ATP was produced by phosphorylation of ADP as follows:
$$18\ ADP + 18\ P_i \star 18\ ATP + 18\ H_2O$$
NADPH was generated by
$$12\ H_2O + 12\ NADP^+ \star 12\ NADPH + 12\ H^+ + 6\ O_2$$
The net reaction is:
$$6\ CO_2 + 6\ H_2O \star glucose + 6\ O_2$$
What the overall reaction doesn't tell us is the source of the O_2. To generate six O_2, 12 H_2Os are split by the light reactions of photosynthesis. However, ATP used in the Calvin cycle accounts for production of 18 H_2Os.

15. The influence of c-subunit stoichiometry on the efficiency of ATP synthesis
The number of c-subunits in F_1F_0-type ATP synthases shows some variation from organism to organism. For example, the yeast ATP synthase contains 10 c-subunits, the spinach CF_1CF_0–ATP synthase has 14, and the cyanobacterium Spirulina platensis enzyme apparently has 15.
*   a. What is the consequence of c-subunit stoichiometry for the H^+/ATP ratio?*
*   b. What is the relationship between c-subunit stoichiometry and the magnitude of Δp (the proton-motive force)?*

Answer: ATP synthase has three binding sites of ATP corresponding to three stages of phosphorylation of ADP to ATP. One turn of the synthase is required to complete the process and the number of c-subunits determines the number of steps taken to complete one turn. Each step involves movement of a proton. As the number of c-subunits increase the ration of H^+/ATP will increase.

The proton-motive force (Δp) can be smaller with increasing numbers of c-subunits.

16. The energetics of hydroquinone reduction by photosynthetic reaction centers
The reduction of membrane-associated quinones, such as coenzyme Q and plastoquinones, is a common feature of photosystems (see Figures 21.15, 21.20, and 21.22). Assume E_o' for PQ/PQH_2 = 0.07 V and the potential of the ground-state chlorophyll molecule = 0.5 V, calculate ΔG for the reduction of plastoquinone by
*   a. 870-nm light.*
*   b. 700-nm light.*
*   c. 680-nm light.*

Answer: The energy associated with a mole of a quantum of light is given by:
$$E = N\frac{hc}{\lambda},$$
h= Planck's constant=6.626×10^{-34}J $\cdot$ sec

c= speed of light in a vacuum=$3 \times 10^8\ \dfrac{m}{sec}$

λ = wavelength of light

N = Avagodro's number = $6.02\ \times 10^{23}$

Using this equation we can calculate the energy associated with light of specific wavelengths.

Wavelength	kJ	kJ per 2 moles
870	137.5	275.0
700	171.0	342.0
680	176.1	352.2

Reduction of plastoquinone involves two electrons, each of which are activated by a quantum of light. Reduction of plastoquinone using a ground-state chlorophyll is unfavorable as shown below.

$$\Delta G^{o\prime} = -n\mathcal{F}\Delta\mathcal{E}_o\prime, \text{ where n is the number of electrons, and}$$

$$\mathcal{F} = \text{Faraday's constant} = 96.485\frac{kJ}{V\cdot mol}$$

$$\Delta\mathcal{E}_o\prime = \mathcal{E}_o\prime(\text{acceptor}) - \mathcal{E}_o\prime(\text{donor})$$

$$\Delta\mathcal{E}_o\prime = \mathcal{E}_o\prime(\text{plastoquinone}) - \mathcal{E}_o\prime(\text{chlorophyll})$$

$$\Delta\mathcal{E}_o\prime = 0.07 - (0.5) = -0.43 \text{ V}$$

$$\Delta G^{o\prime} = -2\times 96.485\frac{kJ}{V\cdot mol}\times(-0.43) = 83.0 \text{ kJ}$$

The overall ΔG for each wavelength is

Wavelength	ΔG (kJ)
870	-275.0 + 83.0 = -192
700	-342.0 + 83.0 = -259
680	-352.2 + 83.0 = -269

17. The energetics of proton translocation by the cytochrome b_6/f complex
Plastoquinone oxidation by cytochrome bc_1 and cytochrome b_6f complexes apparently leads to the translocation of 4 $H^+/2e^-$. If $E_o\prime$ for cytochrome f = 0.365 V (Table 3.5) and $E_o\prime$ for PQ/PQH_2 = 0.07 V, calculate ΔG for the coupled reaction:

$$2\ h\nu + 4\ H^+_{in} \rightarrow 4\ H^+_{out}$$

(Assume a value of 23 kJ/mol for the free energy change (ΔG) associated with moving protons from inside to outside.)

Answer: ΔG for movement of protons from inside to outside is given as 23 kJ/mol. However, 4 protons are moved so $\Delta G = 4\times 23 = 92$ kJ
ΔG for reduction of plastoquinone using cytochrome f is calculated as follows:

$$\Delta G^{o\prime} = -n\mathcal{F}\Delta\mathcal{E}_o\prime, \text{ where n is the number of electrons, and}$$

$$\mathcal{F} = \text{Faraday's constant} = 96.485\frac{kJ}{V\cdot mol}$$

$$\Delta\mathcal{E}_o\prime = \mathcal{E}_o\prime(\text{acceptor}) - \mathcal{E}_o\prime(\text{donor})$$

$$\Delta\mathcal{E}_o\prime = \mathcal{E}_o\prime(\text{cytochrome f}) - \mathcal{E}_o\prime(\text{plastoquinone})$$

$$\Delta\mathcal{E}_o\prime = 0.365 - (0.07) = 0.295 \text{ V}$$

$$\Delta G^{o\prime} = -2\times 96.485\frac{kJ}{V\cdot mol}\times(0.295) = -56.9 \text{ kJ}$$

The ΔG for the coupled reaction is 92 − 56.9 = +35.1 kJ.

18. The overall free energy change for photosynthetic $NADP^+$ reduction
What is the overall free energy change (ΔG) for noncyclic photosynthetic electron transport?
$$4\ (700\text{-nm photons}) + 4\ (680\text{-nm photons}) + 2\ H_2O + 2\ NADP^+ \rightarrow O_2 + 2\ NADPH + 2\ H^+$$

Answer: In problem 16 we calculated the energy associated with a mole of quanta of various wavelengths using the following formula.

$$E = N\frac{hc}{\lambda},$$

h = Planck's constant = $6.626\times10^{-34} J\cdot sec$

c = speed of light in a vacuum = $3\times10^8 \frac{m}{sec}$

λ = wavelength of light

N = Avagodro's number = 6.02×10^{23}

The energy for 700 nm and 680 nm is 171 kJ/mol and 176 kJ/mol. The equation presented in problem 18 involves 4 photons thus the total amount of light energy absorbed is (171×4 + 176×4) = 1388 kJ.

For the reduction of $NADP^+$ by H_2O:

$$\Delta G^{\circ\prime} = -n\mathcal{F}\Delta\mathcal{E}_0\prime, \text{ where n is the number of electrons, and}$$

$$\mathcal{F} = \text{Faraday's constant} = 96.485\frac{kJ}{V\cdot mol}$$

$$\Delta\mathcal{E}_0\prime = \mathcal{E}_0\prime(\text{acceptor}) - \mathcal{E}_0\prime(\text{donor})$$

$$\Delta\mathcal{E}_0\prime = \mathcal{E}_0\prime(NADP^+) - \mathcal{E}_0\prime(H_2O)$$

$$\Delta\mathcal{E}_0\prime = -0.320 - 0.816 = -1.136 \text{ V (values from Table 3.5)}$$

$$\Delta G^{\circ\prime} = -4 \times 96.485\frac{kJ}{V\cdot mol} \times (-1.136) = 438 \text{ kJ}$$

The overall ΔG is -1388 kJ + 438 kJ = -950 kJ.

Preparing for the MCAT® Exam

19. From Figure 21.5, predict the spectral properties of accessory light-harvesting pigments found in plants.

Answer: Figure 21.5 gives absorption spectra of chlorophylls showing that in the ranges 400 nm to 470 nm and 620 nm to 680 nm, light is absorbed. We might expect plants to have evolved to take advantage of the full spectrum of visible light from 400 nm to 700 nm, and so we might expect accessory light-harvesting pigments to cover the range from 470 nm to 620 nm. Many green plants use beta-carotene as an accessory pigment. The absorption spectrum covers up to about 520 nm. Brown algae use fucoxanthin, which covers the spectrum up to about 620 nm. Red algae employ phycoerythrin and phycocyanin whose absorption spectra cover the range from about 440 nm to about 680 nm.

20. Draw a figure analogous to Figure 21.27, plotting [Mg²⁺] in the stroma and thylakoid lumen on the y-axis and dark-light-dark on the x-axis.

Answer: Figure 21.27 shows pH in stroma and thylakoid space in response to dark and light cycles. In the light, light-driven proton pumping moves protons from the stroma into the thylakoid space thereby increasing the pH of the stroma and decreasing the pH of the thylakoid space. When light-driven proton pumping occurs, Mg^{2+} is moved in the opposite direction. There is nearly an equal movement of charge due to magnesium and proton movement such that the electrical potential is unchanged. Thus, we expect a plot of $[Mg^{2+}]$ versus light and dark and light cycles to look similar to what is presented in Figure 21.27 but with some important differences. The y-axis scale is pH, which is a log scale. To substitute $[Mg^{2+}]$ we would have to adjust the scale greatly. Otherwise, the plot should look identical. In light, we expect the concentration of magnesium in the stroma to increase because $[H^+]$ is decreasing, evidenced by the increase in pH (= $-\log[H^+]$).

If we wanted to maintain the log scale we would have to plot pMg^{2+}, which is defined as $-\log[Mg^{2+}]$. The plot would look nearly identical to Figure 21.27 except the labels "Stroma" and "Thylakoid space" would be substituted. And, the location of the lines would be slightly different because magnesium has twice the charge of a proton.

Questions for Self Study

1. Match an item in the first column with an item in the second.
   a. Plastid
   b. Thylakoid vesicle
   c. Granum
   d. Inner membrane
   e. Outer membrane
   f. Lamella

   1. Highly permeable
   2. Stack of thylakoid vesicles
   3. Chloroplast
   4. Flattened sac
   5. Connects grana
   6. Highly impermeable

2. What are the four possible fates of the energy of a quantum of light after it is absorbed by photosynthetic pigments?

3. Two experiments helped to characterize eukaryotic photosystems: (1) measurements of oxygen evolution, normalized to pigment content, as a function of the amount of incident light energy; and, (2) oxygen evolution as a function of wavelength. For each of the statements below, indicate the appropriate experiment.

    a. Showed the red drop or Emerson effect.
    b. Used flashes of white light.
    c. Evidence for photosynthetic units.
    d. Required two light flashes to get maximum oxygen evolution.
    e. Evidence for the existence of two separate photosystems.

4. Fill in the blanks. Eukaryotic photosystems use light energy to move electrons from ___ to ___. During this process ___ is produced as a by-product in a reaction known as ____. Also, protons are moved from the ____ of the chloroplast into the ____. This movement produces a proton gradient that is subsequently used to phosphorylate ADP to produce ATP in a process known as ___. The electrons follow a path known as the ___, which requires ___ quanta of light per electron to function. Light-dependent electron flow with the production of ATP can occur without by-product formation in a process known as ___ ___. This process involves only one of the photosystems, ___.

5. For the reactions, a and b, shown below, identify the reactants and products and name the enzymatic activities.

a.

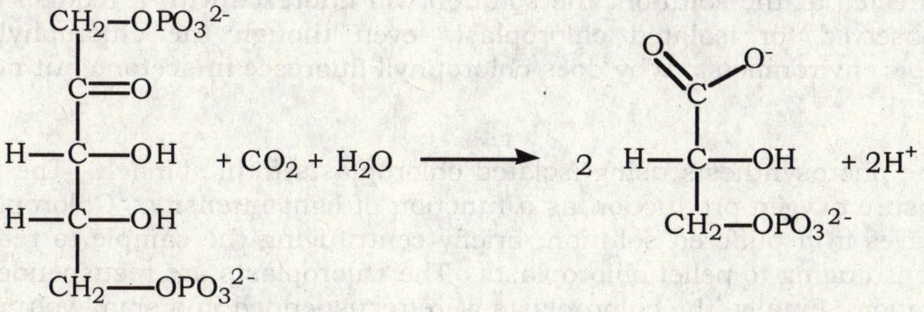

$$
\begin{array}{c}
CH_2-OPO_3^{2-} \\
| \\
C=O \\
| \\
H-C-OH \\
| \\
H-C-OH \\
| \\
CH_2-OPO_3^{2-}
\end{array}
+ CO_2 + H_2O \longrightarrow 2
\begin{array}{c}
COO^- \\
| \\
H-C-OH \\
| \\
CH_2-OPO_3^{2-}
\end{array}
+ 2H^+
$$

b.

$$
\begin{array}{c}
CH_2-OPO_3^{2-} \\
| \\
C=O \\
| \\
H-C-OH \\
| \\
H-C-OH \\
| \\
CH_2-OPO_3^{2-}
\end{array}
+ O_2 \longrightarrow
\begin{array}{c}
COO^- \\
| \\
H-C-OH \\
| \\
CH_2-OPO_3^{2-}
\end{array}
+
\begin{array}{c}
CH_2-OPO_3^{2-} \\
| \\
COO^-
\end{array}
+ 2H^+
$$

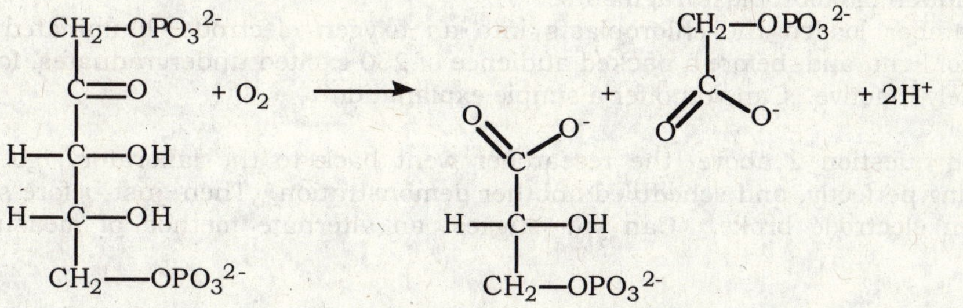

6. The Calvin cycle can be thought of as consisting of two phases, a reductive phase and a nonreductive phase. What are the inputs and outputs of each phase?

7. What is the role of malate production from PEP in C-4 plants? In CAM plants?

8. Carbon dioxide fixation is regulated by what three light-induced signals?

Answers

1. a. 3; b. 4; c. 2.; d. 6; e. 1; f. 5.

2. Lost as heat, lost as light (fluorescence), resonance energy transfer, energy transduction.

3. a. 2; b. 1; c. 1; d. 2; e. 2.

4. Water; NADP⁺; oxygen; photolysis; stroma; thylakoid lumen; photophosphorylation; Z scheme; 2; cyclic photophosphorylation; photosystem I (or P700).

5.   a. Ribulose-1,5-bisphosphate, carbon dioxide and water react to form two molecules of 3-phosphoglycerate. The activity is the carboxylase activity of rubisco.
b. Ribulose-1,5-bisphosphate and oxygen combine to produce 3-phosphoglycerate and phosphoglycolate. The activity is oxygenase activity of rubisco.

6. In the reductive phase, ribulose-1,5-bisphosphate, carbon dioxide, water, ATP, and reducing equivalents in the form of NADPH are used to produce glyceraldehyde-3-phosphate. In the nonreductive phase, triose phosphates are used to produce glucose and to regenerate ribulose-1,5-bisphosphate consumed in the first phase.

7. In C-4 plants, malate serves as a carbon dioxide delivery system from mesophyll cells to bundle sheath cells where the Calvin cycle occurs. This allows carbon dioxide fixation to be carried out in a relatively oxygen-poor environment lessening the possibility of photorespiration. In CAM plants, malate production allows the plants to accumulate carbon dioxide at night when loss of water is minimized.

8. Decrease in stromal pH, increase in reducing power, and Mg^{2+} efflux from thylakoid vesicles.

Additional Problems

1. Chlorophyll can be extracted from plant tissue with acetone. The acetone extract is a dark green solution. If an intense beam of visible light is directed at the solution, the solution will fluoresce with a reddish hue. However, this fluorescence is not observed for isolated chloroplasts even though the chlorophyll in chloroplasts is dissolved in a hydrophobic environment. Why does chlorophyll fluoresce in acetone but not in chloroplasts?

2. A researcher decided to demonstrate photosynthesis using isolated chloroplasts from spinach. The plan was to use an oxygen electrode to measure oxygen production as a function of light intensity. Chloroplasts were isolated by disrupting spinach leaves in a buffered solution, briefly centrifuging the sample to remove cell walls and other debris, and then centrifuging to pellet chloroplasts. The chloroplasts are resuspended in fresh buffer and repelleted by centrifugation. Finally, the chloroplasts were resuspended in a small volume of buffer to make a concentrated solution of chloroplast fragments.

   With great fanfare, the researcher loaded the chloroplasts into an oxygen electrode, illuminated the electrode with an intense source of light, and, before a packed audience of 250 excited undergraduates, found that the preparation was completely inactive. Can you offer a simple explanation?

3. In the scenario described in question 2 above, the researcher went back to the laboratory, got the photosynthesis experiment working perfectly, and scheduled another demonstration. Then, just before show time disaster struck, the oxygen electrode broke. Can you suggest an alternate method of measuring photosynthesis?

4. What do the so-called dark reactions of photosynthesis accomplish?

5. The Hatch-Slack pathway is involved in carbon dioxide fixation in C₄ plants. Describe this pathway (i.e. that portion of the pathway unique to C₄ plants) and identify a similar set of enzymes in animal cells that would accomplish this same process.

6. During electron transport in chloroplasts, hydrogen ions are pumped from one side of a membrane to another. Verbally or in a diagram describe where protons are concentrated in chloroplasts. How is this topologically analogous to proton pumping in mitochondria?

7. The metal ion in chlorophyll is magnesium. In hemoglobin and myoglobin and in the cytochromes, the metal ion is iron. In one sense the magnesium in chlorophyll behaves much like the iron in both hemoglobin and myoglobin, yet chlorophyll itself functions like the cytochrome porphyrins. Explain.

8. The membrane-bond ferredoxins found in the electron transport chain of photosystem I can pass electrons to either soluble ferredoxin or to cytochrome bf complex. What are the outcomes of each path?

9. What experimental evidence supports the hypothesis that pigments are organized into photosynthetic units?

Abbreviated Answers

1. Fluorescence occurs when an electron is excited to a higher energy state, makes a transition to a slightly lower energy state, and then falls back to the unexcited ground state with loss of energy as a photon of light. Isolated chlorophyll fluoresces because the excited state has a long half-life, making the transition to the slightly lower energy level more probable. In chloroplasts, the excited state is effectively quenched by resonance energy transfer so no florescence is observed.

2. For oxygen production to occur a terminal electron acceptor is necessary. Normally, $NADP^+$ serves this role and this small molecule cofactor was probably lost during chloroplast isolation.

3. One possibility is to use a pH meter to measure light-induced proton gradient formation. Oxygen production is accompanied by movement of protons into the thylakoid lumen. In a weakly buffered solution, this will cause a drop in pH that can be measured with a pH meter.

4. The dark reactions constitute the Calvin cycle in which CO_2 is fixed to ribulose-1,5-bisphosphate to produce 2 molecules of 3-phosphoglycerate, which are subsequently reduced by NADPH-dependent glyceraldehyde-3-phosphate dehydrogenase to glyceraldehyde-3-phosphate. Glyceraldehyde-3-phosphate is used to synthesize glucose and to resynthesize ribulose-1,5-bisphosphate.

5. In mesophyll cells, the enzyme pyruvate-phosphate dikinase catalyzes the conversion of pyruvate to PEP. PEP carboxylase produces oxaloacetate using PEP and bicarbonate (derived from CO_2) which is converted to either malate by an NADPH-specific malate dehydrogenase or to aspartate by transamination, and these four-carbon compounds are exported to bundle-sheath cells. In the bundle-sheath cells, aspartate is converted to oxaloacetate by transamination and oxaloacetate is converted to pyruvate and CO_2 by a $NADP^+$-dependent oxidation catalyzed by malic enzyme. The CO_2 is subsequently used in the Calvin cycle whereas the pyruvate is transported back to the mesophyll cells for recycling.

   In animal cells, the gluconeogenic enzyme pyruvate carboxylase produces oxaloacetate from pyruvate and CO_2 in the mitochondria. Oxaloacetate is converted to either malate or aspartate and exported to the cytosol. In the cytosol, malate and aspartate are converted back to oxaloacetate; however, PEP carboxykinase degrades oxaloacetate to PEP and CO_2.

6. During electron transport in chloroplasts, protons are moved from the stroma into the thylakoid lumen. The lumen pH decreases and the stromal pH increases. In mitochondria, protons are pumped from the matrix to the intramembrane space, which is connected to the cytosol. Thus, the cytosolic pH decreases and the mitochondrial matrix pH increases. It appears that the topology is quite different: the stroma of the chloroplast and the matrix of the mitochondria increase in pH upon proton pumping, and the intramembrane space of mitochondria and the thylakoid lumen of chloroplast decrease in pH. However, the biosynthesis of thylakoids is thought to occur by an invagination of the chloroplast outer membrane. Thus, the thylakoid lumen is topologically equivalent to the cytosol.

7. The irons in hemoglobin and myoglobin do not change oxidation state during normal physiological function and remain in the ferrous (2+) oxidation state. The magnesium ion in chlorophyll behaves in a similar manner in that it does not change its oxidation state yet chlorophyll itself in oxidized and reduced during photosynthesis. The cytochromes are similarly oxidized and reduced during electron transport. However, the iron atom undergoes a change in oxidation state during the process.

8. Reduction of soluble ferredoxin leads to reduction of $NADP^+$, whereas reduction of cytochrome *bf* complex leads to cyclic flow of electrons, proton-gradient formation, and ATP synthesis.

9. By measuring O_2 evolved per flash of light at varying intensities of light flashes using the alga *Chlorella* the following data were obtained.

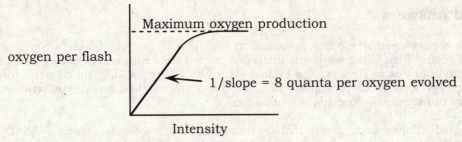

oxygen per flash

Maximum oxygen production

1/slope = 8 quanta per oxygen evolved

Intensity

At saturating intensities of light, the maximum oxygen production corresponds to one oxygen evolved per 2400 chlorophylls in this organism. Thus, the chlorophylls must be acting in units of about this size.

Summary

Photosynthesis is the transformation of light energy into the chemical energy necessary to sustain life. Traditionally, synthesis of carbohydrate ($C_6H_{12}O_6$) has been viewed as the end product of the photosynthetic process, but, in reality, the primary products of photosynthesis are ATP and NADPH, the major energy sources for biosynthesis. These agents, as suppliers of phosphorylation potential and reducing potential, respectively, power many cellular processes, and therefore aptly reflect the role of photosynthesis as the ultimate and essentially universal source of biological energy.

The photosynthetic reactions are intimately associated with membranes, and in photosynthetic eukaryotes, a specialized organelle, the chloroplast, is the site of the photosynthetic process. Photosynthesis can be divided into the "light" reactions (the photochemical generation of energy) and the "dark" reactions (the utilization of chemical energy to drive biosynthesis of carbohydrate in the dark). These different reactions are localized in different chloroplast compartments: the light-dependent processes occur in membrane vesicles (the thylakoids), whereas the dark reactions occur in the stroma (the cytosol-like compartment of the chloroplast).

The transformation of light energy into chemical energy is a manifestation of the photoreactivity of chlorophyll. This π-electron-rich macrocycle readily absorbs photons (quanta of light energy), which promote π-electrons to higher energy orbitals, thereby rendering the Chl molecule a much more effective electron donor. Thus, excited-state chlorophyll (Chl*) initiates a series of electron transfer (oxidation-reduction) reactions whereby light energy is ultimately transduced into reducing power (NADPH) and phosphorylation potential (ATP).

The chlorophyll of photosynthetic organisms is organized into photosynthetic units in order to more efficiently harvest the incident light energy. Within such units, an array of several hundred Chl molecules serve as an antenna to collect photons and funnel quanta of energy to specialized Chl molecules, reaction centers, where the photochemical event occurs. All Chl is in specific Chl-protein complexes which reflect the roles of Chl in the photosynthetic process: a light harvesting complex (LHC), a photosystem I (PSI) complex functioning in NADP+ reduction, and a photosystem II (PSII) complex involved in oxygen evolution. These three complexes are integral components of the thylakoid membrane. PSI and PSII contain unique reaction center chlorophylls, designated P700 in PSI and P680 in PSII.

PSII and PSI act in concert to mediate the light-driven transfer of electrons from H_2O to reduce NADP+. The photoexcitation of P680 in PSII leads to the generation of a weak reductant, QH_2 (a reduced form of plastoquinone), and a strong oxidant, P680+, which, via accepting electrons from water, leads to oxygen evolution. The reductant, QH_2, is coupled to PSI by an integral membrane complex containing Fe/S centers and cytochromes b_6 and f. Electrons transferred from Q via the cytochrome f complex serve to re-reduce the PSI reaction center Chl, P700+. P700 upon photoexcitation generates a strong reductant, P700*, capable of reducing Fe/S centers whose E'_0 are more negative than -0.6 V. The P700+ thus formed has its electron hole filled by an electron from plastocyanin, the blue copper protein serving as e- acceptor to the cytochrome b_6/cytochrome f complex. The Fe/S centers on the reducing side of P700* provide electrons from water to reduce NADP+. Their arrangement according to the redox potential scale and their organization into the unique PSI, PSII complexes bridged by the cytochrome b_6/cytochrome f complex gives rise to a Z-shaped pathway, the so-called "Z" scheme of photosynthesis.

The primary events in photosynthesis are light-induced electron transfers which of necessity result in a situation where oxidized donor and reduced acceptor, for example, P700+ and Chl a-, are in close proximity. These neighboring, oppositely charged species invite charge recombination and consequent dissipation of the photochemical energy. To prevent this, the electron transfer reactions following photo-excitation of reactive centers must be very rapid so that the energized electron is quickly conducted away from the oxidized form of its original donor.

ATP produced by photosynthesis arises because photosynthetic electron transfer is accompanied by transmembrane proton translocations, which establish a proton-motive force across the membrane. This electrochemical H^+ gradient is tapped by the chloroplast ATP synthase, CF_1CF_0 ATP synthase. Quantitative measurements indicate that 1 H^+ is translocated per photon entering the photosynthetic system and that the translocation of 3 H^+ generates the necessary electrochemical gradient to drive the synthesis of 1 ATP under conditions of noncyclic photophosphorylation. The other photophosphorylation pattern, cyclic photophosphorylation, involves a cyclic electron transfer reaction wherein light energy incident at P700 drives an electron through a series of acceptors back to refill $P700^+$. As a consequence of the e^- transfers, proton translocations ensue that can drive ATP formation.

Carbon dioxide fixation, the synthesis of carbohydrate from CO_2, while not uniquely a photosynthetic process, is nevertheless the major investment of the chemical energy derived from photosynthesis. Carbon dioxide is fixed in organic linkage through reaction with ribulose-1,5-bisphosphate to generate 2 molecules of 3-phosphoglycerate. The enzyme catalyzing this reaction is ribulose bisphosphate carboxylase or rubisco. Rubisco also has the capacity to use O_2 in place of CO_2 in the condensation with ribulose-1,5-bisphosphate, a reaction leading to a wasteful loss of this pentose. As an oxygen-consuming, CO_2-evolving reaction, this process is aptly termed photorespiration and is viewed as a wasteful dissipation of RuBP, a crucial metabolic intermediate. The remaining reactions of Calvin cycle CO_2 fixation are similar to reactions in the pentose phosphate pathway and glycolysis. The aim of the Calvin cycle is to account for the synthesis of one hexose molecule ($C_6H_{12}O_6$) from 6 equivalents of carbon dioxide (6 CO_2). Key Calvin cycle enzymes are regulated by light-induced effects on chloroplast reducing power, pH and relative Mg^{2+} concentrations so that CO_2 fixation occurs only when light energy is available. An alternate means of CO_2 capture occurs in so-called C4 plants. These plants form C4 compounds (instead of C3-phosphoglycerate) as the initial products of CO_2 uptake. The C4 pathway is not an alternative process to the Calvin cycle but instead is an effective means for the transport and concentration of CO_2 within cells having high potential for Calvin cycle carbon dioxide fixation. C4 plants achieve greater photosynthetic efficiency by diminishing photorespiration. Typical C4 plants include tropical grasses such as sugar cane and maize.

Chapter 22

Gluconeogenesis, Glycogen Metabolism, and the Pentose Phosphate Pathway

• •

Chapter Outline

❖ Gluconeogenesis
  ⅄ Substrates: Pyruvate, lactate, glycerol, amino acids (except Lys and Leu)
  ⅄ Active in liver and kidney: Supplies glucose to brain and muscle
  ⅄ Seven of ten glycolytic steps in reverse
  ⅄ Three steps not reversal of glycolysis for two reasons
    · Energetics
    · Regulation
❖ Three unique gluconeogenesis steps
  ⅄ Pyruvate to PEP
    · Pyruvate carboxylase converts pyruvate to oxaloacetate
      • Biotin-dependent enzyme
      • ATP + bicarbonate to carbonyl phosphate to carboxybiotin intermediate
      • Allosteric activation by acetyl-CoA
      • Localized in mitochondria
    · Oxaloacetate to malate and back
    · PEP carboxylase converts oxaloacetate to PEP
      • Decarboxylation drives reaction
      • GTP or ATP supplies P and drives reaction
  ⅄ Fructose-1,6-bisphosphate to fructose-6-phosphate
    · Hydrolysis
    · Fructose-1,6-bisphosphatase
      • Citrate stimulates
      • Fructose-2,6-bisphosphate inhibits
      • AMP inhibits
  ⅄ Glucose-6-phosphate to glucose
    · Hydrolysis
    · Glucose-6-phosphatase
      • Releases product into ER lumen
      • Enzyme expressed in liver and kidney (not in muscle or brain)
      • High Km for G6P: Substrate-level control
❖ Reciprocal regulation: Avoidance of substrate cycles
  ⅄ Glucose to glucose-6-phosphate and back
    · Glucokinase high Km and not inhibited by G6P
    · Glucose-6-phosphatase high Km
  ⅄ Fructose-6-phosphate to fructose-1,6-bisphosphate and back
    · Phosphofructokinase
      • AMP stimulates
      • Fructose-2,6-bisphosphate stimulates

- Citrate inhibits
  - Fructose-1,6-bisphosphatase
    - AMP inhibits
    - Fructose-2,6-bisphosphate inhibits
    - Citrate stimulates
  - Fructose-2,6-bisphosphate levels regulated by bifunctional (tandem) enzyme
    - Phosphofructokinase 2 produces F-2,6-BP
      - c-AMP protein kinase phosphorylation inhibits PFK 2
      - F6P activates PFKase 2
    - Fructose-2,6-bisphosphatase hydrolyzes F-2,6-BP
      - c-AMP protein kinase phosphorylation activates phosphatase
      - F6P inhibits FBPase 2
- ❖ Dietary glycogen (and starch) catabolism
  - ⅄ α-Amylase (saliva and pancreas) endoglycosidase
    - Hydrolyzes α(1→4) amylopectin and glycogen at random: Products
      - Maltose, maltotriose, oligosaccharides
      - Limit dextrin: Branches limit
  - ⅄ Debranching enzyme: Two activities
    - Oligo (α1,4 → α1,4) glucanotransferase: Moves three residues to new end
    - α(1 → 6) Glucosidase: Hydrolyzes branch
- ❖ Cellular glycogen
  - ⅄ Glycogen phosphorylase: Uses phosphate to split off glucose-1-P
  - ⅄ Debranching enzyme
- ❖ Glycogen synthesis
  - ⅄ Base: Glycogenin with glucose in acetal linkage to Tyr-OH
  - ⅄ Glycogen synthase: α(1 → 4) linkage of glucose from UDP-glucose
  - ⅄ UDP-glucose: UDP-glucose pyrophosphorylase: UTP + G-1-P to UDP-glucose + PP$_i$
  - ⅄ Branching enzyme: Amylo (1,4 → 1,6) transglycosylase
    - Chain moved is 6-7 residues long and from 11-residue chain
    - Branch points: α(1→ 6) linkage every 8-12 residues
- ❖ Glycogen synthase regulation
  - ⅄ D-form: Less active: Dependent on G6P: Phosphorylated by protein kinase
  - ⅄ I-form: Active (independent of G6P) : No phosphate (phosphate removed by phosphoprotein phosphatase)
- ❖ Hormonal regulation
  - ⅄ Insulin: Released from β–cells islets of Langerhans of pancreas in response to high blood glucose
    - Increase uptake of glucose
    - Increase synthesis of glycolytic enzymes
    - Inhibits gluconeogenesis
  - ⅄ Glucagon α–cells islets of Langerhans of pancreas in response to low blood glucose
    - Stimulates liver to release glucose
    - Involved in long term glucose maintenance
  - ⅄ Epinephrine (adrenaline) released by adrenal glands
    - Active on liver and muscle
    - Stimulates glycogen breakdown and glycolysis
    - Inhibits gluconeogenesis
  - ⅄ Glucocorticoids: Cortisol
    - Steroid hormone active on liver, muscle, adipose tissue
    - In liver: Stimulates gluconeogenesis and glycogen synthesis
    - Catabolic: Promotes protein degradation
- ❖ Pentose phosphate pathway (hexose monophosphate shunt, phosphogluconate pathway)
  - ⅄ Supplies NADPH or ribose
  - ⅄ Two phases
    - Oxidative phase: Glucose-6-P to ribulose-5-P
    - Nonoxidative phase: Ribulose-5-P to 3-, 4-, 5-, 6-, and 7-carbons ketoses and aldoses

❖ Glucose to ribulose-5-P: Oxidative
    ⚘ Glucose-6-phosphate dehydrogenase
        • NADPH produced
        • Gluconolactone produced
        • Enzyme inhibitors
            ● NADPH
            ● Intermediates of fatty acid biosynthesis
    ⚘ Gluconolactonase: Gluconolactone to 6-phosphogluconate
    ⚘ 6-Phosphogluconate dehydrogenase
        • NADPH produced
        • Ribulose-5-P and carbon dioxide produced
❖ Ribulose-5-P to various sugars: Nonreductive: Enzyme types
    ⚘ Phosphopentose isomerase: Ketose to aldose conversion
    ⚘ Phosphopentose epimerase: Sugar interconversion
    ⚘ Transketolase: 2-Carbon unit from keto donor to aldehyde acceptor
    ⚘ Transaldolase: Ketose and aldose formation

Chapter Objectives

Gluconeogenesis

Although gluconeogenesis is not simply the reverse of glycolysis, the two pathways share many steps and it would help to review glycolysis. Briefly, glucose is phosphorylated by hexokinase to glucose-6-phosphate, which is isomerized to fructose-6-phosphate. A second kinase reaction, catalyzed by phosphofructokinase, produces fructose-1,6-bisphosphate, which is cleaved by aldolase to produce dihydroxyacetone phosphate and glyceraldehyde-3-phosphate, two triose phosphates that are interconvertible. Glyceraldehyde-3-P is oxidized and phosphorylated to 1,3-bisphosphoglycerate, a high-energy compound used to support a substrate level phosphorylation of ADP leaving 3-phosphoglycerate, which is converted to 2-phosphoglycerate, dehydrated to phosphoenolpyruvate, and converted to pyruvate with accompanying ATP synthesis.

Gluconeogenesis starts with pyruvate and ends with glucose. Pyruvate can derive from lactate, amino acids, and citric acid cycle intermediates but not from acetyl-CoA units. Gluconeogenesis and glycolysis differ at three steps (see Figure 22.1), glucose to glucose-6-P, fructose-6-P to fructose-1,6-bisphosphate, and PEP to pyruvate. Understand how these three steps are bypassed and you will understand gluconeogenesis. The first two steps are catalyzed by phosphatases that hydrolyze phosphate from the substrates. The PEP to pyruvate step is a two-step process with pyruvate being converted to oxaloacetate by pyruvate carboxylase, a biotin-containing enzyme. Oxaloacetate is converted to PEP by PEP-carboxykinase, which uses ATP or GTP hydrolysis, depending on the enzyme source, to drive the reaction.

Regulation

Glycolysis and gluconeogenesis are reciprocally regulated. You should remember the following regulatory pairs: glucose-6-phosphate inhibition of hexokinase and stimulation of glucose-6-phosphatase; fructose-2,6-bisphosphate stimulation of phosphofructokinase (PFK) and inhibition of fructose-1,6-bisphosphatase (FBPase); AMP stimulation of PFK and inhibition of FBPase; and, acetyl-CoA inhibition of pyruvate kinase and activation of pyruvate carboxylase. Additionally, ATP inhibits PFK and pyruvate kinase, citrate inhibits PFK, alanine inhibits pyruvate kinase, and cyclic-AMP dependent phosphorylation inhibits pyruvate kinase. Fructose-2,6-bisphosphate can be thought of as a signal to stimulate glycolysis. This regulator is not a component of a metabolic pathway but is produced by a bifunctional enzyme regulated by cAMP-dependent phosphorylation.

Glycogen

Dietary glycogen and starch breakdown are catalyzed by α-amylase (β-amylase in plants), and debranching enzymes. Tissue glycogen is degraded by glycogen phosphorylase and synthesized by glycogen synthase. The latter enzyme is supplied with activated glucosyl residues in the form of UDP-glucose produced by UDP-glucose pyrophosphorylase. Understand the role of glycogenin and branching enzyme, which initiate glycogen synthesis and introduce branches. Branches lead to multiple sites for rapid elongation or degradation of glycogen.

Regulation

Glycogen metabolism is carefully controlled by regulating glycogen phosphorylase and glycogen synthase. Glycogen phosphorylase is allosterically activated by AMP and inhibited by ATP and glucose-6-phosphate, whereas glycogen synthase is stimulated by glucose-6-phosphate. Both enzymes are subject to control by covalent modification. Glycogen phosphorylase exists in two forms, a less active *b* form and an active *a* form. Phosphorylase *a* is a phosphorylated form of phosphorylase *b*, catalyzed by a specific kinase, glycogen phosphorylase kinase which is itself under covalent regulation by cAMP-dependent protein kinase. cAMP levels are regulated hormonally by hormone binding to surface receptors, leading to release of G-protein, and stimulation of adenylyl cyclase. Know this cascade. Glycogen synthase also exists in two forms that differ in both activity and in the state of phosphorylation. The less active phosphorylated glycogen synthase D depends on glucose-6-phosphate for activity, whereas the active, unphosphorylated, glycogen synthase I is independent of glucose-6-phosphate levels.

Pentose Phosphate Pathway

Admittedly, this is an impressive pathway but it may simplify matters to remember that it has an oxidative phase and a nonoxidative phase (see Figure 22.22). In the oxidative phase, glucose is converted to ribulose-5-phosphate and CO_2. Clearly, this must involve a phosphorylation (by hexokinase) and oxidations (by two NADP-dependent dehydrogenases). (With reflection, we might have anticipated that two oxidations are required. The CO_2 must derive from one of the alcoholic carbons of glucose via an aldehyde intermediate.)

The nonoxidative phase of the pentose phosphate pathway converts ribulose-5-phosphate into glucose-6-phosphate via 3-, 4-, 5-, 6- and 7- carbon sugar intermediates. Some of the key enzymes involved include the following: phosphopentose isomerase catalyzes a ketose to aldose conversion (We encountered similar biochemistry in glycolysis between glucose-6-P and fructose-6-P and between DHAP and glyceraldehyde-3-P.); phosphopentose epimerase (an epimerase changes the configuration about an asymmetric carbon (we encountered one in galactose catabolism.); transketolases (move two-carbon units between aldoses); and, a transaldolase (with a reaction mechanism similar to the aldolase encountered in glycolysis).

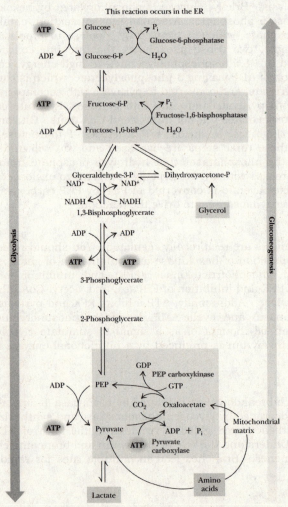

Figure 22.1 The pathways of gluconeogenesis and glycolysis. Species in the boxes indicate other entry points for gluconeogenesis (in addition to pyruvate).

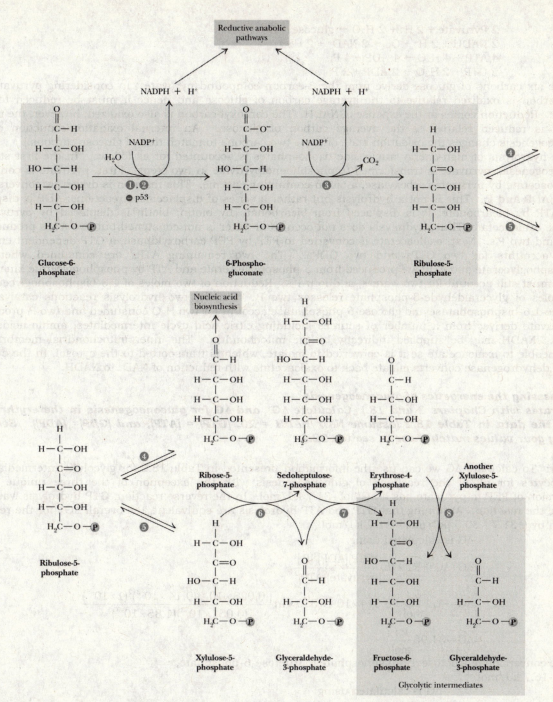

Figure 22.22 The pentose phosphate pathway. The numerals in the circles indicate the steps discussed in the Garrett and Grisham text.

Problems and Solutions

1. Assessing the overall equation for gluconeogenesis
Consider the balanced equation for gluconeogenesis in Section 22.1. Account for each of the components of this equation and the indicated stoichiometry.

Answer: The net reaction for the synthesis of glucose from pyruvate is:

2 Pyruvate + 4 ATP + 2 GTP + 2 NADH + 2 H⁺ + 6 H₂O →
glucose + 4 ADP + 2 GDP + 6 Pᵢ + 2 NAD⁺

The following reactions contribute to this equation:

409

$$2 \text{ Pyruvate} + 2 \text{ H}^+ + 2 \text{ H}_2\text{O} \rightarrow \text{glucose} + \text{O}_2$$
$$2 \text{ NADH} + 2 \text{ H}^+ + \text{O}_2 \rightarrow 2 \text{ NAD}^+ + 2 \text{ H}_2\text{O}$$
$$4 \text{ ATP} + 4 \text{ H}_2\text{O} \rightarrow 4 \text{ ADP} + 4 \text{ P}_i$$
$$2 \text{ GTP} + 2 \text{ H}_2\text{O} \rightarrow 2 \text{ ADP} + 2 \text{ P}_i$$

The six carbons of glucose derive from the 3-carbon compound pyruvate. In considering pyruvate, the keto carbon is oxidized relative to the average carbon of glucose and hence it must be reduced to form glucose. Reduction comes at the expense of NADH. (The carboxyl carbon is also oxidized; however, the methyl carbon is reduced relative to the average carbon of glucose. An internal oxidation-reduction during gluconeogenesis changes the oxidation state of these two carbons to match that of glucose carbons.)

The hydrolysis of high-energy nucleoside triphosphates is accounted for as follows. In the first stage of gluconeogenesis, pyruvate is transformed to phosphoenolpyruvate in two steps. First, pyruvate is converted to oxaloacetate by pyruvate carboxylase, a biotin-containing enzyme. This reaction is driven by conversion of ATP to ADP and P_i. This is not a hydrolysis but rather a series of displacement reactions: ADP is displaced from ATP by bicarbonate; P_i is displaced from bicarbonate by biotin; biotin is displaced by pyruvate to produce oxaloacetate. Thus, hydrolysis does not occur and water is not consumed but two ATPs produce two ADPs and two P_is. Next, oxaloacetate is converted to PEP by PEP-carboxykinase, a GTP-dependent enzyme, which accounts for two GTPs and two GDPs. The two remaining ATPs are consumed when 1,3-bisphosphoglycerate and ADP are produced from 3-phosphoglycerate and ATP by phosphoglycerate kinase.

We must still account for two waters, and four P_is. Reduction of two moles of 1,3-bisphosphoglycerate to two moles of glyceraldehyde-3-phosphate releases two P_i. Finally, two hydrolysis reactions catalyzed by fructose-1,6-bisphosphatase and glucose-6-phosphatase account for two H_2O consumed and two P_i produced.

Pyruvate derives from a number of sources including citric acid cycle intermediates, amino acids, and lactate. NADH may be supplied indirectly by the mitochondria. The inner mitochondrial membrane is impermeable to oxaloacetate so it is converted to malate, which is transported to the cytosol. In the cytosol, malate dehydrogenase converts malate back to oxaloacetate with reduction of NAD^+ to NADH.

2. Assessing the energetics of gluconeogenesis

(Integrates with Chapters 3 and 18.) Calculate $\Delta G^{\circ\prime}$ and ΔG for gluconeogenesis in the erythrocyte, using the data in Table 18.2 (assume $NAD^+/NADH = 20$, $[GTP] = [ATP]$, and $[GDP] = [ADP]$). See how closely your values match those in section 22.1.

Answer: To calculate ΔG we can use the information presented in Table 18.2 for glycolytic intermediates in erythrocytes for all of the reactions of gluconeogenesis with the exception of the three unique steps. Conversion of PEP to pyruvate has a $\Delta G^{\circ\prime}$ of -31.7 kJ/mol. In the reverse reaction, GTP hydrolysis was used to drive the reaction. Assuming that GTP and ATP hydrolysis are equivalent, the overall $\Delta G^{\circ\prime}$ for the reaction is given by $(+31.7 - 30.5)$ kJ/mol $= +1.2$ kJ/mol.

ΔG is calculated using:

$$\Delta G = \Delta G^{\circ\prime} + RT \times \ln \frac{[\text{PEP}][\text{ADP}]^2[\text{P}_i]}{[\text{pyruvate}][\text{ATP}]^2}$$

$$\Delta G = +1.2 \frac{\text{kJ}}{\text{mol}} + 8.314 \times 10^{-3} \times 310 \times \ln \frac{[(0.023 \times 10^{-3})(0.14 \times 10^{-3})^2(1 \times 10^{-3})]}{(0.051 \times 10^{-3})(1.85 \times 10^{-3})^2}$$

$$\Delta G = -31.96 \frac{\text{kJ}}{\text{mol}}$$

For the conversion of fructose-1,6-bisphosphate to fructose-6-phosphate, $\Delta G^{\circ\prime} = -16.7$ kJ/mol.

ΔG is calculated using:

$$\Delta G = \Delta G^{\circ\prime} + RT \times \ln \frac{[\text{F6P}][\text{P}_i]}{[\text{F1,6BP}]}$$

$$\Delta G = -16.7 \frac{\text{kJ}}{\text{mol}} + 8.314 \times 10^{-3} \times 310 \times \ln \frac{[(0.014 \times 10^{-3})(1 \times 10^{-3})]}{(0.031 \times 10^{-3})}$$

$$\Delta G = -36.6 \frac{\text{kJ}}{\text{mol}}$$

For conversion of glucose-6-phosphate to glucose, using the value of $\Delta G^{\circ\prime} = -13.9$ kJ/mol given in Table 3.3, we have:

$$\Delta G = \Delta G^{\circ\prime} + RT \times \ln \frac{[\text{glucose}][P_i]}{[\text{G6P}]}$$

$$\Delta G = -13.9 \frac{kJ}{mol} + 8.314 \times 10^{-3} \times 310 \times \ln \frac{[(5.0 \times 10^{-3})(1 \times 10^{-3})]}{(0.083 \times 10^{-3})}$$

$$\Delta G = -21.1 \frac{kJ}{mol}$$

The overall ΔG is calculated by summing the ΔG values given in Table 18.1 (with signs reversed) for all the reactions with the exception of hexokinase, phosphofructokinase, and pyruvate kinase. These are replaced by the ΔG values calculated above.

$$\Delta G_{\text{gluconeogenesis}} = -21.1 + 2.92 - 36.6 + 0.23 - 2.41 + 1.29 - 0.1 - 0.83 - 1.1 - 31.9$$

$$\Delta G_{\text{gluconeogenesis}} = -89.6 \frac{kJ}{mol}$$

3. Understanding the regulation of the fructose-1,6-bisphosphatase reaction
Use the data of Figure 22.9 to calculate the percent inhibition of fructose-1,6-bisphosphatase by 25 μM fructose-2,6-bisphosphate when fructose-1,6-bisphosphate is (a) 25 μM and (b) 100 μM.

Answer: From Figure 22.9 (a) we can estimate that at 25 μM fructose-1,6-bisphosphate the activity of fructose-1,6-bisphosphatase in the absence of fructose-2,6-bisphosphate is approximately 15 units. In the presence of 25 μM fructose-2,6-bisphosphate, the activity is only 1 unit. The percent inhibition is given by

$$\frac{15-1}{15} \times 100\% = 93.3\%$$

At 100 μM fructose-1,6-bisphosphate the activities of fructose-1,6-bisphosphatase without and with fructose-2,6-bisphosphate are 14 and 8. The percent inhibition in this case is

$$\frac{14-8}{14} \times 100\% = 42.9\%$$

4. Understanding the energetics of the glycogen synthase reaction
(Integrates with Chapters 3.) Suggest an explanation for the exergonic nature of the glycogen synthase reaction ($\Delta G^{\circ\prime}$ = -13.3 kJ/mol). Consult Chapter 3 to review the energetics of high-energy phosphate compounds if necessary.

Answer: Glycogen synthase adds glucose units in α(1★4) linkage to glycogen using UDP-glucose as a substrate and releasing UDP. In comparing UDP-glucose to UDP, we expect very little difference in destabilization due to electrostatic repulsion and no significant entropy factors. The presence of a glucose moiety does interfere with phosphorous resonance states of UDP. The standard free energy of hydrolysis of UDP-glucose is -31.9 kJ/mol whereas the $\Delta G^{\circ\prime}$ for glycogen synthase reaction is -13.3 kJ/mol.

5. Determining the rate of energy consumption by muscles during exercise
Using the values in Table 23.1 for body glycogen content and the data in part b of the illustration for A Deeper Look (page 740), calculate the rate of energy consumption by muscles in heavy exercise (in J/sec). Use the data for fast-twitch muscle.

Answer: From Table 23.1 we are informed that a 70-kg person has about 120 g (dry weight) of glycogen, equivalent to 1,920 kJ of available energy. From *A Deeper Look* we see glycogen supplies are exhausted after approximately 60 minutes of heavy exercise. The rate of energy consumption is given by:

$$\frac{1,920 \text{ kJ}}{60 \text{ min}} = 32 \frac{kJ}{min}$$

$$32 \frac{kJ}{min} = 32 \frac{kJ}{min} \times \frac{1000 \text{ J}}{kJ} \times \frac{1 \text{ min}}{60 \text{ sec}} = 533 \frac{J}{sec}$$

6. Assessing the action of sodium borohydride on the pentose phosphate pathway
Which reactions of the pentose phosphate pathway would be inhibited by NaBH₄? Why?

Answer: The transaldolase reaction that converts sedoheptulose-7-phosphate and glyceraldehyde-3-phosphate to erythrose-4-phosphate and fructose-6-phosphate should be sensitive to NaBH₄. This is because catalysis involves formation of Schiff base between the amino side chain of a lysine residue on the enzyme with the carbonyl carbon of sedoheptulose-7-phosphate (carbon 2). NaBH$_4$ will reduce the Schiff base resulting in irreversible, covalent modification of the enzyme.

7. Determining the extent of branching in glycogen
(Integrates with Chapters 7.) *Imagine a glycogen molecule with 8000 glucose residues. If branches occur every eight residues, how many reducing ends does the molecule have? If branches occur every 12 residues, how many reducing ends does it have? How many nonreducing ends does it have? How many nonreducing ends does it have in each of these cases?*

Answer: The reducing end of a polysaccharide contains a free aldehyde or keto group. In glycogen, the linkage between glucose residues is α(1★4) for nonbranch residues and α(1★6) for branch residues. Thus, both branch residues and nonbranch residues have their anomeric carbons, C-1, in glycosidic bonds. Further, C-1 of the first glucose residue is tied up in linkage to glycogenin at the core so there are no free reducing ends, regardless of the size of the glycogen molecule.

The number of nonreducing ends is very much dependent on the frequency of branches. In the diagram below, let n = number of residues in a branch, b = a branch residue, e = nonreducing end, and x = number of layers of branches. The diagram below shows x = 1, 2, and 3.

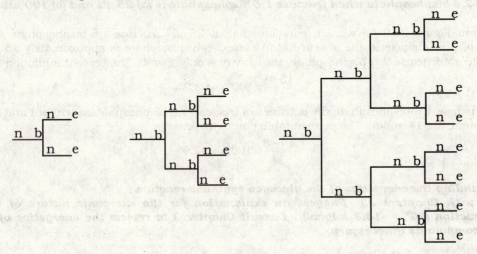

From this diagram the following table may be constructed.

Number of residues	Number of branch points (b)	Number of ends (e)
3n	1	2
7n	3	4
15n	7	8
.	.	.
.	.	.
$[(2^x - 1) + 2^x]n$	$2^x - 1$	2^x

For a glycogen molecule of 8,000 residues with 8 residues per branch (n = 8) we have:

$$[(2^x - 1) + 2^x]n = 8,000 \text{ residues}$$

For n=8.

$$[(2^x - 1) + 2^x] \times 8 = 8,000 \text{ or}$$

$$2^x + 2^x = \frac{8,000}{8} + 1$$

$$2 \times 2^x = 1,001$$

$$2^x = \frac{1,001}{2}$$

$$x = \frac{\log \dfrac{1,001}{2}}{\log 2} = 8.967$$

$$x = 8.967$$

The number of nonreducing ends is given by

$$2^x = 2^{8.967} = 500$$

For a glycogen molecule of 8,000 residues with 12 residues per branch (n = 12), x = 8.383, the number of nonreducing ends is 334 and the number of branch points is 333.
Note: 8,000 residues forms 8,000/12 = 666 12-residue branches. There are 334 terminal branches, 333 internal branches.

8. Describing the regulation of gluconeogenesis and glycogen metabolism
Explain the effects of each of the following on the rates of gluconeogenesis and glycogen metabolism:
   a. Increasing the concentration of tissue fructose-1,6-bisphosphate
   b. Increasing the concentration of blood glucose
   c. Increasing the concentration of blood insulin
   d. Increasing the amount of blood glucagon
   e. Decreasing levels of tissue ATP
   f. Increasing the concentration of tissue AMP
   g. Decreasing the concentration of fructose-6-phosphate.

Answer: Gluconeogenesis and glycolysis are reciprocally controlled. In glycolysis, control is exerted at the strongly exergonic steps catalyzed by hexokinase, phosphofructokinase, and pyruvate kinase. In gluconeogenesis, these steps are bypassed with different reactions catalyzed by glucose-6-phosphatase, fructose-1,6-bisphosphatase, and pyruvate carboxylase and PEP carboxykinase. Glucose-6-phosphatase is under substrate-level control by glucose-6-phosphate. Fructose-1,6-bisphosphatase is allosterically inhibited by AMP and activated by citrate. This enzyme is also inhibited by fructose-2,6-bisphosphate, which is produced from fructose-6-phosphate by phosphofructokinase 2. Phosphofructokinase 2 is allosterically activated by fructose-6-phosphate and inhibited by phosphorylation of a single serine residue by cAMP-dependent protein kinase. Finally, acetyl-CoA stimulates pyruvate carboxylase.

a. Increasing fructose-1,6-bisphosphate will stimulate glycolysis by stimulating pyruvate kinase (see Figure 22.8) and, therefore, gluconeogenesis will be inhibited indirectly.
b. High blood glucose will stimulate glycolysis and glycogen synthesis and decrease gluconeogenesis.
c. Insulin is a signal that blood glucose levels are high. Since additional glucose is not required, gluconeogenesis is inhibited and glycogen synthesis would be stimulated.
d. Glucagon signals the liver to release glucose. This hormone sets off a cascade of signals leading to increase in cAMP and stimulation of cAMP-dependent protein kinase. Pyruvate kinase is inhibited by phosphorylation leading to inhibition of glycolysis. Phosphofructokinase 2 is also inactivated by phosphorylation leading to a drop in the level of fructose-2,6-bisphosphate, which causes relief of inhibition of fructose-1,6-bisphosphatase and stimulation of gluconeogenesis. Glycogen synthesis is inhibited.
e. A decrease in ATP will stimulate glycolysis and inhibit gluconeogenesis.
f. AMP inhibits fructose-1,6-bisphosphatase and inhibits gluconeogenesis.
g. Fructose-6-phosphate is not a regulatory molecule but it is rapidly equilibrated with glucose-6-phosphate, a substrate for glucose-6-phosphatase. Decreased levels of glucose-6-phosphate will lead to inhibition of gluconeogenesis.

Glycogen metabolism is regulated as follows. Glycogen phosphorylase is allosterically activated by AMP and is inhibited by ATP and glucose-6-phosphate. Further, phosphorylation of serine-14 activates glycogen

413

phosphorylase. Glycogen synthase is reciprocally regulated by phosphorylation: It is inactive when phosphorylated. The synthase is also allosterically activated by glucose-6-phosphate.

a. Fructose-1,6-bisphosphate will inhibit glycogen catabolism.
b. High blood glucose leads to an increase in glycogen anabolism.
c. Insulin stimulates glycogen anabolism by increasing glucose intracellularly, which leads to an increase in glucose-6-phosphate, an allosteric activator of glycogen synthase.
d. Glucagon will release glucose from glycogen by setting off a signal cascade ending with phosphorylation, and stimulation, of glycogen phosphorylase and inhibition of glycogen synthase.
e. ATP inhibits glycogen phosphorylase and inhibits glycogen catabolism. A drop in ATP will stimulate glycogen breakdown.
f. AMP activates glycogen phosphorylase leading to stimulation of glycogen catabolism.
g. Fructose-6-phosphate, through glucose-6-phosphate, will activate glycogen synthase leading to stimulation of glycogen formation.

9. Assessing the energetics of the glycogen phosphorylase reaction
(Integrates with Chapters 3 and 15.) *The free-energy change of the glycogen phosphorylase reaction is $\Delta G^{\circ\prime} = + 3.1$ kJ/mol. If $[P_i] = 1$ mM, what is the concentration of glucose-1-P when this reaction is at equilibrium.*

Answer: For the reaction: glycogen$_n$ + P$_i$ →   glycogen$_{n-1}$ + glucose-1-P

$$\Delta G^{\circ\prime} = -RT \times \ln K_{eq} = -RT \times \ln \frac{[\text{glucose-1-P}][\text{glycogen}_{n-1}]}{[P_i][\text{glycogen}_n]}$$

Since [glycogen$_n$]=[glycogen$_{n-1}$] this simplifies to :

$$\Delta G^{\circ\prime} = -RT \times \ln \frac{[\text{glucose-1-P}]}{[P_i]}$$

Solving for [glucose-1-P] we find

$$[\text{glucose-1-P}]=[P_i] \times e^{\frac{-\Delta G^{\circ\prime}}{RT}} = [(1 \times 10^{-3}) \times e^{\frac{-3.1}{8.314 \times 10^{-3} \times 298}}$$

[glucose-1-P]=0.286 mM

10. Understanding enzyme mechanisms related to pyruvate carboxylase
Based on the mechanism for pyruvate carboxylase (Figure 22.3), write reasonable mechanisms for the reactions shown below:

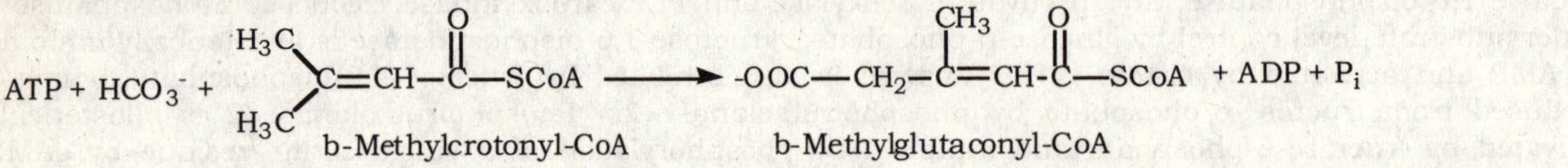

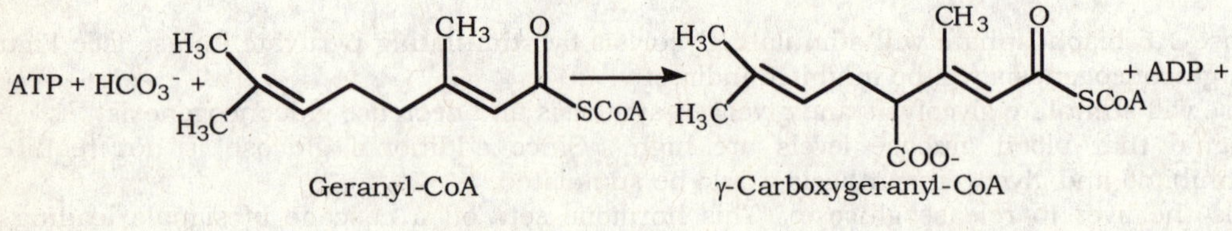

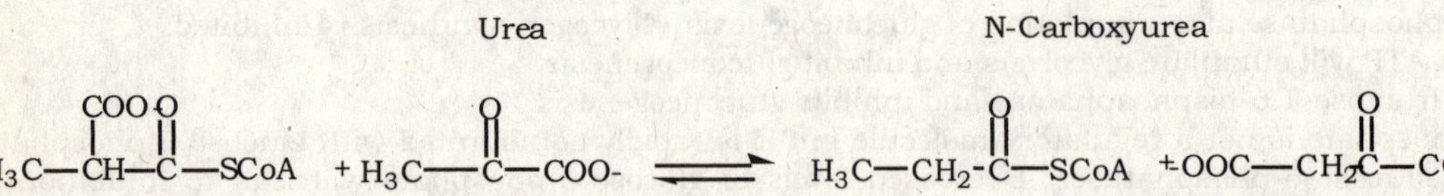

Answer: The first reaction, catalyzed by β-methylcrontonyl-CoA carboxylase, is a step in the pathway for leucine degradation. The reaction is a carboxylation of the γ-carbon of β-methylcrotonyl-CoA. A similar reaction, namely γ-carboxylation, occurs in the second reaction as well. Both are biotin-dependent reactions. The formation of N-carboxyurea by urea carboxylase is also a biotin-dependent reaction. (N-carboxyurea is subsequently hydrolyzed to bicarbonate and ammonium by allophanate amidolyase.) In the last reaction we see a transcarboxylation or movement of a carboxyl group from methylmalonyl-CoA to pyruvate to produce oxaloacetate (and propionyl-CoA). The first three reactions proceed by formation of carboxy-biotin via a carbonyl-phosphate intermediate whose formation is dependent on ATP. In the last reaction, carboxy-biotin formation occurs by transcarboxylation. Thus, in all four reactions we are dealing with biotin-dependent enzymes.

11. *Understanding the mechanisms of reactions related to transketolase*
The mechanistic chemistry of the acetolactate synthase and phosphoketolase reactions (shown below) is similar to that of the transketolase reaction (Figure 22.30). Write suitable mechanisms for these reactions.

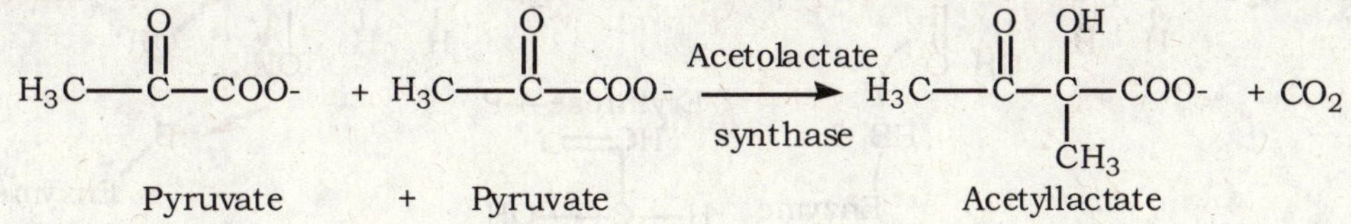

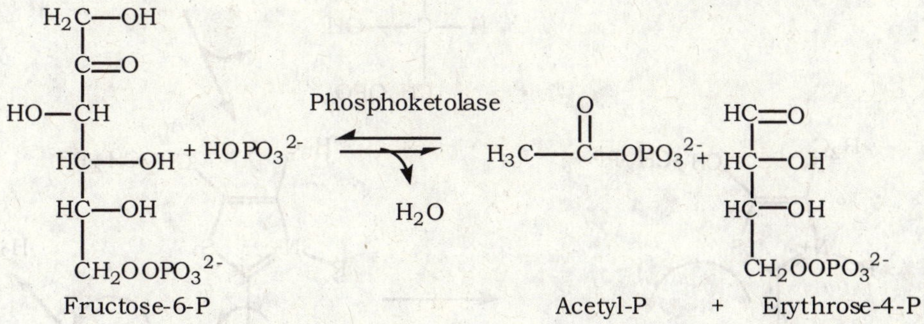

Answer: The enzyme, acetolactate synthase, is involved in valine biosynthesis in bacteria. It is a thiamine pyrophosphate-containing enzyme. A reasonable reaction mechanism might be as follows:

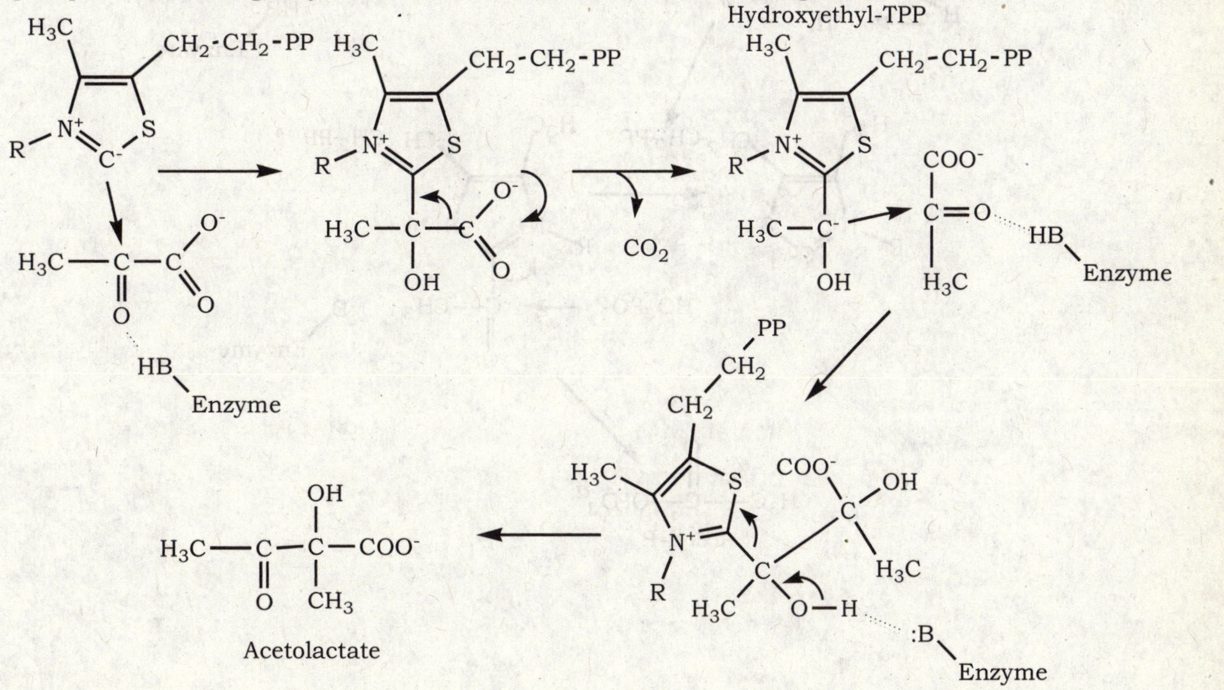

Phosphoketolase is also a thiamine pyrophosphate-containing enzyme. It is used by certain bacteria to ferment glucose into lactic acid and ethanol (using xylulose-5-P in place of fructose-6-P) by the phosphoketolase pathway. (Bacteria capable of running this pathway are used to produce sauerkraut, buttermilk, and certain cheeses.)

The reaction mechanism is shown below:

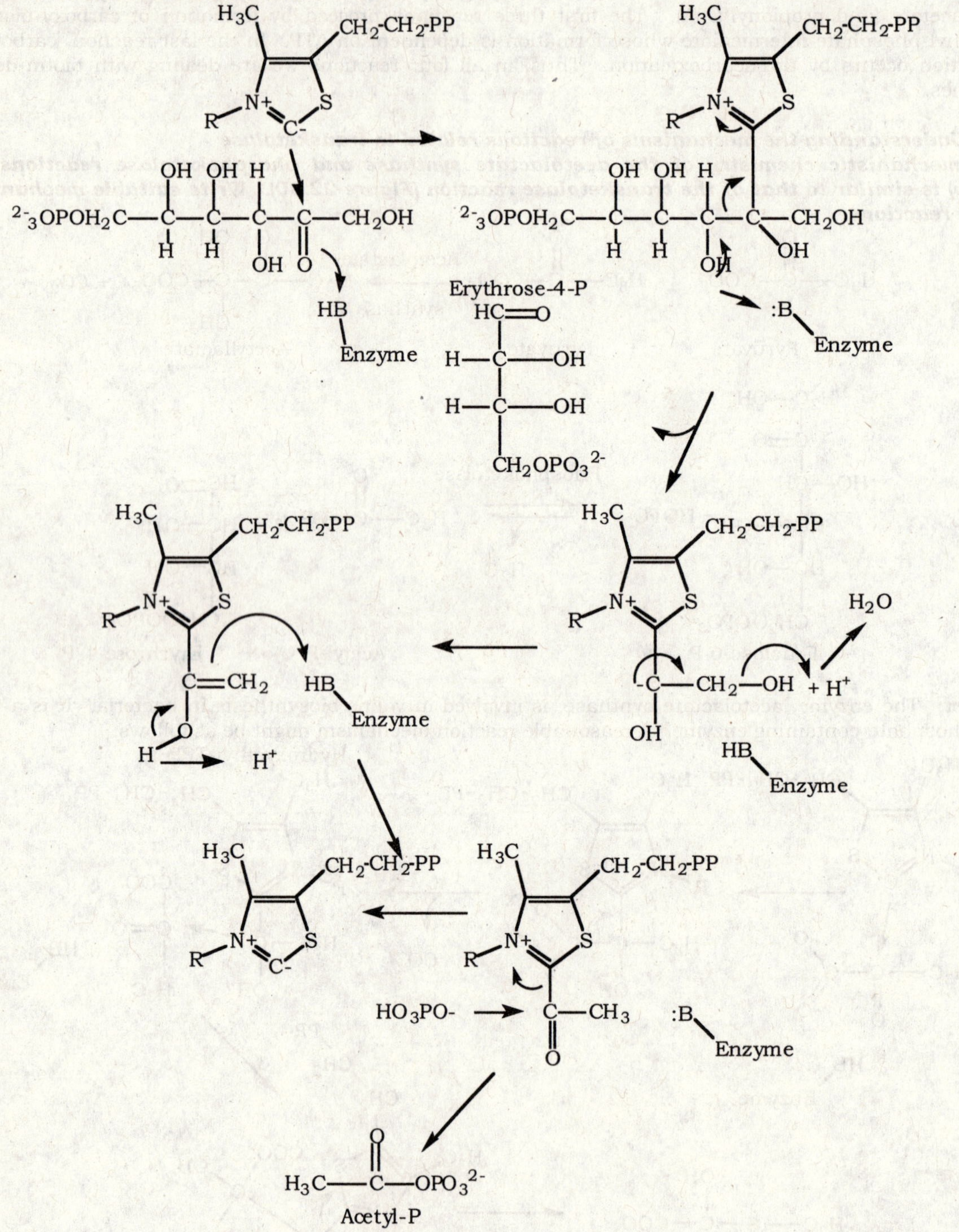

12. Understanding the mechanisms of action of a popular diabetes medication
Metaglip is a prescribed preparation (from Bristol-Myers Squibb) for treatment of type 2 diabetes. It consists of metformin (see Human Biochemistry, page 726) together with glipizide. The actions of metformin and glipizide are said to be complementary. Suggest a mechanism of action of glipizide.

Answer: The structures of both compounds are shown below. Metformin should look somewhat familiar if you recall the structure of the amino acids. Metformin is a biguanide. (The side chain of arginine contains a guanidine group.) Guanidine was recognized as an agent causing hypoglycemia. It is found in extracts of a plant (*Galega officinalis*) used since medieval times to treat diabetes. Metformin seems to have a number of effects on glucose levels. In liver it decreases gluconeogenesis whereas in muscle it increases glucose uptake leading to increased levels of glycogen synthesis. Metformin also sensitizes muscle to insulin and may also decrease glucose uptake by the digestive system.

Glipizide is a sulfonylurea, which stimulates insulin release from beta cells of the pancreas. There is some evidence that the drug may decrease hepatic clearance of insulin.

Metaglip is a combination of <u>met</u>formin and <u>glip</u>izide. Both compounds lead to decrease blood glucose levels. Glipizide has the potential to cause hypoglycemia. Metformin may, in some cases, result in lactic acidosis. (Liver uses lactic acid in gluconeogenesis. Since metformin blocks liver gluconeogenesis, lactic acid metabolism is blocked.)

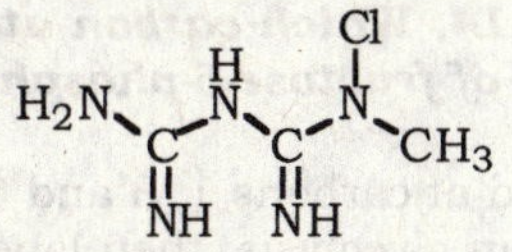
Metformin

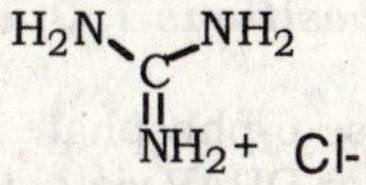
Guanidinium HCl

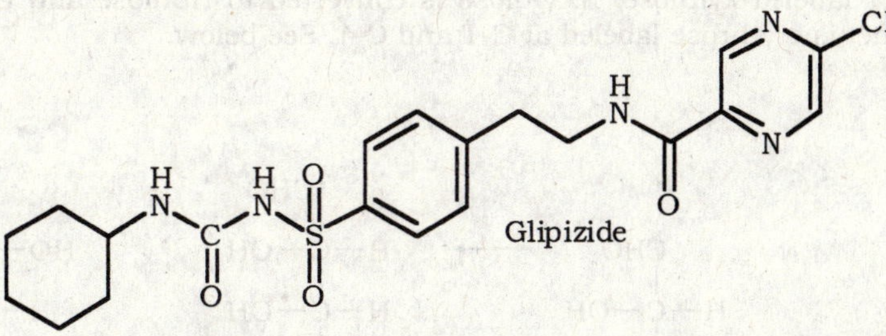
Glipizide

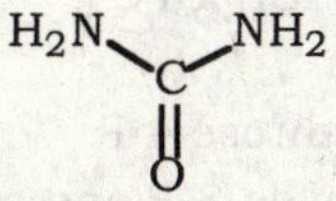
Urea

13. Assessing the mechanism of action of action of aldose reductase inhibitors
Study the structures of tolrestat and epalrestat in the Human Biochemistry box on page 748 and suggest a mechanism of action for these inhibitors of aldose reductase.

Answer: Tolrestat and epalrestat are both inhibitors of aldose reductase, an enzyme in the polyol pathway. Aldose reductase converts glucose to the sugar alcohol sorbitol and uses NADPH as a source of electrons. The structures of tolrestat and epalrestat do not seem to have any similarities to any of the substrates of the enzyme. Studies by Prendergast and co-workers (Ehrig T., Bohren K. M., Prendergast F. G., and Gabbay K. H. (1994) Mechanism of Aldose Reductase Inhibition: Binding of NADP+/NADPH and Alrestatin-like Inhibitors. Biochemistry 33, 7157-7165) show that these inhibitors probably bind to the enzyme:$NADP^+$ complex at a site other than the active site or substrate binding sites.

14. Understanding a version of the pentose phosphate pathway that produces mainly ribose-5-P
Based on the discussion on page 752, draw a diagram to show how several steps in the pentose phosphate pathway can be bypassed to produce large amounts of ribose-5-phosphate. Begin your diagram with fructose-6-phosphate.

Answer: The key here is to avoid the oxidative steps from glucose-6-phosphate to ribulose-5-phosphate and to run fructose-6-phosphate through nonoxidative reactions. Fructose-6-phosphate can be used in the first stage of glycolysis as a source of glyceraldehyde-3-phosphate. So, the 6-carbon fructose-6-phosphate and glyceraldehyde-3-phosphate are available substrates. In order to produce ribose-5-phosphate we are going to need a transketolase. Look at the lower half of Figure 22.22. If we were to run step 8 in reverse, this transketolase would convert the 6-carbon fructose and the 3-carbon glyceraldehyde to 4-carbon erythrose and 5-carbon xylulose. Xylulose is an epimerase and an isomerase away from ribose (via steps 5 and 4). The 4-carbon erythrose and another fructose can be used to produce sedoheptulose and glyceraldehyde by reversal of the aldolase reaction (step 7). Sedoheptulose and glyceraldehyde can be used to produce ribose and xylulose by reversal of step 6. The basic idea is to run reactions that produce 5-carbon sugars that then can be converted into ribose. In Figure 22.22, moving from fructose (6C) and erythrose (4C) to sedoheptulose (7C) and glyceraldehyde (3C) to ribose (5C) and xylulose (5C) will make the conversions. This process depends of formation of erythrose, a 4-carbon sugar produced by reversal of trans ketolase step 8.

15. Assessing the labelling patterns in versions of the pentose phosphate pathway
Consider the diagram you constructed in problem 14. Which carbon atoms in ribose-5-phosphate are derived from carbon atoms in positions 1, 3, and 6 of fructose-6-phosphate?

Answer: If we started with fructose-6-phosphate labeled at carbons 1, 3 and 6 and did not allow label to move into glyceraldehyde-3-phosphate or DHAP via aldolase in glycolysis, then label would immediately show up in the aldehyde carbon of ribose from carbon 1 of fructose. This happens as a consequence of reversal of step 8 (page 749) to produce C-1 labeled xylulose. (Xylulose is converted to ribulose and then to ribose with C-1 labeled.) The other product is erythrose labeled at C-1 and C-4. See below.

```
   *CH2OH
     |
    C=O                                                              *CH2OH
     |                                                                 |
HO—*C-H                                    *CHO                       C=O
     |                                       |                          |
  H—C—OH            CHO         ──▶      H—C—OH              HO—C—H
     |               |                       |                          |
  H—C—OH         H—C—OH                  H—C—OH                H—C—OH
     |               |                       |                          |
  *CH2OPO3⁻       CH2OPO3⁻              *CH2OPO3⁻              CH2OPO3⁻

  Fructose-6-P    Glyceraldehyde-3-P      Erythrose-4-P          Xylulose-5-P
```

Erythrose and another fructose-6-P are converted to glyceraldehyde-3-P and sedoheptulose-7-P by reversal of step 7 (page 750). This produces glyceraldehyde with label on C-3 (from C-6 of fructose) and sedoheptulose-7-P with label on C-1, 3, 4 and 7. (Label at 1 and 3 from fructose and label at 4 and 7 from xylulose.)

```
                                                                 *CH2OH
                                                                   |
   *CH2OH                                                         C=O
     |                                                             |
    C=O                                                       HO—*C-H
     |                                                             |
HO—*C-H              *CHO                                    H—*C-OH
     |                 |                                           |
  H—C—OH          H—C—OH       ──▶        CHO               H—C—OH
     |                 |                    |                      |
  H—C—OH          H—C—OH                H—C—OH               H—C—OH
     |                 |                    |                      |
  *CH2OPO3⁻       *CH2OPO3⁻            *CH2OPO3⁻             *CH2OPO3⁻

  Fructose-6-P     Erythrose-4-P       Glyceraldehyde-3-P     Sedoheptulose-7-P
```

Reversal of step 6 (page 749) will produce ribose labeled on C-1, 2 and 5 from sedoheptulose and xylulose labeled at carbon 1 from sedoheptulose and carbon 5 from carbon 3. When these are converted to ribose, carbons 1, 2 and 5 are labeled. See below.

```
        *CH2OH
         |
         C=O
         |
  HO——*C-H                                              *CHO              *CH2OH
         |                                                |                 |
   H——*C-OH                                        H——*C-OH               C=O
         |                            ——→                 |                 |
   H——C-OH                  CHO               H——C-OH            HO——C-H
         |                    |                            |                 |
   H——C-OH           H——C-OH            H——C-OH            H——C-OH
         |                    |                            |                 |
      *CH2OPO3⁻         *CH2OPO3⁻              *CH2OPO3⁻          *CH2OPO3⁻

  Sedoheptulose-7-P    Glyceraldehyde-3-P       Ribose-5-P        Xylulose-5-P
```

16. Understanding a version of the pentose phosphate pathway that produces mainly NADPH
As described on pages 751 and 752, the pentose phosphate pathway may be used to produce large amounts of NADPH without significant net production of ribose-5-phosphate. Draw a diagram, beginning with glucose-6-phosphate, to show how this may be accomplished

Answer: NADPH production requires the oxidative stage of the pentose phosphate pathway and so steps 1, 2 and 3 must be taken (see Figure 22.22). This will convert glucose-6-phosphate into ribulose-5-phosphate. Ribulose is then passed through the non-oxidative stage of the pathway converting it into glucose. In this stage, ribulose is converted to xylulose and ribose, which are used by transketolase and transaldolase to produce fructose-6-phosphate that can be converted to glucose-6-phosphate.

17. Understanding a version of the pentose phosphate pathway that produces mainly NADPH and ATP
The discussion on page 754 explains that the pentose phosphate pathway and the glycolytic pathway can be combined to provide both NADPH and ATP (as well as some NADH) without net ribose-5-phosphate synthesis. Draw a diagram to show how this may be accomplished.

Answer: To produce NADPH requires the oxidative stage of the pentose phosphate pathway or steps 1, 2 and 3 in Figure 22.22. To get ATP glycolysis needs to be followed and so fructose-6-phosphate produced by consuming ribulose-5-phosphate in the non-oxidative stage of the pathway will have to be metabolized into pyruvate in glycolysis.

18. Assessing the labelling patterns in versions of the pentose phosphate pathway
Consider the pathway diagram you constructed in problem 17. What is the fate of carbon from positions 2 and 4 of glucose-6-phosphate after one pass through the pathway?

Answer: Glucose-6-phosphate labeled at carbons 2 and 4 is converted to ribulose-5-phosphate labeled at carbons 1 and 3. Using Figure 22.22 as a guide the same two carbons remain labeled in ribose-5-phosphate and xylulose-5-phosphate.

```
          CH2OH                            
           |                               
    H——*C-OH                        *CH2OH
           |                               |
   HO——C-H                           C=O
           |                               |
    H——*C-OH                   H——*C-OH
           |                               |
    H——C-OH                    H——C-OH
           |                               |
        CH2OPO3⁻                    CH2OPO3⁻

     Glucose-6-P                    Ribulose-5-P
```

Step 6 converts these to sedoheptulose-7-P labeled at carbons 1, 3 and 5 and glyceraldehyde-3-phosphate labeled at carbon 1.

Xylulose-5-P   Ribose-5-P   Glyceraldehyde-3-P   Sedoheptulose-7-P

Glyceraldehyde-3-P can be converted into pyruvate labeled at carbon 1. And, glyceraldehyde-3-P and sedoheptulose-7-P can be used to produce erythrose-4-P and fructose-6-P by transaldolase (step 7).

Glyceraldehyde-3-P   Sedoheptulose-7-P   Fructose-6-P   Erythrose-4-P

The fructose-6-P formed is labeled at carbons 1, 3 and 4 and these become carbons 1, 3 and 1 of pyruvate. So, label starting on carbons 2 and 4 of glucose-6-P will show up on carbons 1 and 3 of pyruvate.

19. Understanding the initial reaction in formation of glycogen
Glycogenin catalyzes the first reaction in the synthesis of a glycogen particle, with Tyr194 of glycogenin combining with a glucose unit (provided by UDP-glucose) to produce a tyrosyl glucose. Write a mechanism to show how this reaction could occur.

Answer: The mechanism proposed by Hurley and coworkers (Hurley et al., 2005, Requirements for catalysis in mammalian glycogenin. Journal of Biological Chemistry 280:23892-23899) is presented below.

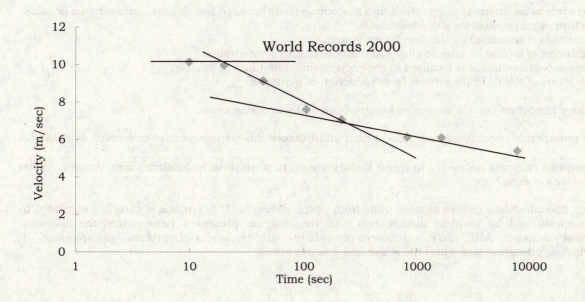

Preparing for the MCAT® Exam

20. (Assessing the biochemistry of exercise
Study the graphs in the Deeper Look box (page 740) and explain the timing of the provision of energy from different metabolic sources during heavy exercise.

Answer: From Graph (a) we see that ATP stores decrease rapidly, within 5 to 10 seconds, followed by phosphocreatine stores, which last 40 seconds or so. Anaerobic metabolism begins after about 10 seconds but peaks at about 40 seconds then declines. As anaerobic metabolism begins to decline, aerobic metabolism begins to increase. In heavy exercise, Graph (b) shows that glycogen levels drop to low levels and are nearly exhausted after 60 minutes of heavy exercise.

In the first 30 seconds or so, energy is being derived from ATP and phosphocreatine. After this time, lactic acid production begins to increase as anaerobic metabolism becomes very active, peaking at about 45 seconds or so. By about 75 seconds, aerobic respiration begins to account for a majority of energy. The following figure shows the velocity of wold records in 2000 against the log of the time. The short races, 100 and 200 m, last less than 30 seconds and are performed at nearly the same rate. These races are being run solely on ATP and phosphocreatine. Runners of the intermediate races (400, 800 and 1500) likely suffer the most because they deplete ATP and phosphocreatine and build up lactic acid from anaerobic metabolism. The long-distance races rely mainly on aerobic respiration and good glycogen stores.

421

21. Understanding the chemistry and energetics of creating phosphate
(Integrates with Chapters 3 and 14.) What is the structure of creatine phosphate? Write reactions to indicate how it stores and provides energy for exercise.

Answer: The structure of creatine phosphate is show below. A little reflection may help you identify this compound's structure as being very similar to the side chain of arginine. (Recall the side chain of arginine, N-methylate it, and attach a phosphate group in P-N linkage.) Creatine phosphate is a phosphagen, a high-energy phosphate compound used to restore ATP levels. Some organisms actually use phosphoarginine as a phosphagen. The P-N bond is highly energetic and can be used to phosphorylate ADP.

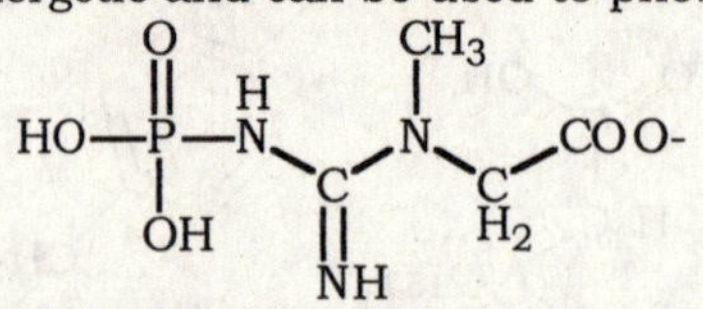

$$HO-\overset{\overset{\displaystyle O}{\|}}{\underset{\underset{\displaystyle OH}{|}}{P}}-\overset{\overset{\displaystyle H}{|}}{N}-\overset{\overset{\displaystyle \|}{C}}{\underset{\underset{\displaystyle NH}{\|}}{C}}-\overset{\overset{\displaystyle CH_3}{|}}{N}-\overset{\overset{}{C}}{\underset{\underset{\displaystyle H_2}{|}}{C}}-COO^-$$

Organisms that use creatine phosphate as a phosphagen have two creatine kinases. Kinases phosphorylate substrates using the gamma phosphate of ATP as the phosphate source. Thus, creatine kinase phosphorylates creatine producing creatine phosphate in the following reaction.

Creatine + ATP → creatine phosphate + ADP

One creatine kinase is located in the mitochondria where high ATP levels allow the reaction to proceed as written. This mitochondrial enzyme is used to re-charge creatine to creatine phosphate. The other creatine kinase is located in the cytoplasm where it runs the reverse reaction when ADP levels begin to rise. This cytosolic creatine kinase recharges ADP at the expense of creatine phosphate. In effect, it catalyzes the following reaction, which is simply the reverse of the one presented above.

creatine phosphate + ADP → Creatine + ATP

To link the two kinase reactions, creatine is transported into the mitochondria while creatine phosphate is transported out of the mitochondria.

Questions for Self Study

1. Fill in the blanks. Gluconeogenesis is the production of ____ from pyruvate and other 3-carbon and 4-carbon compounds. In animals it occurs mainly in two organs, the ____ and ____. This pathway utilized several of the steps of glycolysis but three key glycolytic steps are bypassed. They are the reactions catalyzed by ____, ____, and ____. The conversion of pyruvate to PEP is a two step process in which pyruvate is first converted to ____, a citric acid cycle intermediate with the addition of ____. The enzyme catalyzing this step uses the vitamin ___. Production of PEP from pyruvate results in the hydrolysis of ___ high-energy phosphate bonds. The other two steps unique to gluconeogenesis are catalyzed by ___ and ____. These enzymes carry out ___ reactions.

2. The Cori cycle is an interplay of glycolysis and gluconeogenesis between two tissues. Answer true or false.
   a. The liver is responsible for gluconeogenesis. ___
   b. Pyruvate is transported by the circulatory system. ___
   c. Production of lactate by muscle allows ATP production by fermentation. ___
   d. Glucose-6-phosphatase is localized in the endoplasmic reticulum of muscle. ___
   e. High levels of NADH in the muscle favor reduction of pyruvate. ___

3. Name three important control molecules for glycolysis and gluconeogenesis.

4. What is remarkable about phosphofructokinase-2 and fructose-2,6-bisphosphatase enzymatic activities?

5. There are two enzymes necessary to digest dietary glycogen, α-amylase and debranching enzyme. What are the functions of each?

6. Glycogen phosphorylase cleaves glucose units from tissue glycogen. This enzyme is carefully regulated by allosteric controls and by covalent modification. In the diagram presented below place the following: phosphorylase kinase, AMP, ATP, glucose-6-phosphate, glucose, phosphoprotein phosphatase 1, phosphorylase a, phosphorylase b and the active and inactive states.

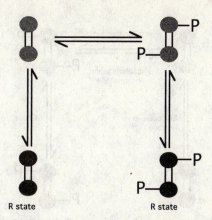

7.   Describe the cascade of events from hormone stimulation to inactivation of glycogen synthase and activation of glycogen phosphorylase.

8. In the oxidative steps of the pentose phosphate pathway, glucose is converted to ribulose-5-phosphate and carbon dioxide in five steps. This short metabolic sequence serves two immediate purposes. What are they?

9. In the nonoxidative steps of the pentose phosphate pathway, 5-carbon sugars are metabolized back to glucose.   The enzymes involved include an isomerase (I), an epimerase (E), transketolases (TK) and a transaldolase (TA). Which enzyme best fits each of the statements below?
    a. Thiamine pyrophosphate-dependent enzyme. ___
    b. Causes conversion of configuration about asymmetric carbon. ___
    c. Ketose-aldose conversion. ___
    d. Forms Schiff base intermediate. ___
    e. Interconverts ribulose-5-phosphate and ribose-5-phosphate. ___
    f. Forms enediolate intermediate, reduction of which may occur on either side of the
       carbon. ___
    g. Reduction of adduct by sodium borohydride leads to inactivation. ___
    h. Moves two carbons from a ketose to an aldose. ___

Answers

1. Glucose; liver; kidney; hexokinase (or glucokinase); phosphofructokinase; pyruvate kinase; oxaloacetate; carbon dioxide; biotin; two; fructose-1,6-bisphosphatase; glucose-6-phosphatase; hydrolysis.

2. a. T; b. F; c. T; d. F; e. T.

3. Fructose-2,6-bisphosphate, AMP, ATP, citrate, acetyl-CoA, alanine and glucose-6-phosphate.

4. Both activities are properties of the same protein molecule, a bifunctional or tandem enzyme.

5. α-Amylase is an endoglycosidase that hydrolyzes α-(1★4) linkages to produce monosaccharides, disaccharides, and short oligosaccharides in addition to limit dextrins, highly branched polysaccharides. Debranching enzyme degrades limit dextrins by removing branches.

6.

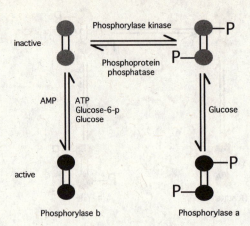

7. Hormone binding to surface receptors leads to stimulation of adenylyl cyclase, which raises the levels of cellular cAMP. This second messenger binds to and activates protein kinase, which phosphorylates and inactivates glycogen synthase. cAMP-dependent protein kinase also phosphorylates glycogen phosphorylase kinase leading to activation. Finally, phosphorylation of glycogen phosphorylase by glycogen phosphorylase kinase leads to activation.

8. The oxidative steps produce 5-carbon sugars and reducing equivalents in the form of NADPH.

9. a. TK; b. E; c. I; d. TA; e. I; f. E; g. TA; h. TK.

Additional Problems

1. What are the three potential substrate cycles in glycolysis/gluconeogenesis and how is each regulated?

2. Pyruvate carboxylase reacts bicarbonate and pyruvate to produce oxaloacetate. It is a vitamin-dependent enzyme. Which vitamin does it use? Describe the reaction the enzyme catalyzes.

3. Alanine inhibits pyruvate kinase. What relationship does alanine have to this enzyme?

4. How are glycogen phosphorylase and glycogen synthase reciprocally regulated?

5. The oxidative phase of the pentose phosphate pathway is the following sequence:
glucose ★ glucose-6-phosphate ★ 6-phosphogluconolactone ★ 6-phosphogluconate ★ ribulose-5-phosphate. Describe the biochemistry of this sequence.

6. Can triacylglycerol (a.k.a. fat) be used in gluconeogenesis?

Abbreviated Answers

1. The conversion of glucose and glucose-6-phosphate catalyzed by glucokinase (in liver) and glucose-6-phosphatase is a substrate cycle confined to cells expressing glucose-6-phosphatase, such as kidney and liver cells. A substrate cycle exists between fructose-6-phosphate and fructose-1,6-bisphosphate catalyzed by phosphofructokinase and fructose-1,6-bisphosphatase. This cycle is regulated by reciprocal control of the two enzymes by fructose-2,6-bisphosphate and AMP. The conversion of PEP to pyruvate in glycolysis and of pyruvate to oxaloacetate to PEP in gluconeogenesis is the final substrate cycle. Acetyl-CoA levels play a key role in regulating this cycle.

2. Pyruvate carboxylase is a biotin-dependent reaction. Biotin participates in one-carbon reactions in which the carbon is fully oxidized. The enzyme catalyzes a two-step reaction. In the first step, carboxyphosphate, an activated form of CO_2, is produced by nucleophilic attack on the γ-phosphate of ATP by a bicarbonate oxygen. Carboxyphosphate is used to transfer CO_2 to biotin, which carboxylates the methyl carbon of pyruvate to produce oxaloacetate.

3. Deamination of alanine produces pyruvate, a product of pyruvate kinase. The presence of alanine is a signal that an alternative carbon source for the production of pyruvate is present. In general, proteins are not stored to the extent that carbohydrates are, and they are therefore broken down into amino acids, which are subsequently metabolized.

4. Glycogen phosphorylase is allosterically activated by AMP and inhibited by glucose-6-phosphate, whereas glycogen synthase is stimulated by glucose-6-phosphate. Both enzymes are also regulated by covalent modification. There are two forms of glycogen phosphorylase. The active form, phosphorylase *a*, differs from the less active form, phosphorylase *b*, by a phosphate group on Ser[14]. Ser[14] is phosphorylated by glycogen phosphorylase kinase, a kinase that also exists in an active, phosphorylated form, and an inactive form. Phosphorylation of glycogen phosphorylase kinase is a result of cAMP-dependent protein kinase. Glycogen synthase also has two forms, an active, glucose-6-phosphate-independent form (glycogen synthase I) and a less active, glucose-6-phosphate-dependent form (glycogen synthase D). Glycogen synthase I is converted to glycogen synthase D by phosphorylation at a number of sites catalyzed by several kinases.

5. The first step is a phosphorylation catalyzed by hexokinase. The next step is an oxidation of glucose-6-phosphate by NADP-dependent glucose-6-phosphate dehydrogenase. 6-Phosphogluconolactone is converted to phosphogluconate by hydrolysis catalyzed by gluconolactonase. Phosphogluconate is oxidatively decarboxylated to ribulose-5-phosphate and CO_2 by NADP-dependent 6-phosphogluconate dehydrogenase. So, the steps are phosphorylation, oxidation, hydrolysis, and oxidative decarboxylation.

6. A majority of the carbons in triacylglycerol is unavailable for incorporation into glucose because they are converted into acetyl-CoA units. However, the glycerol backbone can be used as a carbon source for gluconeogenesis. (For triacylglycerols with fatty acids with an odd number of carbons, a three-carbon unit is metabolized into succinate, a citric acid cycle intermediate and therefore a substrate for gluconeogenesis.

Summary

In order to maintain normal levels of metabolic activity, cells must sustain an appropriate level of sugars, especially glucose. Glucose is synthesized from noncarbohydrate precursors by a process known as gluconeogenesis, and it can also be supplied via breakdown of glycogen (in animals) or starch (in plants or in the diet of animals). The pentose phosphate pathway is the primary source of NADPH, the reduced coenzyme essential to most reductive biosynthetic processes, including fatty acid and amino acid biosynthesis. The pentose phosphate pathway also produces ribose-5-P, an important component of ATP, $NAD(P)^+$, FAD, coenzyme A, DNA and RNA.

Substrates for gluconeogenesis include pyruvate, lactate, most of the amino acids (except leucine and lysine), glycerol and all the TCA cycle intermediates. Fatty acids are not substrates for gluconeogenesis, since animals cannot carry out net synthesis of sugars from acetyl-CoA. Major sites of gluconeogenesis include the liver (90%) and kidneys (10%). Glucose produced by these organs is utilized in brain, heart, muscle and red blood cells to meet their metabolic needs. Gluconeogenesis is not merely the reverse of glycolysis. In order for the net pathway to be exergonic and for the pathway to be suitably regulated, three key enzymatic reactions replace hexokinase (glucokinase), phosphofructokinase, and pyruvate kinase from glycolysis. These reactions are catalyzed by glucose-6-phosphatase, fructose-1,6-bisphosphatase, and the pair of pyruvate carboxylase/PEP carboxykinase. The overall conversion of pyruvate to glucose is made exergonic by these substitutions, with a $\Delta G^{\circ\prime}$ of -30.5 kJ/mole.

The pyruvate carboxylase reaction, which occurs in the mitochondrial matrix, is biotin-dependent and utilizes ATP and bicarbonate as substrates. Pyruvate carboxylase is allosterically activated by acyl coenzyme A. When ATP and acetyl-CoA are high, gluconeogenesis is favored, but when ATP and acetyl-CoA are low, acetyl-CoA is consumed by the TCA cycle for the net production of energy and ATP. Whereas pyruvate carboxylase consumes an ATP to drive a carboxylation, the PEP carboxykinase uses decarboxylation to facilitate formation of PEP. AMP inhibits fructose-1,6-bisphosphatase, and the inhibition by AMP is enhanced by fructose-2,6-bisphosphate. Glucose-6-phosphatase, the final step in gluconeogenesis, is located in the membrane of the endoplasmic reticulum. Glucose formed in liver ER by this reaction is transported to the plasma membrane and secreted directly into the bloodstream for distribution to other organs and cells.

The gluconeogenic pathway is driven by hydrolysis of ATP and GTP - the net free energy change for the pathway from pyruvate to glucose is -37.7 kJ/mole. The pathway is controlled by substrate cycles involving pairs of reactions such as phosphofructokinase (glycolysis) and fructose-1,6-bisphosphatase (gluconeogenesis). Moreover, reciprocal regulatory control ensures that glycolysis is active when gluconeogenesis is inactive and *vice versa*. Acetyl-CoA is a potent allosteric effector of these two pathways, inhibiting pyruvate kinase (glycolysis) and activating pyruvate carboxylase. Hydrolysis of fructose-1,6-bisphosphate by fructose-1,6-bisphosphatase is allosterically inhibited by fructose-2,6-bisphosphate, which

425

also activates phosphofructokinase in another example of reciprocal control. Cellular levels of fructose-2,6-bisphosphate are controlled by phosphofructokinase-2 and fructose-2,6-bisphosphatase, two activities of a bifunctional enzyme that are regulated by phosphorylation and dephosphorylation of a single Ser residue of the 49,000 Dalton subunit.

Adult human beings typically metabolize about 160 g of carbohydrates daily, either from digestion of dietary carbohydrates or from breakdown of tissue glycogen. Dietary carbohydrates are degraded by α-amylase to produce limit dextrins, which can be further degraded by a debranching enzyme. Digestion is a highly efficient (and unregulated) process in which 100% of ingested food is absorbed and metabolized. By contrast, tissue glycogen is metabolized in a highly regulated process. Glycogen breakdown is stimulated by glucagon and epinephrine, whereas glycogen synthesis is activated by insulin. In addition to stimulating glycogen synthesis in liver and muscle, insulin also stimulates transport of glucose across cell membranes and activates several key glycolytic enzymes. Binding of glucagon to liver cell membranes activates intracellular adenylyl cyclase, producing cyclic AMP, a second messenger, which activates a protein kinase. This kinase in turn phosphorylates and activates phosphorylase kinase, which phosphorylates phosphorylase, the enzyme that catalyzes the first step in glycogen breakdown. Such kinase cascades facilitate the amplification of hormonal signals.

Glycogen synthesis involves the transfer of activated glucose units (in the form of sugar nucleotides) to a growing chain in the glycogen synthase reaction. The glycogen chain is initiated on a protein core, glycogenin, with the first glucose residue linked covalently to a Tyr-OH. Glycogen synthase catalyzes the transfer of glucosyl units from UDP-glucose to the C-4 hydroxyl groups of non-reducing ends of glycogen strands. Glycogen branching occurs by transfer of six- or seven-residue segments from the nonreducing end of a linear chain at least eleven residues in length to the C-6 hydroxyl of a glucose residue further up the chain.

The pentose phosphate pathway converts glucose-6-phosphate in two successive oxidations and five non-oxidative reactions to a variety of carbohydrate intermediates, producing NADPH and ribose-5-phosphate as well. Oxidation of glucose-6-phosphate by glucose-6-phosphate dehydrogenase, followed by ring opening and further oxidation by 6-phosphogluconate dehydrogenase yields ribulose-5-phosphate, whereas phosphopentose epimerase produces xylulose-5-phosphate. Further rearrangement to produce other carbohydrate intermediates is carried out by transketolase and transaldolase. Transketolase is a thiamine pyrophosphate-dependent enzyme, and transaldolase functions via formation of an active site Schiff base with a lysine residue on the protein. Utilization of glucose-6-phosphate in glycolysis and the pentose phosphate pathway depends on the cell's needs for ATP, NADPH and ribose-5-P. Depending on the cell's needs for these metabolites, three different metabolic routes produce needed NADPH or ribose-5-P, either selectively or in relatively balanced amounts. An alternative pathway yields large amounts of NADPH and ATP, with relatively small amounts of ribose-5-P.

Chapter 23

Fatty Acid Catabolism

• •

Chapter Outline

❖ Triacylglycerols: Energy storage molecules
   ⅄ Highly reduced carbon source
   ⅄ Stored without large amounts of water
   ⅄ Stored in specialized cells: Adipocytes (adipose or fat cells)
❖ Hormonal regulation
   ⅄ Adrenaline, glucagon, ACTH mobilize fatty acids from adipocytes
   ⅄ Hormone binding stimulates cAMP-dependent protein kinase
   ⅄ Triacylglycerol lipase (hormone-sensitive lipase) activated by phosphorylation
      • Triacylglycerol lipase produces fatty acid and diacylglycerol
      • Diacylglycerol lipase produces fatty acid and monoacylglycerol
      • Monoacylglycerol lipases produces fatty acid and glycerol
         • Fatty acids metabolized by β-oxidation
         • Glycerol metabolized by glycolysis
❖ Dietary triacylglycerols
   ⅄ Pancreatic lipases secreted into duodenum release fatty acids and monoacylglycerol
      • Lipase activity depends on bile salts functioning as detergents
      • Short-chain fatty acids directly absorbed
      • Long-chain fatty acids recondensed onto glycerol in epithelial cells
         • Resynthesized triacylglycerol complexed to proteins
         • Chylomicrons released into blood
❖ β-Oxidation
   ⅄ Oxidation of β-carbon, cleavage of C_α- C_β bond
   ⅄ Occurs in mitochondria
   ⅄ Two-carbon acetate units split off as acetyl-CoA
❖ Fatty acid activation and transport into mitochondria
   ⅄ Short-chain fatty acids
      • Diffuse into mitochondria
      • Acyl-CoA synthetase (Acyl-CoA ligase or fatty acid thiokinase) produces acyl-CoA derivatives
      • ATP to AMP + PP_i ($PP_i \rightarrow 2P_i$) drives synthesis
   ⅄ Long-chain fatty acids: Activated in cytosol
      • Acyl-CoA synthetase produces acyl-CoA derivatives
      • Carnitine acyltransferase I produces acylcarnitine
         • Outside surface of inner mitochondrial membrane
      • Carnitine acyltransferase II reforms acyl-CoA
         • Matrix surface of inner mitochondrial membrane
❖ Enzymology of β-oxidation: Four steps
   ⅄ Acyl-CoA dehydrogenase
      • Produces C-C double bond
      • Noncovalent (but tightly bound) FAD reduced

427

- FADH$_2$ passes electrons through electron transfer protein to coenzyme Q
- Three enzymes: Chain-length specific
- Site of inhibition of hypoglycin
- Enoyl-CoA hydrase (crotonase)
  - Adds water across double bond
  - Three kinds of enzymes
    - Trans double bond hydrated to L-β-hydroxyacyl-CoA
    - Cis double bond hydrated to D-β-hydroxyacyl-CoA
    - Trans double bond hydrated to D-β-hydroxyacyl-CoA
- L-Hydroxyacyl-CoA dehydrogenase
  - Oxidizes OH
  - NAD$^+$ reduced
- Thiolase (β-ketothiolase)
  - Cleaves off acetyl-CoA unit
  - Acyl-CoA two carbons shorter
  - Last cleavage releases
    - Two acetyl-CoA for even carbon numbered chains
    - One propionyl-CoA and one acetyl-CoA for odd carbon numbered chains
- ❖ Odd-chain fatty acids: From plant and marine organism triacylglycerols
  - Propionyl-CoA converted to succinyl-CoA
    - Propionyl-CoA also from Met, Val, Ile degradation
    - Succinyl-CoA TCA cycle intermediate: To metabolize completely
      - Succinyl-CoA to malate
      - Malate to cytosol
      - Cytosolic malate to pyruvate via malic enzyme
    - Propionyl-CoA to succinyl-CoA: Three steps
      - Propionyl-CoA carboxylase produces D-methylmalonyl-CoA
        - ○ Biotin-dependent reaction
        - ○ ATP hydrolysis drives reaction
      - Methylmalonyl-CoA epimerase
        - ○ D-methylmalonyl CoA to L-methylmalonyl-CoA
        - ○ Epimerization not racemization
      - Methylmalonyl-CoA mutase
        - ○ Carbonyl-CoA moiety moved from between carbons
        - ○ B$_{12}$-dependent reaction
- ❖ Unsaturated fatty acids: β-Oxidation and additional reactions
  - Monounsaturated fatty acids: Enoyl-CoA isomerase: cis-Δ^3 to trans-Δ^2 isomerization of double bond
  - Polyunsaturated fatty acids: 2,4-Dienoyl-CoA reductase
- ❖ Peroxisomal β-oxidation: Acyl-CoA oxidase: Oxygen accepts electrons: H$_2$O$_2$ produced
- ❖ α-Oxidation: Branched-chain fatty acids: Phytol: Chlorophyll breakdown product
- ❖ ω-Oxidation (ω, last letter of Greek alphabet)
  - Dicarboxylic acid produced
  - Cytochrome P450 requires NADPH and oxygen
  - ω-Carbon oxidized to carboxylic acid
- ❖ Ketone bodies: Acetoacetate, hydroxybutyrate, acetone
  - Synthesis occurs in liver
  - Converts acetyl units into ketone bodies
  - Ketone bodies serve as fuel for brain, heart, muscle
- ❖ Ketogenesis: Four reactions
  - Thiolase
    - Reversal of last step in β-oxidation
    - Acetoacetyl-CoA produced from two acetyl-CoA
  - HMG-CoA synthase: β-Hydroxy-β-methylglutaryl-CoA
    - This and thiolase reaction in mitochondria lead to ketone body formation
    - In cytosol, HMG-CoA production fuels cholesterol synthesis

⼂ HMG-CoA lyase: Produces acetoacetate
⼂ β-Hydroxybutyrate dehydrogenase
  • NADH-dependent reduction
  • Acetoacetate will spontaneous decarboxylate to produce acetone
❖ Diabetes Mellitus
  ⼂ Type I: Insulin secretion deficient
  ⼂ Type II: Shortage of insulin receptors

Chapter Objectives

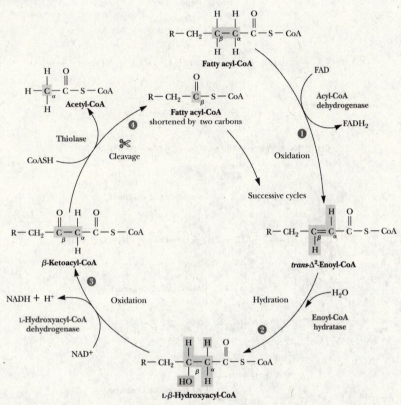

Figure 23.9 The β-oxidation of saturated fatty acids involves a cycle of four enzyme-catalyzed reactions. Each cycle produces single molecules of $FADH_2$, NADH, and acetyl-CoA and yields a fatty acid shortened by two carbons. (The delta [Δ] symbol connotes a double bond, and its superscript indicates the lowest-numbered carbon involved.)

β-Oxidation

β-Oxidation is a simple series of four steps leading to degradation of fatty acyl-coenzyme A derivatives. You should know how fatty acids are supplied to β-oxidation. Lipases hydrolyze triacylglycerols. The fatty acids released are converted to acyl-CoA derivatives by synthetases that use two high-energy phosphoanhydride bonds to drive synthesis. β-Oxidation is reminiscent of the steps in the citric acid cycle leading from succinate to oxaloacetate. You might recall that they included oxidation by an FAD-dependent enzyme to produce a double bond, hydration of the double bond, and reduction of the alcoholic carbon to a ketone by an NAD^+-dependent enzyme. In β-oxidation (Figure 23.9), a similar series operates but with an acyl-CoA as substrate and thiolysis of the product to release acetyl-CoA and a fatty acyl-CoA, two carbons shorter. The FAD-dependent enzyme, acyl-CoA dehydrogenase, will move electrons into the electron transport chain at the level of coenzyme Q and each electron pair will support production of 1.5 ATP. Odd-carbon fatty acids are metabolized by β-oxidation to yield several acetyl-CoAs and one propionyl-CoA, a three-carbon acyl-CoA. Propionyl-CoA is metabolized by a vitamin B_{12}-dependent pathway to succinyl-CoA, a citric acid cycle intermediate. Oxidation of unsaturated fatty acids (Figure 23.22) requires additional enzymes. Keep this in mind by remembering that typical double bonds in fatty acids are in *cis* configuration whereas the double bond formed during β-oxidation is *trans*. Double bonds starting at an odd-numbered carbon are simply isomerized to the *trans* configuration and moved, by one carbon, closer to the carboxyl end. Double bonds at even-numbered carbons are ultimately reduced.

Ketone Bodies

Acetyl units can be joined to form the four-carbon compound acetoacetate which together with β-hydroxybutyrate, a reduced form of acetoacetate, are used as a source of fuel for certain organs of the body and in times of glucose shortage. You should know the metabolic sequence leading to acetoacetate production. An intermediate in the pathway, β-hydroxy-β-methylglutaryl-CoA (HMG-CoA) is an important intermediate in cholesterol synthesis.

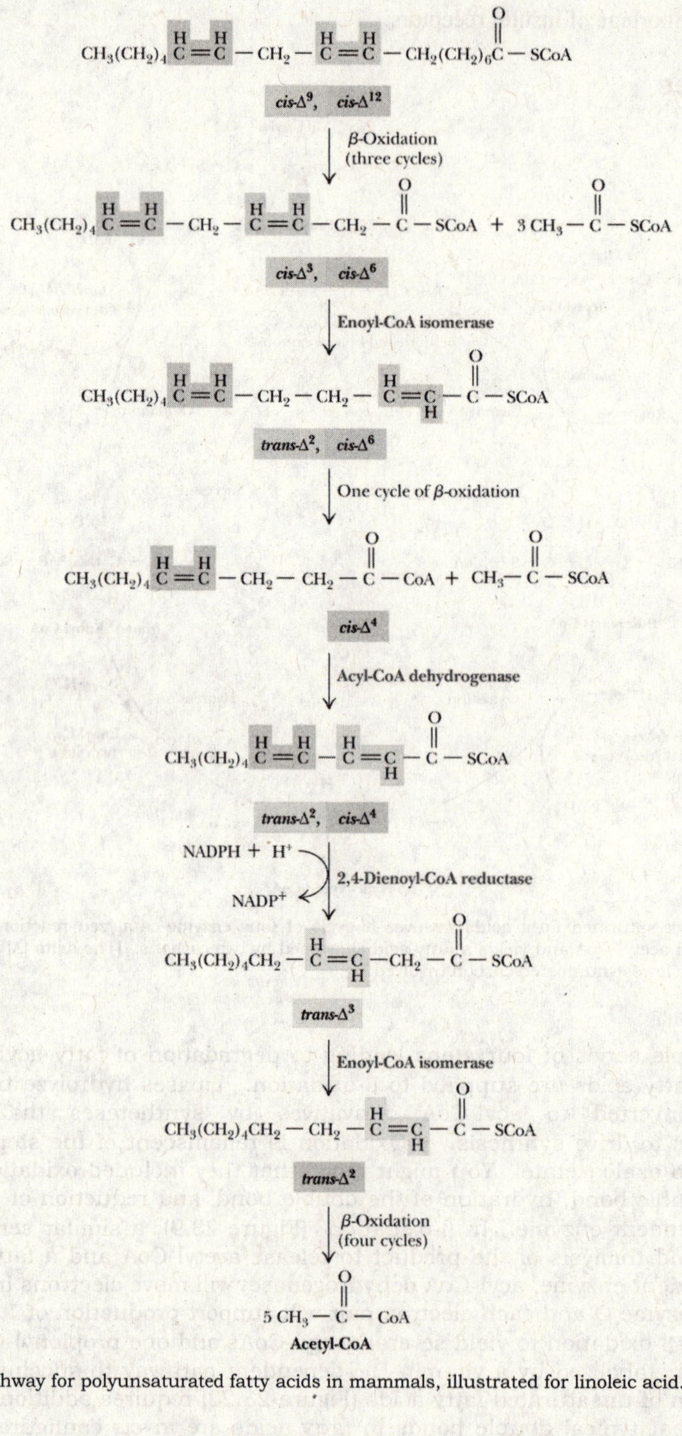

Figure 23.22 The oxidation pathway for polyunsaturated fatty acids in mammals, illustrated for linoleic acid.

Problems and Solutions

1. Determining the amount of water produced from fatty acid oxidation
Calculate the volume of metabolic water available to a camel through fatty acid oxidation if it carries 30 lb of triacylglycerol in its hump.

Answer: If we assume that the fatty acid chains in the triacylglycerol are all palmitic acid, the fatty acid content of a triacylglycerol as a percent of the total molecular weight is calculated as follows:

$$\% \text{ Palmitic acid} = \frac{3 \times M_{r,\text{palmitic acid}}}{3 \times M_{r,\text{palmitic acid}} + M_{r,\text{glycerol}} - 3 \times M_{r,\text{water}}} \times 100\%$$

$$M_{r,\text{palmitic acid}} = 256.4 \ (C_{16}H_{32}O_2), \ M_{r,\text{glycerol}} = 92.09 \ (C_3H_8O_3), \ M_{r,\text{water}} = 18$$

$$\% \text{ Palmitic acid} = \frac{3 \times 256.4}{3 \times 256.4 + 92.09 - 3 \times 18} \times 100\% = 95\%$$

Thirty pounds of triacylglycerol corresponds to:

$$30 \text{ lb} \times \frac{453.6 \text{ g}}{\text{lb}} = 13,608 \text{ g of triacylglycerol}$$

If 95% of this is composed of palmitic acid we have:

$$0.95 \times 13,608 \text{ g} = 12,928 \text{ g of palmitic acid which corresponds to:}$$

$$\frac{12,928 \text{ g}}{256.4 \ \frac{\text{g}}{\text{mol}}} = 50.4 \text{ mol of palmitic acid}$$

β-oxidation of palmitate produces 130 moles of H_2O per mole of palmitate. Therefore, 50.4 moles of palmitate metabolized gives:

$$50.4 \text{ mol} \times 130 \ \frac{\text{mol water}}{\text{mol}} = 6,554 \text{ mol water}$$

At $18 \ \frac{\text{g}}{\text{mol}}$ this represents $6,554 \text{ mol} \times 18 \ \frac{\text{g}}{\text{mol}} = 117,972 \text{ g water or 118 liters}$

2. Determining the amount of ATP produced from fatty acid oxidation
Calculate the approximate number of ATP molecules that can be obtained from the oxidation of cis-11-heptadecenoic acid to CO₂ and water.

Answer: *cis*-11-Heptadecenoic acid is a 17-carbon fatty acid with a double bond between carbons 11 and 12. β-oxidation, ignoring for the moment the presence of the double bond, will produce 7 acetyl-CoA units and one propionyl-CoA. The 7 acetyl-CoA units are metabolized in the citric acid cycle with the following stoichiometry:

$$7 \text{ Acetyl-CoA} + 14 \ O_2 + 70 \text{ ADP} + 70 \ P_i \star 7 \text{ CoA} + 77 \ H_2O + 14 \ CO_2 + 70 \text{ ATP}$$

In producing the 7 acetyl-CoAs, 7 β-carbons had to be oxidized, and only 6 of these by the complete β-oxidation pathway. Thus, 6 FAD ★ 6 FADH₂, and 6 NAD⁺ ★ 6 NADH. And, if electrons from FADH₂ produce 1.5 ATP, whereas electrons from NADH produce 2.5 ATPs, we have 9 ATPs from FADH₂ and 15 ATPs from NADH. Now we will deal with the double bond. The presence of a double bond means that a carbon is already partially reduced, so, the FAD-dependent reduction step is bypassed and only NADH is generated. Thus, one acetyl-CoA unit is generated along with only 2.5 ATPs. Finally, we consider the propionyl-CoA unit that is metabolized into succinyl CoA. In this pathway, propionyl-CoA is converted to methylmalonyl-CoA, a process driven in part by ATP hydrolysis. Succinyl-CoA is a citric acid cycle intermediate that is metabolized to oxaloacetate. This process yields 1 GTP, one FAD-dependent oxidation, and one NAD⁺-dependent reaction. Thus a net of (1 + 1.5 + 2.5 - 1) 4 ATPs are produced. Succinyl-CoA cannot be consumed by the citric acid cycle and we have only converted it to oxaloacetate. One way of oxidizing succinate completely is to remove the carbons from the mitochondria as malate, and convert them to CO₂ and pyruvate using malic enzyme. This reaction is an oxidative-decarboxylation; NADP⁺ is reduced in the cytosol. There is a slight reduction in ATP production potential for electrons on NADPH in the cytosol but we will ignore this and assume that the reduction is equivalent energetically to reduction of malate to oxaloacetate (which we have already taken into account). Pyruvate is metabolized to acetyl-CoA with production of NADH, which supports 2.5 ATPs synthesized. The acetyl-CoA unit results in 10 ATP. To summarize: Oxidation of 7 β-carbons produces 24 ATPs; oxidation of an additional β-carbon (the one involved in a double bond) produces 2.5 ATPs; oxidation of

431

7 acetyl-CoAs contributes 70 ATPs; oxidation of an additional acetyl-CoA (derived from propionyl-CoA) yields 16.5 ATPs. The total is 113.

3. Assessing the mechanism of oxidation of phytanic acid

Phytanic acid, the product of chlorophyll that causes problems for individuals with Refsum's disease, is 3,7,11,15-tetramethylhexadecanoic acid. Suggest a route for its oxidation that is consistent with what you have learned in this chapter. (Hint: The methyl group at C-3 effectively blocks hydroxylation and normal β-oxidation. You may wish to initiate breakdown in some other way.)

Answer: The structure of phytanic acid is:

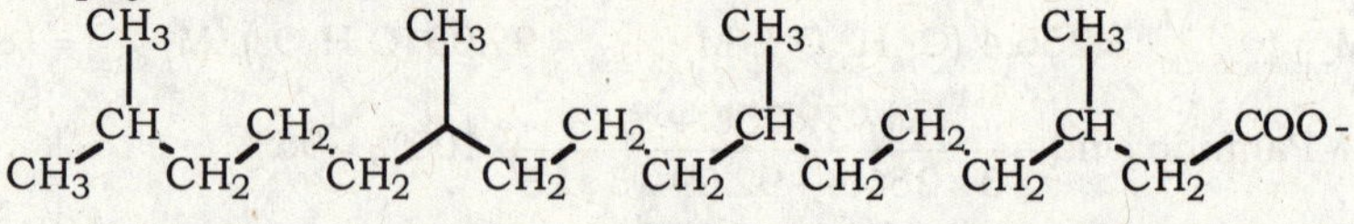

Normally it is metabolized by α-oxidation. The enzyme phytanic acid α-oxidase hydroxylates the α-carbon to produce phytanic acid, which is decarboxylated by phytanate α-oxidase to produce pristanic acid. Pristanic acid can form a coenzyme A ester which is metabolized by β-oxidation to yield 3 propionyl-CoAs, 3 acetyl-CoAs and 2-methyl-propionyl-CoA. The reaction sequence is shown below.

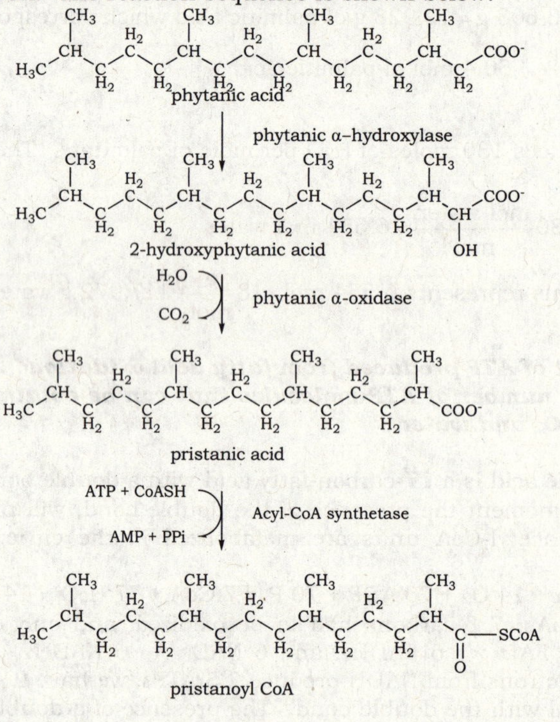

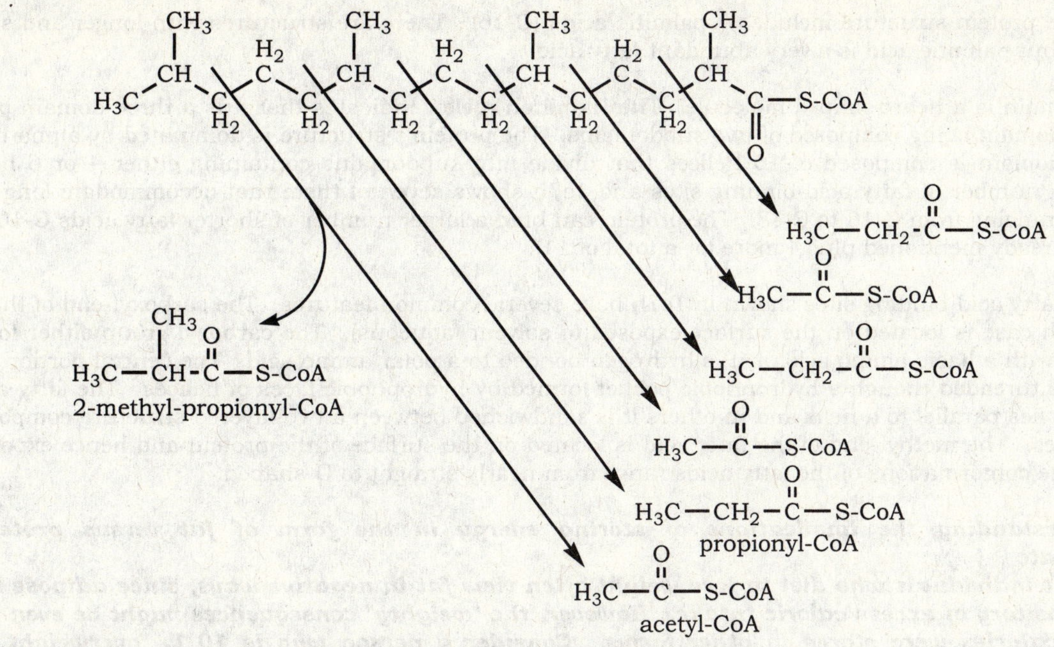

$$CH_3\text{-}CH(CH_3)\text{-}CH_2\text{-}CH_2\text{-}CH_2\text{-}CH_2\text{-}CH_2\text{-}CH_2\text{-}C(=O)\text{-}S\text{-}CoA$$

$$H_3C-CH_2\text{-}C(=O)-S\text{-}CoA$$

$$H_3C-C(=O)-S\text{-}CoA$$

$$H_3C-CH_2\text{-}C(=O)-S\text{-}CoA$$

$$CH_3$$
$$H_3C-CH\text{-}C(=O)-S\text{-}CoA$$
2-methyl-propionyl-CoA

$$H_3C-C(=O)-S\text{-}CoA$$

$$H_3C-CH_2\text{-}C(=O)-S\text{-}CoA$$
propionyl-CoA

$$H_3C-C(=O)-S\text{-}CoA$$
acetyl-CoA

4. Examining the labelling of glucose from ¹⁴C-labeled acetate

Even though acetate units, such as those obtained from fatty acid oxidation, cannot be used for net synthesis of carbohydrate in animals, labeled carbon from ¹⁴C-labeled acetate can be found in newly synthesized glucose (for example in liver glycogen) in animal tracer studies. Explain how this can be. Which carbons of glucose would you expect to be the first to be labeled by ¹⁴C-labeled acetate?

Answer: Acetate, as acetyl-CoA, enters the citric acid cycle where its two carbons show up as carbons 1 and 2 or carbons 3 and 4 of oxaloacetate after one turn of the cycle. C-1 and C-4 label derive from acetate labeled at the carboxyl carbon, whereas C-2 and C-3 label derive from label at the methyl carbon of acetate. (See Figure 19.15 in the textbook.) Oxaloacetate is converted to PEP with carbons 1, 2, and 3 becoming carbons 1, 2, and 3 of PEP, and so label is expected at carbon 1 if carboxy-labeled acetate is used and at carbons 2 or 3 if methyl-labeled acetate is used. Conversion of PEP to glyceraldehyde-3-phosphate results in label either at carbon 1 or at carbon 2 or 3. Isomerization to dihydroxyacetone phosphate labels the same carbons. Carbons 1 of DHAP and glyceraldehyde-3-phosphate become carbons 3 and 4 of fructose-1,6-bisphosphate so aldolase will label fructose-1,6-bisphosphate at carbons 3 and 4 from carboxy-labeled acetate and carbons 1, 2, 5, and 6 from methyl-labeled acetate.

5. Assessing the fatty acid binding capability of serum albumin

Human serum albumin (66.4 kD) is a soluble protein present in blood at 0.75 mM or so. Among other functions, albumin acts as the major transport vehicle for fatty acids in the circulation, carrying fatty acids from storage sites in adipose tissue to their sites of oxidation in liver and muscle. The albumin molecule has up to 11 distinct binding sites. Consult the biochemical literature to learn about the fatty acid–binding sites of albumin. Where are they located on the protein? What are their relative affinities? (Two suitable references with which to begin your study are Bhattacharya, A. A., Grüne, T., and Curry, S., 2000. Crystallographic analysis reveals common modes of binding of medium- and long-chain fatty acids in human serum albumin. Journal of Molecular Biology 303:721–732; and Simard, J. R., Zunszain, P. A., Ha, C-E., et al., 2005. Locating high-affinity fatty acid-binding sites on albumin by X-ray crystallography and NMR spectroscopy. Proceedings of the National Academy of Sciences U.S.A. 102:17958–17963.)

Answer: In addition to the references provided another good source of information is the following review: Hamilton, J.A., (2004) Fatty acid interactions with proteins: what X-ray crystal and NMR solution structures tell us. Progress in Lipid Research 43:177-199. To make sense out of any of this information it helps to view the structure of serum albumin with fatty acids bound to it. To do this, visit the RSCB Protein Data Bank at http://www.rcsb.org/pdb/home/home.do and do a search for human serum albumin. Over 80 hits are returned and the problem is now to decide which to focus on. This task is made easier by adding "fatty acid" to the search. The combined search returned 20 hits. The record 1e7h seemed particularly interesting because the protein structure includes 7 palmitic acids (C-16). There are structures with longer and shorter fatty acids but palmitic acid is a very abundant fatty acid.

Serum albumin is a heart-shaped molecule. The Hamilton review indicates that it is a three-domain protein with each domain being composed of two subdomains. The protein's structure is dominated by alpha helices and each domain is composed of 10 helices that divide into subdomains containing either 4 or 6 helices. There are a number of fatty acid binding sites and 1e7h shows seven of them that accommodate long chain fatty acids ranging from C-16 to C-18. The protein can bind a larger number of shorter fatty acids C-10 to C-14 (the 7 already mentioned plus 4 more for a total of 11).

The seven fatty acid binding sites shown in 1e7h have several common features. The carboxyl-end of the fatty acid in each case is located on the surface exposed to solvent (aqueous). The carboxyl group either forms a salt bridge with a basic amino acid or it is hydrogen bonded to a polar amino acid. The central portion of the fatty acid is threaded though a hydrophobic pocket formed by hydrophobic faces of helices. The fatty acid in some cases lies parallel to a helix and in others it is sandwiched between a two-layered structure composed of alpha helices. The methyl-end of the fatty acid is located on the surface of the protein and hence exposed to solvent. The conformations of the fatty acids range from nearly straight to U-shaped.

6.   *Understanding the implications of storing energy in the form of fat versus protein or carbohydrate*
Overweight individuals who diet to lose weight often view fat in negative ways, since adipose tissue is the repository of excess caloric intake. However, the "weighty" consequences might be even worse if excess calories were stored in other forms. Consider a person who is 10 lb "overweight," and estimate how much more he or she would weigh if excess energy were stored in the form of carbohydrate instead of fat.

Answer: There are two problems to consider: carbohydrates are a more oxidized form of fuel than fats; and, storage of carbohydrates requires large amounts of water, which will add greatly to its weight. From Table 23.1 we see that the energy content of fat and carbohydrate are 37 kJ/g and 16 kJ/g, respectively. Fat has 2.3 times (37 kJ/g ÷ 16 kJ/g) the energy content of carbohydrate on a per weight basis. Therefore, ten pounds of fat is equivalent to 23 pounds of carbohydrate. The information given in Table 23.1 is per gram dry weight. Fat is essentially anhydrous but in vivo carbohydrate is not. The 23 pounds of carbohydrate will be accompanied by a large amount of water adding to its overall weight.

7. *Understanding the consequences of vitamin B$_{12}$ deficiency*
What would be the consequences of a deficiency of vitamin B$_{12}$ for fatty acid oxidation? What metabolic intermediates might accumulate?

Answer: Vitamin B$_{12}$ is a coenzyme prosthetic group for methylmalonyl-CoA mutase. This enzyme is active in metabolism of odd-chain fatty acids. The last round of β–oxidation releases acetyl-CoA and propionyl-CoA from odd-chain fatty acids. Propionyl-CoA is further metabolized by being converted to (S)-methylmalonyl-CoA by propionyl-CoA carboxylase, a biotin-containing enzyme. (S)-Methylmalonyl-CoA racemase then converts (S)-methylmalonyl-CoA to (R)-methylmalonyl-CoA, which is metabolized to succinyl CoA by methylmalonyl-CoA mutase. This enzyme is a vitamin B$_{12}$-dependent enzyme, and a deficiency in vitamin B$_{12}$ will cause a build-up of (R)-methylmalonyl-CoA. Depending on the amount of odd chain fatty acids (and propionic acid) in the diet, a deficiency of B$_{12}$ is expected to slowly accumulating coenzyme A as methylmalonyl-CoA. Methylmalonyl-CoA is also formed during catabolism of valine and isoleucine.

8. *Balancing the equations for fatty acid oxidation*
Write a properly balanced chemical equation for the oxidation to CO$_2$ and water of (a) myristic acid, (b) stearic acid, (c) α-linolenic acid, and (d) arachidonic acid.

Answer: a. Myristic acid is a saturated 14:0 fatty acid, $CH_3(CH_2)_{12}COOH$. To metabolize myristic acid, it is first activated to a coenzyme A derivative in the following reaction:

$$CH_3(CH_2)_{12}COOH + ATP + CoA\text{-}SH \star CH_3(CH_2)_{12}CO\text{-}S\text{-}CoA + AMP + PP_i$$

Six cycles of β-oxidation convert myristoyl-CoA to 7 acetyl-CoA units. The balanced equation for β-oxidation is:

$$CH_3(CH_2)_{12}CO\text{-}S\text{-}CoA + 6\ CoA\text{-}SH + 6\ H_2O + 6\ NAD^+ + 6\ FAD \star$$
$$7\ CH_3CO\text{-}S\text{-}CoA + 6\ NADH + 6\ H^+ + 6\ FADH_2$$

The citric acid cycle metabolizes acetyl-CoA units as follows:

$$7\ CH_3CO\text{-}S\text{-}CoA + 21\ H_2O + 21\ NAD^+ + 7\ FAD \star$$

434

$$14\ CO_2 + 7\ CoA\text{-}SH + 21\ NADH + 21\ H^+ + 7\ FADH_2$$

This is accompanied by substrate-level GDP phosphorylation, which is equivalent to:

$$7\ ADP + 7\ P_i \quad \star \quad 7\ ATP + 7\ H_2O$$

Electron transport recycles NAD^+ as follows:

$$27\ NADH + 27\ H^+ + 13.5\ O_2 \quad \star \quad 27\ NAD^+ + 27\ H_2O$$

that supports the production of 2.5×27 (ADP + P_i $\star$ ATP + H_2O) or

$$67.5\ ADP + 67.5\ P_i \quad \star \quad 67.5\ ATP + 67.5\ H_2O$$

FAD is recycled by:

$$13\ FADH_2 + 6.5\ O_2 \quad \star \quad 13\ FAD + 13\ H_2O$$

that supports the production of 1.5×13 (ADP + P_i $\star$ ATP + H_2O) or

$$19.5\ ADP + 19.5\ P_i \quad \star \quad 19.5\ ATP + 19.5\ H_2O$$

The AMP and PP_i produced in the very first reaction can be metabolized by hydrolysis of PP_i and phosphorylation of AMP to ADP using ATP. Thus, $PP_i + H_2O$ $\star$ $2\ P_i$ and AMP + ATP $\star$ $2\ ADP$:

$$AMP + ATP + PP_i + H_2O \quad \star \quad 2\ ADP + 2\ P_i$$

If we sum these equations we find:

$$CH_3(CH_2)_{12}COOH + 92\ ADP + 92\ P_i + 20\ O_2 \quad \star \quad 92\ ATP + 14\ CO_2 + 106\ H_2O$$

b. Stearic acid or octadecanoic acid is 18:0 and its oxidation is given by:

$$CH_3(CH_2)_{16}COOH + 120\ ADP + 120\ P_i + 26\ O_2 \quad \star \quad 120\ ATP + 18\ CO_2 + 138\ H_2O$$

c. α-Linolenic acid is a polyunsaturated 18-carbon fatty acid with double bonds at carbons 9, 12, and 15. In two out of a total of 8 rounds of β-oxidation, reduction by FAD is bypassed. Therefore 2 fewer moles of $FADH_2$ are reduced in the electron transport chain than for a fully saturated fatty acid and 1.0 fewer moles of O_2 are consumed. The amount of ATP is reduced by $1.5 \times 2 = 3.0$. In addition, a NADH was actually consumed after the fifth cycle of β-oxidation to resolve a conjugated double bond. This accounts for 2.5 fewer ATP and 0.5 fewer O_2 that in b. The balanced equation is:

$$CH_3CH_2(CH=CHCH_2)_3(CH_2)_6COOH + 114.5\ ADP + 114.5\ P_i + 24.5\ O_2 \quad \star$$
$$114.5\ ATP + 18\ CO_2 + 129.5\ H_2O$$

d. Arachidonic acid is 5,8,11,14-eicosatetraenoic acid, a 20-carbon fatty acid with four double bonds. It undergoes a total of 9 cycles of β–oxidation, with 2 cycles not producing $FADH_2$ and consumption of two NADH to resolve two cases of conjugated double bonds. The balanced equation is:

$$CH_3(CH_2)_4(CH=CHCH_2)_4(CH_2)_2COOH + 126\ ADP + 126\ P_i + 27\ O_2 \quad \star$$
$$126\ ADP + 20\ CO_2 + 142\ H_2O$$

9. Tracing tritium incorporation into acetate from fatty acid
How many tritium atoms are incorporated in acetate if a molecule of palmitic acid is oxidized in 100% tritiated water?

Answer: For palmitic acid to be converted to acetate, it must pass through β-oxidation, each cycle, except the last, will result in incorporation of tritium at the α-carbon, which is destined to become carbon 2 of acetate. You might recall that the four steps of β-oxidation are dehydrogenation (oxidation), hydration, dehydrogenation (oxidation), and thiolysis. During the hydration step, the elements of water are added across a carbon-carbon double bond. In the subsequent dehydrogenation, the hydroxyacyl group, is oxidized to a keto group, with loss of two hydrogens, one of which is derived from water. The other water hydrogen remains on the α-carbon. In the thiolase reaction, a proton is abstracted to form a carbanion that is subsequently reprotonated. The proton is donated by a group on the enzyme but it is likely to be an exchangeable proton and thus be tritiated. Thus, a C-2 carbon will have two tritium atoms, one from the hydration step and a second from the thiolase reaction.

10. Understanding the consequences of carnitine deficiency
What would be the consequences of a carnitine deficiency for fatty acid oxidation?

Answer: Carnitine functions as a carrier of activated fatty acids across the inner mitochondrial membrane. A deficiency of carnitine is expected to effectively block fatty acid metabolism by preventing mitochondrial uptake of fatty acids.

11. Assessing energy consumption by a migrating bird
The ruby-throated hummingbird flies 500 miles nonstop across the Gulf of Mexico. The flight takes 10 hours at 50 mph. The hummingbird weighs about 4 grams at the start of the flight and about 2.7 grams at the end. Assuming that all the lost weight is fat burned for the flight, calculate the total energy required by the hummingbird in this prodigious flight. Does anything about the results of this calculation strike you as unusual?

Answer: The hummingbird uses 4 – 2.7 = 1.3 grams of fat for the flight across the Gulf of Mexico. Is we assume 37 kJ/gram of fat this amounts to

$$1.3 \text{ grams} \times 37 \frac{\text{kJ}}{\text{gram}} = 48.1 \text{ kJ}$$

Per mile this works out to about

$$\frac{48.1 \text{ kJ}}{500 \text{ mi}} = 0.096 \text{ kJ per mile}$$

Let's see how this might compare against getting energy out of gasoline. That is, what kind of gas mileage might we expect from the hummingbird. Gasoline has about 45 kJ per gram. The density of gasoline is about 0.8 g per ml.

$$0.096 \frac{\text{kJ}}{\text{mile}} \times \frac{\text{grams}}{45 \text{ kJ}} \times \frac{\text{mL}}{0.8 \text{ grams}} \times \frac{0.264 \text{ gal}}{1000 \text{ mL}} = 7.04 \times 10^{-7} \frac{\text{gal}}{\text{mile}}$$

The inverse of this is miles per gallon, which is 1,417,000 miles per gallon! This sure beats the mileage my automobile gets. (The calculation assumes 100% efficiency, which is unrealistic.) Of course, the hummingbird is moving a lot less mass (4 grams) than does my car's engine.

12. Understanding human energy consumption during exercise
Energy production in animals is related to oxygen consumption. The ruby-throated hummingbird consumes about 250 mL of oxygen per hour during its migration across the Gulf of Mexico. Use this number and the data in problem 11 to determine a conversion factor for energy expended per liter of oxygen consumed. If a human being consumes 12.7 kcal/min while running 8-minute miles, how long could a human run on the energy that the hummingbird consumes in its trans-Gulf flight? How many 8-minute miles would a person have to run to lose 1 pound of body fat?

Answer: The hummingbird uses 48.1 kJ in the 10-hour flight. The energy per L O_2 is:

$$\frac{48.1 \text{ kJ}}{10 \text{ hrs}} \times \frac{\text{hr}}{250 \text{ ml } O_2} \times \frac{1000 \text{ ml}}{\text{L } O_2} = \frac{19.24 \text{ kJ}}{\text{L } O_2}$$

To determine how long a human would run on 48.1 kJ of energy, we must convert 12.7 kcal/min to kJ/min and divide this into 48.1 kJ.

$$\frac{12.7 \text{ kcal}}{\text{min}} \times \frac{1000 \text{ cal}}{\text{kcal}} \times \frac{4.184 \text{ J}}{\text{cal}} \times \frac{\text{kJ}}{1000 \text{ J}} = 53.1 \frac{\text{kJ}}{\text{min}}$$

Using energy at this rate, 48.1 kJ would last:

$$\frac{48.1 \text{ kJ}}{53.1 \frac{\text{k J}}{\text{min}}} = 0.906 \text{ min or } \left(\times \frac{60 \text{ sec}}{\text{min}} \right) \text{ about 54 seconds}$$

To consume a pound of fat (37 kJ/gram) at a consumption rate of 12.7 kcal/min (53.1 kJ/min).

$$1 \text{ lb} \times \frac{453.6 \text{ g}}{\text{lb}} = 453.6 \text{ g}$$

$$453.6 \text{ g} \times 37 \frac{\text{kJ}}{\text{g}} = 16,783 \text{ kJ}$$

$$\frac{16,783 \text{ kJ}}{53.1 \frac{\text{kJ}}{\text{min}}} = 316.1 \text{ min}$$

So, you would have to run 316 minutes or about 5 and one half hours, not a light workout. Running 8-minute miles, you would have to run (316 ÷ 8 =) 39.5 miles. When you take your morning jog or play a few games of squash, if you happen to weigh yourself before and after, it is likely that you will lose more than a pound. Most of this is, however, water loss due to sweating. (A pint's a pound the world around: for every pint of perspiration you lose a pound, which you gain back by drinking fluids.)

13. Understanding the mechanism of the HMG-CoA synthase
Write a reasonable mechanism for the HMG-CoA synthase reaction shown in Figure 23.27.

Answer:

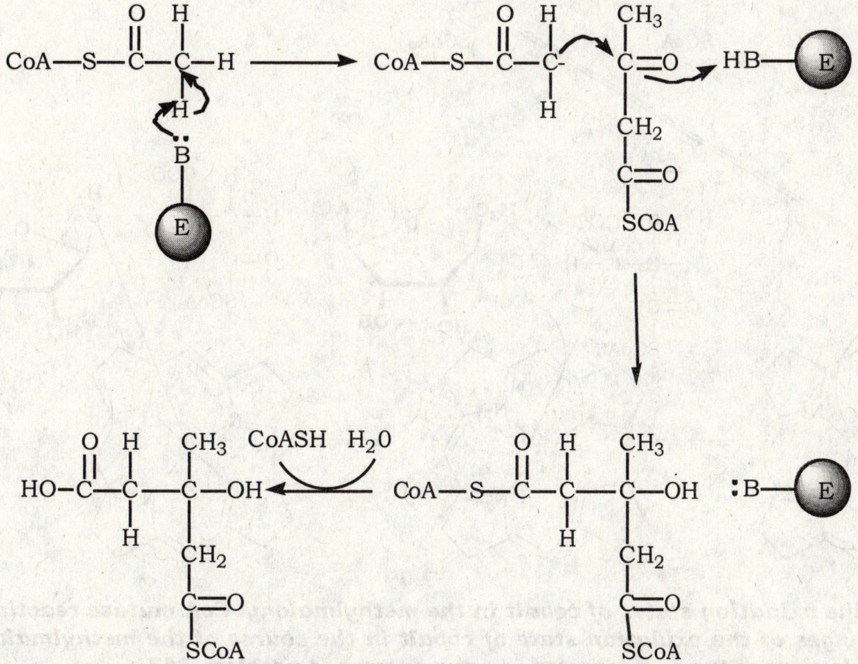

14. Understanding the mechanism of the methylmalonyl-CoA mutase reaction
The methylmalonyl-CoA mutase reaction (see Figure 23.18) involves vitamin B₁₂ as a coenzyme. Write a reasonable mechanism for this reaction. (Hint: The reaction begins with abstraction of a hydrogen atom—that is, a proton plus an electron—from the substrate by vitamin B₁₂. Consider the chemistry shown in A Deeper Look: The Activation of Vitamin B₁₂ as you write your mechanism.)

Answer: The mechanism below produces a radical on L-methylmalonyl-CoA using vitamin B₁₂.

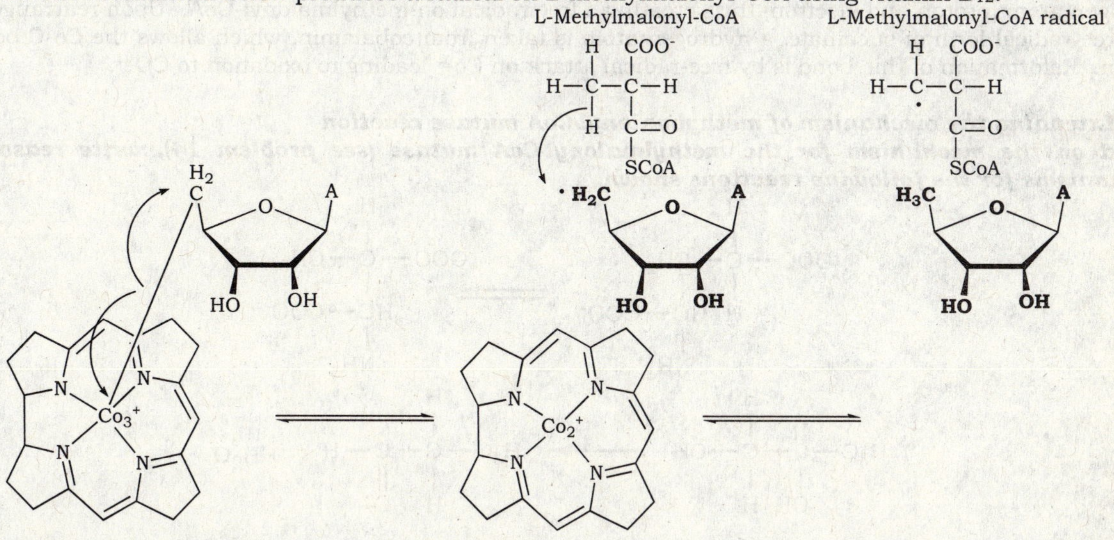

This is followed by rearrangement of the radical with eventual reformation of vitamin B₁₂ and release of succinyl-CoA as follows:

L-Methylmalonyl-CoA radical rearrangement

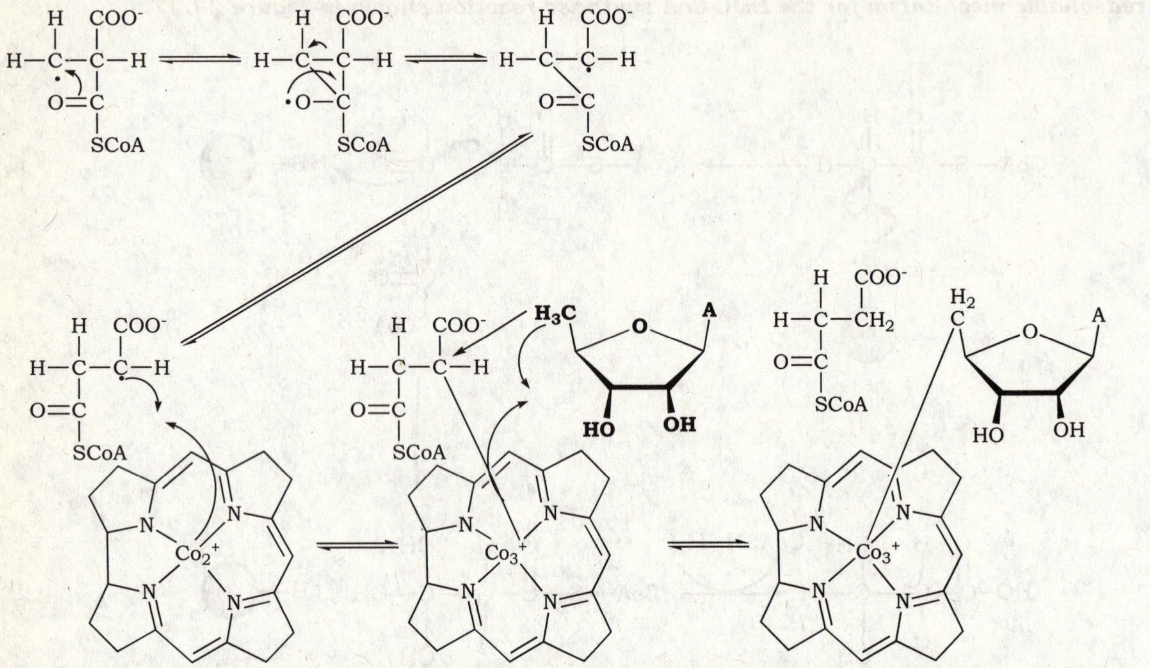

15. Assessing the oxidation states of cobalt in the methylmalonyl-CoA mutase reaction

Discuss the changes of the oxidation state of cobalt in the course of the methylmalonyl-CoA mutase reaction. Why do they occur as suggested in your answer to problem 14?

Answer: The mechanism shown above involves free-radical formation on methylmalonyl-CoA, which results in rearrangement to succinyl CoA. Free-radical formation is initiated by cleavage of the Co-C bond in cobalamin by hemolytic cleavage. Homolytic cleavage is decomposing of a compound into two uncharged atoms or radicals. The electron pair in the Co-C bond is in effect split with one electron going to cobalt and the other to the methylene carbon. Cobalt is thus reduced from Co^{3+} to Co^{2+}. The methylene carbon has an unpaired electron, a free radical, which attacks the methyl carbon of methylmalonyl CoA abstracting a hydrogen atom (proton and electron) thus creating a free radical on methylmalonyl-CoA. Upon rearrangement to a free-radical form of succinate, a hydrogen atom is taken from cobalamin, which allows the Co-C bond to reform. Reformation of this bond is by free-radical attack on Co^{2+} leading to oxidation to CO^{3+}.

16. Extending the mechanism of methylmalonyl-CoA mutase reaction

Based on the mechanism for the methylmalonyl-CoA mutase (see problem 14), write reasonable mechanisms for the following reactions shown.

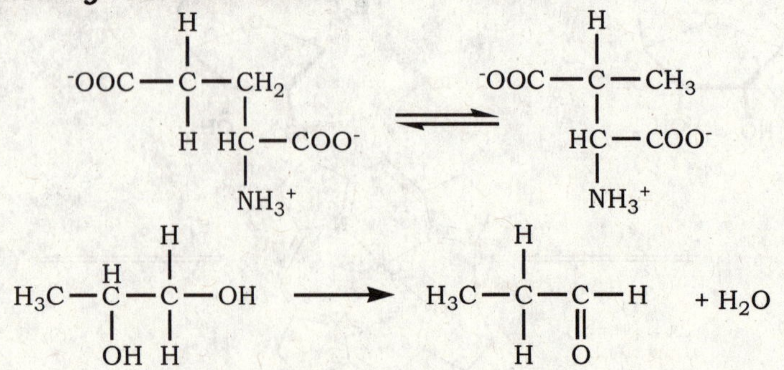

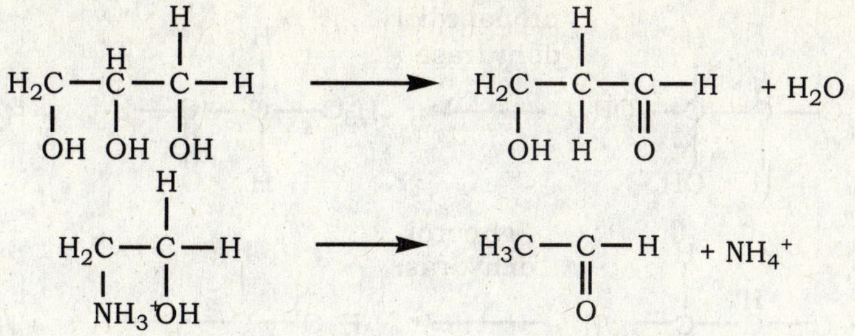

$$\text{H}_2\text{C}-\overset{\text{H}}{\underset{\text{OH}}{\overset{\text{H}}{\underset{}{\text{C}}}}}-\overset{\text{H}}{\underset{\text{OH}}{\overset{}{\text{C}}}}-\text{H} \longrightarrow \text{H}_2\text{C}-\overset{\text{H}}{\underset{\text{OH}}{\text{C}}}-\overset{\text{H}}{\underset{\text{O}}{\text{C}}}-\text{H} \;+\; \text{H}_2\text{O}$$

$$\text{H}_2\text{C}-\overset{\text{H}}{\underset{\text{NH}_3}{\overset{}{\text{C}}}}\!\!\overset{}{\underset{\text{OH}}{}}-\text{H} \longrightarrow \text{H}_3\text{C}-\overset{}{\underset{\text{O}}{\text{C}}}-\text{H} \;+\; \text{NH}_4^+$$

Answer: These reactions are all vitamin B_{12}-dependent reactions. The identity of the enzymes, reactants, and products is shown below.

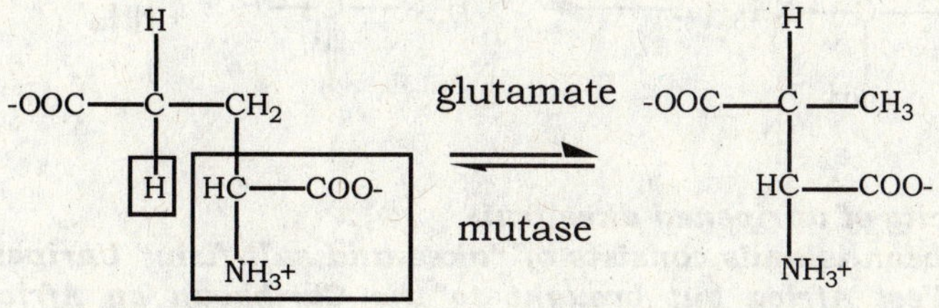

Glutamate $\xrightleftharpoons[\text{mutase}]{\text{glutamate}}$ threo-β-Methyl-L-aspartate

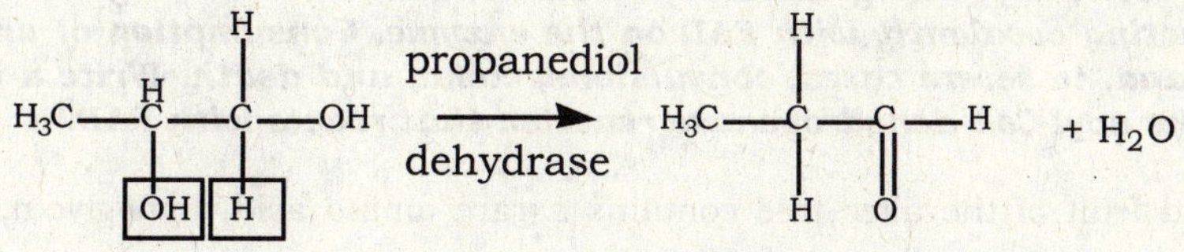

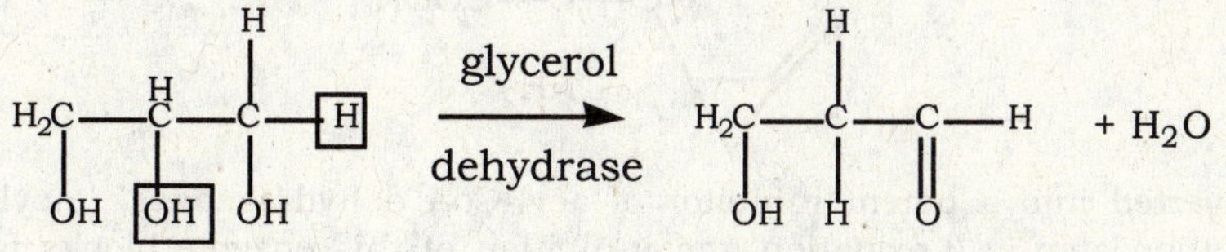

1,2-Propandiol $\xrightarrow{\substack{\text{propanediol}\\\text{dehydrase}}}$ propionaldehyde $+\ \text{H}_2\text{O}$

Glycerol $\xrightarrow{\substack{\text{glycerol}\\\text{dehydrase}}}$ β-Hydroxypropionaldehyde $+\ \text{H}_2\text{O}$

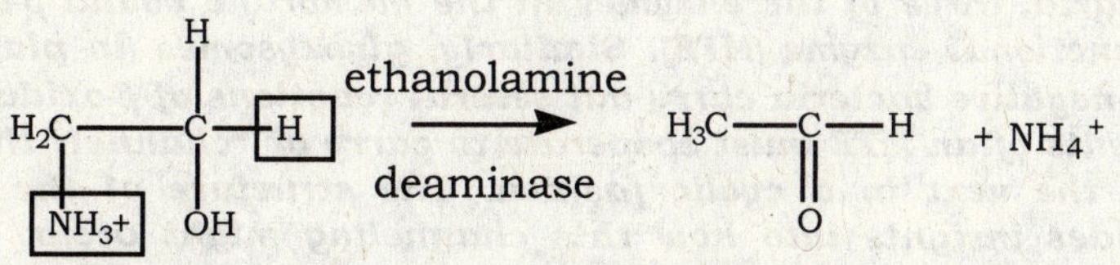

Ethanolamine $\xrightarrow{\substack{\text{ethanolamine}\\\text{deaminase}}}$ Acetaldehyde $+\ \text{NH}_4^+$

In each reaction the positions of the groups in boxes are exchanged. For the last three reactions, rearrangements of hydroxyl groups or an amino group leads to production of unstable intermediates. These are shown below.

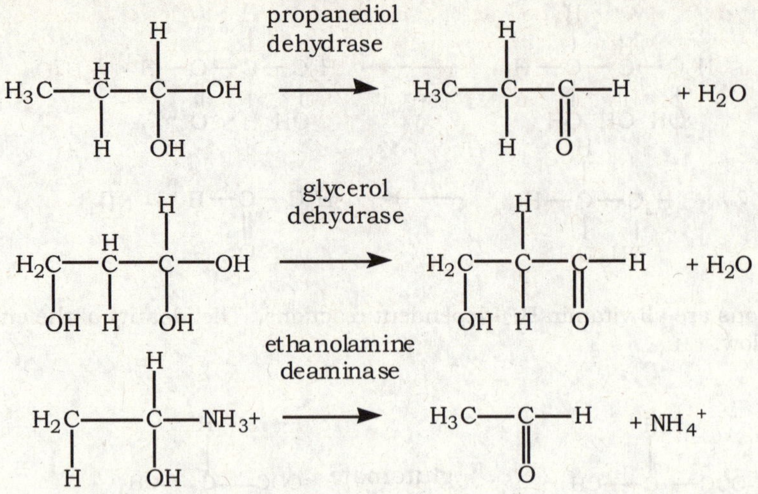

propanediol dehydrase

glycerol dehydrase

ethanolamine deaminase

17. Understanding the toxicity of unripened akee fruit

A popular dish in the Caribbean islands consists of "akee and salt fish." Unripened akee fruit (from the akee tree, native to West Africa but brought to the Caribbean by African slaves) is quite poisonous. The unripened fruit contains hypoglycin, a metabolite that serves as a substrate for acyl-CoA dehydrogenase. However, the product of this reaction irreversibly inhibits the acyl-CoA dehydrogenase by reacting covalently with FAD on the enzyme. Consumption of unripened akee fruit can lead to vomiting and, in severe cases, convulsions, coma, and death. Write a reaction scheme to show the product of the acyl-CoA dehydrogenase reaction that reacts with FAD.

Answer: The unripened fruit of the akee tree contains a rare amino acid, hypoglycin, whose structure is shown below.

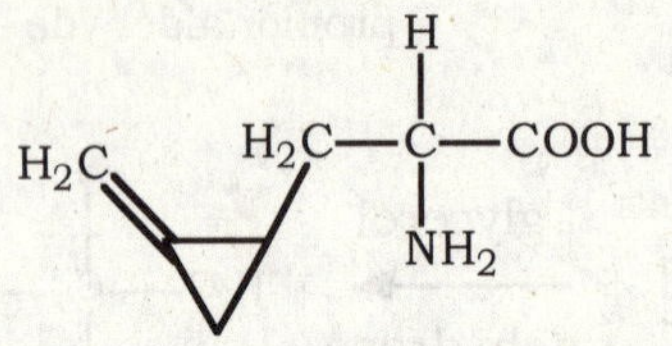

This compound is converted into a potent inhibitor of acyl-CoA dehydrogenase. Acyl-CoA dehydrogenase catalyzes the first oxidation step in β-oxidation and inhibition of this enzyme blocks fatty acid catabolism. Victims of hypoglycin poisoning become severely hypoglycemic because β-oxidation is inhibited and glucose catabolism becomes the primary source of metabolic energy.

18. Exploring substrate channeling in multi-functional enzyme aggregates

In mammalian mitochondria, three of the enzymes in the membrane bound β-oxidation system are components of a multifunctional enzyme (MFE). Similarly, glyoxysomes (in plants), peroxisomes (in most species), and Gram-negative bacteria carry out several reactions of β-oxidation via MFEs. In all these cases, the components of an MFE must cooperate to carry or "channel" the acyl-CoA substrate from one active site to the next in a cyclic fashion. The structure of the complete MFE from Pseudomonas fragi provides insights into how this channeling might occur. Consult the journal article that describes this structure (Ishikawa, M., Tsuchiya, D., et al., 2004. EMBO Journal 23:2745–2754), and explain in your own words how this substrate channeling occurs.

Answer: The enzyme complex in *Psuedomonas fragi* is an α₂β₂ heterotetramer that catalyzes three out of the four reactions in β-oxidation of fatty acyl-CoA. FAD-dependent enzymes that are not part of the complex catalyzes the first step, the formation of a carbon-carbon double bond producing 2-enoyl-CoA. The next three steps, however, are catalyzed by the complex and these include hydration of the double bond to form L-3-hydroxyacyl-CoA, oxidation to 3-ketoacyl-CoA with reduction of NAD⁺ and finally cleavage to release an acetyl-CoA and an acyl-CoA two carbons shorter. These steps are catalyzed by 2-enoyl-CoA hydratase, which hydrates the double bond in enoyl-CoA, by L-3-hydroxyacyl-CoA dehydrogenase, which oxidizes the hydroxyl group to a keto group, and finally 3-keto-CoA thiolase. Acyl-CoA hydratase and L-3-hydroxyacyl-CoA dehydrogenase are properties of the α subunit. This subunit is 715 amino acids long with the hydratase

440

located within the first 300 amino acids on the N-terminus and the dehydrogenase activity on the last 400 amino acids on the C-terminus. Approximately 30 residues separate the two domains. The crystal structure shows that the α subunits dimerize to form a saddle-shaped structure that is mounted on the β_2 homodimer. A large cavity separates the two homodimers. The structure was solved with three ligands: Acetyl-CoA, NAD$^+$, and C$_8$E$_5$ (octyl pentaethylene glycol ether –a surfactant and fatty acid mimic). In addition two conformations were solved for the complex, Form I and Form II. In Form I the two homodimers share a noncrystallographic two-fold axis of symmetry. In Form II the two-fold axis of the alpha dimers and beta dimers are not aligned as a consequence of movement of the upper and lower halves of the structure.

The channeling mechanism between the hydratase and dehydrogenase domains is thought to involve anchored diffusion. The adenine base of the substrate acyl-CoA is shared between the two subunits. Thus, the product of the hydratase is supplied to the dehydrogenase by movement of the fatty acid tail and not by diffusion through solvent. The thiolase domain is the next to receive the subunit and it is barely accessible to the product through its omega end. However, the protein complex is thought to undergo a conformational change (to Form II). The conformational change modifies the adenine-binding site, weakening it. This allows transfer of the ADP (of acyl-CoA) to the thiolase binding site. The acyl chain is maintained within the cavity between the two subunits and thus transfer again is accomplished without release of product to solvent.

Preparing for the MCAT® Exam

19. Comparing the electron transfer characteristics of FAD/FADH₂ and NAD⁺/NADH
Study Figure 23.12 and comment on why nature uses FAD/FADH₂ as a cofactor in the acyl-CoA dehydrogenase reaction rather than NAD⁺/NADH.

Answer: From Figure 23.12 we learn that the acyl-CoA dehydrogenase reaction moves two electrons from fatty acyl CoA to coenzyme Q. The electrons pass from substrate to FAD, which is reduced to FADH$_2$. Electrons are then passed one at a time through an iron-sulfur protein (ETF: UQ oxidoreductase) to coenzyme Q. Electron movement through iron-sulfur centers is one-electron transfers. FAD and coenzyme Q are both capable of accepting one electron at a time. NAD$^+$ can participate on two-electron transfers only.

20. Understanding a ubiquitous series of metabolic reactions
Study Figure 23.9. Where else in metabolism have you seen the chemical strategy and logic of the β-oxidation pathway? Why is it that these two pathways are carrying out the same chemistry?

Answer: Figure 23.9 shows one round of β-oxidation. The steps from acyl-CoA to β-ketoacyl-CoA are the same chemistry as found in the TCA cycle from succinate to oxaloacetate. Succinate to fumarate involves formation of a carbon-carbon double bond, a reduction reaction carried out by succinate dehydrogenase, which used FAD as oxidant. Fumarate to malate is hydration of the double bond. Finally, malate to oxaloacetate is oxidation of a hydroxyl to a keto group. This last step is catalyzed by malate dehydrogenase, which uses NAD$^+$ as electron acceptor. In the case of reduction of FAD to FADH$_2$, both enzymes (acyl-CoA dehydrogenase and succinyl CoA dehydrogenase) move electrons to CoQ in the electron transport chain. In fatty acid synthesis, the reverse of this sequence occurs but for the two oxidation/reduction steps, NADPH is employed by two separate enzymes.

Questions for Self Study

1. What is the difference between initiation of catabolism of short-chain fatty acids and long-chain fatty acids?

2. During β-oxidation a double bond is transiently produced in the fatty acid chain. How is this double bond different than double bonds found in mono- and polyunsaturated fatty acids?

3. Fill in the blanks. Odd-carbon fatty acids are metabolized by β-oxidation until the last three carbons are released as ___. This compound is converted to ___, a citric acid cycle intermediate in a series of three steps. In the first step, a carboxyl group is added by propionyl-CoA carboxylase. This enzyme uses the water-soluble vitamin ___ during catalysis. The third reaction is catalyzed by methylmalonyl-CoA mutase, a vitamin dependent enzyme.

4. Acetoacetate is an example of a ketone body. How is acetoacetate produced from acetyl-CoA? How can acetoacetate be converted to two other ketone bodies?

5. Label the compounds in the catabolic series shown below.

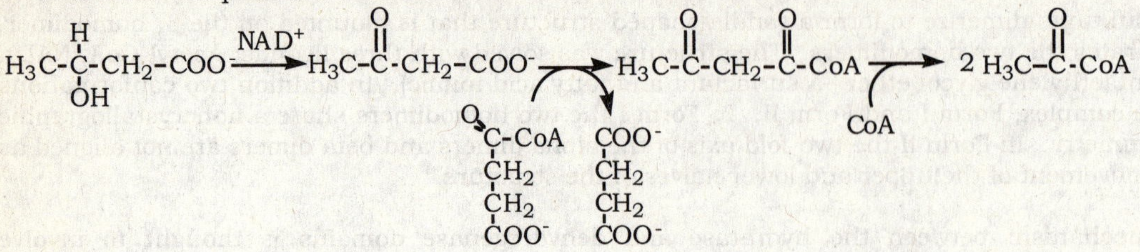

$$H_3C-\underset{OH}{\underset{|}{\overset{H}{\overset{|}{C}}}}-CH_2-COO^- \xrightarrow{NAD^+} H_3C-\overset{O}{\overset{\|}{C}}-CH_2-COO^- \longrightarrow H_3C-\overset{O}{\overset{\|}{C}}-CH_2-\overset{O}{\overset{\|}{C}}-CoA \xrightarrow{CoA} 2\ H_3C-\overset{O}{\overset{\|}{C}}-CoA$$

$$\underset{COO^-}{\underset{|}{\underset{CH_2}{\underset{|}{\underset{CH_2}{\underset{|}{\overset{O}{\overset{\|}{C}}-CoA}}}}}} \qquad \underset{COO^-}{\underset{|}{\underset{CH_2}{\underset{|}{\underset{CH_2}{\underset{|}{COO^-}}}}}}$$

Answers

1. Short-chain fatty acids are transported into the mitochondria as free fatty acids where they are converted to acyl-CoA derivatives. Long-chain fatty acids are attached to carnitine and moved across the inner membrane as O-acylcarnitine. They are then transferred to coenzyme A.

2. The double bond transiently produced during fatty acid catabolism is *trans* whereas double bonds found in naturally occurring fatty acids are usually *cis*. Also, double bonds in naturally occurring fatty acids may start at even-numbered carbons as well as odd-numbered carbons in the fatty acid chain.

3. Propionyl-CoA; succinyl-CoA; biotin, B$_{12}$.

4. Acetoacetate is produced in a three-step metabolic sequence from acetyl-CoA. Acetoacetyl-CoA is produced in the first step and subsequently converted to β-hydroxy-β-methylglutaryl-CoA (HMG-CoA). HMG-CoA is then metabolized to acetoacetate and acetyl-CoA. Decarboxylation of acetoacetate leads to production of acetone. Reduction of acetoacetate leads to production of β-hydroxybutyrate.

5. β-Hydroxybutyrate, acetoacetate, acetoacetyl-CoA, and acetyl-CoA are across the top line. Succinyl-CoA and succinate are shown on the bottom.

Additional Problems

1. It has been suggested that conversion of gasoline-powered automobiles to ethanol will require larger fuel tanks to keep the same effective range. Can you suggest a simple reason why this is the case?

2. Production of free fatty acids from fat deposits involves three separate lipases, triacylglycerol lipase, diacylglycerol lipase, and monoacylglycerol lipase. Fatty acid production is also under hormonal regulation but only one of the lipases is sensitive. Which lipase is hormonally controlled, how is regulation achieved, and why does it make good metabolic sense to regulate this particular lipase?

3. β-Oxidation begins with conversion of a free fatty acid to an acyl-coenzyme A derivative, a reaction catalyzed by acyl-CoA synthetase. Explain why this enzyme is named synthetase.

4. Describe the chemistry involved in the four steps of β-oxidation.

5. Metabolism of fatty acids with *cis* double bonds requires one or two additional enzymes. Explain.

6. In insulin-dependent diabetes mellitus, insulin deficiency can lead to shock as a result of metabolic acidosis. A person in shock may have a detectable odor of acetone on their breath. Explain the source of acetone and the cause of acidosis.

Abbreviated Answers

1. We are informed in Table 23.1 that fat has over twice the energy content of glycogen on a per gram basis. The reason is that the average carbon in fat is less oxidized than the average carbon in glycogen and so the amount of energy released in oxidation of carbon to CO_2 is greater. The same is true in comparing gasoline and ethanol. Gasoline is composed primarily of hydrocarbons whereas the average carbon in ethanol, C_2H_6O, is slightly more oxidized.

2. Adrenaline, glucagon and adrenocorticotropic hormone mobilize fatty acids from adipocytes by stimulation of triacylglycerol lipase or hormone-sensitive lipase. The hormones bind to surface receptors, which results in the stimulation of adenylyl cyclase and an increase in intracellular cyclic AMP levels. cAMP activates protein kinase, which in turn phosphorylates triacylglycerol lipase leading to activation of this enzyme. Diacylglycerol lipase is active against the product of triacylglycerol lipase activity and monoacylglycerol lipase requires the product of diacylglycerol lipase activity. Thus, only the first enzyme in this series of enzymes needs to be regulated.

3. Synthetases use cleavage of high-energy phosphoanhydride bonds to drive product formation. Acyl-CoA synthetase uses ATP to AMP + PP$_i$ conversion to produce acyl-CoA.

4. Starting from acyl-CoA the α- and β-carbons are reduced to form *trans*-Δ^2-enoyl-CoA and hydrated to L-β-hydroxyacyl-CoA. The β-carbon is oxidized to β-ketoacyl-CoA and cleaved to produce acetyl-CoA and an acyl-CoA two carbons shorter.

5. For *cis* double bonds starting at an odd carbon, β-oxidation proceeds normally until the double bond is between the β and γ carbons. An enoyl-CoA isomerase is used to rearrange the *cis*-Δ^3 bond to a *trans*-Δ^2 bond, which is hydrated, oxidized and cleaved normally. For *cis* double bonds starting at an even carbon, β-oxidation cycles eventually produce a *cis*-Δ^4 intermediate, which is converted to a *trans*-Δ^2,*cis*-Δ^4 intermediate by acyl-CoA dehydrogenase. Next, a 2,4-dienoyl-CoA reductase, in mammals, produces a *trans*-Δ^3 enoyl-CoA, which is converted to a *trans*-Δ^2 enoyl-CoA isomer by enoyl-CoA isomerase, a normal β-oxidation cycle intermediate. In *E. coli*, the double bound between carbons 4 and 5 is simply reduced by a 2,4-enoyl-CoA reductase to give *trans*-Δ^2 enoyl-CoA.

6. Acidosis and acetone are a consequence of ketone body formation. In liver mitochondria, acetoacetate and β-hydroxybutyrate are produced and excreted into the blood to be transported to other tissues to be used as fuels. These compounds are acidic with pK_a = 3.5 and in high concentrations causes a drop in blood pH. Acetoacetate, an unstable compound, will nonenzymatically breakdown into acetone and CO_2.

Summary

Fatty acids represent the principal form of energy storage for many organisms. Fatty acids are advantageous forms of stored energy, since 1) the carbon in fatty acids is highly reduced (thus yielding more energy in oxidation than other forms of carbon), and 2) fatty acids are not hydrated and can pack densely in storage tissues. Fatty acids are acquired readily in the diet and can also be synthesized from carbohydrates and amino acids. Whether dietary or synthesized, triacylglycerols are a major source of fatty acids. Release of fatty acids from triacylglycerols is triggered by hormones such as adrenaline and glucagon, which bind to plasma membrane receptors and activate (via cyclic AMP) the lipases, which hydrolyze triacylglycerols to fatty acids and glycerol. A large fraction of dietary triacylglycerols is absorbed in the intestines and passes into the circulatory system as chylomicrons.

Following Knoop's discovery that fatty acids were degraded to two-carbon units, Lynen and Reichart showed that these two-carbon units were acetyl-CoA. The process of β-oxidation of fatty acids begins with the formation of an acyl-CoA derivative by acyl-CoA synthetase, but transport of the acyl chain into the mitochondrial matrix occurs via acyl-carnitine derivatives, which are converted back to acyl-CoA derivatives in the matrix. β-Oxidation involves 1) formation of a Cα-Cβ double bond, 2) hydration across the double bond and 3) oxidation of the β-carbon, followed by 4) a thiolase cleavage, leaving acetyl-CoA and the CoA ester of the fatty acid chain, shorter by two carbons. (A metabolite of hypoglycin, found in unripened akee fruit, inhibits the first of these reactions, resulting in vomiting, convulsions and death.) Repetition of this cycle of reactions degrades fatty acids with even numbers of carbons completely to acetyl-CoA. β-Oxidation of fatty acids with odd numbers of carbon atoms yields acetyl-CoA plus a single molecule of propionyl-CoA, which is subsequently converted to succinyl-CoA in a sequence of three reactions. Complete β-oxidation of a molecule of palmitic acid generates 106 high-energy phosphate bonds, i.e., formation of 106 ATP from ADP and P$_i$. This massive output of energy from β-oxidation permits migratory birds to travel long distances without stopping to eat. The large amounts of water formed in β-oxidation allow certain desert and marine organisms to thrive without frequent ingestion of water.

Unsaturated fatty acids are also catabolized by β-oxidation, but two additional mitochondrial enzymes - an isomerase and a novel reductase - are required to handle the *cis*-double bonds of naturally occurring fatty acids. Degradation of polyunsaturated fatty acids also requires the activity of 2,4-dienoyl-CoA reductase. Although β-oxidation in mitochondria is the principal pathway of fatty acid catabolism, a similar pathway also exists in peroxisomes and glyoxysomes. Fatty acids with branches at odd carbons in the chain are degraded by α-oxidation. Phytol, an isoprene breakdown product of chlorophyll, is degraded via α-oxidation. Defects in this pathway lead to Refsum's disease, which is characterized by poor night vision, tremors and neurologic disorders and which is caused by accumulation in the body of phytanic acid. ω-Oxidation in the ER of eukaryotic cells leads to the synthesis of small amounts of dicarboxylic acids. This is accomplished in part by cytochrome P-450, which hydroxylates the terminal carbon atom of the fatty acid chain. Ketone bodies are a significant source of fuel and energy for certain tissues. Acetoacetate and 3-hydroxybutyrate are the preferred substrates for kidney cortex and heart muscle, and ketone bodies are a major energy source for the brain during periods of starvation. Ketone bodies represent easily transportable forms of fatty acids which move through the circulatory system without the need for complexation with serum albumin and other fatty acid binding proteins.

Chapter 24

Lipid Biosynthesis

• •

Chapter Outline

❖ Fatty acid biosynthesis
  ⌁ Biosynthesis localized in cytosol: Fatty acid degradation in mitochondria
  ⌁ Intermediates held on acyl carrier protein (ACP): Phosphopantetheine group attached to serine (CoA in degradation)
  ⌁ Fatty acid synthase: Multienzyme complex
  ⌁ Carbons derived from acetyl units
    • Acetyl CoA to malonyl CoA by carboxylation
    • Acetyl unit added to fatty acid with decarboxylation of malonyl CoA
  ⌁ Carbonyl carbons of acetyl units reduced using NADPH
❖ Source of acetyl units
  ⌁ Amino acids, glucose
  ⌁ Acetyl CoA used to produce citrate
  ⌁ Citrate exported to cytosol: ATP-citrate lyase forms acetyl-CoA and oxaloacetate
❖ Source of NADPH
  ⌁ Oxaloacetate utilization
    • Oxaloacetate (from citrate) to malate: NADH dependent reaction
    • Malate to pyruvate: Malic enzyme: NADPH produced
  ⌁ Pentose phosphate pathway
❖ Malonyl-CoA production: Acetyl-CoA carboxylase
  ⌁ Biotin-dependent enzyme
  ⌁ ATP drives carboxylation
  ⌁ Enzyme regulation
    • Filamentous polymeric form active
      • Citrate favors active polymer
      • Palmitoyl-CoA favors inactive protomer
      • Citrate/palmitoyl-CoA effects depend on protein phosphorylation
        o Unphosphorylated protein binds citrate with high affinity: Activation
        o Phosphorylated protein binds palmitoyl with high affinity: Inactivation
❖ Acetyl transacetylase: Acetylates acyl carrier protein (ACP): Destined to become methyl end of fatty acid
❖ Malonyl transacetylase: Malonylates ACP
❖ β-Ketoacyl-ACP synthase (acyl-malonyl ACP condensing enzyme): Accepts acetyl group: Transfers acyl group to malonyl-ACP
  ⌁ Malonyl carboxyl group released: Decarboxylation drives synthesis
  ⌁ Malonyl-ACP converted to acetoacetyl-ACP
❖ β-Ketoacyl-ACP reductase
  ⌁ Carbonyl carbon reduced to alcohol
  ⌁ NADPH provides electrons
❖ β-Hydroxyacyl-ACP dehydratase: Elements of water removed: Double bond created
❖ 2,3-trans-Enoyl-ACP reductase
  ⌁ Double bond reduced

       ⅄   NADPH provides electrons
- ❖ Megasynthase complexes
  - ⅄ Bacterial fatty acid synthesis catalyzed by discrete enzymes
  - ⅄ Fungi and higher eukaryotes fatty acid synthesis organized into multiprotein complexes
    - • Mammalian enzyme: Homodimer with each monomer containing domains for multiple enzymatic activities involved in fatty acid synthesis
    - • Fungal enzyme: Multiple subunits with each having one or more domains with enzymatic activity
- ❖ Subsequent cycles produce C-16: Palmitoyl-CoA
- ❖ Additional modifications
  - ⅄ Elongation
    - • Mitochondrial-based system uses reversal of β-oxidation
    - • Endoplasmic reticulum-based system uses malonyl-CoA
  - ⅄ Monounsaturation: One double bond
    - • Bacteria: Oxygen-independent pathway: Chemistry performed on carbonyl carbon
    - • Eukaryotes: Oxygen-dependent pathway
  - ⅄ Polyunsaturation
    - • Plants can add double bonds between C-9 and methyl end
    - • Animals
      - • Add double bonds between C-9 and carboxyl end
      - • Require essential fatty acids to have double bonds closer to methyl end
- ❖ Regulation
  - ⅄ Malonyl-CoA inhibition of carnitine-acyl transferase: Blocks fatty acid uptake
  - ⅄ Citrate/palmitoyl regulation of acetyl-CoA carboxylase
- ❖ Complex lipids
  - ⅄ Glycerolipids: Glycerol backbone
    - • Glycerophospholipids
    - • Triacylglycerols
  - ⅄ Sphingolipids: Sphingosine backbone
  - ⅄ Phospholipids
    - • Sphingolipids
    - • Glycerophospholipids
- ❖ Glycerolipid biosynthesis
  - ⅄ Phosphatidic acid is precursor
    - • Glycerokinase produces glycerol-3-P
    - • Glycerol-3-phosphate acyltransferase acylates C-1 with saturated fatty acid: Monoacylglycerol phosphate
    - • Eukaryotes can produce monoacylglycerol phosphate using DHAP
    - • Acyldihydroxyacetone phosphate reduced by NAPDH to monoacylglycerol phosphate
    - • Acyltransferase acylates C-2: Phosphatidic acid
  - ⅄ Phosphatidic acid used to synthesize two precursors of complex lipids
    - • Diacylglycerol: Precursor of
      - • Triacylglycerol: Diacylglycerol acyltransferase
      - • Phosphatidylethanolamine, phosphatidylcholine
        - o Ethanolamine phosphorylated
        - o CTP and phosphoethanolamine produce CDP-ethanolamine
        - o Transferase moves phosphoethanolamine onto diacylglycerol
        - o (Dietary choline: As per ethanolamine)
        - o (Phosphatidylethanolamine to phosphatidylcholine by methylation)
      - • Phosphatidylserine: Serine for ethanolamine exchange
    - • CDP-diacylglycerol
      - • Phosphatidate cytidylyl transferase produces CDP-diacylglycerol
      - • CDP-diacylglycerol used to produce
        - o Phosphatidyl inositol
        - o Phosphatidyl glycerol

- o Cardiolipin
  - ⋏ Plasmalogens: α,β-unsaturated ether-linked chain at C-1
    - · DHAP acylated
    - · Acyl group exchanged for alcohol
    - · Keto group on DHAP reduced to alcohol and acylated
    - · Head group attached
    - · Desaturase produces double bonds
- ❖ Sphingolipid biosynthesis
  - ⋏ Serine and palmitoyl-CoA condensed with decarboxylation to produce 3-ketosphinganine
  - ⋏ Reduction forms sphingamine
  - ⋏ Sphingamine acylated to form N-acyl sphingamine
  - ⋏ Desaturase produces ceramide
    - · Cerebrosides: Galactose or glucose added
    - · Gangliosides: Sugar polymers: Sugars derive from UDP-monosaccharides
- ❖ Eicosanoids: Derived from 20-C fatty acids: Arachidonate is precursor
  - ⋏ Local hormones: Prostaglandins, thromboxanes, leukotrienes, hydroxyeicosanoic acids
  - ⋏ Prostaglandins
    - · Cyclopentanoic acid formed from arachidonate by prostaglandin endoperoxidase synthase (cyclooxygenase [COX])
    - · Aspirin inhibits enzyme
- ❖ Cholesterol
  - ⋏ Membrane component
  - ⋏ Precursor of important biomolecules
    - · Bile salts
    - · Steroid hormones
    - · Vitamin D
- ❖ Cholesterol biosynthesis: In liver
  - ⋏ Mevalonate biosynthesis
    - · Thiolase condenses two acetyl-CoA to produce acetoacetyl-CoA
    - · HMG-CoA synthase produces HMG-CoA
    - · HMG-CoA reductase produces mevalonate
      - ● Rate limiting step
      - ● Regulation
        - o Inactivated by cAMP-dependent protein kinase
        - o Short half life of enzyme when cholesterol levels high
        - o Gene expression regulated
      - ● Pharmacological target for blood cholesterol regulation
  - ⋏ Isopentenyl pyrophosphate and dimethylallyl pyrophosphate from mevalonate
  - ⋏ Squalene to lanosterol to cholesterol
- ❖ Lipid transport
  - ⋏ Fatty acids complexed to serum albumin
  - ⋏ Phospholipids, triacylglycerol, cholesterol transported as lipoprotein complexes
    - · Lipoprotein complex types: HDL, LDL, IDL, VLDL, Chylomicrons
      - ● Chylomicrons formed in intestine
      - ● HDL, VLDL assembled in liver
        - o Core of triacylglycerol
        - o Single layer of phospholipid
        - o Proteins and cholesterol inserted
        - o VLDL to IDL to LDL to liver for uptake and degradation
        - o HDL: Assembled without cholesterol but picks up cholesterol during circulation
- ❖ Bile salts
  - ⋏ Glycocholic acid
  - ⋏ Taurocholic acid
- ❖ Steroid hormones
  - ⋏ Cholesterol to pregnenolone

447

> λ Pregnenolone to progesterone
> * Hormone
> * Sex hormone precursor
>   * Androgens
>   * Estrogens
> * Corticosteroids precursor
>   * Glucocorticoids
>   * Mineralocorticoids

Chapter Objectives

Fatty Acid Biosynthesis

The steps of fatty acid biosynthesis (Figure 24.7) are similar in chemistry to the reverse of β-oxidation. Two-carbon acetyl units are used to build a fatty acid chain. The carbonyl carbon is reduced to a methylene carbon in three steps: reduction to an alcoholic carbon, dehydration to a carbon-carbon double bond intermediate, and reduction of the double bond. The two reduction steps utilize NADPH as reductant. Two-carbon acetyl units are moved out of the mitochondria as citrate and activated by carboxylation to malonyl-CoA. We have already seen similar carboxylation reactions and should remember that biotin is involved when carbons are added at the oxidation level of a carboxyl group. The enzyme, acetyl-CoA carboxylase, is regulated by polymerization/depolymerization with the filamentous polymeric state being active. You should understand the regulatory effects of citrate (favors polymer formation), palmitoyl-CoA (depolymerizes) and covalent phosphorylation (blocks citrate binding) on acetyl-CoA carboxylase activity. In β-oxidation, we saw that the phosphopantetheine group of coenzyme A functioned as a molecular chauffeur for two-carbon acetyl units. In synthesis, phosphopantetheine, attached to the acyl carrier protein, functions as a molecular carrier by guiding the growth of fatty acid chains.

In plants and bacteria, the steps of fatty acid biosynthesis are catalyzed by individual proteins whereas in animals a large multifunctional protein is involved. Synthesis starts with formation of acetyl-ACP and malonyl-ACP by specific transferases. The carboxyl group of malonyl-ACP departs, leaving a carbanion that attacks the acetyl group of acetyl-ACP to produce a four-carbon β-ketoacyl intermediate, which is subsequently reduced by an NADPH-dependent reductase, dehydrated, and reduced a second time by another NADPH-dependent reductase. To continue the cycle, malonyl-ACP is reformed, decarboxylates, and attacks the acyl-ACP. The original acetyl group is the methyl-end of the fatty acid, whereas the malonyl groups are added at the carboxyl end. NADPH is supplied by the pentose phosphate pathway and by malic enzyme, which converts the oxaloacetate skeleton, used to transport acetyl groups out of the mitochondria as citrate, into pyruvate and CO_2 with $NADP^+$ reduction.

Additional elongation and introduction of double bonds can occur after synthesis of a C_{16} fatty acid. Elongation can occur in the endoplasmic reticulum, where malonyl-CoA is utilized, or in the mitochondria where acetyl-CoA is used. Introduction of double bonds occurs via oxygen-independent mechanisms in bacteria and oxygen-dependent mechanisms in eukaryotes. Be familiar with the reaction catalyzed by stearoyl-CoA desaturase, involving stearoyl-CoA and oxygen as substrates and oleoyl-CoA and water as products.

Complex Lipids

The glycerolipids, including glycerophospholipids and triacylglycerols, are synthesized from glycerol, fatty acids, and head groups. Synthesis starts with the formation of phosphatidic acid from glycerol-3-phosphate and fatty acyl-CoA. C-1 is esterified usually with a saturated fatty acid. Phosphatidic acid may be converted to diacylglycerol and then to triacylglycerol. Alternately, diacylglycerol can be used to synthesize phosphatidylethanolamine and phosphatidylcholine with CDP-derivatized head groups serving as substrates. Phosphatidylserine is produced by exchange of the ethanol head-group from PE with serine. Phosphatidylinositol, phosphatidylglycerol, and cardiolipin (two diacylglycerols linked together by glycerol) are synthesized using CDP-diacylglycerol as an intermediate. Plasmalogens are synthesized from acylated DHAP. The acyl group is exchanged for a long-chain alcohol followed by reduction of the keto carbon of DHAP, acyl group transfer from acyl-CoA to C-2, head group transfer from CDP-ethanolamine and formation of a *cis* double bond between C-1 and C-2 of the long-chain alcohol.

The sphingolipids all derive from ceramide, whose synthesis starts with bond formation between palmitic acid and the α-carbon of serine (with loss of the serine carboxyl carbon as bicarbonate). After a few steps a second fatty acid is attached to serine in amide linkage. Subsequent sugar additions lead to cerebrosides and gangliosides.

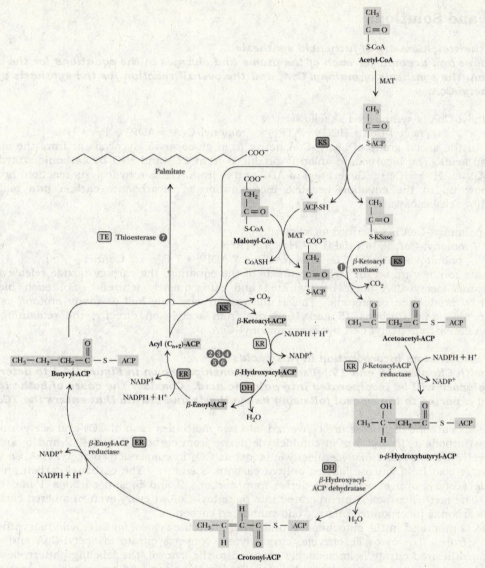

Figure 24.7 The pathway of palmitate synthesis from acetyl-CoA and malonyl-CoA. Acetyl and malonyl building blocks are introduced as acyl carrier protein conjugates. Decarboxylation drives the β-ketoacyl-ACP synthase and results in the addition of two-carbon units to the growing chain

Prostaglandins

The prostaglandins are produced from arachidonic acid released by phospholipase A2 action on phospholipids. Production of these local hormones is blocked by aspirin, and nonsteroid anti-inflammatory agents such as ibuprofen and phenylbutazone.

Cholesterol

Cholesterol derives from HMG-CoA, a product we already encountered in ketone body formation. You might recall that ketone bodies are produced from acetyl-CoA units. HMG-CoA is a six-carbon CoA derivative produced from three acetyl units. The rate-limiting step in cholesterol synthesis is formation of 3R-mevalonate from HMG-CoA by HMG-CoA reductase, which catalyzes two NADPH-dependent reductions. This enzyme is carefully regulated by 1) phosphorylation leading to inactivation, 2) degradation, and 3) gene expression. Mevalonate, a six-carbon intermediate, is converted to isopentenyl pyrophosphate, which is used to synthesize cholesterol. Cholesterol is the precursor of bile salts and the steroid hormones. You should understand how lipoproteins are responsible for movement of cholesterol and other lipids in the body.

Problems and Solutions

1. Explain the stoichiometry of fatty acid synthesis
Carefully count and account for each of the atoms and charges in the equations for the synthesis of palmitoyl-CoA, the synthesis of malonyl-CoA, and the overall reaction for the synthesis of palmitoyl-CoA from acetyl-CoA.

Answer: Malonyl-CoA is synthesized as follows:
$$\text{Acetyl-CoA} + HCO_3^- + ATP^{4-} \; \star \;\; \text{malonyl-CoA}^- + ADP^{3-} + P_i^{2-} + H^+$$
The carbons in the acetyl group of acetyl-CoA derive from glucose via glycolysis or from the side chains of various amino acids. The bicarbonate anion is produced from CO_2 and H_2O by carbonic anhydrase: $CO_2 + H_2O \; \star \;\; H_2CO_3 \; \star \;\; H^+ + HCO_3^-$. Generation of ADP and P_i from ATP is a hydrolysis reaction; however, water does not show up in the equation because incorporation of bicarbonate carbon into malonyl-CoA is accompanied by release of water.

Synthesis of palmitoyl-CoA is described as follows:
Acetyl-CoA + 7 malonyl-CoA$^-$ + 14NADPH + 7 H$^+$ + 7 ATP^{4-} $\star$
$$\text{palmitoyl-CoA} + 7\, HCO_3^- + 14\, NADP^+ + 7\, ADP^{3-} + 7\, P_i^{2-} + 7\, CoASH$$
For bicarbonate to show up on the right hand side of the equation, the carbon dioxide released by reacting malonyl-CoA and acetyl-CoA must be hydrated and subsequently ionized. So, each bicarbonate is accompanied by production of protons. This is the reason why only half as many protons as NADPH are found in the reaction. Carbons 15 and 16 derive from acetyl-CoA directly; the remaining carbons in palmitoyl-CoA derive from acetyl-CoA by way of malonyl-CoA.

2. Tracing carbon atom incorporation in fatty acids
(Integrates with Chapters 18 and 19.) Use the relationships shown in Figure 24.1 to determine which carbons of glucose will be incorporated into palmitic acid. Consider the cases of both citrate that is immediately exported to the cytosol following its synthesis and citrate that enters the TCA cycle.

Answer: The six carbons of glucose are converted into two molecules each of CO_2 and acetyl units of acetyl-coenzyme A. Carbons 1, 2, and 3 of glyceraldehyde derive from carbons 3 and 4, 2 and 5, and 1 and 6 of glucose respectively. Carbon 1 of glyceraldehyde is lost as CO_2 in conversion to acetyl-CoA, so we expect no label in palmitic acid from glucose labeled only at carbons 3 and 4. The carbonyl carbon and the methyl carbon of the acetyl group of acetyl-CoA derive from carbons 2 and 5, and carbons 1 and 6 of glucose, respectively. The methyl carbon is incorporated into palmitoyl-CoA at every even-numbered carbon, whereas the carbonyl carbon is incorporated at every odd-numbered carbon.

Acetyl-CoA is produced in the mitochondria and exported to the cytosol for fatty acid biosynthesis by being converted to citrate. The cytosolic enzyme, citrate lyase, converts citrate to acetyl-CoA and oxaloacetate. When newly synthesized citrate is immediately exported to the cytosol, the labeling pattern described above will result. However, where citrate is instead metabolized in the citric acid cycle, back to oxaloacetate, label derived from acetyl-CoA shows up at carbons 1, 2, 3 and 4 of oxaloacetate. These carbons do not get incorporated into palmitoyl-CoA.

3. Postulating the regulation of acetyl-CoA carboxylase
Based on the information presented in the text and in Figures 24.4 and 24.5, suggest a model for the regulation of acetyl-CoA carboxylase. Consider the possible roles of subunit interaction, phosphorylation, and conformational changes in your model.

Answer: Acetyl-CoA carboxylase catalyzes the formation of malonyl-CoA, the committed step in synthesis of fatty acids. This enzyme is a polymeric protein composed of protomers, or subunits, of 230 kD. In the polymeric form, the enzyme is active whereas in the protomeric form the enzyme is inactive. Polymerization is regulated by citrate and palmitoyl-CoA such that citrate, a metabolic signal for excess acetyl units, favors the polymeric and, therefore, active form of the enzyme whereas palmitoyl-CoA shifts the equilibrium to the inactive form. The activity of acetyl-CoA carboxylase is also under hormonal regulation. Glucagon and epinephrine stimulate cyclic AMP-dependent protein kinase that will phosphorylate a large number of sites on the enzyme. The phosphorylated form of the enzyme binds citrate poorly and citrate binding occurs only at high citrate levels. Citrate is a tricarboxylic acid with three negative charges and its binding site on the enzyme is likely to be composed of positively-charged residues. Phosphorylation introduces negative charges, which may be responsible for the decrease in citrate binding.

In the phosphorylated form, low levels of palmitoyl-CoA will inhibit the enzyme. Thus, the enzyme is sensitive to palmitoyl-CoA binding and to depolymerization in the phosphorylated form. If we assume that the

450

palmitoyl-CoA binding site is located at a subunit-subunit interface, and that phosphorylated, and hence negatively charged subunits interact with lower affinity than do unphosphorylated subunits, we see that it is easier for palmitoyl-CoA to bind to the enzyme.

4. Estimating active site separation in animal fatty acid synthase

Consider the role of the pantothenic acid groups in animal fatty acyl synthase (FAS) and the size of the pantothenic acid group itself, and estimate a maximal separation between the malonyl transferase and the ketoacyl-ACP synthase active sites.

Answer: In fatty acyl synthase, phosphopantetheine, which contains pantothenic acid, is attached to a serine residue as shown below.

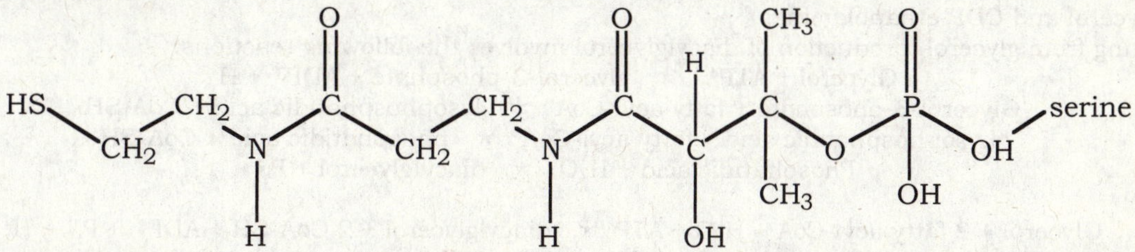

The approximate distance from the sulfur to the α-carbon of serine is calculated as follows. For carbon-carbon single bonds the bond length is approximately 0.15 nm. The distance between carbon atoms is calculated as follows.

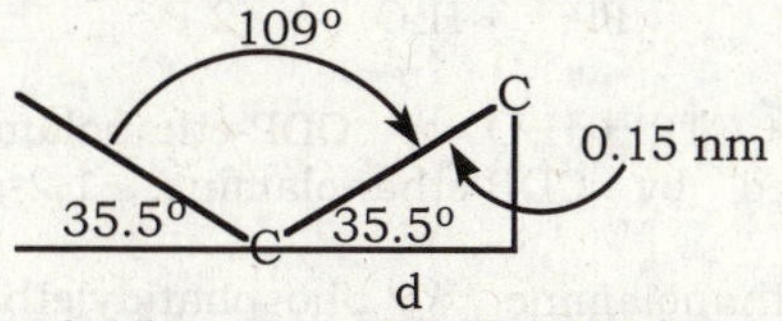

$$d = 0.15 \text{ nm} \times \cos(35.5°) = 0.122 \text{ nm}$$

Let us use this length for carbon-carbon single bonds, carbon-oxygen bonds, oxygen-phosphorous bonds, and carbon-nitrogen bonds exclusive of the amide bond. For the amide bond we will use a distance of 0.132 nm. The overall length is approximately 1.85 nm from the α-carbon of serine to the sulfur. (There are 15 bonds between the sulfur and the alpha carbon of serine and two of them are amide bonds so the distance is $13 \times 0.122 + 2 \times 0.132 = 1.85$ nm.) The maximal separation between malonyl transferase and ketoacyl-ACP is about twice this distance or approximately 3.7 nm. The actual distance between these sites is smaller than this upper limit.

An alternative approach is to visit the pdb database at http://www.pdb.org and do a search for phosphopantetheine. On the returned page on the top activate the Ligand Hits tab and locate 4'-phosphopantetheine. Activating this record will return a page allowing you to view the molecule using Jmol. Right clicking on the Jmol image will open a menu that has an option for making measurements. Only one conformation of the structure is shown but the longest distance measureable is around 1.4 nm. Remembering that phosphopantetheine has to be attached to a serine residue, which will add additional distance of about 0.24 nm, this makes the overall distance approximately 1.64 nm, which is in reasonable agreement with the 1.85 we calculated above.

5. Assessing the stoichiometry of stearoyl-CoA desaturase

Carefully study the reaction mechanism for the stearoyl-CoA desaturase in Figure 24.14, and account for all of the electrons flowing through the reactions shown. Also account for all of the hydrogen and oxygen atoms involved in this reaction, and convince yourself that the stoichiometry is correct as shown.

Answer: Stearoyl-CoA desaturase catalyzes the following reaction

Stearoyl-CoA + NADH + H⁺ + O₂ ⋆ oleoyl-CoA + 2 H₂O

This reaction involves a four-electron reduction of molecular oxygen to produce two water molecules. Two of the electrons come from the desaturation reaction directly, in which desaturase removes two electrons and two protons from stearoyl-CoA to produce the carbon-carbon double bond in oleoyl-CoA. The other two electrons and protons derive from NADH + H⁺. Two electrons from NADH are used by another enzyme, NADH-cytochrome b_5 reductase, to reduce FAD to FADH₂. Electrons are then passed one at a time to cytochrome b_5, which passes electrons to the desaturase to reduce oxygen to water. So, two electrons and two protons come from palmitoyl-CoA and two electrons come from NADH with two protons being supplied by the surrounding solution.

6. Understanding the stoichiometry of glycerophospholipid synthesis
Write a balanced, stoichiometric reaction for the synthesis of phosphatidylethanolamine from glycerol, fatty acyl-CoA, and ethanolamine. Make an estimate for the $\Delta G^{0'}$ for the overall process.

Answer: The synthesis of phosphatidylethanolamine involves the convergence of two separate pathways: A diacylglycerol backbone is synthesized from glycerol and fatty acids; ethanolamine is phosphorylated and activated by transfer to CTP to produce CDP-ethanolamine. CDP-ethanolamine: 1,2-diacylglycerol phosophoethanolamine transferase then catalyzes the formation of phosphatidylethanolamine from diacylglycerol and CDP-ethanolamine.

Starting from glycerol, production of diacylglycerol involves the following reactions:

$$\text{Glycerol} + \text{ATP}^{4-} \;\star\; \text{glycerol-3-phosphate} + \text{ADP}^{3-} + \text{H}^+$$
$$\text{Glycerol-3-phosphate} + \text{fatty acyl-CoA} \;\star\; \text{lysophosphatidic acid} + \text{CoA-SH}$$
$$\text{Lysophosphatidic acid} + \text{fatty acyl-CoA} \;\star\; \text{phosphatidic acid} + \text{CoA-SH}$$
$$\text{Phosphatidic acid} + \text{H}_2\text{O} \;\star\; \text{diacylglycerol} + \text{P}_i^{2-}$$

We have:
$$\text{Glycerol} + 2 \text{ fatty acyl-CoA} + \text{H}_2\text{O} + \text{ATP}^{4-} \star \text{diacylglycerol} + 2 \text{ CoA-SH} + \text{ADP}^{3-} + \text{P}_i^{2-} + \text{H}^+$$

Production of CDP-ethanolamine involves the following:

$$\text{Ethanolamine} + \text{ATP}^{4-} \;\star\; \text{phosphoethanolamine} + \text{ADP}^{3-} + \text{H}^+$$
$$\text{Phosphoethanolamine} + \text{CTP}^{4-} \;\star\; \text{CDP-ethanolamine} + \text{PP}_i^{4-}$$
$$\text{PP}_i^{4-} + \text{H}_2\text{O} \;\star\; 2 \text{ P}_i^{2-}$$

Or,
$$\text{Ethanolamine} + \text{ATP}^{4-} + \text{CTP}^{4-} + \text{H}_2\text{O} \;\star\; \text{CDP-ethanolamine} + \text{ADP}^{3-} + 2 \text{ P}_i^{2-} + \text{H}^+$$

Finally, for the reaction catalyzed by CDP-ethanolamine: 1,2-diacylglycerol phosophoethanolamine transferase we have:

$$\text{Diacylglycerol} + \text{CDP-ethanolamine} \;\star\; \text{phosphatidylethanolamine} + \text{CMP}^{2-} + \text{H}^+$$

The balanced, stoichiometric reaction is:
$$\text{Glycerol} + \text{ethanolamine} + 2 \text{ fatty acyl-CoA} + 2 \text{ ATP}^{4-} + \text{CTP}^{4-} + 2 \text{ H}_2\text{O} \star$$
$$\text{Phosphatidylethanolamine} + 2 \text{ CoA-SH} + 2 \text{ ADP}^{3-} + 2 \text{ H}^+ + 3 \text{ P}_i^{2-} + \text{CMP}^{2-}$$

7. Understanding the stoichiometry of cholesterol synthesis
Write a balanced, stoichiometric reaction for the synthesis of cholesterol from acetyl-CoA.

Answer: The immediate precursors of cholesterol are isopentenyl pyrophosphate and dimethylallyl pyrophosphate, both of which derive from hydroxymethylglutaryl-CoA (HMG-CoA). HMG-CoA is synthesized from acetyl-CoA by the following route:

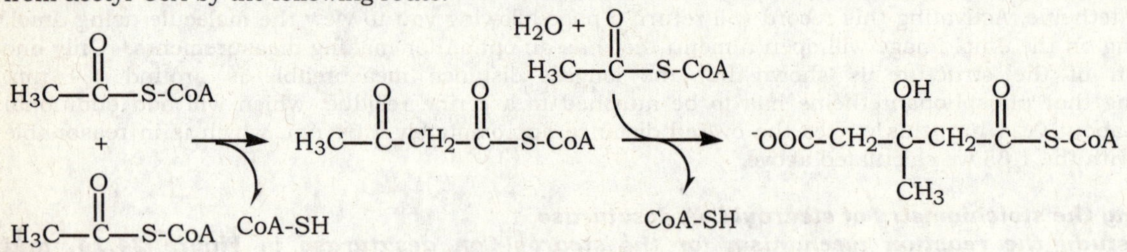

The reaction is:

$$3 \text{ Acetyl-CoA} \star \text{HMG-CoA} + 2 \text{ CoA-SH}$$

HMG-CoA is anabolized into isopentenyl pyrophosphate and dimethylallyl pyrophosphate, both of which are used to synthesize squalene, which is converted by way of lanosterol into cholesterol. (The next question asks us to trace carbons from mevalonate to cholesterol, so it is worthwhile now to look at these reactions in detail).

Synthesis of isopentenyl pyrophosphate from HMG-CoA is as follows:

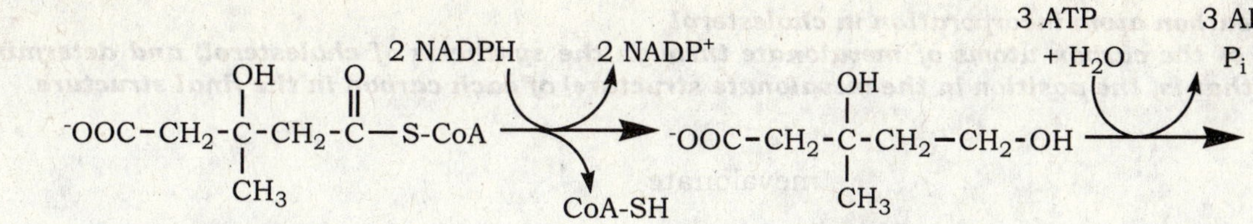

$$^-OOC-CH_2-\underset{\underset{CH_3}{|}}{C}-CH_2-\underset{\underset{}{\|}}{\overset{O}{C}}-S-CoA \xrightarrow[\qquad\qquad]{\text{2 NADPH} \quad \text{2 NADP}^+} {}^-OOC-CH_2-\underset{\underset{CH_3}{|}}{C}-CH_2-CH_2-OH \xrightarrow[\qquad]{\substack{\text{3 ATP} \quad \text{3 ADP +} \\ +\,H_2O \qquad P_i + CO_2}}$$

CoA-SH

HMG-CoA

mevalonate

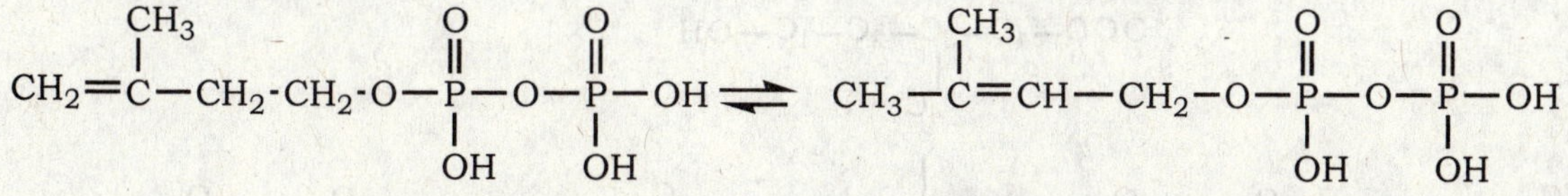

$$CH_2{=}\underset{\underset{CH_3}{|}}{C}-CH_2\cdot CH_2\cdot O-\underset{\underset{OH}{\|}}{\overset{O}{P}}-O-\underset{\underset{OH}{\|}}{\overset{O}{P}}-OH \rightleftharpoons CH_3-\underset{\underset{CH_3}{|}}{C}{=}CH-CH_2-O-\underset{\underset{OH}{\|}}{\overset{O}{P}}-O-\underset{\underset{OH}{\|}}{\overset{O}{P}}-OH$$

isopentenyl pyrophosphate   dimethylallyl pyrophosphate

Overall, the reaction is:

HMG-CoA + 2 NAPDH + 3 ATP ⋆ isopentenyl pyrophosphate (or dimethylallyl
pyrophosphate) + CoA-SH + 2 NADP$^+$ + 3 ADP + P_i + CO_2

Production of squalene using isopentenyl pyrophosphate and dimethylallyl pyrophosphate proceeds as follows. Two farnesyl pyrophosphates are produced from two dimethylallyl pyrophosphates and four isopentenyl pyrophosphates. The farnesyl pyrophosphates are reacted to produce squalene as follows:

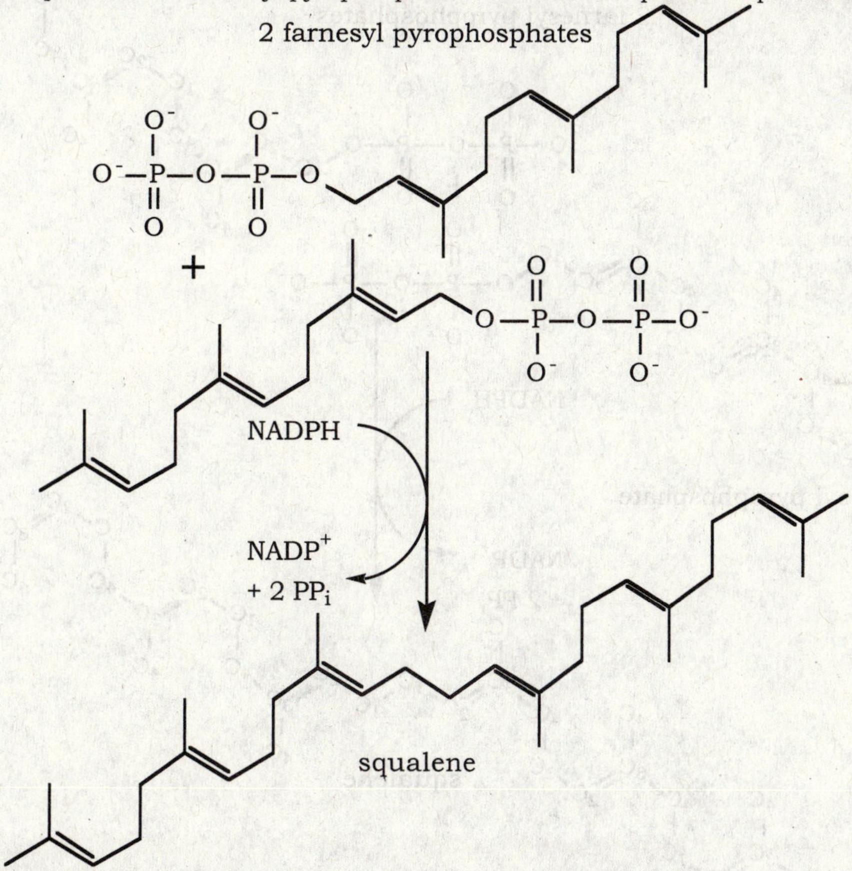

2 farnesyl pyrophosphates

squalene

Squalene is converted to lanosterol in two steps catalyzed by squalene epoxidase and squalene oxidocyclase.

Squalene + 0.5 O_2 + NADPH ⋆ lanosterol

The overall equation for acetyl-CoA to lanosterol is:

18 Acetyl-CoA + 13 NADPH +13 H$^+$ + 18 ATP + 0.5 O_2 ⋆
Lanosterol + 18 CoA-SH + 13 NADP$^+$ + 18 ADP + 6 P_i + 6 PP$_i$ + 6 CO_2

The pathway from lanosterol to cholesterol involves the oxidation and loss of three carbons.

8. Tracing carbon atom incorporation in cholesterol
Trace each of the carbon atoms of mevalonate through the synthesis of cholesterol, and determine the source (that is, the position in the mevalonate structure) of each carbon in the final structure.

Answer:

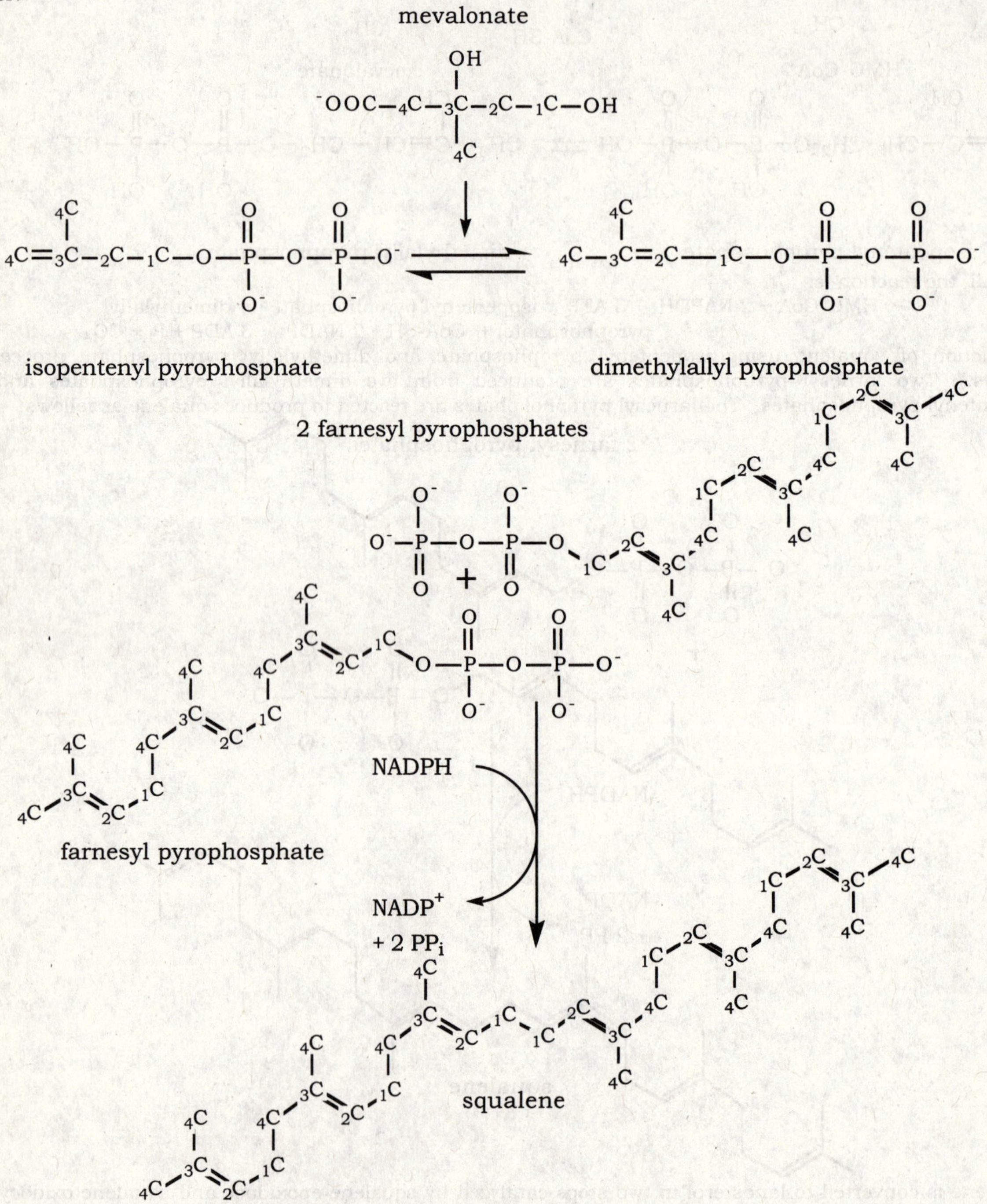

mevalonate

$$^-OOC-{}_4C-{}_3C-{}_2C-{}_1C-OH$$

isopentenyl pyrophosphate

dimethylallyl pyrophosphate

2 farnesyl pyrophosphates

NADPH

NADP$^+$
+ 2 PP$_i$

farnesyl pyrophosphate

squalene

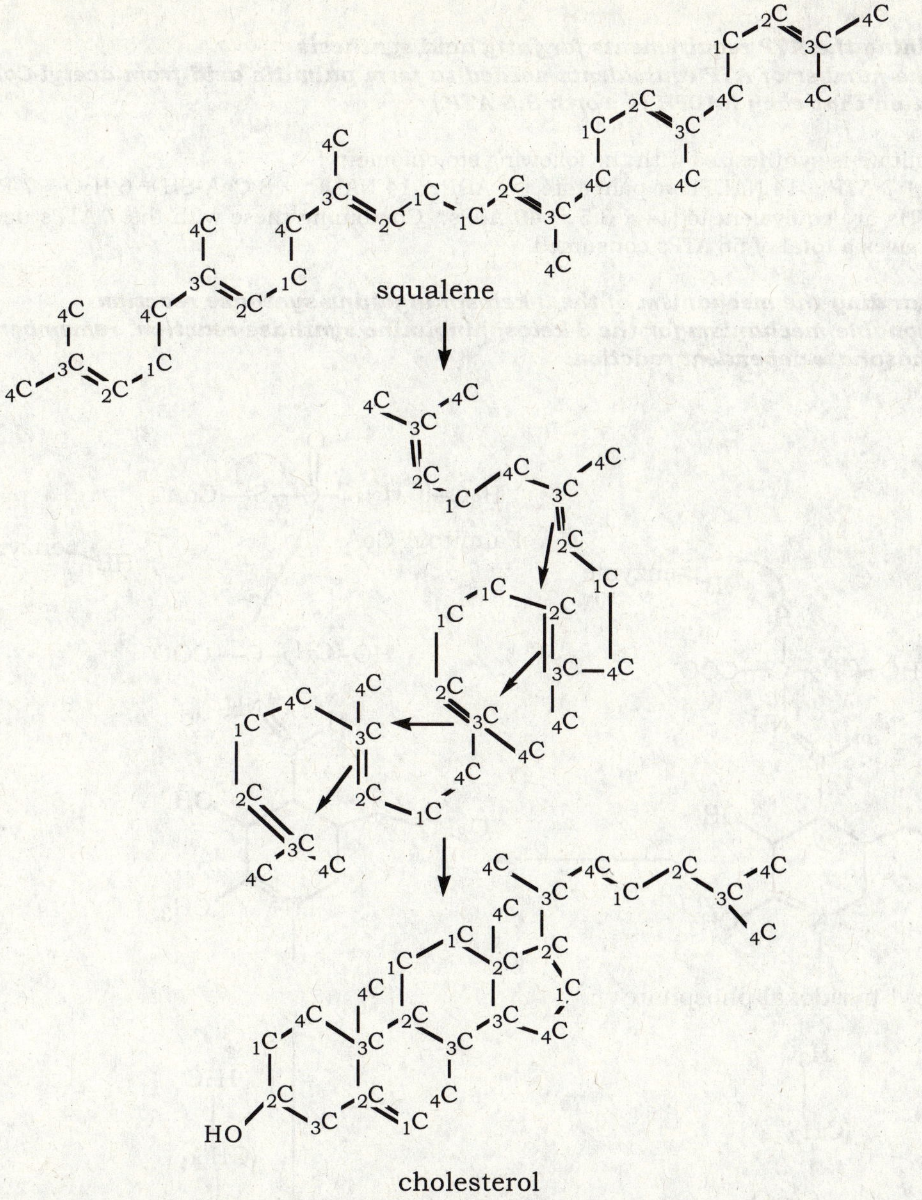

squalene

cholesterol

9. Understanding the functional role of O-linked oligosaccharides in the LDL receptor
Suggest a structural or functional role for the O-linked saccharide domain in the LDL receptor (Figure 24.40).

Answer: LDL receptors are synthesized on the rough endoplasmic reticulum and move through the smooth endoplasmic reticulum and Golgi apparatus before being incorporated into the plasma membrane. On the plasma membrane, LDL receptors bind LDL, aggregate into patches, and are internalized into coated vesicles that fuse with lysosomes where LDL is degraded. The O-linked saccharide domain functions to extend the receptor domain away from the cell surface, above the glycocalyx coat. This allows the receptor to bind circulating lipoproteins.

10. Assessing the consequences of low cellular levels of CTP
Identify the lipid synthetic pathways that would be affected by abnormally low levels of CTP.

Answer: Phosphatidylethanolamine and phosphatidylcholine synthesis depend on the formation of CDP-ethanolamine and CDP-choline respectively. Phosphatidylinositol and phosphatidylglycerol biosynthesis utilize CDP-diacylglycerol. The synthetic pathways of all of these compounds may be affected if the cell experiences low levels of CTP.

11. Determining the ATP requirements for fatty acid synthesis
Determine the number of ATP equivalents needed to form palmitic acid from acetyl-CoA. (Assume for this calculation that each NADPH is worth 3.5 ATP.)

Answer: Palmitate is synthesized with the following stoichiometry:
8 Acetyl-CoA + 7 ATP + 14 NADPH ⋆ palmitate + 7 ADP + 14 NADP$^+$ + 8 CoA-SH + 6 H_2O + 7 P_i
The 14 NADPHs are equivalent to (14 × 3.5 =) 49 ATPs. Combining these with the 7 ATPs used to synthesize malonyl-CoA gives a total of 56 ATPs consumed.

12. Understanding the mechanism of the 3-ketosphinganine synthase reaction
Write a reasonable mechanism for the 3-ketosphinganine synthase reaction, remembering that it is a pyridoxal phosphate-dependent reaction.

Answer:

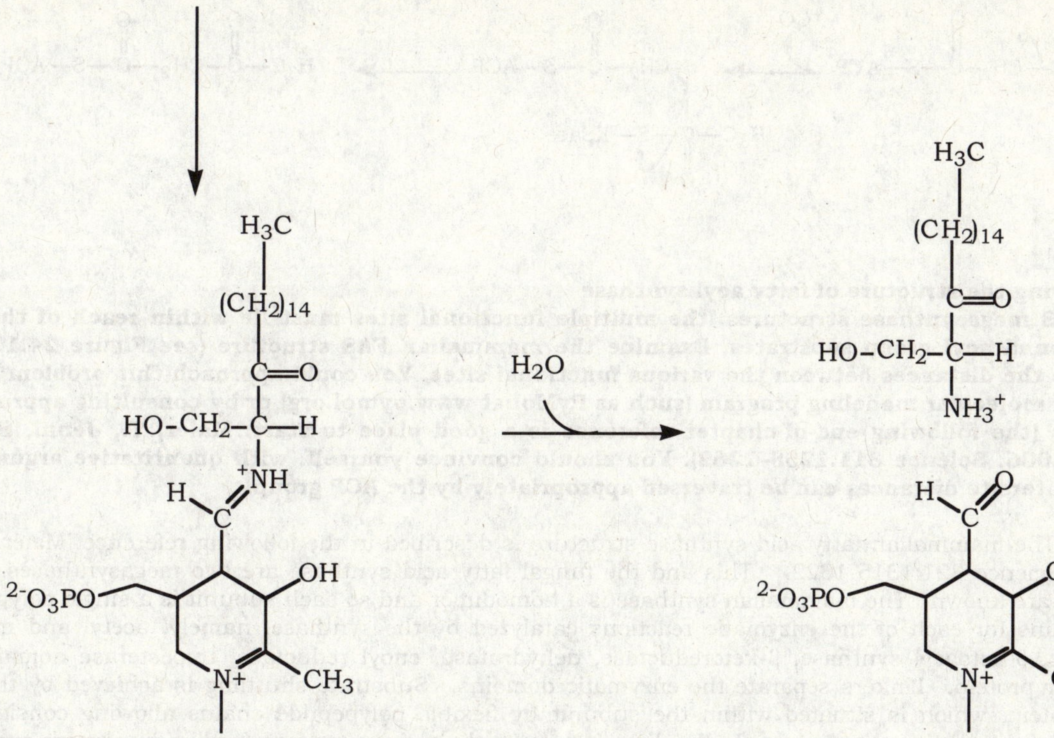

H₃C

(CH₂)₁₄

C=O

HO—CH₂—C—H

NH⁺

H—C=N

C

²⁻O₃PO— —OH

N⁺—CH₃

H₂O →

H₃C

(CH₂)₁₄

C=O

HO—CH₂—C—H

NH₃⁺

H—C=O

C

²⁻O₃PO— —OH

N⁺—CH₃

13. Understanding the involvement of FAD in the conversion of steric acid to oleic acid
Why is the involvement of FAD important in the conversion of stearic acid to oleic acid?

Answer: In eukaryotes, unsaturation reactions are catalyzed by stearoyl-CoA desaturase. This enzyme functions along with cytochrome b₅ reductase and cytochrome b₅ to pass electrons, one at a time to desaturase. Desaturase reduces O_2, a 4-electron reduction for which two electrons come from the fatty acid that is desaturated and two ultimately from NADH. The two electrons from NADH pass through cytochrome b₅ reductase, an FAD-containing enzyme, which must pass electrons one at a time to cytochrome b₅. NAD cannot participate in one-electron transfers whereas FAD can. $FADH_2$ can lose one electron or two.

14. Understanding the mechanism of the HMG-CoA synthase reaction
Write a suitable mechanism for the HMG-CoA synthase reaction. What is the chemistry that drives this condensation reaction.

Answer: The mechanism for HMG-CoA was already discussed in problem 13 of chapter 23. The chemistry is not unlike that of citrate synthase. HMG-CoA synthase produces hydroxymethylglutaryl-CoA from acetyl-CoA and acetoacetyl-CoA. The reaction is accompanied by hydrolysis of a thioester bond linking coenzyme to the acetyl group. We learned in earlier chapters that these thioester bonds are high energy. Production of fatty acyl-CoA by acyl-CoA synthetase, which is used to activate fatty acids for beta oxidation, uses in effect two phosphoanhydride bonds to create the thioester bond. (As an interesting aside, synthase and synthetase both run reactions that join two substrates to form a product. Synthetases, in general, drive these reactions with hydrolysis of high-energy phosphoanhydride bonds. Synthases do not.)

15. Understanding the mechanism of the β-ketoacyl-ACP synthase reaction
Write a suitable reaction mechanism for the β-ketoacyl ACP synthase, showing how the involvement of malonyl-CoA drives this reaction forward.

Answer: β-Ketoacyl ACP synthase links an acetyl group from malonyl-ACP onto an acyl-group (acetyl- in the first round) during fatty acid synthesis. Malonyl-CoA is, in effect, an activated acetyl group produced at the expense of hydrolysis of ATP by acetyl-CoA carboxylase. The mechanism of action involves decarboxylation of malonyl-CoA to produce a carbanion on the beta carbon, which attacks the carbonyl carbon of acyl-ACP producing β-ketoacyl ACP. This mechanism is shown below.

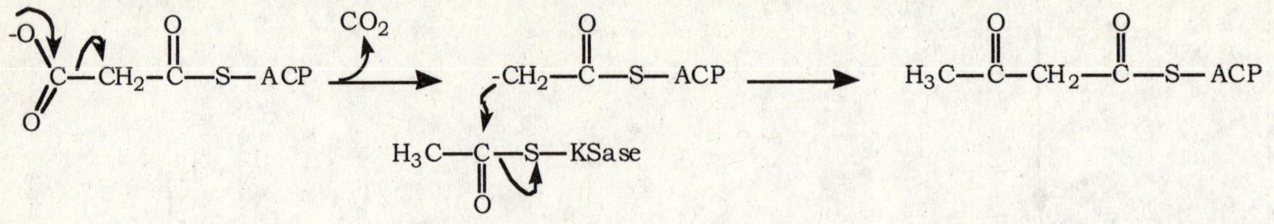

16. Studying the structure of fatty acyl synthase
In the FAS megasynthase structures, the multiple functional sites must lie within reach of the ACP and its bound acyl group substrates. Examine the mammalian FAS structure (see Figure 24.11) and determine the distances between the various functional sites. You could approach this problem either by using a molecular modeling program (such as PyMol at www.pymol.org) or by consulting appropriate references (the following end-of-chapter reference is a good place to start: Maier, T., Jenni, S., and Ban, N., 2006. Science 311:1258–1262). You should convince yourself, with quantitative arguments, that the intersite distances can be traversed appropriately by the ACP group.

Answer: The mammalian fatty acid synthase structure is described in the following reference: Maier, T., et al., 2008 Science 321:1315-1322.   This and the fungal fatty acid synthase are two megasynthases whose structures are known. The mammalian synthase is a homodimer and so each subunit is a single polypeptide with domains for each of the enzymatic reactions catalyzed by the synthase, namely, acetyl and malonyl transferase, β-ketoacyl synthase, β-ketoreductase, dehydratase, enoyl reductase, thioesterase domain and acyl carrier protein. Linkers separate the enzymatic domains. Substrate shuttling is achieved by the acyl carrier protein, which is situated within the subunit by flexible polypeptide chains allowing considerable movement of the acyl carrier protein. In the dimeric enzyme the active sites from both subunits are organized into a lower condensing portion that attaches acetyl groups to the acyl carrier protein and an upper portion that carries out reduction and hydration steps to convert the β-keto intermediate into a fully reduced carbon. The acyl carrier protein is situated between the upper and lower portions of the synthase and thus has access to both the condensing portion and modifying portion of the enzyme complex.

17. Modeling the LDL receptor
In the LDL receptor structure shown in Figure 24.41c, the β-propeller interaction with domains R4 and R5 is partly stabilized by salt bridges between acidic residues on R4 and R5 that also coordinate Ca^{2+} ions. Use a molecular modeling program or consult the literature to identify at least two such interactions. Two suitable references are Beglova, N., and Blacklow, S. C., 2005. Trends in *Biochemical Sciences* 30:309–316; and Rudenko, G., Henry, L., et al., 2002. Science 298:2353–2358.

Answer: From the two references we learn that the LDL receptor is a multi-domain, single-pass transmembrane protein. The N-terminus is extracellular whereas the C-terminus is intracellular. The N-terminus is composed of seven type-A LDL receptor (LA) modules that are responsible for ligand binding. Each LA module contains three disulfide bonds that stabilize two loops. One of the loops contains conserved acid residues that form a calcium-binding site. Calcium binding and the disulfide bonds stabilize the structure. Following the LA modules the protein contains two epidermal growth factor-like (EFG) domains, six YWTD repeats and another EFG domain. This region (the three EFG domains the 6 YWTD repeats – organized into a β-propeller) is important for ligand release. A region with numerous O-linked glycosylation sites follows the ligand release domain and precedes the transmembrane domain, which is followed by a 50-residue C-terminal region located within the cytoplasm. The O-linked sugar domain is responsible for keeping the other extracellular domains away from the membrane surface.

The LDL receptor adopts two conformations, an open conformation at neutral pH and a closed or compact conformation at acidic pH (pH < 6.0). In the closed conformation the ligand release domain interacts with two of the LA domains. Specifically LA4 and 5 interact with the β-propeller structure.

In the open structure the LA domains have a cluster of acidic residues organized by calcium coordination of residues in the loops within the LA domains. These acidic clusters are thought to be responsible for ligand binding. The protein component of LDL has a series of basic residues that is required for LDL binding to the receptor, which likely occurs through electrostatic interactions.

In the closed conformation LA4 and 5 interact with the β-propeller via electrostatic interactions.   Three histidine residues participate in these interactions and are thought to be the pH-sensitive switches. Histidines 562 and 586 are located on the propeller domain. A third histidine, His190, is situated on the tip of one of the loops of LA5. Histidine's pK$_a$ is around 6 and so we expect a dramatic change in histidine's charge in going from neutral pH to pH < 6. There are a number of observations supporting the involvement of these histidines in ligand release. In certain cases of familial hypercholesterolemia mutations are found at either histidine 190 or 562. LDL receptors with these mutations appear to bind LDL normally but show a decrease in LDL-release activity. Site-directed mutagenesis of all three histidines produces a receptor that does not release LDL.

Thus, LDL binds to its receptor through electrostatic interactions between the protein component of LDL and the LA domains on the receptor. When the LDL receptor is internalized in endosomes and the endosomes made acidic, the ligand release region of the LDL receptor becomes positively charged and associates tightly with LA4 and LA5, two domains critical for LDL binding to the receptor. This releases LDL from LA4 and LA5.

18. Understanding the function of the LDL receptor

Insights into the function of LDL receptors in displacing LDL particles in endosomes have come from an unlikely source: a study of LDL receptor binding by a human rhinovirus HRV2 (a common cold virus). Consult suitable references to learn how this study provided support for the model of LDL particle displacement presented in this chapter. Good references are Blacklow, S. C., 2004. Nature Structural and Molecular Biology 11:388–390; Verdaguer, N., Fita, I., et al., 2004. Nature Structural and Molecular Biology 11: 429–434; and Beglova, N., and Blacklow, S. C., 2005. Trends in Biochemical Sciences 30:309–316.

Answer: Human rhinoviruses, many of which are the cause of common colds, are picornaviruses that divide into two groups: a major group and minor group. The minor group infects cells by binding to LDL receptors (and related proteins like VLDL receptor) that are internalized by endocytosis. The picornaviruses are icosahedral viruses composed of a coat of proteins VP1 to VP4. VP1 forms a star-shaped platform at the five-fold vertices of the icosahedron and in the case of the minor group the LDL receptors bind at a five-fold vertex. The LDL receptor structure was described in the answer to problem 17 above. In studies on virus binding to receptors, VLDL receptors were used in place of LDL receptors. Specifically, the virus was crystallized with a two-domain fragment of the VLDL receptor.

The VLDL receptor fragment binds to two adjacent VP1 subunits and what is remarkable about this interaction is that the VP1 subunits present a binding surface that mimics the β-propeller in the closed conformation of the LDL receptor (See answer to previous problem.) The β-propeller associates the LA binding domains, which are responsible for initial LDL binding to its receptor. Thus, as outlined in the answer to the previous question, the β-propeller is likely responsible for dissociation of LDL bound to the receptor. The same interactions are found in both cases and thus viral binding and release appears to follow the same pathway as LDL binding and release. The details include conserved ionic interactions and hydrophobic interactions summarized in the Blacklow review.

Preparing for the MCAT® Exam

19. Comparing the reaction of fatty acid synthesis and the tricarboxylic acid cycle
Consider the synthesis of linoleic acid from palmitic acid and identify a series of three consecutive reactions that embody chemistry similar to three reactions in the tricarboxylic acid cycle.

Answer: Palmitic acid is a saturated 16-carbon fatty acid whereas linoleic acid (shown below) is 18 carbons long with two carbon-carbon double bonds. To convert palmitic acid to linoleic acid it must first be elongated using one cycle of fatty acid synthesis. Elongation of palmitoyl-CoA involves a thiolase reaction using acetyl-CoA, which adds to the carboxyl end. The β-keto acyl CoA derivative is then reduced to β-hydroxy, dehydrated to form a carbon-carbon double bond and then reduced to produce stearyl-CoA. The chemistry of this series of reactions is similar to the chemistry found in the TCA cycle but going in reverse from oxaloacetate to succinate.

To convert stearyl-CoA to linoleilyl-CoA we would have to produce two carbon-carbon double bonds by oxidation of the saturated fatty acid. While plants can produce this polyunsaturated fatty acid, mammals cannot.

459

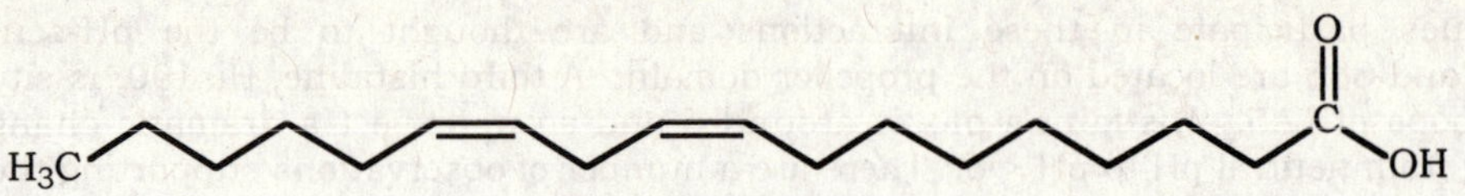

20. Writing an equation to describe the synthesis of behenic acid
Rewrite the equation in Section 24.1 to describe the synthesis of behenic acid (see Table 8.1).

Answer: Behenic acid (a.k.a., docosanoic acid) is a 22-carbon long fatty acid (shown below). On page 799 we are given the equation for synthesis of palmitoyl-CoA, a 16-carbon long fatty acid. To produce behenic acid we need to recognize that we will need to run three more cycles of fatty acid synthesis. Each cycle will consume one acetyl-CoA, 1 ATP and 2 NADPH.

The equation is:
$11\text{-Acetyl-CoA} + 10\ \text{ATP}^{4-} + 20\ \text{NADPH} + 10\ \text{H}^+ \rightarrow \text{behenoyl-CoA} + 20\ \text{NADP}^+ + 10\ \text{CoA-SH} + 10\ \text{ADP}^{3-} + 10\ \text{P}_i^{2-}$.
(There are only 10 H^+ consumed, and not 20 (thinking that each NADPH used is actually used with a proton), because hydrolysis of ATP releases a proton. You can see this by counting charge in the equation.)

Questions for Self Study

1. How are acetate units moved from the mitochondria to the cytosol? What other role does the acetate carrier play in regulation of metabolism?

2. Although acetate units are incorporated into fatty acids during synthesis, they derive from three-carbon compounds attached to coenzyme A. What is this three-carbon coenzyme A derivative? What enzyme is responsible for its formation? How many high-energy phosphate bonds are cleaved to drive its synthesis?

3. Describe how, in animals, the activity of acetyl-CoA carboxylase is regulated by citrate and palmitoyl-CoA and how this regulation is sensitive to covalent modification of the enzyme.

4. Match an enzyme with an activity.
   a. Acetyltransferase
   b. Dehydrase
   c. Malonyltransferase
   d. Enoyl reductase
   e. β-Ketoacyl reductase
   f. β-Ketoacyl synthase

   1. Keto carbon converted to alcoholic carbon.
   2. Attaches malonyl group to fatty acid synthase.
   3. Attaches acetyl group to acyl carrier protein.
   4. Carbon-carbon double bond reduced.
   5. Condensation of acetyl group and malonyl group.
   6. Enoyl intermediate formed.

5. From the following list of compounds identify those that are cholesterol derivatives and appropriately identify each cholesterol derivative as a hormone (H), bile salt (B), or vitamin (V).
   a. Prostaglandin D_2
   b. Glycocholic acid.
   c. Squalene
   d. Testosterone
   e. Arachidonic Acid.
   f. Cortisol
   g. Cholecalciferol
   h. Progesterone
   i. Thromboxanes
   j. Taurocholic acid
   k. Aldosterone

6. What is the rate-limiting step in cholesterol biosynthesis?

7. Match a lipoprotein complex with a function.
   - a. Chylomicrons
   - b. Very low density lipoproteins
   - c. Low-density lipoproteins
   - d. High-density lipoproteins

   1. Formed from very low density lipoproteins.
   2. Formed in the intestine.
   3. Slowly accumulate cholesterol.
   4. Carry lipid from the liver.

8. In eukaryotes, glycerolipids are all derived from phosphatidic acid. Draw the structure of phosphatidic acid and outline its biosynthesis from dihydroxyacetone phosphate and from glycerol-3-phosphate.

9. What is the role of cytidine in lipid biosynthesis?

10. Fill in the blanks. The prostaglandins are ___ that function locally and at very low concentrations. They are synthesized from ___, a 20-carbon polyunsaturated fatty acid. Mammals can produce this fatty acid from $(18:2^{\Delta9,12})$___, but must acquire this polyunsaturated fatty acid from their diet.

Answers

1. Acetate units on acetyl-CoA are used to produce citrate in the mitochondria. Citrate is exported to the cytosol where it is converted to acetyl-CoA and oxaloacetate. Citrate inhibits phosphofructokinase and thus serves as a regulator of glycolysis. It also stimulates fatty acid synthesis.

2. Malonyl-CoA is produced by acetyl-CoA carboxylase at the expense of one high-energy phosphate bond.

3. Acetyl-CoA carboxylase is active in a polymeric state. The equilibrium between active polymer and inactive protomers is affected by citrate and palmitoyl-CoA. Citrate is an allosteric activator of the enzyme and shifts the equilibrium to the polymer. Palmitoyl-CoA shifts the equilibrium to the inactive, protomeric state. The enzyme is phosphorylated by a number of kinases and the phosphorylated state has a low affinity for citrate and a high affinity for palmitate.

4. a. 3; b. 6; c. 2; d. 4; e. 1; f. 5.

5. b. B; d. H; f. H; g. V; h. H; j. B; k. H.

6. The production 3R-mevalonate from HMG-CoA catalyzed by HMG-CoA reductase.

7. a. 2; b. 4; c. 1; d. 3.

8. Dihydroxyacetone phosphate is converted to 1-acyldihydroxyacetone-phosphate by an acyltransferase reaction and reduced to 1-acylglycerol-3-phosphate by a reductase. This compound can also be synthesized from glycerol-3-phosphate by acyltransferase. Transfer of a second acyl group to C-2 produces phosphatidic acid whose structure is shown below.

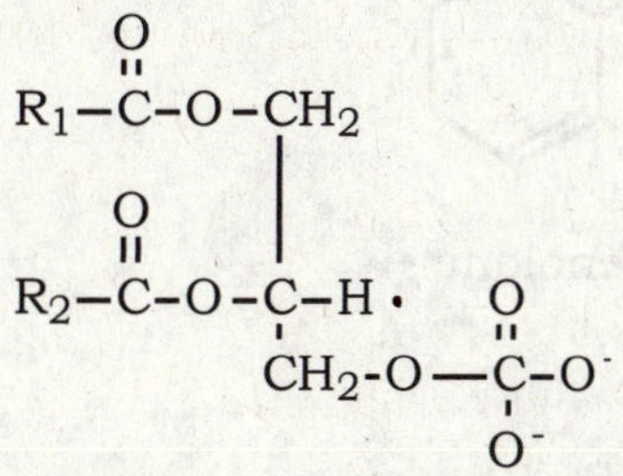

$$
\begin{array}{l}
\quad\quad\quad\quad\; O \\
\quad\quad\quad\quad\; \parallel \\
R_1-C-O-CH_2 \\
\quad\quad\quad\quad\quad | \\
\quad\quad\; O \quad\quad | \\
\quad\quad\; \parallel \quad\quad | \\
R_2-C-O-C-H \cdot \quad O \\
\quad\quad\quad\quad\; CH_2-O-C-O^- \\
\quad\quad\quad\quad\quad\quad\quad\quad\; \parallel \\
\quad\quad\quad\quad\quad\quad\quad\quad\; O^-
\end{array}
$$

9. The head groups of phosphatidylethanolamine and phosphatidylcholine derive from cytidine diphosphate derivatives. CDP-diacylglycerol is a precursor of phosphatidylinositol, phosphatidylglycerol, and cardiolipin.

10. Hormones; arachidonic acid; linoleic acid.

Additional Problems

1. What are the sources of carbons for fatty acid biosynthesis? What is the role of the citrate-malate-pyruvate shuttle in making carbon compounds available for fatty acid biosynthesis?

2. Movement of citrate out of the mitochondria coordinates glycolysis and fatty acid biosynthesis. Explain.

3. Name the three water soluble vitamins that are crucial to fatty acid synthesis and briefly describe the roles they play in this process.

4. Why do mammals require certain essential fatty acids in their diet?

5. Outline the synthesis of glycerophospholipid.

6. Lovastatin lowers serum cholesterol by interfering with HMG-CoA reductase, the enzyme that catalyzes the rate limiting step in cholesterol synthesis. The drug is administered as an inactive lactone that is activated by hydrolysis to mevinolinic acid, a competitive inhibitor of HMG-CoA reductase. Can you recall another lactone hydrolysis reaction encountered in an earlier chapter?

7. What is the role of high-density lipoproteins in regulation of cholesterol levels in the blood?

8. Synthesis of the steroid hormones from cholesterol starts with the reaction catalyzed by desmolase shown below. Why is this a critical reaction for formation of steroid hormones?

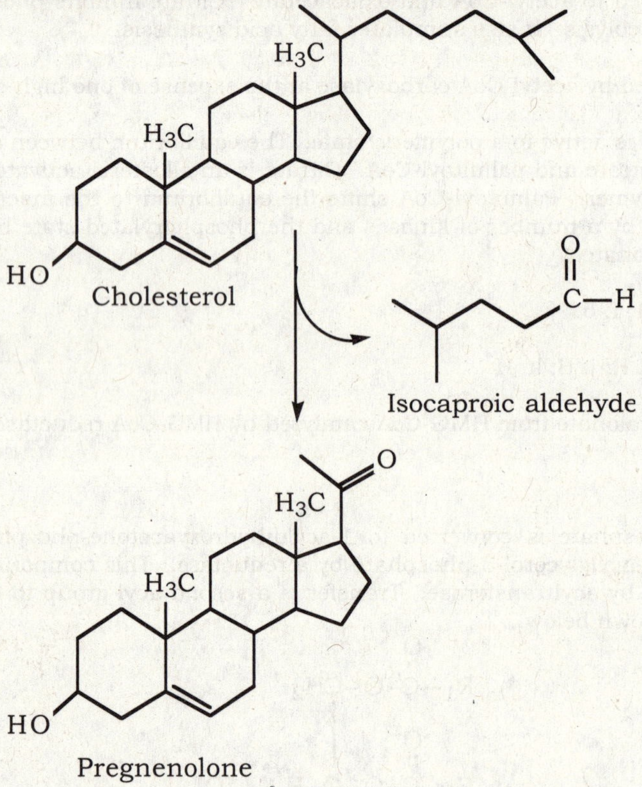

H3C
H3C
HO
Cholesterol
O
C—H
Isocaproic aldehyde
O
H3C
H3C
HO
Pregnenolone

Abbreviated Answers

1. The immediate source of carbons is acetyl-CoAs, which are produced from carbohydrates, amino acids, and lipids. Acetyl-CoA is produced in the mitochondria but fatty acid biosynthesis occurs in the cytosol. To move acetyl units out of the mitochondria, they are condensed onto oxaloacetate to form citrate, in a citric acid cycle reaction. Citrate is then transported out of the mitochondria to the cytosol, where ATP-citrate lyase catabolizes citrate to acetyl-CoA and oxaloacetate. This cytosolic acetyl-CoA is used to synthesize fatty acids. So, the citrate-malate-pyruvate shuttle is responsible for moving two-carbon units from the mitochondria to the cytosol. However, it has another purpose: it supplies some of the NADPH needed for fatty acid synthesis. Cytosolic oxaloacetate is reduced to malate and then oxidatively decarboxylated to CO_2 and pyruvate by malic enzyme, in an $NADP^+$-dependent reaction. The NADPH thus formed is consumed during the reduction steps of fatty acid biosynthesis.

2.   When glycolysis was covered, it was pointed out that phosphofructokinase activity is inhibited by citrate. In this chapter, we saw how citrate is used to move two-carbon units from the mitochondria to the cytosol for fatty acid biosynthesis. An increase in the concentration of citrate is a signal that the citric acid cycle is backing up, either because energy stores are satisfactory or because there is an abundance of acetyl units. In either case, there is not reason to continue glycolysis. Movement of citrate out of the mitochondria shifts acetyl units from degradation via the citric acid to storage via cytosolic fatty acid synthesis and serves to turn down glycolysis at phosphofructokinase.

3. Biotin is a component of acetyl-CoA carboxylase. Nicotinamide is found in NADPH. Phosphopantetheine is covalently attached to acyl carrier protein. Biotin functions as an intermediate carrier of activated carboxyl groups in malonyl-CoA biosynthesis by acetyl-CoA carboxylase. NADPH is required at two reduction steps in each round of chain elongation in fatty acid biosynthesis. Phosphopantotheine serves as a carrier of the growing fatty acid. This group is covalently attached to a serine residue in acyl carrier protein and serves to carry acetyl groups, malonyl groups, and acyl groups during various stages of fatty acid biosynthesis.

4. Mammals cannot introduce a double bond beyond C-9 in a given fatty acid. The prostaglandins are synthesized from linoleic acid, $\Delta^{9,12}$-octadecadienoic acid, which cannot be produced by mammals and is therefore an essential fatty acid.

5. The components of glycerophospholipids are glycerol, phosphate, fatty acids, and an alcoholic head group. Synthesis starts with either glycerol (via reduction of glyceraldehyde) or DHAP being converted to glycerol-3-phosphate by glycerokinase or glycerol-3-phosphate dehydrogenase, respectively. Glycerol-3-phosphate is converted to 1-acylglycerol-3-P and then to phosphatidic acid (1,2-diacylglycerol-3-P) by two acyltransferase reactions. Phosphatidic acid serves as the precursor for triacylglycerol and the glycerophospholipids phosphatidylethanolamine (PE), phosphatidylcholine (PC), phosphatidylserine (PS), phosphatidylinositol (PI), phosphatidylglycerol (PG), and cardiolipin (diphosphatidylglycerol). Phosphatidic acid is converted to diacylglycerol, which is converted to PE or PC by transferases using CDP-derivatized ethanolamine or choline. Alternatively, phosphatidic acid can be converted to its CDP derivative, CDP-diacylglycerol, which is metabolized to PI or PG. PS is produced from PE using serine to displace ethanolamine.

6.   In the pentose phosphate pathway, conversion of 6-phosphogluconolactone to 6-phosphogluconate involves hydrolysis of a lactone.

7.   HDL is assembled in the endoplasmic reticulum of liver cells and secreted into the blood. Newly synthesized HDL contains very little cholesterol but with time it accumulates cholesterol as both free cholesterol and as cholesterol esters. HDL then returns to the liver where cholesterol is either stored or converted to bile salts and excreted.

8. The steroid hormones are transported in the blood to target tissues and must therefore be slightly more soluble than cholesterol. The reaction catalyzed by desmolase removes the hydrocarbon tail of cholesterol, making the product more soluble.

Summary

The biosynthesis of lipid molecules proceeds via mechanisms and pathways, which are different from those of their degradation. In the synthesis of fatty acids, for example, 1) intermediates are linked covalently to the -SH groups of acyl carrier proteins instead of coenzyme A, 2) synthesis occurs in the cytosol instead of the mitochondria, 3) the nicotinamide coenzyme used is NADPH instead of NADH, and 4) in eukaryotes, the enzymes of fatty acid synthesis are associated in one large polypeptide chain instead of being separate enzymes. Fatty acids are synthesized by the addition of two carbon acetate units, which have been activated by the formation of malonyl-CoA, decarboxylation of which drives the reaction forward. Once the growing fatty acid chain reaches 16 carbons in length, it dissociates from the fatty acid synthase and is subject to the introduction of unsaturations or additional elongation. Acetyl-CoA needed for fatty acid synthesis is provided in the cytosol by citrate that is transported across the mitochondrial membrane and converted to acetyl-CoA and oxaloacetate by ATP-citrate lyase. Formation of malonyl-CoA by acetyl-CoA carboxylase (ACC), a biotin-dependent enzyme, commits acetate units to fatty acid synthesis. In animals, ACC is a multifunctional protein, which forms long, filamentous polymers. It is allosterically activated (and polymerized) by citrate and inhibited (and depolymerized) by palmitoyl-CoA. Affinities for both these regulators are decreased by phosphorylation of the enzyme at up to 8 to 10 separate sites. The fatty acid synthesis reactions involve formation of O-acetyl and O-malonyl enzyme intermediates, followed by transfer of the acetyl group to the -SH of an acyl carrier protein (ACP) and then to the β-ketoacyl-ACP synthase. Transfer of the malonyl group to the ACP is followed by decarboxylation of the malonyl group and condensation of the remaining two-carbon unit with the carbonyl carbon of the acetate group on the synthase. This is followed by reduction of the β-carbonyl to an alcohol, dehydration to yield a trans-α,β double bond and reduction to yield a saturated bond. Introduction of unsaturations in the nascent chain occurs by O_2-dependent and O_2-independent pathways and may be followed by further chain elongation. Several mechanisms are utilized to introduce multiple unsaturations in a fatty acid chain. Regulation of fatty acid synthesis is related to regulation of fatty acid breakdown and the activity of the TCA cycle, because of the importance of acetyl-CoA in all these processes. Malonyl-CoA inhibits carnitine transport, blocking fatty acid oxidation. Citrate activates ACC and palmitoyl-CoA inhibits, both in chain-length-dependent fashion. The enzymes of fatty acid synthesis are also under hormonal control.

Glycerolipid synthesis is built around the synthesis of phosphatidic acid from glycerol-3-phosphate or dihydroxyacetone phosphate. Specific acyltransferases add acyl chains to these glycerol derivatives. Other glycerolipids, such as phosphatidylcholine (PC) and phosphatidylethanolamine (PE) are synthesized from phosphatidic acid via CDP-diacylglycerol and diacylglycerol. Base exchange converts phosphatidylethanolamine to phosphatidylserine. Other phospholipids, such as phosphatidylinositol, phosphatidylglycerol and cardiolipin are synthesized from CDP-diacylglycerol. Dihydroxyacetone phosphate is a precursor to the plasmalogens. Platelet activating factor (PAF), an ether lipid, dilates blood vessels, reduces blood pressure and aggregates platelets. Sphingolipids are produced via condensation of serine and palmitoyl-CoA by 3-ketosphinganine synthase and reduction of the ketone product to form sphingamine. Acylation followed by desaturation yields ceramide, the precursor to other sphingolipids and cerebrosides.

Eicosanoids, derived from arachidonic acid by oxidation and cyclization, are ubiquitous local hormones. They include the prostaglandins, thromboxanes, leukotrienes and other hydroxyeicosanoic acids. A variety of stimuli, including histamine, epinephrine, bradykinin, proteases and other agents associated with tissue injury and inflammation, can stimulate the release of eicosanoids, which have short half-lives and are rapidly degraded. Aspirin acetylates endoperoxide synthase on its cyclooxygenase subunit, irreversibly inhibiting the synthesis of prostaglandins.

Cholesterol biosynthesis begins with mevalonic acid, which is formed from acetyl-CoA by thiolase, HMG-CoA synthase and HMG-CoA reductase. The HMG-CoA reductase reaction is the rate-limiting step in cholesterol biosynthesis. Inhibition of this enzyme by lovastatin blocks cholesterol biosynthesis and can significantly lower serum cholesterol. Mevalonate is converted to squalene via isopentenyl pyrophosphate and dimethylallyl pyrophosphate, which join to yield farnesyl pyrophosphate and then squalene. Squalene is cyclized in two steps and converted to lanosterol. The conversion of lanosterol to cholesterol requires another 20 steps.

Lipids circulate in the body in lipoprotein complexes, including high density lipoproteins, low density lipoproteins, very low density lipoproteins and chylomicrons. Lipoproteins consist of a core of mobile triacylglycerols and cholesterol esters, surrounded by a single layer of phospholipid, into which is inserted a mixture of cholesterol and proteins. Lipoproteins are bound to lipoprotein receptors at target sites and progressively degraded in circulation by lipoprotein lipases. Defects of lipoprotein metabolism can lead to elevated serum cholesterol.

Steroids such as the bile acids and steroid hormones are synthesized from cholesterol via key intermediates such as pregnenolone and progesterone. The male hormone testosterone is a precursor to the female hormones including estradiol. Steroid hormones modulate transcription of DNA to RNA in the cell nucleus. The corticosteroids, including glucocorticoids and mineralocorticoids, synthesized by the adrenal glands, are important physiological regulators.

Chapter 25

Nitrogen Acquisition and Amino Acid Metabolism

• •

Chapter Outline

❖ Nitrogen cycle
  ↗ Nitrogen movement in the biosphere
    • N_2 in atmosphere changed to ammonium
      • Nitrogen fixation: Bacteria
    • NO_3^- (nitrate) in soil and oceans changed to ammonium
      • Nitrate assimilation: Plants, fungi, bacteria
  ↗ Animals need reduced dietary N and excrete N as ammonium or organic nitrogen
    • Ammonium to NO_3^- (nitrate): Nitrifying bacteria
    • NO_3^- (nitrate) to N_2: Denitrifying bacteria
❖ Nitrate assimilation: Two steps
  ↗ Nitrate reductase: $NO_3^- + 2H^+ + 2e^- \rightarrow NO_2^- + H_2O$
    • NADH source of electrons
    • SH to FAD to cyto b_{557} to MoCo
  ↗ Nitrite reductase: $NO_2^- + 8H^+ + 6e^- \rightarrow NH_4^+ + 2H_2O$
    • Plants
      • Light-generated reduced ferredoxin source of electrons
      • 4Fe-4S to siroheme
    • Microorganisms
      • NADPH source of electrons
      • Similar to nitrate reductase
❖ Nitrogen fixation: $N_2 + 10H^+ + 8e^- \rightarrow 2NH_4^+ + H_2$: Four requirements
  ↗ Strong reductant: Reduced ferredoxin
  ↗ ATP
  ↗ Oxygen-free environment
  ↗ Nitrogenase (complex)
    • Nitrogen reductase (Fe-protein)
      • 4Fe-4S cluster
      • 2ATP hydrolyzed per electron
    • Nitrogenase (MoFe protein)
      • P-cluster (8Fe-7S)
      • FeMo-cofactor (7Fe-1Mo-9S): Femoco
  ↗ Regulation
    • ADP inhibits
    • Ammonium represses expression of *nif* genes
    • ADP-ribosylation in some systems inactivates
❖ Ammonium entry into organic linkage
  ↗ Carbamoyl-phosphate synthetase I
    • Urea cycle

- $NH_4^+ + HCO_3^- + 2ATP \rightarrow$ carbamoyl-phosphate $+ 2ADP + P_i$
- N-Acetylglutamate activator
- Glutamate dehydrogenase (GDH)
  - $NH_4^+ + \alpha$-ketoglutarate $+ NADPH + H^+ \rightarrow$ glutamate $+ NADP^+ + 2H_2O$
    - Produces glutamate using NADPH
    - Degrades glutamate using NAD^+
  - Catabolic activity
    - ADP allosteric activator
    - GTP allosteric inhibitor
- Glutamine synthetase (GS)
  - $Glu + NH_4^+ \rightarrow Gln$: ATP driven
- ❖ Ammonium assimilation pathways
  - GDH and GS when ammonium high
  - GS and GOGAT when ammonium low
    - GOGAT: Glutamate:oxoglutarate amino transferase
      - Reductant $+ \alpha$-ketoglutarate $+ Gln \rightarrow 2 Glu +$ oxidized reductant
      - Reductant: NADH, NADPH, Reduced ferredoxin
- ❖ Glutamine Synthetase regulation: Dodecameric complex
  - Feedback inhibition
    - Gly, Ala, Ser inhibit by competing with Glu
    - AMP competes with ATP
    - His, Trp, CTP, carbamoyl-P, glucosamine-6-P inhibit
  - Covalent modification
    - Adenylylation of Tyr inactivates
    - Degree of inactivation proportional to number of subunits adenylylated
    - Adenylyl transferase (AT): Adenylylates and deadenylylates
      - $AT \cdot P_{IIA}$ adenylylates
      - $AT \cdot P_{IID}$ deadenylylates
        - Gln activates P_{IIA} inhibits P_{IID}
        - α-Ketoglutarate inhibits P_{IIA} activates P_{IID}
  - Gene expression: GS subunit encoded by Gln A
    - Transcriptional enhancer: NR_I: Must be phosphorylated
    - NR_{II}: NR_I kinase
    - NR_{II}/P_{IID}-complex: NR_I-phosphatase
- ❖ Amino acid biosynthesis: Transamination of appropriate α-keto acid
  - Essential amino acids: Required in diet
  - Nonessential amino acids: Synthesized by organism
  - Transamination by aminotransferases
    - Glutamate aminotransferase most common
- ❖ Carbon skeletons
  - α-Ketoglutarate family
    - Glu, Gln
    - Pro from Glu
    - Arg from Glu via ornithine
      - Ornithine's metabolic fates
        - Arg precursor
        - Urea cycle intermediate
        - Arg breakdown product
    - Lys
  - Aspartate family
    - Asp, Asn
    - Thr, Met, Lys
    - Ile
  - Pyruvate family
    - Ala, Val, Leu

- 3-Phosphoglycerate family
  - Ser, Gly, Cys
- Aromatic amino acids
  - Shikimate path produces chorismate
  - Chorismate branch point
    - Phe, Tyr
    - Trp
  - His
- ❖ Amino acid degradation: 20 Amino acids converted to 7 metabolic intermediates
  - Pyruvate
    - Ala, Ser, Cys, Gly, Trp (carboxyl-, α–, β-carbons), Thr
  - Oxaloacetate
    - Asp, Asn
  - α-Ketoglutarate
    - Glu, Gln, Pro, Arg, His
  - Succinyl-CoA
    - Val, Ile, Met
  - Acetoacetate
    - Leu (+ acetyl-CoA)
    - Lys, Trp ring
    - Phe, Tyr (+ fumarate)
  - Acetyl-CoA
    - Leu (+ fumarate)
  - Fumarate
    - Phe, Tyr (+ acetoacetate)

Chapter Objectives

Nitrogen Acquisition

All organisms produce important compounds containing reduced nitrogens. However, the principal forms of nitrogen in the environment, N_2 (dinitrogen) and NO_3^- (nitrate), contain nitrogen in an oxidized state. Nitrogen is abstracted from the environment and converted to NH_4^+ (ammonium) by two pathways: nitrogen assimilation and nitrogen fixation. Know the basic chemistry involved and the important enzymes in these two processes.

Nitrate Assimilation

Nitrate assimilation converts nitrate to ammonium in a two-step process
$$NO_3^- + 2\ H^+ + 2\ e^- \rightarrow NO_2^-\ (nitrite) + H_2O$$
$$NO_2^- + 8\ H^+ + 6\ e^- \rightarrow NH_4^+ + 2\ H_2O$$
The first reaction is catalyzed by nitrate reductase, a NAD(P)H-dependent enzyme that moves electrons through a protein-SH, protein-bound FAD, cytochrome b_{557} and a molybdenum cofactor (MoCo) to reduce nitrate to nitrite. The second reaction is catalyzed by nitrite reductase. Plants utilize reduced ferredoxin, a product of photosynthesis, as a source of electrons for this reaction and direct them through a 4Fe-4S cluster to a siroheme to nitrite. Fungi and bacteria use NADPH as an electron source.

Nitrogen Fixation

Only certain prokaryotes are capable of nitrogen fixation:
$$N_2 + 8\ H^+ + 8\ e^- \rightarrow 2\ NH_3 + H_2$$
Nitrogen fixation is catalyzed by nitrogenase using reduced ferredoxin or flavodoxin as a source of electrons and ATP hydrolysis to drive the reaction. Nitrogenase is a complex of nitrogenase reductase, an O_2-sensitive enzyme that binds ATP and contains one (4F2-4S) cluster, and nitrogenase proper, an O_2-sensitive, $\alpha_2\beta_2$ heterodimer containing an iron-molybdenum cofactor (FeMoCo). Electron pair movement through nitrogenase is driven by hydrolysis of 4 ATP, so the complete reaction consumes 16 ATP.

Ammonium Utilization

Ammonium is incorporated into organic compounds in the following ways:

1. $NH_4^+ + HCO_3^- + 2\ ATP \longrightarrow H_2N-\overset{\overset{\displaystyle O}{\displaystyle \|}}{C}-O-PO_3^{2-} + 2\ ADP + P_i + H^+$

catalyzed by carbamoyl-phosphate synthetase I, a urea cycle enzyme.

2. $NH_4^+ + \alpha$-ketoglutarate + NADPH $\star$ glutamate + NADP$^+$ + H$_2$O

catalyzed by glutamate dehydrogenase.

3. $NH_4^+ +$ glutamate + ATP $\star$ glutamine + ADP + P$_i$

catalyzed by glutamine synthetase.
Glutamate dehydrogenase has a higher K_m for ammonium than does glutamine synthetase and in conditions in which ammonium is not abundant, reaction 3 above predominates and glutamate is consumed. To replenish glutamate, glutamate synthase or GOGAT (glutamate:oxo-glutarate amino-transferase) catalyzes

NADPH + α-ketoglutarate + glutamine $\star$ 2 glutamate + NADP$^+$

Regulation of *E. coli* Glutamine Synthetase

Regulation of this key enzyme in nitrogen metabolism is achieved by three mechanisms. The enzyme is allosterically regulated by nine inhibitors including Gly, Ala, Ser, His, Trp, CTP, AMP, carbamoyl-phosphate, and glucosamine-6-phosphate. Covalent modification of glutamine synthetase by adenylylation of a tyrosine residue leads to inactivation. Gene expression is regulated to control enzyme levels.

The level of adenylylation of glutamine synthetase is regulated by adenylyl transferase. This enzyme both adds and removes adenylyl groups but activity in general is dependent on a complex between adenylyl transferase and a regulatory protein P$_{II}$. P$_{II}$ is itself a target for covalent modification by uridylyl transferase. Uridylylated P$_{II}$ complexed with adenylyl transferase results in deadenylylation activity, deadenylylation of glutamine synthetase, and glutamine synthetase activity. Unuridylylated P$_{II}$ complexed to adenylyl transferase stimulates deadenylylation activity leading to inhibition of glutamine synthetase. Uridylyl transferase activity is dependent on glutamine.

Amino Acid Biosynthesis

The key point to remember about amino acid biosynthesis is that an α-keto acid is first produced and then an amino group is introduced by transamination from glutamate, catalyzed by an aminotransferase. Some α-keto acids important in this regard include α–ketoglutarate, oxaloacetate, and pyruvate. α–Ketoglutarate, a citric acid cycle intermediate, is used to produce glutamic acid, glutamine, proline, and ornithine, an amino acid intermediate in the urea cycle, which leads to arginine synthesis. Oxaloacetate, another citric acid cycle intermediate, is the precursor of aspartic acid, asparagine, threonine, methionine and lysine. Four of the carbons of isoleucine derive from oxaloacetate via threonine, with two carbons coming from pyruvate. Pyruvate is used to form alanine, valine and leucine. The glycolytic intermediate, 3-phosphoglycerate, is used to synthesize serine, glycine, and cysteine. (You might recall a relationship between serine and glycine in photorespiration.) The aromatic amino acids, phenylalanine, tyrosine, and tryptophan, all have chorismate as a common intermediate. Chorismate is produced from erythrose-4-phosphate, a pentose phosphate intermediate, and PEP via shikimate. Finally, histidine biosynthesis shares metabolic intermediates with purine synthesis. ATP and phosphoribosyl pyrophosphate are used for all of the carbons and one of the nitrogens. The remaining two nitrogens come from glutamine and glutamate.

Amino Acid Degradation

The α-amino group is removed from amino acids by transamination and excess nitrogen is excreted in one of three ways depending on the organism. It may be released as free ammonia, as urea (Figure 25.23), or as uric acid. The carbon skeletons are metabolized into seven metabolic intermediates including pyruvate, oxaloacetate, α–ketoglutarate, acetyl-CoA, fumarate, and acetoacetate. Alanine, serine, glycine, cysteine, and parts of threonine and tryptophan are converted to pyruvate. Aspartate and asparagine produce oxaloacetate, and, glutamate, glutamine, proline, arginine, and histidine produce α-ketoglutarate. Valine, isoleucine, and methionine are degraded to succinyl-CoA. Leucine is converted to acetyl-CoA and acetoacetate, a ketone body. Lysine is decarboxylated twice to yield acetoacetate. Phenylalanine is converted to tyrosine by hydroxylation and tyrosine is degraded into acetoacetate and fumarate. Amino acids that can support gluconeogenesis are termed glucogenic amino acids whereas amino acids producing only acetyl-CoA or acetoacetate are ketogenic.

468

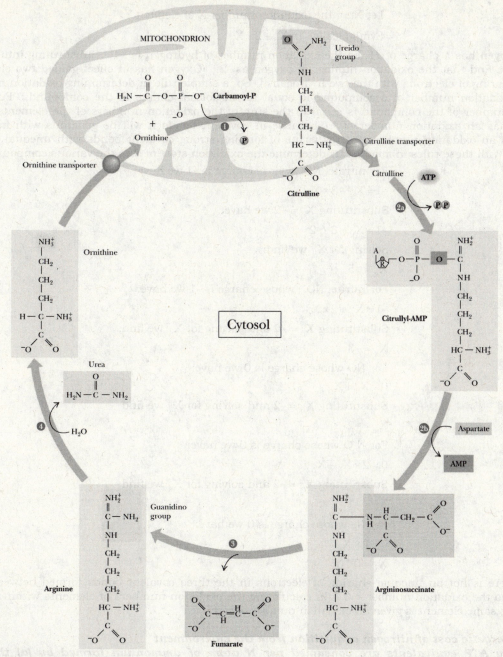

Figure 25.23 The urea cycle.

Problems and Solutions

1. The oxidation states of nitrogen
What is the oxidation number of N in nitrate, nitrite, NO, N₂O and N₂?

Answer: The oxidation number is in effect a measure of the number of electrons being shared unequally in covalent bonds. An easy compound to see this in is water. In water each hydrogen is involved in a covalent bond with oxygen, an electronegative element. In these bonds, there is an unequal sharing of the electron pair such that the oxygen becomes partially negatively charged whereas the hydrogens are partially positively charged. The oxidation number of a compound is equal to the formal charge of the compound and it is the sum of the oxidation numbers of the elements of the compound. In water for example, the formal charge is zero, which is equal to the sum of the oxidation state of oxygen plus twice the oxidation state of hydrogen.

Let X_Y = the oxidation number X of element Y.

Symbolically, $0 = X_O + 2 \times X_H$

Each hydrogen has a charge of +1, so the oxidation number of hydrogen is +1. Substituting into the above equation we find that the oxidation number of oxygen as -2. (Oxygen has in effect gained two electrons by unequal sharing of electrons with the two hydrogens.) So, a simple rule in determining oxidation numbers is that the oxidation number of a compound is equal to the formal charge of the compound. Further, the oxidation number of the compound is equal to the sum of the oxidation numbers of the elements. Finally, hydrogen has an oxidation number of +1 because in covalent bonds it will be generous with its electron; oxygen has an oxidation number of -2 by virtue of forming strong covalent bonds with unequal sharing of electrons. With these rules in mind let us determine the oxidation state of N in the various compounds.

For nitrate, NO_3^-, whose charge is -1 we have:

$-1 = X_N + 3 \times X_O$

Substituting $X_O = -2$ we have:

$-1 = X_N + 3 \times -2$

Solving for X_N we find:

$X_N = +5$

For nitrite, NO_2^-, whose charge is -1 we have:

$-1 = X_N + 2 \times X_O$

Substituting $X_O = -2$ and solving for X_N we find:

$X_N = +3$

For NO whose charge is 0 we have:

$0 = X_N + X_O$

Substituting $X_O = -2$ and solving for X_N we find:

$X_N = +2$

For N_2O whose charge is 0 we have:

$0 = 2 \times X_N + X_O$

Substituting $X_O = -2$ and solving for X_N we find:

$X_N = +1$

For N_2 whose charge is 0 we have:

$0 = 2 \times X_N$

$X_N = 0$

For N_2, there is not an unequal sharing of electrons in the three covalent bonds formed between the two nitrogens so the oxidation number = 0. (In calculating the oxidation numbers of elements within a group, a bond to the same element is given an oxidation number of zero.)

2. The energetic cost of nitrogen acquisition from the environment
How many ATP equivalents are consumed per N atom of ammonium formed by (a) the nitrate assimilation pathway and (b) the nitrogen fixation pathway? (Assume NADH, NADPH, and reduced ferrodoxin are each worth 3 ATPs)

Answer: a. In nitrate assimilation, nitrate is converted to ammonium in two steps: nitrate to nitrite; nitrite to ammonium. From the answer to problem 1 we know that the oxidation number of nitrogen in nitrate is +5. We can calculate the oxidation number of nitrogen in ammonium as follows:

For ammonium, NH_4^+, whose charge is +1 we have:

$+1 = X_N + 4 \times X_H$

Substituting $X_H = +1$ and solving for X_N we find:

$X_N = -3$

Therefore, in going from nitrate to ammonium there is a change in oxidation of (-3 - 5 =) - 8, a reduction caused by acceptance of 8 electrons or 4 electron pairs. If we consider the 4 pairs of electrons deriving from

NADPH then 4 NADPH equivalents are required by this reaction. The ATP equivalent of NADPH is 4 ATP per NADPH, therefore, a 4 NADPH reduction is equivalent to 12 ATP.

b. For the nitrogen fixation pathway, nitrogen gas or dinitrogen, N_2, is converted to two molecules of ammonia. In this case the oxidation number of each nitrogen is changed from zero to -3, a change in oxidation number of -6. This pathway also produces hydrogen gas, H_2, from two protons and two electrons and consumes 16 ATP per N_2 reduced (Figure 25.6). Thus, the overall reaction sequence involves 8 electrons, or 4 electron pairs, which, if derived from NADH, are equivalent to 12 ATP, plus 16 ATP for a total of 28 ATP.

3. Regulation of glutamine synthetase by covalent modification
Suppose at certain specific metabolite concentrations in vivo the cyclic cascade regulating E. coli glutamine synthetase has reached a dynamic equilibrium where the average state of GS adenylylation is poised at n = 6. Predict what change in n will occur if:
> **a. [ATP] increase.**
> **b. P_{IIA}/P_{IID} increases.**
> **c. [α-KG]/[Gln] increases.**
> **d. [P_i] decreases.**

Answer: Glutamine synthetase is a key enzyme in nitrogen metabolism. In *E. coli* the enzyme is a 600-kD dodecamer that is sensitive to a number of metabolic regulators. It catalyzes the following reaction

$$\text{glutamate} + \text{ATP} + NH_3 \star \text{glutamine} + \text{ADP} + P_i$$

The activity of the enzyme is further regulated by covalent modification; specifically the enzyme is inhibited by adenylylation catalyzed by ATP:GS:adenylyl transferase or AT. Thus, the activity of AT determines *n*, the average state of GS adenylylation. AT is itself regulated by interaction with a regulatory protein, P_{II}, which also can exist it two states, P_{IIA} and P_{IID}. Interaction with P_{IIA} results in adenylyl transferase activity whereas interaction with P_{IID} results in deadenylylation activity. The interconversion of P_{II} between its two states is a result of uridylylation/deuridylylation catalyzed by a converter enzyme, UT. Uridylylated P_{II} is in the P_{IID} state whereas deuridylylated P_{II} is in the P_{IIA} state. UT catalyzes both uridylylation and deuridylylation, the exact activity dependent on the ratio of [Gln]/[α-ketoglutarate]. Excess glutamine is a signal that additional glutamine is not required. Therefore, GS must be inactivated by adenylylation, a reaction itself dependent on P_{IIA}, which is formed by deuridylylation of P_{II}.
a. An increase in [ATP] is expected to favor adenylylation of GS causing *n* to increase leading to increased inhibition.
b. An increase in P_{IIA} levels, leading to deadenylylation, will cause an increase in *n* and increase in inhibition.
c. An increase in the ratio of [α-KG]/[Gln] is a signal that Gln is needed. Thus, GS activity is increased by deadenylylation and *n* will decrease.
d. A decrease in [P_i], the substrate of the deadenylylation reaction, will lead to a decrease in deadenylylation, and an increase in *n*.

4. The energetic cost of nitrogen excretion via the urea cycle
How many ATP equivalents are consumed in the production of one equivalent of urea by the urea cycle?

Answer: In the urea cycle, the amino acid ornithine serves as a skeleton to synthesize arginine, which is hydrolyzed to form urea and ornithine. The two nitrogens in urea derive immediately from ammonia and the amino group of aspartic acid. The nitrogen from ammonia is introduced during the synthesis of citrulline from ornithine and carbamoyl phosphate. Carbamoyl phosphate is produced from ammonia and bicarbonate by carbamoyl phosphate synthetase, in a reaction driven by hydrolysis of two 2 ATPs. Citrulline and aspartate are combined to form argininosuccinate in a reaction catalyzed by argininosuccinate synthetase and driven by hydrolysis of ATP to AMP and PP_i. This step is equivalent to hydrolysis of two high-energy phosphate bonds because PP_i is hydrolyzed by pyrophosphatase into 2 P_i. So far we have accounted for 4 ATPs consumed to form one urea and one fumarate from one CO_2, one NH_3, and one aspartate.

However, there are some hidden energetic benefits to the urea cycle. Argininosuccinate is resolved into arginine and fumarate by argininosuccinase. In order to convert fumarate back into aspartate it is hydrated to malate and oxidized to oxaloacetate (two citric acid cycle steps). Malate to oxaloacetate is accompanied by NADH production, which is equivalent to approximately 2.5 ATPs. (An aminotransferase reaction converts oxaloacetate to aspartate.) Finally, ammonia is generated by glutamate dehydrogenase in an oxidative deamination of glutamate to α-ketoglutarate with accompanying reduction of NAD$^+$ to NADH accounting for another 2.5 ATP equivalents. Therefore, if we account for ammonia and aspartate, we see that as long as there is a source of amino acids, the urea cycle actually accounts for the net production of 1 ATP.

5. Why a high-protein diet requires greater water intake
Why are persons on a high-protein diet (such as the Atkins diet) advised to drink lots of water?

Answer: Humans do not store excess proteins or amino acids, and a dietary excess of either will result in degradation of amino acids and production of urea by the urea cycle. The urea cycle is carried out by liver cells, which release urea to the bloodstream to be carried to the kidney where it is excreted in urine. Excess water will be required to meet this increase in urea synthesis and urine production.

6. The energetic cost of lysine synthesis
How many ATP equivalents are consumed in the biosynthesis of lysine from aspartate by the pathway shown in Figure 25.27?

Answer: Lysine, methionine, and threonine are all members of the aspartate family of amino acids. Synthesis of lysine begins with phosphorylation and reduction of aspartate to β-aspartate-semialdehyde in two steps: Phosphorylation is at the expense of ATP; reduction uses electrons from NADPH. This accounts for 5 ATP equivalents (assuming that NADPH is worth 4 ATP). Next, dihydropicolinate is produced from β-aspartate-semialdehyde and pyruvate and converted by reduction to Δ^1-piperidine-2,6-dicarboxylate with electrons from NADPH accounting for 4 more ATPs. Finally, L,L-α,ε-diaminopimelate is produced from Δ^1-piperidine-2,6-dicarboxylate, which is converted in two steps to lysine. The production of L,L-α,ε-diaminopimelate is made favorable by hydrolysis of succinyl-CoA to succinate and CoASH, a step equivalent in free energy to one ATP. In sum, lysine production requires 10 ATP equivalents.

7. PEP as a precursor for aromatic amino acid biosynthesis
If PEP labeled with ^{14}C in the 2-position serves as precursor to chorismate synthesis, which C atom in chorismate is radioactive?

Answer: Chorismate, which is synthesized from PEP and erythrose-4-phosphate, is a precursor to the aromatic amino acids, tyrosine, phenylalanine, and tryptophan. Its synthesis is as follows

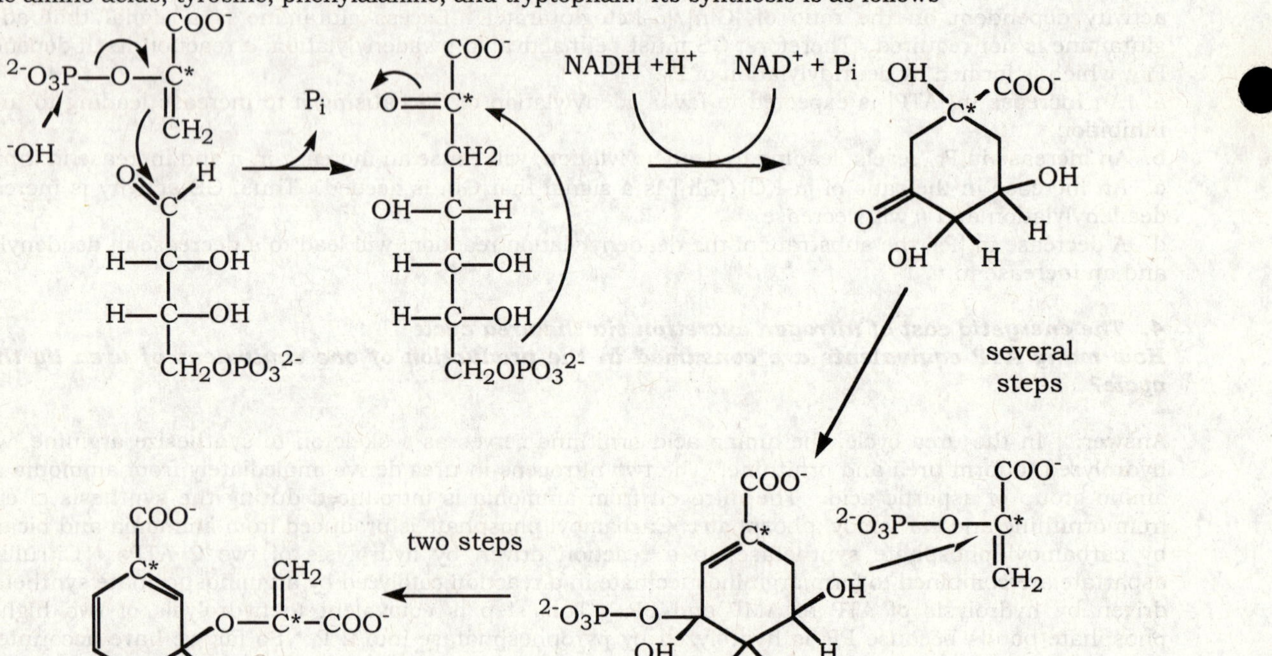

8. Gluconeogenesis from aspartate
(Integrates with Chapter 22.) Write a balanced equation for the synthesis of glucose (by gluconeogenesis) from aspartate.

Answer: In gluconeogenesis, pyruvate is converted to glucose with the following stoichiometry
2 pyruvate + 2 NADH + 4 H+ + 4 ATP + 2 GTP + 6 H$_2$O ⋆

$$\text{glucose} + 2 \text{ NAD}^+ + 4 \text{ ADP} + 2 \text{ GDP} + 6 \text{ P}_i$$

ATP is consumed at two steps: Conversion of pyruvate to oxaloacetate and production of 1,3-bisphosphoglycerate from 3-phosphoglycerate. In addition, GTP is consumed by PEP carboxykinase in producing PEP from oxaloacetate.

With aspartate as a source of carbons for gluconeogenesis, oxaloacetate is produced by transamination of aspartate to α-ketoglutarate. Thus, one energy-utilizing step in gluconeogenesis starting from pyruvate is bypassed when aspartate is metabolized. Further, CO_2 shows up as a product. The balanced equation is

2 aspartate + 2 α-ketoglutarate + 2 NADH + 6 H$^+$ + 2 ATP + 2 GTP + 4 H$_2$O ⋆

$$\text{glucose} + 2 \text{ glutamate} + 2 \text{ NAD}^+ + 2 \text{ ADP} + 2 \text{ GDP} + 4 \text{ P}_i + 2 \text{ CO}_2$$

(Note: 4 of the 6 protons are necessary to balance the charge on the 4 carboxylate groups on 2 aspartates.)
We may wish to metabolize glutamate with glutamate dehydrogenase as follows

Glutamate + NAD(P)$^+$ + H$_2$O ⋆ α-ketoglutarate + NH$_3$ + NAD(P)H + H$^+$

If we then consider this reaction, the balanced equation is

2 Asp + 2 H$^+$ + 2 ATP + 2 GTP + 6 H$_2$O ⋆ glucose + 2 ADP + 2 GDP + 4 P$_i$ + 2 CO$_2$ + 2 NH$_3$

9. Synthesis of amino acids from α-keto acids
For each of the 20 common amino acids, give the name of the enzyme that catalyzes the reaction providing its α-amino group.

Answer: For alanine, pyruvate is the α-keto acceptor for glutamate: pyruvate aminotransferase.

Arginine is synthesized from glutamate via ornithine so its α–amino group is glutamate's α-amino group, which derives from α-ketoglutarate and ammonia by glutamate dehydrogenase.

For aspartic acid, oxaloacetate is the α-keto acceptor for glutamate:oxaloacetate aminotransferase. Aspartic acid may be converted to asparagine by asparagine synthetase.

Cysteine is derived from serine, which derives from 3-phosphoserine via glutamate:3-phosphoserine aminotransferase.

Glutamate is from glutamate dehydrogenase as discussed under arginine.

Glutamine may derive from glutamate.

Glycine derives from serine, which is also used to produce cysteine. So, its α-amino group is from glutamate:3-phosphoserine aminotransferase.

Histidine's α-amino group is introduced by l-histidinol phosphate aminotransferase, which uses glutamate as amino donor.

Isoleucine's α-amino group is from the reaction catalyzed by glutamate:α-keto-β-methylvalerate aminotransferase.

Leucine's α-amino group is from the reaction catalyzed by glutamate:α-ketoisocaproate aminotransferase.

Lysine's α-amino group comes from glutamate via N-6-(L-1,3-dicarboxypropyl)-L-Lysine: NADP$^+$ oxidoreductase.

Methionine is produced from aspartate whose α-amino group comes from glutamate:oxaloacetate aminotransferase.

Phenylalanine is produced from phenylpyruvate via aminotransferase using glutamate as amino donor. The enzyme is glutamate:phenylpyruvate aminotransferase.

Proline is synthesized from glutamate so glutamate dehydrogenase incorporates its α–amino group.

Serine was already discussed under cysteine (glutamate:3-phosphoserine aminotransferase).

Threonine is synthesized from aspartate so glutamate:oxaloacetate aminotransferase is responsible for threonine's α-amino group.

Tryptophan's α–amino group comes from serine via glutamate:3-phosphoserine aminotransferase.

Tyrosine may derive from phenylalanine. Alternatively, it may be produced from 4-hydroxyphenylpyruvate by aminotransfer from glutamate by 4-hydroxyphenylalanine:glutamate aminotransferase.

Valine gets its α-amino group from glutamate:α-ketoisovalerate aminotransferase.

10. The essential coenzyme in amino acid metabolism
Which vitamin is central in amino acid metabolism? Why?

Answer: The amino acids, as metabolites, receive a lot of attention from pyridoxal-phosphate containing enzymes. Pyridoxal phosphate is produce from pyridoxal (a.k.a., vitamin B6). Pyridoxal phosphate is used as coenzyme in aminotransferase reactions in addition to other reactions involving amino acids. (See the answer to the next question for an example.)

11. Vitamins and coenzymes in homocysteine metabolism
Vitamins B6, B12, and folate may be recommended for individuals with high blood serum levels of homocysteine (a condition called hyperhomocysteinemia). How might these vitamins ameliorate homocysteinemia?

Answer: Homocysteine comes from the compound S-adenosylmethionine (SAM). The sulfur of methionine is attached to carbon 5 of adenosine. SAM is a very important compound in the so-call one-carbon pool because it is the source of methyl groups for methyltransferases (methylase). Methyltransferases, as their name implies, transfer a methyl group from SAM to a target. (There are a large number of methyltransferases whose targets range from proteins, to nucleic acids to small molecule compounds and metabolites). When the terminal methyl group is removed from SAM, it is converted to S-adenosylhomocysteine. Two pathways are used to deal with homocysteine. S-adenosylhomocysteine itself may be methylated back to SAM and in this pathway the methyl group comes from tetrahydrofolate, a key player in the one-carbon pool. (Tetrahydrofolate moves carbons at several oxidation states.) Tetrahydrofolate, in turn, acquires a methyl group in a vitamin B12-dependent reaction. In the other pathway, S-adenosylhomocysteine is hydrolyzed to release homocysteine, which is converted to cystathionine. This conversion involves a pyridoxal-phosphate-dependent reaction in which serine's β-carbon is attached to the sulfur of homocysteine. Metabolism of cystathionine can lead to production of cysteine or other metabolic intermediates. In any case, all of these reactions are catalyzed by pyridoxal-phosphate-dependent reactions. High blood levels of homocysteine may be corrected by vitamins.

12. Vitamins and coenzymes in branched-chain amino acid degradation
(Integrates with Chapter 19.) On the basis of the following information, predict a reaction mechanism for the mammalian branched-chain α-keto acid dehydrogenase complex (the BCKAD complex). This complex carries out the oxidative decarboxylation of the α-keto acids derived from valine, leucine, and isoleucine.
   a. *One form of maple syrup urine disease responds well to administration of thiamine.*
   b. *Lipoic acid is an essential coenzyme.*
   c. *The enzyme complex contains a flavoprotein.*

Answer: Mention of thiamine, lipoic acid and flavoprotein brings to mind pyruvate dehydrogenase complex and α-ketoglutarate dehydrogenase complex. BCKAD complex is similar to these two multi-protein complexes. Oxidative decarboxylation is accomplished by decarboxylation of the keto acid followed by transfer to thiamine then to lipoic acid and finally to coenzyme A. In the process lipoic acid is reduced. It is deoxidized by NAD+ via FAD. The complete reaction mechanism was discussed in Chapter 19.

13. Dietary concerns in phenylketonuria
People with phenylketonuria must avoid foods containing the low-calorie sweetener Aspartame, also known as NutraSweet. Find the structure of the low-calorie sweetener Aspartame in the Merck Index (or other scientific source) and state why these people must avoid this substance.

Answer: To locate the structure use chemfinder (http://chemfinder.cambridgesoft.com/), DBGET (http://www.genome.jp/dbget/), or NCBI's PubChem Compound database (http://www.ncbi.nlm.nih.gov/). DBGET had aspartame but it did not list its brand name NutraSweet. Chemfinder listed both as did NCBI. The structure is shown below. Aspartame is aspartic acid linked in peptide bond to phenylalanine whose carboxyl group is methyl esterified. Phenylketonurics need to avoid this compound because they lack the

474

enzyme phenylalanine 4-hydroxylase, which converts phenylalanine into tyrosine. The problem is not that tyrosine cannot be synthesized (phe is an essential amino acid because humans cannot synthesize it whereas tyrosine is not because phenylalanine can normally be converted to tyrosine). Rather, the problem is that phenylalanine cannot be degraded. It is deaminated by an aminotransferase producing the α-keto acid phenylpyruvate, a phenylketone that builds up in the blood. The structures of aspartame and phenylpyruvate are shown below.

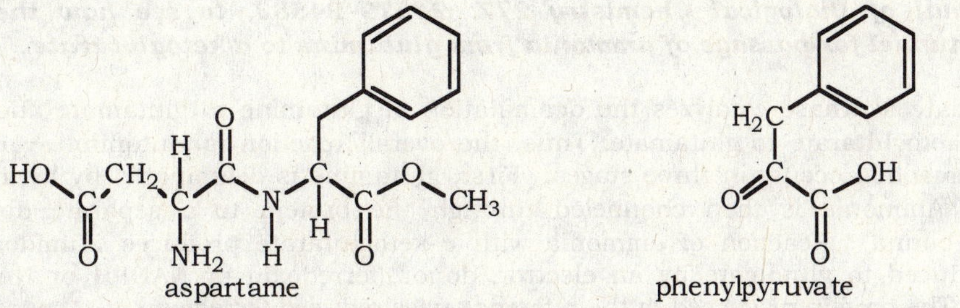

aspartame   phenylpyruvate

14. How the herbicide RoundUp affects plant metabolism

Glyphosate (otherwise known as RoundUp) is an analog of PEP. It acts as a noncompetitive inhibitor of 3-enolpyruvylshikimate-5-P synthase; it has the following structure in its fully protonated state:

$$HOOC\text{-}CH_2\text{-}NH\text{-}CH_2\text{-}PO_3H_2$$

Consult Figures 25.35 and 25.36 and construct a list of the diverse metabolic consequences that might be experienced by a plant cell exposed to glyphosate.

Answer: Glyphosate inhibits 3-enolpyruvylshikimate-5-phosphate synthase, which produces a compound that is converted to chorismate. Chorismate is a key metabolic intermediate that is used to synthesize the aromatic amino acids, phenylalanine, tyrosine and tryptophan, in addition to being an intermediate in ubiquinone and folate biosynthesis. Ubiquinone is an intermediate in synthetic pathways for vitamins K and E. Phenylalanine is used in plants to produce alkaloids, lignin (a plant cell wall component) and flavonoids. A great way to see these connections is to search the KEGG LIGAND database (http://www.genome.jp/kegg/ligand.html). A keyword search for chorismate returns several kinds of links including pathway links, which list multiple pathways (in addition to chorismate's own biosynthetic pathway) in which chorismate is involved.

15. Glycine synthesis from glucose

(Integrates with Chapter 18.) When cells convert glucose to glycine, which carbons of glucose are represented in glycine?

Answer: Glycine is made from serine by serine hydroxymethyltransferase, which removes the β-carbon of serine to produce glycine. Serine is synthesized from 3-phosphoglycerate. So, if we can recall glycolysis, we can figure out the labeling pattern in glycine from label in glucose. In glycolysis, 3-phosphoglycerate is made by spiting fructose-1,6-bisphosphate. The phosphorylated carbons become the phosphorylated carbon of 3-phosphoglycerate. Thus, carbons 1 and 6 of glucose appear in carbon 3 of 3-phosphoglycerate, which becomes the β–carbon of serine and is lost when producing glycine. Carbon 2 of 3-phosphoglycerate derives from carbons 2 and 5 of glucose. This carbon becomes glycine's α-carbon. Finally, carbon 1 of 3-phosophoglycerate comes from carbons 3 and 4 of glucose. This carbon becomes the carboxyl carbon of 3-phosphoglycerate and the carboxyl carbon of glycine.

16. A deficiency in 3-phosphoglycerate dehydrogenase can affect amino acid metabolism

Although serine is a nonessential amino acid, serine deficiency syndrome has been observed in humans. One such form of the syndrome is traceable to a deficiency in 3-phosphoglycerate dehydrogenase (see Figure 25.31). Individuals with this syndrome not only are serine-deficient, but also are impaired in their ability to synthesize another common amino acid, as well as a class of lipids. Describe why.

Answer: Serine is an important precursor for a number of biologically active compounds including the amino acid glycine. A defect in phosphoglycerate dehydrogenase will cause a decrease in glycine synthesis from serine. Serine is a precursor for nucleotides and phospholipids as well. Sphingolipids are synthesized from serine and some glycerophospholipids have serine as a head group. Finally, L-serine is converted to D-serine by a racemase active in astrocytic glia cells. D-serine appears to bind and activate NMDA receptors. These receptors are activated by glutamate. However, glycine functions as a co-agonist as does D-serine.

17. Using the Protein Data Bank to explore the structure and function of glutamate synthase
Go to www.pdb.org and examine the PDB file 1LM1 for glutamate synthase. Find its iron–sulfur cluster and FMN prosthetic group. Discover how this enzyme is organized into an N-terminal domain that functions in ammonia removal from glutamine (the glutaminase domain) and the α-ketoglutarate–binding site near the Fe/S and flavin prosthetic groups. Consult van den Heuvel, R. H. H., et al., 2002. Structural studies on the synchronization of catalytic centers in glutamate synthase. Journal of Biological Chemistry 277: 24579–24583, to see how these two sites are connected by a tunnel for passage of ammonia from glutamine to α-ketoglutarate.

Answer: Glutamate synthase catalyzes the deamination of glutamine to glutamate and couples this with conversion of α-ketoglutarate to glutamate. Thus, the overall reaction is glutamine + α-ketoglutarate to 2 glutamate. The reaction occurs in three stages. First, glutamine is deaminated by hydrolysis to glutamate and ammonia. Ammonia is then channeled through the protein to a separate domain at which α-ketoglutarate is bound. Reaction of ammonia with α-ketoglutarate produces 2-imidoglutarate, which is subsequently reduced to glutamate by an electron donor (ferredoxin or NADPH or NADH, depending on enzyme source). The protein discussed in the reference uses reduced ferredoxin as a source of electrons. The enzyme is an iron-sulfur flavoprotein. Electrons are moved from ferredoxin, which binds non-covalently to the enzyme, to a 3Fe-4S cluster to a protein bound FMN cofactor. The protein is composed of four domains that are functionally and structurally distinct regions of a 1523 amino acid residue single polypeptide chain. The N-terminal domain, residues 1-422, is responsible for deamination of glutamine with its N-terminal cysteine responsible for hydrolysis of ammonia. Electron transport from ferredoxin to FMN via the iron/sulfur cluster is located in the FMN-bonding domain, residues 787-1239. The N-terminal domain and the FMN-binding domain are connected by the central domain, residues 423-786. Finally, the C-terminal domain, which includes the region from residues 1240 to 1507, makes numerous contacts with the N-terminal domain and the FMN-binding domain and these contacts are thought to stabilize the domains. In viewing the protein (1LM1) it should be an easy task to identify and highlight the N-terminal cysteine, which marks the deamination domain. The protein contains two heteromolecules, the iron/sulfur cluster and FMN, and these can also be highlighted using a visualization program of your choice. (I am using iMol but I use others as well.) In looking at the structure the tunnel must run between these two domains and through a small portion of the central domain. (The central domain mainly contacts the outer surface of the N-terminal domain and the FMN-binding domain.) A striking feature of the C-terminal domain is a barrel like structure dominated by β–sheets.

18. The energy cost of using dietary protein as food
The thermic effect of food is a term used to describe the energy cost of processing the food we eat, digesting it, and either turning it into precursors for needed biosynthesis, usable energy in the form of ATP, or storing the excess intake as fat. The thermic effect is usually approximated at 10% of the total calories consumed, but the thermic effect of fat is only 2% to 3% of total fat calories and the thermic effect of protein is 30% or more of calories consumed as protein. Why do you suppose dietary protein has a much higher thermic effect than either dietary carbohydrate or fat?

Answer: Proteins present numerous challenges to the process of food digestion. Proteins are first degraded into their amino acid building blocks in a complicated process that starts in the stomach with HCl and pepsin fragmenting proteins into small peptides. This is followed by conversion of peptides into amino acids in the small intestine, which involves a number of proteases including trypsin, chymotrypsin, aminopeptidases and carboxypeptidases. Once amino acids are liberated from proteins they must be transported across biological membranes. Uptake occurs through specific transporters that move relatively polar amino acids across hydrophobic membrane barriers. Finally, when amino acid supplies exceed demands for protein synthesis, the amino acids are broken down. The first step in this breakdown process is removal of nitrogen and incorporation of it into urea.

Each stage in amino acid utilization presents energy demands. For example, the use of proteases to convert proteins into amino acids requires synthesis of proteolytic enzymes as inactive precursors, which are activated by proteolysis and are ultimately consumed as any other protein during digestion. Metabolism of amino acids requires formation of urea to deal with nitrogen and this is an energy demanding process. Finally, some amino acids are converted into precursors that can be used in gluconeogenesis but even this is energy demanding. Fats, on the other hand, ultimately have one metabolic use: formation of acetyl units to be consumed in the citric acid cycle.

Preparing for the MCAT® Exam

19. From the dodecameric (α₁₂) structure of glutamine synthetase shown in Figure 25.14, predict the relative enzymatic activity of GS monomers (isolated α-subunits).

Answer: The structure of glutamine synthetase was reviewed in the Protein Data Bank, Molecule of the Month feature in 2002. (URL is below.) The protein forms an impressive structure that may be explored using visualization programs like RasMol and others. GS monomers are inactive in part because some active sites occur at the interface between two subunits.

http://www.rcsb.org/pdb/101/motm.do?momID=30

20. Consider the synthesis and degradation of tyrosine as shown in Figures 25.37, 25.38, and 25.48 to determine where the carbon atoms in PEP and erythrose-4-P would end up in acetoacetate and fumarate.

Answer: Figure 25.37 shows conversion of chorismate to phenylalanine and pyruvate. Chorismate is synthesized from PEP and erythrose-4-P, which is shown in Figure 25.36. Therefore, we should start there. PEP is actually used twice, once to form 5-dehydroshikimate and again to form 3-enolpyruvylshikimate-5-P. The α-carboxyl group, α-carbon and β-carbon of both phenylalanine and tyrosine derive from carbons 1, 2 and 3 of PEP. The ring carbons derive from PEP and erythrose-4-P. Starting with Figure 25.36, you can see that ring closure involves carbon 2 of pyruvate. This carbon becomes the γ-carbon of phenylalanine and tyrosine. Moving to Figure 25.37 you see that formation of both amino acids involves loss of carbon 1 of PEP by decarboxylation. Carbons of phenylalanine that derive from erythrose-4-P are shown as bold numerals. Carbons derived from PEP are numbered and labeled as carbons.

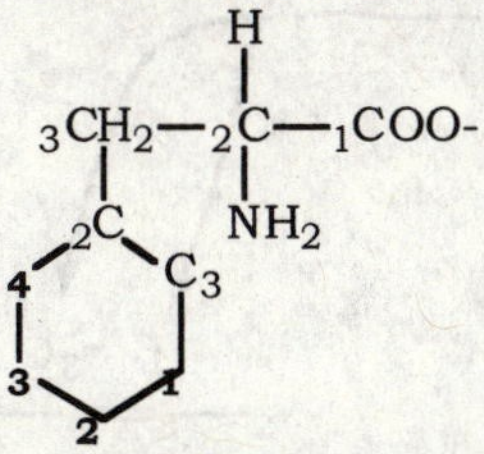

$$\underset{2C}{\overset{3CH_2 - 2C - 1COO^-}{|}}\quad \begin{array}{c} H \\ | \\ \\ NH_2 \end{array}$$

Degradation of phenylalanine and tyrosine is shown in Figure 25.48 (Phe is hydroxylated to Tyr, which is subsequently degraded.) The α-carbon is lost as CO_2. Formation of homogentisate is a rather involved reaction that moves the attachment of the β-carbon to a δ-carbon (labeled C₃ or **4** in the structure shown above.

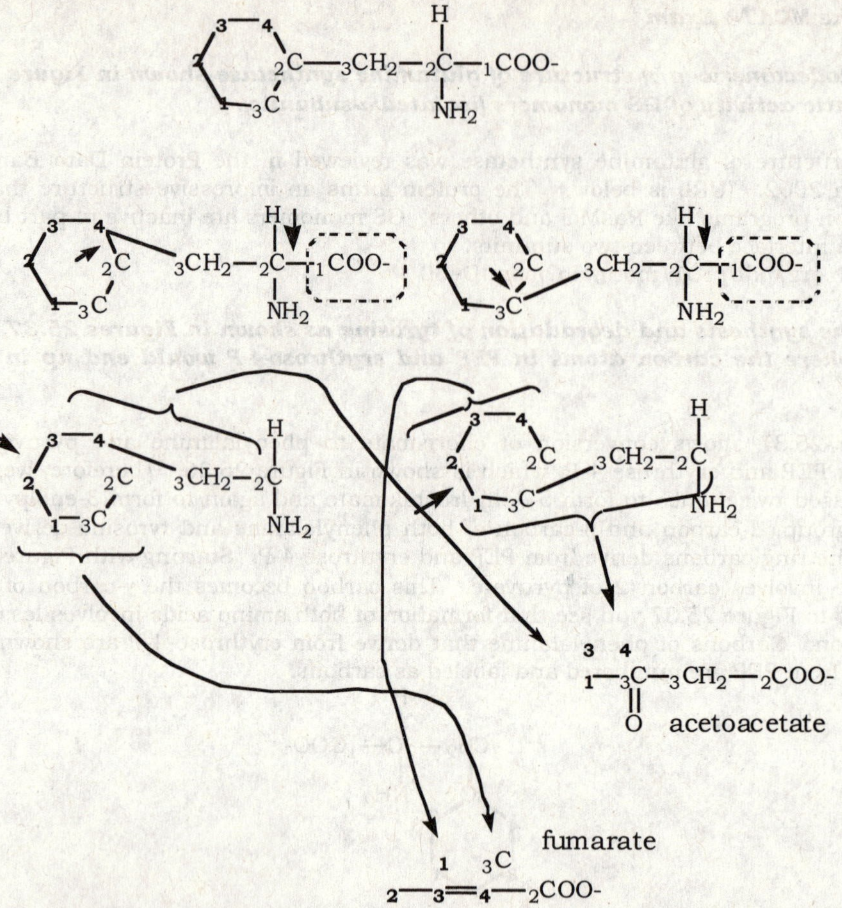

The bottom line is that acetoacetate is labeled on its carboxyl group from carbon 2 of PEP and on its β-carbon from carbon 3 of PEP. The carbonyl carbon receives label from either carbon 3 of PEP or carbon 4 of erythrose-4-P. Fumarate's carboxyl groups are labeled from carbon 2 of PEP and carbon 2 of erythrose-4-P. The two middle carbons receive label from carbon 1 or 3 of erthryose-4-P and carbon 4 of erythrose-4-P or carbon 3 of PEP as shown above.

Questions for Self Study

1. The two reactions below occur in the process of ___. In the first step nitrate is reduced to nitrite while in the second step nitrite is converted to ammonium. Step (1) is catalyzed by ___, and step (2) is catalyzed by ___.

$$NO_3^- + 2 H^+ + 2 e^- \star NO_2^- + H_2O$$
$$NO_2^- + 8 H^+ + 6 e^- \star NH_4^+ + 2H_2O$$

The enzyme, ___ , is responsible for catalysis of the following reaction:
$$N_2 + 8 H^+ + 8 e^- \star 2 NH_3 + H_2$$
in a process known as ___.

2. Ammonium is incorporated into organic compounds as a result of action of which three enzymes?

3. Write a two step reaction sequence leading from the citric acid cycle intermediate, α-ketoglutarate, to the amino acid glutamine and name the two enzymes involved.

4. Complete the following reaction, name the general enzyme type that catalyzes this kind of reaction, and identify the products.

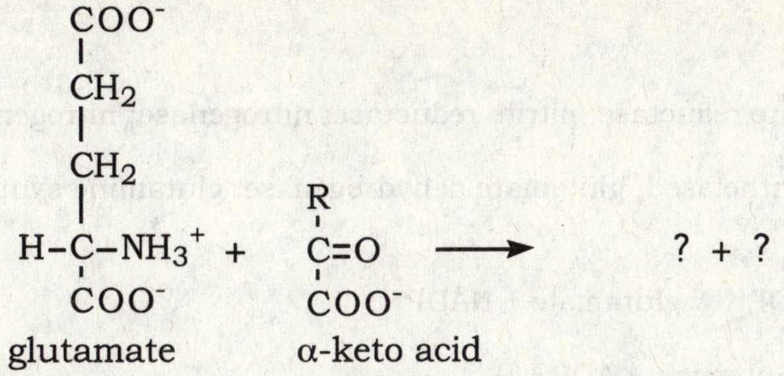

$$\underset{\text{glutamate}}{\overset{\displaystyle COO^-}{\underset{\displaystyle COO^-}{\overset{\displaystyle |}{\underset{\displaystyle |}{\overset{\displaystyle CH_2}{\underset{\displaystyle H-C-NH_3^+}{\overset{\displaystyle |}{\overset{\displaystyle CH_2}{\overset{\displaystyle |}{}}}}}}}}} \quad + \quad \underset{\text{α-keto acid}}{\overset{\displaystyle R}{\underset{\displaystyle COO^-}{\overset{\displaystyle |}{\overset{\displaystyle C=O}{\overset{\displaystyle |}{}}}}}} \quad \longrightarrow \quad ? \ + \ ?$$

5. Match the nonessential amino acids with their precursor molecules.

a. Glutamate   1. serine and H_2S
b. Glutamine   2. serine (one step)
c. Aspartate   3. phenylalanine
d. Asparagine   4. glutamate (one step)
e. Alanine   5. glutamate (several steps)
f. Serine   6. 3-phosphoglycerate
g. Glycine   7. pyruvate
h. Proline   8. aspartate
i. Tyrosine   9. α-ketoglutarate
j. Cysteine   10. oxaloacetate

6. Fill in the blanks. The 20 common amino acids are degraded into 7 metabolic intermediates. These include the citric acid cycle fuel ___, four citric acid cycle intermediates ___, ___, ___, and___, a ketone body___, and the end product of glycolysis___.

7. Identify the compound drawn below. Its synthesis and degradation convert an amino acid into fumarate and produce one of the 20 common amino acids. Name the amino acid consumed and produced. This compound is part of which metabolic cycle?

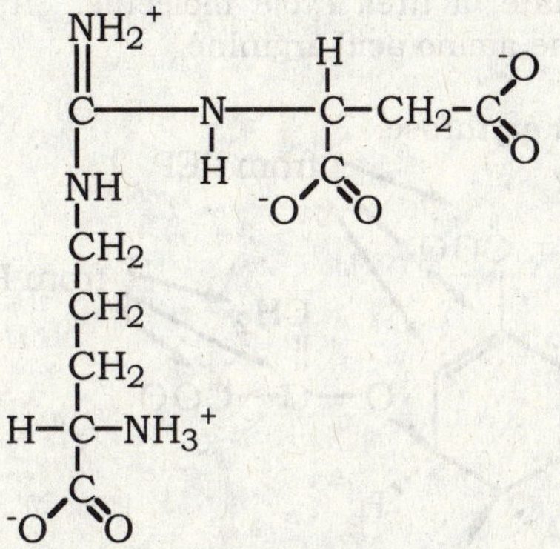

8. Chorismate is a key metabolic intermediate that is derived from erythrose-4-phosphate and two PEP in the shikimate pathway. Given that the pathway does not contain oxidation steps, indicate the carbons in chorismate that derive from PEP and from erythrose-4-phosphate.

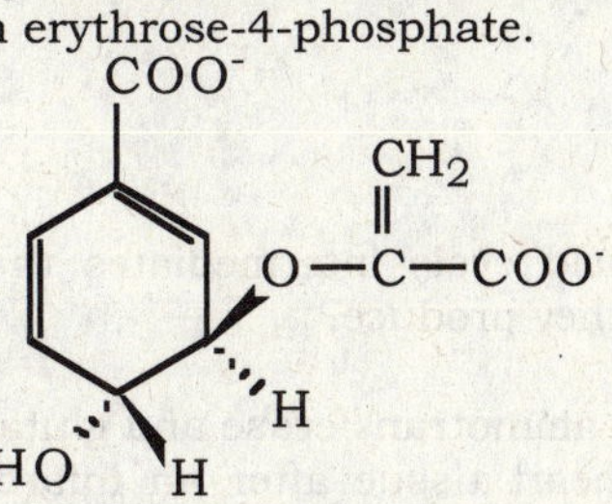

9. List four compounds that derive from chorismate.

Answers

1. Nitrate assimilation; nitrate reductase; nitrite reductase; nitrogenase; nitrogen fixation.

2. Carbamoyl-phosphate synthetase I, glutamate dehydrogenase, glutamine synthetase.

3. Glutamate dehydrogenase:
$NH_4^+ + \alpha$-ketoglutarate $+ NADPH \star$ glutamate $+ NADP^+$
Glutamine synthetase :
Glutamate $+ NH_4^+ + ATP \star$ glutamine $+ ADP + P_i$

4. Aminotransferases catalyze the following reaction.

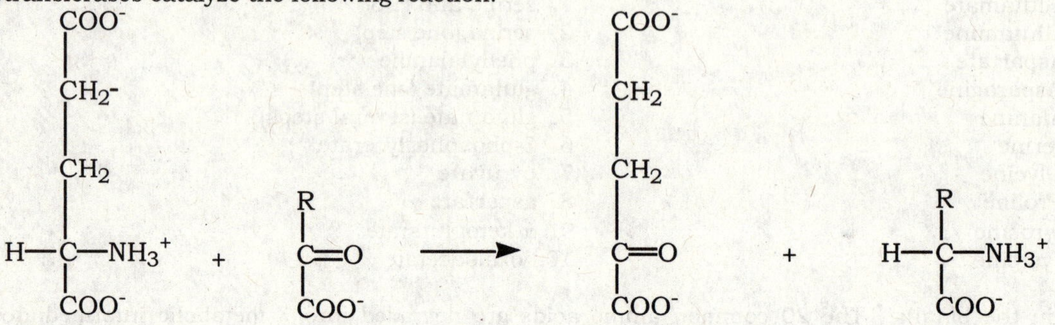

glutamate   α-keto acid   α-ketoglutatate   amino acid

5. a. 9; b. 4; c. 10; d. 8; e. 7; f. 6; g. 2; h. 5; i. 3; j. 1.

6. Acetyl-CoA; α-ketoglutarate; succinyl-CoA; fumarate; oxaloacetate; acetoacetate; pyruvate.

7. The compound is argininosuccinate, a urea cycle molecule. Its formation and degradation converts aspartate to fumarate and produces the amino acid arginine.

8. The unlabelled carbons derive from erythrose.

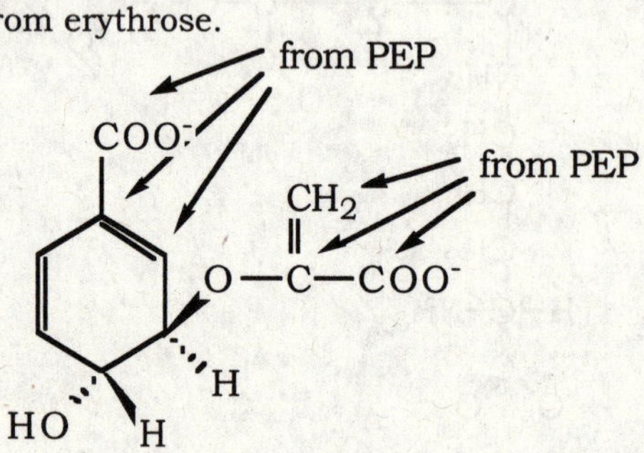

9. Any four of tryptophan, phenylalanine, tyrosine, Vitamins E and K, Coenzyme Q, plastoquinone, and folic acid.

Additional Problems

1. Draw the structure of the citric acid cycle intermediates that can be converted to amino acids by transamination. Which amino acids do they produce?

2. The two enzymes glutamate-pyruvate aminotransferase and glutamate-oxaloacetate aminotransferase were formerly used to diagnose damage to heart tissue after an infarction. By comparing the levels of these enzymes in serum drawn at various times after an incident, an estimate of the extent of heart muscle damage could be made. Can you suggest a simple coupled assays to measure the activity of these two enzymes?

3. The polyamine spermidine is produced from two amino acids. Given the structure of spermidine, can you suggest which two amino acids are involved?

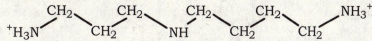

4. Describe the reaction catalyzed by glutamine synthetase, suggest a reaction mechanism, and explain why this reaction is favorable even in low concentrations of ammonium.

5. The essential amino acids in mammals include phenylalanine but not tyrosine. Draw the side chains of these two amino acids to convince yourself that tyrosine is in fact more complicated than phenylalanine in structure and explain why phenylalanine is required but tyrosine is not.

6. What amino acids are involved in the urea cycle?

7. Excrement from birds makes an excellent fertilizer. Why?

8. Would you expect aminotransferase activity to be sensitive to sodium borohydride ($NaBH_4$)? Explain.

Abbreviated Answers

1. Oxaloacetate is converted to aspartate and α–ketoglutarate is converted to glutamate. The structures of the α-keto acids are shown below.

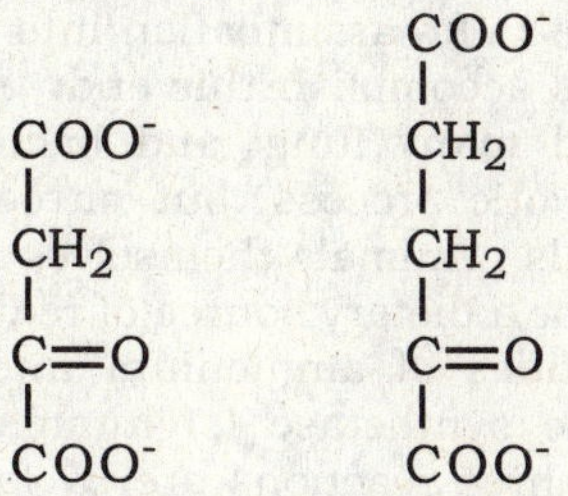

$$
\begin{array}{cc}
& \text{COO}^- \\
& | \\
\text{COO}^- & \text{CH}_2 \\
| & | \\
\text{CH}_2 & \text{CH}_2 \\
| & | \\
\text{C=O} & \text{C=O} \\
| & | \\
\text{COO}^- & \text{COO}^-
\end{array}
$$

oxaloacetate   α-ketoglutarate

2. Glutamate-pyruvate aminotransferase and glutamate-oxaloacetate aminotransferase catalyze the following reactions.

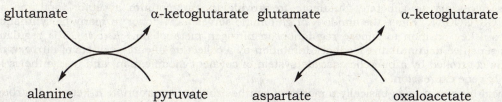

For glutamate-pyruvate aminotransferase, lactate dehydrogenase can be used to produce lactate from pyruvate with oxidation of NADH, which can be followed spectrophotometrically at 340 nm. The aminotransferase reaction can be forced toward pyruvate production by adding excess glutamate and alanine. In a similar manner, glutamate-oxaloacetate aminotransferase activity can be measured using malate dehydrogenase.

3. The butanediamine group, known as putrescine, is produced by decarboxylation of ornithine, a urea cycle intermediate. The aminopropane group derives from methionine via S-adenosylmethionine by decarboxylation and loss of 5'-methylthioadenosine.

4. Glutamine synthetase converts glutamate and ammonium to glutamine. The name synthetase implies that the enzyme uses ATP to drive the reaction. The reaction mechanism involves nucleophilic attack of a γ-carboxylate oxyanion on the γ-phosphorus of ATP to produce γ-glutamyl-phosphate intermediate. The phosphate group is subsequently displaced by ammonium to produce glutamine.

5. Phenylalanine can be converted to tyrosine by phenylalanine hydroxylase in an irreversible oxidation reaction involving molecular oxygen. The side chains of phenylalanine and tyrosine are:

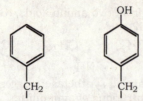

6. Arginine, ornithine, citrulline, and aspartate are involved in the urea cycle. Only arginine and aspartate are common amino acids found in proteins. Ornithine and citrulline are not incorporated into proteins during protein synthesis. (Citrulline was first isolated from watermelon, *Citrullus vulgaris*.)

7. Bird droppings contain nitrogen compounds like uric acid. Birds deal with excess nitrogen by converting it into highly water insoluble compounds that are excreted as a paste requiring minimum amounts of water.

8. The mechanism of action of aminotransferases involves a Schiff base intermediate on pyridoxal phosphate or Vitamin B6. The incoming amino acid is deaminated with the amino group being transferred to an enzyme-bound pyridoxal phosphate. This step involves Schiff base formation and sodium borohydride can reduce the Schiff base and inactivate the enzyme.

Summary

Nitrogen is a crucial macronutrient. In the common inorganic forms of nitrogen prevalent in the environment, nitrogen is in an oxidized state. Its assimilation into organic compounds requires reduction to the level of NH^+_4. Two metabolic pathways accomplish this end: nitrate assimilation and nitrogen fixation. Virtually all higher plants and algae and many fungi and bacteria are capable of nitrate assimilation. Nitrogen fixation is an exclusively prokaryotic process, but nitrogen-fixing bacteria do occur in symbiotic association with selected plants and animals. Animals themselves lack both of these pathways and thus are dependent on plants and microorganisms for a dietary source of reduced-N, principally in the form of protein.

The incorporation of significant quantities of ammonium into organic linkage occurs via only three enzymatic reactions, carbamoyl phosphate synthetase I, glutamate dehydrogenase (GDH) and glutamine synthetase (GS). The latter two of these three reactions are by far the most important. Two pathways of ammonium assimilation are prevalent. If $[NH^+_4]$ levels are adequate, the GDH/GS pathway operates so that one equivalent of α-ketoglutarate picks up two equivalents of NH^+_4 and glutamine is formed. If ammonium is limiting, GDH is inefficient, and thus, GS becomes the only assimilating reaction. Then, glutamate synthase (GOGAT) is necessary to generate glutamate to provide the ammonium acceptor, and the assimilatory pathway is GS/GOGAT. Since the amide-N of glutamine serves as N-donor in many biosynthetic pathways, GS occupies a key position to impose regulation on nitrogen metabolism. *E. coli* GS is regulated at three levels: it is sensitive to cumulative feedback inhibition by a collection of end products of nitrogen metabolism, its activity is controlled by a bi-cyclic cascade system of covalent modification, and its synthesis is regulated at the level of gene expression.

Amino acid biosynthesis is basically a matter of synthesizing the appropriate α-keto acid carbon skeletons for the various amino acids, followed by transamination of the α-keto acid by a glutamate-dependent aminotransferase. Plants and most micro-organisms can synthesize all of the 20 amino acids commonly found in proteins, but mammals lack the metabolic pathways for synthesis of 10 of these. It is a generality of nature that the longer the metabolic pathway, the more likely it is that the pathway has been lost over evolutionary time. ("Longer" here means "the greater the number of unique reactions intervening between central metabolic intermediates and the final end product"). The amino acids that mammals cannot synthesize are called "essential amino acids" to denote that it is essential for the mammal to obtain these amino acids in their diet. Amino acid biosynthetic pathways can be grouped according to shared precursors, as in the "α-ketoglutarate family of amino acids" (Glu, Gln, Pro, Arg and in some organisms, Lys) or "the 3-phosphoglycerate family" (Ser, Gly and Cys). The aromatic amino acids (Phe, Tyr and Trp) are synthesized from chorismate, a metabolic intermediate central to the biosynthesis of practically all common aromatic compounds. In addition, the histidine and the aromatic amino acid biosynthetic pathways share intermediates in common with the pathways for synthesis of purine and pyrimidine nucleotides.

Chapter 26

The Synthesis and Degradation of Nucleotides

• •

Chapter Outline

- ❖ Nucleotides
    - ⚊ NTPs: RNA biosynthesis
        - • ATP: Energy currency
        - • UTP: Carbohydrate metabolism
        - • CTP: Phospholipid metabolism
        - • GTP: Protein synthesis, signal transduction
    - ⚊ dNTPs: DNA biosynthesis
- ❖ Formation: Two options
    - ⚊ *De novo* synthesis
    - ⚊ Salvage pathway
- ❖ One-carbon pool
    - ⚊ Biotin: Used to move carboxylate carbons
    - ⚊ Folate: Used to move carbons at other oxidation states
        - • Active form tetrahydrofolate (THF): Produced by dihydrofolate reductase
        - • Carbon attached to N_5 N^{10} or both
        - • Carbon enters via serine hydroxymethyltransferase reaction
- ❖ Purine biosynthesis: IMP biosynthesis
    - ⚊ Synthesized on ribose-5-P
    - ⚊ Ribose-5-phosphate pyrophosphokinase reaction rate limiting: PRPP formed
        - • ADP, GDP inhibitors
    - ⚊ Glutamine phosphoribosyl pyrophosphate amidotransferase
        - • Committed step
        - • GMP, GDP, GTP inhibitory site
        - • AMP, ADP, ATP inhibitory site
        - • Azaserine inhibition: Antitumor agent
    - ⚊ GAR synthetase: Condensation of glycine via carboxyl group
    - ⚊ GAR transformylase: Uses formyl-THF
        - • Methotrexate inhibits
    - ⚊ FGAR amidotransferase (FGAM synthetase)
    - ⚊ AIR synthetase
        - • Formation of imidazole ring
        - • Activation of formyl group for second ring closure
    - ⚊ AIR carboxylase
    - ⚊ SAICAR synthetase
        - • This and previous enzyme on bifunctional polypeptide
    - ⚊ Adenylosuccinase (adenylosuccinate lyase) produces AICAR
        - • AICAR: Histidine biosynthesis
    - ⚊ AICAR transformylase: Uses formyl-THF: Produces FAICAR
        - • Methotrexate inhibits
    - ⚊ IMP cyclohydrolase: Dehydration and ring closure
        - • This and AICAR transformylase on bifunctional polypeptide

❖ IMP precursor of AMP and GMP
   ⚲ AMP production
      • Adenylosuccinate synthetase: GTP-dependent reaction
      • Adenylosuccinase: AMP and fumarate
   ⚲ GMP production
      • IMP dehydrogenase
      • GMP synthetase: ATP-dependent reaction
❖ NTP production
   ⚲ Adenylate kinase: (d)AMP + ATP → (d)ADP + ADP
   ⚲ Guanylate kinase (d)GMP + ATP → (d)GDP + ADP
   ⚲ Nucleoside diphosphate kinase: (d)NDP + ATP → (d)NTP + ADP
❖ Purine salvage
   ⚲ Adenosine phosphoribosyltransferase: APRT
   ⚲ Hypoxanthine-guanine phosphoribosyltransferase: HGPRT
      • Lesch-Nyhan syndrome defect
❖ Purine nucleotide degradation
   ⚲ Dietary nucleotides
      • Nucleotidases and phosphatases produce nucleosides
      • Nucleosidases or nucleoside phosphorylases produce bases
   ⚲ Intracellular nucleotides
      • Nucleotidase produces nucleosides
      • A and dA to inosine by adenosine deaminase
      • Purine nucleoside phosphorylase releases base as hypoxanthine
      • Hypoxanthine to xanthine by xanthine oxidase
      • Xanthine
         ● Hypoxanthine (from inosine) to xanthine by xanthine oxidase
         ● Guanine to xanthine by guanine deaminase
❖ Anaplerotic cycle: Fumarate production
   ⚲ AMP to IMP via adenosine deaminase: Degradation
   ⚲ IMP to AMP via adenylosuccinate synthetase and adenylosuccinase: Fumarate produced
❖ Uric acid: Metabolism/excretion
   ⚲ Primates: End product (but urea production accounts for nitrogen excretion also)
      • Excess levels leads to crystallization and gout
      • Allopurinol blocks xanthine oxidase
   ⚲ Birds, terrestrial reptiles, insects: Uric acid only pathway of nitrogen excretion
   ⚲ Mollusks and other mammals: Uric acid to allantoin by urate oxidase
   ⚲ Bony fish: Allantoin to allantoic acid by allantoinase
   ⚲ Sharks, rays, amphibians: Allantoic acid to glyoxylic acid and urea by allantoicase
   ⚲ Marine invertebrates: Convert urea to ammonia and carbon dioxide
❖ Pyrimidine biosynthesis: Precursors are carbamoyl-P and aspartate
   ⚲ Carbamoyl-P synthetase II
      • CPS-II and following two enzymes on multifunctional polyprotein: In mammals
      • UDP/UTP inhibitory, PRPP and ATP stimulatory
   ⚲ ATCase: Carbamoyl-P condensed with aspartate
      • Regulation
         ● *E. coli* enzyme inhibited by CTP, activated by ATP
   ⚲ Dihydroorotase: Ring closure
   ⚲ Dihydroorotate dehydrogenase
      • NAD^+-linked in bacteria
      • Quinone-linked in eukaryotes
   ⚲ Orotate phosphoribosyltransferase
      • Transferase and decarboxylase (below) on UMP synthase: In mammals single polypeptide
   ⚲ OMP decarboxylase to form UMP
❖ UTP and CTP formation
   ⚲ Nucleoside monophosphate kinase: UMP + ATP → UDP + ADP
   ⚲ Nucleoside diphosphate kinase: UDP + ATP → UTP + ADP

- CTP synthase: Animation of UTP to give CTP
- ❖ Pyrimidine degradation
    - U and C: β-Alanine, ammonium, carbon dioxide
    - T: β-Aminoisobutyric acid
- ❖ Deoxyribonucleotide biosynthesis
    - Ribonucleotide reductase
        - $(R1)_2(R2)_2$ subunit structure
            - R1: Two regulatory sites
                - o Overall activity site: ATP activates, dATP inhibits
                - o Substrate specificity site
                    - Bound ATP specifies UDP or CDP
                    - Bound dTTP specifies GDP
                    - Bound dGTP specifies ADP
        - NADPH source of electrons
            - Electrons moved to thioredoxin by thioredoxin reductase
- ❖ Thymine nucleotides
    - dUDP to dUTP to dUMP: dUTPase catalyzes last step; Or, dCMP to dUMP by dCMP deaminase
    - dUMP to dTMP by thymidylate synthase
        - 5-Fluorouracil converted to inhibitor or thymidylate synthase

Chapter Objectives

Purine Biosynthesis

Purine biosynthesis starts with 5-phosphoribosyl pyrophosphate and the complete pathway involves formation of nucleotide derivatives. N-9 is the first component of the purine ring to be added from the side chain amino group of glutamine. The five-membered, imidazole ring components include N-9 and a formylglycine residue introduced by amide linkage of glycine to N-9 followed by formylation. Before imidazole ring formation is completed, N-3 is incorporated from glutamine. Imidazole ring formation produces an amino derivative of imidazole ribotide or AIR. To complete synthesis of the purine inosine monophosphate, the precursor of ATP and GTP, two carbons and one nitrogen are needed. The first carbon is introduced as a carboxyl group, the remaining nitrogen derives from the α-amino group of aspartic acid. Aspartic acid is attached by an amide linkage to carboxylated AIR and the carbon skeleton is eliminated as fumarate. (A similar reaction occurs in the urea cycle.) Formylation of N-3, incorporated prior to imidazole-ring closure, and subsequent ring closure produces inosine monophosphate. Amination of IMP using the α-amino group of aspartate and releasing the carbon skeleton of aspartate as fumarate produces AMP. GMP is produced by oxidation of C-2 followed by amination using the side chain nitrogen of glutamine.

Purine Degradation

The common pathway for purine degradation is to produce free bases that are metabolized to xanthine, then uric acid, and depending on the organism, allantoin, allantoic acid, urea, or CO_2, H_2O and ammonia.

Purine Points to Consider

Know why methotrexate inhibits purine formation and why the sulfonamides, structural analogs of p-aminobenzoic acid (PABA), function as folic acid antagonists. Understand the role of phosphoribosyltransferases in purine salvage pathways and the consequence of HGPRT deficiency (Lesch-Nyhan syndrome). Gout is a consequence of excess uric acid production. Know why allopurinol is used to treat gout.

Pyrimidine Biosynthesis

In contrast to purine synthesis, pyrimidine ring formation is completed before being attached to ribose. Synthesis starts with production of carbamoyl-phosphate from bicarbonate and the side chain amino group of glutamine. (We encountered carbamoyl-phosphate synthesis in the urea cycle where it was produced in the mitochondria. Carbamoyl-P synthesis leading to pyrimidine biosynthesis is catalyzed by a second synthetase located in the cytosol.) The nitrogen and carbon of carbamoyl-phosphate account for two of the ring components of the six-membered pyrimidine ring. The remaining components derive from aspartate. Aspartic acid is carbamoylated by amide linkage of a carbamoyl group to the α-amino group of aspartic acid. This is followed by ring closure and oxidation to produce orotate (uracil-6-carboxylic acid, with the carboxyl

485

group from the α-carboxyl group of aspartic acid). Orotate is attached to a phosphoribosyl group and decarboxylated to form UMP.

Deoxyribonucleotide Biosynthesis

Ribonucleotide reductase catalyzes the conversion of NDP to dNDP without regard to base (see problem 6 for details on regulation of this important enzyme). The reducing power derives from thioredoxin, a small protein with reactive sulfhydryl groups that undergo oxidation-reduction cycles. The reducing power of thioredoxin is replenished by NADPH-dependent thioredoxin reductase. Thymine nucleotides are synthesized from dUMP by methylation with methylene-tetrahydrofolate as methyl-group donor.

Problems and Solutions

1. Metabolic origin of the atoms in purine and pyrimidine rings
Draw the purine and pyrimidine ring structures, indicate the metabolic source of each atom in the rings.

Answer: For purine biosynthesis:

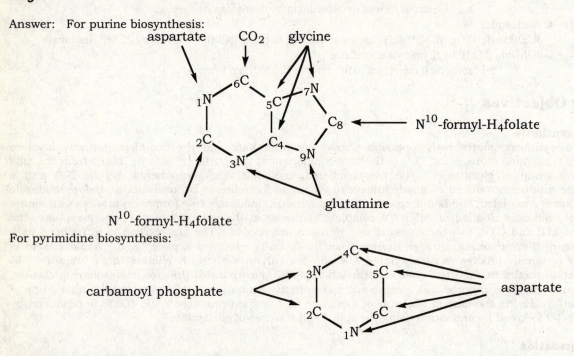

For pyrimidine biosynthesis:

2. The energy cost of nucleotide biosynthesis
Starting from glutamine, aspartate, glycine, CO_2 and N-formyl THF, how many ATP equivalents are expended in the synthesis of (a) ATP, (b) GTP, (c) UTP, and (d) CTP?

Answer: Synthesis of purines begins with pyrophosphorylation of ribose-5-phosphate using ATP to produce 5-phosphoribosyl-α-pyrophosphate (PRPP) and AMP, which means that two ATP equivalents have been expended. PRPP is then converted to glycinamide ribotide (GAR) in two steps, the first of which incorporates N-9 from glutamine with release of PP$_i$ and the second adds glycine (see answer to problem 1) with hydrolysis of ATP driving the reaction. Thus, these three steps account for 2 ATP ★ AMP + ADP + PP$_i$ + P$_i$ or, 3 ATP ★ 3 ADP + 3 P$_i$. GAR is converted to formylglycinamide ribotide (FGAR), which in two steps is converted to 5-aminoimidazole ribotide (AIR), with each step driven by ATP hydrolysis. So far we have accounted for 5 ATPs. AIR is then carboxylated and subsequently reacted with aspartate to produce 5-aminoimidazole-4-(N-succinylocarboxamide) ribotide (SACAIR), both reactions again dependent on ATP hydrolysis and accounting for the sixth and seventh ATP utilized. SACAIR is three steps away from inosine monophosphate (IMP), the precursor of both AMP and GMP.

a. To produce ATP from IMP, the amino group at C-6 is incorporated from aspartate with GTP hydrolysis driving the reaction and subsequent elimination of fumarate to produce AMP. Finally, AMP is phosphorylated to ATP at the expense of two high energy phosphate bonds. So, to produce ATP a total of 10 high energy phosphate bonds are hydrolyzed, 7 for production of IMP, 1 to produce AMP, and 2 to produce ATP. (If we start from ribose instead of ribose-5-phosphate we must account for an additional ATP.)

b. Production of GTP from IMP requires an oxidation step to convert IMP to xanthosine monophosphate (XMP), the direct precursor of AMP. This step generates reduced NAD⁺, which is equivalent to production of 3.0 high energy phosphate bonds.   Next, XMP is converted to GMP with ATP hydrolysis to AMP and PP$_i$ driving the reaction. This step is equivalent to hydrolysis of 2 ATPs. Finally GMP is converted to GTP with 2 more ATPs used. The overall synthesis of GTP consumes 8 ATP.

Pyrimidine synthesis begins with formation of carbamoyl phosphate from bicarbonate and the amino group of glutamine with hydrolysis of 2 ATP. Carbamoyl phosphate is reacted with aspartate to produce carbamoyl aspartate, which is subsequently converted to dihydroorate. Dihydroorotate is oxidized to orotate with accompanying reduction of NAD⁺, a step equivalent to +3.0 ATP. Orotate is converted to a nucleotide monophosphate, orotidine monophosphate (OMP). The sugar moiety derives from 5-phospho-α-D-ribosyl-1-pyrophosphate (PRPP). The pyrophosphate group of PRPP is from ATP; it is released upon OMP production. This then accounts for a net of 2 ATP equivalents. Decarboxylation converts OMP to UMP. The total ATPs consumed to produce UMP is 1.

c. To produce UTP from UMP, two ATPs are consumed. Thus, a total of 3 ATP are needed to synthesize UTP.

d. CTP formation requires an additional ATP when UTP is aminated, using glutamine and hydrolysis of ATP to drive the reaction. Thus, CTP production requires 4 ATP.

3. Allosteric regulation of purine and pyrimidine biosynthesis
Illustrate the key points of regulation in (a) the biosynthesis of IMP, AMP, and GMP; (b) E. coli pyrimidine biosynthesis; and (c) mammalian pyrimidine biosynthesis.

Answer: a. IMP is the precursor of both AMP and GMP and its production is regulated by both purine nucleotides. Regulation occurs in the first two steps catalyzed by ribose-5-phosphate pyrophosphorylase and glutamine phosphoribosyl-pyrophosphate amidotransferase.   Ribose-5-phosphate pyrophosphorylase is inhibited by both ADP and GDP. Glutamine phosphoribosyl-pyrophosphate amidotransferase regulation is similar, but more complex. This enzyme is sensitive to inhibition by AMP, ADP, and ATP at one site, and inhibition by GMP, GDP, and GTP at another site, and the enzyme is activated by PRPP. IMP is a branch point to either AMP or GMP. AMP production from IMP initiates with adenylosuccinate synthetase, an enzyme that is inhibited by AMP, the final product of the branch. In an analogous fashion, GMP production from IMP initiates with IMP dehydrogenase subject to GMP inhibition.

b. The first step in pyrimidine biosynthesis is the formation of carbamoyl phosphate in a reaction catalyzed by aspartate transcarbamoylase. In *E. coli*, ATCase is feedback inhibited by CTP and activated by ATP.

c. In contrast, regulation of pyrimidine biosynthesis in mammals occurs at the level of carbamoyl phosphate synthesis.   Both UTP and CTP inhibit carbamoyl phosphate synthetase II, whereas PRPP and ATP are allosteric activators.

4. Inhibition of purine and pyrimidine metabolism by pharmacological agents
Indicate which reactions of purine or pyrimidine metabolism are affected by the inhibitors (a) azaserine, (b) methotrexate, (c) sulfonamides, (d) allopurinol, and (e) 5-fluorouracil.

Answer: a. Azaserine (O-diazoacetyl-L-serine) and a related compound DON, 6-diaxo-5-oxo-L-norleucine are glutamine analogs. Their structures are shown on the next page.

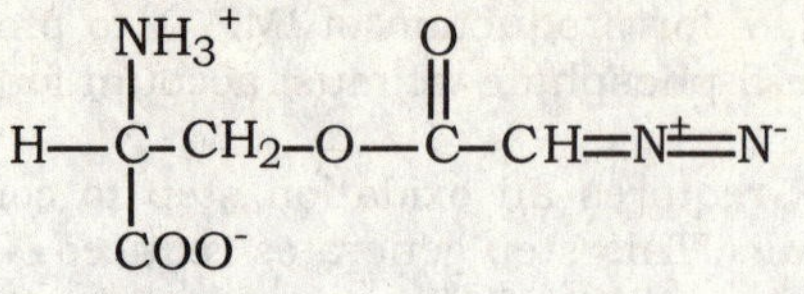

$$H-\overset{\overset{\displaystyle NH_3^+}{|}}{\underset{\underset{\displaystyle COO^-}{|}}{C}}-CH_2-O-\overset{\overset{\displaystyle O}{\|}}{C}-CH=N^{\pm}=N^-$$

Azaserine

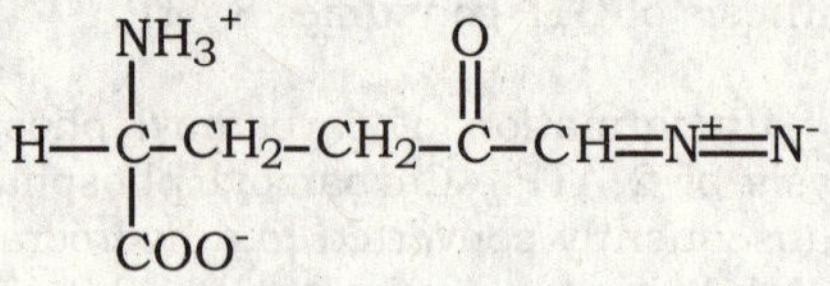

$$H-\overset{\overset{\displaystyle NH_3^+}{|}}{\underset{\underset{\displaystyle COO^-}{|}}{C}}-CH_2-CH_2-\overset{\overset{\displaystyle O}{\|}}{C}-CH=N^{\pm}=N^-$$

DON

$$H-\overset{\overset{\displaystyle NH_3^+}{|}}{\underset{\underset{\displaystyle COO^-}{|}}{C}}-CH_2-CH_2-\overset{\overset{\displaystyle O}{\|}}{C}-NH_2$$

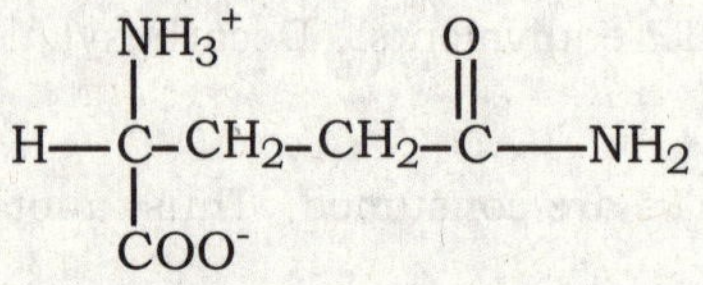

Glutamine

Azaserine and DON bind to glutamine-binding proteins and are capable of reacting with nucleophiles, leading to covalent modification and inactivation. The following reactions in purine synthesis are sensitive to these inhibitors

1. PRPP + glutamine + H_2O ⋆ β-5-phosphoribosylamine + glutamate + PP$_i$
catalyzed by glutamine:PRPP amidophosphoribosyl transferase.
2. FGAR + ATP + glutamine + H_2O ⋆ FGAM + ADP + glutamate + P$_i$
catalyzed by FGAM synthetase.
3. XMP + glutamine + ATP + H_2O ⋆ GMP + glutamate + AMP + PP$_i$
catalyzed by GMP synthetase.

For pyrimidine synthesis, there one reaction sensitive to these inhibitors.

1. UTP + glutamine + ATP + H_2O ⋆ CTP + glutamate + ADP + P$_i$
catalyzed by CTP synthetase.

b. Methotrexate is an analog of dihydrofolate that will compete, with extremely high affinity, for folate binding sites on enzymes. The following reactions in purine biosynthesis will be affected.

1. GAR + N^{10}-formyl-THF ⋆ FGAR + THF
catalyzed by GAR transformylase.
2. AICAR + N^{10}-formyl-THF ⋆ FAICAR + THF
catalyzed by AICAR transformylase.

In pyrimidine synthesis, methotrexate will block the formation of dTMP produced in the following reaction:

1. dUMP + N^5,N^{10}-methylene-THF ⋆ dTMP + DHF
catalyzed by thymidylate synthase.

c. Sulfonamides are structural analogs of para-aminobenzoic acid and they competitively inhibit production of folic acid in bacteria. Folic acid is a precursor of THF and the same steps indicated above for methotrexate inhibition will be affected by sulfonamides. However, inhibition will occur because of an inability of bacteria to produce folic acid; inhibition is a result of a lack of substrate as opposed to the presence of an inhibitor of the enzymes for these steps. Since folic acid is a dietary requirement for animals, sulfonamides will not have an effect on nucleotide biosynthesis in animals.

d. Allopurinol is a suicide inhibitor of xanthine oxidase. Xanthine oxidase will hydroxylate allopurinol to form alloxanthine, which remains tightly bound to the enzyme and thus inactivates it. This enzyme plays a key role in degradation of purines. Both adenosine and guanosine are metabolized to xanthine, which is subsequently converted to uric acid by xanthine oxidase. Uric acid formation is blocked.

e. 5-Fluorouracil in and of itself is not an important inhibitor of nucleotide metabolism but it is capable of being converted to 5-fluorodeoxyuridylate (FdUMP), which is a potent inhibitor of thymidylate synthase.

5. The metabolic role of dUDP
Since dUTP is not a normal component of DNA, why do you suppose ribonucleotide reductase has the capacity to convert UDP to dUDP?

Answer: Thymidine, a component of DNA, is produced by methylation of dUMP by the enzyme thymidylate synthase. dUMP is derived from dUTP through the action of dUTP diphosphohydrolase in the following reaction:

$$dUTP + H_2O \star dUMP + PP_i$$

Thus, to produce dUMP a source of dUTP is required. Ribonucleotide reductase converts NDP to dNDP without regard to base. Thus, dUDP is produced and can be phosphorylated to dUTP. Ultimately UDP is the precursor to dTTP, which is one of the deoxyribonucleotide triphosphates required for DNA synthesis.

6. Allosteric regulation of ribonucleotide reductase by ATP and deoxynucleotides
Describe the underlying rationale for the regulatory effects exerted on ribonucleotide reductase by ATP, dATP, dTTP, and dGTP.

Answer: The diagram below shows the various levels of nucleotide regulation of ribonucleotide reductase. By starting with ATP bound to the overall specificity site and following the arrows we see that the enzyme first produces dCDP and dUTP which leads to increased dTTP which binds to the substrate specificity site by competing with ATP. dTTP leads to production of dGTP, which in turn leads to dATP and enzyme inactivation.
ATP thus sets off this cascade of changing activity.

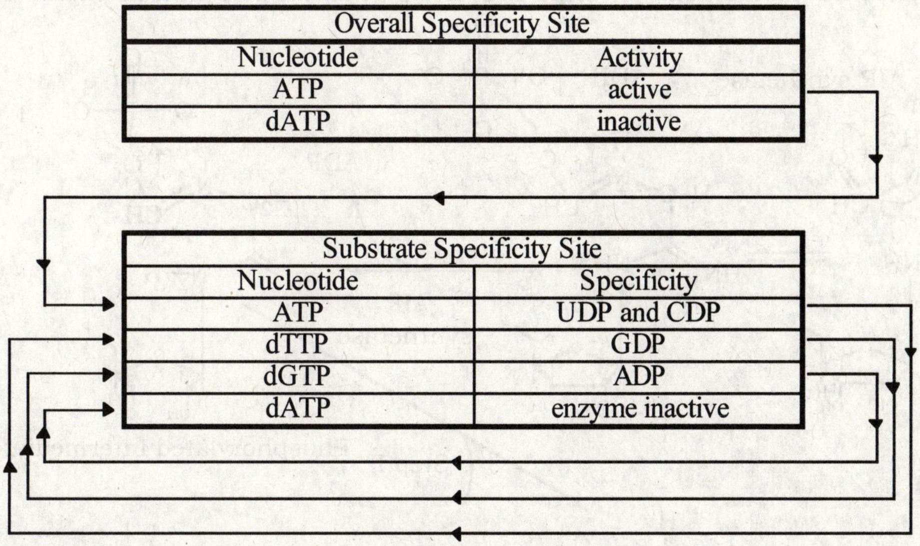

Overall Specificity Site	
Nucleotide	Activity
ATP	active
dATP	inactive

Substrate Specificity Site	
Nucleotide	Specificity
ATP	UDP and CDP
dTTP	GDP
dGTP	ADP
dATP	enzyme inactive

7. Nucleotide catabolism as a source of metabolic energy
(Integrates with Chapters 18-20 and 22.) By what pathway(s) does the ribose released upon nucleotide degradation enter intermediary metabolism and become converted to cellular energy? How many ATP equivalents can be recovered from one equivalent of ribose?

Answer: Ribose is metabolized in the pentose phosphate pathway, glycolysis, and the citric acid cycle. To enter the pentose phosphate pathway, ribose is phosphorylated at the expense of ATP and converted to ribulose-5-phosphate, which is metabolized as follows. Ribulose-5-phosphate is used to produce triose phosphate equivalents that are fed through glycolysis into the citric acid cycle. Thus, 3 ribose units follow this route and will produce 5 triose phosphates. Starting with 3 ribose (or a total of 15 carbons), the nonoxidative phase of the pentose phosphate pathway is used to produce a total of 5 triose phosphates. For example, using ribulose-5-phosphate isomerase, ribulose-5-phosphate epimerase, transketolase, and transaldolase, three ribose-5-phosphates are converted to two fructose-6-phosphate and glyceraldehyde-3-phosphate. To convert fructose-6-phosphate to glyceraldehyde-3-phosphate, ATP is consumed. Thus, to this stage, a total of 5 ATP are consumed; three to phosphorylate ribose initially* and two additional ATP to metabolize fructose-6-phosphate.
Net to this stage: -5 ATPs

The 5 glyceraldehyde-3-phosphates are metabolized to pyruvate with production of 5 NADH, equivalent to 15 ATP, and 10 ATP directly for a total of 25 ATP. Pyruvate metabolism accounts for 1 NADH to form acetyl CoA, and 1 GTP, 3 NADH, and 1 FADH$_2$ in the citric acid cycle for a total of 15 ATP per pyruvate. So, 5 pyruvates yield 75 ATP.

The total is -5 + 25 + 75 = 95 ATP per 3 ribose or 31.7 ATP per ribose.

* Starting with ribose-1-phosphate the net would be -2 ATP instead of -5 ATP. The extra expense of 3 ATPs is avoided by rearranging ribose-1-phosphate to ribose-5-phosphate.

8. Convergence of purine and histidine biosynthesis
(Integrates with Chapter 25.) **At which steps does the purine biosynthetic pathway resemble the pathway for biosynthesis of the amino acid histidine?**

Answer: The biosynthesis of histidine begins with 5-phosphoribosyl-α-pyrophosphate (PRPP), which is also used for purine biosynthesis. Thus, the very first step in both biosynthetic pathways is formation of PRPP by ribose phosphate pyrophosphokinase. After this reaction, the pathways diverge. However, a byproduct of histidine biosynthesis is also an intermediate in purine biosynthesis. In histidine biosynthesis, imidazole glycerol phosphate is produced from N^1-5'-phosphoribulosylformimino-5-aminoimidazole-4-carboxamide ribonucleotide and glutamine by glutamine amidotransferase. A byproduct of this reaction is 5-aminoimidazole-4-carboxamide ribonucleotide or AICAR, an intermediate in IMP biosynthesis.

9. The mechanism of action of purine biosynthetic enzymes
Write reasonable chemical mechanisms for steps 6, 8, and 9 in purine biosynthesis (Figure 26.3).

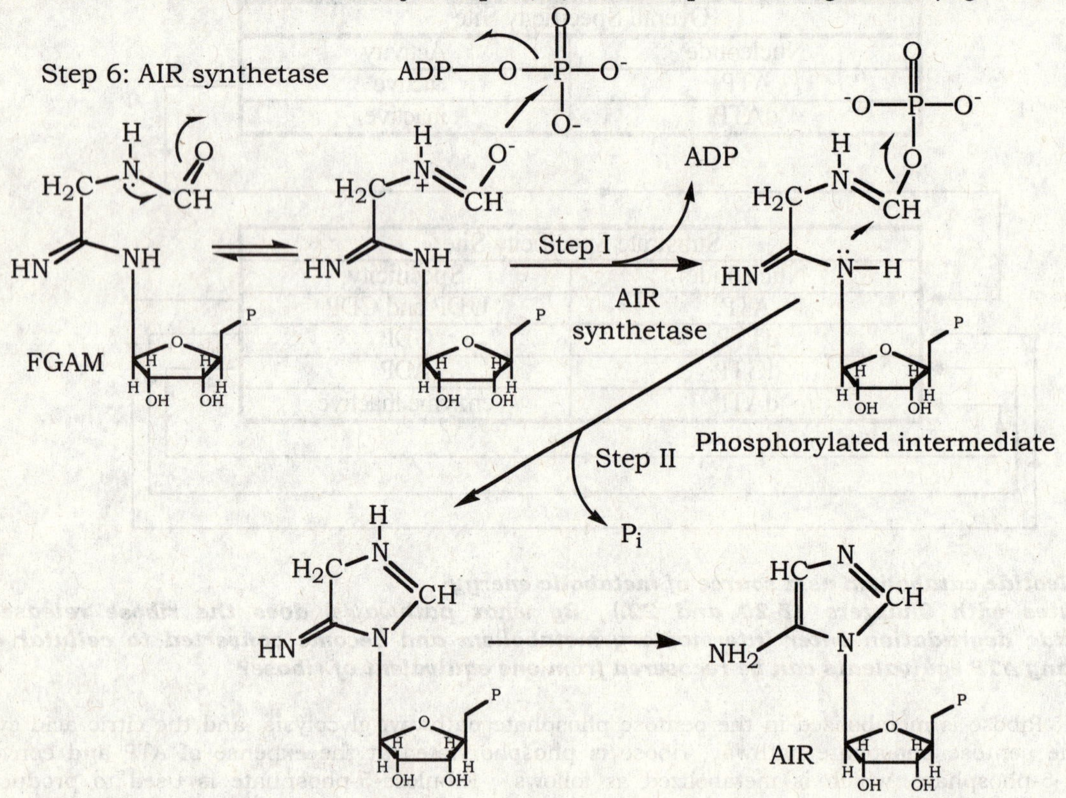

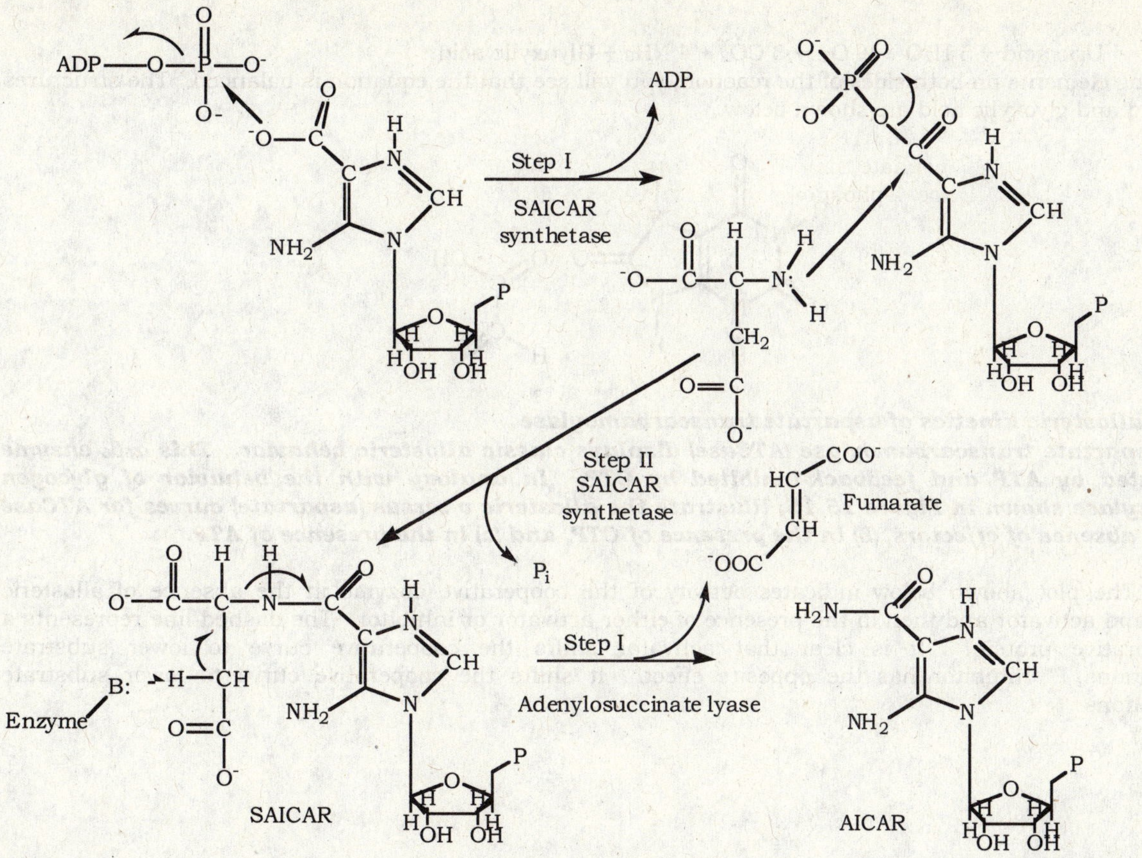

10. The anaplerotic purine nucleoside cycle of skeletal muscle
Write a balanced equation for the conversion of aspartate to fumarate by the purine nucleoside cycle in skeletal muscle.

Answer: The sequence of reactions that converts aspartate to fumarate is shown in Figure 26.9. The amino group of aspartate is used to convert IMP to AMP, a transamination reaction with IMP serving as the keto acceptor. AMP is subsequently deaminated by AMP deaminase to release ammonium and IMP. The reactions are shown below.

$$\text{IMP} + \text{Aspartate} + \text{GTP}^{4-} \rightarrow \text{Adenylosuccinate} + \text{GDP}^{3-} + \text{P}_i^{2-} + \text{H}^+ \text{ (Adenylosuccinate synthetase)}$$
$$\text{Adenylosuccinate} \rightarrow \text{Fumarate} + \text{AMP} \text{ (Adenylosuccinase)}$$
$$\text{AMP} + \text{H}_2\text{O} \rightarrow \text{IMP} + \text{NH}_3 \text{ (AMP deaminase)}$$
$$\text{NH}_3 + \text{H}^+ \rightarrow \text{NH}_4^+$$

Net:
$$\text{Asparate} + \text{GTP}^{4-} + \text{H}_2\text{O} \rightarrow \text{Fumarate} + \text{GDP}^{3-} + \text{P}_i^{2-} + \text{NH}_4^+$$

11. The catabolism of uric acid
Write a balanced equation for the oxidation of uric acid to glycolic acid, CO_2, and NH_3, showing each step in the process and naming all of the enzymes involved.

Answer: The conversion of uric acid to glycolic acid and ammonia is shown in Figure 26.12. Uric acid is converted to allantoin, which is converted to allantoic acid. Allantoic acid is then converted to glyoxylic acid and urea, which is broken down to carbon dioxide and water. The sequence is shown below.

$$\text{Uric acid} + 2\ \text{H}_2\text{O} + \text{O}_2 \rightarrow \text{Allantoin} + \text{C}_2\text{O} + \text{H}_2\text{O}_2 \text{ (Urate oxidase)}$$
$$\text{Allantoin} + \text{H}_2\text{O} \rightarrow \text{Allantoic acid} \text{ (Allantoinase)}$$
$$\text{Allantoic acid} + \text{H}_2\text{O} \rightarrow \text{Glyoxylic acid} + 2\ \text{Urea} \text{ (Allantoicase)}$$
$$2\ \text{Urea} + 2\ \text{H}_2\text{O} \rightarrow 2\ \text{CO}_2 + 4\ \text{NH}_3 \text{ (Urease)}$$

To get rid of hydrogen peroxide, we will use hydrogen peroxide oxidoreductase, which reacts two moles of hydrogen peroxide to produce water and oxygen as shown below.

$$\text{H}_2\text{O}_2 \rightarrow \text{H}_2\text{O} + {}^1/_2\ \text{O}_2 \text{ (Hydrogen peroxidase oxidoreductase)}$$

Net: Uric acid + 5 H_2O + ½ O_2 → 3 CO_2 + 4 NH_3 + Glyoxylic acid

By counting elements on both sides of the reaction, you will see that the equation is balanced. The structures of uric acid and glyoxylic acid are shown below.

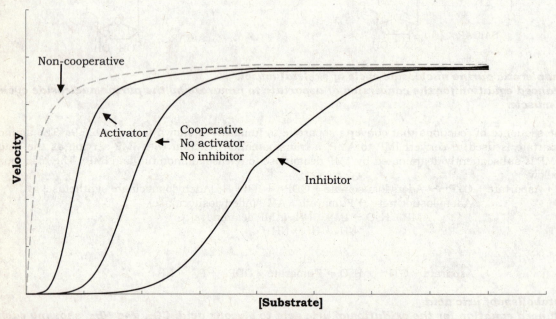

12. The allosteric kinetics of aspartate transcarbamoylase

E. coli aspartate transcarbamoylase (ATCase) displays classic allosteric behavior. This $\alpha_6\beta_6$ enzyme is activated by ATP and feedback-inhibited by CTP. In analogy with the behavior of glycogen phosphorylase shown in Figure 15.15, illustrate the allosteric v versus [aspartate] curves for ATCase (a) in the absence of effectors, (b) in the presence of CTP, and (c) in the presence of ATP.

Answer: The plot shown below indicates activity of the cooperative enzyme in the absence of allosteric inhibitor and activator and then in the presence of either activator or inhibitor. The dashed line represents a non-cooperative protein.   It is clear that activator shifts the cooperative curve to lower substrate concentrations.   Inhibitor has the opposite effect.   It shifts the cooperative curve to lower substrate concentrations.

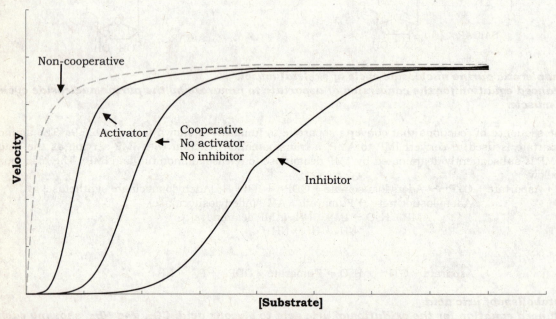

13. The functional organization of a heteromeric allosteric enzyme

(Integrates with Chapter 15.)   Unlike its allosteric counterpart glycogen phosphorylase (an α_2 enzyme), E. coli ATCase has a heteromeric ($\alpha_6\beta_6$) organization. The α-subunits bind aspartate and are considered catalytic subunits, whereas the β-subunits bind CTP or ATP and are considered regulatory subunits. How would you describe the subunit organization of ATCase from a functional point of view?

Answer: Since each catalytic subunit should be associated with a regulatory subunit we could consider ATCase as a hexamer of αβ heterodimers or ($\alpha\beta$)$_6$. When we discussed hemoglobin its structure suggests it is a dimer of heterodimers or ($\alpha\beta$)$_2$.

14. The energy cost of dTTP synthesis
(Integrates with Chapter 20.) Starting with HCO₃⁻, glutamine, aspartate, and ribose-5-P, how many ATP equivalents are consumed in the synthesis of dTTP in a eukaryotic cell, assuming dihydroorotate oxidation is coupled to oxidative phosphorylation? How does this result compare with the ATP costs of purine nucleotide biosynthesis calculated in problem 2?

Answer: Starting with Figure 26.15, **two ATP** are consumed by carbamoyl phosphate synthetase II to produce carbamoyl-P from bicarbonate and glutamine. From carbamoyl-P to dihydroorotate no additional ATP is needed. Oxidation of dihydroorotate to orotate is at the expense of reduction of CoQ to CoQH₂. When reoxidized, CoQH₂ should support **1.5 ATP** produced. To convert orotate, a pyrimidine, into a nucleoside 5-phosphoribosyl-α-pyrophosphate is needed.   To produce this from ribose-5-P Ribose-5-phosphate pyrophosphokinase is needed, which consumes ATP and produces PRPP and AMP. Production of OMP by oratate phosphoribosyltransferase releases PPᵢ, which we can anticipate being hydrolyzed to 2 Pᵢ. Thus, **two ATP** are used to produce OMP. Decarboxylation of OMP produces UMP.

UMP could be converted to UTP at the expense of **two ATP**. UTP to CTP using CTP synthetase consumes **one ATP**. To make dTTP from CTP, CDP is produced from CTP regaining **one ATP**. CDP is then converted to dCDP using NADPH, which is equivalent to **4 ATP**. dCDP could then produce **one ATP** by being converted to dCMP, which may be converted to dUMP by dCMP deaminase and finally dTMP by thymidylate synthase. Finally, dTMP is phosphorylated twice to dTTP using **two ATP**. The net is **7.5 ATP**. Purine biosynthesis accounts for 10 ATP.

15. The chemistry of dTMP synthesis and the substrate-binding site of thymidylate synthase
Write a balanced equation for the synthesis of dTMP from UMP and N⁵,N¹⁰-methylene-THF. Thymidylate synthase has four active-site arginine residues (Arg²³, Arg¹⁷⁸ʹ, Arg¹⁷⁹ʹ, and Arg²¹⁸) involved in substrate binding. Postulate a role for the side chains of these Arg residues.

Answer: UMP is first converted to UDP by nucleoside monophosphate kinase. UDP then serves as substrate for ribonucleotide reductase, which converts UDP to dUDP. To produce dTMP, dUDP is phosphorylated to dUTP, dUTP converted to dUMP by pyrophosphorylase and finally, dUMP to dTMP by thymidylate synthase, which uses N⁵,N¹⁰-methylene-THF. The reactions are shown below:

$$\text{UMP} + \text{ATP} \rightarrow \text{UDP} + \text{ADP}$$
$$\text{UDP} + \text{NADPH} + \text{H}^+ \rightarrow \text{dUDP} + \text{NADP}^+ + \text{H}_2\text{O}$$
$$\text{dUDP} + \text{ATP} \rightarrow \text{dUTP} + \text{ADP}$$
$$\text{dUTP} + \text{H}_2\text{O} \rightarrow \text{dUMP} + \text{PP}_i$$
$$\text{PP}_i + \text{H}_2\text{O} \rightarrow 2\ \text{P}_i$$
$$\text{dUMP} + \text{N}^5,\text{N}^{10}\text{-methylene-THF} \rightarrow \text{dTMP} + \text{DHF}$$

Net:
$$\text{UMP} + 2\ \text{ATP} + \text{H}_2\text{O} + \text{NADPH} + \text{H}^+ + \text{N}^5,\text{N}^{10}\text{-methylene-THF} \rightarrow$$
$$\text{dTMP} + 2\ \text{ADP} +\ 2\ \text{P}_i + \text{NADP}^+ + \text{DHF}$$

One might guess that the role of positively-charged arginine is to coordinate negatively-charged groups and this is apparently what happens in thymidylate synthase. The enzyme functions as a homodimer and the phosphate moiety of dUMP seems to be coordinated by Arg²³ and Arg²¹⁸ on one protomer (monomer) and Arg¹⁷⁸ and Arg¹⁷⁹ from the other protomer. Notice that in the question 178 and 179 are primed i.e., 178' and 179'. This indicates that they are not on the same subunit as arginine 23 and 218. For a discussion of the role of these arginine residues see "Replacement set mutagenesis of the four phosphate-binding arginine residues of thymidylate synthase" by Kawase, S. et al. in Protein Engineering 13:557-563.

16. Representatives of the PRT family of enzymes in purine synthesis
Enzymes that bind 5-phosphoribosyl-1α-pyrophosphate (PRPP) have a common structural fold, the PRT fold, which unites them as a structural family. PRT here refers to the phosphoribosyl transferase activity displayed by some family members. Typically, in such reactions, PPᵢ is displaced from PRPP by a nitrogen-containing nucleophile. Several such reactions occur in purine metabolism. Identify two such reactions where the enzyme involved is likely to be a PRT family member.

Answer: A convenient way of identifying enzymes that bind PRPP is to use KEGG: Kyoto Encyclopedia of Genes Genomes at http://www.genome.jp where you can do a search against all databases for "PRPP". On the returned page focus on hits in the KEGG Compound database, which should include C00119. Activating this will return an information sheet on PRPP including structure, reactions in which it participates, pathways in which it is found and enzymes with which it reacts. Activating E.C. numbers will identify each enzyme.   Adenine phosphoribosyl transferase (E.C. 2.4.2.7) and hypoxanthine-guanine phosphoribosyl transferase (E.C. 2.4.2.8) are both involved in purine metabolism as is glutamine phosphoribosyl pyrophosphate amidotransferase (E.C. 2.4.2.14).   Ribose-5-phosphate pyrophosphokinase (E.C. 2.7.6.1) is also involved in purine metabolism (and the pentose phosphate pathway).   In pyrimidine metabolism, orotate phosphoribosyltransferase uses PRPP as a substrate.   Additionally there are enzymes involved in histidine metabolism, aromatic amino acid biosynthesis and nicotinamide metabolism that use PRPP as substrate.

17. Explore the structure of DHFR reductase
The crystal structure of E. coli dihydrofolate reductase (DFR) with NADP⁺ and folate bound can be found in the Protein Data Bank (www.rcsb.org/pdb) as ID 7DFR. Go to this website, enter "7DFR" in the search line, and click on "KiNG" under "Display options" when the 7DFR page comes up. Explore the KiNG graphic of the DFR structure to visualize how its substrates are bound. (If you hold down the left button on your mouse and move the cursor over the image, you can rotate the structure to view it from different perspectives.) Note in particular the spatial relationship between the nicotinamide ring of NADP⁺ and the pterin ring of folate. Do you now have a better appreciation for how this enzyme works? Note also the location of polar groups on the two substrates in relation to the DFR structure.

Answer: The binding sites for dihydrofolate is a hydrophobic pocket whose bottom is formed by a parallel pleated sheet. The protein's overall structure is dominated by a eight-stranded sheet structure, with the last two strands being antiparallel and the rest parallel. In the region forming the dihydrofolate binding pocket the parallel beta sheet strands are joined by crossover structures with helices and random coils that form an opening to the pocket. Dihydrofolate is buried in this pocket with its glutamic acid residue on the surface of the protein and exposed to solvent. The binding site for NADP⁺ is located below the sheet and an opening between two strands of the sheet allows access by the nicotinamide ring of NADP⁺ to the hydrophobic interior of the dihydrofolate binding pocket. The rest of the NADP⁺ (essentially ADP) is bound on the surface of the enzyme on the side opposite of the glutamic acid residue of dihydrofolate. The nicotinamide ring is in close proximity to dihydrofolate's two fused 6-membered rings and this places the hydride donor of the nicotinamide ring close to the 7-position of folate, which is the hydride acceptor.

18. Explore the structure of aspartate transcarbamoylase, an allosteric enzyme
E. coli aspartate transcarbamoylase is an allosteric enzyme (see problem 12) composed of six catalytic (C) subunits and six regulatory (R) subunits. Protein Data Bank file 1RAA shows one-third of the ATCase holoenzyme (two C subunits and two R subunits; CTP molecules are bound to the R subunits). Explore this structure using the KiNG display option. What secondary structural motif dominates the R subunit structure? Protein Data Bank file 2IPO also shows one-third of the ATCase holoenzyme (two C subunits and two R subunits), but in this structure molecules of the substrate analog N-(2-phosphonoacetyl)-L-asparagine are bound to the C subunits. Explore this structure using the KiNG display option. Note the distance separating the ATCase active site from its allosteric site. Interpret what you see in terms of the Monod–Wyman–Changeux model for allosteric regulation (see Chapter 15). Which of these structures corresponds to the MWC R-state, and which corresponds to the T-state?

Answer: The portion of aspartate transcarbamoylase shown in 1RAA includes two regulatory subunits and two catalytic subunits. The regulatory subunits (chains B and D) each have a zinc atom and CTP bound at opposite ends of the protein. The CTP binding site is located on one end of the subunit, the end facing away from the catalytic subunit interface. The zinc atom is located on the end of the regulatory subunit that is in contact with the catalytic subunit. The CTP binding site is a beta sheet structure composed of two parallel strands and one antiparallel strand. These are joined by two more parallel strands to form a five-stranded sheet with CTP binding to one edge of the sheet and to the face of the sheet exposed to solvent. The other side of the sheet has two α-helices and on top of them is a four-stranded sheet. The outer surface of this sheet has the zinc binding site. The structure appears to have three layers: sheet/helix/sheet.

In 2IPO we are also shown a structure with two catalytic and two regulatory subunits. In addition to protein this structure contains zinc and N-2-phosphonoacetyl-L-asparagine, a substrate analog bound to the active site of the catalytic domain. The active site at which the substrate analog is bound is located on the surface of the subunit on the opposite side of the regulatory subunit interface. From the description of the regulatory subunit given above it should be clear that the CTP binding site and the substrate binding site are on opposite ends of the protein and thus communication between these sites must involve conformational changes transmitted through the whole protein, from one end of a regulatory subunit to one end of a catalytic subunit.

The structure 2IPO contains substrate (actually an analog) and so this must represent the R (relaxed) state. The structure shown in 1RAA is the T state.

Preparing for the MCAT® Exam

19. Examine Figure 26.6 and predict the relative rates of the regulated reactions in the purine biosynthetic pathway from ribose-5-P to GMP and AMP under conditions in which GMP levels are very high.

Answer: Since the pathway produces both GMP and AMP we would expect it to respond to high GMP levels by blocking production of GMP and allowing a shift of resources into AMP production. The obvious place to accomplish this is at IMP, which may be converted into either GMP or AMP. High GMP, however, inhibits IMP dehydrogenase, the enzyme that converts IMP to XMP, which leads to GMP production. High GMP also has an effect on the production of IMP at the step catalyzed by Gln-PRPP amidotransferase. This enzyme is partially inhibited by GMP. Complete inhibition is achieved by both GMP and AMP, so GMP alone inhibits but not enough to shut down the pathway completely. This insures that IMP continues to be made until AMP levels rise.

20. Decide from Figures 18.1, 25.31, 26.26, and the Deeper Look box on page 913 which carbon atom(s) in glucose would be most likely to end up as the 5-CH₃ carbon in dTMP.

Answer: From the Deeper Look we learn that tetrahydrofolate (THF) is converted to N^5, N^{10},-methylene THF from either serine or glycine. Glycine, as we learned in the last chapter, derives from serine. Serine is produced from 3-phosphoglycerate, a glycolytic intermediate.

When glycine is used to charge THF, it is the α-carbon that becomes the methylene group of N^5, N^{10},-methylene THF. When serine is used, serine's β-carbon becomes the methylene carbon. Thymidylate synthase uses N^5, N^{10},-methylene THF to convert dUMP to dTMP. Thus, the methyl group of dTMP derives from either the α-carbon of glycine or the β-carbon of serine. How are these carbons related to glucose?

Serine is formed from 3-phosphoglycerate with carbon 3 destined to become the β-carbon of serine. Carbon 3 derives from carbon 1 or carbon 6 of glucose. (Remember: Glycolysis begins with phosphorylation of these two carbons.) So, the methyl carbon of dTMP is labeled by carbons 1 and 6 of glucose via serine.

The methyl carbon may also be labeled using glycine directly. When glycine is produced from serine, its α-carbon derives from carbons 2 and 5 of glucose.

Questions for Self Study

1. The de novo synthesis of AMP and GMP starts with IMP.
   a. Synthesis of IMP starts with ___ .
   b. Two of the carbons of the purine ring of IMP are supplied by the one-carbon pool as formyl groups attached to ___ .
   c. Two of the nitrogens derive from the side chain of ___, a common amino acid.
   d. One carbon derives from _____ in a reaction that interestingly does not require biotin (nor ATP hydrolysis).
   e. The amino acid ___ is completely consumed during IMP synthesis accounting for two carbons and one nitrogen.
   f. One nitrogen derives from ___, an amino acid whose carbon skeleton is released as fumarate.

2. The intermediate 5-phosphoribosyl-α-pyrophosphate plays key roles in both de novo purine biosynthesis and in purine salvage pathways. Explain.

3. Match the disease or clinical condition with its appropriate enzyme.
   a. Lesch-Nyhan Syndrome   1. Xanthine Oxidase
   b. Severe Combined Immunodeficiency   2. Hypoxanthine-guanine
   Syndrome (SCID)   phosphoribosyltransferase(HGPRT)
   c. Gout   3. Adenosine deaminase

4. In mammals there are two enzymes that synthesize carbamoyl phosphate. The enzymes are localized in different compartments of the cell and participate in different metabolic pathways. Draw the structure of carbamoyl phosphate. Name the two enzymes that synthesize this compound, identify their cellular locations, and state in which metabolic pathways they function.

5. Fill in the blanks.
The enzyme ___ produces deoxyribonucleotide diphosphates from ribonucleotide diphosphates. The reducing equivalents for this reaction come from NADPH but are supplied to the enzyme through ___, a small protein with two cysteine residues that undergo reversible oxidation-reduction cycles between cysteine sulfhydryls and ___ disulfide.

Answers

1. a. ribose-5-phosphate; b. tetrahydrofolate; c. glutamine; d. carbon dioxide; e. glycine; f. aspartate.

2. The purine ring of IMP is synthesized on a ribose moiety and the first reaction involves 5-phosphoribosyl-α-pyrophosphate. This derivatized sugar is used as a substrate for adenine phosphoribosyltransferase and hypoxanthine-guanine phosphoribosyltransferase to salvage free bases.

3. a. 2; b. 3; c. 1

4. Carbamoyl phosphate synthetase II is a cytosolic enzyme involved in pyrimidine biosynthesis. Carbamoyl phosphate synthetase I is a mitochondrial enzyme of the urea cycle (and arginine biosynthesis). The structure of carbamoyl phosphate is shown.

$$H_2N-\overset{\overset{\displaystyle O}{\|}}{C}-O-\overset{\overset{\displaystyle O}{\|}}{\underset{\underset{\displaystyle O^-}{|}}{P}}-O^-$$

5. Ribonucleotide reductase; thioredoxin; cystine.

Additional Problems

1. Why are the sulfonamides effective against bacterial but not animal cells?

2. Describe the anaplerotic pathway involving AMP deaminase, adenylosuccinate synthetase and adenylosuccinase.

3. Some patients with gout suffer from a defect in PRPP amidotransferase. What reactions does this enzyme catalyze? Can you suggest a defect in the enzyme that might lead to gout?

4. How does purine biosynthesis differ fundamentally from pyrimidine biosynthesis?

Abbreviated Answers

1. Sulfonamides are structural analogs of p-aminobenzoic acid, a component of folate. Animals do not synthesize folate. However, bacteria synthesize it *de novo* or using precursors including p-aminobenzoic acid. Thus, sulfonamides will compete with p-aminobenzoic acid and lead to inhibition of folic acid biosynthesis.

2. AMP deaminase converts AMP to IMP. Adenylosuccinate synthetase attaches aspartate to IMP to form adenylosuccinate, which is degraded into AMP and fumarate, a citric acid cycle intermediate.

3. PRPP amido transferase produces 5-phosphoribosylamine from 5-phosphoribosyl pyrophosphate and glutamine. It is the first step in purine biosynthesis. The defect in this enzyme cannot lead to enzyme inactivation as this would have the effect of lowering uric acid levels (and perhaps being lethal). The enzyme is subject to feedback inhibition by guanine nucleotides and adenine nucleotides. A defect in allosteric inhibitor binding might lead to a loss of feedback inhibition of purine production leading to increased uric acid production and gout.

4. Purines are synthesized as nucleotide derivatives whereas pyrimidines are produced as the free base orotate, which is subsequently converted to a nucleotide.

Summary

Nearly all organisms have the ability to synthesize purine and pyrimidine nucleotides *de novo*. Since these nucleotides are the direct precursors for RNA and DNA synthesis, their formation is a particularly active process in rapidly proliferating cells, as seen in bacterial infections or cancerous malignancies. Consequently, enzymes of nucleotide biosynthesis are often the targets of drugs designed to halt uncontrolled cell division.

The nine atoms of the purine ring are contributed by five separate precursors: CO_2, the amino acids glycine, aspartate and glutamine, and one-carbon units contributed by THF derivatives. The purine ring is constructed on a ribose-5-phosphate scaffold. PRPP, the "active form" of ribose-5-P, is the limiting substance in purine biosynthesis. The *de novo* purine synthetic pathway has as its first purine product, IMP. IMP synthesis is regulated at the first two steps: 1) Ribose-5-P pyrophosphokinase is allosterically inhibited by ADP and GDP; 2) Glutamine-PRPP amidotransferase is subject to cumulative feedback inhibition by adenine and guanine nucleotides. This latter enzyme has two sites for allosteric inhibition, one where AMP, ADP and ATP act, and a second where GMP, GDP and GTP act. The two prominent purines, AMP and GMP, are formed from IMP via a bifurcated pathway with two reactions in each branch. The first enzyme in the AMP branch, adenylosuccinate synthetase, has GTP as a substrate and is feedback inhibited by AMP. The initial step in the GMP branch, IMP dehydrogenase, is feedback inhibited by GMP, while the second step, GMP synthetase, uses ATP as its energy source. This product inhibition and reciprocity in energy donor provides an effective regulatory mechanism for balancing the formation of AMP and GMP to meet cellular needs.

Purine salvage pathways furnish a means to recover free purine bases and nucleosides released as a consequence of nucleic acid degradation and turnover. Though these salvage enzymes seem peripheral to the major utilization, deficiencies in them, as in the absence of HGPRT (hypoxanthine-guanine phosphoribosyltransferase) in Lesch-Nyhan Syndrome, have tragic consequences. Further, aberrations in purine catabolism, as in ADA (adenosine deaminase) deficiency, the underlying cause of Severe Combined Immunodeficiency Syndrome or "SCID", can also be devastating. The end product of purine catabolism is uric acid. An excess accumulation of uric acid leads to the clinical disorder known as gout. Allopurinol, an analog of hypoxanthine, is an effective treatment. Allopurinol binds tightly to xanthine oxidase, inhibiting its activity and preventing uric acid build-up. Purine metabolism even fulfills an anaplerotic role. In skeletal muscle, a cyclic series of purine transformations known as the purine nucleoside cycle has the net effect of converting aspartate to fumarate to replenish the level of citric acid cycle intermediates.

De novo pyrimidine biosynthesis first yields the six-membered pyrimidine ring; only then is the ribose-5-moiety added to create a nucleotide. Just two precursors, carbamoyl-P and aspartate, provide the requisite six atoms of the pyrimidine. In *E. coli*, pyrimidine synthesis is regulated at aspartate transcarbamoylase (ATCase), the step where carbamoyl-P and aspartate condense to form carbamoyl-aspartate. ATCase is the paradigm of allosteric enzymes (Chapter 12); it is feedback inhibited by the ultimate ribonucleotide end product of the pyrimidine pathway, CTP, and is activated by ATP. In animals, carbamoyl phosphate synthetase II (CPS-II) is the committed step in pyrimidine biosynthesis. Animal CPS-II is feedback inhibited by UDP and UTP, while ATP and PRPP are allosteric activators. Pyrimidine synthesis in mammals provides an example of "metabolic channeling". The first three reactions are catalyzed by a multi-functional polypeptide, which contains the active sites for CPS-II, ATCase and DHOase. A second multi-functional polypeptide carries both OPRTase and OMP decarboxylase activities (reactions 5 and 6 of the pathway). Such multi-functional proteins achieve metabolic channeling: the direct transfer of substrate from one active site to the next, avoiding dissociation from the protein followed by diffusion (and dilution) in free solution to the next enzyme.

Deoxyribonucleotides have only one metabolic purpose - to serve as building blocks in DNA synthesis. Ribonucleoside diphosphates (NDPs) are the substrates for deoxyribonucleotide formation; NDP reduction at the 2'-position forms the corresponding dNDP. The enzyme mediating this reaction is ribonucleotide reductase, and NADPH is the ultimate source of reducing power for dNDP synthesis. Ribonucleotide reductase has two distinct allosteric sites in addition to its active site, and its activity is regulated by an elegant feedback control circuit that balances the supply of dNTPs. One of the active sites determines the overall activity of the enzyme depending on whether ATP or dATP is bound. ATP activates; dATP inhibits. The second allosteric site is the substrate specificity site. Occupation of this site by its effectors, ATP, dTTP, dGTP or dATP, determines which NDP substrate (CDP, UDP, GDP or ADP) is bound and reduced in the active site.

Synthesis of thymine deoxynucleotides begins with the deoxyribonucleotides, dUDP and dCDP, both of which can lead to dUMP: dUDP★dUTP★dUMP★or dCDP★dCMP★dUMP. The conversion of dCDP to dUMP is catalyzed by dCMP deaminase, an allosteric enzyme that provides a control point for regulation of dTTP formation. Thymidylate synthase catalyzes the synthesis of dTMP from dUMP and N^5, N^{10}-methylene THF. Because of its pivotal role in the pathway of dTTP formation, thymidylate synthase has become an attractive target for chemotherapeutic agents designed to selectively inhibit cell division through denial of adequate amounts of DNA precursors.

Chapter 27

Metabolic Integration and Organ Specialization

• •

Chapter Outline

❖ Metabolism: Five functional blocks
   ⋏ Catabolic activities
      • Foods oxidized to carbon dioxide and water
      • ATP and NADPH produced
   ⋏ Anabolic activities
      • Metabolic intermediates from catabolism converted to variety of molecules
      • ATP and NADPH consumed
   ⋏ Macromolecular synthesis
      • Anabolic products used to synthesize biopolymers
      • ATP principle source of energy
         ● GTP: Protein synthesis
         ● CTP: Phospholipid synthesis
         ● UTP: Polysaccharide synthesis
   ⋏ Photochemical activities
      • Light energy used to produce ATP and NADPH
   ⋏ Carbon dioxide fixation
      • ATP and NADPH used to fix carbon dioxide and convert to intermediate
❖ Ten key intermediates
   ⋏ Carbohydrates
      • Triose-P, tetrose-P, pentose-P, hexose-P
   ⋏ α-Keto acids
      • Pyruvate, oxaloacetate, α-ketoglutarate
   ⋏ CoA derivatives
      • Acetyl-CoA, succinyl-CoA
   ⋏ PEP
❖ ADP/ATP and NAD/NADPH couple catabolism to anabolism
❖ Stoichiometries of ATP utilization
   ⋏ Reaction stoichiometry: Chemical stoichiometry: Reactants and products balanced
   ⋏ Obligate coupling stoichiometry: Initial substrate of one pathway and final product of second pathway determine stoichiometry
   ⋏ Evolved coupling stoichiometry: Stoichiometry of ATP coupled to pathway not fixed
❖ Metabolic roles of ATP
   ⋏ Stoichiometry establishes large K_{eq}: Unidirectional process
   ⋏ ATP (AMP and ADP) allosteric effectors
   ⋏ Adenylate kinase interconverts ATP, ADP, and AMP
   ⋏ Energy charge: $E.C. = \dfrac{1}{2} \times \left(\dfrac{2 \times [ATP] + [ADP]}{[ATP] + [ADP] + [AMP]} \right)$ E.C. varies from 0 to 1
      • R response to E.C.
         ● Active when E.C. low; Activity decreases as E.C. approaches 1
         ● Enzymes in catabolic pathways show R response

499

- U response
  - Active when E.C. close to 1: Activity decreases as E.C. decreases
  - Enzymes in anabolic pathways show U response
- Phosphorylation potential: $\Gamma = \dfrac{[ATP]}{[ADP][P_i]}$ Ranges from 200 to 800 M^{-1}

❖ Energy balance in cells: AMP-activated protein kinase (AMPK) is cellular sensor
  - Inactive when ATP is high, active when ATP is low: AMP stimulates
  - Serine/threonine kinase
  - Regulated by hormone (e.g., leptin) and muscle activity
  - Activated by diabetes drug metformin
  - Key targets
    - Phosphofructokinase-2 (to stimulate glycolysis)
    - Glycogen synthase, acetyl-CoA carboxylase, HMG-CoA reductase (inhibited)
    - Glycogen: Liver and muscle
❖ Human metabolism
  - Fuel stores
    - Glycogen: Liver and muscle
    - Triacylglycerol: Adipose tissue
    - Protein: Muscle
  - Fuel utilization preferences: Glycogen > triacylglycerol > protein
  - Organ metabolism and interplay
    - Brain
      - High respiratory metabolism
      - No fuel reserves
      - Glucose preferred fuel: From diet or from liver via gluconeogenesis
      - β-Hydroxybutyrate during starvation: From liver
    - Muscle
      - Fatty acid, glucose, and ketone body metabolism at rest
      - P-creatine, and glycogen utilization during intense activity
      - Fatigue caused by decrease in pH, not by depletion of reserves
      - Fasting muscle utilizes amino acids from protein
    - Heart
      - Preferred fuel fatty acids
      - Minimal reserves: Fatty acids, glucose, and ketone bodies must be supplied
    - Adipose tissue
      - Glucose converted to acetyl-CoA and fatty acids
      - Fatty acids also supplied by liver
      - Triacylglycerol production relies on glycerol-3-P from glucose
      - Glucose plays pivotal role
        o Source of glycerol-3-P
        o Fuel for pentose phosphate pathway
    - Liver
      - Buffers blood glucose levels
      - Fatty acid metabolism and ketogenesis
      - Cori cycle
      - Glucose-6-P plays key role
        o Glycogen metabolism
        o Gluconeogenesis/glycolysis
        o Production of NADPH from pentose phosphate
        o Production of acetyl-CoA
❖ Regulation of eating
  - Neuroendocrine signals from stomach, small intestines, pancreas, adipose tissue and central nervous system (hypothalamus mainly)
  - Shot term and long term responses

- Ghrelin and cholecystokinin are short term
  - Ghrelin from stomach stimulates appetite when stomach is empty
  - Cholecystokinin from GI track inhibits eating
- Insulin and leptin are long term
  - Insulin from pancreas stimulates glucose uptake
  - Leptin from adipocytes (released in response to insulin)
- Leptin
  - Blocks release of NPY (see below): Suppresses appetite
- Neurons involved: Two subsets
  - NPY/AgRP-producing neurons release neuropeptide Y: Stimulates eating
    - PYY_{3-36} from small intestine, leptin block NPY release
  - Melanocortin-producing neurons release melanocyte-stimulating hormones: Inhibit eating: AgRP (agouti-related peptide) blocks these
- AMPK
  - Leptin inhibits AMPK via melanocortin-4 receptor
  - Ghrelin and NPY activate AMPK
  - Malonyl-CoA serves as indicator molecule
- Longevity
  - Caloric restriction diet leads to longevity
  - Genetic control of longevity via SIRT1
    - NAD^+-dependent protein deacetylase
    - Inactivates PPARγ (Active PPARγ stimulates genes involved in fat storage
    - Stimulates PGC-1 leading to stimulation of gluconeogenesis
    - Resveratrol: Component of red wines that stimulates SIRT1

Chapter Objectives

Metabolism

So far, we have developed a complicated picture of intermediary metabolism and it is time to attempt to simplify and unify. There are a small number of intermediates that serve crucial roles in intermediary metabolism. These include sugar phosphates, pyruvate, oxaloacetate, α-ketoglutarate, acetyl-CoA, succinyl-CoA and PEP. The sugar phosphates are found in glycolysis, gluconeogenesis, the pentose phosphate pathway, and the Calvin cycle. Pyruvate, oxaloacetate and α-ketoglutarate are keto acids. Pyruvate derives from a number of sources including glycolysis and amino acids and is the port of entry into the citric acid cycle for glucose-derived carbons. Oxaloacetate and α-ketoglutarate are citric acid cycle intermediates and both can be produced from amino acids by deamination. Acetyl-CoA is consumed in the citric acid cycle and is a common denominator between fatty acids, sugars, and amino acid. Succinyl-CoA, a citric cycle intermediate, is the place of entry of propionate from dietary sources and odd-chain fatty acid catabolism, is a product of amino acid catabolism, and is used in heme biosynthesis.

ATP and NADPH serve critical roles in coupling catabolism and anabolism. Catabolism is largely oxidative in nature, leading to reduction of cofactors NAD^+ and FAD. Anabolic pathways are reductive with NADPH usually serving as the immediate source of electrons. This coenzyme is reduced in the pentose phosphate pathway. Additionally, cycles exist to move electrons from NADH to $NADP^+$. Catabolic pathways are exergonic and lead to synthesis of ATP. ATP is then consumed in anabolic, energy requiring pathways. The coupling of ATP production to a complex metabolic pathway such as aerobic oxidation of glucose to CO_2 and H_2O has a stoichiometry that is not defined by simple chemical considerations. Rather ATP coupling stoichiometry is an evolved quantity. Nevertheless, under physiological conditions, the complete oxidation of glucose gives high yields of ATP. Furthermore, the process is always far from equilibrium making regulation by kinetic controls possible.

ATP Equivalents

The metabolic unit of energy exchange is the ATP equivalent defined as the amount of energy released upon hydrolysis of ATP to ADP. The ATP equivalent of key metabolic reactions is given below.

ATP hydrolysis, 1.0
PP_i hydrolysis, 1.0
ATP to AMP and 2 P_i, 2.0
NADH oxidation, 3.0 ATP (2.5 in mitochondria)
$FADH_2$ oxidation, 2.0 ATP (1.5 in mitochondria)

NADPH oxidation, 3.5 to 4 ATP

Metabolic Integration

The major organs specialize in metabolism of particular fuels and there is interplay among the liver, muscles, heart, adipose tissue, and brain to insure that the energetic demands are met. For example, glucose can be supplied to other tissues by the liver by gluconeogenesis and glycogenolysis, muscle can produce lactic acid during times of intense energy demands and this lactic acid is sent to the liver for reprocessing into glucose. Energy demands are ultimately met by diet and animals have evolved a complex system of hormonal regulation to regulate energy storage and appetite. Brain, stomach, small intestines, pancreas, and adipose tissue all play roles in stimulating or suppressing appetite.

Problems and Solutions

1. The energetics of opposing metabolic sequences
(Integrates with Chapters 3, 18, and 22.) The conversion of PEP to pyruvate by pyruvate kinase (glycolysis) and the reverse reaction to form PEP from pyruvate by pyruvate carboxylase and PEP carboxykinase (gluconeogenesis) represents a so -called substrate cycle. The direction of net conversion is determined by the relative concentrations of allosteric regulators that exert kinetic control over pyruvate kinase, pyruvate carboxylase, and PEP carboxykinase. Recall that the last step in glycolysis is catalyzed by pyruvate kinase:

$$PEP + ADP \leftrightarrows pyruvate + ATP$$

The standard free energy change is -31.7 kJ/mol.
a. Calculate the equilibrium constant for this reaction.
b. If [ATP] = [ADP], by what factor must [pyruvate] exceed [PEP] for this reaction to proceed in the reverse direction?
The reversal of this reaction in eukaryotic cells is essential to gluconeogenesis and proceeds in two steps, each requiring an equivalent of nucleoside triphosphate energy:

Pyruvate carboxylase
$$Pyruvate + CO_2 + ATP \rightarrow oxaloacetate + ADP + P_i$$

PEP carboxykinase
$$Oxaloacetate + GTP \rightarrow PEP + CO_2 + GDP$$
Net: Pyruvate + ATP + GTP → PEP + ADP + GDP + P_i

c. The $\Delta G°'$ for the overall reaction is +0.8 kJ/mol. What is the value of K_{eq}?
d. Assuming [ATP] = [ADP], [GTP] = [GDP], and P_i = 1 mM when this reaction reaches equilibrium, what is the ration of [PEP]/[pyruvate]?
e. Are both directions in the substrate cycle likely to be strongly favored under physiological conditions?

Answer: a. From $\Delta G° = -RT \ln K_{eq}$ we can write:

$$K_{eq} = e^{\frac{-\Delta°'G}{RT}}$$

The value of RT at room temperature is:
$$(8.314 \times 10^{-3})(298) = 2.48$$
$$K_{eq} = e^{\frac{-(-31.7)}{2.48}} = 360,333$$

b. Given [ATP] = [ADP], and K_{eq} is 360,333, the reaction will proceed in reverse when $\Delta G > 0$. ΔG is given by:

$$\Delta G = \Delta G°' + RT\ln \frac{[ATP][pyruvate]}{[ADP][PEP]}$$

But, $\Delta G°' = -RT\ln K_{eq}$

Thus, $\Delta G = RT\ln \frac{[ATP][pyruvate]}{[ADP][PEP]K_{eq}}$

$K_{eq} = 360,333$

For the reaction to go in reverse and $\Delta G > 0$, term in the ln must be > 1. Thus,
[ATP][pyruvate] > [ADP][PEP] 360,333. When [ATP] = [ADP], [pyruvate] > 360,333[PEP].

502

c. The following formula is used to determine K_{eq}: $\Delta G^{o\prime} = -RT\ln K_{eq}$
Solving for K_{eq} we find that

$$K_{eq} = e^{\frac{\Delta G^{o\prime}}{RT}}$$

Given R = 8.3145 J/mol, T = 298 (i.e., 25°C)

$\Delta G^{o\prime}$ = =0.8 kJ/mol or 800 J/mol

$$K_{eq} = e^{-\frac{800}{8.3145 \times 298}} = 0.724$$

d. When the reaction reaches equilibrium, the ratio of [PEP] to [pyruvate] is:

$$Keq = 0.724 = \frac{[PEP][ADP][GDP][P_i]}{[pyruvate][ATP][GTP]}$$

$$0.724 = \frac{[PEP] \times 1 \times 10^{-3}}{[pyruvate]} \text{ or}$$

$$\frac{[PEP]}{[pyruvate]} = 724$$

Answer: For the reaction: PEP + ADP $\star$ pyruvate + ATP, from b we calculated:

$$\frac{[pyruvate]}{[PEP]} = 360,333 \text{ at equilibrium}$$

Ratios of [pyruvate] to [PEP] smaller than this value will react in the direction of pyruvate production until the equilibrium value is reached. Thus,

$$\frac{[pyruvate]}{[PEP]} < 360,333 \text{ proceeds spontaneously, or}$$

$$\frac{[PEP]}{[pyruvate]} > \frac{1}{360,333} > 0.0000028 \text{ proceeds spontaneously}$$

From d, we found that as long as the ratio of [PEP] to [pyruvate] does not exceed 724, the reaction will proceed as written. Thus, both reactions will be favorable as long as [PEP]/[pyruvate] is between 0.0000028 and 724.

2. Calculating energy charge and phosphorylation potential
(Integrates with Chapter 3.) Assume the following intracellular concentrations in muscle tissue: ATP = 8 mM, ADP = 0.9 mM, AMP = 0.04 mM, P_i = 8 mM. What is the energy charge in muscle? What is the phosphorylation potential?

Answer: The energy charge is given by:

$$\text{Energy Charge} = \frac{\frac{1}{2}(2 \times [ATP] + [ADP])}{[ATP] + [ADP] + [AMP]}$$

$$\text{Energy Charge} = \frac{\frac{1}{2}(2 \times 8 \times 10^{-3} + 0.9 \times 10^{-3})}{8 \times 10^{-3} + 0.9 \times 10^{-3} + 0.04 \times 10^{-3}}$$

Energy Charge = 0.945

The phosphorylation potential, Γ, is given by:

$$\Gamma = \frac{[ATP]}{[ADP][P_i]}$$

$$\Gamma = \frac{8 \times 10^{-3}}{0.9 \times 10^{-3} \times 8 \times 10^{-3}} = 1,111 \text{ M}^{-1}$$

3. How long do ATP and phosphocreatine supplies last during muscle contraction?
Strenuous muscle exertion (as in the 100-meter dash) rapidly depletes ATP levels. How long will 8 mM ATP last if 1 gram of muscle consumes 300 μmol of ATP per minute? (Assume muscle is 70% water). Muscle contains phosphocreatine as a reserve of phosphorylation potential. Assuming [phosphocreatine] = 40 mM, [creatine] = 4 mM, and $\Delta G°'$ (phosphocreatine + H_2O ✶ creatine + P_i) = -43.3 kJ/mol, how low must [ATP] become before it can be replenished by the reaction phosphocreatine + ADP ✶ ATP + creatine. [Remember, $\Delta G°'$ (ATP hydrolysis) = -30.5 kJ/mol.]

Answer: One gram of muscle contains approximately 0.7 g of H_2O, or 0.7 mL. If the [ATP] = 8 mM, 0.7 mL contains

$$0.7 \times 10^{-3} L \times 8 \times 10^{-3} \frac{mol}{L} = 5.60 \times 10^{-6} mol \text{ or } 5.6 \text{ μmol ATP}$$

If ATP is consumed at the rate of 300 μmol per min it will last

$$\frac{5.6 \text{ μmol}}{300 \frac{\text{μmol}}{\text{min}}} = 0.019 \text{ min or } 1.12 \text{ sec}$$

Phosphocreatine and ATP are coupled by the following reaction

$$\text{phosphocreatine} + ADP \text{ ✶ } ATP + \text{creatine}$$

This is the sum of two reactions

$$\text{phosphocreatine} + H_2O \text{ ✶ } \text{creatine} + P_i \qquad \Delta G°' = -43.3 \text{ kJ/mol}$$

and

$$P_i + ADP \text{ ✶ } ATP + H_2O \qquad \Delta G°' = 30.5 \text{ kJ/mol}$$

Thus, the overall $\Delta G°' = -12.8$ kJ/mol. For the reaction to be favorable, ΔG must be less than zero or

$$\Delta G = \Delta G°' + RT \ln \frac{[\text{creatine}][ATP]}{[\text{phosphocreatine}][ADP]} < 0, \text{ or}$$

$$\Delta G°' + RT \ln \frac{[\text{creatine}][ATP]}{[\text{phosphocreatine}][ADP]} < 0$$

$$RT \ln \frac{[\text{creatine}][ATP]}{[\text{phosphocreatine}][ADP]} < -\Delta G°'$$

$$\frac{[ATP]}{[ADP]} < \frac{[\text{phosphocreatine}]}{[\text{creatine}]} \times e^{\frac{-\Delta G°'}{RT}}$$

$$\frac{[ATP]}{[ADP]} < \frac{40 \text{ mM}}{4 \text{ mM}} \times e^{\frac{-(-12.8)}{2.48}}$$

$$\frac{[ATP]}{[ADP]} < 1,750$$

The reaction is favorable and ADP is phosphorylated at the expense of phosphocreatine when [ATP] < 1,750 [ADP] and [Cr-P] = 40 mM and [Cr] = 4 mM.

4. What is the value of NADPH in ATP equivalents?
(Integrates with Chapter 20.) The standard reduction potentials for the (NAD^+/NADH) and the ($NADP^+$/NADPH) couples are identical, namely -320 mV. Assuming the in vivo concentration ratios NAD^+/NADH = 20 and $NADP^+$/NADPH = 0.1, what is ΔG for the following reaction?

$$NADPH + NAD^+ \rightleftharpoons NADP^+ + NADH$$

Calculate how many ATP equivalents can be formed from ADP + P_i by the energy released in this reaction.

Answer: From $\Delta G = -nF\Delta\mathcal{E}_o$, where n is the number of electrons transferred, F is Faraday's constant (96,494 J/V· mol), and $\Delta\mathcal{E}_o'$ is the change in redox potential. We can calculate ΔG given that:

$$\Delta\mathcal{E} = \Delta\mathcal{E}^{\circ'} - \frac{RT}{n\mathcal{F}}\ln\frac{[\text{NADH}][\text{NADP}^+]}{[\text{NAD}^+][\text{NADPH}]}$$

$$\Delta\mathcal{E}^{\circ'} = \mathcal{E}^{\circ'}_{\text{acceptor}} - \mathcal{E}^{\circ'}_{\text{donor}} = -320 \text{ mV} - (-320 \text{ mV}) = 0$$

$$\Delta\mathcal{E} = 0 - \frac{8.314 \times 298}{2 \times 96,494}\ln\frac{0.1}{20}$$

$$\Delta\mathcal{E} = 0.068 \text{ V}$$

$$\Delta G = -n\mathcal{F}\Delta\mathcal{E} = -2 \times 96,494 \times 0.068\text{V} = -13126 \frac{\text{J}}{\text{mol}}$$

$$\Delta G = -13.1 \frac{\text{kJ}}{\text{mol}}$$

Under standard conditions and under physiological conditions hydrolysis of ATP is a very favorable reaction with large negative ΔG values (on the order of -30 kJ/mol and -50 kJ/mol respectively). Assuming ΔG = -50 kJ/mol, -13.1 kJ/mol will support (-13.4/-50 =) 0.26 mol of ATP.

5. Adenylate kinase is an amplifier
(Integrates with Chapter 3.) Assume the total intracellular pool of adenylates (ATP + ADP + AMP) = 8 mM, 90% of which is ATP. What are [ADP] and [AMP] if the adenylate kinase reaction is at equilibrium? Suppose [ATP] drops suddenly by 10%. What are the concentrations now for ADP and AMP, assuming adenylate kinase reaction is at equilibrium? By what factor has the AMP concentration changed?

Answer: Adenylate kinase catalyzes the following reaction
$$\text{ATP} + \text{AMP} \star \text{ADP} + \text{ADP}$$
The equilibrium constant for the reaction K_{eq} = 1.2 (given in Figure 27.2).

$$K_{eq} = \frac{[\text{ADP}]^2}{[\text{ATP}][\text{AMP}]}$$

Let T=[ATP]+[ADP]+[AMP], the total concentration of adenylates

If [ATP]=90% $\times$ T=0.9 $\times$ 8 mM=7.2 mM

[ADP]+[AMP]=8 mM-7.2 mM=0.8 mM, or

[AMP]=0.8 mM-[ADP]

Substituting this expession and the value of 7.2 mM for [ATP]

into the equilibrium equation we find:

$$K_{eq} = \frac{[\text{ADP}]^2}{7.2 \text{ mM} \times (0.8 \text{ mM-[ADP]})}, \text{ or}$$

$[\text{ADP}]^2 + 7.2 \text{ mM} \times K_{eq} \times [\text{ADP}] - 7.2 \text{ mM} \times 0.8 \text{ mM} \times K_{eq} = 0$

By substitution K_{eq} =1.2 we find:

$[\text{ADP}]^2 + 8.64 \times 10^{-3} \times [\text{ADP}] - 6.91 \times 10^{-6} = 0$

A quadratic equation whose solution is [ADP]=0.737 mM

[AMP]=0.8 mM-[ADP]=0.8 mM-0.737 mM=0.063 mM

Thus, we have:

[AMP]=0.063 mM, [ADP]=0.737 mM, and [ATP]=7.2 mM

If the [ATP] suddenly falls by 10%, we have:

ATP=7.2 mM-10% $\times$ 7.2 mM=6.48 mM

[ADP]+[AMP]=8 mM-6.48 mM=1.52 mM, or

[AMP]=1.52 mM-[ADP]

Substituting this expression and the value of 6.48 mM for [ATP] into the equilibrium equation we find:

$$[K_{eq} = \frac{[ADP]^2}{6.48 \text{ mM} \times (1.52 \text{ mM}-[ADP])} , \text{ or}$$

$$[ADP]^2 + 6.48 \text{ mM} \times K_{eq} \times [ADP] - 6.48 \text{ mM} \times 1.52 \text{ mM} \times K_{eq} = 0$$

By substitution $K_{eq} = 1.2$ we find:

$$[ADP]^2 + 7.78 \times 10^{-3} \times [ADP] - 1.18 \times 10^{-5} = 0$$

A quadratic equation whose solution is [ADP]=1.30 mM

[AMP]=1.52 mM-[ADP]=0.22 mM, [ATP]=6.48 mM

This relatively modest change in [ATP], causes almost a doubling of [ADP] levels and a 3.5-fold increase in [AMP].

6. Regulating the flux through metabolic pathways
(Integrates with Chapters 18 and 22.) The reactions catalyzed by PFK and FBPase constitute another substrate cycle. PFK is AMP-activated; FBPase is AMP-inhibited. In muscle, the maximal activity of PFK (µmoles of substrate transformed per minute) is ten times greater than FBPase activity. If the increase in [AMP] described in Problem 5 raised PFK activity from 10% to 90% of its maximal value but lowered FBPase activity from 90% to 10% of its maximal value, by what factor is the flux of fructose-6-P through the glycolytic pathway changed? (Hint: Let PFK maximal activity = 10, FBPase maximal activity = 1; calculate the relative activities of the two enzymes at low [AMP] and at high [AMP]; let J, the flux of F-6-P through the substrate cycle under any condition, equal the velocity of PFK reaction minus the velocity of the FBPase reaction.)

Answer:

Let $J = \text{Velocity}_{PFK} - \text{Velocity}_{FBP}$

At low [AMP],

$\text{Velocity}_{PFK} = 0.1 \times \text{Velocity}_{max, PFK}$, and

$\text{Velocity}_{FBP} = 0.9 \times \text{Velocity}_{max, FBP}$

So, $J_{low \, AMP} = 0.1 \times \text{Velocity}_{max, PFK} - 0.9 \times \text{Velocity}_{max, FBP}$

So, $J_{low \, AMP} = 0.1 \times 10 - 0.9 \times 1 = 0.1$

At high [AMP],

$\text{Velocity}_{PFK} = 0.9 \times \text{Velocity}_{max, PFK}$, and

$\text{Velocity}_{FBP} = 0.1 \times \text{Velocity}_{max, FBP}$

So, $J_{high \, AMP} = 0.9 \times \text{Velocity}_{max, PFK} - 0.1 \times \text{Velocity}_{max, FBP}$

So, $J_{high \, AMP} = 0.9 \times 10 - 0.1 \times 1 = 8.9$

The ratio $\dfrac{J_{high \, AMP}}{J_{low \, AMP}} = \dfrac{8.9}{0.1} = 89$

7. The effects of leptin on fat metabolism
(Integrates with Chapters 23 and 24.) Leptin not only induces synthesis of fatty acid oxidation enzymes and uncoupling protein-2 in adipocytes, but it also causes inhibition of acetyl-CoA carboxylase, resulting in a decline in fatty acid biosynthesis. This effect on acetyl-CoA carboxylase, as an additional consequence, enhances fatty acid oxidation. Explain how leptin-induced inhibition of acetyl-CoA carboxylase might promote fatty acid oxidation.

Answer: Acetyl-CoA carboxylase catalyzes the production of malonyl-CoA and inhibition of this enzyme will immediately inhibit fatty acid biosynthesis because malonyl-CoA is a substrate for fatty acid synthase. Malonyl has another regulatory role in fatty acid metabolism: it inhibits carnitine acyltransferase the enzyme responsible for fatty acid uptake by the mitochondria. As malonyl-CoA levels fall, carnitine acyltransferase will cause an increased uptake of fatty acids into the mitochondria where they are metabolized by β-oxidation

8. Ethanol as a source of metabolic energy
(Integrates with Chapters 19 and 20.) Acetate produced in ethanol metabolism can be transformed into acetyl-CoA by the acetyl thiokinase reaction:

$$Acetate + ATP + CoASH \rightarrow acetyl\text{-}CoA + AMP + PP_i$$

Acetyl-CoA then can enter the citric acid cycle and undergo oxidation to 2 CO_2. How many ATP equivalents can be generated in a liver cell from the oxidation of one molecule of ethanol to 2 CO_2 by this route, assuming oxidative phosphorylation is part of the process? (Assume all reactions prior to acetyl-CoA entering the citric acid cycle occur outside the mitochondrion.) Per carbon atom, which is a better metabolic fuel, ethanol or glucose? That is, how many ATP equivalents per carbon atom are generated by combustion of glucose versus ethanol to CO_2?

Answer: Let's try to answer the second to last question first: namely, which might be a better fuel, ethanol or glucose? The formulas for glucose and ethanol are ($C_6H_6O_6$) and (C_3H_6O). Clearly, glucose is more oxidized than is ethanol and so if we are to metabolize these oxidatively we should anticipate ethanol to be the better fuel. Let's look into the biochemistry to see if we are correct.

Ethanol is converted to acetaldehyde and then to acetate by action of alcohol dehydrogenase and acetaldehyde dehydrogenase, both use NAD^+ and oxidant. Thus, two NAD are generated in the cytosol. Depending on how electrons are moved into the mitochondria, we may recover 1.5 ATP per NADH. So, ethanol to acetate is accompanied by 3 ATP produced.

The acetyl thiokinase reaction costs two phosphoanhydride bonds bring the conversion of ethanol to acetyl-CoA and CO_2 to one ATP. Complete combustion of acetyl-CoA in the citric acid cycle generates 10 ATP (from 3 NADH at 2.5 ATP each, one $FADH_2$ at 1.5 ATP and 1 GTP). Ethanol combustion thus accounts for 11 ATP or 11 ATP ÷ 2 carbons or 5.5 ATP/carbon.

Conversion of glucose to two pyruvate supports 2 ATP directly and two NADH that account for 1.5 ATP each. Thus, glucose to two pyruvate supports 5 ATP. Pyruvate to acetyl-CoA and subsequent oxidation of the acetyl unit in the citric acid cycle supports 12.5 ATP (from 3 NADH at 2.5 ATP each, one $FADH_2$ at 1.5 ATP and one GTP). For two pyruvates there is a gain of 25 ATP for a total of 30 ATP per glucose. This accounts for (30÷6) 5 ATP/carbon.

To summarize, ethanol accounts for 5.5 ATP/carbon whereas glucose accounts for 5.0 ATP/carbon. (This is not an endorsement for the use of ethanol as a source of calories. Excessive consumption of alcohol leads to numerous problems.)

9. ATP coupling coefficients and the thermodynamics of opposing metabolic sequences I
(Integrates with Chapter 19.) Assuming each NADH is worth 3 ATP, each $FADH_2$ is worth 2 ATP, and each NADPH is worth 4 ATP: How many ATP equivalents are produced when one molecule of palmitoyl-CoA is oxidized to 8 molecules of acetyl-CoA by the fatty acid β-oxidation pathway? How many ATP equivalents are consumed when 8 molecules of acetyl-CoA are transformed into one molecule of palmitoyl-CoA by the fatty acid biosynthetic pathway? Can both of these metabolic sequences be metabolically favorable at the same time if ΔG for ATP synthesis is +50 kJ/mol?

Answer: To convert palmitoyl-CoA to 8 acetyl-CoAs we must run β-oxidation 7 times. This produces 7 NADH and 7 $FADH_2$, which accounts for (7 x 3 = 21 and 7 x 2 = 14) 35 ATP.

To synthesize palmitoyl-CoA starting with 8 acetyl-CoAs we would have to convert seven acetyl-CoA to malonyl-CoA and then run 7 cycles of fatty acid biosynthesize with each cycle consuming 2 NADPH. Production of 7 malonyl-CoA would consume 7 ATP. Consumption of 14 NADPH would account for 56 ATP. The total ATP cost for production is 63 ATP.

For the reactions to be favorable at the same time each of their ΔGs would have to be negative. Since palmitoyl-CoA to 8 acetyl-CoA supports production of 35 ATP, this "costs" 35 x 50 = 1,750 kJ/mol. Provided ΔG for the reaction is more negative than this value, the reaction will be favorable. For the reaction going from acetyl-CoA to palmitoyl-CoA, it consumes 63 ATP at a cost of 63 x 50 = 3,150 kJ/mol. (That is, hydrolysis of 63 ATP releases –3,150 kJ/mol.)

10. ATP coupling coefficients and the thermodynamics of opposing metabolic sequences II
(Integrates with Chapters18-21.) If each NADH is worth 3 ATP, each $FADH_2$ is worth 2 ATP, and each NADPH is worth 4 ATP, calculate the equilibrium constant for cellular respiration, assuming synthesis of each ATP costs 50 kJ/mol of energy. Calculate the equilibrium constant for CO_2 fixation under the same conditions, except here ATP will hydrolyze to ADP + P_i with the release of 50 kJ/mol. Comment on whether these reactions are thermodynamically favorable under such conditions.

Answer: Because the question talks about carbon dioxide fixation, we can assume that cellular respiration is from glucose to CO_2 and H_2O. For complete oxidation of glucose, we generate 2 ATP to pyruvate and two

507

NADH, which account for 6 ATP, for a total of 8 ATP. From 2 pyruvate to 2 acetyl-CoA and through the citric acid cycle, we produce 2 GTP, 2 FADH$_2$, which account for 6 ATP, and 8 NADH, which account for 24 ATP. The total ATP is 38 ATP. At 50 kJ/mol, this accounts for 1,900 kJ per mol glucose.

For CO$_2$ fixation, the metabolic cost is 12 NAPDH and 18 ATP (See equation 21.3). This is equivalent to (12 x 4 = 48 + 18 =) 66 ATP. At 50 kJ/mol, this accounts for 3,300 kJ per mol per 6 CO$_2$.

To calculate the equilibrium constants we would have to know the $\Delta G^{o'}$ for both reactions. It would then be a simple task to calculate K$_{eq}$ given $\Delta G^{o'}$ = -RTln K$_{eq}$. Let us assume that the energy calculations above correspond to ΔG^o values for ATP hydrolysis or ATP production. Further, let us assume the following $\Delta G^{o'}$ for oxidation of glucose to carbon dioxide and water:

$$C_6H_{12}O_6 + 6\ O_2 \rightarrow 6\ CO_2 + 6\ H_2O, \Delta G^{o'} = -2870\ kJ/mol$$

This reaction is, in effect, cellular respiration but without coupling to ATP synthesis. Since ΔG is a state function, $\Delta G^{o'}$glycolysis = $\Delta G^{o'}$glucose oxidation + $\Delta G^{o'}$ATP hydrolysis. Therefore, $\Delta G^{o'}$glycolysis = +1,900 −2870 = -970 kJ/mol. And, K$_{eq}$ = e$^{(970/RT)}$ = e^{391}. My Hewlett-Packard calculator had a problem when I tried to evaluate this. To help, I took the log base 10 of e^{391}, which is 391 x log e (391 x .434 =169) and then raised 10 to this power. The final answer is 10^{169}.

To calculate the equilibrium constant for carbon dioxide fixation we would go through a similar analysis using now +2870 kJ/mol for formation of glucose from carbon dioxide and water and −3,300 kJ/mol for ATP hydrolysis. Therefore, $\Delta G^{o'}$gluconeogenesis = -3,300 + 2870 = -430 kJ/mol. And, K$_{eq}$ = e$^{(430/RT)}$ = e^{173}, which equals to 10^{75}.

11. Metabolic inhibitors as a diabetes treatment strategy
(Integrates with Chapters 22.) In type 2 diabetes, glucose production in the liver is not appropriately regulated, so glucose is over produced. One strategy to treat this disease focuses on the development of drugs targeted against regulated steps in glycogenolysis and gluconeogenesis, the pathways by which liver produces glucose for release into the blood. Which enzymes would you select for as potential targets for such drugs?

Answer: Type II diabetes is characterized by hyperglycemia. Since the liver is a source of blood glucose via gluconeogenesis or glycogenolysis, lowering hepatic glucose release might address the problem. It would make sense to stop glycogen breakdown and stimulate glycogen synthesis by targeting glycogen phosphorylase to inhibit it or glycogen synthase to stimulate it. Another target you could consider is glycogen phosphorylase kinase, the enzyme that regulates both glycogen phosphorylase and glycogen synthetase. For glucose to exit liver cells and enter the blood stream, glucose-6-phosphatase must be active. This enzyme might also be a good target.

Finally, it might be useful to regulate glycolysis/gluconeogenesis at the interconversion between fructose-6-phosphate and fructose-1,6-bisphosphate. This interconversion is accomplished by phosphofructokinase I and fructose-1,6-bisphosphatase. Finally, fructose-2,6-bisphosphate plays a role in regulation of glycolysis and gluconeogenesis so it might be useful to target phosphofructokinase II, a bifunctional enzyme that also has fructose-2,6-bisphosphatase activity.

12. Drug targets to regulate eating behavior
As chief scientist for drug development at PhatFarmaceuticals, Inc., you want to create a series of new diet drugs. You have a grand plan to design drugs that might limit production of some hormones or promote the production of others. Which hormones are on your "limit production" list and which are on your "raise levels" list?

Answer: The hormones you would likely want to target involve hunger. Hormones you might want to up regulate include α-melanocyte-stimulating hormone, which is derived from proopiomelanocortin. Stimulation of melanocortin 1 receptors leads to depression of appetite and increase in energy expenditure. Appetite-stimulating (orexigenic) hormones like neuropeptide Y and agouti-related protein, are antagonists of the appetite-suppressing (anorexigenic) hormone melanocortin by suppressing production of proopiomelanocortin, the precursor of α-melanocyte-stimulating hormone. So, you might want to down regulate neuropeptide Y (NPY) and agouti-related protein (AgRP). Leptin and insulin are long-term regulators of eating behavior. Insulin is responsive to blood glucose, which when high, leads to production of insulin. Insulin, in turn, stimulates fat cells to release leptin, which leads to appetite suppression. So, both insulin and leptin may be targeted to increase. Short-term appetite signals include ghrelin and cholecystokinin. Ghrelin might be down regulated as its release by stomach leads to appetite stimulation. Cholecystokinin, released by the gastrointestinal tract, causes appetite suppression. Finally, the gut hormone PYY$_{3-36}$ suppresses appetite and should be scheduled for up-regulation.

13. Leptin injections as an obesity therapy
The existence of leptin was revealed when the ob/ob genetically obese strain of mice was discovered. These mice have a defect leptin gene. Predict the effects of daily leptin injections on ob/ob mice on food intake, fatty acid oxidation, and body weight. Similar clinical trials have been conducted on humans, with limited success. Suggest a reason why this therapy might not be a miracle cure for overweight individuals.

Answer: Leptin, a 146-amino acid peptide is produced by adipocytes in response to insulin. High leptin levels in blood lead to appetite suppression. Fat metabolism is also regulated by leptin. Triacylglycerol production is inhibited and fat metabolism increased. Further, leptin induces synthesis of uncoupling protein 2, which leads to energy loss as heat. Uncoupling protein 2, in effect, short-circuits the proton gradient in the mitochondria. The energy of the gradient is dissipated as heat and thus not used for ATP production. Leptin also blocks NPY release by the hypothalamus. Since NYP is orexigenic, down regulation of it by leptin leads to loss of appetite. Given all this we might expect the ob/ob mice to have lower body weight as a consequence of increase fat metabolism in adipocytes, inefficiency of electron transport in adipocytes and decrease in food intake.

In obese individuals, leptin may not be the problem. Since leptin is produced by adipocytes, obese individuals may already have high blood levels of leptin and so increasing levels by injection may not have big effects.

14. Hormones that regulate eating behavior
Would it be appropriate to call neuropeptide Y (NPY) the obesity-promoting hormone? What would be the phenotype of a mouse whose melanocortin-producing neurons failed to produce melanocortin? What would be the phenotype of a mouse lacking a functional MC3R gene? What would be the phenotype of a mouse lacking a functional leptin receptor gene?

Answer: Since neuropeptide Y stimulates eating behavior it might be reasonable to think that it would cause obesity. A review by Levens and Della-Zuana (in Current Opinion in Investigative Drugs (2003) 1198-204) discusses research on NPY and NPY receptors. As it turns out decreasing NPY levels with antisense oligonucleotides, antibodies or NPY knockout mice has not consistently supported a role for NPY in obesity. Drugs targeted towards NPY receptors have been equally disappointing. The evidence to date seems to indicate that NPY is not an important obesity-promoting hormone. Lack of melanocortin would lead to appetite stimulation. Lack of a MC3R gene would have a phenotype similar to lack of melanocortin. Finally, lack of leptin receptors would also be expected to lead to obesity.

15. Consequences of alcohol consumption on the NAD⁺/NAHD ratio
The Human Biochemistry box The Metabolic Effects of Alcohol Consumption, points out that ethanol is metabolized to acetate in the liver by alcohol dehydrogenase and aldehyde dehydrogenase:
$$CH_3CH_2OH + NAD^+ \leftrightarrows CH_3CHO + NADH + H^+$$
$$CH_3CHO + NAD^+ \leftrightarrows CH_3COO^- + NADH + 2\,H^+$$
These reactions alter the NAD⁺/NADH ratio in liver cells. From your knowledge of glycolysis, gluconeogenesis, and fatty acid oxidation, what might be the effect of an altered NAD⁺/NADH ratio on these pathways? What is the basis of this effect?

Answer: With these two reactions going rapidly, NAD⁺ supplies might be expected to be limiting. This would turn down glycolysis at glyceraldehydes-3-phosphate dehydrogenase. Fatty acid oxidation would also suffer because NAD⁺ is required in one step of β-oxidation. The citric acid cycle would also be affected because NADH is a negative regulator of several steps. Lastly, we might expect in increase in gluconeogenesis because the glyceraldehyde-3-phosphate step would be stimulated by excess NADH to run in the direction of gluconeogenesis.

16. The phenotype of a T172D mutation in AMPK
A T172D mutant of the AMPK is locked in a permanently active state. Explain.

Answer: T172D indicates that the mutation is at position 172 in AMPK changing a threonine to an aspartic acid. The effect of this mutation is to replace a polar neutral amino acid (threonine) for a negatively charged amino acid (aspartic acid). It is often the case that proteins are regulated by phosphorylation most commonly of serines and threonines but in some cases tyrosines (but by different kinases). The consequence of phosphorylation is either activation or inhibition of the protein. The T172D mutation was likely engineered to mimic the phosphorylated form of the enzyme.

17. Malonyl-CoA as a key indicator of nutrient availability
a. Some scientists support the "malonyl-CoA hypothesis," which suggests that malonyl-CoA is a key indicator of nutrient availability and the brain uses its abundance to assess whole-body energy homeostasis. Others have pointed out that malonyl-CoA is a significant inhibitor of carnitine acyltransferase-1 (see Figure 24.16). Thus, malonyl-CoA may be influencing the levels of another metabolite whose concentration is more important as a signal of energy status. What metabolite might that be?
b. Another test of the malonyl-CoA hypothesis was conducted through the creation of a transgenic strain of mice that lacked functional hypothalamic fatty acid synthase (see Chapter 24). Predict the effect of this genetic modification on cellular malonyl-CoA levels in the hypothalamus, the eating behavior of these transgenic mice, their body fat content, and their physical activity levels. Defend your predictions.

Answer: Malonyl-CoA derives from acetyl-CoA by carboxylation and it is the immediate source of two-carbon acetyl units for fatty acid biosynthesis. High levels of malonyl-CoA would likely be an indicator of abundant metabolites that are converted to acetyl-CoA (like sugars or amino acids). Additionally, high malonyl-CoA levels would occur when fatty acids are abundant and there is little need for synthesis of fatty acids. In this case fatty acids are activated by addition of coenzyme A. Thus, malonyl-CoA levels would also be expected to reflect fatty acyl-CoA levels.

For b, mice missing fatty acid synthase could not consume malonyl-CoA leading to elevated levels of this metabolic intermediate. This would be expected to lead to a decrease in eating behavior and an increase in weight loss. In studies on mice using inhibitors of fatty acid synthase, food intake was suppressed and skeletal muscles increased oxidation of fatty acids leading to weight loss. (See Lane, M.D. et al, 2008, Regulation of food intake and energy expenditure by hypothalamic malonyl-CoA *International Journal of Obesity* 32 (suppl.4) S49-S54.) All of these responses were reversed when a viral vector expressing malonyl-CoA decarboxylase was introduced into the hypothalamus (stereotactically) indicating that malonyl-CoA levels were driving the responses.

18. Exploring leptin's mechanism of action
a. Leptin was discovered when a congenitally obese strain of mice (ob/ob mice) was found to lack both copies of a gene encoding a peptide hormone produced mainly by adipose tissue. The peptide hormone was named leptin. Leptin is an anorexic (appetite suppressing) agent; its absence leads to obesity. Propose an experiment to test these ideas.
b. A second strain of obese mice (db/db mice) produces leptin in abundance but fails to respond to it. Assuming the db mutation leads to loss of function in a protein, what protein is likely to be nonfunctional or absent? How might you test your idea?

Answer: Leptin is a peptide hormone secreted by adipocytes. In humans the mature peptide is 146 amino acids long and its structure is dominated by four antiparallel alpha helices. Leptin binds to leptin receptors, which a single-pass membrane spanning protein that is expressed in nervous tissue and peripheral tissue. Leptin binding to its receptor sets off a complicated signal transduction pathway with connections to insulin transduction pathways. (For a detailed discussion see: Anubhuti and Arora, S., 2008, Leptin and its metabolic interactions –an update *Diabetes, Obesity and Metabolism* 10:973-993.) Leptin seems to play two physiological roles. During periods of caloric balance (when intake balances output), leptin levels correlate with total body fat mass. So individuals with more body fat have higher leptin levels. The second role of leptin comes into play during periods of negative or positive energy balance. Under conditions of caloric excess, leptin levels rise and this results in signals that decrease appetite and increase energy expenditure. To test this in mice one could study the effects of leptin injections in ob/ob mice, which are genetically obese. Regular leptin injections would be expected to decrease appetite and weight.

The db/db mice lack leptin receptors. Testing this would be a bit tricky since injecting leptin receptors would not have the same effect in these mice as injecting leptin in ob/ob mice. (Leptin is a soluble protein whereas the receptor is a membrane-protein.) One could conduct binding assays to see if db/db mice lack leptin receptors but experiments with negative results must be carefully controlled.

Preparing for the MCAT® Exam

19. Consult Figure 27.7 and answer the following questions: Which organs use both fatty acids and glucose as a fuel in the well-fed state, which rely mostly on glucose, which rely mostly on fatty acids, which one never uses fatty acids, and which one produces lactate.

Answer: The organ that uses both fatty acids and glucose is the heart. The brain relies mostly on glucose but in times of low glucose it will use ketone bodies. Fatty acids are metabolized by adipose tissue, which can also use glucose to produce glycogen. Muscle relies heavily on fatty acids. Lactate is produced by skeletal muscle.

20. Figure 27.3 illustrates the response of R (ATP-regenerating) and U (ATP-utilizing) enzymes to energy charge.
a. Would hexokinase be an R enzyme or a U enzyme? Would glutamine:PRPP amidotransferase, the second enzyme in purine biosynthesis, be an R enzyme or a U enzyme?
b. If energy charge = 0.5: Is the activity of hexokinase high or low? Is ribose-5-P pyrophosphokinase activity high or low?
c. If energy charge = 0.95: Is the activity of hexokinase high or low? Is ribose-5-P pyrophosphokinase activity high or low?

Answer: a. Hexokinase is an R enzyme because it is in glycolysis, an ATP-producing pathway. Despite being an ATP consuming enzyme, it is active when energy charge is low because its activity eventually leads to ATP production.

Glutamine:PRP amidotransferase reacts glutamine with PRPP to produce 5-phosphoribosylamine, an intermediate in purine biosynthesis. It is regulated by adenine nucleotide pools through feedback inhibition by AMP, ADP and ATP. It is not, however, an ATP-dependent enzyme. Therefore, it is neither an R nor a U enzyme.

b. When the energy charge is 0.5, the activity of hexokinase is high. Despite the enzyme consuming ATP, it is stimulated by low ATP because its reaction is a gateway to glycolysis, an ATP producing pathway. Ribose-5-phosphate pyrophosphorylase converts ribose-5-phosphate to 5-phosphoribose-1-pyrophosphate (PRPP). This enzyme is active in purine metabolism. It is likely that this enzyme is a U enzyme and thus inhibited when energy charge is low. At EC of 0.5, we should expect the activity of this enzyme to be low.

c. When the energy charge is 0.95, we expect hexokinase, an R enzyme, to have low activity and ribose-5-phosphate pyrophosphokinase, a U enzyme, to have high activity.

Questions for Self Study

1. Describe the interplay between β-oxidation, a catabolic pathway, and fatty acid biosynthesis, an anabolic pathway. How are the two connected? Does degradation of a fatty acid provide sufficient resources to support resynthesis of a fatty acid of the same length?

2. What two important compounds couple anabolism and catabolism?

3. Metabolism in a typical aerobic heterotropic cell (a chemoheterotroph) can be described as an interaction of three functional blocks: catabolism, anabolism, and macromolecular synthesis and growth. How is this picture altered for a chemoautotroph? A photoheterotroph? A photoautotroph?

4. The balanced equation for oxidation of glucose via glycolysis, the citric acid cycle, and electron transport is $C_6H_{12}O_6 + 6 O_2 + 38 ADP + 38 P_i \star 6 CO_2 + 38 ATP + 44 H_2O$. Of the ATPs produced, how many derive from simple chemical stoichiometry and how many from evolved coupling stoichiometry?

5. Describe the reaction catalyzed by adenylate kinase. What important allosteric regulator is produced by this reaction?

6. Table 27.1 is reproduced below with the last three columns randomly ordered. Rearrange the entries in these columns to correct Table 27.1

Organ	Energy Reservoir	Preferred Substrate	Energy Source Exported
Brain	Glycogen, triacylglycerol	Fatty acids	Fatty acids, glycerol
Skeletal muscle (resting)	Triacylglycerol	Glucose(ketone bodies during starvation)	Fatty acids, glucose, ketone bodies
Skeletal muscle (prolonged exercise)	Glycogen	Fatty acids	None
Heart muscle	None	Glucose	Lactate
Adipose tissue	Glycogen	Amino acids, glucose, fatty acids	None
Liver	None	Fatty acids	None

Answers

1. In β-oxidation, fatty acids are converted into acetyl-CoA units. During the process, coenzymes are reduced. Fatty acid biosynthesis involves joining of acetyl units and oxidation of reduced coenzymes. Per acetyl unit, the number of reduced coenzymes produced in β-oxidation balances the number of reduced coenzymes consumed during fatty acid synthesis. However, fatty acid synthesis is driven by hydrolysis of high-energy phosphate bonds whereas β-oxidation, once primed, is thermodynamically spontaneous. To support anabolism of fatty acids from a catabolic series, some of the acetyl units of the catabolic series would have to be consumed in the citric acid cycle to provide energy for fatty acid biosynthesis.

2. ATP and NADPH.

3. For a photoautotroph an additional block is added in which light energy is converted to ATP and reducing equivalents which are used, in turn, to fix carbon dioxide. A photoheterotroph can also harvest light but cannot fix carbon dioxide. Thus, its block lacks the ability to produce reducing equivalents using light energy and to fix carbon dioxide. The chemoautotroph can fix carbon dioxide with reducing equivalents derived from inorganic compounds.

4. The conversion of glucose to two pyruvates produces two ATP. Two additional ATPs are generated by substrate level phosphorylation in the citric acid cycle in which two pyruvates are metabolized to carbon dioxide and water. Thus, the simple chemical reaction stoichiometry is 4 ATP.

5. The reaction catalyzed by adenylate kinase is ATP + AMP ⋆ 2 ADP. In the reverse reaction, AMP, an important allosteric regulator, is produced from ADP.

6. Table 27.1

Organ	Energy Reservoir	Preferred Substrate	Energy Source Exported
Brain	None	Glucose(ketone bodies during starvation)	None
Skeletal muscle (resting)	Glycogen	Fatty acids	None
Skeletal muscle (prolonged exercise)	None	Glucose	Lactate
Heart muscle	Glycogen	Fatty acids	None
Adipose tissue	Triacylglycerol	Fatty acids	Fatty acids, glycerol
Liver	Glycogen, triacylglycerol	Amino acids, glucose, fatty acids	Fatty acids, glucose, ketone bodies

Additional Problems

1. A plot of the rate of running versus distance for recent world records of several races is shown below. It appears that the data can be divided into three linear regions. For each region, describe the source of energy being utilized by muscle.

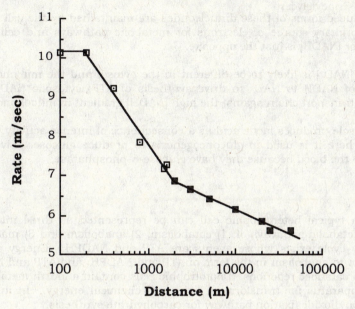

2. In brown fat, the protein thermogenin functions in a substrate cycle. Describe this cycle and state its purpose.

3. The energy of ATP hydrolysis is a highly exergonic process with a large negative free energy change. We have discussed in this chapter how coupling of ATP hydrolysis is used to drive pathways. Give an example of a reaction driven by ATP hydrolysis and describe how ATP hydrolysis is coupled to the reaction.

4. Explain why NADPH and NADH have different energy equivalents?

5. Why is there a difference in energy between cytosolic NADH and mitochondrial NADH?

6. Carl and Gerti Cori were the first to describe the metabolic interplay between muscle and liver involving lactic acid formation in muscle and glucose formation in liver. This metabolic couple is known as the Cori cycle. Describe it.

Abbreviated Answers

1. The first two points on the graph represent the winning performances for the 100- and 200-meter dash. These races are run almost completely on high-energy phosphate stores of ATP and phosphocreatine, which limit the muscle to about 20 sec of intense activity. The group of races constituting the line of high slope is using a combination of aerobic and anaerobic respiration. At the end of these races high-energy phosphate stores are exhausted, lactic acid levels have risen, and aerobic respiration is functioning maximally. The long-distance races rely completely on aerobic respiration. (It is of interest to note that the 2000-meter race, a grueling contest, is at the break-point between the long-distance races and the anaerobic/aerobic-dependent races.)

2. Thermogenin functions as a proton channel in the inner mitochondrial membrane. The substrate cycle involves protons as substrates. Protons are pumped out of the mitochondria by electron transport but are allowed back in via thermogenin. Thus, protons move from one side of the membrane to the other. The purpose is to generate heat.

3. Any one of the biotin containing enzymes is a good example. Biotin is involved in metabolism of one-carbon units at the oxidation level of a carboxylate. The reaction mechanism involves formation of a phosphorylated bicarbonate intermediate by nucleophilic attack of bicarbonate on the γ-phosphate of ATP, releasing ADP. This activated carboxylate is then transferred to enzyme-bound biotin and from there to substrate to produce the carboxylated product. Carboxylation of biotin thus depends on ATP hydrolysis.

4. The oxidized and reduced forms of these dinucleotides are maintained by the cell at very different levels. [NADPH] is used as a primary source of electrons for metabolic pathways and cells maintain [NADPH] > [NADP+]. The situation for [NADH] is just the opposite.

5. The ratio [NADH] to [NAD+] is likely to be different in the cytosol and the mitochondria. In addition, in order to use oxidation of NADH to NAD+ to drive synthesis of ATP, cytosolic NADH will have to donate electrons to the electron transport chain against the high [NADH] gradient in mitochondria.

6. In the Cori cycle, muscle produces lactic acid as a consequence of intense activity. Lactic acid is sent via the blood to the liver where it is used in gluconeogenesis to produce glucose. Liver cells are capable of releasing glucose back to the blood because they have glucose-6-phosphatase.

Summary

The metabolism of a typical heterotrophic cell can be represented as three interconnected functional blocks composing the metabolic pathways of: 1) catabolism, 2) anabolism and 3) macromolecular synthesis and growth. An energy cycle exists whose agents are ATP and NADPH. Energy and reducing power is delivered from catabolism to anabolism in the form of ATP and NADPH, and ATP and NADPH are regenerated from ADP and NADP+ via catabolic reactions. Phototrophic cells contain a fourth metabolic system consisting of the photochemical apparatus for transforming light into chemical energy. In autotrophic cells, a fifth system occurs, the carbon dioxide fixation pathway for carbohydrate synthesis.

Three levels of stoichiometry can be recognized in metabolism: 1) simple reaction stoichiometry, 2) obligate coupling stoichiometry, and 3) evolved coupling stoichiometry, particularly as represented in ATP coupling. The net yield of 38 ATP/ glucose in cellular respiration is a biological adaptation, not the consequence of inviolable chemical laws. At 38 ATP/glucose, the K_{eq} for cellular respiration is 10^{170}. The ATP coupling coefficient is defined as the number of ATP phosphoric anhydride bonds formed or broken in a given metabolic sequence. The ATP equivalent is the metabolic unit of energy exchange. All substrates, metabolites and coenzymes can be assigned an ATP value reflecting the net cost of their metabolic conversion in ATP equivalents. NADH has an ATP value of 3; FADH$_2$ of 2; and NADPH of 4. Considering the thermodynamics of a generally defined conversion, A⋆B, the nature and magnitude of the ATP equivalent can be illustrated. Coupling the A⋆B conversion to ATP hydrolysis raises the equilibrium ratio, $[B]_{eq}/[A]_{eq}$, by 1.4 x 10^8-fold. The energetics of ATP is crucial to the solvent capacity of the cell: The thermodynamic favorability of phosphoryl transfer by ATP makes it possible for the cell to carry out reactions with great efficiency, even though [reactants] are low.

If the ATP coupling coefficient for a particular metabolic sequence in one direction differs by 1 or more from the ATP coupling coefficient for the counterpart sequence running in the opposite direction, a substrate cycle is possible. In the absence of regulation, such substrate cycles could result in the net hydrolysis of ATP and wasteful dissipation of cellular energy. Phosphofructokinase (PFK) and fructose bisphosphatase (FBPase) constitute an example of a substrate cycle; both reactions are thermodynamically favorable under physiological conditions. It turns out that ATP coupling coefficients for opposing sequences always differ, and this difference allows both sequences to be thermodynamically favorable.

The concept of unidirectionality reveals an important feature of metabolic relationships: Nearly all metabolic pathways are unidirectional and fulfill only one purpose, either synthesis or degradation. For this to be possible, both pathways in pairs of opposing sequences must be thermodynamically favorable at essentially the same time and under the same conditions. The ATP coupling coefficient for each metabolic sequence has evolved so that the overall equilibrium for the conversion is highly favorable. This function of ATP in metabolism can be represented as the stoichiometric role of ATP. The most vivid illustration of ATP and unidirectionality is given by the grandest pair of opposing metabolic sequences in the biosphere - cellular respiration and photosynthetic CO$_2$ fixation. Cellular respiration has a coupling coefficient of +38 ATP and proceeds with a K_{eq} of 10^{170}; carbon dioxide fixation has a coupling coefficient of -66 and a K_{eq} of 10^{60}.

ATP also acts as an important allosteric effector in the kinetic regulation of metabolism, thereby serving a second distinctive role in metabolism. Energy charge is a concept that provides a measure of how fully charged the adenylate system is with high-energy phosphoryl groups. Enzymes can be classified as **R** or **U** in terms of their response to energy charge. **R**-type enzymes are members of ATP-regenerating metabolic pathways, while **U** enzymes are characteristically in biosynthetic, or ATP-utilizing sequences. Thus, **R** and **U** pathways are diametrically opposite with regard to ATP involvement. The reciprocal relationship between **R** and **U** systems means that energy charge reaches a point of metabolic steady-state at an E.C. value of 0.85 - 0.88. Phosphorylation potential is a better index of the cell's momentary capacity to drive phosphorylation reactions.

In multicellular organisms, organ systems have arisen to carry out specific physiological functions. Essentially all cells in these organisms have the same set of enzymes in the central pathways of intermediary metabolism, but the various organs do differ significantly in the metabolic fuels - glucose, glycogen, fatty acids, amino acids - that they prefer to use for energy production. Brain has a very high respiratory metabolism, essentially no fuel reserves, and is dependent on a supply of blood glucose as its principal fuel. Muscle is intermittently active as muscle contraction and relaxation takes place on demand. ATP is necessary to drive contraction in response to an increased $[Ca^{2+}]$ pulse as the metabolic signal. The Ca^{2+}-ATPase operating during muscle relaxation uses almost as much ATP as the acto-myosin contractile system. Muscle is a major storage site of glycogen. During strenuous exercise, the rate of glycolysis in muscle may increase 2,000-fold. Muscle fatigue is the result of a drop in cytosolic pH due to the accumulation of H^+ released during glycolysis. Heart, a rhythmically active muscle system, has minimal reserves of fuel such as glycogen or triacylglycerols, and prefers free fatty acids as fuel. Adipose tissue is a metabolically active, amorphous tissue widely distributed throughout the body. A 70-kg person has enough triacylglycerols stored in adipose tissue to fuel 3 months' worth of modest activity. Glucose is the signal for release of fatty acids from adipocytes to the blood: When glucose levels are sufficient, adipocytes can form glycerol 3-phosphate allowing triacylglycerol synthesis from fatty acids generated by triacylglycerol turnover. Brown fat is a form of adipose tissue containing lots of mitochondria. In brown fat, oxidative phosphorylation is uncoupled by thermogenin and the energy of fatty acid oxidation is released as heat. The liver is a major metabolic processing organ. It serves an essential role in buffering [blood glucose]. It is also a center for fatty acid turnover, ketone body formation and conversion of amino acids into other metabolic fuels.

Chapter 28

DNA Metabolism: Replication, Recombination, and Repair

• •

Chapter Outline

❖ Characteristics of DNA replication
  ⅄ Semiconservative
    • Meselson and Stahl (1958)
      • Cells grown in 15-N: Shifted to 14-N
      • DNA analyzed by CsCl gradient centrifugation after 1 or 2 generations
  ⅄ Initiation at oriC in *E. coli*
    • DnaA binds to 245 bp region
    • Strands separate at three AT-rich repeats on 5' end of oriC
    • DnaB binds to oriC forming prepriming complex: Helicase activity
    • Addition of primase completes primosome
  ⅄ dsDNA unwinding
    • Helicases unwind dsDNA
    • DNA gyrase relieves torsional stress from unwinding
    • Single-stranded DNA-binding protein prevents reannealing
  ⅄ Semidiscontinuous
    • Leading strand: Continuous
    • Lagging strand: Discontinuous: Okazaki fragments
❖ Properties of DNA polymerases
  ⅄ Template strand and primer required
  ⅄ Chain growth 5' to 3'
❖ Enzymology of DNA replication
  ⅄ DNA polymerase I: Major DNA polymerases in *E. coli*
    • First characterized by Kornberg (1959): Repair and removal of primers
      • 5' Polymerase activity
        ○ Low processivity
      • 3' Exonuclease
        ○ Proofreading
        ○ Exonuclease and polymerase on large (Klenow fragment)
      • 5' Exonuclease: Nick translation
  ⅄ Polymerase III: Replication
    • Core enzyme: Three subunits
      ○ α = polymerase
      ○ ε = 3' exonuclease: Proofreading
      ○ θ = ε subunit stabilization
    • DNA polymerase III*
      ○ Two cores
      ○ γ = Clamp loader
    • DNA polymerase III holoenzyme

517

- ○ Two β dimers (one dimer per core): Sliding DNA clamp
  - ⅄ Primase: Adds RNA primers
  - ⅄ DNA ligase: Joins 3' hydroxyl with 5' phosphate
  - ⅄ Termination
    - Ter locus contains oppositely oriented ter sequences
- ❖ Eukaryotic DNA replication
  - ⅄ Multiple replicators (ori's)
  - ⅄ Pre-Replication Complex
    - ORC (origin recognition complex) binds replicators
    - Cdc6p binds to ORC
    - RLFs (replication licensing factors) bind to DNA: Two required
      - RLF-B: Confined to cytosol: Binds when nuclear envelop is lost
      - RLF-M: MCM protein complex
  - ⅄ pre-RC substrate for two protein kinases
    - Cyclin-dependent protein kinase/cyclin B complex: Cyclin B-CDK
      - Phosphorylates ORC, Cdc6p, MCM, Cdc7p-Dbf4p
    - Cdc7p-Dbf4p kinase activated by cyclin B-CDK
      - Activity serves as replication switch
- ❖ Eukaryotic DNA polymerases
  - ⅄ DNA polymerase α: Initiator of nuclear DNA replication
  - ⅄ DNA polymerase δ: Principle enzyme
    - Forms complex with PCNA (DNA clamp)
    - High processivity
  - ⅄ DNA polymerase ε: Role in replication
  - ⅄ DNA polymerase β: DNA repair
  - ⅄ DNA polymerase γ: Mitochondrial enzyme
- ❖ DNA ends: Telomeres
  - ⅄ Tandem repeats of G-rich 5 to 8 bp segments
  - ⅄ Telomerase maintains telomeres
    - RNA-dependent DNA polymerase
    - Ribonucleoprotein particle where RNA serves as template for telomere
- ❖ Reverse transcriptase: Three activities
  - ⅄ RNA-dependent DNA polymerase
  - ⅄ RNase H: Removes RNA of RNA/DNA hybrid
  - ⅄ DNA-directed DNA polymerase
- ❖ General (homologous) recombination
  - ⅄ Meselson and Weigle: Recombination in λ
    - Density label two strains: Recombinants had intermediate density
  - ⅄ Holliday model
    - Chromosomes pair (synapse)
    - DNA unwinds
    - Strand invasion
    - Holliday junction
    - Branch migration
    - Resolution of junctions
      - Patch recombination
      - Splice recombination
- ❖ Enzymology of recombination
  - ⅄ RecBCD
    - Helicase and nuclease activities
    - Chi site: GCTGGTGG: Induces RecBCD nuclease activity
    - Product is ssDNA tail with chi site at 3'-end
  - ⅄ ssDNA binding protein: Coats DNA as RecBCD helicase unwinds dsDNA
  - ⅄ RecA
    - ATP-dependent DNA strand exchange

- Binds ssDNA and dsDNA: Forms extended filaments
- High-affinity primary DNA-binding site binds ssDNA
- Second dsDNA binds at secondary DNA-binding site
- When homologous DNA encountered dsDNA converted to ssDNA
- Strand exchange occurs
- RuvA, RuvB, RuvC proteins
    - RuvA and RuvB: Holliday-junction specific helicase: Catalyzes branch migration
    - RuvC endonuclease that resolves Holliday junctions
- ❖ Transposons
    - McClintock identified activator gene in maize in 1950 (Nobel prize for work in 1983): Mobile genetic element
    - Insertion sequences: Simplest transposon
- ❖ DNA repair
    - Mismatch repair: Methyl-directed pathway determines template strand
    - Photoreaction of pyrimidine dimers: Photolyases use light energy to reverse pyrimidine dimers
    - Excision repair
        - Base excision: Removal of single altered base
            - DNA glycosidase removes damaged base
            - AP endonuclease cleaves at AP site
            - Exonuclease removes AP site
        - Nucleotide excision: Removes damaged base by removing short ssDNA fragment
- ❖ Molecular nature of mutations
    - Point mutations: Base substitutions
        - Transitions: Purine to purine or pyrimidine to pyrimidine
        - Transversions: Purine to pyrimidine or pyrimidine to purine
    - Insertions or deletions
        - Cause frame shifting
        - Produced by intercalating agents
    - Base analog mutations
        - 5-Bromouracil: Thymine analog whose tautomeric form pairs with G (not A)
        - 2-Aminopurine: Incorporates as A but pairs with C
    - Chemical mutagens
        - Nitrous acid: Deamination of bases: C changed to U gives G to A mutation
        - Hydroxylamine: Modifies C: Modified C pairs with A not G
        - Alkylating agents: Modified bases mispair
- ❖ Immunoglobulin genes
    - Antibodies: Immunoglobulin proteins: Immune response agents that bind foreign substances called antigens
    - IgG: Major class of circulating antibodies
        - H_2L_2 heterotetramer: H = heavy chain, L = light chain
            - L chain: 214 amino acids
                - V_L: Residues 1-108 variable
                    - Hypervariable regions (complementarity-determining regions)
                - C_L: Residues 109-214 constant
            - H chain: 446 amino acids
                - V_H: Residues 1-108 variable
                    - Hypervariable regions
                - C_H: Residues 109-446 constant
                    - C_H1, C_H2, C_H3
        - Structure: Immunoglobulin fold
    - Immunoglobulin genes
        - DNA rearrangement during B lymphocyte formation produces novel genes
        - L-chain genes
            - Kappa genes and Lambda genes
            - Each has V and J (**j**oining) regions

- o Multiple J regions
- o Multiple V regions
- o V and J joined in different combinations
- • H-chain genes
  - • Multiple V regions
  - • Multiple D genes (diversity)
  - • Multiple J genes
  - • Multiple C genes
  - • V, D, and J joined in different combinations
- • V-J and V-D-J joining
  - • In germ line cells, V and D genes followed by CACAGTGN$_{23}$ACAAAAACC
  - • In germ line cells, D and J genes preceded by GGTTTTTGTN$_{12}$CACTGTG
  - • Sequences are complementary to each other: Recombination recognition signals
  - • RAG1 and RAG2 bind sequences: V(D)J recombinase

Chapter Objectives

DNA Replication

Review the structure of DNA before starting this chapter. You should already know that dsDNA is composed of two strands that run antiparallel to each other and that a single stand of DNA has a 5'-end and a 3'-end. Each strand is a polymer of deoxyribonucleoside monophosphates held together in phosphodiester linkage between the 5'-carbon of one nucleotide and the 3'-carbon of the next nucleotide. The strands are held to each other by complementary hydrogen bonds between pairs of bases AT, TA, GC, and CG.

DNA is replicated semiconservatively. Understand what this means and how the experiments of Meselson and Stahl proved it. These experiments employed the stable heavy isotope of nitrogen ^{15}N to grow a culture of *E. coli* to contain dense DNA. The culture was then shifted to ^{14}N-containing media, grown for a number of generations, and after each generation sampled to determine the density of DNA.

Replication initiates at specific locations termed origins of replication or ori. Typically, replication from an ori proceeds in both directions.

Enzymology

All DNA polymerases require a ssDNA template to direct the synthesis of a complementary strand. The complementary strand is produced by addition of deoxyribonucleoside monophosphate groups derived from dNTPs to the 3'-end of the growing complementary strand. Thus, the polymerase moves 3'★5' along the template and extends the newly synthesized strand 5'★3'.

The Kornberg polymerase DNA pol I, a 109-kD, single-chain polypeptide, was the first DNA polymerase to be characterized. It plays a role in replication and repair. You should know the various activities exhibited by this DNA-dependent DNA polymerase including polymerase, 3' exonuclease, and 5' exonuclease. The 3' exonuclease functions in proofreading, whereas the 5' exonuclease, acting concurrently with polymerase activity, can cause nick translation or movement of a DNA gap by removing DNA (or RNA) on the 5'-end of the gap and adding DNA to the 3'-end of the gap. DNA polymerase I can cause strand displacement and has low processivity.

The principal polymerase in DNA replication in *E. coli* is DNA polymerase III, a complex of several proteins with separate functions. The α subunit has polymerase activity, the ε subunit has 3'★5' exonuclease activity or proofreading, and several other protein components are active in assembly of polymerase III, binding of polymerase III to template DNA, and high processivity. The ssDNA used as template by DNA polymerase III is supplied by a combination of actions by DNA gyrase and DNA helicase. DNA gyrase introduces negative supercoils into dsDNA in an ATP-dependent process. Negative supercoils overcome torsional stress caused by helicase-catalyzed unwinding of dsDNA, a process also driven by ATP hydrolysis. This action at an origin of replication leads to a replication bubble consisting of two replication forks. At each fork there is leading strand synthesis and lagging strand synthesis. Leading strand synthesis occurs in the direction of replication fork movement whereas lagging strand replication is away from the replication fork. At a replication fork, DNA is replicated semidiscontinuously. Synthesis is primed on both strands at a replication fork by primase, a DNA-dependent RNA polymerase. As the replication fork moves to expose unreplicated DNA, the lagging strand is periodically reprimed. Later the RNA primers are removed by RNase H, an RNase specific for RNA:DNA hybrids, and by nick translation by DNA polymerase I. Any gaps are filled in leaving nicks that are sealed by DNA ligase.

RNA-dependent DNA polymerase or reverse transcriptase can convert an RNA template into DNA. Reverse transcriptases have, in addition to DNA polymerase activity, RNase H activity, and DNA-dependent DNA polymerase activity.

DNA Repair

Mismatch repair removes mispaired bases on DNA by scanning duplex DNA for mispairs, excising the incorrect member of the pair, and replacing it with the correct base. Mispairs in newly synthesized DNA are corrected because the template DNA is methylated while the nascent DNA strand is unmethylated. UV-induced pyrimidine dimers are corrected by a light-dependent cleavage catalyzed by photolyase. Excision repair is active against several kinds of damaged DNA, including apurinic and apyrimidinic sites created by base removal. The general approach to excision repair is to create flanking single-strand breaks on the same stand in order to remove the damaged DNA.

Recombination

General recombination involves exchange of DNA between two molecules that have similar sequences. DNAs with similar sequences that are derived from a common sequence are homologous. Thus, the process is known as homologous recombination. In homologous recombination the two DNAs pair, single-stranded nicks are produced, and the ssDNAs displace each other and ligate to form a branched structure. The branch migrates, causing strand exchange, and is finally resolved by strand cleavage into either patch recombinants or splice recombinants (see answer to problem 12). The enzymology of general recombination is best understood in *E. coli* in which Rec A protein, RecBCD protein complex, and ssDNA binding protein or SSB play key roles. The RecBCD protein complex initiates recombination by attaching to the end of DNA and unwinding it. As DNA is unwound it begins to rewind but at a slower rate giving rise to a bubble of ssDNA. When a specific sequence, GCTGGTGG, known as a Chi site is encountered, RecBCD causes a single strand nick producing a ssDNA tail, which is coated with SSB. RecA binds to SSB-coated ssDNA to form a nucleoprotein filament that binds to dsDNA and causes strand displacement. Ligation leads to branch formation, branch migration, and resolution of the Holliday junction.

Alteration in DNA sequence

DNA sequences can be altered by mutation, recombination, and transposon insertion. Mutations include insertions and deletions of one or more base pairs and changes in a single base (point mutations) leading to either a transition (a purine-purine change or a pyrimidine-pyrimidine change) or a transversion (purine-pyrimidine of vice versa). Small insertions and deletions can be produced by intercalating agents such as acridine orange whereas large deletions and insertions can be produced by recombination or by transposon mutagenesis. Transposons, first described by McClintock, are genetic elements that can change their location within DNA.

Immunoglobulin Genes

Understand how somatic recombination leads to cells with unique immunoglobulin genes. Recombination involves joining, in combinatorial fashion, segments of genetic information to produce a virtually unique recombinant.

Problems and Solutions

1. Semiconservative or conservative DNA replication
If ^{15}N-labeled E. coli DNA has a density of 1.724 g/mL, ^{14}N-labeled DNA has a density of 1.710 g/mL, and E. coli cells grown for many generations on $^{14}NH_4^+$ as nitrogen source are transferred to media containing $^{15}NH_4^+$ as sole N-source, (a) what will be the density of the DNA after one generation, assuming replication is semiconservative? (b) Suppose replication took place by a conservative mechanism in which the parental strands remained together and the two progeny strands were paired. Design an experiment that could distinguish between semiconservative and conservative modes of replication.

Answer: After one generation for both semiconservative and dispersive modes of replication, DNA of an intermediate density, $(1.724 + 1.710)/2 = 1.717$ g/mL, is expected. In order to distinguish between the two models, DNA is denatured into single-stranded DNA and subsequently analyzed by CsCl density gradient centrifugation to determine the density of the single strands. For the dispersive model, single-stranded DNA of only a single density is expected whereas for the semiconservative model, two DNA populations with slightly different densities will be observed, one corresponding to ^{15}N DNA and the other to ^{14}N DNA.

After one round of conservative replication one would expect to see two bands, one representing the template dsDNA with density corresponding to ^{14}N DNA and the other representing the newly synthesized DNA with density corresponding to ^{15}N DNA. Heat denaturation would not be expected to result in a change in density of either band. Additional rounds of replication would show an increase in the amount of ^{15}N DNA while ^{14}N DNA would remain constant.

2. The enzymatic activities of DNA polymerase I
(a) What are the respective roles of the 5'-exonuclease and 3'-exonuclease activities of DNA polymerase I? (b) What would be a feature of an E. coli strain that lacked DNA polymerase I 3'-exonuclease activity?

Answer: DNA polymerase I functions during replication to replace RNA primers with DNA and during DNA repair to fill in gaps produced by DNA repair endonucleases. The 5'-exonuclease activity is largely responsible for removing RNA primers and for the process known as nick translation. RNA primers are abundant in the lagging strand of newly synthesized DNA because this strand is reprimed frequently. Frequent repriming is necessary because the replication fork moves in a direction opposite to primer elongation on the lagging strand. During elongation, whenever DNA polymerase III encounters the 5'-end of a RNA primer, elongation stops, resulting in a gap between the 3'-end of newly synthesized DNA and the 5'-end of the previously used RNA primer. DNA polymerase I will extend these 3'-ends, using its DNA polymerase activity, while simultaneously removing the 5'-end of the RNA primer. The result is movement of the gap in a process known as nick translation and replacement of the RNA primer with DNA.

The 3'-exonuclease activity of DNA polymerase I is a proofreading activity. During polymerization, when the 3'-end is being extended, if an incorrect base is incorporated into the 3'-end causing a mismatch, DNA polymerase I pauses at that site and removes the mismatched base on the 3'-end. Cells with DNA polymerase I lacking this 3'-exonuclease proofreading capability will acquire random mutations at high frequency.

3. Multiple replication forks in E. coli I
Assuming DNA replication proceeds at a rate of 750 base pairs per second, calculate how long it will take to replicate the entire E. coli genome. Under optimal conditions, E. coli cells divide every 20 minutes. What is the minimal number of replication forks per E. coli chromosome in order to sustain such a rate of cell division?

Answer: The *E. coli* genome is 4.64×10^6 bp. Assuming that DNA replication proceeds at a rate of 750 base pairs per second, the time required to replicate this amount of DNA assuming bidirectional replication (i.e., two replication forks starting at a single origin of replication and moving in opposite directions) is given by:

$$\frac{4.64 \times 10^6 \text{bp}}{2 \times 750 \dfrac{\text{bp}}{\text{sec}}} = 3{,}093 \text{ sec} = 51.6 \text{ min} = 0.86 \text{ hr}$$

For the genome to be replicated within 20 min, the cell must be replicating DNA at the rate of:

$$\frac{4.64 \times 10^6 \text{bp}}{20 \text{ min}} = 2.32 \times 10^5 \frac{\text{bp}}{\text{min}}$$

At a replication bubble, DNA is being synthesized at the rate of

$$2 \times 750 \frac{\text{bp}}{\text{sec}} = 1{,}500 \frac{\text{bp}}{\text{sec}} = 90{,}000 \frac{\text{bp}}{\text{min}}$$

There are $\dfrac{2.32 \times 10^5 \dfrac{\text{bp}}{\text{min}}}{90{,}000 \dfrac{\text{bp}}{\text{min}}} = 2.58$ replication bubbles or 5.16 replication forks

4. Multiple replication forks in E. coli II
On the basis of Figure 28.2, draw a simple diagram illustrating replication of the circular E. coli chromosome (a) at an early stage, (b) when one-third completed, (c) when two-thirds completed, and (d) when almost finished, assuming the initiation of replication at oriC has occurred only once. Then, draw a diagram showing the E. coli chromosome in problem 3 where the E. coli cell is dividing every 20 minutes.

Answer: Replication in *E. coli* begins at a single origin of replication (ori C) and proceeds bidirectional at equal rates. At an early stage one expects to see a replication bubble extended from the origin equally to both sides of it. When one-third complete, the replication bubble will cover one-third of the circular genome. When nearly finished, the two daughter DNA's will appear as concatenated double-stranded DNA circles. The origin of replication of the daughter DNA's will initiate replication well before there own synthesis is complete. This gives rise to concatenated dsDNA molecules that are, themselves, beginning replication. For genes near the origin of replication, rapidly growing *E. coli* cells are effectively polyploidy for these genes.

In the figure below horizontal arrows show the progression of replication forks eventually resulting in two fully replicated chromosomes that are still joined. The vertical arrow shows the results of a second round of replication beginning before the previous round is complete.

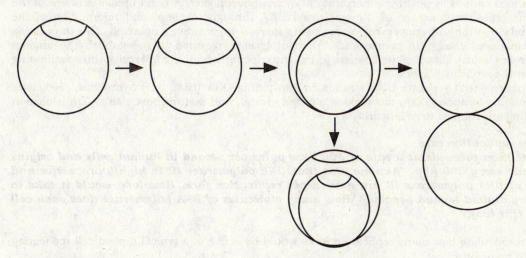

5. Molecules of DNA polymerase III per cell vs. growth rate
It is estimated that there are forty molecules of DNA polymerase III per E. coli cell. Is it likely that E. coli growth rate is limited by DNA polymerase III availability?

Answer: For maximum growth rate approximately 5 replication forks are active. (See problem 3.) If each replication fork has two copies of DNA polymerase III then 10 molecules of the enzyme should be sufficient to support maximum growth.

6. Number of Okazaki fragments in E. coli and human DNA replication
Approximately how many Okazaki fragments are synthesized in the course of replicating an E. coli chromosome? How many in replicating an "average" human chromosome?

Answer: Okazaki fragments are on the order of 1,000 to 2,000 nucleotides in length and are produced during lagging strand synthesis. The result of DNA replication is production of two duplex molecules. As a consequence of bidirectional replication, half of each newly synthesized strand is produced in lagging strand synthesis. Thus, the equivalent of a single strand of chromosomal DNA is produced as Okazaki fragments. The number of Okazaki fragments, assuming a length of 1000 nucleotides, is given by:

$$\frac{4.64 \times 10^6 \text{nt}}{1,000 \dfrac{\text{nt}}{\text{Okazaki fragment}}} = 4,640 \text{ Okazaki fragments, or}$$

2,320 Okazaki fragments 2,000 nucleotides long.

For a human chromosome:

$$\frac{3 \times 10^9 \dfrac{\text{nt}}{\text{hapoid genome}} \times 2 \dfrac{\text{haploid genome}}{\text{diploid genome}}}{1,000 \dfrac{\text{nt}}{\text{Okazaki fragment}}} = 6 \times 10^6 \frac{\text{Okazaki fragment}}{\text{diploid genome}}$$

Using the numbers given for Okazaki fragments we expect around 3 to 6 million Okazaki fragments. In eukaryotes, Okazaki fragments tend to be only 100 to 200 nucleotides in length and thus we would see ten-times more than determined by our calculation.

7. The roles of helicases and gyrases
How do DNA gyrases and helicases differ in their respective functions and modes of action?

Answer: DNA gyrases introduce negative supercoils into DNA. These enzymes act by binding to DNA at crossover configurations caused by positive supercoils, catalyzing phosphodiester bond breakage of one of the duplex DNAs at the crossover, passage of the uncleaved DNA through the gap, and reformation of the phosphodiester bonds resulting in conversion of positive supercoil to negative supercoil. The process is driven by a conformational change in gyrase and ATP hydrolysis is required to re-establish the original conformation. Gyrases move ahead of replication forks, introducing negative supercoils thus facilitating strand unwinding at the replication fork.

Helicases are enzymes that separate DNA strands by unwinding DNA using ATP hydrolysis. Helicases bind to regions of double-stranded DNA, move along a single strand, and disrupt base pairs. Translocation and strand unwinding are coupled to ATP hydrolysis.

8. Human genome replication rate
Assume DNA replication proceeds at a rate of 100 base pairs per second in human cells and origins of replication occur every 300 kbp. Assume also that DNA polymerase III is highly processive and only 2 molecules of DNA polymerase III are needed per replication fork. How long would it take to replicate the entire diploid human genome? How many molecules of DNA polymerase does each cell need to carry out this task?

Answer: First let us calculate how many replication forks would be active in a typical diploid cell replicating approximately 6×10^9 bp of DNA.

$$\frac{6 \times 10^9 \, \dfrac{\text{nt}}{\text{genome}}}{300 \times 10^3 \, \dfrac{\text{bp}}{\text{origin}}} = 2 \times 10^4 \, \text{origins or } 4 \times 10^4 \, \text{replication forks.}$$

If each replication fork is producing 100 base pairs per sec it would take:

$$\frac{6 \times 10^9 \, \text{bp}}{4 \times 10^4 \, \text{replication forks} \times 100 \, \dfrac{\text{bp}}{\text{sec}}} = 1,500 \, \text{sec or 25 min}$$

If there are two molecules of DNA polymerase per replication fork the cell would require 80,000 DNA polymerase molecules.

9. Heteroduplex DNA formation in recombination
From the information in Figure 28.17 and 28.18, diagram the recombinational event leading to the formation of a heteroduplex DNA region within a bacteriophage chromosome.

Answer: Assume that the dsDNA illustrated below represents the same gene from two bacteriophages. The genes are not identical but are homologous. To make heteroduplex DNA, single strand nicking, followed by strand invasion, ligation, branch migration, and resolution of the Holliday junction produces two heteroduplex molecules. Subsequent DNA mismatch repair or replication will produce recombinant bacteriophage.

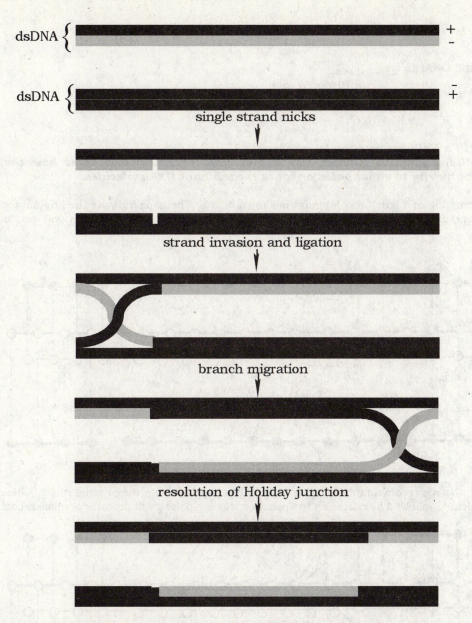

dsDNA { +−

dsDNA { −+

single strand nicks

strand invasion and ligation

branch migration

resolution of Holiday junction

10. Homologous recombination, heteroduplex DNA, and mismatch repair
Homologous recombination in E. coli leads to the formation of regions of heteroduplex DNA. By definition, such regions contain mismatched bases. Why doesn't the mismatch repair system of E. coli eliminate these mismatches?

Answer: Mismatch repair systems in *E. coli* rely on DNA methylation in order to distinguish between two strands to determine which will serve as a template. In homologous recombination, methylated DNA duplexes may be joined. In this case, both strands are methylated and mismatch repair will not be able to distinguish between them.

11. RecA:DNA nucleoprotein helix/B-DNA comparison
If RecA protein unwinds duplex DNA so that there are about 18.6 bp per turn, what is the change in Δφ the helical twist of DNA, compared to its value in B-DNA?

Answer: B-DNA has 10 bp per turn. The Δφ is calculated as follows:

$$\Delta\phi = \frac{360°}{10 \text{ bp}} = 36° \text{per bp}$$

For RecA-unwound DNA, $\Delta\phi$ is:

$$\Delta\phi = \frac{360°}{18.6 \text{ bp}} = 19.4° \text{per bp}$$

The change in $\Delta\phi = 36° - 19.4° = 16.6°$

12. Resolution of a Holliday junction by resolvase
Diagram a Holliday junction between two duplex DNA molecules and show how the action of resolvase might give rise to either patch or splice recombinant DNA molecules.

Answer: The formation of a patch recombinant is shown below. The gaps indicate the original nicks. Strand cleavage at the points indicated followed by religation to opposite cleavage points will produce a patch recombinant.

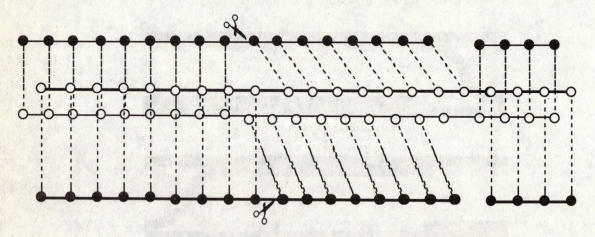

The formation of a splice recombinant is shown below. The gaps indicate the original nicks. Strand cleavage at the points indicated followed by religation to opposite cleavage points will produce a splice recombinant.

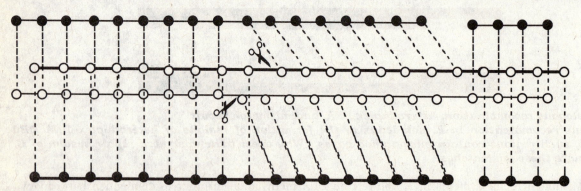

13. Chemical mutagenesis of DNA bases
Show the nucleotide sequence changes that might arise in a dsDNA (coding strand segment GCTA) upon mutagenesis with (a) HNO₂, (b) bromouracil, and (c) 2-aminopurine.

Answer: (a.) Nitrous acid (HNO_2) converts cytosine to uracil and adenine to hypoxanthine by deamination. Uracil base pairs with adenine, which subsequently pairs with thymine. Thus, cytosine to uracil results in C to T transitions. For the sequence GCTA, C to T transitions on the top strand will produce:

G**T**TA

Modification of C on the complementary strand (complementary to GCTA) will produce:

ACTA

It is possible, although much less probably, to get double-mutations:

ATTA

Hypoxanthine, produced by deamination of adenine, can pair with cytosine, which in turn pairs with G. The net result is A to G transitions caused by deamination of A. The sequence GCTA may be converted to:

GCT**G**

Modification of A on the complementary strand (complementary to GCTA) will produce:

GC**C**A

It is possible, although much less probably, to get double-mutations:

GC**CG**

We considered only mutations to C's or A's separately. It is possible to have simultaneous mutations to C's and A's, which would generate combinations of the sequences shown above.

(b.) Bromouracil is a base analog of uracil with a bromine atom at carbon 5. It resembles thymine and can be incorporated into DNA in place of T. However, the presence of the electronegative Br influences the keto to enol tautomerization. In the enol tautomer, 5-bromouracil pairs with guanine. Thus, incorporation of 5-bromouracil for T produces T to C transitions or A to G transitions. In addition, the enol tautomer may be incorporated into DNA in place of C, resulting in C to T transitions and G to A transitions but this might be expected to occur less often.

The sequence GCTA and its complement, TAGC, (both written 5' to 3') would each be synthesized in the presence of bromouracil and during synthesis of each strand the T's could be replaced by bromoU (U_{Br}). Thus, GCTA would be GCU$_{Br}$A and TAGC would be U$_{Br}$AGC. Mutations would occur when these U$_{Br}$-containing DNAs were replicated and mispaired with G. GCU$_{Br}$A would produce TGGC (written 5' to 3') the complement of which is **GCCA**. Thus, GCTA is mutated to GCCA. U$_{Br}$AGC would product GCTG (written 5' to 3'). Thus, GCTA is mutated to **GCTG**. The double mutation **GCCG** would occur but less frequently.

Bromouracil could substitute for C during DNA synthesis if it were in its enol form and paired with a G. The sequence GCTA and its complement, TAGC, (both written 5' to 3') would each be synthesized in the presence of bromouracil and during synthesis of each strand the C's could be replaced by bromoU (U_{Br}). Thus, GCTA would be GU$_{Br}$TA and TAGC would be TAGU$_{Br}$. Mutations would occur when these U$_{Br}$-containing DNAs were replicated and mispaired with G. GU$_{Br}$TA would produce TAAC (written 5' to 3') the complement of which is **GTTA**. Thus, GCTA is mutated to GTTA. TAGU$_{Br}$ would product ACTA (written 5' to 3'). Thus, GCTA is mutated to **ACTA**. The double mutation **ATTA** would occur but less frequently.

(c.) 2-Aminopurine (A_2) is an adenine analog that substitutes for A but can pair with cytosine. The sequence GCTA and its complement TAGC (both written 5' to 3') serve as templates during which A_2 is incorporated in place of A. Thus, GCTA produces TA$_2$GC and when this is replicated it produces **GCCA**. TAGC produces GCTA$_2$ and when this is replicated it produces CAGC the complement of which is **GCTG**. GCCG would be produced very rarely.

14. Transposons as mutagens
Transposons are mutagenic agents. Why?

Answers: Transposons are mobile genetic elements capable of relocating within genomes. They are also capable of moving DNA sequences to new locations via homologous recombination. This has several consequences. If a transposon moves into a gene it may inactivate the gene by insertional inactivation. If transposons, upon relocating, take with them genomic sequences this can cause problems depending on the nature of the genomic sequences moved. For example, if a poorly expressed gene is relocated nearby an enhancer element, expression of the gene may be greatly amplified. Alternatively, if a gene is separated from an enhancer it will decrease expression. Transposon relocation can therefore inactivate genes or alter the level of gene expression by either increasing or decreasing transcription. Simply changing the location of DNA within a genome may have consequences for gene expression.

15. Recombination in immunoglobulin genes
If recombination between V_K and J_K gene formed a CCA codon at codon 95 (Figure 28.39), which amino acid would appear at this position?

Answer: The recombination position is shown below.

```
      94      95      96      97
      Val     Pro     Ser     Leu
V•    GTTC A T C T T C G A
J•    A T G G C A A G C T T G
```

16. Helicase unwinding of the E. coli chromosome
Hexameric helicases, such as DnaB, the MCM proteins, and papilloma virus E1 helicase (illustrated in Figures 16.23–16.25), unwind DNA by passing one strand of the DNA duplex through the central pore, using a mechanism based on ATP-dependent binding interactions with the bases of that strand. The genome of E. coli K12 consists of 4,686,137 nucleotides. Assuming that DnaB functions like papilloma virus E1 helicase, from the information given in Chapter 16 on ATP-coupled DNA unwinding, calculate how many molecules of ATP would be needed to completely unwind the E. coli K12 chromosome.

Answer: Hexameric helicases use ATP hydrolysis to unwind DNA with approximately 1 ATP consumed per base pair. To completely unwind the genome of E. coli K12 it would require 4,686,137 molecules of ATP. Or, converting to moles:

$$4,686,137 \text{ molecules ATP} \times \frac{1 \text{ mol}}{6.02 \times 10^{23} \text{molecules}} = 7.8 \times 10^{-18} \text{mole}$$

17. Translesion DNA polymerase IV experiment
Asako Furukohri, Myron F. Goodman, and Hisaji Maki wanted to discover how the translesion DNA polymerase IV takes over from DNA polymerase III at a stalled replication fork (see Journal of Biological Chemistry 283:11260–11269, 2008). They showed that DNA polymerase IV could displace DNA polymerase III from a stalled replication fork formed in an in vitro system containing DNA, DNA polymerase III, the β-clamp, and SSB. Devise your own experiment to show how such displacement might be demonstrated. (Hint: Assume that you have protein identification tools that allow you to distinguish easily between DNA polymerase III and DNA polymerase IV.)

Answer: It would be useful to conduct an experiment that could identify proteins that are released from replication forks and those that are bound to them. Identification could be made using monoclonal antibodies specific for DNA polymerase III and antibodies specific for DNA polymerase IV. A stalled replication fork is a large macromolecular complex that should be separable from the DNA polymerase complexes using a technique like centrifugation, perhaps gradient centrifugation. Stalled replication fork complexes could be treated with increasing amounts of DNA polymerase IV and then separated by centrifugation. Fractions across a sucrose gradient, for example, could then be quantitated with respect to DNA polymerase III and IV. One would expect less DNA polymerase III associated with replication forks as the concentration of competing DNA polymerase IV is increased.

18. Functional consequences of Y-family DNA polymerase structure
The eukaryotic translesion DNA polymerases fall into the Y family of DNA polymerases. Structural studies reveal that their fingers and thumb domains are small and stubby (see Figure 28.10). In addition, Y-family polymerase active sites are more open and less constrained where base pairing leads to selection of a dNTP substrate for the polymerase reaction. Discuss the relevance of these structural differences. Would you expect Y-family polymerases to have 3'-exonuclease activity? Explain your answer.

Answer: The Y-family of DNA polymerases are involved in translesion DNA synthesis and lack 3'-5' proofreading exonuclease activity. In general DNA polymerases contain three structural regions, finger, thumb and palm. The palm region contains conserved acidic amino acids that coordinate the incoming dNTP. The finger domain also contacts the dNTP whereas the thumb is involved in duplex DNA binding. In high-fidelity polymerases these structural features create a very restricted active site that accommodates only a single unpaired template base. The Y-family polymerases contain a palm structure but the finger and thumb domains are short and stubby. This results in a less restricted active site because fewer contacts are made with both the incoming dNTP and the DNA template. The template-binding region is in fact more open allowing the polymerase to accommodate a lesion on the DNA.

Preparing for the MCAT® Exam

19. *Figure 28.11 depicts the eukaryotic cell cycle. Many cell types "exit" the cell cycle and don't divide for prolonged periods, a state termed G$_0$; some, for example, neurons, never divide again.*
a. What stage of the cell cycle do you suppose a cell might be in when it exits the cell cycle and enters G$_0$?
b. The cell cycle is controlled by checkpoints, cyclins, and CDKs. Describe how biochemical events involving cyclins and CDKs might control passage of a dividing cell through the cell cycle.

Answer: a. The cell cycle consists of four stages, mitosis, G$_1$, S and G$_2$. Cells not actively cycling enter a state termed G$_0$. Cells enter G$_0$ from G$_1$. A cell in G$_1$ has a normal complement of DNA because G$_1$ precedes DNA synthesis. Thus, activities in G$_0$, such as transcription and translation, are done with a "normal" complement of DNA. Some cells, such as vertebrate and invertebrate oocytes, stop their cell cycle in G$_2$ (G$_2$ arrest) and wait for events such as fertilization to continue the cell cycle.

b. In progressing through the cell cycle, a cell needs to be sure one phase is complete before entering the next. Thus, there are a number of "check-points" that control entry into the next phase of the cell cycle. These checkpoints ensure that a cell has completed one stage before entering the next phase. For example, mitosis is completed only after DNA is replicated and any necessary repairs to DNA are made to insure that daughter cells get identical copies of DNA. Controlling the use of origins of replication carefully regulates DNA synthesis. Origins are used (fired) once during DNA synthesis to insure that multiple rounds of replication do not occur, which would lead to polyploidy. To regulate usage of an origin, a protein complex is assembled at the origin. The complex, the Origin Recognition Complex (ORC), is joined by a number of proteins whose levels or activities are regulated during the cell cycle. These include Cdc6 (cell division cycle), various Mcm (mini chromosome maintenance) proteins and other proteins. They join ORC to form the pre-replication complex. The activity of the complex is to function as a helicase making ssDNA available to serve as template for replication. This activity depends, however, on removal of Cdc6 and modification of other proteins including some of the Mcm proteins by phosphorylation by a specific protein kinase termed CDK2 or cyclin-dependent kinase. As its name implies, this kinase is dependent on another protein, cyclin E, whose levels fluctuate during the cell cycle. Just before S-phase cyclin E accumulates and binds to CDK2. This complex then binds to the pre-RC and phosphorylates a number of proteins. Phosphorylation of Cdc6 leads to its degradation while phosphorylation of other proteins causes the origin or replication to fire.

20. *Figure 28.39 gives some examples of recombination in IgG codons 95 and 96, as specified by the V$_\kappa$ and J$_\kappa$ genes. List the codon possibilities and the amino acids encoded if recombination occurred in codon 97. Which of these possibilities is least desirable?*

Answer: The sequences to be joined are GTT CAT CTT CGA from V$_\kappa$ and ATG GCA AGC TTG from J$_\kappa$ codon for positions 94 through 97. During V/J joining position 97 may be produced as a composite of one to two bases from V$_\kappa$ and the remaining bases from J$_\kappa$. The first three codons from V$_\kappa$ are combined with a hybrid codon produced from V$_\kappa$ and J$_\kappa$. This is shown below. Recombinations at 97 give:
```
GTT CAT CTT C
+              TG = GTT CAT CTT CTG = Val-His-Leu-Leu

GTT CAT CTT CG
+               G = GTT CAT CTT CGG = Val-His-Leu-Arg
```
How could we know the benefits to Leu or Arg at codon 97? We could only say that the two amino acids are quite different and could cause profound consequences to protein structure. But we cannot say which would be better.

Questions for Self Study

1. The experiments of Meselson and Stahl showed that DNA is replicated semiconservatively. The experiments ruled out conservative and dispersive models. Describe the experiments and explain how they favored semiconservative replication over the other two models.

2. DNA polymerase I and DNA polymerase III in *E. coli* play key roles in DNA replication. Determine which polymerase best fits each of the statements below.
   a. Proof-reading on a separate subunit.
   b. Highly processive enzyme.
   c. Used to generate Klenow fragment.

    d. Multisubunit enzyme.
    e. Active in DNA repair.
    f. Repairs gaps left after RNA primers removed.
    g. Single polypeptide chain.
    h. Functions at a replication fork.
    i. Moderately processive enzyme
    j. Carries out nick translation.

3. Match the columns.

a. DNA gyrase	1. Rep protein.
b. Helicase	2. Joins 3' hydroxyl to 5' phosphate.
c. Leading strand	3. ATP-dependent dsDNA unwinder.
d. Lagging strand	4. RNA polymerase.
e. Primase	5. Okazaki fragments.
f. RNase H	6. Attaches holoenzyme to DNA.
g. DNA ligase	7. 3' end toward replication fork.
h. SSB	8. Type II topoisomerase.
i. Localizes helicase to replication fork	9. Blocks DNA 2° structure formation.
j. β-subunit of DNA pol III	10. Removes primers from Okazaki fragments.

4. What are the three enzymatic activities of reverse transcriptase?

5. If an error occurs during replication resulting in a mismatched base pair, how does the repair mechanism distinguish between the correct, template base and the incorrect base?

6. UV-damaged DNA can be repaired by excision repair or by dimer reversal. Explain.

7. To acquire new genes, cells can be transformed by plasmids. In this case the plasmid has the necessary genetic elements to direct its replication and segregation into daughter cells thus insuring that it is stably maintained. However, cells can also be transformed by DNA devoid of such elements (origins of replication). For example, in the classic experiments of F. Griffith, cells were transformed using sheared, chromosomal DNA. How can linear DNA (lacking an origin of replication) get incorporated stably into cells?

8. General recombination in *E. coli* depends on at least three proteins, RecA protein, RecBCD protein complex, and single-stranded DNA-binding protein (SSB). For each of the statements below indicate which protein fits best.
    a. Promotes RecA binding to ssDNA.
    b. Forms nucleoprotein filaments that unwind DNA to 18.6 bp per turn.
    c. Recognizes the sequence GCTGGTGG (Chi site).
    d. Catalyzes DNA strand exchange.
    e. Binds to single-stranded DNA only.
    f. Unwinds DNA in an ATP-dependent reaction.
    g. Binds to the ends of a DNA duplex.

9. In 1983, Barbara McClintock was awarded the Nobel Prize in physiology or medicine for discoveries she made in the 1950s. What important class of genetic elements did McClintock discover?

10. Briefly explain how antibody diversity is accomplished.

11. Match.

a. Deletion	1. GGGACCC ★ GGGGCCC
b. Insertion	2. GGGGCCC ★ GGGCCCC
c. Transversion	3. GGGACCC ★ GGGAACCC
d. Transition	4. GGGACCC ★ GGGCCC

12. The base analog 5-bromouracil (5-BU) is a thymine analog that causes transitions because it tautomerizes readily from the keto form to the enol form. In the keto form, 5-BU pairs with A but in the enol form it pairs with G. The structure of 5-BU in the keto form is shown below. Draw the enol form and show how this form pairs with G and how the keto form pairs with A.

5-BU   Adenine   Guanine

Answers

1. *E. coli* was grown in ^{15}N for several generations to uniformly label DNA with this heavy nitrogen isotope. Cells were then shifted to ^{14}N for a few generations and DNA was isolated after each generation. The densities of DNA were then compared by density centrifugation. It was found that after one generation DNA of a density intermediate to that of ^{15}N-labeled DNA and ^{14}N-labeled DNA was produced. Subsequent generations produced two bands of density, one corresponding to the band after one generation and a second corresponding to ^{14}N-labeled DNA. The intermediate density DNA, produced after one generation, was further analyzed under denaturing conditions and it was found to be composed of single-stranded ^{15}N DNA and single-stranded ^{14}N DNA.

2. a. III; b. III; c. I; d. III; e. I; f. I; g. I; h. III; i. I; j. I.

3. a. 8; b. 3; c. 7; d. 5; e. 4; f. 10; g. 2; h. 9; i. 1; j. 6.

4. RNA-directed DNA polymerase, RNase H (degrades RNA of RNA/DNA duplex), DNA-directed DNA polymerase.

5. The DNA is scanned for methyl groups located on the parent, template strand.

6. In excision repair, the damaged base is removed along with a few residues flanking the damage site. This action leaves a gap that is subsequently filled. In dimer reversal, photolyase binds to the dimer and, using the energy of visible light, disconnects the bases from each other.

7. The DNA must be incorporated into genomic DNA by either site-specific or general recombination. (Hfr crosses and general transduction also depend on recombination.)

8. a. SSB; b. RecA; c. RecBCD; d. RecA; e. SSB; f. RecBCD; g. RecBCD.

9. Mobile elements (transposable elements or transposons).

10. Genetic rearrangements produce an intact gene by combining several separate elements. Each element derives from a set of similar elements. Further, imprecise joining of elements adds to genetic diversity.

11. a. 4; b. 3; c. 2; d. 1.

12.

5-BU (enol)   Guanine   5-BU (keto)   Adenine

Additional Problems

1. The enzyme uracil-N-glycosylase is part of a DNA repair system that removes deoxyuracil from DNA. The enzyme scans DNA for uracil and hydrolyzes the glycosidic bond between sugar and base to leave an apyrimidinic site. How is this apyrimidinic site dealt with subsequently?

2. Suggest two mechanisms by which deoxyuracil gets into DNA. (Hint: one has to do with deoxynucleotide biosynthesis, the other has to do with instability of a certain base.)

3. Suggest a method of incorporating labeled dNTPs into DNA using DNA polymerase I.

4. Okazaki fragments arise from lagging strand synthesis. Yet in wild-type cells, no long fragments are initially observed. Can you explain why both long and short fragments are expected at early stages of replication? Bonus: Can you explain why long fragments are not observed in wild type cells but are observed in *ung* cells (*ung* codes for uracil-N-glycosylase).

5. The genetic locus of DNA Q mutations was first identified as one of a number of so-called *mut* (mutator) loci. What function of DNA metabolism might this locus be responsible for?

6. How was bidirectional replication first shown?

7. Explain why the base analog 5-bromouracil can give rise to T-A to C-G transitions.

8. The chemical mutagens and base analogs described in this chapter are very effective at inducing transitions. Transversions occur less frequently. Provide an explanation.

9. Based on your knowledge of enzymes that have been encountered in your studies on intermediary metabolism, can you correctly restate the "one-gene, one-enzyme hypothesis"?

Abbreviated Answers

1. An apurinic endonuclease recognizes the site and causes strand cleavage to remove the deoxyribose and several flanking positions, creating a gap that is repaired by DNA polymerase and ligase.

2. Deoxynucleotide biosynthesis involves formation of dNDPs from NDP without regard to base. Thus, dUDP is formed and converted to dUTP. Normally dUTPase hydrolyzes dUTP to dUMP and PP_i but with less than perfect efficiency. Occasionally dUTP is incorporated into DNA in place of dTTP. The second mechanism for formation of uracil in DNA is by nonenzymatic oxidative deamination of Cs.

3. DNA can be labeled by taking advantage of the nick translation activity of DNA polymerase I. DNA is first nicked with limited digestion by DNase I, which creates randomly positioned single-strand breaks. These breaks or nicks are positions at which DNA polymerase I will initiate nick translation. If radioactively labeled dNTPs are employed, the newly synthesized stand will be appropriately labeled.

4. Because the lagging strand is periodically reprimed, it is synthesized as short DNA fragments or Okazaki fragments. With time, Okazaki fragments are converted into long DNA fragments by removal of the RNA primers and ligation. The leading strand is synthesized continuously because its elongation is in the same direction as replication fork migration. No long fragments are observed in wild type cells because of the uracil repair mechanism. Uracil is randomly incorporated into both the leading and lagging strands. Removal of the base followed by chain cleavage excises the uracil residues, leaving gaps to be repaired. The leading strand, although produced continuously, is fragmented by this repair system. Mutant *ung* cells do not repair uracil-containing DNA and produce long and short fragments

5. DNA Q codes for the proof reading subunit of DNA polymerase III. Certain mutations at this locus cause cells to undergo frequent mutagenesis because of error-prone DNA replication.

6. Cells were grown for a number of generations in low amounts ^{3}H-thymidine to uniformly label DNA with a small amount of radioactivity. Cells were then shifted for a brief period of time to high levels of ^{3}H-thymidine. Cells were subsequently fixed and autoradiographed to determine the location of heavily labeled DNA.

7. The base analog 5-bromouracil will more readily undergo keto-enol tautomerization because of the influence of the bromine group. In the enol tautomeric form, 5-bromouracil will pair with G. Thus, a T-A pair will be converted to a 5-bromouracil-G pair then to a C-G pair.

8. Mutagens are effective only because the damage they cause is not efficiently repaired by DNA repair mechanisms. To produce a transition requires a purine-pyrimidine mismatch during mutagenesis and these are less likely to be repaired than purine-purine or pyrimidine-pyrimidine mismatches produced during transversion mutagenesis.

9. On several occasions we have encountered single polypeptide chains that have two enzymatic activities. For example, phosphofructokinase 2 and fructose bisphosphatase 2 activities, responsible for fructose-2,6-bisphosphate metabolism, are found on a single protein. A more correct statement would be "one-gene, one-polypeptide".

Summary

The maintenance of the genetic information encoded in the sequence of bases in DNA is accomplished via replication; the expression of this information is achieved through transcription. DNA replication is semi-conservative: The two strands of the DNA double helix separate and each serves as a template for the synthesis of a new complementary strand. Thus, two daughter DNA double helices identical in every respect are reproduced from a single parental double helix. Each daughter double helix consists of one parental strand and one new strand. Using the techniques of ^{15}N-density labeling of DNA and CsC1 density gradient ultracentrifugation, Meselson and Stahl provided the experimental proof establishing the semi-conservative mechanism of replication.

In 1957, Arthur Kornberg and his colleagues reported the discovery of an enzyme from *E. coli* capable of DNA synthesis. This enzyme, DNA polymerase I, catalyzed the incorporation of the deoxynucleoside 5'-triphosphates dATP, dGTP, dCTP and dTTP into DNA in the presence of a template strand to copy and a primer strand that provided a free 3'-OH end for nucleotide addition. Successive addition of nucleotides at the 3'-end of the primer drove chain elongation in the 5'★3' direction. *E. coli* DNA pol I is a 109 kD polypeptide of 928 amino acid residues possessing three catalytic sites: one responsible for the 5'★3' DNA polymerase activity and two others that catalyzed distinct exonuclease reactions, one in the 3'★5' direction (the "3'-exonuclease") and one in the 5'★3' direction (the "5'-exonuclease"). The 3'-exonuclease activity improves the fidelity of replication by checking the accuracy of base-pairing between the base just incorporated and its complementary base in the template. If this pairing is inappropriate, the 3'-exonuclease excises the offending base, giving the polymerase another chance to insert the correct base. The 5'-exonuclease activity acts only on dsDNA and serves a repair function: It "edits out" sections of damaged DNA or any ribonucleotides incorporated during the initiation of DNA replication (see below).

All DNA polymerases discovered thus far catalyze chain growth in the 5'★3' direction, adding each new base as specified by a template DNA strand according to the A:T/G:C base-pairing rules of Watson and Crick. The template is read in the antiparallel 3'★5' direction and the incoming nucleotide is added to the 3'-end of a primer chain. It turns out that DNA pol III, not DNA pol I, is the principal DNA replicating enzyme of *E. coli*. DNA pol III consists of a 165 kD "core" polymerase of α, ε and θ subunits, where the 120 kD α subunit provides the polymerase active site and the 27.5 kD ε subunit contributes the proofreading 3'-exonuclease function. In vivo, this "core" polymerase is part of an 800 kD complex, the DNA polymerase holoenzyme. The processivity of DNA polymerase III holoenzyme exceeds 5,000; that is, it can associate with a template strand and read along it to synthesize a complementary DNA strand greater than 5,000 nucleotides long without once dissociating.

DNA replication is a complex process. First, the DNA helix must be unwound to expose single-stranded template regions. Unwinding can impose torsional stress and ATP-dependent DNA gyrases act to introduce negative supercoils to counteract this stress. ATP-dependent helicases catalyze the actual disruption of the double helix, breaking the H-bonds between the base pairs as they move along a strand of the DNA duplex. Replication is bidirectional: It begins at unique sites on chromosomes ("origins of replication"). The strands are separated here to form a "bubble", and so-called "replication forks" proceed away from the origin in both directions along the parental dsDNA, growing two daughter DNA duplexes in their wake. Replication is semi-discontinuous because DNA polymerases only work in the 5'★3' direction, reading a template in the 3'★5 sense. This polarity means that DNA polymerases can continuously copy the emerging 3'★5' strand, but the 5'★3' parental strand must be copied discontinuously: Only when a single-stranded stretch of this 5'★3' strand has been exposed can the DNA polymerase move long it in the 3'★5' direction to synthesize its complement. Reiji Okazaki provided the experimental verification for this semi-discontinuous mode of replication when he discovered that the radioactivity immediately incorporated into newly synthesized DNA occurred in short fragments, 1,000-2,000 nucleotides in length. With time, this radioactivity was associated with progressively longer DNA strands as the Okazaki fragments were ligated together to form a covalently contiguous DNA chain. DNA ligase catalyzes this reaction. The continuously synthesized DNA strand is called the "leading strand"; the discontinuously synthesized strand is called the "lagging strand". The primers for DNA synthesis in vivo are RNA oligonucleotides complementary to the DNA template. These RNA primers are synthesized by primase, an RNA polymerase. No RNA is found in mature DNA duplexes because DNA pol I binds at the 3'-OH nicks of Okazaki fragments. Its 5'★3' exonuclease then cuts out the RNA and the gaps this creates are filled in with DNA by the 5'★3' polymerase activity. DNA ligase then seals the junctions.

In *E. coli*, DNA replication begins at a unique 245 bp site, *oriC*, and proceeds bidirectionally. *DnaA protein* first binds at four 9 bp repeats within *oriC* and then additional *DnaA protein* binds to create a nucleosome-like assembly. *DnaA protein* then mediates the opening of the duplex at three A:T-rich tandem repeats at the 5'-side of *oriC*. Other proteins bind to form a "pre-priming complex" and the helix is further unwound. Primase then joins the pre-priming complex to yield the primosome. Two primosomes are formed at *oriC*, one for each replication fork. At each replication fork, two DNA polymerase III holoenzymes are present; one catalyzes leading strand synthesis and the other, lagging strand synthesis.

In eukaryotes, DNA replication occurs during the S phase of the cell cycle. A number of multimeric DNA polymerases are found in eukaryotic cells. DNA polymerase α and DNA polymerase δ are believed to constitute an "asymmetric dimer" replicase located at each replication fork, with DNA polymerase δ synthesizing the leading strand and DNA polymerase α forming the lagging strand.

RNA-directed synthesis of DNA occurs in certain RNA viruses known as retroviruses. The enzyme responsible is reverse transcriptase. Like all DNA polymerases, it incorporates nucleotides in the 5'★3' direction, reading its template in the 3'★5' direction. A tRNA H-bonded to the RNA template serves as primer. Reverse transcriptases display two other enzymatic activities in addition to their RNA-dependent DNA polymerase activity: an RNase H activity and a DNA-dependent DNA polymerase activity. RNase H degrades RNA chains in RNA:DNA hybrid duplexes. Its role in retroviral replication is to digest the genomic RNA chain so that the DNA-dependent DNA polymerase activity of reverse transcriptase can then copy the just-made DNA strand to give a duplex DNA. This DNA duplex then either mediates the subsequent course of the viral infection or becomes inserted into the host genome where it can lie dormant for many years. The reverse transcriptase of human immunodeficiency virus, the etiological agent in AIDS, is inhibited by the triphosphate derivative of AZT.

Genetic recombination involves the breakage and reunion of DNA strands, so that a physical exchange of parts takes place. The Holliday model for general recombination postulates a sequence of events including alignment of homologous sequences on two different duplex DNA molecules, introduction of single-strand nicks at analogous sites on both DNAs, invasion of each of the single strands into the other DNA duplex (strand invasion), and ligation of the free ends from different duplexes to create a cross-stranded intermediate, the Holliday junction. This junction can migrate along the DNA molecules so that considerable lengths of nucleotide sequence from one duplex become associated via base pairing with the other DNA duplex. If strands of the duplexes become nicked and then re-ligated with strands of the other, recombinant duplexes can be formed. The enzymology of general recombination is understood, at least in outline. The *Rec*BCD complex attaches to the end of a DNA duplex and progresses along it, unwinding the helix in an ATP-dependent reaction. *Rec*BCD enzyme also nicks the duplex at Chi "hotspots" of recombination to produce a ssDNA tail, a necessary prelude to the entry of *Rec*A protein into the process. The protein *Rec*A catalyzes the DNA strand exchange reaction, the reaction by which one strand of a duplex is displaced and replaced by an invading single strand from another duplex in an ATP-driven process. SSB (single-stranded DNA-binding protein) facilitates the action of RecA by binding to ssDNA and keeping it from assuming any secondary structure, which might impede strand exchange. The Holliday junction ensuing from RecBCD, RecA action upon two adjacent and homologous DNA duplexes is resolved in *E. coli* by the *ruv*C endonuclease/resolvase into one of the two isomeric recombinant forms, patch or splice. Cleavage of exchanged strands yields patch recombinants; cleavage of parental strands gives splice recombinants.

The immunoglobulin genes represent a system that has evolved for generating a large repertoire of protein diversity from a finite amount of genetic information. DNA rearrangement (gene reorganization) is the biological mechanism that allows a variety of protein isoforms from a limited number of genes. DNA rearrangement occurs only among genes encoding proteins of the immune response, like the immunoglobulin genes. IgG is an α₂β₂ tetramer composed of 2 light (L) chains and 2 heavy (H) chains. The L and H polypeptide chains are remarkable in that the amino acid sequence over the first 108 residues of both is highly variable. It is the variable region of the immunoglobulin polypeptide chains that is ultimately responsible for antigen recognition and binding. Sequence diversity in these regions endows these proteins with the unparalleled repertoires of structure recognition essential to the immune response. The unusual organization of the immunoglobulin genes accounts for this amino acid sequence diversity: the genetic information for an immunoglobulin chain is scattered among multiple gene segments in germline cells.

534

During vertebrate development, these segments are assembled by DNA rearrangement into complete genes. The mechanism for light-chain and heavy-chain gene assembly is based on secondary structures formed between consensus heptamer and nonamer sequences separated by 12-bp and 23-bp spacers. These heptamer/nonamer, 12-23-bp spacers are located adjacent to the rearranging genes and act to bring the genes into juxtaposition for the recombination events that underlie L-chain and H-chain gene assembly. Imprecise joining (the joining of these genes within codes) creates even greater amino acid sequence diversity in the variable regions of κ chains.

Mutations change the nucleotide sequence of DNA, either by substituting one base pair for another, or by inserting or deleting one or more base pairs. Point mutations are those mutations due to the substitution of one base pair for another. They may be either transitions (where a purine replaces a purine, or a pyrimidine, a pyrimidine) or transversions (where a purine replaces a pyrimidine, or vice versa). Point mutations arise from a number of causes. Some are due to mistakes in base pairing between the usual bases of DNA, others appear via the incorporation of base analogs with aberrant base pairing properties into DNA during replication, and still others are a result of chemical modifications altering pairing of bases already present in the DNA. Intercalation of flat aromatic molecules like acridine orange creates insertion/deletion mutations through distortion of the DNA double helix that leads to a misreading during replication. Transposons are discrete DNA segments, about 10 kb in length, which promote their own movement into and out of DNA. Transposons are thus a natural source of insertion mutations.

Chapter 29

Transcription and the Regulation of Gene Expression

• •

Chapter Outline

❖ Central Dogma: Crick: DNA → RNA → protein
  ⋏ Transcription: DNA to RNA
  ⋏ Translation: RNA to protein
❖ Transcription
  ⋏ Enzymology (*E. coli*)
    • DNA-dependent RNA polymerase (RNA polymerase)
      • 3' Polymerase activity
      • NTPs as substrates
    • Subunit structure
      • αβ'β: Core enzyme: Elongation
      • αβ'βσ: Holoenzyme: Initiation
      • β' = DNA binding subunit
      • β = NTP binding sites
      • σ = promoter recognition
  ⋏ Initiation
    • Closed promoter complex
      • σ of holoenzyme binds to promoter
      • Promoter: Two conserved sequences
        ○ Pribnow Box (-10): TATAAT
        ○ -35: TTGACA
        ○ Separation between elements: 17 bp
    • Open promoter complex
      • dsDNA unwound
      • RNA polymerase at start of initiation (+1)
        ○ Negative numbers: Upstream of start site
        ○ Positive numbers: Downstream of start site
    • Several phosphodiester bonds made
      • Initiation site binds ATP or GTP (complementary to template start site)
      • Elongation site binds NTP complementary to template
      • Phosphodiester bond formation
      • After 5 to 9 bonds formed σ dissociates
        ○ Rifamycin: Blocks initiation site NTP binding
        ○ Rifampicin: Blocks translocation
        ○ Cordycepin: Adds to 3' end of RNA but lacks 3'OH: Chain termination
  ⋏ Termination: Two types
    • Rho independent termination: Three terminator elements
      • G/C rich inverted repeat forms stem
      • Nonrepeated sequence separates inverted repeats forms loop

- 6 to 8 Us on end
- RNA polymerase pauses and terminates after element is transcribed
    - Rho dependent termination
        - Rho: ATP-dependent helicase
        - Rho binding site: C-rich regions of transcript
- ❖ Transcriptional regulation in prokaryotes
    - ⋏ Gene organization
        - Genes coding for enzymes with common goals often transcribed together
        - Genes organized into operon
        - Transcription produces polycistronic message
        - Operator near operon promoter regulates initiation of transcription
            - Induction: Increased synthesis in response to specific substrate
                - o Inducer: Small molecule binds to protein
                - o Gratuitous inducer: Not metabolized (e.g., *lac* operon's IPTG)
            - Repression: Decreased synthesis in response to specific metabolite
            - Constitutive expression: Expression not regulated
    - ⋏ *lac* Operon: Jacob and Monod: Operon hypothesis
        - Genes for lactose metabolism expressed when lactose present and glucose absent
        - Operator: Palindromic sequence near promoter recognized by *lac* repressor protein
        - *Lac* repressor protein: *lacI* gene product
            - In absence of inducer *lac* repressor binds to operator and blocks promoter
            - Inducer binding to *lac* repressor causes release of repressor from operator: Promoter available
        - CAP: Positive regulator
            - Catabolite activator protein: Binds near promoter increases promoter strength
                - o Binding is cAMP dependent
                - o cAMP levels determined by adenylyl cyclase levels
                - o Glucose leads to low cAMP levels
    - ⋏ *araBAD* operon: Positive and negative control by *AraC*
        - CAP protein: Catabolite repression
        - *araC* gene product: Induction via arabinose
            - *AraC* gene product binds to operator: Blocks transcription
            - Arabinose binding relocates *araC* gene product on operator and binds CRP-cAMP: Activation
    - ⋏ *trp* Operon
        - Trp repressor: *trpR* Gene product
            - When complexed to tryptophan: Binds to operator and blocks transcription
            - Low tryptophan cause trp repressor to release from operator
            - Trp repressor regulates two other operons
                - o *aroH*: Aromatic amino acid synthesis: Related metabolism
                - o *trpR*: Example of autogenous regulation (autoregulation)
        - Attenuation: Any regulatory mechanism that manipulates transcription termination or transcription pausing to regulate gene transcription downstream
            - Leader peptide mRNA properties
                - o Two tryptophan codons (High percentage of tryptophan in 14 amino acid polypeptide)
                - o Three secondary structures:
                    - ▪ Pause structure and terminator: Compatible
                    - ▪ Antiterminator: Incompatible with pause and terminator
                - o Low trp
                    - ▪ Slow progression of ribosome
                    - ▪ Formation of antiterminator
                - o High trp
                    - ▪ Rapid progression of ribosome
                    - ▪ Formation of pause/terminator structure

538

- ❖ Eukaryotic transcription: Three RNA polymerases: Require transcription factors to bind to promoter
  - ↟ RNA polymerase I
    - • Transcribes rRNA genes
    - • α−Amanitin resistant
  - ↟ RNA polymerase II
    - • Transcribes mRNA genes
    - • α−Amanitin sensitive
  - ↟ RNA polymerase III
    - • Transcribes tRNA and 5S rRNA genes
    - • α−Amanitin sensitive but less than RNA polymerase II
- ❖ RNA polymerase II
  - ↟ Multiple subunits (12 in yeast)
    - • RPB1: DNA binding (β'-like)
      - • C-terminal domain (CDT): PTSPSYS repeats
        - ○ Unphosphorylated CDT: Initiation competent
        - ○ Phosphorylated CDT: Elongation competent
    - • RPB2: NTP binding (β−like)
    - • RPB3 (α-like), RPB4 (σ-like)
  - ↟ Promoters: General transcription factors bind to common promoter elements
    - • TATA box at -25
    - • Initiator element (Inr) at start site
    - • DPE at +30
    - • Proximal elements
      - • CAAT box: Around -80: Strong promoter
      - • GC box: Housekeeping genes
    - • Enhancers (upstream activation sequences)
      - • Distance and orientation insensitive
    - • Response elements: Promoter modules that make genes responsive to common regulation
  - ↟ Initiation: Core promoter bound by basal apparatus: Components
    - • RNA polymerase II
    - • General transcription factors: TFIIB, TFIID, TFIIE, TFIIF, TFIIH, TFIIS
      - • TFIID
        - ○ TATA-binding protein
        - ○ TAFs: TBP-associated factors
    - • Mediator: Connects enhancers to general transcription factors
- ❖ Transcriptional regulation in eukaryotes
  - ↟ Nucleosome structure
    - • Swi/Snf complex disrupts nucleosomal arrays in ATP-dependent manner
    - • Histone Modifications
      - • Histone acetyltransferse (HAT): Acetylate lysine to remove charge
      - • Phosphorylation of ser, methylation of lys, sumoylation of lys
      - • Histone deacetylases complex (HDAC): Remove acetyl groups
      - • Histone code: Patterns of modifications recruit specific proteins
- ❖ DNA-binding motifs
  - ↟ Helix-turn-helix
    - • C-terminal helix recognition helix
      - • Binds to major groove
  - ↟ Zinc finger motifs
    - • C_2H_2 class: Zn coordinated by 2 C and 2 H
    - • C_x class: Zn coordinated by cysteine
  - ↟ bZIP motif: Basic region with leucine zipper
    - • Leucine zipper: Protein dimerization domain: Coiled-coils formed between two zippers
    - • Basic region: Rich in basic amino acids: DNA binding domain
- ❖ Post-transcriptional processing in eukaryotes
  - ↟ Capping: 5'-end activity

539

- Guanylyl transferase adds guanylyl in 5'-5' linkage
- 7 Position of G methylated
- Polyadenylation: 3'-end activity
  - Consensus AAUAAA in mRNA signal for endonuclease and polyA polymerase activity
    - mRNA cut 10 to 30 nt downstream
    - 100 to 200 adenine residues added to 3'-end
- Splicing
  - hnRNA (heterogeneous nuclear RNA) converted into mRNA
  - Introns removed
    - 5' Splice site
    - 3' Splice site
    - Branch site: Lariat formed
  - Exons joined
  - Process catalyzed by spliceosome
    - Small nuclear ribonucleoprotein particles (snRNPs)
    - Constitutive splicing: Formation of single mRNA
    - Alternative splicing: Formation of a variety of RNAs

Chapter Objectives

This chapter covers several important topics including RNA production or transcription, RNA processing, regulation of transcription and gene expression, and DNA-binding proteins.

RNA

RNAs are linear polymers of ribonucleotide monophosphates held together by phosphodiester bonds between the 5'-carbon of one nucleotide and the 3'-carbon of another nucleotide. The polymers have two distinct ends, a 5'-end and a 3'-end. Single-stranded RNAs often can form intrachain hydrogen-bonded structures by complementary base pairing. There are three general classes of RNAs: mRNAs, the agents of gene expression that bring genetic information from DNA to the ribosome to direct assembly of proteins; ribosomal RNAs (rRNA), structural and functional components of ribosomes; and, small stable RNAs like tRNAs, adaptor molecules that participate in protein synthesis.

RNA Polymerase

The principal RNA polymerase in *E. coli* is RNA polymerase I, a DNA-dependent RNA polymerase with $\alpha_2\beta\beta'$ subunit composition. In order to initiate transcription, the core polymerase requires the sigma factor σ which functions to identify specific DNA sequences termed promoters. A typical *E. coli* promoter is composed of a -35 region, with a consensus sequence of TCTTGACAT and a -10 region or Pribnow box with consensus sequence TATAAT. Initiation begins with the RNA polymerase holoenzyme, $\alpha_2\beta\beta'\sigma$, binding to the -35 region, and migrating to the -10 region where strand separation is initiated to expose the start-of-transcription site. Initiation of transcription usually begins with ATP or GTP in part because the initiation site, a nucleotide binding site on the RNA polymerase preferentially binds these nucleotides. Nucleotide binding at this site, located on the β subunit, can be blocked by rifamycin B. (A related compound rifampicin allows the first phosphodiester bond to form but blocks subsequent RNA polymerase movement.) After a short oligonucleotide has been synthesized, the σ factor dissociates from the RNA polymerase leaving the core complex to the task of elongation. Elongation continues until termination signals are encountered of which there are two types, rho-independent and rho-dependent. In rho-independent termination the 3'-end of the RNA is synthesized to include the termination signal, a sequence that forms a short G:C-rich stem and small loop with a run of unpaired Us. Termination sequence folding causes RNA polymerase to pause and terminate transcription. Rho-dependent termination, a less common mechanism, involves ρ factor, an ATP-dependent helicase that unwinds RNA:DNA duplexes in response to C-rich regions of the mRNA unoccupied by ribosomes.

Eukaryotes contain three classes of nuclear RNA polymerases: RNA polymerase I responsible for rRNA transcription, RNA polymerase II which transcribes protein-encoding genes, and RNA polymerase III responsible for 5S rRNA, tRNA, and other small stable RNAs. The three classes of RNA polymerases are distinguishable by their sensitivity to α-amanitin, a mushroom poison. RNA polymerase I is resistant to the compound, RNA polymerase II is sensitive to inhibition of chain elongation by α-amanitin, and RNA polymerase III is less sensitive. All three classes of RNA polymerases interact with specific promoter sequences with the aid of transcription factors.

RNA polymerase II promoters commonly have sequence elements near the start of transcription, examples include the TATA box and an initiator element. The TATA box has a TATAAA consensus sequence and is located 25 bases upstream of the start of transcription. The transcription start site is located within the initiator element.

Regulation of Transcription

Lactose is metabolized by a set of gene products encoded by the *lac* operon. Transcription of the *lac* operon results in a polycistronic mRNA, a mRNA with coding regions for three separate proteins. Regulation of the *lac* operon is achieved by repression and activation. RNA polymerase requires an additional protein to transcribe the *lac* operon because the promoter is a rather weak, inefficient promoter. The catabolite activator protein (CAP) acts as a positive regulator in this regard by binding to the *lac* promoter. However, binding occurs only in the presence of cAMP, which binds to CAP. *E. coli* prefers glucose and will metabolize it in preference to lactose when both sugars are present. Glucose preference results because glucose inhibits cAMP production leading to loss of promoter binding by CAP. Repression of the *lac* operon is achieved by *lac* repressor protein binding to an element termed the operator located near the promoter. Repressor binding blocks RNA polymerase from gaining access to the transcription start site. Release of repression occurs when lactose is present. Lactose is converted to allolactose which functions as an inducer by binding to the *lac* repressor protein, causing it to dissociate from the operon. Derivatives of lactose that are nonmetabolizable yet function as inducers are termed gratuitous inducers. The *lac* operon is under both positive and negative control and is both inducible and repressible.

The arabinose operon is also regulated by repression and induction. Repression is caused by the *araC* gene product binding as a dimer to two sites, one of which is located near the promoter and the other some distance away to produce a DNA loop. Arabinose binding to *araC* protein causes it to vacate the distant site, remain attached to the promoter site and bind to an additional site in the promoter. In this conformation, the *araC* dimer binds two CAP-cAMP, causing the promoter to turn on.

The *trp* operon is regulated by repression and by transcription attenuation. Repression is achieved by *trp* repressor protein binding to the promoter in a trp-dependent manner. The ability to transcribe and translate concurrently makes attenuation possible. The attenuator is a short region near the end of the first gene in the *trp* operon. The attenuator is remarkable in having two tandem Trp codons in a region that can form two different secondary structures. When Trp is plentiful, translation is tightly coupled to transcription and one of the two secondary structures, the terminator secondary structure, is favored. When Trp is low, the ribosome pauses at the tandem Trp codons, allowing the attenuator to assume an antiterminator secondary structure.

Transcription in eukaryotes is regulated at initiation by promoter strength, by enhancers or upstream activation sequences (UAS), and by responsive elements. The function of enhancers is independent of proximity to the promoter and orientation of the enhancer with respect to the promoter. Responsive elements are located near the transcription start site and are recognized by specific proteins that bind to the responsive element.

DNA Binding Motifs

There are three prominent classes of DNA-binding motifs, helix-turn-helix, zinc finger, and leucine zippers. The helix-turn-helix motif is typically 20 residues arranged into a 7-residue helix, a small β-turn, and a second helix of about 7 residues that functions as the recognition helix. The recognition helix interacts with a specific DNA sequence via the major groove. The zinc finger motif contains a zinc-binding site defined by either two cysteines and two histidines (C_2H_2) or by several cysteines (C_x). The domain is at the base of a small loop that interacts with DNA via the major groove. The leucine zipper motif is a periodic repetition of leucine zippers along a helix such that leucines line one face of the helix. Proteins containing leucine zippers do not actually bind DNA via the leucine zipper. Rather DNA binding is a property of a sequence of predominantly basic amino acids N-terminal to the leucine zipper. The zipper functions as a protein dimerization site.

Post-Transcriptional Processing

In eukaryotes, transcription produces a class of primary transcripts in the nucleus known as heterogeneous nuclear RNA. This RNA is processed in three ways to produce mature mRNA. The primary transcripts are modified on the 5'-end by a cap structure, on the 3'-end by a polyA tail, and internally by splicing. The 5' cap consists of a guanine residue joined in 5'-5' phosphodiester linkage that is methylated at position N-7 on the base and may contain additional ribose methylations on the first and second nucleotide of the primary transcript. 3'-Polyadenylation occurs after transcription of a poly (A) processing signal (consensus sequence AAUAAA) has been transcribed in the primary transcript. Transcripts are cleaved downstream of this signal and 100 to 200 adenine residues are attached to the end. Splicing removes introns or intervening sequences from the primary transcript and joins the remaining exons to produce mRNA.

541

Nucleoprotein complexes in the nucleus carry out splicing. The process involves recognition of the two ends of an intron, cleavage and removal of the intron, and ligation of the exon. The splicing reaction passes through an intermediate structure known as a lariat structure or RNA branch with the 5'-end of the intron joined to a specific base within the intron in 5'-2' phosphodiester linkage.

Problems and Solutions

1. Template-transcript relationships
The 5'-end of an mRNA has the sequence:
             ...AGAUCCGUAUGGCGAUCUCGACGAAGACUCCUAGGGAAUCC...
What is the nucleotide sequence of the DNA template strand from which it was transcribed? If this mRNA is translated beginning with the first AUG codon in its sequence, what is the N-terminal amino acid sequence of the protein it encodes? (See Table 30.1 for the genetic code.)

Answer: The nucleotide sequence of the DNA sense strand has the following sequence, written 5' to 3'
             ...GGATTCCCTAGGAGTCTTCGTCGAGATCGCCATACGGATCT...
The first AUG codon, underlined, is shown below
             ...AGAUCCGU<u>AUG</u>GCGAUCUCGACGAAGACUCCUAGGGAAUCC...
The sequence of the protein from the N-terminal amino acid to the C-terminal amino acid is
     ...AGAUCCGUA<u>UG GCG AUC UCG ACG AAG ACU CCU AGG GAA UCC.</u>..
             Met Ala Ile Ser Thr Lys Thr Pro Arg Glu Ser...

2. The events in transcription initiation
Describe the sequence of events involved in the initiation of transcription by E. coli RNA polymerase. Include in your description those features a gene must have for proper recognition and transcription by RNA polymerase.

Answer: RNA polymerase in *E. coli* is a multimeric protein of the form $\alpha_2\beta\beta'\sigma$. Sigma factor ($\sigma$) functions during initiation of transcription to recognize specific DNA regions called promoters. Promoters contain two DNA sequence elements important for initiation. Located approximately 35 bp upstream of the start of transcription is an element with a consensus sequence TTGACA called the -35 region that is recognized by the sigma factor and represents the initial point of binding of RNA polymerase to DNA. Once the RNA polymerase contacts this element it migrates to a second region, known as the Pribnow box or the -10 region, where it forms an open complex. Whereas the initial binding of RNA polymerase occurs on double-stranded DNA, RNA polymerase eventually requires single-stranded DNA to serve as a template for RNA synthesis. Strand separation is initiated at the Pribnow box, an AT-rich region with a consensus sequence of TATAAT.
    Initiation of polymerization starts with a purine nucleotide triphosphate that binds to the initiation site, one of two nucleotide binding sites on RNA polymerase. This nucleotide is destined to become the 5'-end of the RNA. The next nucleotide binds to the elongation site on RNA polymerase and the enzyme catalyzes phosphodiester bond formation between the 3' hydroxyl group of the nucleotide in the initiation site and the 5'-α–phosphate of the nucleotide in the elongation site with elimination of inorganic pyrophosphate (PP$_i$). Next, translocation along the antisense strand occurs moving the RNA polymerase to the next base. The elongation site is filled with a nucleotide triphosphate complementary to this base and a phosphodiester bond is formed. This process, catalyzed by RNA polymerase holoenzyme (i.e., $\alpha_2\beta\beta'\sigma$), is repeated until an oligonucleotide 6 to 10 nucleotide residues long has been formed. The sigma factor dissociates from the holoenzyme leaving the core enzyme to complete elongation, marking the end of initiation.

3. Substrate binding by RNA polymerase
RNA polymerase has two binding sites for ribonucleoside triphosphates, the initiation site and the elongation site. The initiation site has a greater K$_m$ for NTPs than the elongation site. Suggest what possible significance this fact might have for the control of transcription in cells.

Answer: The initiation site has a preference for purine ribonucleoside triphosphates in addition to having a greater K$_m$ for NTPs than the elongation site. This means that at a given concentration of NTPs the initiation site will fill only if the concentration of NTPs is high enough to also fill the elongation site. Thus, initiation occurs only where there are sufficient NTPs to support elongation. Having a preference for ATP or GTP ties initiation of polymerization intimately to energy levels in cells.

4. Comparison of prokaryotic and eukaryotic transcription
Make a list of the ways that transcription in eukaryotes differs from transcription in prokaryotes?

Answer: In eukaryotes, transcription is compartmentalized within the nucleus. Another fundamental difference is the number of RNA polymerases involved in transcription. In prokaryotes, a single RNA polymerase is responsible for all classes of RNAs including mRNA, rRNA, and small, stable RNAs such as tRNAs. (The only other RNA polymerase activity, primase, functions during replication to produce RNAs used as primers.) In contrast, eukaryotes have several RNA polymerases each dedicated to production of one class of RNA. RNA polymerase I is localized in the nucleolus and is responsible for transcription of ribosomal RNA genes, except 5S rRNA. RNA polymerase II is active in the nucleus and is responsible for synthesis of mRNAs. RNA polymerase III also acts in the nucleus to produce small, stable RNAs such as tRNAs, 5S rRNA, and the small RNAs involved in mRNA processing and in protein targeting. Finally, mitochondria and chloroplasts have RNA polymerases responsible for transcription of organellar genes.

Eukaryotic RNA polymerases are themselves generally incapable of initiation of transcription and require additional protein factors or transcriptional factors to initiate protein synthesis. The enzyme complexes are larger than those found in prokaryotes. Eukaryotic RNA polymerases are also more complicated than their counterparts in prokaryotes.

Termination of transcription is different between eukaryotes and prokaryotes. In bacteria, termination involves either rho-dependent terminators or so-called intrinsic (rho-independent) terminators. In eukaryotes, termination signals are less precise. For RNAs destined to be polyadenylylated, termination occurs downstream of the AAUAAA cleavage and polyadenylation signal sequence.

For prokaryotic promoters there are essentially two elements, one at –35 and another, an AT-rich sequence at –10. The –35 element is recognized by sigma factor and while there are several different sigma factors that recognize different –35 sequences, the situation is not nearly as complicated as found in eukaryotes. Eukaryotes have a TATA-box at –25 and a weakly conserved sequence, Inr, around the site of initiation of transcription, and a number of other elements that are found near the site of initiation of transcription. Robust transcription requires additional elements (like enhancers). DNA in a eukaryotic genome is organized into nucleosomes and effective gene expression requires DNA to be available for transcription. Prokaryotic DNA is less structured and hence typically more readily available for transcription.

In prokaryotes, as the transcript is being produced it can also be translated because protein synthesis and transcription occur in the same compartment. As soon as the 5' end of the primary transcript clears the RNA polymerase it is available to the ribosome for translation. In eukaryotes, transcription is localized in the nucleus whereas protein synthesis occurs in the cytoplasm; thus there is no opportunity to couple the two processes. Furthermore, eukaryotic primary transcripts undergo extensive processing, including capping, polyA addition and splicing before being exported from the nucleus to the cytoplasm.

5. DNA recognition by DNA-binding proteins
DNA-binding proteins may recognize specific DNA regions either by reading the base sequence or by "indirect readout." How do these two modes of protein DNA recognition differ?

Answer: When the base sequence is used for recognition, a DNA-binding protein typically binds to dsDNA through the major groove. In the major groove, each base pair displays a characteristic pattern of hydrogen bond donors and acceptors. There is similar information in the minor groove but not as diverse as that presented in the major groove.

Indirect readout refers to conformational variations in the dsDNA helix formed as a consequence of the sequence. Base pairs have slightly different sizes. In addition, neighboring base pairs influence stacking interactions. Thus, the exact shape of a region of dsDNA is influenced by the sequences of bases.

6. The size of gene promoters
(Integrates with Chapter 11.) The metallothionein promoter is illustrated in Figure 29.27. How long is this promoter, in nm? How many turns of B-DNA are found in this length of DNA? How many nucleosomes (approximately) would be bound to this much DNA? (Consult Chapter 11 to review the properties of nucleosomes.)

Answer: The metallothionein promoter is approximately 265 base pairs in size. B-form DNA has a pitch of 3.4 nm and 10 base pairs per turn (or 0.34 nm per base pair). Thus, 265 base pairs makes 26.5 turns, which is (26.5 x 3.4 nm =) 90.1 nm.

Each nucleosome covers about 146 base pairs of DNA and so 265 base pairs are about enough for nearly two nucleosomes. Nucleosomes are separated by about 50 bp of linker DNA and if we include this then the 265 base pairs would only account for a little more than one nucleosome.

7. Heterodimer formation by bZIP proteins
Describe why the ability of bZIP proteins to form heterodimers increases the repertoire of genes whose transcription might be responsive to regulation by these proteins.

Answer: bZIP polypeptides form dimers by interacting through leucine zippers, amphipathic α-helices containing a periodic repetition of leucine residues spaced every seven residues such that they line up on one side of the helix. These proteins are DNA-binding proteins as well, interacting with DNA through a short stretch of basic residues located N-terminal to the leucine zipper motif. The specificity of DNA binding is determined by the sequence of basic residues. Because bZIP proteins function as dimers, two sequence elements on DNA are recognized. In the case of homodimer formation, the sequence elements form a dyad. However, heterodimers can bind to DNA lacking dyad symmetry. In fact, the bZIP-binding site may be composed of two distinct regions, each recognized separately by different bZIP proteins that function together as a heterodimer. The ability to form heterodimers allows for composite bZIP binding sites, lacking dyad symmetry, to be recognized by bZIP proteins.

8. Alternative splicing possibilities
Suppose exon 17 was deleted from the fast skeletal muscle troponin T gene (Figure 29.46). How many different mRNAs could now be generated by alternative splicing? Suppose that exon 7 in a wild-type troponin T gene was duplicated. How many different mRNAs might be generated from a transcript of this new gene by alternative splicing?

Answer: The fast skeletal troponin T gene consists of 18 exons. Exons 1 through 3, 9 through 14, and 18 are constitutive and are thus always found in mature mRNAs. Exons 4 through 8 are combinatorial and may be individually included or excluded in any combination. Finally, exons 16 and 17 are mutually exclusive, either one or the other is found. How many different mRNAs are possible? Let us first determine the number of different combinations of exons 4, 5, 6, 7, and 8. For each of these exons there are only two states or choices: present or not present. The number of combinations is given by $2 \times 2 \times 2 \times 2 \times 2 = 32$. So, considering exons 4 through 8, 32 different combinations are possible. There is also a choice between exons 16 and 17. Thus, 2×32 or 64 different mRNAs may be produced.

If exon 17 is deleted, the number of different mRNAs possible is only 32.

In the event that exon 7 is duplicated the number of combinations of exons 4, 5, 6, 7, duplicated 7, and 8 is given by the number of combinations of exons 4, 5, 6, and 8 ($2 \times 2 \times 2 \times 2 = 16$) multiplied by the number of different combinations of 7 and 7 duplicate which is 3 (either they are both present, both absent, or one or the other is present). This gives $3 \times 16 = 48$ different combinations of exons 4, 5, 6, 7, 7 duplicate, and 8. When we factor in the two possibilities for exon 16 or 17 we have $2 \times 48 = 96$ different mature mRNAs.

9. Histone modifications and histone-DNA interactions
Figure 29.30 illustrates some of the various covalent modifications that occur on histone tails. How might each of these modifications influence DNA histone interactions?

Answer: The modifications mentioned in Figure 29.30 include acetylation and methylation of lysine and phosphorylation of serine in addition to methylation of arginine. Acetylation of the ε-amino group of lysine will effectively remove lysine's positive charge. The acetyl group is attached via amide linkage (in effect, a peptide bond). The lone pairs on the nitrogen are lost to electron delocalization due to the carbonyl group. Thus, acetylation will remove lysine's positive charge. Methylation of lysine will add a more bulky group to lysine's ε-amino group and while this is not expected to change charge, it might be expected to interfere with ionic interactions. (The pK_a's of methyl amine and dimethyl amine are essentially the same. Thus, we should not expect a major change in pK_a when lysine's ε-amino group, a primary amine, is methylated and converted to a secondary amine.)

Phosphorylation of serine will convert a neutral polar group to a negatively charged phosphate group with 2- charge. This modification allows for additional ionic interactions. Methylation of arginine is not expected to change arginine's charge but it may interfere with ionic interactions.

Histones interact with DNA through ionic bonding and the large number of basic amino acids, such as arginine and lysine, are ideal for this purpose. Acetylation of lysine is expected to effectively remove lysine's positive charge thereby reducing ionic interactions between histones and DNA. Phosphorylation of serine will produce a strong negative charge on an otherwise neutral amino acid. This modification is expected to allow certain regions of histones to be repelled by DNA.

Modification by methylation will influence ionic bonding capabilities by adding a bulkier group on a primary nitrogen.

10. Inducer binding to lac repressor
(Integrates with Chapter 15.) Predict from Figure 29.12 whether the interaction of lac repressor with inducer might be cooperative. Would it be advantageous for inducer to show cooperative binding to lac repressor? Why?

Answer: The lac repressor protein is a homotetramer but it is structurally organized as a dimer of dimers. This arrangement allows the protein to bind to two promoter elements separated by several DNA turns. The result is a DNA/protein complex with a large DNA loop. Binding might be expected to be cooperative and it is. When a lac repressor tetramer binds to one site on DNA it increases the probability that another DNA binding site on the bound repressor binds a second DNA site. In effect, it greatly increases the local repressor concentration making it much more likely that a second DNA/protein interaction occurs.

Inducer binding to the lac repressor might also be expected to be cooperative and it is but only when the repressor is bound to DNA (See O'Gorman, R.B. et al., J. Biol. Chem. 255:10107-14 (1980)). The advantage to cooperative binding of inducer is the system becomes very sensitive to inducer concentration resulting in responses over a narrow range of inducer concentration

11. Post-transcriptional modification of eukaryotic mRNAs
What might be the advantages of capping, methylation, and polyadenylylation of eukaryotic mRNAs?

Answer: Capping and polyadenylation effectively protect the 5'- and 3'-end of RNA from exonucleolytic degradation. There is evidence that the cap and polyA tail interact and that this interaction is important for mRNA recognition. This insures that only full length, completely processed mRNAs are translated. A specialized structure on the 5' end helps in mRNA recognition during initiation of translation. Cap formation occurs shortly after the 5' end of a primary transcript is produced and the cap is bound by factors required for further processing of the primary transcript. Finally, export of a processed primary transcript is dependent on polyadenylation of the 3' end. This insures that no RNA is released before its time.

12. The RNA polymerase mechanism of action
(Integrates with Chapter 15.) Although Figure 29.24 shows only one Mg^{2+} ion in the RNA polymerase II active site, more recent studies reveal the presence of two. Why is the presence of two Mg^{2+} ions significant?

Answer: In the last chapter we learned that DNA polymerases follow a mechanism that involves two metal ions. The incoming dNTP is actually dimagnesium/dNTP complex in which two magnesium ions are bound to the triphosphate end of dNTP. These metal ions are further complexed by two conserved aspartic acid residues in the polymerase. This positions one of the metal ions close to the 3' hydroxyl group that will function as a nucleophile. Proximity of the metal ion lowers the pKa of the hydroxyl group making it a better nucleophile, which attacks the α-phosphate of the dNTP. Since RNA polymerases catalyze essentially identical reactions, the finding that RNA polymerase II has two magnesium ions suggests that the mechanism is followed by both RNA and DNA polymerases.

13. Chromatin remodeling exposes DNA
(Integrates with Chapter 11.) The SWI/SNF chromatin-remodeling complex peels about 50 bp from the nucleosome. Assuming B-form DNA, how long is this DNA segment? In forming nucleosomes, DNA is wrapped in turns about the histone core octamer. What fraction of a DNA turn around the core octamer does 50 bp of DNA comprise? How does 50 bp of DNA compare to the typical size of eukaryotic promoter modules and response elements?

Answer: B-form DNA has 3.4 nm per turn and 10 base pairs per turn or 0.34 nm per base pair. Thus, a 50 base pair segment of B-form DNA is (50 bp x 0.34 nm/bp) 17 nm. Approximately 147 bp of DNA are wrapped around a nucleosome core making 1.65 turns. Thus, 50 bp represents (50 bp x 1.65 turns/146 bp =) 0.565 turn. Promotes and response elements are around 20 to 25 bp thus the 50 bp peeled off from a nucleosome by the SWI/SNF complex could easily expose one of these elements

14. The lariat intermediate in RNA splicing
Draw the structures that comprise the lariat branch point formed during mRNA splicing: the invariant A, its 5' -R neighbor, its 3' –Y neighbor, and its 2' -G neighbor.

Answer: The lariat branch point is a 5'-2'-phosphodiester bond joining the 5' end of the intron to the 2'-hydroxyl group of the branch point A. The structure is shown below.

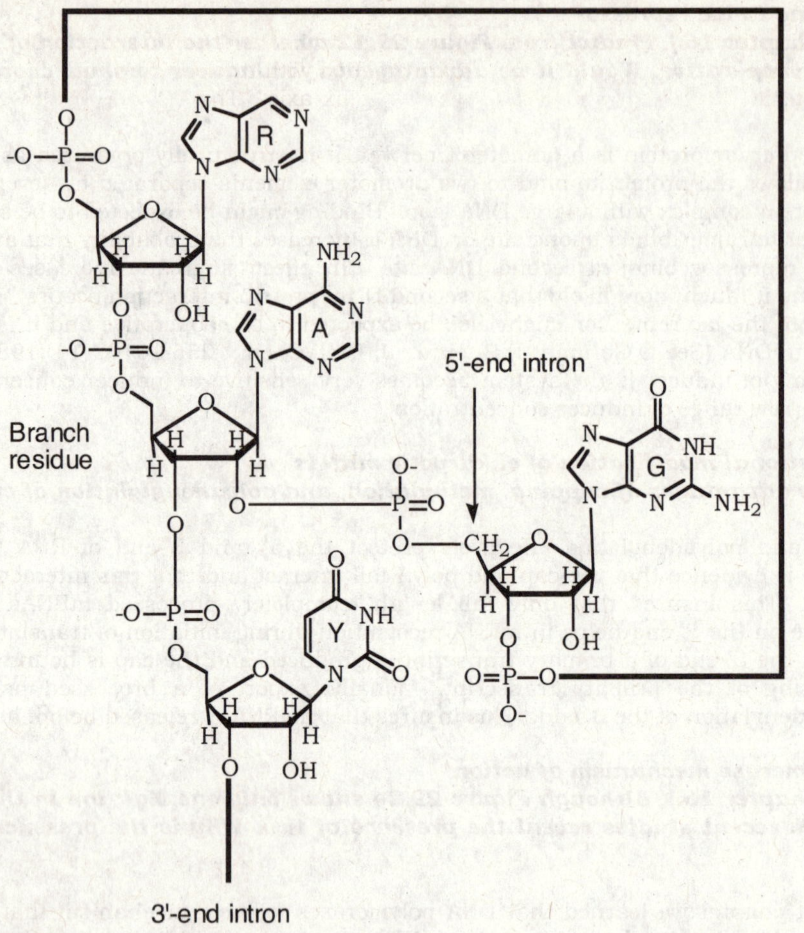

Branch
residue

5'-end intron

3'-end intron

15. HTH DNA-binding proteins and B-DNA dimensions
(Integrates with Chapters 6 and 11.) *The α-helices in HTH (helix-turn-helix motif) DNA-binding proteins are formed from 7– or 8–amino acid residues. What is the overall length of these α-helices? How does their length compare with the diameter of B-form DNA?*

Answer: The Δz for an α-helix is 0.15 nm per amino acid residue. For a helix 7 to 8 residues long, its length ranges from (0.15 nm x [7 or 8]) 1.05 nm to 1.2 nm. The diameter of B-form DNA is about 2 nm but the distance across a base pair is approximately 1.1 nm (between 1'-carbons). The major groove is approximately 1.2 nm wide and 0.8 nm deep. This would easily accommodate an α-helix (whose diameter is around 0.6 nm ignoring side chains). The 1.2 nm of length would cover about 3 base pairs (1.2 nm ÷ 0.34 nm/bp).

16. Exploring the structure of RNA polymerase
Bacteriophage T7 RNA polymerase bound to two DNA strands and an RNA strand, as shown in pdb 1MSW, provides a glimpse of transcription. View this structure at www.pdb.org to visualize how the template DNA strand is separated from the nontemplate strand and transcribed into an RNA strand. Which Phe residue of the enzyme plays a significant role in DNA strand separation? In which domain of the polymerase is this Phe located? (You might wish to consult Yin, Y. W., and Steitz, T .A., 2002. Structural basis for the transition from initiation to elongation transcription in T7 RNA polymerase. Science 298:1387–1395 to confirm your answer.)

Answer: Start out by downloading 1MSW (at http://www.rcsb.org/pdb/home/home.do) and viewing it using Swiss PDB Viewer or your favorite viewing program. Swiss PDB Viewer allows one to view only selected parts of the molecule so start out viewing only the DNA. You will see that the DNA is 10 base pairs of double-stranded DNA and about 9 base pairs that are single stranded. You can use the curser to identify the last base pair (an A:T pair) of the double stranded region. This is the last base pair before the single stranded region. By selecting one of the bases of this pair you can use the Select command (pull down menu option) to restrict the view to a prescribed distance from the base pair. From the select command activating "Neighbors of select aa" will return a text box with options for displaying only groups that are within 5 Å. (If you select a larger distance it will return more residues.) One of the residues that is located within 5 Å of the last base pair is Phe 644. The side chain ring is almost perfectly aligned with the helix axis of the double stranded DNA and the plane of the side chain is perpendicular to the helix axis. This residue functions as a wedge that serves to split the dsDNA into single stranded regions.

17. α-Amanitin inhibition of RNA polymerase II
RNA polymerase II is inhibited by α-amanitin. This mushroom derived toxin has no effect on the enzyme's affinity for NTP substrates, but it dramatically slows polymerase translocation along the DNA. Go to www.pdb.org to view pdb file 1K83, which is the structure of RNA polymerase II with bound α-amanitin. Locate α-amanitin within this structure and discuss why its position is consistent with its mode of inhibition.

Answer: Download the file as outlined in the answer to question 16. To get a quick picture of the situation display the protein as a ribbon and color the ribbons using the chain coloring option. This colors each polypeptide chain a different color. Next locate α-amanitin and display it as a space filled molecule. You should discover that the inhibitor is located at the bottom of a cleft formed between the two largest subunits of RNA polymerase II, chains A and B (Rpb1 and Rbp2 subunits). The bottom of the cleft is formed by two long α-helices from chain A that are antiparallel to each other. While the cleft is formed by chains A and B, the residues that form the α-amanitin binding site come from chain A. The α-amanitin binding site is located at the bottom of the cleft bounded by three helices from the chain A each at right angles to each other and include one of the two helices that form the bottom of the cleft, the "bridge" helix that extends from the main part of chain A across the cleft and another short helix. The bridge helix is closest to the α-amanitin binding site and makes the majority of contact with the inhibitor.

The α-amanitin binding site is approximately 20 Å from a manganese ion that is located at the active site, the nucleotide binding site and the site of phosphodiester bond formation. Thus, α-amanitin is not expected to interfere with either nucleotide binding or bond formation. The inhibitor interferes with DNA translocation and thus elongation is blocked.

18. Exploring the structure of a DNA-bound bZIP transcription factor
C/EBPβ is a bZIP transcription factor in neuronal differentiation, learning and memory process, and other neuronal and glial functions. The structure of the bZIP domain of C/EBPβ bound to DNA is shown in pdb file 1GU4. Explore this structure to discover the leucine zipper dimerization domain and the DNA-binding basic regions. On the left side of the www.pdb.org 1GU4 page under "Display Files," click "pdb file" to see the atom-by-atom coordinates in the three-dimensional structure (scroll down past "Remarks" to find this information). Toward the end of this series, find the amino acid sequence of the C/EBPβ domain used in this study. Within this amino acid sequence, find the leucine residues of the leucine zipper and the basic residues in the DNA-binding basic region.

Answer: An easier approach to getting the amino acid sequence is to activate the FASTA Sequence option on the left side of 1gu4. Depending on how your browser is set up you might be asked either to save the file to a disk or to open it with a program. Do either but eventually get the file opened in a word processing program and use Courier font. The contents of the file are shown below. It is apparent that chains A and B are identical protein chains whereas chains C and D are nucleic acid sequences. The sequence file (below) is in Courier font because it is a non proportional font and so each character occupies the same space.

```
>1GU4:B|PDBID|CHAIN|SEQUENCE
VKSKAKKTVDKHSDEYKIRRERNNIAVRKSRDKAKMRNLETQHKVLELTAENERLQKKVEQLSRELSTLRNLFKQLPE
>1GU4:A|PDBID|CHAIN|SEQUENCE
VKSKAKKTVDKHSDEYKIRRERNNIAVRKSRDKAKMRNLETQHKVLELTAENERLQKKVEQLSRELSTLRNLFKQLPE
>1GU4:D|PDBID|CHAIN|SEQUENCE
AATATTGCGCAATCCT
>1GU4:C|PDBID|CHAIN|SEQUENCE
TAGGATTGCGCAATAT
```

To visualize leucines we can use the replace command (under the edit menu in MSWord) to replace L with **L**. (Activate replace under the edit menu, enter "L" in the Find what text box, and enter "L" in the Replace with text box. Expand the Find and Replace menu and on the bottom of the expanded menu, use the format pull down menu to activate font. From the font menu select Bold and Underline format.) You should get the following.

547

```
>1GU4:B|PDBID|CHAIN|SEQUENCE
VKSKAKKTVDKHSDEYKIRRERNNIAVRKSRDKAKMRNLETQHKVLELTAENERLQKKVEQLSRELSTLRNLFKQLPE
>1GU4:A|PDBID|CHAIN|SEQUENCE
VKSKAKKTVDKHSDEYKIRRERNNIAVRKSRDKAKMRNLETQHKVLELTAENERLQKKVEQLSRELSTLRNLFKQLPE
```

You might notice immediately that the leucines are all located on the C-terminal half of the protein sequences. The first and second leucines are separated by 7 residues. Starting with the third leucine there are leucine residues every 7th position for four cycles. You can show this by simply placing paragraphs before the first, second and third leucines. Then, starting with the third leucine, place a paragraph every 7th residue. This should generate the following.

```
>1GU4:B|PDBID|CHAIN|SEQUENCE
VKSKAKKTVDKHSDEYKIRRERNNIAVRKSRDKAKMRN
LETQHKV
LE
LTAENER
LQKKVEQ
LSRELST
LRNLFKQ
LPE
```

A more sophisticated approach to the question is to use some of the protein analysis tools at ExPASy. Visit the following URL:http://us.expasy.org/tools/#primary. You might consider running ProtParam to get the amino acid composition of the protein. For comparison you could use RandSeq to generate a random protein with an average amino acid composition for comparison. Finally, you could use one of the programs to determine the isoelectric point of the first half of the protein and the second half. These analyses should convince you that the protein's N-terminal half is rich in basic amino acids and that its C-terminal half forms a leucine zipper.

Preparing for the MCAT® Exam

19. Figure 29.15 highlights in red the DNA phosphates that interact with catabolite activator protein (CAP). What kind of interactions do you suppose predominate and what kinds of CAP amino acid side chains might be involved in these interactions?

Answer: In order to interact with the negatively charged phosphate backbone of DNA we would expect a protein to use positively charged lysine and arginine residues to form ionic bonds (electrostatic interactions). Using the directions given in the answer to the previous problem you could download the sequence of 1g6n and analyze its amino acid composition. The CAP protein is not rich in basic amino acids.

20. Chromatin decompaction is a preliminary step in gene expression (Figure 29.48). How is chromatin decompacted?

Answer: The first step is to decrease the interaction between histone protein and DNA and this is accomplished by histone acetyl transferases (HATs). HATs acetylate lysine residues on histones and acetylation effectively removes the positive charge on lysine side chains. This disrupts electrostatic interactions between histones and DNA. Next, chromosome remodeling complexes unwind DNA from the nucleosome using ATP hydrolysis to drive the process.

Questions for Self Study

1. Describe the events leading up to initiation of transcription by *E. coli* RNA polymerase. Be sure to include in your discussion the importance of promoter elements in a typical promoter.

2. Name the two types of transcriptional termination mechanisms found in bacteria.

3. The last two columns of the table shown below are incorrect. Correct them.

RNA polymerase Type	Genes Transcribed	α-Amanitin Sensitivity
RNA polymerase I	protein-encoding genes	intermediate sensitivity
RNA polymerase II	tRNAs, 5S rRNA and others	resistant
RNA polymerase II	ribosomal RNA genes	very sensitive

4. Eukaryotic RNA polymerase II requires what additional class of proteins to function?

5. The *lac* operon is under both positive and negative control. Explain how both the presence of lactose and the absence of glucose are required for initiation of transcription of the *lac* operon.

6. How does the availability of tryptophan regulate gene expression in the *trp* operon?

7. Eukaryotic genes. Match the terms.
    a. A common element found in many promoters.
    b. A promoter element found in "housekeeping " genes.
    c. Orientation and distance independent regulatory sequences
    d. Responsive elements

    1. GC box
    2. Subject promoters to regulation by some common signal.
    3. TATA box
    4. Enhancers (or upstream activation sequences)

8. What are the three major classes of DNA-binding motifs?

9. How are the 5'- and 3'- ends of eukaryotic transcripts typically modified?

10. The primary transcript of an eukaryotic gene is typically longer than its mRNA. Explain.

Answers

1. Initiation of transcription begins when RNA polymerase holoenzyme, through its σ subunit, binds to the -35 region of a promoter to form a closed complex. The polymerase moves along the promoter until it reaches the -10 region or Pribnow box where it unwinds about 12 base pairs of DNA exposing the transcription start site. This arrangement is known as the open promoter complex. Initiation begins with a nucleoside triphosphate, usually ATP or GTP, binding to the initiation site on RNA polymerase. The elongation site binds the next nucleotide and phosphodiester bond formation occurs. The holoenzyme produces a short oligonucleotide chain before its σ factor dissociates leaving the core complex to complete elongation.

2. ρ-Dependent (rho-dependent) and ρ-independent.

3.

RNA polymerase Type	Genes Transcribed	α–Amanitin Sensitivity
RNA polymerase I	ribosomal RNA genes	resistant
RNA polymerase II	protein-encoding genes	very sensitive
RNA polymerase II	tRNAs, 5S rRNA and others	intermediate sensitivity

4. RNA polymerase II requires general transcription factors.

5. Initiation of transcription is blocked by *lac* repressor protein binding to an element in the promoter of the *lac* operon known as the operator. Repressor protein binding occurs only in the absence of an inducer, a galactose derivative. The *lac* promoter is a weak promoter and requires an additional factor for transcription. This factor is a complex of catabolite activator protein (CAP) and cyclic AMP (cAMP). This complex binds to a region upstream of the RNA polymerase binding site in the promoter. cAMP is necessary for CAP binding and the levels of cAMP are sensitive to glucose such that glucose decreases cAMP levels. Thus, in the absence of glucose, cAMP levels are high enough to support CAP binding to the promoter.

6. There are two mechanisms. Initiation of transcription is blocked by the Trp repressor protein. This protein binds tryptophan and associates with the promoter, at a site termed the operator, to block initiation of transcription. The operon is also controlled by transcriptional attenuation. The operon codes for a small leader peptide. The rate at which this leader sequence is translated affects transcriptional termination.

7. a. 3; b. 1; c. 4; d. 2.

8. Helix-turn-helix, zinc finger, and leucine zipper.

9. The 5' end of a transcript is capped by a 7-methylguanylyl residue in 5' to 5' linkage. The 3' end contains a polyA tail 100 to 200 adenine residues in length.

10. The primary transcript contains exons and introns. The exons are sequences destined to be present in the mature mRNA but they are not contiguous in the primary transcript. Rather, they are separated by introns. The process of splicing removes introns and joins exons to form the mature mRNA.

Additional Problems

1. What characteristics are associated with the -10 region (Pribnow box) of RNA polymerase promoters in *E. coli*?

2. What is a consensus sequence?

3. Describe the differences and similarities between RNA polymerase I and DNA polymerase I in *E. coli*.

4. Describe the structure of a rho-independent transcriptional terminator.

5. The operator of the *lac* operon was first identified by *cis-trans* complementation tests. In these tests a partial diploid is produced from two cell lines showing defects in lactose metabolism. The idea behind *cis-trans* complementation is to determine if the defects in lactose metabolism are due to mutations in the same gene or in different genes. If mutations are in different genes then a partial diploid will be capable of metabolizing lactose. For example, the *lac* operon contains two genes necessary for lactose metabolism that code for β-galactosidase and lactose permease. A partial diploid between a cell defective in β-galactosidase but containing a normal permease and a cell defective in permease but containing a normal β-galactosidase will result in a diploid capable of lactose metabolism. How might operator mutations behave in *cis-trans* complementation tests?

6. Why does IPTG cause induction of the *lac* operon? How does induction by IPTG differ from induction by lactose via allolactose?

7. How might splicing increase the versatility of eukaryotic genes?

8. How might the fact that mature mRNAs have 3'-polyA tails be used to isolate mature mRNAs from eukaryotic cells?

9. Modified nucleosides can be identified by reverse phase HPLC. How might you expect ribose-methylated and unmethylated nucleosides to differ on reverse phase HPLC? What about 7-methyl G *versus* unmodified G?

Abbreviated Answers

1. The -10 region or Pribnow box is an A-T rich sequence of about six bases located approximately 10 bases upstream of the start of transcription. The high A-T content allows this region to "melt" so that the strands separate. The RNA polymerase-promoter complex shifts from the closed form to the open form at the -10 region.

2. A consensus sequence is a sequence representing the most commonly found base in each position within a DNA element.

550

3. Both enzymes utilize nucleotide triphosphates to produce nucleic acid polymers, both require a template, read the template 3'★5' direction, and produce the nucleic acid is the 5'★3' direction. RNA polymerase is capable of initiating RNA synthesis by incorporating the very first 5' nucleotide whereas DNA polymerase requires a primer to function. RNA polymerase does not proofread nor does it have exonuclease activity. DNA polymerase has 3'- and 5'-exonuclease activity for proofreading and nick translation.

4. Rho-dependent terminators are DNA elements that when transcribed produce a short sequence of mRNA capable of forming a G:C-rich stem and loop followed by a stretch of Us.

5. Operator mutations may give rise to constitutive expression of the *lac* operon if for example the *lac* repressor protein binding site is altered. The phenotype is constitutive expression of the *lac* operon even in a partial diploid. Operator mutations are examples of *cis* acting mutations because they affect or influence genes only on the same DNA.

6. IPTG (isopropylthiogalactoside) is a galactose derivative capable of binding to the *lac* repressor protein, thereby altering its ability to bind to the operator of the *lac* promoter. Induction by allolactose is transient because allolactose is converted to lactose, which is metabolized to galactose and glucose by β-galactosidase. IPTG is not degraded by β-galactosidase and induction is long-lasting.

7. For certain genes, alternative splicing patterns give rise to different gene products that code for proteins with different properties.

8. A polyT derivatized matrix can be used to separate polyA mRNA from total cellular RNA.

9. In reverse phase HPLC, hydrophobic interactions are responsible for adsorption. Therefore, ribose methylated nucleosides are expected to bind with higher affinity and require higher concentrations of organic solvent to be removed from the reverse phase matrix. The same is true for modified bases, with the exception of 7-methyl modification of G (and A). This modification gives rise to a positive charge on N-7 leading to a decrease in hydrophobicity. 7-Methyl G will elute from a matrix at lower concentrations of organic solvent than unmodified G.

Summary

Transcription is the process whereby RNA chains are synthesized according to a DNA template. The enzymes responsible are DNA-dependent RNA polymerases. Prokaryotes have a single RNA polymerase, which synthesizes rRNA and tRNA as well as mRNA. RNA polymerases, like DNA polymerases, polymerize nucleotides in the 5'★3' direction, reading along templates in the 3'★5' direction. In contrast to DNA polymerases, RNA polymerases can catalyze the *de novo* synthesis of nucleotide chains.

E. coli RNA polymerase consists of a catalytic "core" of subunit composition $\alpha_2\beta\beta'$. This core requires a σ subunit in order to recognize promoter sequences, which signal the sites where genes begin. Addition of a σ subunit to the "core" RNA polymerase yields the RNA polymerase holoenzyme. This holoenzyme binds at promoter sites which are characterized by two features: the Pribnow box, a TATAAT sequence element located at position -10, and a 9 bp consensus sequence TCTTGACAT near position -35. RNA polymerase holoenzyme then "melts" the DNA at the A:T-rich Pribnow box to form an "open complex" adjacent to the +1 site where transcription is initiated. Chain elongation proceeds at the rate of 20 to 50 nucleotides per sec, moving slower in G:C-rich regions where DNA strand separation is more difficult. Chain termination occurs when a G:C-rich inverted repeat sequence near the end of the transcript forms a stable stem-loop structure which halts further translocation of the RNA polymerase. This stem-loop is followed by a series of A residues in the template, which are transcribed, into Us in the RNA. The hydrogen bonds holding these A:U base pairs are exchanged for A:T pairs between the "sense" and "anti-sense" DNA strands to re-anneal the DNA duplex and release the nascent transcript.

Eukaryotes have three classes of DNA-dependent RNA polymerases; all are found in the nucleus. RNA pol I transcribes the major rRNA genes, RNA pol II transcribes the protein-encoding genes and thus leads to mRNA synthesis, and RNA pol III transcribes tRNA genes as well as those genes encoding a number of other small cellular RNAs. All three RNA polymerases are large complex multimeric proteins. Each class of RNA polymerase recognizes a distinct set of promoter elements appropriate to the genes it transcribes. Distinct sets of DNA-binding proteins called transcription factors mediate these interactions, much as the σ subunit of the *E. coli* RNA polymerase directs it to its proper "transcription start sites".

Many genes in bacteria are organized into clusters known as operons. Expression of these genes to produce a single polycistronic mRNA is under the control of promoter and operator regulatory elements lying upstream. Transcription is controlled by induction and repression, processes in which the interaction of regulatory proteins with small metabolites determines whether these regulatory proteins will bind to DNA and facilitate or limit transcription. For example, the *lac* operon is expressed provided the regulatory protein, *lac* repressor, is not bound at the operator site. The presence of the *lac* repressor on the operator blocks transcription, an example of negative control. β-Galactosides induce expression of the *lac* operon by binding to *lac* repressor and promoting its dissociation from the operator. The affinity of *lac* repressor binding to operator DNA binding is 3 orders of magnitude greater than the affinity of the inducer(IPTG):*lac* repressor complex for operator DNA.

The *lac* operon is also under positive control by CAP, the catabolite activator protein. In the presence of cAMP, CAP binds near the *lac* promoter and assists transcription initiation by RNA polymerase. When [cAMP] is low, free CAP cannot associate with RNA polymerase and promote transcription, so the *lac* operon is not expressed. The availability of glucose, *E. coli*'s preferred carbon and energy source, leads to lower cAMP levels and repression of the *lac* operon as well as other operons involved in the metabolism of alternative carbon and energy sources. This phenomenon is called catabolite repression.

The *araBAD* operon, another of the catabolite repressible operons, is also controlled both positively and negatively by the regulatory protein, *AraC*. In the absence of arabinose, *AraC* binds to three regulatory sites ($araO_1$, $araO_2$ and *araI*) in the *araBAD* operon and prevents transcription. However, when arabinose binds to *AraC*, this protein releases from one of the sites and joins with CAP-$(cAMP)_2$ to activate transcription of the *araBAD* operon.

Operon expression may also be regulated after transcription has begun through a mechanism known as attenuation. Attenuation is determined by the availability of charged aminoacyl-tRNAs. Operons encoding enzymes of amino acid biosynthesis often have multiple codons for that particular amino acid early in their polycistronic message. If the particular aminoacyl-tRNA is low in concentration, the ribosome translating these codons stalls, awaiting arrival of the aminoacyl-tRNA. This pausing by the ribosome allows a secondary structure called the antiterminator to form downstream in the mRNA. RNA polymerase perceives this antiterminator and continues transcription of the entire operon. However, if the relevant aminoacyl-tRNA is present in abundance, the ribosome translates these codons quickly and an alternative secondary structure appears further downstream in the mRNA, the terminator. RNA polymerase perceives the terminator and terminates transcription early in the operon, so the operon's genes are not expressed. The antiterminator and terminator are mutually exclusive secondary structures since one of the RNA segments they use is the same.

Transcriptional regulation in eukaryotes is more complicated. The promoters of eukaryotic genes are defined by modules of short conserved sequences, such as the TATA, CAAT and GC boxes, which bind transcription factors essential to RNA polymerase II function. In addition to these promoter elements, eukaryotic genes have enhancers, sequence elements whose presence is necessary for full expression of the gene. Enhancers may be located in either orientation with respect to the gene and may lie either upstream or downstream from it. Enhancers act nonspecifically to enhance transcription from any nearby promoter, but enhancer function is dependent on binding a specific transcription factor. Many genes respond to a common regulation and are characterized by the presence of common sequence elements in their promoters known as response elements. The heat shock element (HSE) and glucocorticoid response element (GRE) are examples. These response elements are recognized by specific transcription factors that coordinate expression of all genes containing the particular response element. DNA looping provides a mechanism for convening a variety of specific regulatory proteins at the transcription initiation site.

Certain structural motifs recur in DNA-binding proteins. In particular, DNA-binding proteins often employ an α-helix as their DNA recognition element. The diameter of an α-helix matches closely the width of the major groove in B-form DNA, and these recognition helices fit within the major groove, identifying specific nucleotide sequences through a variety of atomic interactions, including hydrogen bonding patterns with the edges of the bases, ionic and H-bonding with the sugar-phosphate backbone, as well as van der Waals fits and hydrophobic interactions with projecting nonpolar -CH_3 groups on thymine residues. DNA-binding proteins erect a variety of structural scaffolds serving to deliver the α-helix or other recognition elements to the DNA site, and it is really these scaffolds that have lent their names to the structural motifs: the helix-turn helix, the Zn-finger and the leucine zipper. The great majority of DNA-binding proteins possess one of these three basic motifs.

Chapter 30

Protein Synthesis

● ●

Chapter Outline

❖ Code characteristics
  ⚔ Triplet code: Three bases decoded as one amino acid
    • All 64 triplets used: 61 amino acid codons: 3 stop codons
  ⚔ Nonoverlapping, no commas, unambiguous, universal (or nearly so)
  ⚔ Degenerate: Synonymous codons decode same amino acid
    • Third base degenerate (except in two cases)
    • Second base
      • Pyrimidines code for hydrophobic amino acids
      • Purines code for polar or charged amino acids
  ⚔ Aminoacyl-tRNA synthesis: At least 20 enzymes charge tRNAs that decode 61 codons
    • Cognate tRNAs: All tRNAs charged by particular synthetase
    • Reaction mechanism: Two step process
      • Activation of amino acid by formation of aminoacyl adenylate
      • Transfer of amino acid to 3'OH of tRNA
    • Recognition
      • Amino acid binding site
      • tRNA binding site: Determinants on tRNA
        o At least one of anticodon bases
        o One or more of three base pairs in acceptor stem
        o Discriminator base N of NCCA at 3'-end
      • Two classes of synthetases
        o Class I
          ▪ Typically monomeric
          ▪ Adds amino acid to 2'OH (but it moves to 3'OH)
          ▪ Nucleotide binding site is Rossman fold
            • HIGH sequence
            • KMSKS sequence
          ▪ Binds tRNA acceptor stem from minor groove
        o Class II
          ▪ Typically oligomeric
          ▪ Adds amino acid to 3'OH
          ▪ Three motifs
            • Motif 1: Dimerization motif
            • Motifs 2 and 3: Active site
❖ Wobble Rules: tRNA interactions with mRNA: Explains degeneracy of code
❖ Codon Usage: Certain codons preferred over others
❖ Nonsense suppression: Mutant tRNA that recognizes stop codons (UAA, UAG, UGA)
❖ Ribosomes: Cellular translational machinery
  ⚔ Prokaryotic ribosomes (*E. coli*): 70S particle: Two subunits
    • Small subunit: 30S

- 21 proteins: S-proteins
- 16S rRNA: 1542 nt
    - Large subunit: 50S
        - 31 distinct proteins: L-proteins
        - 23S rRNA: 2904 nt
        - 5S rRNA: 120 nt
    - rRNAs derive from one precursor transcript
- Eukaryotic cytoplasmic ribosomes: 80S particle: Two subunits
    - Small subunit: 40S
        - 33 proteins
        - 18S rRNA: Around 1800 nt
    - Large subunit: 60S
        - 49 proteins
        - 28S rRNA: Around 4700 nt
        - 5S rRNA: 120 nt
        - 5.8S rRNA: 160 nt: Homologous to 5'-end of prokaryotic 23S rRNA
- ❖ Stages of protein synthesis in prokaryotes
    - Initiation: Formylmet-tRNA, mRNA, small subunit, large subunit
        - Formylmet-tRNA: Initiator tRNA
            - 5'-Base not paired
            - CCU sequence in D-loop
            - Three G:C bp in anticodon stem
            - Formyl group added to charged tRNA by methionyl-tRNA formyl transferase
        - mRNA
            - Ribosome binding site (Shine-Dalgano sequence) near 5'-end interacts with 16S rRNA
            - Complementary sequence on 16S rRNA near 3'-end interacts with ribosome binding site
        - Initiation factors
            - IF-1, IF-3: Dissociate 30S subunits
            - IF-2: Binds GTP and fmet-tRNA and deposits fmet-tRNA into P site
                - GTP hydrolysis accompanied by IF release and 50S binding
                - 70S initiation complex formed
    - Elongation
        - Elongation factor Tu (EF-Tu)
            - Binds GTP and charged-tRNA
            - Deposits charged-tRNA into A-site
            - GTP hydrolysis releases EF-Tu·GDP
                - GDP exchanged for GTP by EF-Ts (Guanylyl exchange protein)
        - Peptide bond formation: Peptidyl transferase: 50S activity: 23S rRNA activity
        - Translocation
            - Deacylated tRNA removed from P site
            - Peptidyl-tRNA moved to P site
            - mRNA moved so that new codon in A site
            - EF-G: GTPase catalyzes translocation
    - Termination: Stop codon in A site
        - RF-1 recognizes UAA and UAG
        - RF-2 recognizes UAA and UAG
        - RF-3: GTPase
        - Ribosome recycling factor: Disassembles ribosome
- ❖ Eukaryotic protein synthesis
    - Eukaryotic initiator tRNA$_i$: Charged with Met: Not formylated
    - Three stages of initiation
        - 43S preinitiation complex formation: eIF2, eIF4A, eIF3 and Met-tRNA$_i$

- mRNA binding to 43S preinitiation complex
  - eIF4E: mRNA cap binding protein
- 80S initiation complex formed upon large subunit binding
- Elongation
  - EF-1: Two components
    - EF1A: EF-Tu counterpart
    - EF1B: EF-Ts counterpart
- Termination: Single RF recognizes all stop codons
- Regulation of translation: Phosphorylation
  - S6 phosphorylation stimulated by growth factors
  - eIF2α phosphorylated by variety of kinases leading to inactivation
- Inhibitors of protein synthesis
  - Streptomycin: Low concentrations cause misreading
  - Macrolide antibiotics: Plugs exit tunnel for peptide chain

Chapter Objectives

Ribosomes

The site of cellular protein synthesis is the ribosome, a two-subunit ribonucleoprotein complex. You should be familiar with the ribosome vocabulary including the following terms: 16S rRNA, 18S rRNA, 23S rRNA, 28S rRNA, 5S rRNA, 5.8S rRNA, small subunit, 30S subunit, 40S subunit, large subunit, 50S subunit, 60S subunit, 70S ribosome, 80S ribosome, S-proteins, and L-proteins. The rRNAs derive from operons containing structural genes for the small- and large-subunit rRNAs and depending on cell type, the 5S rRNA. A primary transcript is processed by nucleolytic cleavage and post-transcriptional modifications, typically methylations, to produce the mature rRNAs. Even before processing occurs, certain ribosomal proteins begin to bind to the rRNA. This binding is followed by binding of other ribosomal proteins, leading to an ordered assembly of the ribonucleoprotein complex.

Protein Synthesis

Although the ribosome plays the leading role in protein synthesis, it is assisted at each stage by nonribosomal protein factors specific for stages of protein synthesis including initiation, elongation, and termination. Initiation is a property of the small subunit and various initiation factors. A special tRNA charged with Met is used in initiation of protein synthesis. (In prokaryotes it is termed f-Met-tRNAf^{Met} where f indicates that Met is formylated. In eukaryotes, Met-tRNAi^{Met} is used.) Initiation begins when a ribosome, aided by initiation factors, dissociates into subunits in order to produce free small subunits. The small subunit binds mRNA and a complex of f-Met-tRNAf^{Met} and another initiation factor to form a preinitiation complex that is joined by the large subunit to form the initiation complex. Initiation complex formation is accompanied by GTP hydrolysis and release of ribosome-bound initiation factors. The outcome of initiation is an aminoacylated-tRNA bound in the so-called P-site on the ribosome. In peptide chain elongation, a second aminoacyl-tRNA binding site, the A-site, is filled, in an elongation factor-dependent manner, with an aminoacylated-tRNA appropriate for the codon on the mRNA. In *E. coli*, the elongation factor EF-Tu binds to aminoacyl-tRNA and GTP to form a complex. This complex binds to the A-site on the ribosome and deposits the charged tRNA at that location with subsequent hydrolysis of GDP and release of P$_i$ and EF-Tu:GDP complex. For EF-Tu to participate in another round of elongation, it must be recharged by binding to another elongation factor, EF-Ts, which causes GDP to be released in forming an EF-Tu:EF-Ts complex that is resolved to EF-Tu:GTP and EF-Ts by GTP binding. Peptide bond formation is a property of the ribosome alone. The large subunit catalyzes the transpeptidation or peptidyl transfer of the P-site amino acid to the A-site aminoacylated-tRNA. The uncharged tRNA in the P-site is released and the A-site tRNA, now carrying a dipeptide and hence called a peptidyl-tRNA, is translocated along with the mRNA from the A- to the P-site. Translocation is dependent on elongation factor EF-G in the case of *E. coli*. The vacated A-site is now free to accept another aminoacylated tRNA to continue elongation. In the event that a stop codon enters the A-site, chain termination ensues. Protein release factors bind to the A-site and cause hydrolysis of the peptide from peptidyl-tRNA.

Inhibitors of Protein Synthesis

Protein synthesis is an important target for antibiotics. You should be familiar with the kinds of inhibition caused by various antibiotics including inhibition of initiation, of A-site binding, of peptidyl transferase activity, and of translocation. Puromycin causes premature release of peptides. Diphtheria toxin inhibits EF-

2 in eukaryotes by covalent ADP-ribosylation. Ricin causes chain cleavage of the large subunit rRNA, and Colicin E-3 cleaves prokaryotic small subunit ribosomes.

Post-Translational Processing

The most common form of protein modification is proteolytic cleavage to activate a protein, or to release mature protein products from a larger primary translational product. Proteins are also targeted to specific locations within the cell or are destined for transport out of the cell. In prokaryotes, a signal sequence, located on the N-terminus of a protein, or at a more internal location, may direct the protein to be exported from the cell or to become a membrane-bound protein. Signal recognition particles play a role in halting protein synthesis shortly after a signal peptide is produced and, in conjunction with docking protein, directing the nascent protein to the endoplasmic reticulum membrane where protein synthesis resumes.

Problems and Solutions

1. Translating nucleotide sequences into amino acid sequences
(Integrates with Chapter 12.) The following sequence represents part of the nucleotide sequence of a cloned cDNA:

. . . CAATACGAAGCAATCCCGCGACTAGACCTTAAC. . .

Can you reach an unambiguous conclusion from these data about the partial amino acid sequence of the protein encoded by this cDNA?

Answer: We are faced with the problem of deciding on a reading frame. Assuming that we have the noncoding strand sequence, we can inspect it for stop codons, UGA, UAG, UAA. (It makes no sense to look for AUG start codons because the sequence might have come from any location within the cDNA.) By inspection we find two stop codons as shown below.

. . . CAATACGAAGCAATCCCGCGAC**TAG**ACCT**TAA**C. . .

These two stop codons are not in the same reading frame (because they are not separated by an integral multiple of three nucleotides). Therefore, either of the two different reading frames that include TAG and TAA may represent the region coding for the C-terminal portion of the protein. The third possible reading frame, (the one that starts with CAA) does not contain a stop codon. More information is required in order to decide which is the correct reading frame. One possibility is to compare codon usage in the three frames with codon usage in other genes from the same organism. Alternatively, we might turn to amino acid composition analysis. The amino acid sequence of the third reading frame is:

Gln-Tyr-Glu-Ala-Ile-Pro-Arg-Leu-Asp-Leu-Asn

Certain amino acids are often found in low frequencies in proteins as for example Met (AUG), Trp (UGG), His (CAU/C), Cys (UGU/C), Tyr (UAU/C), and Phe (UUU/C). However, of these relatively rarely used amino acids, we find only the tyrosine codon shown below.

. . . CAA**TAC**GAAGCAATCCCGCGACTAGACCTTAAC. . .

It may be possible to check the protein for chymotrypsin cleavage (chymotrypsin cleaves after tyrosine) and this may help us decide if the tyrosine codon is in the appropriate reading frame.

A similar analysis may be applied to the complementary strand whose sequence, written 5' to 3', is:

GTTAAGGTCTAGTCGCGGGATTGCTTCGTATTG

This sequence has two stop codons in different reading frames as shown below.

GT**TAA**GGTC**TAG**TCGCGGGATTGCTTCGTATTG

The third reading frame, shown below, may be an open reading frame.

G **TTA A**GG TC**T AG**T CGC GGG ATT GCT TCG TAT TG

If translated it would produce the following amino acid sequence:

Leu-Arg-Ser-Ser-Arg-Gly-Ile-Ala-Ser-Tyr-

2. Nucleotide sequences, possible codons, and amino acid specification
A random (AG) copolymer was synthesized with polynucleotide phosphorylase using a mixture of 5 parts adenine nucleotide to one part guanine nucleotide as substrate. If this random copolymer is used in a cell-free protein synthesis system, which amino acids will be incorporated into the polypeptide product? What will be the relative abundances of these amino acids in the product?

Answer: Polynucleotide phosphorylase catalyzes the following reaction *in vitro*

$$NDP + NDP \star RNA + P_i.$$

The mixture of NDPs consists of 5 parts ADP and 1 part GDP. The probabilities of selecting ADP and GDP from the mixture are 5/6 and 1/6 respectively. To determine the probability of any codon we apply the product rule. For example, the probability of an AAA codon is given by (5/6)(5/6)(5/6) = 0.58. The relative probability of a codon is calculated by dividing the probability of a codon by the probability of the most probable codon.

Codon	Probability	Relative Probability	Amino acid
AAA	(5/6)(5/6)(5/6) = 0.58	1.00	Lys
AAG	(5/6)(5/6)(1/6) = 0.12	0.20	Lys
AGA	(5/6)(1/6)(5/6) = 0.12	0.20	Arg
GAA	(1/6)(5/6)(5/6) = 0.12	0.20	Glu
AGG	(5/6)(1/6)(1/6) = 0.023	0.04	Arg
GAG	(1/6)(5/6)(1/6) = 0.023	0.04	Glu
GGA	(1/6)(1/6)(5/6) = 0.023	0.04	Gly
GGG	(1/6)(1/6)(1/6) = 0.005	0.008	Gly

To determine the abundance of amino acids we can apply the sum rule of probability. For example, the probability of a lysine codon is the sum of the probabilities of AAA and AAG. The relative probability is calculated by dividing the probability of an amino acid by the probability of the most probable amino acid.

Amino acid	Probability	Relative Probability	Normalized Probability
Lys	1.00 + 0.20 = 1.20	1.20	100
Arg	0.20 + 0.04 = 0.24	0.24	20
Glu	0.20 + 0.04 = 0.24	0.24	20
Gly	0.04 + 0.008 = 0.048	0.048	4

3. The second genetic code
Review the evidence establishing that aminoacyl-tRNA synthetases bridge the information gap between amino acids and codons. Indicate the various levels of specificity possessed by aminoacyl-tRNA synthetases that are essential for high-fidelity translation of messenger RNA molecules.

Answer: Information flow in biological systems is described in the central dogma first postulated by Francis Crick. Genetic information is stored and transmitted from generation to generation in the form of a sequence of bases in nucleic acids. Nucleic acids are faithfully copied in the process of replication, which is directed by base-pair interactions involving complementary hydrogen bonds. The information content of a nucleic acid is its sequence of bases. A portion of this sequence is used to direct the synthesis of proteins, linear chains of amino acids, in a process known as gene expression. For most genes, gene expression involves at least two steps, transcription and translation. In transcription, an mRNA is produced by a base-pair directed polymerization. In translation, mRNAs are used to direct the synthesis of proteins by ribosomes. Even in this process, complementary base pair interactions between mRNAs and tRNAs are paramount in importance. Yet, translation results in the production of linear chains of amino acids. Thus, during gene expression a change in information content from nucleotide sequences to amino acid sequences occurs. The point at which this information gap is bridged is the production of aminoacyl-tRNAs by aminoacyl-tRNA synthetases. These enzymes must speak two languages, one based on nucleotides, the other based on amino acids.

In order for translation to occur with high fidelity, aminoacyl-tRNA synthetases must accurately charge tRNAs. This is accomplished by a two-step reaction. As the name implies, synthetases produce aminoacylated-tRNAs at the expense of ATP hydrolysis. In the first step in catalysis, an amino acid is adenylated to form an aminoacyl-adenylate that is tightly bound to the enzyme. This reaction is accompanied by elimination of PP$_i$, which is subsequently hydrolyzed by pyrophosphatases rendering the reaction irreversible. In the second stage of catalysis, the amino acid is transferred to either the 2'-OH or the 3'-OH of a tRNA.

Specificity exists at each of these two stages. The aminoacyl adenylation reaction depends on the amino acid specificity of the amino acid binding site. This specificity is not absolute and activation of inappropriate amino acids occurs with low frequency. This may lead to misacylation. Aminoacyl-tRNA synthetases carefully edit this step to insure that misacylated tRNAs are not released. Misacylated tRNAs trigger a deacylase that hydrolyzes inappropriately charged tRNAs.

The "second genetic code" refers to so-called charging determinants, which are features on tRNAs that aminoacyl-tRNA synthetases recognize in order to charge them with the correct amino acid. One might imagine that the anticodon is a charging determinant and in some cases it is but for many tRNAs the synthetase does not use the anticodon as a recognition feature but rather uses a number of other features on tRNAs including single bases, base pairs, and modified bases.

4. *Codon-anticodon recognition: base-pairing possibilities*
(Integrates with Chapter 11.) Draw base-pair structures for (a) a G:C base pair, (b) a C:G base pair, (c) a G:U base pair, and (d) a U:G base pair. Note how these various base pairs differ in the potential hydrogen-bonding patterns they present within the major groove and minor groove of a double-helical nucleic acid.

Answer: For the G:C and C:G pairs, the pattern of hydrogen bond donor and acceptor groups in the major and in the minor groove are opposite to each other. For example, in the major groove of the G:C pair we find, on G, two acceptors, N-7, and the carbonyl oxygen of C-6, and on C, a donor, the amino group at C-4. In the minor groove, G presents an acceptor, N-3, and a donor, the amino group at C-2, whereas C presents an acceptor, the carbonyl oxygen of C-2. For the C:G pair the pattern is in the reverse order.

For the G:U and U:G pair the pattern of hydrogen-bonding groups is symmetrical. This is shown below where, reading from left to right in the major and minor groove, the pattern of hydrogen bond donors and acceptors is presented.

G:C

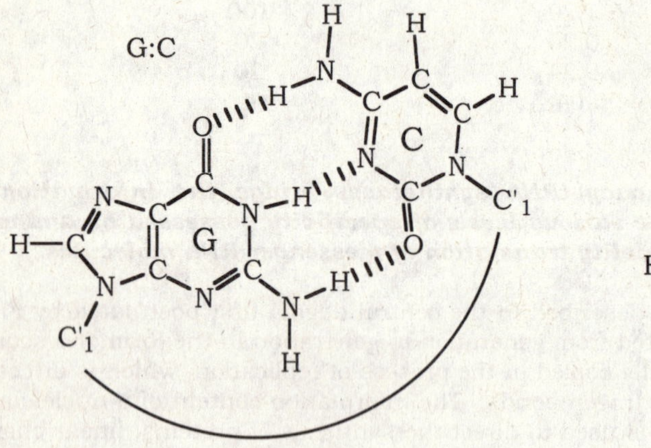

minor groove

C:G

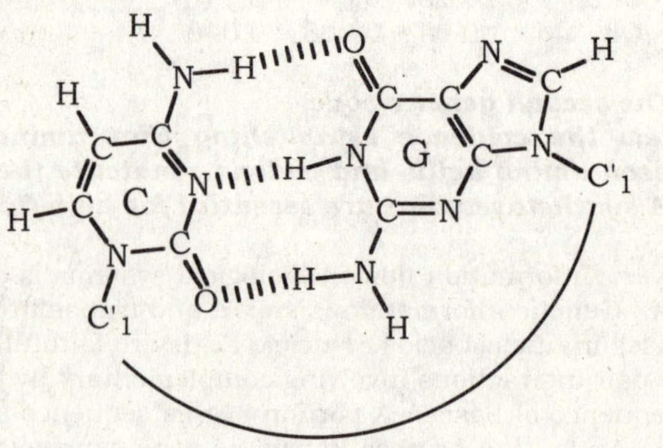

minor groove

G:U

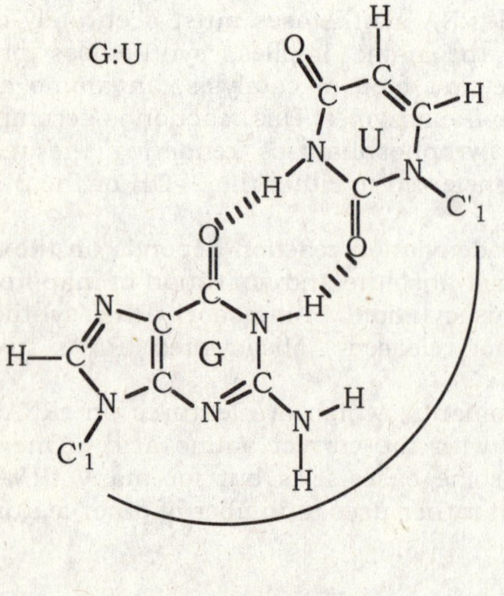

minor groove

U:G

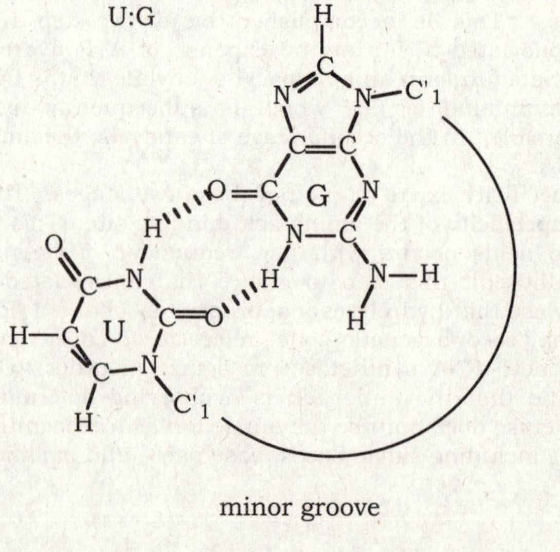

minor groove

Base Pair	Groove	Hydrogen-Bonding Groups
G:C	major	A-A-D
G:C	minor	A-D-A
C:G	major	D-A-A
C:G	minor	A-D-D
G:U	major	A-A-A
G:U	minor	A-D-A
U:G	major	A-A-A
U:G	minor	A-D-A

5. Consequences of the wobble hypothesis
Point out why Crick's wobble hypothesis would allow fewer than 61 anticodons to be used to translate the 61 sense codons. How does "wobble" tend to accelerate the rate of translation?

Answer: The wobble hypothesis allows base pairs to form between the 5' anticodon position and the 3' codon position according to the following rules:

5' Anticodon	3' Codon
C	G (canonical pair)
A	U (canonical pair)
U	A (canonical pair) or G
G	C (canonical pair) or U
I	U, C, or A

To decode a set of four codons that differ only in their 3' base, a minimum of two anticodons is necessary. For example to decode the four codons of threonine, ACU, ACC, ACA, and ACG, the anticodons IGU and CGU (written 5' to 3') are sufficient. Because inosine (I) can base pair with U, C, or A, the IGU anticodon will decode AGU, AGC, and AGA. The anticodon CGU decodes ACG only. In this example we can see that four codons can be decoded by only two anticodons.

The wobble rules explain the degeneracy of the genetic code. In all cases the codons NNU/C (or NNY) are always degenerate. To decode NNU requires a 5' anticodon base that pairs with U; the choices are A, G, or I. However, A is normally converted to I whenever it occurs in the 5' anticodon position. Therefore, only G or I are possible, both of which base pair with C as well. If G is used, then the anticodon will recognize both NNU and NNC. If I is used, this anticodon will recognize NNU, NNC, and NNA.

The codons NNA/G (NNR) are often degenerate because the 5' position of anticodons may contain U that is capable of pairing with both A or G. Anticodons with C for 5' base will only decode a single codon and in fact there are two instances where this occurs in the genetic code. Both methionine and tryptophan are coded for by a single codon each, AUG and UGG respectively. Their anticodons must be CAU and CCA.

Wobble allows non-canonical base pairs to form. This makes codon-anticodon interactions less stable, allowing for an increase in the rate of translation. Wobble also allows decoding of the 61 sense codons with far fewer tRNAs. A bit of reflection will show that the minimum number of tRNAs is 32. The standard genetic code can be represented by 16 blocks of 4 codons with codons in a particular block differing only in their 3' nucleotide. Each block will require a minimum of 2 tRNAs except the block containing TAT, TAC, TAA and TAG. The first two code for tyrosine whereas the last two are stop codons. So this block requires only one tRNA to decode TAT and TAC. The rest of the blocks require two tRNAs each and this sums to 31 tRNAs. Initiation of protein synthesis, however, requires an additional initiator tRNA that is charged with methionine giving a total of 32 tRNAs in a minimal set. (Additional tRNAs would be required to decode selenocysteine and pyrrolysine.)

6. Sense-to-nonsense codon mutations
How many codons can mutate to become nonsense codons through a single base change? Which amino acids do they encode?

Answer:

Termination Codons

	UAA	UAG	UGA
Third Base Change	UA**U** (Tyr) UA**C** (Tyr)	UA**U** (Tyr) UA**C** (Tyr)	UG**U** (Cys) UG**C** (Cys) UG**G** (Trp)
Second Base Change	U**U**A (Leu) U**C**A (Ser)	U**U**G (Leu) U**C**G (Ser) UG**G** (Trp)	U**U**A (Leu) U**C**A (Ser)
First Base Change	**C**AA (Gln) **A**AA (Lys) **G**AA (Glu)	**C**AG (Gln) **A**AG (Lys) **G**AG (Glu)	**C**GA (Arg) **A**GA (Arg) **G**GA (Gly)

7. Amino acid possibilities in nonsense suppression
Nonsense suppression occurs when a suppressor mutant arises that reads a nonsense codon and inserts an amino acid, as if the nonsense codon was actually a sense codon. Which amino acids do you think are most likely to be incorporated by nonsense suppressor mutants?

Answer: The most likely candidates for nonsense suppressors are tRNAs that decode codons that differ from stop codons by one base. For UAA and UAG termination codons, tRNAs that decode the following amino acid codons are candidates: Leu, Ser, Tyr, Trp, Gln, Lys, and Glu. For the UGA termination codon, tRNAs that decode Trp, Cys, Arg, Gly, Ser, and Leu are candidates. However, there are certain constraints on nonsense suppression. The tRNA gene to be mutated must not be an essential gene to the cell. Ideally a tRNA gene whose function is covered by another tRNA is best suited to become a nonsense suppressor. In the above list Leu, Ser, and Arg tRNAs are good candidates because each of these amino acids have six codons. An amino acid coded by four codons, such as glycine, is also a good candidate. The remaining amino acids are coded by only one codon (e.g., Trp) or two codons each. These are the least likely candidates to serve as suppressors. The tRNA that decodes the UGG Trp codon would have to be first duplicated to provide an additional gene copy to mutate to a nonsense suppressor.

8. Prokaryotic vs. eukaryotic protein synthesis
Why do you suppose eukaryotic protein synthesis is only 10% as fast as prokaryotic protein synthesis?

Answer: Protein synthesis, a fundamental reaction to all cell types, is carried out by ribosomes. Ribosomes from prokaryotes, eukaryotes, and organelles such as mitochondria and chloroplasts are all very similar in structure and they carrying out the process of protein synthesis in similar manners. At each stage of protein synthesis, initiation, elongation, and termination, the ribosome is dependent of protein factors, which together with the ribosome are responsible for a specific phase of protein synthesis.

One striking difference between prokaryotic ribosomes and eukaryotic ribosomes is the larger size and hence complexity of eukaryotic ribosomes. Not only are eukaryotic ribosomes considerably larger, they contain more ribosomal proteins and larger rRNAs. In addition, eukaryotic stage-specific protein factors are more complex. Because of this, eukaryotic ribosomes function at slower rates.

The advantage to slower ribosomal activity is an increase in fidelity. Eukaryotic cells produce a wider range of proteins whose proper functions are critical to cell survival. Thus, there is a greater demand for higher fidelity of protein synthesis in eukaryotes.

9. Capacity of the ribosomal peptide exit tunnel
If the tunnel through the large ribosomal subunit is 10 nm long, how many amino acid residues might be contained within it? (Assume that the growing polypeptide chain is in an extended β-sheet–like conformation.)

Answer: For a peptide in extended β-sheet conformation the distance between residues is 0.35 nm. For a length of 10 nm this represents (10 nm/0.35 nm/residue) 28.6 residues.

10. The consequences of ribosome complexity
Eukaryotic ribosomes are larger and more complex than prokaryotic ribosomes. What advantages and disadvantages might this greater ribosomal complexity bring to a eukaryotic cell?

Answer: More complex ribosomes allow for a greater repertoire of control mechanisms to function to regulate protein synthesis. In addition, the ability to produce proteins at a low error rate is improved. The costs of complexity include a lower rate of protein synthesis and a protein synthesis system that requires more components in order to function.

11. Ribosomes as two-subunit structures
What ideas can you suggest to explain why ribosomes invariably exist as two-subunit structures, instead of a larger, single-subunit entity?

Answer: Ribosomes catalyze peptide bond formation using two aminoacyl-tRNAs as substrates that bind to two distinct tRNA binding sites on the ribosome known as the P-site and the A-site. During each round of elongation, the P-site is transiently occupied by a peptidyl-tRNA containing the newly synthesizes N-terminal portion of the polypeptide attached via aminoacyl linkage to the 3'-end of a tRNA. The A-site is filled with an aminoacyl-tRNA appropriate to the codon displayed at the A-site. The ribosome catalyzes peptide bond formation by nucleophilic attack by the amino group of the aminoacyl-tRNA in the A-site on the carbonyl carbon of the acyl bond on the peptidyl-tRNA in the P-site. The result is polypeptide chain transfer from the P-site to the A-site. As a result the P-site is now occupied by an uncharged tRNA whereas the A-site contains a peptidyl-tRNA one amino acid longer. In order for elongation to continue, the P-site must be vacated, the peptidyl-tRNA, now in the A-site, must be moved into the P-site, and the A-site must be refilled. This process, known as translocation, occurs in at least two steps or stages. In the first step, during peptide bond formation, the acceptor end of the A-site aminoacyl-tRNA moves into the P-site to form the peptide bond. The result of this movement is formation of a hybrid tRNA binding site: the 3'-end of the A-site tRNA is now in the P-site, a P/A hybrid state. What is the fate of the now deacylated tRNA? Interestingly, this tRNA is not immediately released from the ribosome. Rather, it is transferred to a third tRNA binding site, the E-site or exit site. However, the transfer is accomplished in two steps. As the 3'-end of the A-site tRNA moves into the P-site and peptide bond formation occurs, the 3'-end of the now deacylated tRNA moves into the E-site forming an E/P hybrid state. To occupy these two hybrid tRNA binding states, E/P and P/A, the tRNAs must move relative to the large subunit, an event that may be accomplished by movement of the small subunit. In order for elongation to continue, the A-site must be vacated. This is accomplished by resolving the P/A hybrid state into P-site binding and the E/P hybrid state into E-site binding. The A-site is now free to bind a new aminoacyl-tRNA, which initially interacts with the ribosome through a third hybrid-binding site, the T/A. The transition of aminoacyl-tRNAs from T/A binding to the A-site may be a proofreading step, and this transition may be influenced by E-site binding such that A-site binding results in release of an uncharged tRNA from the E-site.

The tRNA binding sites involve aspects of both the large and the small subunits. To produce hybrid sites, the subunits may be required to move relative to each other. The functional significance of this movement may be responsible for the high fidelity of protein synthesis.

12. tRNAᵢMet recognition and translation initiation
How do prokaryotic cells determine whether a particular methionyl-tRNA^{Met} is intended to initiate protein synthesis or to deliver a Met residue for internal incorporation into a polypeptide chain? How do the Met codons for these two different purposes differ? How do eukaryotic cells handle these problems?

Answer: Initiator tRNA is a specific tRNA capable of decoding AUG codons responsible for the N-termini of proteins. In prokaryotes, this tRNA, tRNAᵢMet, is charged with methionine, which is subsequently formylated by methionyl-tRNAᵢMet formyl transferase. Compared to tRNAs exclusively involved in elongation, tRNAᵢMet has distinguishing features including an unpaired 5'-terminal base on the acceptor end, a unique CCU sequence it its D loop, and a set of three G:C base pairs in its anticodon stem. These features are essential to tRNAᵢMet functioning during initiation.

There is only one codon for methionine, AUG, however it has two meanings, initiation or elongation, depending on context. Specifically, the context for initiation in prokaryotes is having a purine-rich sequence of 4 to 8 bases located on the mRNA approximately 10 bases upstream of the AUG which will serve to initiate protein synthesis. This purine-rich sequence is known as the ribosome-binding site or the Shine-Dalgarno sequence. It interacts with a complementary, pyrimidine-rich sequence near the 3'-end of the small subunit rRNA. This interaction is responsible for determining the site of initiation of protein synthesis: initiation

561

occurs, in most cases, at an AUG codon appropriately spaced from the ribosome-binding site. Any other AUG codon is decoded as methionine by methionyl-tRNA^{Met}.

In eukaryotes, a special initiator tRNA charged with methionine is used to initiate protein synthesis. However, tRNA$_i^{Met}$ is not formylated as is the case in prokaryotes but it is a special tRNA with features that distinguish it from tRNAs used in elongation. It is clear that initiation in eukaryotes does not involve an interaction between the 3'-end of the small subunit rRNA and the mRNA because eukaryotic small subunit rRNAs lack Shine-Dalgarno sequences. Initiation is dependent on formation of a 40S initiation complex that involves a cap binding protein responsible for recognizing the 5'-end cap modification.

13. The function of the Shine-Dalgarno sequence
What is the Shine-Dalgarno sequence? What does it do? The efficiency of protein synthesis initiation may vary by as much as 100-fold for different mRNAs. How might the Shine-Dalgarno sequence be responsible for this difference?

Answer: The Shine-Dalgarno sequence is a purine-rich sequence of 4 to 8 nucleotides located on prokaryotic mRNAs around 10 nucleotides upstream of an AUG codon used as an initiation codon. The Shine-Dalgarno sequence is often called the ribosome-binding site because mRNAs interact with ribosomes by interacting with a complementary, pyrimidine-rich sequence located near the 3'-end of the small subunit rRNA. This rRNA sequence is fixed for a given prokaryote; there is only one population of ribosomes. Shine-Dalgarno sequences on mRNAs show a range of variability from gene to gene. The degree to which a Shine-Dalgarno sequence is complementary to the pyrimidine-rich sequence on rRNA determines the efficiency with which a particular mRNA is used in initiation complex formation. mRNAs showing a high degree of complementarity are capable of interacting strongly with ribosomal small subunits and are therefore initiated frequently.

14. Reconciling peptide bond formation and tRNA movements
In the protein synthesis elongation events described under the section on translocation, which of the following seems the most apt account of the peptidyl transfer reaction: (a) The peptidyl-tRNA delivers its peptide chain to the newly arrived aminoacyl-tRNA situated in the A site, or (b) the aminoacyl end of the aminoacyl-tRNA moves toward the P site to accept the peptidyl chain? Which of these two scenarios makes most sense to you? Why?

Answer: A description of elongation is given in the answer to question 11 above and choice (b) agrees with this description. Using various conditions that stop or freeze the ribosome at various stages of protein synthesis, the state of occupancy of the P-site, A-site, and E-site have been determined. During elongation, a peptidyl-tRNA appears to bind to a hybrid P/A-site, thus exposing a portion of the A-site. In choice (a), a portion of the P site might be expected to be exposed whereas in choice (b), a portion of the A is exposed. The evidence favors (b).

For economy of effort it would be easier to move the aminoacylated end than the peptidyl group due to size. Finally, because the nascent protein appears to be threaded through the large subunit of the ribosome there may be fewer degrees of freedom for this group.

15. Why translation factors are G proteins
(Integrates with Chapter 15.) Why might you suspect that the elongation factors EF-Tu and EF-Ts are evolutionarily related to the G proteins of membrane signal transduction pathways described in Chapter 15?

Answer: EF-Tu functions during elongation as a ternary complex with GTP and an aminoacyl-tRNA. This ternary complex delivers charged tRNAs to an empty A-site on a ribosome. Upon binding to the ribosome, GTP hydrolysis occurs resulting in deposition of the aminoacyl-tRNA into the A-site and release of an EF-Tu:GDP complex and inorganic phosphate. EF-Tu functions catalytically and in order to participate in another round of elongation, GDP must be exchanged for GTP. This is accomplished indirectly by action of EF-Ts. (Ts and Tu refer to the temperature-stability of these two proteins. EF-Ts activity is stable to high temperatures whereas EF-Tu activity is unstable.) EF-Ts binds to the EF-Tu:GDP complex, displaces GDP, and forms a heterodimeric EF-Tu:EF-Ts protein complex. This complex dissociates upon GTP binding to EF-Tu, resulting in free EF-Ts and EF-Tu:GTP.

Many signal transduction pathways involve G proteins, GTP-binding proteins capable of stimulating or inhibiting another proteins. G proteins are capable of these activities only when they have bound GTP, and, these activities are transient because G proteins have intrinsic GTPase activity, which is responsible for hydrolyzing GTP to inactivate the protein. G proteins consist of three subunits α, β, and γ. The α subunit is a guanine nucleotide-binding protein that binds either GTP or GDP and exhibits GTPase activity. There are two types of α subunits, α$_s$ and α$_i$ that stimulate or inhibit respectively. Following GTP hydrolysis, the α subunit contains bound GDP, which must be replaced with GTP. This is accomplished by the α subunit associating with the β and γ subunits to form a membrane-bound complex. This complex is capable of interacting with membrane-bound receptor proteins but only when the receptor protein is activated by hormone binding. This interaction of G protein complex and hormone-loaded receptor protein results in dissociation of GDP, binding of GTP, and dissociation of an Gα:GTP complex that can now act on adenylate cyclase. Thus, G proteins participate in a GTP-GDP cycle reminiscent of EF-Tu:EF-Ts and it comes as no great surprise that the proteins are related.

16. The energetic cost of peptide elongation
How many ATP equivalents are consumed for each amino acid added to an elongating polypeptide chain during the process of protein synthesis?

Answer: The first step is to activate the amino acid and attach it to an appropriate tRNA using an aminoacyl tRNA synthetase. These enzymes use ATP to first activate the amino acid on its carboxyl end by attaching AMP in a mixed anhydride bond. The inorganic pyrophosphate released from ATP is subsequently hydrolyzed to two inorganic phosphates and this overall process is equivalent to hydrolysis of 2 ATP. The next step is to get the aminoacyl tRNA into the A site using elongation factor EF-Tu, which forms a ternary complex with GTP and the charged tRNA. The charged tRNA is deposited into the A site by EF-Tu when its GTPase activity is stimulated by the ribosome, which accounts for a third ATP equivalent. (The GDP gets released by EF-Ts forming a complex with EF-Tu/GDP and releasing GDP. Then GTP displaces EF-Ts to recharge EF-Tu with GTP.) Finally, translocation or movement of the charged tRNA from the A site to the P site is catalyzed by EF-G/GTP, which also exhibits ribosome-stimulated GTPase activity resulting in another ATP equivalent. The total is four.

17. Exploring the structure of the 30S ribosomal subunit
Go to www.pdb.org and bring up PDB file 1GIX, which shows the 30S ribosomal subunit, the three tRNAs, and mRNA. In the box on the right titled "Biological Assembly", click "More Images…", and then scroll down to look at the Interactive View. By moving your cursor over the image, you can rotate it to view it from any perspective.
  a. How are the ribosomal proteins represented in the image?
  b. How is the 16S rRNA portrayed?
  c. Rotate the image to see how the tRNAs stick out from the structure. Which end of the tRNA is sticking out?
  d. Where will these ends of the tRNAs lie when the 50S subunit binds to this complex?

Answer: The More Images option only shows static views of the molecule. In fact it shows either the asymmetric unit or the biological molecule, which happen to be the same for this file. You can toggle between both views by using the box with arrows just below the heading Images and Visualization. To manipulate the model you must use one of the viewing programs like Jmol. If you use Jmol it should launch right in your web browser and you should see that the proteins are depicted as ribbons of different color to make protein identification easy. The rRNA is a single-stranded ribbon or line. The tRNA is shown as a line for the sugar/phosphate backbone and ring structures for the bases. The aminoacylated 3' ends of the tRNAs are sticking out away from the 30S subunit. When the 50S joins the small subunit to form the ribosome these tRNA ends will be located in the peptidyl transferase center of the large subunit.

18. Exploring the structure of the 50S ribosomal subunit
Go back to www.pdb.org and bring up PDB file 1FFK, which shows the 50S ribosomal subunit. In the box titled "Biological Assembly," click "More Images," and scroll down to look at the Interactive View. Right-click the image to discover more information and tools.
  a. How many atoms are represented in this structure?
  b. Are the bases of the nucleotides visible anywhere in the structure?
  c. Can you find double helical regions of RNA?
  d. Right-click and, under "Select," select all proteins. Right-click again and select "Render," then "Scheme," and then "CPK Spacefill" to highlight the ribosomal proteins. Go back and cancel the protein selection. Then select "Nucleic," and render nucleic acid in "CPK Spacefill." Which macromolecular species seems to predominate the structure?

Answer: Before opening the image study the summary page. Under "Molecular Description" you will see that the structure is that of the large ribosomal subunit with 23S rRNA, 5S rRNA and 27 large-subunit proteins. If you expand the abstract you will get additional details. On the top of the summary page you will see several tabs. Activate the "Methods" tab. This will show details about the experimental conditions used to crystallize the molecule and solve its structure. Under the "Refinement" section (which is below the "Refinement Data" section) there is a section listing the number of non-hydrogen atoms used in the refinement. Protein accounts for 3,656 atoms, RNA 60,612 and heteroatoms 13 for a total of 64,281 atoms.

Open the image using Jmol. The default view shows RNA as a strand or line. Use the "Color" pull down menu to color "By Chain". This will render each individual protein and rRNA in a different color. To visualize double stranded helical regions of RNA the view should be simplified by first removing protein. Right click on the model and from the menu chose "Select" then "protein" then "All". When you reopen the menu by right clicking you should see Select (3,656). Next, expand Style, Structures and then "off". This will remove proteins from the view. Right click on the model and then select "Select" -nucleic –All (or –RNA). When you right click again you should see Select (2,826). To simplify the view, right click then chose Style –Structure - Trace. This will convert the RNA into lines that will make looking for double-stranded regions much easier.

To answer "d" right click, select all proteins then right click again followed by Style –Scheme –CPK Space fill. The proteins will be shown as grey balls whereas the RNA will a white line. It should be apparent that the RNA dominates the molecule. If you are not convinced by this just compare the number of atoms associated with RNA and protein when each is selected. Or, select either protein or RNA and then Style –Structure – Trace. This will changed the selected component to lines and leave the other as space fill. Then, reverse the procedure.

19. Exploring the structure of the peptide chain termination complex
Go again to www.pdb.org and bring up the PDB file 3F1E. Scroll down and view the designations for the various components in the structure, as given in the Molecular Description box (note, for example, that the mRNA is designated "chain V"). In the box on the right click on "view in Jmol." View the overall structure, then scroll down below the Jmol box and under "PDP", click on the various 3F1E sections to highlight RF-2 in the Jmol image. Within the Jmol image, right click to explore the structure further. Clicking on "model 1/2" allows you to select the 30S or 50S subunits. Other possibilities allow you too delve further.

Answer: When you start this exercise and you are looking at the Molecular Description box make a note that RF-2 is chain X and it is 378 residues. When you launch Jmol proteins are colored in blue and RNA in a flesh tone. From the section under PDP activate all records associated with chain X, which are 3F1EXa, b, and c. This will highlight in colors various domains of RF-2. You can then view only the 30S or only the 50S ribosomal subunit or you can view the whole ribosome with RF-2 highlighted. The ribosome was the January 2010 Molecule of the Month, which may be found at http://www.pdb.org/pdb/101/motm.do?momID=121. The structure of release factor 2 is entry 1GQE. You might start out with this structure to get a sense of the structure of RF-2. Since it binds to a tRNA binding site on the ribosome it might make sense that its structure mimics a tRNA and it does. If you highlight acidic amino acids in the structure you will see that they are distributed within a three alpha-helical bundle.

Preparing for the MCAT® Exam

20. Review the list of Shine–Dalgarno sequences in Figure 30.18 and select the one that will interact best with the 3'-end of E. coli 16S rRNA.

Answer: The Shine-Dalgarno is a purine-rich sequence located near the 5'-end of bacterial mRNA's. It binds to a polypyrimidine region near the 3' end of the small subunit rRNA of bacteria. The sequence of the 3'-end of the small subunit rRNA is 3'-AUUCCUCCACUCA. The polypyrimidine region is AUUCCUCCA (written unconventionally as 3' to 5'). The complement of this is U**AAGGAGG**U (written conventionally 5' to 3'). If you allow for GU pairing as well then U**(A/G)(A/G)GG(A/G)GG**U will work but less well. The mRNA that would be expected to interact with the ribosome should have a sequence close to this. The data in Figure 30.18 are shown below. Potential purine-rich sequences are in bold font as is the translational start codon, AUG. The terms in the last column indicate the number of mismatches (MM), GU pairs, and conventional base pairs (BP) that would occur if the region in bold were paired to the polypyrimidine sequence near the 3' end of rRNA. In terms of base pairs, the Qβ phage replicase makes the largest number of base pairs.

araB	UUUGGGAU**GGAG**UGAAACG**AUG**GCGAUU	4BP
galE	AGCCCUAAU**GGAG**CGAAUU**AUG**AGAGUU	4BP
lacI	CAAUUUC**AGGGUGGU**GAUU**AUG**AAACCA	7BP 1MM 1GU
lacZ	UUCAACAC**AGGA**AACAGCU**AUG**ACCAGU	4BP

Qβ phage replicase	UAACC**UAAGGA**UGAAAUGC**AUG**UCUAAG	6BP
X174 phage A protein	AAUCCUU**GGAGG**CUUUUUU**AUG**GUUCGU	4BP
R17 phage coat protein	UCAAACC**GGGGU**UUGAAGC**AUG**GCUUCU	5BP 1GU
ribosomal protein S12	AAAAACC**AGGAG**CUAUUUA**AUG**GCAACA	5BP
ribosomal protein L10	CUACCC**AGGAG**CAAAGCUA**AUG**GCUUUA	5BP
trpE	CAAAAAUUA**GAG**AAUAACA**AUG**CAAACA	3BP
trpL leader	GUAAAA**AAGGG**UAUCGACA**AUG**AAAGCA	5BP 1GU

21. Chloramphenicol (Figure 30.31) inhibits the peptidyl transferase activity of the 50S ribosomal subunit. The 50S peptidyl transferase active site consists solely of functionalities provided by the 23S rRNA. What sorts of interactions do you think take place when chloramphenicol binds to the peptidyl transferase center? Which groups on chloramphenicol might be involved in these interactions?

Answer: The structure of chloramphenicol is shown below. The nitrophenyl group might be expected to intercalate into RNA helical regions. The two hydroxyl groups represent potent hydrogen bond donors and acceptors and could interact with rRNA through hydrogen bonding. There is an amide (peptide) bond in chloramphenicol. We know from protein structure studies that this is a rather polar bond capable of hydrogen bonding as donor and acceptor. The structure of chloramphenicol bound to the 50S large ribosomal subunit has been solved by X-ray diffraction. It may be found in the Protein Data Bank at http://www.rcsb.org/pdb/. From the structure it is quite obvious that the nitrophenyl group stacks on an A residue. In addition, one of the hydroxyl groups makes a hydrogen bond with a G residue. Within 5Å of chloramphenicol are three G residues, two of which seem to be hydrogen bonding with the hydroxyl groups on chloramphenicol. Finally, the dichloromethyl group has a water molecule that may be coordinating it.

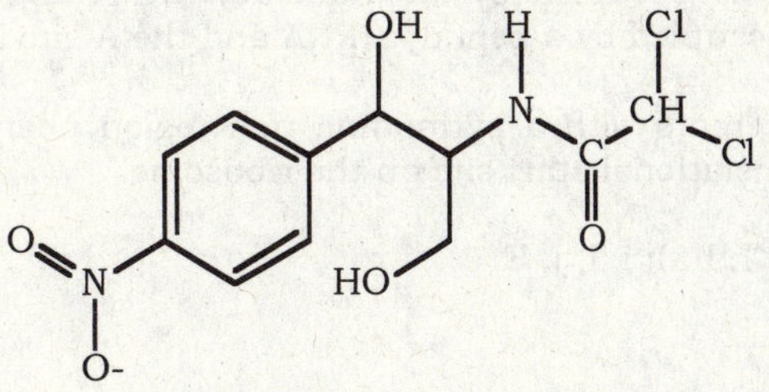

Questions for Self Study

1. Match

a. 30S	1. Large subunit rRNA of eukaryotic ribosome.
b. 50S	2. Small rRNA in large subunit of eukaryotes only.
c. 70S	3. Eukaryotic ribosome.
d. 40S	4. Small subunit rRNA of prokaryotic ribosome.
e. 60S	5. Small subunit rRNA of eukaryotic ribosome.
f. 80S	6. Small subunit of prokaryotic ribosome.
g. 23S	7. Large subunit rRNA of prokaryotic ribosome.
h. 16S	8. Large subunit rRNA in both prokaryotes and eukaryotes.
i. 28S	9. Prokaryotic ribosome.
j. 18S	10. Large subunit of eukaryotic ribosome.
k. 5S	11. Large subunit of prokaryotic ribosome.
l. 5.8S	12. Small subunit of eukaryotic ribosome.

2. Translation can be divided into three phases. What are they? During each phase, two tRNA binding sites on the ribosome, the A site and the P site, are occupied by various modified tRNAs or release factors. For each phase of protein synthesis, state the occupancy of the A site and the P site.

3. What is the role of the Shine-Dalgarno sequence (ribosome-binding site), a purine rich region of prokaryotic mRNAs located upstream of the initiator codon?

4. Identify the following as associated with either initiation (I), elongation (E), or termination (T) of translation.
   a. IF-2
   b. EF-Tu
   c. Cap binding protein
   d. EF-Ts
   e. EF-G
   f. RF-1
   g. IF-3
   h. 40S preinitiation complex
   i. eIF-2
   j. polysomes

5. Match a protein synthesis inhibitor with its mode of action.

Inhibitor	Mode of Action
a. Streptomycin	1. Causes release of peptide in covalent linkage with antibiotic.
b. Puromycin	2. ADP-ribosylated eEF-2.
c. Diphtheria Toxin	3. Blocks peptidyl transferase activity of 50S subunit.
d. Ricin	4. Cleaves prokaryotic small subunit rRNA.
e. Colicin E3	5. Inactivates eukaryotic large subunit rRNA by base removal.
f. Chloramphenicol	6. Codon misreading.

Answers

1. a. 6; b. 11; c. 9; d. 12; e. 10; f. 3; g. 7; h. 4; i. 1; j. 5; k. 8; l. 2.

2. During initiation of protein synthesis the P site is occupied by charged initiator tRNA and the A site is occupied by an aminoacyl-tRNA. During elongation the P site is occupied by a peptidyl tRNA or transiently by an uncharged tRNA. The A site is occupied by an aminoacyl-tRNA and transiently by a peptidyl tRNA. During termination, the P site is occupied by a peptidyl-tRNA and the A site is recognized by a release factor.

3. The ribosome-binding site interacts with a pyrimidine rich region near the 3' end of the small subunit ribosome to properly align the translational start site on the ribosome.

4. a. I; b. E; c. I; d. E; e. E; f. T; g. I; h. I; i. I; j. E.

5. a. 6; b. 1; c. 2; d. 5; e. 4; f. 3.

Additional Problems

1. In addition to ribosomes, *in vitro* protein synthesis systems from *E. coli* require a high-speed supernatant that supplies protein factors required for all of the side reactions necessary for protein synthesis. The reaction must be supplied with ATP, GTP, amino acids and a compound known as citrovorum. Citrovorum is an old name for folinic acid, formyl tetrahydropteroyl-glutamic acid. Explain why ATP, GTP and citrovorum are added. What role do they play in protein synthesis?

2. How are the 5S rRNA and 5.8S rRNA related? How is the 5.8S rRNA related to the large-subunit rRNA?

3. The substrates for protein synthesis are not free amino acids but aminoacylated tRNAs. Explain what an aminoacyl tRNA is and how is it formed.

4. During the elongation phase of protein synthesis, two tRNA binding sites orchestrate the elongation of a protein. Describe this process.

5. The small subunit of the ribosome is responsible for initiation of protein synthesis. It interacts with several proteins, and various RNAs to form a preinitiation complex. Later, the preinitiation complex is joined by the large subunit to form the initiation complex, a complete ribosome lacking an aminoacyl-tRNA at the A site. Describe how a preinitiation complex forms.

6. In eukaryotic ribosomes, initiation of protein synthesis is regulated by phosphorylation of eIF-2. Explain.

7. The elongation factor EF-Tu is a GTPase required for binding of aminoacylated tRNA to the A site. This protein is recycled during protein synthesis by interacting with another protein, EF-Ts. Describe this interaction.

8. In terms of codon recognition, how is termination fundamentally different than elongation in protein synthesis?

Abbreviated Answers

1. ATP is required for aminoacylation of tRNAs by aminoacyl-tRNA synthetases. Hydrolysis of GTP is used at various stages of protein synthesis to drive reactions. Formyl-tetrahydropteroly-glutamic acid serves as a formyl-group donor for methionyl-tRNA$_f$Met formyl transferase to formylate the charged initiator tRNA.

2. The 5S rRNA and the 5.8S rRNA are both components of the large subunit of the ribosome. The 5.8S is specific for eukaryotic cytoplasmic ribosomes whereas the 5S rRNA is found in eukaryotes and prokaryotes. The 5.8S rRNA of eukaryotes has a sequence similar to the 5'-end of the large subunit rRNA of prokaryotes.

3. Aminoacyl-tRNAs represent activated forms of amino acids used by the ribosome as substrates for peptide bond formation during protein synthesis. Aminoacyl-tRNAs are formed by aminoacyl-tRNA synthetases in a two-step reaction driven by hydrolysis of ATP.

4. During a cycle of elongation, the P-site is occupied by a peptidyl-tRNA produced by preceding rounds of elongation and the A-site is occupied by an aminoacyl-tRNA appropriate for the codon in the A-site. Peptide bond formation occurs by nucleophilic attack of the amino group of the A-site aminoacyl-tRNA on the carbonyl carbon of the ester bond in the peptidyl-tRNA resulting in peptide bond formation, deacylation of the P-site tRNA, and formation of a peptidyl-tRNA in the A-site. The P-site tRNA is released and the peptidyl-tRNA in the A-site is translocated to the P-site in a factor-dependent manner. Translocation of the tRNA is accompanied by movement of the mRNA as well, to bring a new codon into the A-site. The A-site is filled with the appropriate aminoacylated-tRNA by a factor-dependent process.

5. A ribosome is dissociated by initiation factors into a large subunit and a small subunit. The small subunit binds mRNA and a complex of a specific initiation factor (IF-2 in prokaryotes), GTP, and charged initiator-tRNA to form a preinitiation complex. The complex is joined by the large subunit with release of initiation factors to produce the initiation complex.

6. In eukaryotes, initiation of protein synthesis begins with formation of a ternary complex with eIF-2, GTP, and Met-tRNA$_i$Met. The complex associates with a small subunit to produce a 40S preinitiation complex. The α-subunit of eIF-2 can be phosphorylated by a specific kinase. Phosphorylated eIF-2 binds eIF-2B tightly and effectively eliminates eIF-2B-dependent recycling of eIF-2.

7. EF-Tu forms a complex with GTP and an aminoacyl-tRNA before binding to the ribosome. Ribosome binding stimulates GTPase activity resulting in release of P$_i$, A-site binding of the aminoacyl-tRNA, and release of EF-Tu:GDP complex. EF-Ts binds to this complex and displaces GDP to produce an EF-Tu:EF-Ts complex. GTP can dissociate EF-Ts from this complex, thus allowing EF-Tu to participate in another round of elongation.

8. For termination codons, codon recognition is achieved by protein release factors instead of tRNAs.

Summary

Translation is the process whereby genetic information is expressed through synthesis of a protein. Ribosomes are the agents of protein synthesis. Ribosomes are self-assembling ribonucleoprotein particles ubiquitous to all cells, invariably consisting of two unequal subunits, each containing more than twenty different proteins and at least one characteristic ribosomal RNA (rRNA) molecule.

Like chemical polymerization processes, protein biosynthesis has three phases: initiation, elongation and termination. In initiation, the small ribosomal subunit joins with an mRNA molecule and a particular methionine-bearing tRNA is used only in peptide chain initiation, never in peptide chain elongation. The large ribosomal subunit then adds to the small subunit:mRNA:Met-tRNA complex and the elongation phase commences.

All the peptide bonds in the protein, from the first to the last, are formed in the elongation phase. Elongation proceeds via the codon-directed association of aminoacyl-tRNAs with the ribosome:mRNA complex, as the ribosome moves along the mRNA and encounters successive codons. The ribosome:mRNA complex has two aminoacyl-tRNA binding sites: the A site where an incoming aminoacyl-tRNA binds and the P site where peptidyl-tRNA (the tRNA carrying the growing polypeptide chain) is held. Peptide bond formation is achieved when the α-carboxyl end of the polypeptide chain is transferred from its tRNA to the free α-NH$_2$ group of the aminoacyl-tRNA in the A site. The deacylated tRNA (the tRNA left after transfer of the polypeptide) leaves the P site. The tRNA in the A site now bears the polypeptidyl chain and it is translocated from the A site to the P site as the ribosome moves one codon further along the mRNA. This codon now directs the binding of the next incoming aminoacyl-tRNA in the A site, and the process repeats itself until all the codons have been translated. Termination of polypeptide synthesis is triggered by the appearance of a nonsense codon (UAA, UAG or UGA) in the A site. Release factor proteins bind at the A site, transforming the peptidyl transferase activity of the ribosome into a peptidyl hydrolase, and the free polypeptide is released. The ribosome:mRNA complex then dissociates.

The reactions of peptide chain initiation, elongation and termination are orchestrated at each step by specific soluble proteins - initiation factors, elongation factors and release factors. GTP hydrolysis provides the energy to drive conformational changes in the protein-synthesizing apparatus. These conformational changes represent the workings of the protein synthesis machinery in its cycle of successive amino acid addition to the growing peptidyl chain. Two GTPs are expended for each amino acid added to the chain: one in aminoacyl-tRNA binding at the A site and one in the coupled movement of the ribosome along the mRNA which translocates the peptidyl-tRNA from the A site to the P site. Ribosomal subunits cycle rapidly through protein synthesis. Polysomes, consisting of an mRNA molecule with several ribosomes simultaneously translating it, serve as the active protein-synthesizing units.

While the translation is similar in prokaryotes and eukaryotes, both the process and the accompanying apparatus are considerably more complex in eukaryotes. Protein synthesis inhibitors have been very useful in probing the mechanism of protein synthesis; some have clinical value as antibiotics.

Chapter 31

Completing the Protein Life Cycle: Folding, Processing, and Degradation

● ●

Chapter Outline

❖ Protein folding: Chaperones
  ⅄ TF (Trigger factor in *E. coli*), NAC (in eukaryotes) bind to nascent chain on ribosome
  ⅄ Hsp70 (DnaK): Binds to nascent polypeptide chain at hydrophobic residues to prevent folding
  ⅄ DnaJ (Hsp40): Binds to unfolded proteins and passes them to Hsp70 (DnaK)
  ⅄ GrpE: ADP/ATP exchange on DnaK
  ⅄ Hsp60: Chaperonins
    ◆ Group I in eubacteria
      ● GroES-GroEL
    ◆ Group II in archaea and eukaryotes
      ● CCT (triC) a GroEL analog
      ● Prefoldin (GimC)
  ⅄ Hsp90: Eukaryotic foldosome that folds signal transduction molecules
❖ Post-translational processing
  ⅄ Post-translational modifications like phosphorylation, acetylation, methylation, etc.
  ⅄ Proteolytic cleavage of pro-enzymes
    ◆ Introduce diversity of protein products
    ◆ Activation of preproprotein
    ◆ Targeting of proteins in protein translocation
    ◆ Protein translocation
    ◆ Characteristics of translocation systems
      ● Proteins made as preproteins with signal peptides
      ● Specific protein receptors exist on target membrane
      ● Movement catalyzed by complex structures: Translocons: ATP (or GTP) driven
      ● Proteins generally maintained in loosely folded conformations for translocation competence
  ⅄ Prokaryotic translocation
    ◆ N-terminal leader sequence
      ● N-terminus of leader sequence: Basic amino acids
      ● Central domain hydrophobic
      ● C-terminus: Nonhelical structure
    ◆ Leader peptidase: Removes leader sequence
  ⅄ Eukaryotic translocation and protein sorting
    ◆ Secreted and membrane proteins synthesized on ER-localized ribosomes
      ● Cytoplasmic ribosome initiates translation
      ● N-terminal signal sequence detected by signal recognition particle (SRP)
      ● SRP/ribosome complex binds to docking protein: ER membrane protein
      ● Ribosome delivers peptide to translocon
      ● Signal peptidase cleaves leader sequence

- Membrane proteins carry 20-residue stop transfer sequence
    - Retrograde transport: Sec61p: Moves protein from ER back to cytosol
    - Mitochondrial protein import
        - N-terminal sequence 10 to 70 residues long
            - Form amphiphilic α-helix
            - Binds to TOM (mitochondria outer membrane translocase)
        - Outer membrane protein
            - TOM passes protein to SAM (sorting and assembly complex)
        - Inner membrane protein
            - TOM to TIM22
        - Matrix protein
            - TOM to TIM23
    - Chloroplasts
        - TOCs and TICs
- ❖ Protein degradation: Ubiquitination most common pathway in eukaryotes
    - ⋏ Ubiquitin: Conserved 76-residue protein
        - E1: Ubiquitin-activation protein: Attaches to C-terminal Gly of ubiquitin
        - E2: Ubiquitin-carrier protein: Accepts ubiquitin from activator protein: Carried on cysteine residue
        - E3: Ubiquitin-protein ligase: Binds target protein: Protein ubiquitinated on amino groups
    - ⋏ Proteins with acidic N-termini
        - N-termini altered by Arg-tRNA
    - ⋏ PEST sequences: Target proteins for degradation
    - ⋏ Proteasomes
        - 20S proteasomes
        - 26S proteasomes
    - ⋏ HtrA protease: DegP in *E. coli*
        - Functions as chaperone at low temperature (20°C)
        - Switch from chaperone to protease function as temperature increases
- ❖ Sumoylation: Addition of small ubiquintin-like modifiers
    - ⋏ Reaction requires three-enzyme conjugation system
    - ⋏ Consequences of SUMO addition
        - Blocking protein binding site: Inhibition of protein/protein interaction
        - Creation of binding site for interacting partner protein
        - Induction of protein conformational change

Chapter Objectives

Protein Folding

Protein folding starts before the polypeptide is released from the ribosome. In many cases, folding is assisted by molecular chaperones. You should be familiar with the names, order of action and mechanism of a few of them: TF, NAC, HSP70, HSP60, GroES-GroEL, CCT (or TriC) and Hsp90. Chaperones act via cycles of binding and hydrolysis of ATP. If hydrophobic patches are exposed and not buried within the protein structure on their surface, they have the potential for incorrect associations with other like patches, ultimately leading to precipitation. Hsp70s recognize these exposed hydrophobic regions as signals that the protein is not folded correctly, and bind to them, thereby blocking damaging associations. Both GroES-GroEL and CCT are large structures into which proteins are sequestered for ATP-dependent folding. Hsp90s are chaperones that act in concert with Hsp70 to assist in the folding of proteins operating in signal transduction pathways.

Post-Translational Processing

The most common form of protein modification is proteolytic cleavage to activate a protein, or to release mature protein products from a larger primary translational product. Proteins are also targeted to specific locations within the cell or are destined for transport out of the cell. In prokaryotes, a signal sequence, located on the N-terminus of a protein, or at a more internal location, may direct the protein to be exported from the cell or to become a membrane-bound protein. Signal recognition particles play a role in halting protein synthesis shortly after a signal peptide is produced and, in conjunction with docking protein, direct the nascent protein to the endoplasmic reticulum membrane where protein synthesis resumes. Proteins cross into the endoplasmic reticulum lumen, or become integrated into the membrane, via an aqueous tunnel called a translocon. The translocon is aligned with the exit tunnel from the 50S ribosomal subunit, and together they form a continuous channel from the ribosome through the membrane. Once the N-terminus reaches the endoplasmic reticulum lumen, the signal sequence is proteolytically removed by leader peptidase.

Most mitochondrial proteins are encoded in the nucleus and are post-translationally imported from the cytoplasm. Proteins are directed to the mitochondria by N-terminal targeting presequences that have the ability to form amphipathic α-helices. Proteins enter the mitochondria through the TOM and TIM translocons, located in the outer and inner mitochondrial membranes, respectively.

Protein Degradation

Ubiquitin plays a role in degradation of proteins in eukaryotes. This pathway is specific and efficient in removing defective proteins from the cell by attaching multiple copies of ubiquitin to the protein to be degraded, which in turn targets the protein for destruction in the proteosome. Proteins having Arg, Lys, His, Phe, Tyr, Trp, Leu, Asn, Gln, Asp or Glu at their N-termini are particularly susceptible to ubiquitination and proteosome-mediated degradation. These proteins have half-lives of between 2 and 30 minutes. In contrast, proteins with Met, Ser, Ala, Thr, Val, Gly or Cys at their N-termini are resistant to ubiquitin-mediated degradation.

Proteins marked for destruction are degraded by one of two related proteosomes, differentiated on the basis of their size. These are large oligomeric structures enclosing a central cavity wherein degradation takes place. HtrA proteases have a dual function in prokaryotic cells. At low temperatures HtrA acts as a chaperone, preventing aggregation and promoting folding. At high temperatures, however, HtrA develops protease activity, and instead of correcting a misfolded protein, destroys it.

Problems and Solutions

1. Assessing the ATP cost of protein synthesis
(Integrates with Chapter 30.) Human rhodanese (33kD) consists of 296 amino acid residues. Approximately how many ATP equivalents are consumed in the synthesis of the rhodanese polypeptide chain from its constituent amino acids and the folding of this chain into an active tertiary structure?

Answer: The energetics of protein synthesis requires at least 4 ATP equivalents per amino acid residue incorporated into protein. During elongation, the A-site is filled at the expense of GTP that occurs during recycling of EF-1, a eukaryotic elongation factor. A second GTP hydrolysis drives translocation catalyzed by EF-2, the eukaryotic translocation factor. Peptide bond formation is driven by the high-energy aminoacyl bond on the aminoacyl-tRNA substrate. Formation of aminoacyl-tRNAs is catalyzed by aminoacyl-tRNA synthetases, enzymes that consume ATP and produce AMP and PP$_i$. Pyrophosphatase hydrolysis of PP$_i$ accounts for an additional high-energy phosphate bond. For synthesis of human rhodanese with 296 amino acid residues, the A-site will have to be filled 295 times, accounting for (295 × 4 =) 1,180 ATP equivalents. Met-tRNA$_i$ formation consumes two ATPs. Initiation requires two ATP equivalents, one in the form of GTP during eIF-2 mediated Met-tRNA$_i$ binding to form the 40S pre-initiation complex, and one in the form of ATP during 40S initiation complex formation. Peptide chain termination in eukaryotes requires GTP hydrolysis thus accounting for an additional ATP equivalent. Thus, 1,180 + 4 + 1 = 1,185 ATP equivalents are consumed to synthesize rhodanese. In order to properly fold the polypeptide chain, an additional 130 equivalents of ATP are consumed during the folding cycle catalyzed by molecular chaperones. Active rhodanese production requires approximately 1,315 equivalents of ATP.

2. Understanding the consequences of proteolysis
A single proteolytic break in a polypeptide chain of a native protein is often sufficient to initiate its total degradation. What does this fact suggest to you regarding the structural consequences of proteolytic nicks in proteins?

Answer: Although peptide bonds in proteins are quite stable, proteins are not made to last forever. During their lifetime they are subjected to a number of insults including oxidation of side chains, chemical modifications, and cleavage of peptide bonds. The consequences of these events may be inactivation of a protein or alteration of its activity to a new, undesirable form. To avoid accumulating inactive protein, cells must either correct the defect or degrade the protein and replace it through gene expression. In the case of peptide bond breakage, cells have evolved the ability to recognize and degrade nicked protein. In eukaryotic cells this process involves the protein ubiquitin, a highly conserved, 76 amino acid polypeptide. Ubiquitin is

571

ligated to free amino groups on proteins and serves as a molecular tag, directing the protein's degradation by proteolysis. The ubiquitin pathway is apparently rapid and efficient and, because of this, cells never accumulate breakdown products of protein degradation.

3. Understanding the relevance of chaperones in protein folding
Protein molecules, like all molecules, can be characterized in terms of general properties such as size, shape, charge, solubility/hydrophobicity. Consider the influence of each of these general features on the likelihood of whether folding of a particular protein will require chaperone assistance or not. Be specific regarding Hsp70 chaperones or Hsp70 chaperones and Hsp60 chaperonins.

Answer: Hsp70 binds to exposed hydrophobic portions of target polypeptides via an 18 kDa domain residing in the central portion of its sequence. Since hydrophobic patches are not exposed in proteins that have attained their final conformation, Hsp70s interact only with polypeptides that have not yet folded. It would be expected that since the occurrence of charged residues within a sequence decreases its overall hydrophobicity, charge would negatively impact the binding of the chaperones. Size, on the other hand, would not be expected to play a role in Hsp70 binding, as multiple chaperones could bind to larger polypeptides, or to multiple hydrophobic regions on a single polypeptide.

For Hsp 60 and HSP 70, the Hsp60 chaperonin acts in concert with Hsp70 to achieve the final folded conformation of certain proteins. Unlike the Hsp70, however, Hsp60 action requires that the polypeptide fit inside the chaperonin cavity. This sets an upper limit on the size of the polypeptide that can be folded by Hsp60. Since the cavity has a diameter of 5 nm, we can calculate the molecular weight of a protein that can be folded as follows: The volume of a protein with a radius of 2.5 nm is

$$V = \frac{4}{3}\pi r^3 = \frac{4}{3}\pi(2.5)^3 = 65.4 \text{ nm}^3$$

Using 1.25 g/ml as the density of a typical protein, we calculate the molecular weight of a protein with a radius of 2.5 nm as

$$M_r(kD) = V(nm^3) \times \frac{1.25 \dfrac{g}{cm^3 \cdot molec} \times 6.02 \times 10^{23} \dfrac{molec}{mol} \times 0.001 \dfrac{kD}{g/mol}}{10^{21} \dfrac{nm^3}{cm^3}}$$

$$= V(nm^3) \times 0.7525 \frac{kD}{nm^3} = 65.4 \text{ nm}^3 \times 0.7525 \frac{kD}{nm^3} = 49.2 \text{ kD}.$$

Thus, a protein larger than approximately 50 kDa will be too large to fit into the Hsp60 cavity and cannot be folded by this chaperonin.

4. Assessing the chaperone-assisted folding of multi-subunit proteins
Many multidomain proteins apparently do not require chaperones to attain the fully folded conformations. Suggest a rational scenario for chaperone-independent folding of such proteins.

Answer: While many proteins renature under certain conditions after experimental denaturing, the question asks us specifically to consider multidomain proteins. Many such proteins contain structural domains residing in regions of contiguous sequence that fold independently, and which are often linked by flexible hinge regions. It is likely that the domains fold as they emerge from the ribosome, and that the polypeptide never exposes large regions of unfolded sequence that could then be bound by chaperones. Once such a protein is released from the ribosome, its pre-folded domains would adopt a spatial arrangement resulting in the protein's final tertiary structure.

5. Assessing the dimensions and capacity of GroEL
The GroEL ring has a 5-nm central cavity. Calculate the maximum molecular weight for a spherical protein that can just fit into this cavity, assuming the density of the protein is 1.25 g/mL.

Answer: This calculation was made for problem 3 above. In that problem, the following relationship was developed between the volume of a spherical protein and its molecular weight:

$$Mr(kD) = V(nm^3) \times 0.7525 \frac{kD}{nm^3}$$

This can be further modified to give the relationship between molecular weight and a protein's radius:

$$Mr\ (kD) = V\ (nm^3) \times 0.7525 \frac{kD}{nm^3}$$

$$= \frac{4}{3}\pi r^3 \times 0.7525 \frac{kD}{nm^3} = 3.15 \frac{kD}{nm^3} \times r^3$$

Accordingly, a spherical protein with a radius of 2.5 nm has a molecular weight of approximately 50 kDa.

6. Understanding the possible phosphorylated states of acetyl-CoA carboxylase
(Integrates with Chapter 24.) Acetyl-CoA carboxylase has at least seven possible phosphorylation sites (residues 23, 25, 29, 76, 77, 95 and 1200) in its 2345-residue polypeptide (see Figure 24.4). How many different covalently modified forms of acetyl-CoA carboxylase protein are possible if there are seven phosphorylation sites?

Answer: Each phosphorylation site had two possibilities for covalent modification; it either is or it isn't phosphorylated. Combinatorial statistics dictates that the number of unique combinations that can be formed from seven amino acids, each of which can exist in one of two states is $2^7 = 128$. One of these configurations is the state in which none of the potential sites is phosphorylated. So, the number of covalently modified forms of acetyl-CoA carboxylase is $2^7 - 1 = 127$.

7. Comparing the mechanisms of action of EF-Tu/EF-Ts and DnaK/GrpE
(Integrates with Chapter 30.) In what ways are the mechanisms of action of EF-Tu/EF-Ts and DnaK/GrpE similar? What mechanistic functions do the ribosome A-site and DnaJ have in common?

Answer: GrpE and EF-Ts are nucleotide exchange factors working on DnaK and EF-Tu, respectively. Their reaction mechanisms can be summarized as follows:

GrpE catalyzes replacement of ADP with ATP on DnaK → DnaK converted to a form having low affinity for substrate → Causes release of substrate → DnaK changes conformation to a form having a high affinity for a DnaJ/unfolded protein substrate → DnaJ stimulates ATPase activity of DnaK → DnaK:ATP is converted to DnaK:ADP, and the reaction cycle starts again.

EF-Ts catalyzes replacement of GDP with GTP in EF-Tu → EF-Tu converted to a form having high affinity for aminoacyl-t-RNA → Binding of aminoacyl-t-RNA:EF-Tu to the A site of the ribosome stimulates ET-Tu GTPase activity → EF-Tu:GTP is converted to EF-Tu:GDP → aminoacyl-t-RNA:EF-Tu adjusts conformationally to the A site, allowing codon:anticodon recognition, and the reaction cycle starts again.

DnaJ and the A site of the ribosome have similar functions in that they both stimulate the NTP hydrolysis activity of their respective partner NTPases, thereby causing in them stabilizing conformational changes.

8. Understanding the structure and function of signal sequences and signal peptidases
The amino acid sequence deduced from the nucleotide sequence of a newly discovered human gene begins: MRSLLILVLCFLPAALGK.... Is this a signal sequence? If so, where does the signal peptidase act on it? What can you surmise about the intended destination of this protein?

Answer: A signal sequence is an amino-terminal sequence that contains the information necessary for the protein to be targeted to and transported into the lumen of the ER. Signal peptides are characterized by a sequence consisting of one or more basic residues followed by a stretch of 6 to 12 hydrophobic amino acids. Leader peptidase within the ER lumen removes the signal peptide at a position immediately following the motif A-X-A, with X representing any amino acid and allowing conservative substitutions of the alanine. In the sequence MRSLLILVLCFLPAALGK..., a basic amino acid is found at position 2, which is followed immediately by a hydrophobic sequence of the length appropriate for a signal peptide. Thus, this sequence could function as a signal peptide, causing the passenger protein to be translocated into the ER lumen. The motif A-X-A is roughly met in the sequence A-L-G in this peptide, and so leader peptidase would be expected to cleave the signal sequence between the G and K. This protein may pass through the endomembrane system and ultimately be secreted.

There are Internet resources to analyze sequences for potential signal sequences. The Swissprot database has a link to one of them in their "Tools" page at http://expasy.org/tools/. On that page locate SignalP and activate it. This should bring you to SignalP 3.0 Server page. I submitted the short sequence and got a result that suggested it contained a signal sequence. The sequence, however, seemed to be too short to analyze properly so I attached to its C-terminus a random protein sequence of 200 amino acid residues I generated at using a program at http://www.bioinformatics.org/sms/index.html. Analysis indicated that the N-terminal sequence was indeed a signal sequence with a potential cleavage site between G and K in the N-terminal sequence.

9. Understanding the mechanism of transport of integral membrane proteins
Not only is the Sec61p translocon complex essential for translocation of proteins into the ER lumen, it also mediates the incorporation of integral membrane proteins into the ER membrane. The mechanism for integration is triggered by stop-transfer signals that cause a pause in translocation. Figure 31.5 shows the translocon as a closed cylinder spanning the membrane. Suggest a mechanism for lateral transfer of an integral membrane protein from the protein-conducting channel of the translocon into the hydrophobic phase of the ER membrane.

Answer: Translocation of the nascent polypeptide continues through the Sec61p translocon until a transmembrane-spanning region of hydrophobic amino acids is encountered. This stop-transfer sequence causes translocation to pause and a conformational change takes place within the translocon that opens the channel to the core of the membrane bilayer. The transmembrane region then slips laterally out of the translocon into the membrane, and translation at the ribosome continues.

10. Assessing the dimensions and capacity of Sec61p
The Sec61p core complex of the translocon has a highly dynamic pore whose internal diameter varies from 0.6 to 6 nm. In post-translational translocation, folded proteins can move across the ER membrane through this pore. What is the molecular weight of a spherical protein that would just fit through a 6-nm pore? (Adopt the same assumptions used in problem 5.)

Answer: From the equations developed in problems 3 and 5, we have

$$M_r(kD) = 3.15 \frac{kD}{nm^3} \times r^3$$

with r being the radius of the spherical protein in units of nm. Using r = 3 nm for a spherical protein with a 6 nm diameter,

$$M_r(kD) = 3.15 \frac{kD}{nm^3} \times 3^3 = 85.1 \ kD$$

Thus the Sec61p translocon channel can accommodate a protein of about 85 kD.

11. Assessing the dimensions and capacity of the protein conducting translocon
(Integrates with Chapters 6, 9, and 30.) During co-translational translocation, the peptide tunnel running from the peptidyl transferase center of the large ribosomal subunit and the protein conducting-channel are aligned. If the tunnel through the ribosomal subunit is 10 nm and the translocon channel has the same length as the thickness of a phospholipid bilayer, what is the minimum number of amino acid residues sequestered in this common conduit?

Answer: The thickness of a bilayer is approximately 5 nm, so the combined channel through the Sec61p translocon and the ribosome is 15 nm. In the most extended β-sheet-like conformation, each residue spans a distance of 0.347 nm. In this conformation the overall channel would contain:

$$\frac{15nm}{0.347 \frac{nm}{residue}} \approx 43 \text{ residues}$$

By using the distance values for the most extended β-sheet-like conformation, this represents the minimum number of residues that can be sequestered in the ribosome/translocon channel. In fact, the experimentally determined number is 70 residues, indicating that the nascent polypeptide is in a less extended conformation with an average distance per residue of approximately 0.21 nm.

12. Understanding the nature of the ubiquitin-ubiquitin linkage
Draw the structure of the isopeptide bond formed between Gly[76] of one ubiquitin molecule and Lys[48] of another ubiquitin molecule.

Answer: As with a true peptide bond, the isopeptide bond is formed between a carboxylic acid and an amine with the removal of water. In ubiquitin, the C-terminal Gly[76] contributes the carboxylic group, and the amine is contributed by the side chain group (ε-amine) of Lys[48]. The isopeptide bond can then be represented as

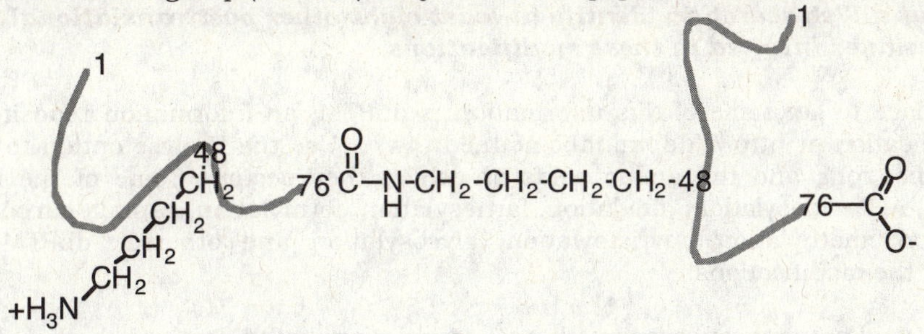

$$\overset{1}{\big|}$$

$$\overset{48}{\underset{\underset{+H_3N}{\overset{\displaystyle CH_2}{\overset{\displaystyle CH_2}{\overset{\displaystyle CH_2}{\overset{\displaystyle CH_2}{|}}}}}}{} \qquad 76\;\overset{O}{\overset{\|}{C}}-\underset{H}{N}-CH_2\text{-}CH_2\text{-}CH_2\text{-}CH_2\text{-}48 \qquad 76\text{-}\overset{O}{\overset{\|}{C}}\overset{}{\underset{O^-}{}}$$

$$\overset{1}{\big|}$$

13. Understanding the nature of E3 ubiquitin protein ligation
Assign the 20 amino acids to either of two groups based on their susceptibility to ubiquitin ligation by E3 ubiquitin protein ligase. Can you discern any common attributes among the amino acids in the less susceptible versus the more susceptible group?

Answer: E3 ubiquitin protein ligase selects proteins by the nature of the N-terminal amino acid. Since susceptible proteins must have a free α-amino terminus to be susceptible to degradation, proteins with Pro at their N-terminus are not degraded through the ubiquitin pathway. Proteins with Met, Ser, Ala, Thr, Val, Gly or Cys at the amino terminus are resistant to selection by E3 ubiquitin protein ligase, whereas those having Arg, Lys, His, Phe, Tyr, Trp, Leu, Ile, Asn, Gln, Asp or Glu at their N-terminus are susceptible.

It is difficult to recognize common chemical attributes that apply to all amino acids residing in one group or another. For instance, while the charged amino acids are in the susceptible group, other hydrophilic amino acids, i.e., Ser and Thr, are resistant. Similarly, both groups contain hydrophobic amino acids.

14. Assessing the consequences of proteasome inhibition
Lactacystin is a Streptomyces natural product that acts as an irreversible inhibitor of 26S proteosome β-subunit catalytic activity by covalent attachment to N-terminal threonine –OH groups. Predict the effects of lactacystin on cell cycle progression.

Answer: Progression through the cell cycle requires the cyclic accumulation and degradation of various cellular proteins. This requires the continuous operation of both the protein synthesis and degradation machineries. Inhibition of either one of these pathways would be expected to disrupt numerous cellular functions among which would be progression through the cell cycle.

15. Understanding the dual functionality of the HtrA protease
HtrA proteases are dual-function chaperone-protease protein quality control systems. The protease activity of HtrA proteases depends on a proper spatial relationship between the Asp-His-Ser catalytic triad. Propose a mechanism for the temperature-induced switch of HtrA proteases from chaperone function to protease function.

Answer: You should remember this triad from material presented in chapter 14 on mechanisms of enzyme action. This triad is the catalytic triad of so-called serine proteases. Figure 14.15 shows the location of these amino acid residues in the primary structure of several serine protease and they are far apart in the primary structure. Thus, to function the protein needs to fold such that the triad is close together. A temperature-induced change in protein function implies that the protein undergoes a change from one conformation at low temperature to another conformation at the higher temperature. In the case of HtrA, it is likely that the high-temperature conformation brings the Asp-His-Ser triad into a spatial configuration that allows them to act catalytically.

16. Understanding the scope and diversity of posttranslational protein modification
As described in this chapter, the most common post-translational modifications of proteins are proteolysis, phosphorylation, methylation, acetylation, and linkage with ubiquitin and SUMO proteins. Carry out a Web search to identify at least eight other posttranslational modifications and the amino acid residues involved in these modifications.

Answer: A good place to get some of this information is dbPTM, an information repository of protein post-translational modification at http://dbptm.mbc.nctu.edu.tw/. Use the Browse option to get a table of post-translational modifications and the amino acids on which they occur. Some of the modifications listed include acetylation, ADP-ribosylation, amidation, farnesylation, formylation, gamma-carboxylation of glutamic acids, hydroxylation, methylation, myristoylation, glycosylation and others. dbPTM provides links to structure pages for the modifications.

Mass spectrometry is often very concerned with post-translational modifications, both natural and synthetic, of amino acids and there are several databases related to mass spectrometry work that list post-translational modifications.

17. Understanding the nature and functional utility of fluorescence resonance energy transfer (FRET)
Fluorescence resonance energy transfer (FRET) is a spectroscopic technique that can be used to provide certain details of the conformation of biomolecules. Look up FRET on the Web or in an introductory text on FRET uses in biochemistry, and explain how FRET could be used to observe conformational changes in proteins bound to chaperonins such as GroEL. A good article on FRET in protein folding and dynamics can be found here: Haas, E., 2005. The study of protein folding and dynamics by determination of intramolecular distance distributions and their fluctuations using ensemble and single-molecule FRET measurements. ChemPhysChem 6:858–870. Studies of GroEL using FRET analysis include the following: Sharma, S., et al., 2008. Monitoring protein conformation along the pathway of chaperonin-assisted folding. Cell 133:142–153; and Lin, Z., et al., 2008. GroEL stimulates protein folding through forced unfolding. Nature Structural and Molecular Biology 15:303–311.

Answer: Fluorescence resonance energy transfer or FRET is a process by which energy is transferred from one fluorescent molecule to another. This process can be used to measure distance between fluorescent molecules in the range of 1 nm to 10 nm, dimensions useful in studying proteins and other macromolecules. The energy transfer is between two fluorescent dyes. For transfer to occur the emission spectrum of the so-called donor dye must overlap with the absorption (or excitation) spectrum of the acceptor dye. Furthermore, the distance between the two dyes must be within approximately 10 nm or less. Measurements of efficiency of energy transfer are used to measure the distance between two dyes. Techniques employing FRET are widely used in biochemistry and molecular biology. In the particular experimental problem under consideration, protein folding, the protein under study must be covalently modified with two fluorescent dyes on specific amino acid residues. Considering two such dyes on a polypeptide chain, the efficiency with which FRET occurs between them will depend on the distance separating them. This distance will obviously depend on the separation distance along the polypeptide chain. That is, dyes located on different ends will transfer energy less efficiently than dyes located on amino acids close together on the chain. But, proteins do not remain as extended polypeptide chains. Rather, they fold up into a conformation dictated by the protein's amino acid sequence and solvent conditions. As a polypeptide folds, regions along the chain that are far apart in primary structure may be close together in the final conformation. The folding process follows a pathway or a series of pathways that start off with local hydrophobic interactions. GroEL participates in protein folding by preventing protein aggregation (among protein molecules) during the folding process by sequestering the protein. GroEL also presents the folding protein a changing environment ranging from hydrophobic to hydrophilic. In the hydrophobic environment, proteins relinquish their local hydrophobic interactions in favor of interaction with GroEL. As the environment changes, proteins move though conformations that are stabilized by a hydrophobic core characteristic of many globular proteins.

18. Understanding the nature and functions of the phosphodegron
The cross-talk between phosphorylation and ubiquitination in protein degradation processes is encapsulated in the concept of the "phosphodegron." What is a phosphodegron, and how does phosphorylation serve as a recognition signal for protein degradation? (A good reference on the phosphodegron and crosstalk between phosphorylation and ubiquitination is Hunter, T., 2007. The age of crosstalk: Phosphorylation, ubiquitination, and beyond. Molecular Cell 28:730–738.)

Answer: Phosphodegrons are short sequence motifs that are substrates for E3 ubiquitin ligase when the sequence is phosphorylated by a kinase. The amino acid to be phosphorylated is localized in a short consensus sequence that is recognized by the kinase. Kinases phosphorylate serine, threonine or tyrosine and the phosphorylated amino acid residue serves as a recognition site for an E3 ubiquitin ligase. Ubiquitination typically targets the protein for degradation. In some cases phosphorylation leads to inhibition of E3 ligase binding, which results in stabilization of the protein from ubiquitin-mediated decay.

Preparing for the MCAT® Exam

19. Assessing the consequences of a common post-translational modification
A common post-translational modification is the removal of the universal N-terminal methionine in many proteins by Met-aminopeptidase. How might Met removal affect the half-life of the protein?

Answer: Whether or not Met removal form the N-terminus of a protein affects its half-life depends on the penultimate amino acid, since this will become the new N-terminal residue. In problem 13 above the amino acids that render a protein susceptible to ubiquitination and degradation when found in the N-terminal position are listed. From this list it can be seen that if the penultimate amino acid is Arg, Lys, His, Phe, Tyr, Trp, Leu, Ile, Asn, Gln, Asp or Glu, the protein resulting after Met-aminopeptidase action will be expected to have a shortened half-life. No change in the protein's half-life would be expected after removal of the N-terminal Met if the penultimate residue is Met, Ser, Ala, Thr, Val, Gly or Cys.

20. Designing a sequence for a peptide with amphipathic α-helical secondary structure
Figure 31.6 shows the generalized structure of an amphipathic α-helix found as an N-terminal presequence on a nuclear-encoded mitochondrial protein. Write out a 20-residue-long amino acid sequence that would give rise to such an amphipathic α-helical secondary structure.

Answer: An amphipathic α-helix is one in which hydrophobic amino acids are predominantly arrayed on one side of the helical axis and hydrophilic resides are arrayed on the other. In addition, mitochondrial targeting sequences are generally devoid of acidic residues. Thus, a 20-residue mitochondrial targeting sequence might contain basic amino acids at positions 1, 4, 7, 11, 14 and 18, and hydrophobic residues at the remaining positions.

Questions for Self Study

1. What is the dominant driving force for protein folding?

2. How does an Hsp70 chaperone recognize a protein as unfolded?

3. Answer True or False.
   a. All chaperones bind and release from their target proteins via cycles of ATP binding and hydrolysis. ___
   b. All chaperons contain central cavities in which proteins are sequestered to allow them to fold. ___
   c. All proteins follow the same pathway through the various chaperones during folding. ___
   d. Hsp90 and Hsp70 act in conjunction to fold proteins involved in signal transduction. ___
   e. Chaperones operate independently and never use "helper" proteins or "co-chaperones" to mediate protein folding. ___

4. Fill in the blanks. The endoplasmic reticulum is an internal compartment typically composed of two regions, the smooth ER and the _____. The later region is so named because ___ stud its surface. These macromolecular complexes are translating proteins destined to be processed in the endoplasmic reticulum. Translation of these proteins actually starts in the cytosol. When a short N-terminal sequence of the nascent polypeptide, termed the ___, emerges from the ribosome it is recognized by the ___. This ribonucleoprotein complex binds to the ribosome and blocks further protein synthesis until the ribosome is escorted to the cytosolic surface of the endoplasmic reticulum. The ___ is responsible for docking the ribosome to the membrane. Protein synthesis then continues with the newly synthesized proteins being deposited into the lumen of the endoplasmic reticulum. There the N-terminal sequence is cleaved by ___.

5. What are the roles of ubiquitin and the large 26S ATP-dependent protease complex in degrading proteins?

Answers

1. The burial of hydrophobic residues away from the aqueous solvent.

2. Hsp70s bind to exposed hydrophobic patches on proteins. Because the protein's structure would never tolerate the placement of a significant number of hydrophobic amino acids on the exposed surface, the fact that hydrophobic patches are exposed at all signals that the protein is not in its native conformation, and may be in need of a chaperone to assist in folding.

3. a. T; b. F; c. F; d. T; e. F.

4. Rough endoplasmic reticulum or RER; ribosomes; signal sequence; signal recognition particle; docking protein or SRP receptor; signal peptidase.

5. Proteins destined for degradation are covalently modified by conjugation to ubiquitin. Ubiquitinated proteins are degraded by the protease complex.

Additional Problems

1. The textbook tells us that Christian Anfinsen pointed out 40 years ago that all the information necessary for a protein to fold is contained within its primary structure. What made him say this?

2. Describe the mechanism of action of chaperonins in protein folding.

3. What does retrograde transport from the ER mean?

4. Describe the routes followed by proteins that reside in different mitochondrial compartments.

5. It is known from experiments that mature mitochondria and chloroplast will still actively import proteins. Speculate as to why this might be the case.

Abbreviated Answers

1. Anfinsen knew that the amino acid sequence contained the information required for proper folding because some proteins could be denatured and subsequently renatured all on their own. This generally required the protein to be in a very dilute solution, conditions never encountered in the living cell. Accordingly, many proteins fold with the assistance of molecular chaperones.

2. Chaperonins are members of the Hsp60 family. An unfolded protein with surface-exposed hydrophobic regions binds to hydrophobic patches present on the apical domain of the interior surface of the GroEL cavity. Upon binding ATP, GroES is recruited to cover the apical domain and conformational changes occur within GroEL that bury the hydrophobic patches. This causes the release of the unfolded or partially folded polypeptide, where is exists in the central cavity essentially free from other similarly partially folded polypeptides with which it could potentially form aggregates. After a short time the bound ATP is hydrolyzed, causing release of GroES and repositioning of the GroEL apical domain hydrophobic patches. If the polypeptide has folded successfully, it will escape from the GroEL cavity. If not, it can be re-captured by GroEL for another round of ATP-induced folding reactions.

3. If a protein is translocated into the ER lumen, but fails to fold correctly or assemble as a subunit of a multimeric protein, it can be transported back out of the ER through the Sec61p translocon. Upon reaching the cytoplasm, the offending protein can then be destroyed by the proteosome. The movement of a protein out of the ER through the Sec61p translocon is called retrograde transport..

4. Mitochondrial precursors all start their journey into the organelle by interacting with the TOM complex, a series of receptors and translocons in the outer membrane. A protein destined to integrate into the outer membrane is transferred from the TOM complex to a SAM complex, also in the outer membrane, from which it is inserted across the hydrophobic core of the membrane. Proteins that will take up residence in the inner membrane pass through the TOM complex and engage the TIM22 complex. The TIM22 complex mediates the insert in of the protein into the membrane. Proteins residing in the mitochondria matrix pass from the TOM complex to a different inner membrane complex, TIM23. The TIM23 complex mediates the passage of the

precursor protein across the inner mitochondria membrane into the matrix, where it is acted upon by the matrix processing protease to remove the targeting peptide.

5.   Although the organelles are mature, they must still import proteins to replace those lost to normal turnover. Many active proteolytic enzymes are present in these organelles, and proteins can have a short half-life here as well as in the cytoplasm.

Summary

A great variety of modifications are introduced post-translationally into proteins, proteolytic cleavage being a particularly prevalent change. Proteolytic cleavage can serve a number of ends: Generating diversity in the protein, activating its biological function, or facilitating the sorting and dispatching of proteins to their proper destinations in the cell. In the latter role, a leader peptide located at the N-terminus of the protein acts as a signal sequence, which tags the protein as belonging in a particular compartment or organelle. These signals are recognized by protein translocation systems.

Some of these systems identify the protein before its synthesis is complete. For example, the eukaryotic "signal recognition particle" (SRP) binds to the leader sequence of proteins destined for processing in the endoplasmic reticulum, just as this signal emerges from the ribosome. The SRP then chauffeurs the ribosome to the cytosolic face of the ER and docks it there so that the growing polypeptide chain is threaded through the ER membrane into the ER lumen. Within the lumen, the leader sequence is removed by signal peptidase. Additional modifications (such as glycosylation) may be introduced before the nascent polypeptide becomes a mature protein and gains its final destination.

Other protein translocation systems, such as those serving in mitochondrial protein import or prokaryotic protein export, employ molecular chaperones to keep the completed polypeptide in a partially unfolded state as it is shepherded through the cytoplasm from the ribosome to the membrane-embedded translocation apparatus. Once there, this apparatus catalyzes the energy-dependent translocation of the protein across the membrane. Typically, signal sequences are "cleavable presequences" clipped from the polypeptide once they have reached the other side of the membrane.

Proteins are in a dynamic state of turnover, with protein degradation serving an important role in determining the half-life of a protein in the cell. Protein degradation may be non-selective, as in lysosomal degradation, or selective, as in the ubiquitin-mediated pathway. The later system targets proteins for degradation by a large multi subunit catalytic ATP-dependent protease complex by labeling them with ubiquitin. Ubiquitin is a 76 residue polypeptide that becomes covalently attached to proteins via linking its C-terminal glycine α-carboxyl group to free -NH_2 groups in the protein. Ubiquitinylation of a protein is determined by the nature of the amino acid residue at the protein's N-terminus.

Chapter 32

The Reception and Transmission of Extracellular Information

• •

Chapter Outline

❖ Hormones: Three types
   ↳ Steroid hormones: Cholesterol derivatives
      • Regulate metabolism, salt and water balance, inflammatory process, sexual function
      • Two modes of action
         • Pass through cell membrane and localize on receptors on DNA
         • Bind to cell membrane to regulate ion channels
   ↳ Amino acid derivatives
      • Epinephrine and norepinephrine
         • Regulate smooth muscle contraction, cardiac rate, lipolysis, glycogenolysis
      • Thyroid hormone
         • Stimulates metabolism
   ↳ Peptide hormones
      • Secreted hormones initially synthesized with signal sequence
      • Synthesized as preprohormones
      • Cleavage of preprohormone may produce several peptide hormones
❖ Signal Transduction Systems: Features
   ↳ Receptor proteins
   ↳ Protein interaction domains used to form signalsomes
   ↳ Messages include covalent modifications and small molecule second messengers
   ↳ Signaling pathways of proteins
❖ Transmembrane signal processing hormone receptors: Three families
   ↳ G-protein-coupled receptors
      • Extracellular hormone-binding domain
      • 7 Transmembrane segments
      • Intracellular G protein-binding site
   ↳ Single transmembrane segment
      • Extracellular ligand recognition site
      • Intracellular catalytic domain: Either of two types
         • Tyrosine kinase
         • Guanylyl cyclase
   ↳ Oligomeric ion channels: Multi subunit proteins
      • Ligand-gated channels: Ligands are neurotransmitters
❖ Receptor tyrosine kinases (RTK) and receptor guanylyl cyclase (RGC)
   ↳ Extracellular ligand-binding, transmembrane helix, intracellular catalytic domains
      • EFG receptor: RTK that dimerizes when EFG binds
      • Insulin receptor: RTK that dimerizes when insulin binds
      • Natriuretic hormone receptors: RGC: Dimers that activate with ligand binding
      • pp60^{v-src}:Nonreceptor tyrosine kinase involved in transforming cells

581

- Soluble guanylyl cyclases: Activated by NO
- ❖ Signal Transduction events
  - ⅄ G-protein coupled receptors activate G-protein
  - ⅄ Hormone binding to receptor induces GDP for GTP exchange on G protein heterotrimer
  - ⅄ G_α-GTP dissociates from trimer
  - ⅄ GTPase activity of G_α hydrolyzes GTP: Adenylyl cyclase stimulation stops
  - ⅄ G_α-GDP reassociates with $G_\beta G_\gamma$ to reform trimer
  - ⅄ Two kinds of G_α
    - Stimulatory
    - Inhibitory
- ❖ Cyclic AMP: A second messenger
  - ⅄ Adenylyl cyclase produces cAMP from ATP
  - ⅄ Phosphodiesterase hydrolyzes cAMP to AMP
  - ⅄ cAMP activates protein kinase A
- ❖ Phospholipase activation produces second messengers
  - ⅄ Hydrolysis of phosphatidylinositol produces two products
    - DAG (diacylglycerol)
      - Remains in membrane: Activates protein kinase C
    - Inositol phosphate (IP) or IP_2 (which is converted to IP_3)
      - IP_3: Water soluble stimulates intracellular calcium release
        - ○ Catabolized by phosphatases into inositol-1,4-bisphosphate, I-4-P, and myo-inositol
        - ○ Metabolized by kinases into inositol-1,2,4,5-tetraphosphate
  - ⅄ Activation of phospholipase C
    - Phospholipase C-β, -γ, -δ calcium stimulated
    - Phospholipase C-β also stimulated by G protein
    - Phospholipase C-γ also stimulated by receptor tyrosine kinase
  - ⅄ Hydrolysis of phosphatidylcholine produce diacylglycerol or phosphatidic acid
  - ⅄ Hydrolysis of sphingomyelin produces ceramide: Stimulates ceramide-activated protein kinase
- ❖ Calcium as second messenger
  - ⅄ Two sources
    - Extracellular: Entry via plasma membrane channels in response to cAMP
    - Intracellular: Endoplasmic reticulum and calciosomes release calcium in response to IP_3
      - Calcium released from ER opens plasma membrane channels and together with DAG stimulates protein kinase C
      - Calcium oscillations often induced by excitation of phosphoinositide pathway
  - ⅄ Calcium-binding proteins: Three groups
    - Calcium-modulated proteins: Calmodulin, parvalbumin, troponin-C: Contain EF-hands
      - Calcium binding allows protein to bind to target
      - Target is basic amphiphilic alpha (Baa) helix
    - Annexin proteins: Peripheral membrane proteins
    - Protein kinase C
      - Soluble protein inactive
      - Target-bound protein active
        - ○ DAG binding to soluble protein opens substrate binding domain
          - ▪ Phorbol esters mimic DAG binding
        - ○ Calcium binding activates kinase
- ❖ Nervous system cells
  - ⅄ Neurons: Transmit nerve impulses
    - Axons: Cell processes that transmit impulse away from cell body
      - Synaptic terminal: Specialized ends of axon
    - Dendrite: Receives input
  - ⅄ Glial cells: Protect and support neurons
- ❖ Electrical signals: Change in electrical potential: Potential generated by ion gradients
  - ⅄ Nernst equation: Membrane potential based on unequal distribution of single ion

582

- Goldman equation: Membrane potential based on unequal distribution of several ions
- Action potential
  - Activated by depolarization
  - Voltage-gated sodium channels open allow Na^+ in
  - Sodium causes depolarization until channels close
  - Potassium channels open to allow K^+ out
  - Potassium causes repolarization (hyperpolarization)
- Synapses: Junctions
  - Electrical synapse: Small gaps: Pass depolarization electrically
  - Chemical synapse: Axon releases neurotransmitter into synaptic cleft
    - Acetylcholine
      - Cholinergic synapse receptors
        - Nicotinic receptors: Ligand-gated cation channel
        - Muscarinic receptors: Interact with G proteins
          - 7-TMS proteins: Stimulation causes
            - Inhibition of adenylyl cyclase
            - Stimulation of phospholipase C
            - Potassium channel opening
    - Amino acid neurotransmitters
      - Glu, Asp: Excitatory
      - GABA, Gly: Inhibitory
    - Catecholamines: Epinephrine, norepinephrine, dopamine, L-dopa
    - Peptides: Endorphins, enkephalins
    - Gaseous neurotransmitters

Chapter Objectives

Hormones

There are three major classes of hormones, steroid hormones, which are derived from cholesterol, amino acid derived hormones, and peptide hormones. The steroid hormones either interact with surface receptors or bind to intracellular proteins. Amino acid-derived hormones and peptide hormones bind to surface receptors and lead to stimulation of a signal transduction pathway leading to second messenger production. There are three families of membrane receptors. The 7-transmembrane segment (7-TMS) receptors have seven helical segments that span the membrane, an intracellular recognition site for G-proteins, and an extracellular domain for hormone binding. The single-transmembrane segment catalytic receptors have a single transmembrane segment, an extracellular domain containing the hormone binding site, and an internal domain with catalytic activity, either tyrosine kinase or guanylyl cyclase. The third family is the oligomeric ion channels. These are ligand-gated ion channels that open in response to neurotransmitters.

Second Messages

cAMP is produced by adenylyl cyclase and degraded by phosphodiesterase. cGMP is produced by guanylyl cyclase. Ca^{2+} is released from internal stores or moves across the membrane via a calcium channel from the extracellular environment. Inositol trisphosphate, diacylglycerol, and phosphatidic acid are released by metabolism of membrane components. Nitric oxide is produced by action of NO synthase.

G-Proteins

G-proteins function to transduce a cell surface signal into an internal signal. Hormone binding to a hormone receptor protein on the cell surface results in release of G-protein. The G-proteins are heterotrimers with the α-subunit functioning as the messenger between hormone binding to the receptor and production of an intracellular signal. This subunit is a guanine nucleotide binding protein. GTP bound the to G_α leads to release of a G_α-GTP complex that activates adenylyl cyclases. G_α has intrinsic GTPase activity and hydrolysis of GTP leads to inactivation of G_α. There are many G-proteins exhibiting a range of activities from stimulatory to inhibitory.

Membrane-Derived Second Messages

Specific phospholipases function to produce a host of second messages including inositol trisphosphate (IP_3), diacylglycerol, free arachidonic acid, which serves as the precursor of eicosanoids, and ceramide. IP_3

stimulates release of calcium from internal stores and calcium in concert with diacylglycerol stimulates protein kinase C.

Calcium

The importance of calcium as a second message has already been covered in muscle contraction, where troponin C functions as the key calcium binding protein. Calcium can be released from internal stores by IP_3 or from the plasma membrane by other signals. Calcium binds to calcium-modulated proteins, including calmodulin, parvalbumin and troponin C or to annexin proteins. The first three proteins all have virtually identical calcium binding domains, an EF-hand, consisting of a two helices separated by a calcium-binding loop. Calmodulin can bind to a number of other proteins and modulate their activity in a calcium-dependent manner.

Tyrosine Kinases

The first tyrosine kinase to be discovered was an oncogene carried by Rous sarcoma virus. The gene product, pp60[v-src], is capable of phosphorylating a number of cellular target proteins. This activity is responsible for transformation. Epidermal growth factor, insulin, and platelet-derived growth factor all function by binding to surface receptor proteins that have tyrosine kinase activity that is dependent on hormone binding.

Peptide Hormones, NO, and Steroids

The peptide hormones, a diverse group of hormones, derive from a primary translation product by post-translational processing events including: proteolytic cleavage of a hydrophobic N-terminal signal peptide, proteolytic cleavage internally, and post-translational covalent modifications. NO is a small, gaseous hormone produced from arginine. It functions as a hormone by stimulating guanylyl cyclase. The steroid hormones bind to internal receptor proteins that function as transcriptional activators leading to synthesis of new proteins.

Neurons

Understand the basic anatomy of a neuron and the role of dendrites, axons, nodes of Ranvier, and synaptic knobs. Sensory neurons are the input gateways to the nervous system, interneurons are the connections between neurons, and motor neurons pass information to muscle cells. All cells have an electrical potential across their membrane. You should understand the Nernst equation and the Goldman equation. The sequence of events responsible for an action potential includes opening and closing of voltage-sensitive sodium channels followed by opening and closing of potassium channels. Action potentials are uniform in size and thus their number and frequency are important in carrying information. The ability to affect specific channels by various agents has led to our understanding of channel functioning. Tetrodotoxin and saxitoxin lock sodium channels in a closed conformation, veratridine and batrachotoxin lock channels in an open conformation. Potassium channels are blocked by 4-aminopyridine, tetraethylammonium ion, dendrotoxins, mast cell degranulation peptide, and charybdotoxin. The anatomy of synaptic clefts should be understood. Be familiar with these terms: presynaptic cell, postsynaptic cell, synaptic cleft, synaptic vesicle. Migration and fusion of synaptic vesicles is a calcium-mediated event. Voltage-gated calcium channels open in response to the arrival of an action potential, allowing calcium to enter the presynaptic cell followed by synaptic membrane fusion and release of neurotransmitter into the synaptic cleft. Neurotransmitter receptors on the postsynaptic cell are often voltage-gated ion channels. Acetylcholine is the neurotransmitter of cholinergic synapses. There are muscarinic receptors and nicotinic receptors for acetylcholine. Nicotinic receptors are ligand-gated ion channels. Antagonists of nicotinic receptors include tubocurarine, cobratoxin and α-bungarotoxin. Muscarinic receptors cause the opening of potassium channels in a G-protein mediated process. Agents active against muscarinic receptors include atropine and muscarine. Succinylcholine and decamethonium are acetylcholine analogs that over-stimulate receptors. DIFP, malathion, parathion, sarin, and tabun all inhibit acetylcholinesterase. Other important neurotransmitters include the amino acids glutamate, aspartate, and glycine; γ-aminobutyric acid; the catecholamines, epinephrine, norepinephrine, dopamine, and L-dopa; and various peptide neurotransmitters.

Problems and Solutions

1. Assessing the features and physiology of hormones
Compare and contrast the features and physiological advantages of each of the major classes of hormones, including the steroid hormones, polypeptide hormones, and the amino acid-derived hormones.

Answer: The steroid hormones have two modes of action. They can bind to surface receptors leading to regulation of ion channels. Alternately, steroid hormones can diffuse into the cell and interact with specific internal receptor proteins leading to stimulation of transcription and gene expression. The time course and duration of stimulation is quite different for these two modes of action. Ion channel stimulation is a rapid though brief response. Alteration in gene expression is slower but long-lasting. The polypeptide hormones function by receptor binding at the cell surface resulting in production or release of a second messenger. Production of the peptide hormones requires gene expression and several post-translational modifications including proteolytic cleavages and covalent modifications. Peptide hormones are the most diverse group of hormones and as a result can participate in a wide variety of specific hormone-receptor interactions. The amino-acid derived hormones include two groups. The catecholamine hormones, including epinephrine and norepinephrine, are derived from the amino acid tyrosine and function by surface receptor binding and second message production. The second group of amino-acid derived hormones includes the thyroid hormones thyroxine and triiodothyronine, which bind to intracellular receptors that in turn regulate gene expression.

2. Understanding the features and physiology of the second messengers
Compare and contrast the features and physiological advantages of each of the known classes of second messengers.

Answer: Second messengers include Ca^{2+}, the cyclic nucleotides cAMP and cGMP, diacylglycerol (DAG), inositol trisphosphate (IP_3), phosphatidic acid, ceramide, nitric oxide (NO), and cyclic ADP-ribose. At rest, cells maintain low concentrations of intracellular calcium but maintain calcium stores in internal compartments, such as the endoplasmic reticulum and other vesicles, and cells are typically bathed in an extracellular solution of high calcium concentration. The internal calcium concentration can be rapidly changed by opening calcium channels either in the plasma membrane or in an intracellular calcium storage compartment, or both. Calcium is a small molecule that can rapidly diffuse to intracellular targets that include calcium binding proteins. Calcium-dependent signals are fast-acting but require the action of calcium pumps and sequestering agents to be switched off.

The cyclic nucleotides derive from ATP and GTP by adenylyl and guanylyl cyclase. They bind to target proteins in a highly specific manner and the levels of cyclic nucleotides are regulated by controlling both activity of the cyclases and the phosphodiesterases (that degrade cyclic nucleotides into nucleoside monophosphates by hydrolysis). The cyclic nucleotides and inositol trisphosphate and its derivatives are strictly intracellular agents. They are impermeable to membranes and remain within the cytosol. Inositol trisphosphate can be converted to a large number of phosphorylated intermediates that can serve as second messengers. Diacylglycerol, phosphatidic acid, and ceramide are second messages that act at membrane surfaces. Nitric oxide is the smallest second message. (Calcium is smaller but its hydrate is larger). It can diffuse rapidly to target sites and requires no special uptake mechanism because it is freely permeable across the membrane. In addition it is rapidly degraded or inactivated by non-enzymatic reactions. Thus, nitric oxide gives rise to rapid but localized response.

3. Are there gaseous second messengers besides NO?
Nitric oxide may be merely the first of a new class of gaseous second messenger/neurotransmitter molecules. Based on your knowledge of the molecular action of nitric oxide, suggest another gaseous molecule that might act as a second messenger and propose a molecular function for it.

Answer: Nitric oxide is synthesized from arginine by NO synthase. This gaseous compound is capable of rapidly diffusing across cell membranes. The target of NO is the heme prosthetic group of soluble guanylyl cyclase. NO binding results in an increase in enzyme activity leading to increased concentrations of cGMP. In studying hemoglobin (Chapter 12) we discovered that carbon monoxide has a high affinity for heme groups. Carbon monoxide is produced during heme catabolism when one of the methylene carbons of heme is converted to CO while the remainder of the heme group is converted to bilirubin. The enzyme that produces CO from heme is heme oxygenase.

NO is known to activate guanylyl cyclase, which leads to production of cGMP. cGMP is involved in a number of physiological responses including gating of ion channels in phototransduction, activation of myosin phosphatase to relax smooth muscle and roles in platelet inhibition and cell growth and differentiation. CO also activates guanylyl cyclase but it is much less effective than is NO. CO, however, is a comparatively stable molecule relative to NO. There is evidence that NO synthase activity depends on normal oxygen concentrations and so in anaerobic conditions, NO production may decline. In contrast, heme oxygenase remains active at low oxygen levels. NO and CO signaling may be acting through the same pathways but at different levels of oxygen tension.

The first gaseous hormone to be discovered was ethylene used by plants to stimulate a number of processes including fruit ripening, flower senescence and others.

4. Assessing the action of an antibiotic
Herbimycin A is an antibiotic that inhibits tyrosine kinase activity by binding to SH groups of cysteine in the src gene tyrosine kinase and other similar tyrosine kinases. What effect might it have on normal rat kidney cells that have been transformed by Rous sarcoma virus? Can you think of other effects you might expect for this interesting antibiotic?

Answer: Transformed cells exhibit a number of properties distinct from nontransformed cells, including loss of contact inhibition, continuous division, increased glucose uptake and glycolysis, loss of cytoskeletal elements leading to cell shape changes, alteration in surface antigens, loss of differentiation, and growth in liquid culture (anchorage-independent growth). Rous sarcoma virus carries the src oncogene, which codes for a tyrosine kinase that is responsible for transformation by the sarcoma virus. By inhibiting the tyrosine kinase with herbimycin A, the characteristics of transformed cells are all reversed.

5. Proposing uses for monoclonal antibodies against phosphotyrosine
Monoclonal antibodies that recognize phosphotyrosine are commercially available. How could such a monoclonal antibody be used in studies of cell signaling pathways and mechanisms?

Answer: The ability to measure changes of phosphotyrosine levels in response to a hormonal signal is necessary to establish a role for tyrosine kinases in a particular signaling pathway. It is an easy task to use the radioactive isotope of phosphorous, ^{32}P, to establish that a protein is phosphorylated but it is difficult to distinguish between phosphotyrosine, phosphoserine and phosphothreonine. The protein would have to be converted to amino acids but this presents a problem because phosphotyrosine is not particularly stable. Further, phosphoproteins are not abundant and therefore difficult to isolate. Monoclonal antibodies against protein phosphotyrosine would allow rapid quantitation of phosphotyrosine levels, and using Western blots, identification of phosphotyrosine-containing proteins.

6. Are there functions of hormone receptors other than signal amplification?
Explain and comment upon this statement: The main function of hormone receptors is that of signal amplification.

Answer: In the case of a hormone receptor that functions enzymatically, the statement is true and the receptor tyrosine kinases are good examples. Hormone binding leads to stimulation of protein kinase activity, which leads to phosphorylation of multiple protein targets. In G-protein-mediated pathways, hormone receptors serve the purpose of transmitting information from the surface to the cell interior. Amplification occurs at two stages. A hormone receptor with bound hormone can stimulate several G-proteins. The G-proteins stimulate adenylyl cyclase, leading to an increase in cAMP, stimulation of cAMP-dependent protein kinase, and phosphorylation of multiple targets. When hormone receptors function to stimulate transcription, the hormone signal is converted to a hormone-hormone receptor complex with a fixed stoichiometry. The complex then stimulates transcription and only at this stage is the signal amplified.

7. Determining the concentration of a neurotransmitter in a synaptic vesicle
Synaptic vesicles are approximately 40 nm in outside diameter, and each vesicle contains about 10,000 acetylcholine molecules. Calculate the concentration of acetylcholine in a synaptic vesicle.

Answer: A synaptic vesicle with a 40 nm outside diameter has an inside diameter of approximately 36 nm. The volume occupied by one synaptic vesicle is given by:

$$V=\frac{4}{3}\pi r^3 = \frac{4}{3}\times 3.1416\times(\frac{36\times10^{-9}m}{2})^3 = 2.44\times10^{-23}m^3$$

$$V=2.44\times10^{-23}m^3\times(\frac{100\text{ cm}}{1\text{ m}})^3\times(\frac{1\text{ L}}{1000\text{ cm}^3})$$

$$V=2.44\times10^{-20}L$$

The concentration of acetylcholine in a vesicle is given by:

$$[Acetylcholine]=\frac{\frac{10{,}000\text{ molecules}}{6.02\times10^{23}\text{ molecules/mol}}}{2.44\times10^{-20}L}=0.68\text{ M}$$

8. Imagining a model for neurotransmitter release

GTPγS is a nonhydrolyzable analog of GTP. Experiments with squid giant synapses reveal that injection of GTPγS into the presynaptic end (terminal) of the neuron inhibits neurotransmitter release (slowly and irreversibly). The calcium signals produced by presynaptic action potentials and the number of synaptic vesicles docking on the presynaptic membrane are unchanged by GTPγS. Propose a model for neurotransmitter release that accounts for all of these observations.

Answer: The fact that the nonhydrolyzable analog of GTP leads to inhibition of neurotransmitter release implicates a G-protein. Since calcium release occurs normally and the number of synaptic vesicles that dock on the presynaptic membrane is unchanged by GTPγS, the defect may be in fusion of the synaptic vesicle to the membrane. A cAMP-dependent vesicle fusion step that is blocked by an inhibitory G-protein would account for the observations.

9. Assessing the binding of a hormone to its receptor

A typical hormone binds to its receptor with an affinity (K_D) of approximately 1×10^{-9} M. Consider an in vitro (test-tube) system in which the total hormone concentration is approximately 1 nM and the total concentration of receptor sites is 0.1 nM. What fraction of the receptor sites is bound with hormone? If the concentration of receptors is decreased to 0.033 nM, what fraction of receptor is bound with the hormone?

Answer: For simple equilibrium binding of hormone to receptor the following equilibrium applies:

$$[HR] \leftrightharpoons [H] + [R]$$

Where [HR], [H] and [R] refer to the concentrations of hormone receptor complex, free hormone (i.e., unbound hormone), and uncomplexed receptor. The equilibrium dissociation constant for this reaction is:

$$K_D = \frac{[H][R]}{[HR]}$$

We are given the total concentration of receptor sites. In the presence of hormone, total receptor partitions into free receptor, R, and HR. Thus, $[R_{total}] = [R] + [HR]$. The total hormone concentration is $[H_{total}] = [H] + [HR]$. We are asked to calculate the fraction of receptor sites bound so we looking for $f_{bound} = [HR]/[R_{total}]$ using these three equations. Using K_D, we can write [HR] as a function of $[R_{total}]$ and $[H_{total}]$ as follows:

$$K_D[HR] = [H][R]$$
$$[H] = [H_{total}] - [HR] \text{ and } [R] = [R_{total}] - [HR]$$
$$K_D[HR] = \{[H_{total}] - [HR]\}\{[R_{total}] - [HR]\}$$
$$[HR]^2 - [HR]\{[H_{total}] + [R_{total}] + K_D\}\} + [H_{total}][R_{total}] = 0$$

The solution to this quadratic equation is:

$$[HR] = \frac{[H_{Total}]+[R_{Total}]+K_D \pm \sqrt{([H_{Total}]+[R_{Total}]+K_D)^2 - 4[H_{Total}][R_{Total}]}}{2}$$

The correct solution is the minus root. (When there are no hormones present, [HR] must be zero. For the above equation to equal to zero in this situation, the minus root must apply.)

The answers are $f_{bound} = 0.488$ at 0.1 nM receptor and $f_{bound} = 0.496$ at 0.033 nM receptor. This seems reasonable. The K_D is the concentration of free hormone that would half-saturate receptors. The free hormone concentration in both cases will be very close to 1 nM because the receptor concentrations are smaller than this number. If receptors were completely saturated, the free hormone concentration would be 1 nM – 0.1 nM or 1 nM – 0.033 nM, which are both close to 1 nM.

10. Assessing the effects of statins on steroid hormones
(Integrates with Chapter 24.) All steroid hormones are synthesized in the human body from cholesterol. What is the consequence for steroid hormones and their action from taking a "statin" drug, such as Zocor, which blocks the synthesis of cholesterol in the body? Are steroid hormone functions compromised by statin action?

Answer: There are a couple of considerations that suggest that steroid hormone functions will not be drastically altered in individuals taking statin drugs. First, it is unlikely that the cells most heavily involved in steroid hormone synthesis make cholesterol themselves de novo. Instead, the required cholesterol is taken up from the store circulating in the blood. Second, a considerable fraction of the body's total cholesterol is supplied from consumed foods and is not derived from de novo cholesterol biosynthesis. Thus, under normal circumstances shutting down the cholesterol biosynthetic pathway with a statin drug would not eliminate cholesterol from the bloodstream, as those trying to lower their cholesterol with Zocor will attest. One could imagine this situation might be more acute in strict vegetarians since cholesterol is not synthesized in plants. However, it would be unlikely that those individuals would be taking such drugs to begin with. Finally, the levels of most steroid hormones decline with age whereas concerns about high blood cholesterol levels increase with age. Accordingly, those most likely to be receiving statin drugs for medical problems are already experiencing declining levels of steroid hormones.

11. Reconciling the apparent preference for helices in transmembrane proteins
Given that β-strands provide a more genetically economical way for a polypeptide to cross a membrane, why has nature chosen α-helices as the membrane-spanning motif for G-protein–coupled receptors? That is, what other reasons can you imagine to justify the use of α-helices?

Answer: The statement that β-strands are a more genetically economical way for a polypeptide to cross a membrane refers to the fact that a β-strand is in a more extended conformation than an α-helix. Thus, fewer amino acids would be required to span the distance across the membrane. However, recall that the peptide bond is polar, and that secondary structural elements are stabilized by hydrogen bonds to the backbone atoms. In an α-helix, the hydrogen bonds are between groups on the same strand of polypeptide chain, whereas β-structures require stabilizing hydrogen bonds to be formed between adjacent runs of β-conformation. Thus, for single-spanning proteins there would be no possibility for shielding the polarity of the polypeptide backbone from the hydrophobic core of the membrane if a β-conformation were adopted. This is not to say that β-conformations do not exist in integral membrane proteins. Indeed, there is a class of pores known as β-barrel proteins in which β-strands are perpendicular with respect to the membrane and hydrogen bonded to adjacent β-strands. However, the majority of membrane-spanning regions of integral membrane proteins are α-helical, thereby satisfying hydrogen bonding requirements of the polar backbone.

12. Understanding reaction mechanisms for adenylyl cyclase and phosphodiesterase
Write simple reaction mechanisms for the formation of cAMP from ATP by adenylyl cyclase and for the breakdown of cAMP to 5'–AMP by phosphodiesterases.

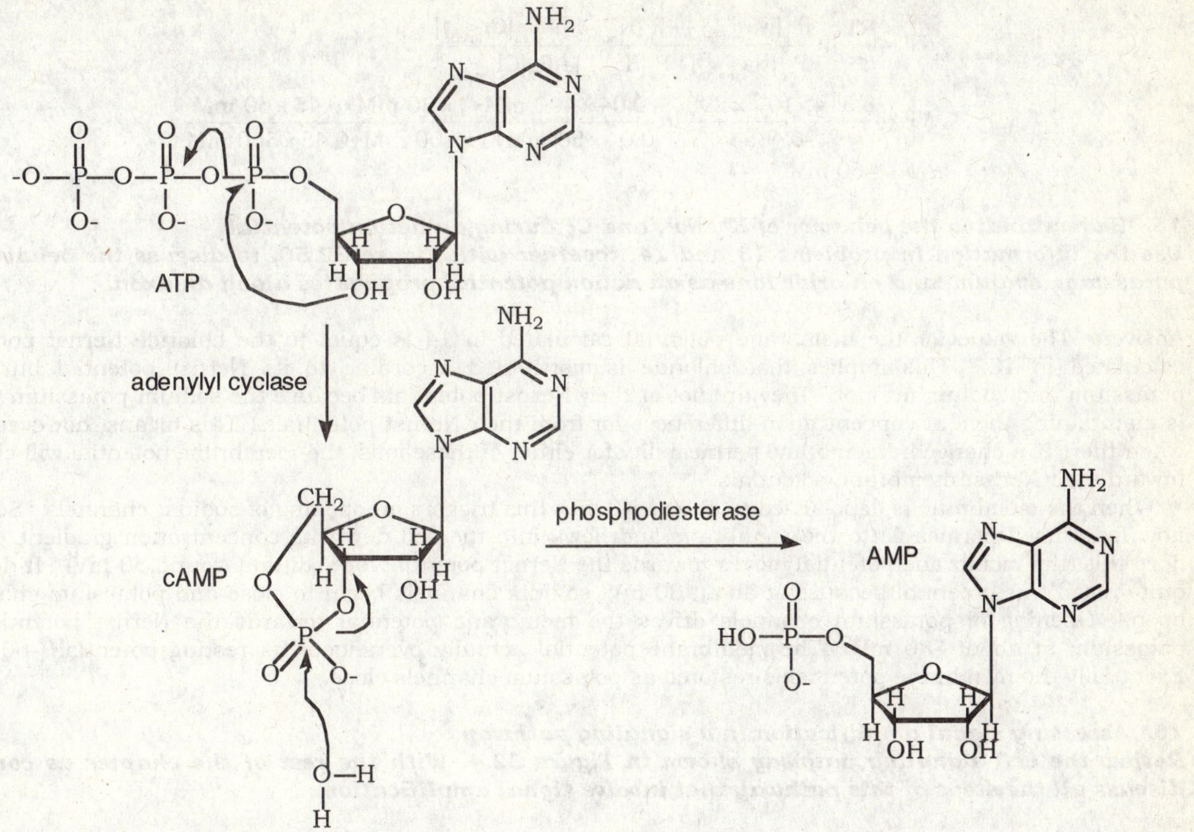

13. Determining the equilibrium transmembrane potential for K⁺ and Na⁺
(Integrates with Chapter 9.) Consider the data in Figure 32.49a. Recast Equation 9.2 to derive a form from which you could calculate the equilibrium electrochemical potential at which no net flow of potassium would occur. This is the Nernst equation. Calculate the equilibrium potential for K⁺ and also for Na⁺, assuming T=37°C.

Answer: Equation 9.2 is $\Delta G = RT\ln([C_2]/[C_1]) + Z\mathcal{F}\Delta\psi$. For no net flow of potassium to occur the system must be at equilibrium in which case $\Delta G = 0$. Making this substitution and solving for $\Delta\Psi$ we find that

$$\Delta\psi = -(RT/Z\mathcal{F}) \times \ln([C_2]/[C_1])$$

Given R = 8.3145 J/mol, $\mathcal{F}$ = 96485 J/V·mol and T = 37°C = (273 + 37 =) 310 K and from Figure 32.49a K⁺ inside and outside is 400 mM and 20 mM, respectively. Substituting these numbers into the equation above gives $\Delta\psi$ = -80 mV. Since there is more potassium inside, the inside is at a potential –80 mV relative to the outside. For sodium (50 mM inside and 400 mM outside) $\Delta\psi$ = +55.6 mV (with the inside more positive than the outside). For chloride, using 60 mM inside and 560 mM inside the potential is approximately -60 mV. (Note: The values at 25°C are –77 mV and +53.4 mV for potassium and sodium respectively.)

14. Calculating transmembrane potential using the Goldman equation
The calculation of the actual transmembrane potential difference for a neuron is accomplished with the Goldman equation:

$$\Delta\Psi = \frac{RT}{\mathcal{F}}\ln\frac{\sum P_C[C_{outside}] + \sum P_A[A_{inside}]}{\sum P_C[C_{inside}] + \sum P_A[A_{outside}]}$$

where [C] and [A] are the cation and anion concentrations, respectively, and P_C and P_A are the respective permeability coefficients of cations and anions. Assume relative permeabilities for K⁺, Na⁺, and Cl⁻ of 1, 0.04, and 0.45, respectively, and use this equation to calculate the actual transmembrane potential difference for the neuron whose ionic concentrations are those given in Figure 32.49a.

Answer: The concentrations of sodium, potassium and chloride on the inside and outside are 50 mM, 400 mM, 60 mM, 400mM, 20 mM, 560 mM. Substitution into the Goldman equation is shown below.

589

$$\Delta\Psi = \frac{RT}{\mathscr{F}} \ln \frac{P_{Na}[Na_{outside}] + P_K[K_{outside}] + P_{Cl}[Cl_{inside}]}{P_{Na}[Na_{inside}] + P_K[K_{inside}] + P_{Cl}[Cl_{outside}]}$$

$$\Delta\Psi = \frac{8.314 \times 10^{-3} \times 298}{96.485} \ln \frac{0.04 \times 400 \text{ mM} + 1 \times 20 \text{ mM} + 0.45 \times 60 \text{ mM}}{0.04 \times 50 \text{ mM} + 1 \times 400 \text{ mM} + 0.45 \times 560 \text{ mM}}$$

$$\Delta\Psi = -60 \text{ mV}$$

15. Understanding the behavior of K⁺, Na⁺, and Cl⁻ during an action potential
Use the information in problems 13 and 14, together with Figure 32.50, to discuss the behavior of potassium, sodium, and chloride ions as an action potential propagates along an axon.

Answer: The value for the membrane potential calculated in 14 is equal to the chloride Nernst potential calculated in 13.   This implies that chloride is distributed according to its Nernst potential but both potassium and sodium are not. They are not at their Nernst potentials because the sodium-potassium pump is maintaining them at concentration differences far from their Nernst potentials. This means, however, that when there is a change in membrane permeability for either of these ions, the membrane potential will change toward their Nernst membrane potentials.

When the membrane is depolarized to about –40 mV, this triggers an opening of sodium channels. Sodium now finds itself permeable to the membrane and flows into the cell down its concentration gradient. As it does this, the membrane potential moves towards the Nernst potential for sodium (around 50 mV). It doesn't quite reach this potential because at about 30 mV, sodium channels begin to close and potassium channels open.   Opening of potassium channels drives the membrane potential towards the Nernst potential for potassium at about –80 mV. The membrane potential actually overshoots its resting potential (-60 mV). Eventually the membrane potential is restored as potassium channels close.

16. Assessing signal amplification in a signaling pathway
Review the cell signaling pathway shown in Figure 32.4. With the rest of the chapter as context, discuss all the steps of this pathway that involve signal amplification.

Answer: The obvious steps in signal transduction that could result in amplification involve steps catalyzed by enzymes.   In the pathway shown in Figure 32.4 hormone binding to a tyrosine kinase receptor leads to autophosphorylation of the receptor. This step functions more like an on/off switch conveying the hormone-binding event into the cell. Phosphorylation leads to activation of Ras by GTP exchanging for GDP. Activated Ras then activates Raf, a protein kinase, which phosphorylates MAP kinase kinase, which phosphorylates MAP kinase, MAP kinase phosphorylates various target proteins that function in regulation of gene expression, which is the cell's ultimate response to the initial hormone binding signal.

The steps involving activated Raf, activated MAP kinase kinase and MAP kinase are all steps that involve signal amplification. Phosphorylation of tyrosine kinase might be considered to be an activation step rather than amplification. The phosphorylated receptor then recruits SOS, which functions as a guanylyl exchange factor, which catalyzes GTP/GDP exchange on Ras. This step could provide a modest amplification as could Ras activation of Raf.   But, these two steps are confined to the plasma membrane and involve a two-dimensional diffusion of target protein in the plane of the membrane. Their purpose may be for rapid transmission of information about hormone binding rather than to amplify the signal. Starting with Raf, amplification should be significant for the steps described in the first paragraph.

17. Understanding the consequences of a Ras mutation on a signaling pathway
In the cell signaling pathway shown in Figure 32.4, what would be the effect if Ras were mutated so that it had no GTPase activity?

Answer: Signal transduction pathways include both on and off switches to ensure that signals last for appropriate lengths of time. The consequence of activation of a signal transduction pathway is to elicit a physiological response to some environmental or developmental condition.   The response will somehow address the condition and examples of this might be responses to blood glucose or serum calcium levels. Once the response to the signal is made the condition setting off the signal is reassessed. If the condition is still uncorrected an additional signal is required.

On and off switches in signal transduction pathways often include kinases and phosphatases that phosphorylate and dephosphorylate a target protein. In the case of G proteins, they are activated by GTP binding but they inactivate themselves by intrinsic GTPase activity. If Ras lacked GTPase activity but could still bind GTP and become active (as an inhibitor or activator of some downstream protein) then the signal would be stuck in the on position. In this case Ras would be capable of signaling in the absence of hormone stimulation.   As it turns out mutant forms of Ras are commonly associated with human cancers. The mutants lead to constitutively active Ras as a consequence of inactivation of GTPase activity.

18. Understanding the action of GPCRs in G-protein-independent pathways
One of the topics discussed in this chapter is the ability of GPCRs to exert signaling effects without the involvement of G proteins. Using the pathway shown in Figure 32.40, and considering everything you have learned in this chapter, suggest some reasons that would explain why this G-protein–independent signaling was difficult to verify experimentally.

Answer: G-protein coupled pathways are complex and involve multiple components and teasing out the components from a complex mixture is a difficult task experimentally. G-protein-dependent and G-protein-independent pathways converge and many components are shared. Additionally, there is evidence that dependence and independence may be set off by different concentrations of the ligand. A given concentration of ligand (hormone) may result in stimulation of both pathways. GPCRs can interact with downstream effectors either directly or through adapter proteins. Identifying these interactions would require receptor mutations that block particular interactions or mutations in adapter proteins like arrestin.

Preparing for the MCAT® Exam

19. *Assessing the actin of a popular pesticide*
Malathion (Figure 32.58) is one of the secrets behind the near complete eradication of the boll weevil from cotton fields in the United States. For most of the 20th century, boll weevils wreaked havoc on the economy of states from Texas to the Carolinas. When boll weevils attacked cotton fields in a farming community, the destruction of cotton plants meant loss of jobs for farm workers, bankruptcies for farm owners, and resulting hardship for the entire community. Relentless application of malathion to cotton crops and fields has turned the tide, however, and agriculture experts expect that boll weevils will be completely gone from cotton fields within a few years. Remarkably, malathion-resistant boll weevils have not emerged despite years of this pesticide's use. Consider the structure and chemistry of malathion and suggest what you would expect to be the ecological consequences of chronic malathion application to cotton fields.

Answer: The structure of malathion is shown below along with acetylcholine. Malathion is an acetylcholinesterase inhibitor. It is very difficult to access ecological consequences to chronic malathion application just from the structure. One important property of malathion that has contributed to it being widely used is that it is labile and breaks down under a number of ecologically-relevant conditions. While it may be very good against boll weevils it is unlikely that it is specific to these insects. So, any non-target insects may be affected by malathion.

20. *Assessing the structural requirements for a regulatory protein in a signaling network*
Consult the excellent review article "Assembly of Cell Regulatory Systems Through Protein Interaction Domains" (Science 300:445–452, 2003, by Pawson and Nash) and discuss the structural requirements for a regulatory protein operating in a signaling network.

Answer: Proteins operating in regulatory systems often possess two independent domains, one containing a protein-protein interaction domain through which target pathways are selected, and the second imparting catalytic activity required for signal transduction itself. These domains can be joined in modular fashion, allowing relatively few catalytic activities to be used repeatedly in different signaling pathways. A well-characterized example of this is the use of kinases in signal cascades, wherein the activity of signaling proteins is modulated by phosphorylation. Often, the phosphorylated proteins are themselves kinases, which when activated go on to phosphorylate other downstream targets, greatly amplifying the original signal. The specificity of these kinases for their immediate targets is enhanced by protein interaction domains located in separate regions of the enzyme. Because catalytic and interaction domains can often fold independently, the activity of one is often independent of the activity of the other. The modular nature of these signaling proteins thus allows for use of common and recognizable protein interaction and catalytic domains to be expressed in different contexts and provide for the high specificity required of signal transduction pathways.

Questions for Self Study

1. Name three classes of chemical species that act as hormones.

2. List the three receptor superfamilies that mediate transmembrane signal processing and give a brief description how signal processing occurs.

3. What two enzymatic activities are responsible for regulating cAMP levels?

4. cAMP is an example of a second messenger. Name five other second messengers.

5. A large family of second messengers can be derived from phospholipase C activity on phosphatidylinositol and its derivatives. Explain.

6. Protein kinase C is sensitive to what two intercellular signals?

7. Membrane-bound guanylyl cyclases and soluble guanylyl cyclases are stimulated by very different signals. What are they?

8. Match a term with its definition.
   a. Node of Ranvier   1. Connects to sensory receptor.
   b. Synapse   2. Carries nerve impulses away from cell body.
   c. Schwann cell   3. Moves nerve impulses to the cell body.
   d. Interneuron   4. Insulating layer around axons.
   e. Sensory neuron   5. Gap between an axon and a dentrite.
   f. Dendrite   6. Moves signals from one neuron to another.
   g. Axon   7. Gap between Schwann cells along an axon's length.

9. Answer True of False.
a. Typically, action potentials are induced by a hyperpolerization of the membrane voltage.___
b. Resting potential of an axon is determined by the concentration gradients of all impermeable ions.___
c. During an action potential, the membrane voltage changes from approximately -60 mV to +40 mV because of changes in sodium permeability. ___
d. Potassium permeability changes alone are responsible for returning the action potential back to resting values. ___
e. Action potentials attenuate with distance. ___

10. Fill in the blanks. In cholinergic synpases, small vesicles termed ____ are localized on the inside of the synaptic knob and contain large amounts of the neurotransmitter ____. When an action potential arrives at the synaptic knob, voltage-gated ____ channels open. This is followed by fusion of vesicles with the plasma membrane and release of neurotransmitter into the synaptic cleft. Neurotransmitter induces action potentials in postsynaptic cells by binding to receptors. There are two kinds of receptors ____ and ____ distinguished by there responses to a toxic alkaloid in toadstools or to nicotine. Neurotransmitter action is usually short lived because it is rapidly hydrolyzed by the enzyme ___.

11. Match
   a. Endorphin   1. Binds to nicotinic receptors and blocks their opening.
   b. Catecholamine   2. Insecticide that blocks muscarinic receptor.
   c. Glycine   3. Excitatory amino acid transmitter.
   d. Glutamate   4. Epinephrine.
   e. Malathion   5. Peptide neurotransmitter.
   f. d-tubocurarine   6. Inhibitory amino acid transmitter.

Answers

1. Steroids, amino acid derivatives, and peptides.

2. 7-Transmembrane segmented receptors are integral membrane proteins composed of seven helical segments, an extracellular hormone-binding domain, and an intracellular G-protein binding domain. GTP-binding proteins or G-proteins bind GTP and are released from the transmembrane receptor upon hormone binding. Depending on the type of G-protein, the G protein-GTP complex activates or inhibits adenylyl cyclase leading to changes in cyclic AMP levels. Single-transmembrane segment catalytic receptors are integral membrane proteins with an extracellular hormone receptor domain, a transmembrane segment, and an intracellular domain with either tyrosine kinase or guanylyl cyclase activity. Ligand gated oligomeric ion channels open or close in response to ligand binding. Many such ion channels are regulated by neurotransmitters.

3. Adenylyl cyclase and phosphodiesterase.

4. cGMP, inositol phosphates, diacylglycerol (DAG), calcium, phosphatidic acid, ceramide, nitric oxide, cyclic ADP-ribose.

5. Phospholipase C activity produces diacylglycerol (DAG) and phosphoinositol. Depending on the phosphorylation state of phosphatidylinositol, inositol-1-P, inositol-1,4-P, or inositol-1,4,5-P may be released. Inositol-1,4,5-P leads to inositol-1,3,4,5-P, inositol-1,3,4-P, inositol-pentaphosphate, and inositol hexaphosphate. DAG may also be converted to phosphatidic acid.

6. Calcium and diacylglycerol.

7. Membrane-bound guanylyl cyclases are stimulated by a variety of peptide hormones. Soluble guanylyl cyclases are stimulated by nitric oxide.

8. a. 7; b. 5; c. 4; d. 6; e. 1; f. 3; g. 2.

9. a. F; b. F; c. T; d. F; e. F.

10. Synaptic vesicles; acetylcholine; calcium; muscarinic; nicotinic; acetylcholinesterase.

11. a. 5; b. 4; c. 6; d. 3; e. 2; f. 1.

Additional Problems

1. Caffeine (1,3,7-trimethylxanthine), theophylline (1,3-dimethylxanthine found in teas), and theobromine (3,7-dimethylxanthine found in chocolate) mimic the action of hormones that cause an increase in cAMP. Why?

2. Anabolic steroids have been used to increase athletic performance because these hormones lead to increase in mass and strength of muscle. However, their use for this purpose is banned in many competitions and random tests are often conducted to determine if an athlete has taken anabolic steroids. (Notable examples of athletes who have tested positive for drug use include Ben Johnson, Diego Maradona, and Hulk Hogan.) The tests are conducted on urine samples and steroid use can be detected even if active steroid use had been discontinued at some time prior to testing. From the properties of steroid hormones, can you suggest why synthetic steroids might have long biological half-lives?

3. Explain why the GTP analogs GMP-PNP and GMP-PCP are activators of adenylyl cyclase.

4. Draw the structure of inositol and explain why it can be metabolized into a large number of active compounds.

5. Based on the description of calmodulin target proteins and the Baa helix and referring to a wheel plot of melittin, suggest a possible function for melittin.

6. The cellular membrane potential is established by ion pumping mechanisms that are electrogenic. How can an electrical potential across a synthetic membrane be established in the absence of electrogenic pumps?

7. How does hyperpolarization lead to inhibition of action potentials?

8. Electrical recordings from an unstimulated neuromuscular junction reveal small, transient fluctuations in the membrane voltage termed miniature end-plate potentials. The magnitude of these end-plate potentials is reduced in size with curare. Further, their frequency is altered with changes in the membrane potential with hyperpolarization of the presynaptic membrane decreasing frequency and depolarization increasing frequency. Based on these observations, explain how miniature end-plate potentials arise.

9. In myelinated axons, sodium channels are clustered in the nodes of Ranvier where gaps in the myelin sheath leave the axon exposed. How does a nerve impulse propagate along a myelinated axon and is the spacing of nodes of Ranvier at all important for this process?

10. What mechanism insures that nerve stimuli move in one direction along a neuron?

Abbreviated Answers

1. These agents inhibit phosphodiesterase, the enzyme responsible for cAMP metabolism to AMP. With phosphodiesterase inhibited, levels of cAMP increase and remain high for prolonged periods of time. (They also interfere with adenosine receptors.)

2. Steroids are fat-soluble substances that can partition into hydrophobic substances like fats and be retained there for long periods of time. Further, the steroid ring is very stable and only slowly metabolized by reduction to more water-soluble forms that are excreted by the kidney.

3. These GTP analogs bind to the GTP binding site on G-proteins; however, because they lack the normal phosphate ester linkage they are not hydrolyzed, resulting in prolonged stimulation of adenylyl cyclase.

4. Inositol can be converted to a large number of phosphorylated derivatives that have different biological functions. The structure of inositol is shown below.

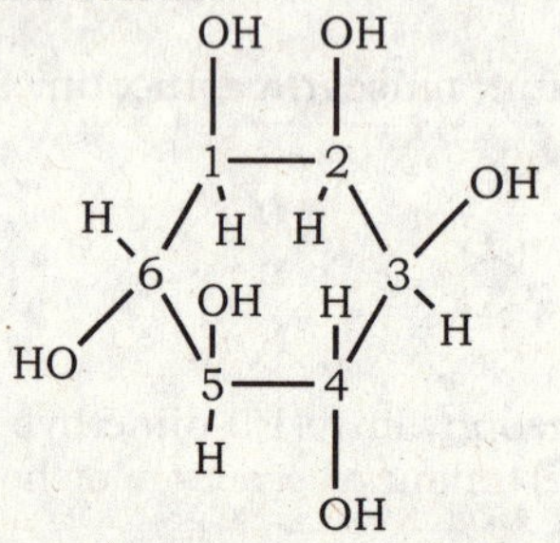

5. The α-helical wheel plot of melittin shows a helix with hydrophobic residues on one face and a collection of hydrophilic residues on another face with several basic amino acids at one end of the peptide. Melittin is a component of bee venom and shows hemolytic activity. In addition, it binds to calmodulin and can block calmodulin stimulation of calmodulin-binding proteins.

6. An electrical potential across a semipermeable membrane can be formed by placing a salt, composed of a permeable cation or anion and an impermeable counterion, on one side of the membrane. For example, consider a salt composed of a negatively-charged, large, impermeable protein and a small cation like sodium. Upon dissolving the salt in solution on one side of the membrane, the salt dissociates into the large protein anion and Na^+. Sodium freely diffuses across the membrane because it moves down its concentration gradient. But because the impermeable anion cannot follow, sodium exiting from the membrane produces a charge imbalance across the membrane creating a voltage gradient.

7. An action potential is initiated above a certain threshold voltage for sodium channels. This threshold is slightly less negative than the normal resting potential of the cell. Hyperpolarization drives the membrane to a more negative value and a greater voltage change is required to reach the threshold voltage.

8. Miniature end-plate potentials arise from random release of packets of acetylcholine as synaptic vesicles fuse with the presynaptic membrane and release their contents into the synaptic cleft. Curare blocks end-plate potentials by blocking acetylcholine receptors. Hyperpolarization lowers the frequency of their occurrence whereas depolarization increases it by altering the presynaptic calcium flux and thus reducing or increasing spontaneous synaptic vesicle release.

9. For an action potential to be propagated along an axon, a change in the membrane potential must be large enough to exceed the threshold voltage for sodium-channel opening. Once a voltage change due to an action potential occurs in a local region of the axon, it must be conducted to nearby regions of the axon to cause a depolarization of the membrane to a value above the threshold value. The myelin sheath greatly increases the transverse resistance, causing current, induced by an action potential, to be carried down the axon. As the current moves down the axon it is attenuated; however, the myelin sheath reduces attenuation and the nodes of Ranvier are spaced such that the attenuated current is still large enough to cause a local depolarization of the membrane exceeding the threshold voltage. In effect, action potentials jump from node to node. Between nodes the stimulus is carried as a passive electrical current.

10. If an axon is stimulated electrically to initiate an action potential, the action potential will be conducted in both directions along an axon. However, the transmission of information across a synaptic cleft occurs only in one direction.

Summary

Hormones are chemical signal molecules that control and coordinate the many and diverse processes that occur in different parts of an organism. Steroid hormones regulate metabolism, salt and water balances, inflammatory processes and sexual function. Epinephrine and norepinephrine regulate smooth muscle contraction and relaxation, blood pressure, cardiac rate and other processes. Peptide hormones likewise regulate processes in all body tissues, as well as the release of other hormones. Hormones bind with very high affinity to their receptors.

Non-steroid hormones bind exclusively to outward-facing membrane receptors and activate signal transduction pathways which mobilize second messengers, including cyclic AMP, cyclic GMP, Ca^{2+}, inositol-1,4,5-trisphosphate, diacylglycerol and nitric oxide. The receptors that mediate transmembrane signaling include three receptor superfamilies, including 1) the 7-transmembrane segment receptors, which transmit their signals via GTP-binding proteins, 2) the single transmembrane segment catalytic receptors, which possess tyrosine kinase or guanylyl cyclase activity, and 3) oligomeric ion channels.

The first second messengers to be discovered were calcium and cAMP. cAMP is produced by adenylyl cyclase and hydrolyzed by phosphodiesterase. The activity of cAMP is mediated by G proteins, which exchange GDP for GTP upon binding of hormone to the associated receptor protein. G_{α}-GTP dissociates from $G_{\beta\gamma}$ and binds to adenylyl cyclase to activate the synthesis of cAMP. Inhibitory G proteins act in a similar manner to inhibit adenylyl cyclase. The toxic effects of cholera toxin are due to ADP-ribosylation and consequently activation of G_s, whereas pertussis toxin ADP-ribosylates and inactivates G_i. The small G proteins, typified by the protein product of the *ras* gene, function in a manner similar to the heterotrimeric G proteins. Mutant forms of the *ras* protein have been implicated in the tumorigenic activity of certain tumor viruses.

A diverse array of second messengers is generated by breakdown of membrane phospholipids. Phospholipase C hydrolyzes phosphatidylinositol-4,5-bisphosphate to produce inositol-1,4,5-trisphosphate (IP_3) and diacylglycerol (DAG). IP_3 stimulates an increase in intracellular $[Ca^{2+}]$, whereas DAG activates protein kinase C. Activation of phospholipase C enzymes is mediated by G proteins or by receptor tyrosine kinases. IP_3 has a half-life of only a few seconds. It is catabolized by phosphatases and/or kinases to produce a series of other inositol-phosphate metabolites which may themselves be second messengers in certain cellular processes. Second messengers may also be produced by breakdown of phosphatidylcholine. Action of phospholipase A_2 on PC produces arachidonic acid, which can be metabolized to eicosanoid compounds, and phospholipases C and D can produce DAG and phosphatidic acid, respectively. Sphingomyelinase action on sphingomyelin produces ceramide, which stimulates a protein kinase, and gangliosides and their breakdown products modulate the activity of protein kinases and G protein-coupled receptors.

Calcium ion is an important second messenger, which is released by action of IP_3 into the cytoplasm from intracellular stores including calciosomes and ER. Extracellular calcium can enter the cell via ligand-gated, voltage-gated, or second-messenger-regulated channels. Ca^{2+} binds to a series of calcium-binding proteins, such as calmodulin, which act to modulate enzyme and ion channel activity within the cell. Calmodulin target proteins possess a basic amphiphilic α-helix, to which calmodulin binds specifically and with high affinity. Ca^{2+} is also an activator of protein kinase C, which elicits a variety of cellular responses by phosphorylation of target proteins at Ser and Thr residues. Protein kinase C contains a pseudosubstrate sequence that masks the active site to inhibit PKC activity in the absence of DAG and Ca^{2+}. Many protein kinases and protein phosphatases are regulated by such intrasteric control. Protein phosphatases may either be specific for Ser- and Thr-phosphates or for Tyr-phosphates. The tumor promoting activity of okadaic acid, the major cause of diarrhetic shellfish poisoning, arises from its potent inhibition of protein phosphatases.

The single transmembrane segment receptor proteins display either a tyrosine kinase or guanylyl cyclase activity. Additionally, soluble tyrosine kinases and guanylyl cyclases also participate in cellular signaling processes. Receptor tyrosine kinases are integral transmembrane proteins, whereas non-receptor tyrosine kinases, which are related to retroviral transforming proteins, are peripheral, lipid-anchored proteins. Receptor tyrosine kinases are membrane-associated allosteric enzymes, since hormone binding induces oligomerization of receptors in the membrane, activating phosphorylation of the cytoplasmic domains and stimulating tyrosine kinase activity by the receptors.

Polypeptide hormones are synthesized as inactive precursors and undergo cellular processing which produces the active hormone. Processing events common to all polypeptide hormones include: 1) proteolytic cleavage of a hydrophobic N-terminal signal sequence, 2) proteolytic cleavage at sites defined by pairs of basic amino acids, 3) proteolytic cleavage at the site of a single Arg residue and 4) post-translational modification of individual amino acids, including C-terminal α-amidation, phosphorylation, glycosylation or acetylation of the N-terminal residue.

Membrane-bound guanylyl cyclases are single-TMS receptors, which synthesize cGMP in response to hormone binding. Peptides, which activate membrane-bound guanylyl cyclases, include atrial natriuretic peptides. Soluble guanylyl cyclases are receptors for nitric oxide, which is synthesized by NO· synthase. NO· acts as a neurotransmitter and as a second messenger, inducing relaxation of vascular smooth muscle and mediating penile erection. It also enables macrophages to kill tumor cells and bacteria. NO· is unique among second messengers, because, as a dissolved gas, it is capable of rapid diffusion across cell membranes without the assistance of carriers. NO· has a half-life of 1-5 seconds and is degraded by nonenzymatic pathways. NO· activates guanylyl cyclase by binding to the heme prosthetic group to form a nitrosoheme.

Steroid hormones exert their effects in two ways: 1) in the nucleus, steroids act as transcription regulators, modulating gene expression. Steroids also can act at the cell membrane, directly regulating ligand-gated ion channels and other processes. Specialized receptor proteins carry steroids to the nucleus. These receptor proteins, members of a gene superfamily, each contain a hydrophobic steroid binding domain near the C-terminus and a central, hydrophilic DNA binding domain. Receptor proteins for thyroid hormones are highly homologous with steroid receptors.

The nervous system is composed of neurons and neuroglia or glial cells. There are several types of glial cells that serve various purposes in the nervous system. Schwann cells envelope neurons and form an insulating layer. Phagocytic glial cells remove cellular debris from nervous tissue and other glial cells form linings in the cavities of the brain and in the spinal column. Neurons contain three distinct regions: the cell body, which contains the nucleus, mitochondria, ribosomes, endoplasmic reticulum and other organelles; the axon, a long extension of the cell body whose primary function is to carry nerve impulses from the cell body to the cell periphery; and dendrites, short, branched structures that carry impulses to the cell body from the periphery. Certain axons have myelin sheath derived from Schwann cells. The Schwann cells wrap around the axon to produce an insulating layer with periodic gaps called the nodes of Ranvier where the axon is exposed. The distal end of the axon ends in a synaptic knob or synaptic bulb.

There are three kinds of neurons: sensory neurons that transduce sensory signals into nervous signals; interneurons that pass signals from one neuron to another; and motor neurons that pass signals to muscles. The signals carried by neurons are electrical signals that arise from changes in the electrical potential across the neuron membrane as a result of changes in permeability of the membrane to various ions. The Nernst equation predicts the contribution to the membrane electrical potential for a single permeable ion unequally distributed across a membrane. The sign and magnitude of the electrical potential are given by $\Delta\psi = (RT/\mathcal{F})\times\ln[C_2]/[C_1]$ where $[C_2]$ and $[C_1]$ are the molar concentrations of the ion on both sides of the membrane, R is the gas constant, T the temperature in degrees Kelvin, and $\mathcal{F}$ Faraday's constant. For multiple ions, the Goldman equation is used where $\Delta\psi = (RT/\mathcal{F})\ln\{([\Sigma P_C[C_{outside}] + \Sigma P_A[A_{inside}])/(\Sigma P_C[C_{inside}] + \Sigma P_A[A_{outside}])\}$ and P_A and P_C are the permeabilities of anions and cations to the membrane. Nerve impulses are called action potentials and they arise when the membrane becomes permeable to specific ions. A resting cell maintains a membrane potential of about -60 mV (with the inside negative relative to the outside), due in large part to the Na^+-K^+ pump. When a cell is depolarized, the membrane potential is made more positive (the opposite, to hyperpolarize, is to change the membrane potential to a more negative value) by about 20 mV, voltage-gated ion channels specific for sodium are opened. Because sodium is in high concentration external to the cell and the inside of the cell is at a negative potential, sodium rushes into the cell causing the cell to depolarize to a membrane potential close to the value predicted for the Nernst potential for sodium. At this point, the sodium channels close and potassium specific voltage-gated channels open allowing potassium to rush out of the cell down its electrochemical gradient to hyperpolarize the cell to a value below the resting membrane potential. Eventually, potassium channels close and the voltage returns to the resting potential. Action potentials are propagated very rapidly, are not attenuated with distance, and show a fixed amplitude. Because the amplitude is fixed, the information content is the number and frequency of action potentials. The voltage-gated sodium channels are integral membrane proteins. Tetrodotoxin from the puffer fish, and saxitoxin from two species of marine dinoflagellates responsible for "red tides", bind to the sodium channels and prevent their opening. Veratridine from a plant of the lily family and batrachotoxin, a compound from the skin secretions of a Colombian frog, can block sodium channels in the open position. Voltage-gated potassium channels are also intrinsic membrane proteins that can be specifically blocked by 4-aminopyridine, tetraethylammonium ion, dendrotoxin I (a 60-residue peptide from the black mamba snake), mast cell degranulating peptide (a 22-residue peptide from bee venom), and charybdotoxin (a 37-residue peptide from scorpion venom).

Transmission from one neuron to another involves information transfer across the synaptic cleft. There are two kinds of synapses, electrical synapses with a very small gap, approximately 2 nm, separating the presynaptic cell from the postsynaptic cell, and chemical synapses, with 20 nm to 50 nm gaps. Chemical substances called neurotransmitters are released by the presynaptic cell, diffuse across the synaptic cleft, and bind to receptors in the postsynaptic cell. Neurotransmitters include acetylcholine, various amino acids, catecholamines, peptides, and gaseous compounds. The cholinergic synapse employs acetylcholine as a neurotransmitter. The presynaptic cell contains synaptic vesicles filled with acetylcholine. A voltage-gated calcium channel in the presynaptic membrane opens in response to the arrival of an action potential and allows calcium to enter the cell. Calcium, complexed to calmodulin, binds to a 75-kD protein, synapsin-I, allowing synaptic vesicle fusion to the plasma membrane and acetylcholine release into the synaptic cleft. Acetylcholine release can be specifically blocked by several toxins produced by *Clostridium botulinum*, the bacteria responsible for botulism poisoning. α-Latrotoxin, a black widow venom protein, stimulates acetylcholine release.

Acetylcholine released into the synaptic cleft binds to acetylcholine receptors in the postsynaptic membrane, leading to membrane depolarization, and initiation of an action potential in the postsynaptic cell. There are two classes of acetylcholine receptors that are distinguished by their responses to muscarine, a toxic toadstool alkaloid, and nicotine. Nicotinic receptors are K^+-Na^+ channels located in motor endplates of skeletal muscles. Nicotine binding causes the receptor to lock into the open conformation. The nicotinic acetylcholine receptors are ligand-gated ion channels that undergo a conformational change upon acetylcholine binding that results in the opening of a channel equally permeable to both K^+ and Na^+. The effects of sodium on the membrane voltage are greater than those of potassium and as a result the postsynaptic cell depolarizes and an action potential is initiated. The channel remains open for only a few milliseconds, then closes and as long as acetylcholine remains bound to the receptor it will not reopen. To resensitize the receptor, acetylcholine must be removed. This occurs when the concentration of acetylcholine in the synaptic cleft drops to below 10 nM in response to acetylcholinesterase-dependent degradation of acetylcholine. Repackaging of acetylcholine into synaptic vesicles occurs in the presynaptic cell. Empty synaptic vesicles are produced by endocytosis of the presynaptic membrane. A proton pump, a V-type ATPase, produces a proton gradient across the vesicle membrane by pumping protons into the vesicle and acetylcholine transport protein uses this gradient to concentrate acetylcholine in the vesicles. Agents that prevent nicotinic acetylcholine receptors from opening include d-tubocurarine, cobratoxin, and α-bungarotoxin.

The muscarinic receptors, located in smooth muscle and glands, are stimulated by muscarine. Binding of acetylcholine to muscarinic receptors results in 1) inhibition of adenylyl cyclase, 2) stimulation of phospholipase C, and 3) opening of potassium channels. All of these effects are mediated by G-proteins. As mentioned, muscarine stimulates these receptors. A potent antagonist is atropine from the deadly nightshade plant.

Acetylcholine receptors can be activated by the acetylcholine analogs, decamethonium and succinylcholine. These agents lead to long term stimulation of acetylcholine receptors because they are broken down only slowly. Organophosphorus compounds are potent inhibitors of acetylcholinesterase, a serine esterase with an active site serine that is a target for DIFP, malathion and parathion (commonly used insecticides), and sarin and tabun (nerve gases). Physostigmine and neostigmine are mild inhibitors of acetylcholinesterase that have been used in the treatment of myasthenia gravis.

The amino acids glutamate and aspartate are also excitatory neurotransmitters that are stored in presynaptic vesicles and released into the presynaptic cleft by calcium-dependent exocytosis. Synaptic vesicles are filled with glutamate in a manner similar to acetylcholine-filled vesicles. A proton pump creates a proton gradient that provides the driving force for glutamate accumulation. To remove glutamate from the synaptic cleft, glutamate is taken into glial cells and converted to glutamine. Glutamine is transported to the presynaptic neuron and recycled by the mitochondria back to glutamate. Glutamate receptors include N-methyl-D-aspartate (NMDA), kainate, and AMPA receptors, examples of ligand-gated ion channels, and metabotropic receptors, which are G-protein mediated receptors that are coupled to phosphotidylinositol metabolism. The NMDA receptors the best characterized glutamate receptor. It functions as a Ca^{2+}, Na^+, and K^+ channel that is closed by Mg^{2+} in a voltage-dependent manner. The hallucinogenic drug, phencyclidine (PCP, a.k.a angel dust) is a specific antagonist of the NMDA receptor.

γ-Aminobutyric acid (GABA) and glycine are inhibitory neurotransmitters that function by hyperpolarizing the postsynaptic neuron. The receptors are ligand-gated chloride channels. GABA receptors are localized in the brain and opened by ethanol. Glycine receptors are found in the spinal cord and are blocked specifically by strychnine.

The catecholamine neurotransmitters include L-dopa, dopamine, norepinephrine, and epinephrine, all derived from the amino acid tyrosine. Norepinephrine is the neurotransmitter in junctions between sympathetic nerves and smooth muscle. Excessive production of dopamine or hypersensitivity of dopamine receptors is responsible for schizophrenia while lowered production of dopamine is found in Parkinson's disease. Inhibitors of catecholamine uptake and vesicular repackaging have been used to treat clinical depression. Monoamine oxidase inhibitors block degradation of norepinephrine. The tricyclics act on norepinephrine transporters, reserpine inhibits proton-monoamine exchange protein from reloading synaptic vesicles with norepinephrine, and cocaine blocks the reuptake transporters of monoamine neurotransmitters in the presynaptic membrane.

Peptide neurotransmitters include endorphins and enkephalins, natural opioid substances that act as pain relievers, endothelins, and regulatory peptides that act on smooth muscle and connective tissue, vasoactive intestinal peptide, and many more. Vasoactive intestinal peptide causes an increase in cAMP via a G-protein-mediated pathway.

Glossary

. . . .

AAA+ ATPase family A group of proteins with ATPase activity associated with a variety of cellular activities.

ABC transporters Membrane transport proteins that move xenobiotics across biological membranes. Often problematic in cancer chemotherapy because they expel therapeutic drugs.

abzymes Catalytic antibodies.

accessory pigments Visible light-absorbing pigments in plants and photosynthetic bacteria that, along with chlorophylls, absorb sunlight.

acetylcholinesterase An enzyme in the synaptic cleft that hydrolyzes acetylcholine, a neurotransmitter.

acetylcholine transport protein A membrane protein used to restock synaptic vesicles with acetylcholine. Uses a proton gradient formed by ATP-driven proton pump to concentrate acetylcholine.

active site The region of the enzyme that binds the substrate and catalytically transforms it.

active transport Movement of molecules across a membrane that uses an energy source, typically hydrolysis of ATP, to produce a concentration gradient of the molecules across the membrane.

acyl carrier protein A protein that binds activated intermediates of fatty acid synthesis through a thioester linkage.

acyl-CoA dehydrogenases Enzymes involved in catabolism of fatty acids by β-oxidation including very long-chain (VLCAD), long-chain (LCAD), medium-chain (MDAC) and short-chain acyl-CoA dehydrogenases (SCAD)

ADAR Adenosine deaminases that act on RNA in RNA editing to convert A to I.

adipocytes (adipose cells) Animal cells where fats (triacylglycerols) are stored.

aerobes Organisms that use oxygen as an electron acceptor in energy-producing pathways.

agarose and agaropectin Agarose is a chain of alternating D-galactose and 3,6-anhydro-L- galactose with side chains of 6-methyl-galactose. Agaropectin is similar to agarose but with sulfate ester side chains and D-glucuronic acid. Agarose is the neutral component of agar whereas agaropectin is the negatively charged component.

agouti-related peptide Peptide hormone that blocks activity of melanocortin-producing neurons. Melanocortin blocks eating behavior.

alcaptonuria A genetic disease in which homogentisate collects in the urine because of a deficiency in homogentisate dioxygenase.

aldaric acid A sugar acid with two carboxylic acid groups often at C-1 and C-6 (for hexoses).

aldonic acid A sugar acid formed from by oxidation of the aldehyde carbon of an aldose sugar to a carboxyl group.

aldose A monosaccharide containing an aldehyde group.

alginates Polysaccharides that bind metal ions, particularly calcium.

alleles Alternative forms of a gene at a particular location or locus.

allosteric regulation The activation or inhibition of enzymatic activity through noncovalent interaction of the enzyme with small molecules other than the substrate.

α-helix A common secondary structure in proteins, which is almost always right handed.

α-amanitin A mushroom poison that differentially inhibits eukaryotic RNA polymerases. RNA polymerase I is resistant, II is highly sensitive and III intermediate.

amino acids Building blocks of proteins with a tetrahedral α-carbon covalently linked to an amino group and a carboxyl group.

aminoacyl-tRNA synthetases Enzymes that attach amino acids to appropriate tRNAs in aminoacyl linkage.

aminotransferases Pyridoxal phosphate containing enzymes that exchange nitrogen between amino acids and α-keto acids.

ammonotelic organisms Aquatic animals that excrete excess nitrogen as ammonia.

amphibolic pathway A metabolic pathway that can be both catabolic and anabolic.

amphipathic Possessing both polar and nonpolar groups.

amphiphilic Having both strongly polar and strongly nonpolar groups on a single molecule. Also known as amphipathic.

anabolism Metabolism in which complex biomolecules, such as carbohydrates, lipids and proteins, are generated from simpler precursors.

anaerobes Organisms that can subsist without O_2.

anaplerotic reaction A reaction that replenishes metabolites that are removed from a central metabolic pathway like the citric acid cycle.

androgens Male sex hormone steroids.

annexins Family of calcium binding proteins that interact with membranes and phospholipids in a calcium dependent manner.

antibody An immune system protein capable of recognizing and binding a "foreign substance" known as an antigen.

anticodon The three-nucleotide unit in tRNA that recognizes and base pairs with a particular mRNA codon.

antigen A molecule specifically recognized and bound by an antibody.

antiparallel α-helix proteins Proteins whose structures are heavily dominated by α-helices.

antiparallel β-sheet A secondary protein conformation in which the adjacent strands of a β-strand run in opposite directions.

apoenzyme The protein portion of an enzyme devoid of its cofactors or prosthetic groups.

apoptosis Programmed cell death.

archaea Prokaryotes that are capable of living in extreme conditions like temperature, pH or osmotic pressure

aspartic proteases A family of proteolytic enzymes that are active at acidic pH and possess two aspartic acid residues at the active (catalytic) site.

ATP equivalent A metabolic unit of exchange; the conversion of ATP to ADP.

autogenous regulation of a gene (autoregulation of a gene) The regulation of expression of a gene by the product of the gene.

autophosphorylation The ability of a kinase to phosphorylate residues in its own amino acid sequence.

autotroph An organism that uses carbon dioxide as its sole source of carbon.

• •

baa helix *B*asic *a*mphiphilic *a*lpha helix to which calmodulin binds.

bacteriophage Viruses that infect bacteria.

ballistic method A technique of transformation in which microprojectiles coated with DNA are fired into recipient cells by a gas-powered gun.

base pair Two nucleotides in nucleic acid chains that are paired because of hydrogen bonding between their bases, such as A with T and C with G.

β-bulge A structure in which one residue of a β-strand is not H-bonded but residues on either side of it are H-bonded to contiguous residues in the adjacent strand.

β-oxidation The oxidative degradation of a fatty acid into acetyl-CoA by repeated oxidations and cleavages at the β-carbon of a fatty acid.

β-pleated sheet A common secondary structure in proteins in which the polypeptide chain is arranged in an extended, pleated fashion. Several such strands side-by-side are typically joined by H-bonds to form a sheet. Adjacent strands may be either parallel or antiparallel.

β-turn A simple structure observed in many proteins in which the peptide chain forms a tight loop with the carbonyl oxygen of one residue hydrogen-bonded with the amide proton of the residue three positions down the chain.

bifunctional enzyme (tandem enzyme) A protein possessing two distinct enzymatic functions.

bile acids Polar carboxylic acid derivatives of cholesterol that are important in the digestion of food, especially the solubilization of ingested fats.

binuclear center A closely-associated pair of metal ions.

BLOSUM A scoring matrix used in protein sequence comparisons to quantitate the degree of sequence similarity between two proteins.

Bohr effect When the oxygen affinity of hemoglobin in red blood cells decreases due to a lowering of the pH.

branch migration The procession of base-pairing between an invading DNA strand and one strand of a DNA duplex during recombination.

branching and debranching enzymes Enzymes involved in metabolism of glucose polymers by adding or removing branches.

bromodomains Protein domains that bind acetylated lysine.

brown fat A specialized type of adipose tissue found in newborns and animals involved in heat production.

buffer A solution component that tends to resist changes in pH as acid or base is added. Typically, a buffer is composed of a weak acid and its conjugate base.

bZIP A DNA binding motif characterized by an α-helical structure with basic amino acids on one end and a leucine zipper at the other. Leucine zipper have leucines spaced every seven residues along the helix.

• •

Calmodulin Calcium binding protein with EF hand calcium binding motif that imparts calcium regulation to a number of proteins.

caloric homeostasis The constant availability of fuels in the blood.

Calvin-Benson cycle (Calvin cycle) The set of reactions that transform 3-phospho-glycerate into hexose during photosynthesis.

CAP Catabolite activator protein –a positive regulator of transcription that binds cAMP and sometimes called CRP (cAMP receptor protein). The term "cap" also refers to modification of the 5' end of mRNAs in eukaryotes by 7-methylG.

carbamoyl-phosphate synthetase Enzymatic activity that produces carbamoyl phosphate from ammonia and bicarbonate using ATP hydrolysis.

carbohydrate A class of molecules with molecular formula $(CH_2O)_n$, where n is three or more.

carbon dioxide fixation The synthesis of carbohydrates from carbon dioxide and water.

cardiac glycosides Potent inhibitors of the Na^+/K^+ pump. Inhibition of the pump results in increases in sodium that is exchanged for calcium, which increases the strength of contraction of cardiac muscle.

carnitine acyltransferase Enzyme that moves an acyl group from acyl CoA to carnitine. Acyl carnitine is carried across the inner mitochondrial membrane.

caspases Proteases that are activated by apoptosome during apoptosis.

catabolism Energy-yielding metabolism, which involves the oxidative degradation of complex nutrient molecules such as carbohydrates, lipids, and proteins.

catalytic power The ability of enzymes to accelerate reaction rates.

CCT The eukaryotic analog of GroEL, a chaperonin that functions in protein folding.

cDNA library A DNA library constructed by synthesizing cDNA from purified cellular mRNA.

cellulose A structural polysaccharide and linear homopolymer of D-glucose units linked by $\beta(1\rightarrow4)$ glycosidic bonds.

cerebroside A sphingolipid containing one sugar residue as the head group.

chaperones Proteins or protein complexes that aid in protein folding by prevention inappropriate hydrophobic interactions.

chemical mutagens Agents that chemically modify bases so that their base-pairing characteristics are altered.

chemotrophic organism An organism that obtains energy by metabolizing organic compounds.

chiral compound A compound with one or more atoms that have four different groups attached. Variations in the arrangement of groups form stereoisomers.

chitin A structural polysaccharide with repeating units of N-acetyl-D-glucosamine in $\beta(1\rightarrow4)$ linkage.

chloroplasts Chlorophyll-containing photosynthetic organelles that harvest light and transform it into metabolically useful chemical forms. Found in plants and algae.

cholecystokinin A peptide hormone released by the GI tract during eating. Signals satiety and curtails eating.

chromatin A complex of DNA and protein in a eukaryotic cell nucleus.

chromatography Experimental techniques for the separation of mixtures. Methods include ion exchange chromatography, high-performance liquid chromatography and gas chromatography.

chromodomains Protein domains that bind methylated lysine.

chromosome A single large DNA molecule, and its associated proteins, containing many genes.

chromosome walking A technique for ordering DNA fragments in a genomic library. It involves hybridization, restriction mapping, and isolation of overlapping DNA molecules.

chromatin dsDNA associated with nucleosomes, protein complexes composed of histones. Chromatin remodeling complexes mediate ATP-dependent reorganization of nucleosomes along DNA.

clamp loader A protein complex that loads a β-subunit dimer around DNA. The β-subunit dimer is a protein ring that tethers a polymerase to DNA.

cloning The amplification of identical DNA molecules or cells from a single parental DNA molecule or cell.

cloning vector A plasmid used to carry a segment of a foreign DNA and to introduce it into a cell where it can be replicated.

codon A sequence of three adjacent nucleotides in mRNA that codes for a specific amino acid.

coenzyme An organic molecule that helps an enzyme to carry out its catalytic function.

coenzyme Q Ubiquinone –a compound that participates in one- and two-electron transfer reactions. A mobile lipid-soluble carrier of electrons.

cofactor An inorganic ion or organic molecule required by an enzyme for catalytic activity.

collagen A fibrous protein that is a principal constituent of connective tissue in animals.

colligative properties Properties of solutions that depend only on the concentration of solute species and not on their nature.

competitive inhibitor A reversible inhibitor of an enzyme that competes with the substrate for binding at the active site of the enzyme.

complementarity-determining region (CDR) In an immunoglobulin, regions that form the structural site that is complementary to some part of an antigen's structure, providing the basis for the antibody:antigen recognition.

complementary DNA (cDNA) A DNA copy of a mRNA molecule.

configuration A spatial arrangement of covalently-bonded atoms, which cannot be altered without breaking and reforming covalent bonds.

conformation A three-dimensional structure or spatial arrangement of a molecule that results from free rotation of substituent groups around single bonds.

constitutive expression The continuous expression of genes independently of regulation.

constitutive splicing A type of splicing in which every intron is removed and every exon is incorporated into the mature RNA without exception.

converter enzyme (modifying enzyme) An enzyme that catalyzes the covalent modification of another enzyme, thereby changing its catalytic activity.

cooperativity A behavior in which the binding of one ligand to a protein influences the affinity of the protein for additional molecules of the same ligand. Cooperativity can be positive or negative.

corepressors A metabolite that binds to and activates a protein that functions to repress gene expression.

corticosteroids Steroid hormones formed by the adrenal cortex.

coupled reactions Two chemical reactions that have a common intermediate to transfer energy to one another.

covalent modification The reversible covalent attachment of a chemical group to an enzyme in order to regulate the activity of the enzyme.

cristae The folds of the inner mitochondrial membrane, which provide the inner membrane with a large surface area in a small volume.

critical micelle concentration (CMC) Concentration of an amphipathic molecule that form micelles. Below this value molecules do not aggregate into micelles.

cruciform A cross-shaped tertiary structure of DNA that arises when inverted repeats form a structure that involves intrastrand base pairing.

cyclic photophosphorylation Phosphorylation of ADP as a consequence of light-driven movement of electrons in a cyclic pathway leading to proton gradient formation.

cytochromes Proteins containing heme prosthetic groups; they serve as electron carriers in oxidation-reduction reactions such as respiration and photosynthesis.

cytochrome P-450 A cytochrome found in liver and involved in hydroxylation of xenobiotics to detoxify them by making them more water soluble.

• •

dark reactions of photosynthesis The reactions of photosynthesis that do not require light, notably carbon dioxide fixation.

debranching enzyme An enzyme that can degrade a branched polymer, such as a limit dextrin, by repositioning or removing branches.

degenerate codons (synonymous codons) Codons that specify the same amino acid.

denaturation The loss of structural order in a macromolecule; it is accompanied by loss of function.

denitrifying bacteria Bacteria capable of using NO_3^- and similar forms of oxidized inorganic nitrogen as electron acceptors in place of oxygen in energy-producing pathways.

density gradient ultracentrifugation A centrifugation technique for separating macromolecules based on their densities.

deoxyribonucleotide Nucleotides containing 2-deoxy-D-ribose as the pentose. **designer enzymes** Synthetic enzymes tailored to carry out specific catalytic processes. **dextrorotatory isomer** A stereoisomer that rotates the plane of plane-polarized light clockwise.

diabetes mellitus A common endocrine disease characterized by an abnormally high level of glucose in the blood.

dialysis Removal of small molecules and ions from macromolecules in solution by allowing them to pass through a semipermeable membrane.

diastereomer For molecules with more than one asymmetric center, one of a pair of non-mirror image stereoisomers that differ in configuration at only one of the asymmetric centers.

diploid cells Cells containing two sets of chromosomes.

direct reversal repair DNA repair process that directly corrects the DNA defect.

discriminator base The N of NCCA the last four bases in a tRNA. This base is invariant in the tRNAs for a particular amino acid.

DNA polymerase An enzymatic activity that uses dNTPs to elongate the 3' end of a DNA or RNA using a complementary ssDNA template. DNA polymerase I, II, III, IV and V are bacterial enzymes, DNA polymerase α, ϵ, ϕ, τ, β, γ, δ are eukaryotic polymerases.

docking protein (SRP receptor) A heterodimeric transmembrane protein that binds to the SRP-ribosome complex to stimulate the ribosome to resume polypeptide synthesis.

domains Compact structures found in proteins as distinct regions that are often stable by themselves. Also referred to as modules.

double-reciprocal plot (Lineweaver-Burk plot) A graph of the reciprocal of the initial velocity versus the reciprocal of the substrate concentration for an enzyme-catalyzed reaction. It allows an accurate graphical determination of the K_m and the V_{max}.

• •

EC **numbers** Systematic classification of enzymes into one of six major classes. Major classes are further divided into subclasses, sub-sub classes and sub-sub-subclasses. For example, EC 1.1.1.1 is alcohol dehydrogenase that uses NAD+.

Edman degradation N-terminal chemistry used to sequence proteins. The chemistry modifies and cleaves only the N-terminal residue creating a new N-terminus that is subjected to an other round of Edman degradation.

EF-Tu and EF-Ts Proteins involved in elongation of protein synthesis in bacteria. EF-Tu is a thermal unstable protein; EF-Ts is a thermal stable protein.

eicosanoids (prostaglandins, thromboxanes, and leukotrienes) Breakdown products of phospholipids. Compounds with hormone-like activities that are derived from 20-carbon polyunsaturated fatty acid arachidonic acid used to synthesize prostaglandins, thromboxanes, and leukotrienes.

electrolytes Substances capable of generating ions in solution, thereby causing an increase in the electrical conductivity of the solution.

electron transport pathway A metabolic pathway in which electrons from the tricarboxylic acid cycle are passed through membrane-associated proteins to reduce O_2 to H_2O.

electroporation A transformation procedure in which the membranes of cells exposed to pulses of high voltage are rendered momentarily permeable to DNA molecules.

elongation In protein synthesis, the successive addition of amino acids to a polypeptide chain by ribosomes.

elongation factors Proteins required for the ribosome-catalyzed growth of a polypeptide chain. Examples include EF-Tu, EF-Ts and EF-G. EF-G catalyzes translocation during protein synthesis during which the peptidyl-tRNA is moved from the A site to the P site.

Embden-Meyerhof pathway Glycolysis —conversion of glucose into two pyruvate.

enantiomer Mirror image, non-superimposable stereoisomers.

endergonic A process that has a positive ΔG value (energy absorbing).

energy charge An index of the capacity of the adenylate system (ATP/ADP/AMP system) to provide high-energy phosphoryl groups in order to drive thermodynamically unfavorable reactions.

energy transduction The transformation of energy from one form to another.

enhancer (upstream activation sequence; UAS) Nucleotide sequence involved in regulating the transcription of a gene.

enthalpy (H) A thermodynamic quantity, defined as the heat of a reaction at constant pressure.

entropy A measure of disorder and randomness in a system or its surroundings.

enzyme A protein that catalyzes a specific chemical reaction. Catalytic RNAs are ribozymes.

enzyme kinetics The study of rates of enzyme-catalyzed reactions.

enzyme substrate complex (ES) A complex formed when substrate molecule(s) bind noncovalently to the active site of an enzyme.

episomes Plasmids capable of chromosomal integration.

essential amino acids Amino acids that cannot be synthesized by the body and must be obtained from dietary sources.

estrogens Female sex hormone steroids.

eukaryote A single-celled or multi-celled organism whose cells have a membrane-bound nucleus, organelles and many chromosomes.

excision repair The replacement of damaged or modified bases in DNA by endonucleolytic cleavage.

exergonic A process with a negative ΔG value (energy releasing).

exon The region of a eukaryotic gene that codes for sequences joined during the process of splicing.

exon insertion The process whereby exons encoding particular functional or structural domains of proteins from one gene are inserted into another.

exon shuffling The exchange of exons between genes via recombination events.

exosome A protein complex involved in degradation of RNA.

expressed sequence tags (ESTs) Relatively short (< 200 bases) sequences obtained by determining the nucleotide sequence of randomly selected cDNAs.

expression vector A cloning vector that allows a cloned inserted DNA to be transcribed into RNA and often translated into protein.

• •

facultative anaerobes Organisms that can adapt to anaerobic conditions by substituting other electron acceptors for O_2 in their energy-producing pathways.

familial hypercholesterolemia A variety of inherited metabolic defects that lead to greatly elevated levels of serum cholesterol, much of it in the form of LDL particles.

fasting state A situation in which food intake by an organism is zero.

fatty acid synthase A single polypeptide chain that contains all the enzymatic activities necessary to synthesize fatty acids. Found in animals.

fermentation The anaerobic catabolism of organic molecules for the production of energy. In alcoholic fermentation, pyruvate is converted to ethanol and carbon dioxide.

fibrous proteins Proteins that have relatively simple, regular linear structures that often serve structural roles in biological systems.

filamentous phage Single-stranded DNA-containing bacterial viruses that package their genome into long, thin tubes constructed from thousands of protein monomers.

first law of thermodynamics A principle stating that the total energy of an isolated system is conserved.

fitness The ability of an organism to survive and reproduce.

flagellum A bacterial filament made out of the protein flagellin. Also, long projection of the plasma membrane of a eukaryotic cell filled with microtubules and involved in motility. Cilium is a short version of a flagellum. Cells might have one or two flagella but many cilia.

flavoprotein Protein conjugated with a flavin coenzyme.

flippase ATP-dependent enzyme that transports phosphatidyl serine (and to a lesser extend PE) from the outer leaflet to the inner leaflet of a plasma membrane.

floppase ATP-dependent enzyme that transports cholesterol, phosphatidyl choline and sphingomyelin from the outer leaflet to the inner leaflet of a plasma membrane.

fluid mosaic model A model proposed for the structure of biological membranes. In this model, membranes are dynamic structures composed of protein and phospholipid molecules that move laterally.

freeze fracture electron microscopy A technique used to visualize the structure of biological membranes.

furanose A five-membered ring monosaccharide formed by intramolecular hemiacetal formation.

• •

gaba Gamma amino butyric acid, an inhibitory neurotransmitter that regulates chloride channels.

galactosemia A disease in which toxic levels of galactose accumulate in the body, causing cataracts and permanent neurological disorders.

gene conversion A recombination process whereby two different sequences in a genome interact in such a way that one is converted to the other.

gene pool The sum of all the alleles of a particular gene found in a population.

gene The element or unit carrying and transferring inherited characteristics from parent to offspring. Genes are contained in chromosomes.

gene replacement therapy A technique in which cells are removed from a patient, genetically manipulated and returned to the patient in order to correct a human genetic disorder.

gene sharing When the product of a single gene gains a second function without losing its primary function.

gene therapy The repair of a genetic deficiency by introduction of a functional version of the gene.

general genetic recombination The process whereby DNA sequences are exchanged between homologous chromosomes, resulting in the arrangement of genes into new combinations.

general transcription factors Proteins required for transcription by RNA polymerase II that bind to elements near the site of initiation of transcription.

genetic recombination The formation of chromosomes with combinations of gene types different from those found in the parental chromosomes.

genomic library A DNA library prepared by isolating the total DNA from an organism, digesting it into fragments and cloning the fragments into an appropriate vector.

genomics The study of the whole genome of organisms.

genotype The genetic make-up of an organism, as opposed to its phenotype or outward characteristics.

germ cells The sperm and the eggs of multicellular organisms.

ghrelin An appetite-stimulating peptide hormone produced in the stomach.

Gibbs free energy (G) A thermodynamic state function relating enthalpy, entropy and temperature, G = H - TS.

globular protein A water-soluble protein that is roughly spherical in shape. It consists of one or more compactly-folded polypeptide chains.

glucogenic compound A compound that can be used for gluconeogenesis in animals. Most amino acids are glucogenic.

gluconeogenesis An anabolic pathway for the production of glucose from noncarbohydrate precursors.

glutamate (glutamic acid) An acidic amino acid with a side chain containing a carboxyl group. Also used as an excitatory neurotransmitter amino acid.

glutamate dehydrogenase An enzyme that catalyzes the reductive amination of α-ketoglutarate to produce glutamate.

glutamine synthetase An enzyme that catalyzes the ATP-dependent amidation of the γ-carboxyl group of glutamate to form glutamine.

glycerophosphate shuttle A process whereby two different glycerophosphate dehydrogenases, one in the cytoplasm and one in the outer face of the mitochondrial inner membrane, work together to deliver electrons to the mitochondrial electron transport system.

glycoconjugate A carbohydrate in which one or more sugar residues is covalently linked to a peptide chain, a lipid or a protein.

glycogen The major form of storage polysaccharide in animals.

glycogen phosphorylase Enzyme that releases glucose from glycogen as glucose-1-phosphate. Activated by glycogen phosphorylase kinase.

glycogen synthase Enzyme that uses UDP-glucose to synthesize glycogen.

glycolipid A carbohydrate covalently linked to a lipid molecule.

glycolysis The metabolic pathway responsible for the stepwise degradation of glucose and other simple sugars to pyruvate during which energy is released from the sugar and captured in the form of ATP under anaerobic conditions.

glycoprotein A protein containing a covalently linked carbohydrate.

glycosaminoglycans A class of polysaccharides consisting of linear chains of repeating disaccharides in which one of the monosaccharide units is an amino sugar and one (or both) of the monosaccharides contains at least one negatively charged sulfate or carboxylate group.

glycosidic bond A bond between the anomeric carbon of a sugar and the oxygen atom of an alcohol.

glyoxylate cycle A modification of the tricarboxylic acid cycle, used by plants and bacteria to produce four-carbon dicarboxylic acids (and eventually sugars) from two-carbon acetate units.

glyoxysomes A plant peroxisome that contains the enzymes for the glyoxylate cycle.

G-protein Protein capable of binding and hydrolyzing GTP. Found in signal transduction pathways and protein translation.

G-protein-coupled receptors Integral membrane proteins with several membrane-spanning regions and an intracellular domain that binds G-protein

grana Stacks of flattened vesicles formed from the thylakoid membrane in chloroplasts.

gratuitous inducers Nonmetabolizable substrate analogs capable of activating the transcription of catabolic enzymes.

GroES-GroEL complex A bacterial protein complex that functions as a chaperonin to help proteins fold.

group transfer potential The measure of the ability of a compound to donate a functional group to a specific receptor molecule or to water.

growth medium The nutrients needed for a microorganism to grow in a laboratory.

GTPase-activating protein (GAP) Protein capable of increasing GTPase activity of a G-protein.

gyrase A type II topoisomerase that acts to overcome torsional stress imposed upon DNA unwinding by adding negative supercoils to DNA using ATP hydrolysis.

• •

halophiles Prokaryotes that can live in water with high salt levels

haploid cells Cells containing only one set of chromosomes.

Hanes-Woolf plot Plots of [S]/v versus [S] for enzyme kinetic data.

HAT Histone acetyltransferase –an enzyme that acetylates lysines on histone proteins using acetyl-CoA.

Hatch-Slack pathway A reaction pathway in plants to attach carbon dioxide to form a 4-carbon compound that serves as a CO_2 delivery system.

HDACs Histone deacetylase complexes –enzyme activity that removes acetyl groups from histones

heat shock proteins   Proteins synthesized by cells in response to a sudden increase in temperature. Many heat shock proteins are chaperonins.

helix-turn-helix A description of a structural motif found in DNA-binding proteins.

heme The iron-porphyrin prosthetic group of heme proteins.

hemiacetal The product formed when an alcohol reacts with an aldehyde.

hemiketal The product formed when an alcohol reacts with a ketone.

hemoproteins Heme-containing proteins.

Henderson-Hasselbach equation An equation relating the pH, the pKa and the concentrations of the proton-acceptor [A⁻] and proton-donor species [HA] in a solution.

heredity   Loosely defined as the tendency of an organism to possess the characteristics of its parent(s).

heterogeneous nuclear RNA (hnRNA)   A large mRNA precursor with stretches of nucleotide sequence that have no protein-coding capacity.

heterotroph An organism that requires an organic form of carbon, such as glucose, in order to generate metabolic energy and to synthesize other essential carbon compounds.

HGPRT (hypoxanthine-guanine phosphoribosyltransferase) An enzyme in the purine salvage pathway that produces IMP. Defects in this enzyme cause Lesch-Nyhan syndrome.

high performance liquid chromatography (HPLC) A chromatographic technique for separating components of a mixture by forcing the liquid mixture though a column under high pressure.

high-density lipoprotein (HDL)   A plasma lipoprotein that contains a significant amount of protein and that transports cholesterol and cholesterol esters from tissues to the liver.

histones A class of arginine- and lysine-rich basic proteins that interact ionically with anionic phosphate groups in the DNA backbone to form nucleosomes.

HIV-1 Human immunodeficiency retrovirus that causes AIDS.

HMG-CoA reductase Hydroxymethylglutaryl-CoA reductase catalyzes production of 3R- mevalonate from HMG CoA. This is the rate limiting step in cholesterol synthesis and the site of action of statin drugs.

Holliday junction A cross-strand intermediate in genetic recombination in which two double-stranded DNA molecules are joined by the reciprocal crossover involving one strand of each molecule.

holoenzyme A catalytically-active complex of protein and cofactor.

homeobox domain   A DNA motif that encodes a related 60-amino acid sequence found in a number of proteins that act as a sequence specific transcriptional factor.

homeostasis   Maintaining a constant internal environment with respect to parameters like temperature, osmotic pressure and pH.

homologous genetic recombination The recombination of two DNA molecules of similar sequences.

612

homologous proteins Homologous proteins are proteins that share a common evolutionary ancestor, which is often evidenced by a significant degree of sequence or structural similarity.

HTH motif A DNA binding motif characterized by two successive α-helices separated by a β-turn. Proteins with HTH motifs function as dimers.

hydration A state in which a molecule or ion is surrounded by water molecules.

hydrogen bonds Weak chemical forces between a hydrogen atom covalently bonded to an electronegative atom such as oxygen or nitrogen and a second electronegative atom that serves as the hydrogen bond acceptor.

hydropathy plot A technique to detect hydrophobic and hydrophilic regions within a polypeptide chain by plotting the average hydrophobicity of a window of amino acids versus location as the window is moved across the amino acid sequence.

hydrophobic interactions The association of nonpolar groups or molecules with each other in aqueous environments.

hydroxyapatite Calcium/phosphate complex found in microcrystals in bone tissue.

hydroxyurea A free-radical quenching agent and inhibitor of ribonucleotide reductase.

hyperuricemia Chronic elevation of blood uric acid levels, which occurs in about 3% of the population as a consequence of impaired excretion of uric acid or overproduction of purines.

• •

illegitimate recombination A rare recombination that occurs between nonhomologous DNA independently of any unique sequence element.

immunoglobulin A major class of antibody molecules found circulating in the bloodstream.

immunoglobulin fold A collapsed β-barrel domain consisting of two layers of a total of seven β-strands.

indirect readout The ability of a protein to indirectly recognize a particular nucleotide sequence by recognizing local conformational variations in double-helical DNA resulting from the effects that base sequences have on DNA structure.

inducers Substrates capable of activating the synthesis of enzymes that metabolize them.

induction of enzyme synthesis An increase in enzyme synthesis by the activation of transcription of the gene encoding the enzyme.

inhibitor A compound that decreases the velocity of an enzymatic reaction.

initiation In protein synthesis, the first phase that primes the ribosome for polypeptide formation.

initiation complex A complex consisting of 70S ribosome with GTP, N-formyl-Met-tRNA$_f$^{Met} and mRNA ready for the elongation steps in protein synthesis.

initiation factors (IFs) Proteins necessary for the initiation of protein synthesis.

initiator tRNA A special tRNA charged with methionine and used to start protein synthesis.

insulin Polypeptide hormone of the β-cells of the islets of Langerhans in the pancreas of all vertebrates that regulates carbohydrate metabolism.

integral membrane protein (intrinsic membrane protein) A membrane protein with hydrophobic surfaces that penetrate or extend all the way across the lipid bilayer.

intercalation The insertion of a molecule between stacked structural elements such as the insertion of a planar molecule between two successive bases in a nucleic acid.

intermembrane space The space between an inner and outer membrane.

international unit (IU) One IU is the amount of enzyme that catalyzes the formation of one micromole of product in one minute at a specified pH, temperature and ionic strength.

intrinsic termination Transcription termination in bacteria that occurs because the transcript forms a RNA hairpin with a GC rich stem followed by a run of Us that cause the RNA polymerase to pause and terminate transcription.

intron (intervening sequence) A sequence of nucleotides in a gene that is transcribed but is not found in the mature mRNA.

invertase An enzyme that hydrolyzes sucrose to fructose and glucose. The optical rotation changes sign during digestion.

ion channels Membrane proteins through which ions can pass. Channels can be gated (regulated to open and close) by ligands (**ligand-gated**) or membrane voltage (**voltage-gated**).

ion exchange chromatography A method for separating substances on the basis on their charge.

ionic bonds Weak bonds that are the result of attractive forces between oppositely charged polar functions, such as negative carboxyl groups and positive amino groups.

islets of Langerhans Cells in the pancreas that secrete insulin.

isoacceptor tRNAs The set of tRNAs specific for a particular amino acid.

isoelectric focusing An electrophoretic technique for separating proteins according to their isoelectric point or pI.

isoelectric point The pH at which a compound is electrically neutral.

isomerases A group of enzymes that interconvert isomers.

• •

katal One katal is the amount of enzyme catalyzing the conversion of one mole of substrate to product per second. One katal equals 6×10^7 international units.

keratins Insoluble structural proteins with parallel polypeptide chains in the α-helical or β-sheet conformations.

ketogenesis The synthesis of acetone, acetoacetate and β-hydroxybutyrate generated from acetyl-CoA.

ketogenic compounds A compound such as an amino acid that can be degraded to yield acetyl-CoA, thereby contributing to the synthesis of fatty acids or ketone bodies.

ketone bodies Fuel molecules (acetone, acetoacetate and β-hydroxybutyrate), synthesized in the liver from acetyl-CoA. During starvation and in untreated diabetes mellitus, ketone bodies become a major source of fuel.

ketose A monosaccharide containing a ketone group.

kinases Enzymes that transfer the terminal (gamma) phosphate of ATP to some target molecule, usually on a hydroxyl group.

Klenow fragment The larger fragment of DNA polymerase I that contains the polymerase and 3'-exonuclease activity.

Krebs cycle The tricarboxylic acid cycle.

• •

Lactose intolerance A metabolic disorder caused by the inability to digest lactose due to the absence of the enzyme lactase in the intestine.

lamellae The paired folds of the thylakoid membrane of chloroplasts.

lariat A branched, covalently closed loop of RNA formed as an intermediate during RNA splicing.

leader peptide An N-terminal sequence of amino acids in a protein that is synthesized on the RER and is destined to be a membrane protein or to be secreted.

leading strand/lagging strand During DNA replication the leading strand is the newly synthesized strand that is elongated in a continuous manner in the direction of movement of a replication fork. The lagging strand is synthesized as Okazaki fragments because it is on the opposite strand from the lagging strand.

Leloir pathway The catabolism of galactose.

lethal synthesis The metabolic transformation of an otherwise innocuous compound into a poisonous derivative.

leucine zipper A dimerization structural motif consisting of 7-residue repeats of leucine in an alpha helix and found in DNA-binding proteins.

leukotrienes A class of molecules derived from arachidonic acid; they stimulate contractions in vascular, respiratory, and intestinal smooth muscle.

levorotatory isomer A stereoisomer that rotates the plane of plane-polarized light counterclockwise.

ligand A molecule that is bound to another, usually larger molecule.

light reactions of photosynthesis The reactions of photosynthesis that require light and cannot occur in the dark.

limit dextrins Highly-branched oligosaccharides that are left over after extensive removal of glucose units from starch or glycogen by α-amylase, glycogen phosphorylase or starch phosphorylase.

Lineweaver-Burk double reciprocal plot Plot of $1/v$ versus $1/[S]$ for enzyme kinetic data.

linking number (L) The basic parameter characterizing supercoiled DNA. The number of times the two strands of DNA are intertwined.

lipid A class of biological molecules defined by low solubility in water and high solubility in nonpolar solvents.

lipid anchored protein Protein to which lipid is covalently attached causing the protein to localize in a biological membrane

lipid bilayer A back-to-back arrangement of lipid monolayers in which the hydrophobic tails aggregate in the interior of the bilayer and the polar head groups face outward into the aqueous environment.

lipidomics Global analysis of lipids and their interacting protein partners in organs, cells and organelles.

lipopolysaccharide A lipid group joined to a polysaccharide made up of long chains with many different sugars.

lipoproteins Proteins conjugated with lipids including high-density (HDL), low-density (LDL), intermediate-density (IDL), very-low-density lipoproteins (VLDL) and chylomicrons.

liposome A closed, spherical structure formed from a single phospholipid bilayer, which can be used as a drug and enzyme delivery system in therapeutic applications.

locus The chromosomal location of a gene.

low density lipoproteins (LDLs) Plasma lipoproteins with a low protein-to-lipid ratio.

lymphocytes White blood cells involved in the immune response. B lymphocytes, from bone marrow, synthesize and secrete antibodies. T lymphocytes, from the thymus gland, play a regulatory role in the immune response or act as a killer of foreign and virus-infected cells.

• •

macromolecules Proteins, polysaccharides, and polynucleotides (DNA and RNA).

major groove The larger of the two grooves created on the surface when DNA forms a double helix.

malate-aspartate shuttle An electron-transport shuttle system in which malate and aspartate are carried across the inner mitochondrial membrane.

MAPKs Mitogen-activated protein kinases are protein kinases activated by mitogens, agents that induce cell division.

maple syrup urine disease A heredity defect in the oxidative decarboxylation of branched chain α-keto acids leading to elevated levels of valine, leucine and isoleucine in blood and urine. The urine of these individuals smells like maple syrup; the disease is fatal unless dietary intake of these amino acids is restricted in early life.

matrix The space enclosed by the inner mitochondrial membrane.

mediator A protein complex active in eukaryotic transcription that links specific transcription activators bound to enhancers and general transcription factors/RNA polymerase II located at promoters.

melanocortin Peptide hormones that inhibit neurons that initiate eating behavior.

membrane rafts Microdomains within biological membranes that are aggregates of cholesterol and glycosphingolipids.

messenger RNA (mRNA) A ribonucleic acid, which serves to carry the information or "message" that is encoded in genes to the sites of protein synthesis in the cell where this information is translated into a polypeptide.

met aminopeptidase A protein that removes the invariant Met initiating all polypeptide chains.

metabolic channeling A process whereby the product of an enzymatic reaction in a pathway is delivered directly to the next enzyme in the pathway for which it serves as the substrate.

metabolism The sum of the chemical changes that convert nutrients into energy and finished products in an organism.

metabolites Simple organic compounds that are substrates, intermediates or products in cellular energy transformation and in the biosynthesis or degradation of biological molecules, such as amino acids, sugars, fatty acids, and nucleotides.

metabolomics The study of small molecule profile of a biological system under a particular condition.

metalloenzyme An enzyme that binds a metal tightly or that requires a metal ion to maintain activity or stability.

metalloproteins Proteins conjugated with metals.

methanogens Archaea that are obligate anaerobes and that generate methane from CO_2 and H_2 or from organic compounds.

methylmalonyl-CoA mutase A vitamin B_{12}-dependent enzyme used in catabolism of odd chain fatty acids.

micelle A structure formed by amphipathic molecules in aqueous solution in which the hydrophobic portions aggregate in the interior of the structure and the hydrophilic portions of the molecules project into the aqueous environment.

Michaelis constant (K_m) The concentration of substrate at which an enzyme-catalyzed reaction proceeds at one-half its maximum velocity.

Michaelis Menten equation A rate equation relating the initial velocity (v_o) of an enzymatic reaction to the substrate concentration [S]. $v_o = V_{max}$ [S]/ (K_m + [S]) where V_{max} is the maximum velocity and K_m is the Michaelis constant.

microtubules Hollow, cylindrical structures, approximately 30 nm in diameter, formed from tubulin, a heterodimeric protein.

mineralocorticoids Steroid hormones that regulate water and salt balance like aldosterone.

minor groove The smaller of the two grooves created on the surface when DNA forms a double helix.

mismatch repair DNA repair that corrects base pair errors introduced during DNA replication

Mitchell's chemiosmotic hypothesis A proposal stating that the energy stored in a proton gradient across a membrane by electron transport drives the synthesis of ATP in cells.

mitochondria The power plants of cells, which carry out the energy-releasing aerobic metabolism of carbohydrates and fatty acids with the concomitant capture of energy in metabolically useful forms such as ATP.

modulator proteins Proteins that bind to enzymes and by binding influence the activity of the enzyme.

monosaccharide A simple sugar of three or more carbon atoms with the formula $(CH_2O)_n$ where n is 3 or larger.

mosaic proteins Proteins composed of several different structural motifs.

motor proteins Proteins that function as molecular motors to produce movement by transferring chemical energy (ATP hydrolysis or ion gradients) into mechanical energy of motion, examples include dyneins, kinesins, myosins and DNA helicase.

multifunctional polypeptides Single polypeptide chains having two or more enzymatic centers.

mutant An organism with a change in its genetic information.

mutase An enzyme that moves the location of some functional group on a molecule.

myofibrils Actin/myosin fibers that are found in striated muscle cells known as muscle fibers (long multinucleated skeletal muscle cells).

• •

natriuretic hormones Atrial (ANP) or brain (BNP) peptide hormones that regulate sodium levels.

natural selection The differential reproduction of genetically distinct individuals within a population.

618

neo-Darwinian theory of evolution The view that natural selection determines the frequency of alleles in a population and ultimately, the genetic makeup of populations.

neutral theory of molecular evolution A postulate stating that the majority of molecular changes in evolution are due to the random fixation of neutral mutations within the population.

neuropeptide Y A protein that stimulates neurons that initiate eating behavior.

neurotransmitter Chemical substance used to transmit a signal across a synaptic cleft.

nitrate assimilation A two-step metabolic pathway, where NO_3^- is reduced to NH_4^+ in green plants, various fungi and certain bacteria.

nitrate reductase The enzyme that catalyzes the first step in nitrate assimilation: reduction of nitrate to nitrite.

nitrifying bacteria A group of chemoautotrophs, which oxidize NH_4^+ to NO_3^-.

nitrite reductase The enzyme that catalyzes the second step in nitrate assimilation: reduction of nitrite to ammonium.

nitrogen fixation The formation of NH_4^+ from N_2 gas; this reduction occurs only in certain prokaryotic cells.

nitrogenase The enzyme that catalyzes nitrogen fixation, an 8 electron reduction of dinitrogen to ammonium.

noncyclic photophosphorylation Phosphorylation of ADP as a consequence of movement of electrons in a photosystem from water to $NADP^+$ resulting in a proton gradient.

nonessential amino acids Amino acids that can be synthesized by an organism and thus are not required in the diet.

nonsense codons Codons UAA, UAG and UGA that do not specify any amino acids but serve as termination codons.

nonsense suppressors Mutations in tRNA genes that alter the tRNA so that the mutant tRNA can now read a stop codon and insert an amino acid.

nuclear magnetic resonance (NMR) A spectroscopic technique used to study the structures of molecules in solution. In NMR, the absorption of electromagnetic energy by molecules in magnetic fields provides information about molecular structure and dynamics.

nucleases Enzymes that hydrolyze nucleic acids.

nucleic acid Linear polymers of nucleotides linked in a 3' to 5' fashion by phosphodiester bridges. (DNA and RNA)

nucleolus A specialized region of the nucleus at which ribosome subunits are assembled. The nucleolus is centered on a chromosome in a region at which ribosomal RNA genes are in a tandem cluster.

nucleoproteins Proteins conjugated with nucleic acids.

nucleoside A compound formed by the linkage of pentose sugar to a purine or pyrimidine base.

nucleoside diphosphate kinase An enzyme that phosphorylates NDPs using ATP thus producing GTP, CTP and UTP.

nucleosomes Structures in which the DNA double helix is wound around a protein core composed of pairs of four different histone proteins.

nucleotide A nucleoside with a phosphoric acid esterified to a sugar hydroxyl group.

nucleus The organelle that is the repository of genetic information in the form of linear sequences of nucleotides in the DNA of eukaryotes.

• •

Obligate aerobes Organisms, like humans, for which O_2 is essential to sustain life

obligate anaerobes Organisms that cannot use O_2 at all and are even poisoned by it.

Okazaki fragments Short ssDNA chains of about 1000 residues in length, which are formed during the discontinuous synthesis of the lagging strand of DNA.

oligopeptides Peptide chains of more than 12 and less than 20 amino acid residues.

oligosaccharide A polymer of approximately 2 to 10 monosaccharide units linked by glycosidic bonds.

operon A cluster of genes encoding the enzymes of a particular metabolic pathway, along with the regulatory sequences that control their transcription. Inducible operons are expressed only in the absence of their corepressors or only in the presence of small molecule inducers.

operon hypothesis A theory that accounts for the coordinate regulation of related metabolic enzymes.

optical activity The ability of a substance to rotate the plane of plane-polarized light.

ORC Origin recognition complex – a protein complex that binds to replication origins.

organelles Membrane-bound structures found in eukaryotic cells; they perform specialized cell functions. Examples include nucleus, mitochondria, chloroplasts, endoplasmic reticulum, Golgi apparatus, vacuoles, peroxisomes, lysosomes, and chromoplasts.

ORI origin of replication A DNA sequence site for the beginning of DNA replication.

orthologous proteins Proteins from different species that have an evolutionary ancestor that existed before the species split. Orthologous proteins often have similar functions in different species.

ouabain A highly toxic cardiac glycoside used to inhibit the sodium pump.

oxidative phosphorylation The enzymatic phosphorylation of ADP to ATP coupled to the transfer of electrons from a substrate to O_2.

oxygen-evolving complex The PSII complex of photosynthesis that is responsible for the photolysis of water.

● ●

Palindromes (inverted repeats) A segment of duplex DNA in which the base sequences of the two strands exhibit two-fold rotational symmetry.

paralogous proteins Homologous proteins that evolved from an ancestral gene by gene duplication.

parvalbumin Calcium binding protein in neurons with high firing rate. First calcium binding protein discovered to have calcium-binding EF hand.

patch recombinants Recombinant heteroduplexes that are formed by (-) strand cleavage at a Holliday junction.

PCR (polymerase chain reaction) An in vitro technique for amplifying a specific region of DNA using a thermostabile DNA polymerase and primers that flank the sequence to be amplified.

pentose phosphate pathway (hexose monophosphate shunt; phosphogluconate pathway) A metabolic pathway that interconverts pentoses and hexoses and is a source of NADPH.

peptide bond An amide bond between amino acids.

peptidyl transferase The catalytic activity of the large RNA of the large subunit of a ribosome that produces peptide bonds.

peripheral membrane protein (extrinsic membrane protein) A membrane protein which does not penetrate the bilayer to any significant extent but which is associated with the membrane by ionic interactions and hydrogen bonds.

peroxisomes Eukaryotic, cytoplasmic, membrane-bound organelles which carry out a variety of flavin-dependent oxidation reactions, regenerating oxidized flavins by reaction with oxygen to produce hydrogen peroxide, H_2O_2.

phenotype The observable characteristics of an organism.

phenylketonuria A genetic disease in which phenylpyruvate collects in the urine, causing severe mental retardation unless a newborn is placed on a diet low in phenylalanine.

phosphatase An enzyme that removes phosphates from substrates by hydrolysis.

phosphofructokinase-2 Along with fructose-2,6-bisphosphatase, regulates levels of fructose-2,6 bisphosphate, a regulator of glycolysis and gluconeogenesis.

phospholipids A class of lipids, each member of which contains a phosphate moiety.

phosphoproteins Proteins that have phosphate groups typically esterified to the hydroxyls of serine, threonine, or tyrosine residues.

phosphoprotein phosphatase A protein that hydrolyzes phosphate from phosphorylated proteins.

photophosphorylation Light-driven ATP synthesis from ADP and P_i catalyzed by ATP synthase.

photorespiration The light-dependent uptake of O_2 accompanied by the release of CO_2 and metabolism of phosphoglycolate.

photosynthesis The use of light energy to produce ATP and NADPH. These energy-rich substances can drive carbohydrate synthesis from carbon dioxide and water.

photosystem (photosynthetic unit) In photosynthetic cells, a membrane-bound reaction center with an antenna of several hundred light-harvesting chlorophyll molecules.

phototrophic organisms Organisms that grow by transforming light energy into chemical energy

phylogeny The origin and evolution of the many types and species of organisms.

phytol A breakdown product of chlorophyll that occurs in the fat of ruminant animals.

PIDs (protein interaction domains) Protein domains that mediate protein-protein interactions.

pitch The axial distance required to complete one turn of a helix.

plasmalogens Ether glycerophospholipids in which the alkyl moiety is *cis*-α,β-unsaturated.

plasmids Circular, extrachromosomal DNA.

plastids Self-replicating organelles in plants.

platelet activating factor The compound, 1-alkyl-2-acetylglycerophosphocholine, which has the ability to dilate blood vessels in order to reduce blood pressure in hypertensive animals and to aggregate platelets.

P/O ratio The ratio of the number of molecules of ATP synthesized to the number of atoms of oxygen reduced in oxidative phosphorylation.

point mutations A class of mutations in which one base pair is substituted for another. Two possible types are transitions (a purine-purine or pyrimidine-pyrimidine replacement) and transversions (a purine-pyrimidine or pyrimidine-purine replacement).

polyA tail The 3' end of eukaryotic mRNAs with from 100 to 200 A's. Polyadenylation signal (AAUAAA) is a sequence on a RNA transcript that directs cleavage of the RNA downstream of the signal. The new 3' end is then polyadenylated by poly(A) polymerase.

polycistronic mRNA An mRNA containing more than one open reading frame each coding for a protein.

polylinker A short region of DNA sequence bearing numerous restriction sites.

polymerase chain reaction (PCR) A repetitive polymerization technique for dramatically amplifying the amount of a specific DNA segment.

polypeptide A long chain of amino acids linked by amide bonds.

polyribosomes (polysomes) Multiple ribosomes attached to mRNA.

polysaccharide A polymer of many monosaccharide units linked by glycosidic bonds. These polymers can be either linear or branched.

porins Proteins found in the outer membranes of Gram-negative bacteria that span the membranes forming a β-barrel.

post-translational modification Enzymatic processing of a polypeptide chain after genetic information from DNA has been translated into newly formed protein.

preprohormones Peptide hormones that must be processed by proteolysis to become active.

Pribnow box A nucleotide sequence involved in the initiation of transcription of prokaryotic genes.

primary structure The amino acid sequence of a protein.

primer A short nucleotide oligomer to which a polymerase adds monomers.

processing Alterations that convert a newly synthesized RNA into mature messenger RNA.

processivity In DNA synthesis, the degree to which an enzyme remains associated with the template during nucleotide addition.

prochiral An achiral center that can be made chiral by replacement of a single group about the achiral center.

prokaryotes Single-celled organisms that lack nuclei and other organelles.

promoter A nucleotide sequence at which transcription initiation is regulated and initiated.

proproteins Larger inactive protein precursors that are activated through proteolysis.

prostaglandins Compounds that have hormone-like properties and are derived from arachidonic acid.

prosthetic group A molecule that is tightly bound to an enzyme.

proteasome A protein complex involved in degradation of intracellular proteins.

protein A molecule composed of one or more polypeptide chains, each with a characteristic sequence of amino acids linked by peptide bonds.

protein isoforms A set of related polypeptides derived from a common gene by differential RNA splicing.

protein kinase A Cyclic AMP-dependent protein kinase that phosphorylates Ser or Thr on a number of target proteins.

protein module A tertiary structural motif or domain that may exist in several different proteins or that may occur two or more times in the same protein.

protein translocation The targeting of proteins to their proper destinations in cells.

proteoglycans A family of glycoproteins whose carbohydrate moieties are predominately glycosaminoglycans.

proteomics The study of the totality of proteins expressed by an organism under a particular condition.

pseudosubstrate sequence A sequence used to regulate protein kinases by intrasteric control. Pseudosubstrate sequences mimic the target sequence but lack the OH-bearing side chain.

purines A structure containing a pyrimidine ring fused to a five-membered imidazole ring.

pyranose A six-membered ring monosaccharide formed by intramolecular hemiacetal formation.

pyrimidine A six-membered heterocyclic aromatic ring containing two nitrogen atoms.

pyruvate dehydrogenase complex Enzyme complex that metabolizes pyruvate to acetyl-CoA and CO_2. Catalysis is a multistep process involving water soluble vitamins niacin, riboflavin, thiamine, lipoic acid and pantothenic acid.

$PYY_{3\text{-}36}$ Peptide hormone produced in small intestines. Inhibits eating by acting on NPY/AgRP-producing neurons in the brain.

• •

Qcycle Movement of electrons from coenzyme QH_2 to cytochrome c during electron transport in complex III.

quaternary structure The three-dimensional organization of two or more polypeptide chains in a multisubunit protein.

• •

Rstate The active conformation of an allosteric enzyme.

Ramachandran plot Plots of the two dihedral angles about the alpha carbon of amino acid residues within a protein chain.

random genetic drift A random change in allelic frequency.

rate limiting step The slowest step in a chemical reaction. This step has the highest activation energy among the steps leading to the formation of a product from a reactant.

rate of nucleotide substitution The number of nucleotide substitutions per site per year.

rate of product formation In an enzyme-catalyzed reaction, the rate of the formation of product with time: $v = d[P]/dt$.

reaction center A complex of a pair of photochemically-reactive chlorophyll a molecules that forms the core of a photosystem. The reaction center is the site of conversion of photochemical energy into electrochemical energy during photosynthesis.

receptor guanylyl cyclase Single transmembrane segment protein with guanylyl cyclase activity on intracellular domain.

receptor tyrosine kinase Single transmembrane segment protein with tyrosine kinase activity on intracellular domain.

recombinant DNA A DNA molecule that includes DNA from different sources.

recombinant DNA technology (genetic engineering) The process of isolating, manipulating and cloning a sequence of DNA.

recombinant plasmids Hybrid DNA molecules consisting of plasmid DNA sequences plus inserted DNA elements (called inserts). Also known as chimeric plasmids.

reducing sugar A carbohydrate whose free anomeric OH group can reduce oxidizing agents.

Refsum's disease An inherited metabolic disorder that results in defective night vision, tremors, and other neurological abnormalities, caused by an accumulation of phytanic acid in the body.

regulatory proteins Proteins that do not perform any obvious chemical transformation, but can regulate the ability of other proteins to carry out their physiological functions.

release factors Proteins that promote polypeptide release from the ribosome.

replication In vivo DNA synthesis. Replication is always semiconservative meaning the new dsDNA is composed of one strand from the original DNA and one newly synthesized complementary strand.

replication forks During DNA replication, it is the Y-shaped junction where the double-stranded DNA template is unwound and the new DNA strands are synthesized.

replicons The units of DNA replication.

repression A decrease in protein synthesis arising from cessation of transcription of the gene encoding the protein.

response elements Promoter modules in genes that serve as binding sites for proteins that activate transcription.

restriction endonucleases Enzymes, isolated chiefly from bacteria, that have the ability to cleave double-stranded DNA, typically at short, defined sequences.

resveratrol A phytoalexin found in red wines that is a potent activator of sirtuin, an enzyme involved in longevity.

retroviruses A class of eukaryotic RNA viruses that has single-stranded RNA genomes that replicate through a double-stranded DNA intermediate using reverse transcriptase.

rho termination factor A bacterial protein that functions during termination of transcription.

ribonucleotide A nucleotide containing D-ribose as its pentose.

ribosomal RNA (rRNA) Ribonucleic acid, which serves as a component of ribosomes.

ribosomes Compact ribonucleoprotein particles responsible for protein synthesis and found in the cytosol of all cells, as well as in the matrix of mitochondria and the stroma of chloroplasts.

ribozymes Catalytic RNAs.

RNA polymerase An enzyme that uses NTPs to produce RNA. RNA polymerase holoenzyme refers to a protein complex in bacteria capable of initiating transcription as specific locations termed promoters. The core complex cannot initiate but does elongate. RNA polymerases I, II and III are eukaryotic enzymes.

rubisco Ribulose bisphosphate carboxylase/oxygenase, an enzyme in photosynthesis that attaches carbon dioxide to an organic compound.

• •

Saponification The process of hydrolysis of acylglycerols with alkali to yield glycerol and salts of free fatty acids.

sarcomeres, sarcoplasmic reticulum, sarcolemma Sarcomeres are the functional structural units of muscle contraction. Linear arrays of sarcomeres are myofibrils. Sarcoplasmic reticulum is a specialized endoplasmic reticulum surrounding sarcomeres. Sarcolemma is a specialized plasma membrane of a muscle fiber.

saturated fatty acid A fatty acid that does not contain a carbon-carbon double bond.

scaffolding proteins Proteins that bind on either side of a plasma membrane and influence membrane curvature.

scramblase Calcium-activated enzyme that catalyze bidirectional movement of lipids between inner and outer leaflets of the plasma membrane.

second law of thermodynamics A law stating in part that the total entropy of the universe always increases in a spontaneous process.

second messenger An intracellular agent synthesized in response to an external signal or first messenger, such as a hormone.

secondary structure The arrangement in space of atoms in the backbone of a polypeptide chain or a nucleic acid stabilized by hydrogen bonds.

selenocysteine An amino acid coded by special UGA stop codons and derived from serine by replacement of serine's hydroxyl group with selenium. One letter and three letter codes: U and Sec.

semiconservative model for DNA replication The process for duplicating DNA in which the nucleotide sequence in one strand dictates the sequence in the other, complementary strand, resulting in two daughter molecules of double-stranded DNA, each of which contains one of the parent strands.

sequence alignments A comparison of two sequences by juxtaposing two nucleotide or amino acid sequences and analyzing their similarities.

sequential model for allosteric behavior A model for the cooperative binding of identical ligands to an oligomeric protein.

serine proteases A family of proteolytic enzymes that have serine at their active site. Serine is a member of a catalytic triad that includes histidine and aspartic acid.

severe combined immunodeficiency syndrome (SCID) A group of related inherited disorders characterized by the lack of an immune response to infectious disease.

Shine-Dalgarno sequence A mRNA sequence, rich in purine, which is required for mRNA binding to prokaryotic ribosomes.

shuttle vectors Plasmids capable of propagating and transferring genes between two different organisms, one of which is typically a prokaryote and the other a eukaryote.

sickle-cell anemia A human disease characterized by crescent-shaped red blood cells.

signal recognition particle (SRP) A nucleoprotein assembly that binds to the signal sequence to halt further protein synthesis by the ribosome.

signal sequence (signal peptide) The N-terminal sequence of residues in a newly synthesized polypeptide that targets the protein for translocation across a membrane.

signalsome Complex formed by a protein with several protein interaction domains during signal transduction.

single-stranded assimilation (single-stranded uptake) A process driven by branch migration, which displaces the homologous DNA strand from the DNA duplex and replaces it with the invading ssDNA strand.

siroheme A novel heme prosthetic group found in photosynthetic nitrite reductases.

siRNA Small interfering RNAs 20 to 25 nucleotides long that are involved in RNA interference, a process by which RNA blocks gene expression.

sirtuin 1 NAD^+-dependent protein deacetylases involved in longevity.

site-specific genetic recombination Genetic recombination occurring only at specific sequences.

small nuclear RNA (snRNA) A class of RNA molecules found only in the nucleus of eukaryotic cells.

snRNPs ("snurps") Small nuclear ribonucleoprotein particles.

solid phase synthesis Synthesis during which the product is produced while attached to a solid support matrix. Used in chemical synthesis of proteins and nucleic acids.

solvent capacity The capacity of the cell to keep all of its essential metabolites and macromolecules in an appropriate state of solvation.

somatic cells All body cells except for the germ cells (sperm and eggs).

SOS response A repair system that allows DNA replication to continue at lesion in DNA.

SNAREs A family of proteins that "snare" vesicles to initiate fusion of a vesicle with the plasma membrane.

specific activity The number of micromoles of a substrate transformed by an enzyme per minute per milligram of protein at 25°C; it is a measure of enzyme activity and purity.

specificity The ability of an enzyme to discriminate among competing substrates.

spectroscopic methods Techniques which measure the absorption and emission of energy of different frequencies by molecules and atoms.

sphingolipids A class of lipids that contains an 18-carbon amino alcohol backbone, called a sphingosine, joined to a fatty acid.

splice recombinants Recombinant heteroduplexes that are formed by (+) strand cleavage at a Holliday junction.

spliceosome A multicomponent complex formed from the association of various snRNPs with pre-RNA.

standard reduction potential A quantity (voltage) that indicates the tendency of a chemical species to donate or accept electrons.

starch A storage polysaccharide in plants.

steady state A state in which the rate of disappearance of a compound is equal to its rate of synthesis. That is, the change in concentration of the compound with time is equal to zero.

steady state (cellular) A state of apparent constancy, which is actually very dynamic. In this state, energy and material are consumed by the cell and used to maintain the harmonious stability and order of the cell.

steroid A lipid containing 18 or more carbon atoms and a fused four-ring polyprenyl structure.

steroid hormones Hormones that derive from cholesterol.

stoichiometry The measurement of the amounts of chemical elements involved in chemical reactions.

stomata Microscopic pores on a leaf through which carbon dioxide diffuses directly into photosynthetic cells.

stop-transfer sequence A hydrophobic 20-residue peptide that stops the passage of a growing polypeptide chain through the ER membrane.

storage proteins A class of proteins whose biological function is to provide a reservoir of amino acids.

stroma The soluble portion of a chloroplast.

structural complementarity Having a molecular surface with chemical groups arranged to interact specifically with complementary chemical groups on another molecule.

substrate level phosphorylation Phosphorylation of a nucleoside diphosphate to a nucleoside triphosphate by transfer of a phosphoryl group from a non-nucleoside substrate.

structural polarity Arrangement of monomers in a polymer such that the polymer ends are distinct.

substrate The substance upon which an enzyme acts.

suicide substrate A substrate analog that is transformed by an enzyme into a substance that irreversibly inactivates the enzyme.

sumoylation Post translational modification of a protein by a small *ubiquitin-like mo*difer (sumo).

supercoil Underwound (negative supercoil) or overwound (positive supercoil) double-stranded DNA.

supramolecular complexes The combination of various members of one or more classes of macromolecules. Examples include multifunctional enzyme complexes, ribosomes, chromosomes, and cytoskeletal elements.

synapsis A process of chromosome pairing in which two homologous DNA duplexes are juxtaposed so that their sequences are aligned.

synaptic cleft Space between two neurons across which a signal must be transmitted to pass a signal from one neuron to the next.

synonymous codons Different codons that specify the same amino acid.

system That portion of the universe with which we are concerned; the rest of the universe is called the surroundings. Systems may be closed, isolated, or open to the surroundings.

• •

T **state** The less active conformation of an allosteric enzyme.

telomeres Special tandem repetitive DNA 5 to 8 bp long on the ends of eukaryotic chromosomes. Telomeres are formed by telomerase, an RNA-dependent DNA polymerase that extends the 3' end of chromosomes using an RNA template that is part of the telomerase complex.

template A strand of DNA or RNA whose sequence of nucleotide residues acts as a pattern for the synthesis of a complementary strand.

termination The final step in a process such as protein synthesis.

terpenes A class of lipids formed from combinations of two or more molecules of 2-methyl-1,3-butadiene, also known as isoprene.

tertiary structure The compact three-dimensional folded shape of a polymer.

thermodynamics A collection of laws and principles describing the flows and interchanges of heat, energy and matter in systems of interest.

thermogenin An inner mitochondrial protein that short circuits the proton gradient converting the energy to heat.

third base degeneracy The irrelevance of the third base in a codon.

third law of thermodynamics A law which states that entropy of any crystalline, perfectly ordered substance must approach zero as the temperature approaches 0 K, and at T = 0 K, entropy is exactly zero.

thromboxanes Molecules derived from arachidonic acid which are involved in platelet aggregation during blood clotting.

thylakoid membrane The inner-membrane system of chloroplasts, which is organized into paired folds that extend throughout the organelle. It is the site of light-dependent reactions of photosynthesis, leading to the formation of NADPH and ATP.

thylakoid space (thylakoid lumen) The interior of the thylakoid vesicles.

thylakoid vesicles Flattened sacs or disks arising from the paired folds (lamellae) in chloroplasts.

titration curve A graph of the pH versus the equivalents of base added during the titration of an acid.

tobacco mosaic virus An RNA virus infecting plants.

topoisomerase An enzyme capable of changing the linking number of a DNA molecule by breaking one or both strands of DNA, and rejoining the ends.

transcription An enzymatic process in which an RNA copy is made of the sequence of bases along one strand of DNA.

transcription attenuation A regulatory mechanism that manipulates transcription termination or transcription pausing to regulate gene transcription downstream.

transcriptomics The study of global messenger RNA expression of an organism.

transduction The transfer of genetic information from one cell to another by means of a viral vector.

transfection The uptake of viral DNA by competent cells.

transfer RNA (tRNA) Ribonucleic acids that serve as carriers of amino acid residues for protein synthesis.

transformation The uptake, integration and expression of naked DNA by competent cells.

transgenic animals Animals that have acquired new genetic information as a consequence of the introduction of foreign genes.

transition state A high-energy, unstable arrangement of atoms, in which bonds are being broken and formed. It represents the transition between reactants and products in a chemical reaction.

transition state analogs Stable molecules that are chemically and structurally similar to the transition state and that bind more strongly than a substrate or a product to the active site of an enzyme.

transitions DNA mutations that change one purine for another or one pyrimidine for another (c.f. transversions).

translation The process that converts genetic information embodied in the base sequence of a messenger RNA molecule into the amino acid sequence of a polypeptide chain.

translocons A selective permeable protein-conducting channel that catalyzes movement of proteins across membranes.

transpeptidation The peptide bond-forming reaction in protein synthesis.

transport proteins A class of proteins whose job is to transport specific substances from one place to another.

transposons Segments of DNA that are moved enzymatically from place to place in the genome, the smallest of which is called an insertion sequence or IS.

transversions DNA mutations that replace a purine for a pyrimidine or vice versa.

triacylglycerols (triglycerides) Molecules consisting of glycerol esterified with three fatty acids.

tricarboxylic acid cycle (TCA cycle; citric acid cycle; Krebs cycle) A metabolic pathway in which an acetyl group is oxidized to CO_2, and the energy released is captured as ATP, NADH, and $FADH_2$.

troponin-C Calcium binding protein found in skeletal muscle.

turnover number (k_{cat}) The measure of an enzyme's maximal catalytic activity, or the number of substrate molecules converted into product per unit of time when the enzyme is saturated with substrate.

tyrosine kinases A group of enzymes that phosphorylate targets on tyrosine residues using ATP.

● ●

Ubiquitin A small 76 amino acid protein often attached to lysines on proteins destined to be degraded. Ubiquitination is involved in a large number of cellular processes in addition to protein degradation.

UDP-glucose An activated form of glucose used during glycogen synthesis.

ultrafiltration A technique for removing small molecules from solutions of macromolecules and/or for concentrating dilute solutions of macromolecules.

uncoupler A compound that disrupts the normally tight coupling between electron transport and the phosphorylation of ADP.

unsaturated fatty acid A fatty acid that contains one or more carbon-carbon double bonds.

upstream/downstream Refers to directions along a DNA. Upstream is in the 5' direction and downstream is in the 3' direction. Upstream locations are indicated by negative numbers whereas downstream locations are positive numbers. In transcription, +1 is defined as the start of transcript.

ureotelic organisms Organisms, such as terrestrial vertebrates, which excrete excess nitrogen as urea.

uricotelic organisms Organisms, such as birds and reptiles, that excrete nitrogen as uric acid.

uronic acid A acid sugar formed by oxidation of a hydroxyl group to a carboxyl group. Uronic acid sugars contain both a carbonyl carbon (ketose or aldose) and a carboxyl group.

• •

Van der Waals forces Weak, chemical forces induced by electrical interactions between approaching atoms or molecules as their electron clouds fluctuate instantaneously in time.

viruses Supramolecular complexes of nucleic acid (DNA or RNA), encapsulated in a protein coat and in some instances, surrounded by a membrane envelope. Within a host cell, they become self-replicating.

• •

Waxes Esters of long-chain alcohols with long-chain fatty acids.

weak chemical forces Forces ranging from 4 to 30 kJ/mol, including hydrogen bonds, van der Waals forces, ionic bonds and hydrophobic interactions.

wild-type The reference phenotype.

wobble position The first base of the anticodon. It pairs with the third base of a codon.

• •

YACs Yeast artificial chromosomes –cloning vectors that are used to clone megabase pairs of DNA.

• •

Zinc finger A description of a structural motif found in DNA-binding proteins.

zwitterion A neutral molecule with both a positive and a negative charge.

zymogen (proenzyme) An enzyme that exists as an inactive precursor until one or several of its peptide bonds are cleaved. At that point the enzyme acquires full enzymatic activity.